# An Introduction to Electrodynamics

P. A. DAVIDSON

*University of Cambridge*

OXFORD
UNIVERSITY PRESS

OXFORD
UNIVERSITY PRESS

Great Clarendon Street, Oxford, OX2 6DP,
United Kingdom

Oxford University Press is a department of the University of Oxford.
It furthers the University's objective of excellence in research, scholarship,
and education by publishing worldwide. Oxford is a registered trade mark of
Oxford University Press in the UK and in certain other countries



First Edition published in 2019

Impression: 1



Published in the United States of America by Oxford University Press
198 Madison Avenue, New York, NY 10016, United States of America

British Library Cataloguing in Publication Data
Data available

Library of Congress Control Number: 2018963890

ISBN 978–0–19–879812–5 (hbk.)
ISBN 978–0–19–879813–2 (pbk.)

DOI: 10.1093/oso/9780198798125.001.0001

Printed and bound by
CPI Group (UK) Ltd, Croydon, CR0 4YY

*For Catherine*

# PREFACE

Prefaces are, by and large, rather dull affairs, usually starting out with a declaration of the intended readership, followed by an outline of the contents of the book. Perhaps the most useful section is the list of acknowledgements towards the end, where the author has the opportunity to apologise to a long-suffering spouse who has been sadly neglected during the writing of the book. I shall, of course, conform to this convention, if only in the interests of marital harmony. However, before doing so, I shall use this preface for quite another purpose: that of providing the reader with a few historical snippets that might help motivate what is often seen by undergraduates as a rather daunting subject. The hope, of course, is to help bring the subject alive.

Towards the end of the nineteenth century two controversies were sweeping through the theoretical physics and applied mathematics communities, provoking animated debate amongst many of the great and the good. The issues at stake were far from trivial and the outcome was to have a crucial influence on the development of physics in the first part of the twentieth century, especially relativity. On the one hand, a small group of European physicists, commonly known as the Maxwellians, were trying to convince their fellow physicists of the merits of the Faraday–Maxwell interpretation of the laws of electricity and magnetism. On the other hand, physicists and mathematicians were trying to agree on the appropriate mathematical apparatus for manipulating the various vector fields that lay at the centre of Maxwell's new theory.

Let us start with the Maxwellians. Maxwell had embraced and formalised Faraday's firmly held position that *vector fields* (magnetic fields and electric fields) lay at the heart of all observed electromagnetic phenomena and that the traditional notion of action at a distance was inadequate. Moreover, in the process of translating Faraday's modes of thought into mathematical form, Maxwell introduced the concept of the displacement current, which provides a direct connection between electromagnetism and light. As Maxwell put it in 1865 in a letter to a cousin: 'I also have a paper afloat, containing an electromagnetic theory of light, which, till I am convinced to the contrary, I hold to be great guns.'

However, the Maxwellians had their work cut out, with Maxwell's revolutionary theory receiving little attention and often being poorly understood by those who did engage with it. Given that Maxwell's equations are now seen as one of the crowning glories of nineteenth-century science, it is interesting to look back and reflect on how slow the scientific community was to first comprehend, and then endorse, the theory of electromagnetism as articulated by Faraday and Maxwell. Nevertheless, the Maxwellians had a powerful advocate in the form of Oliver Heaviside, the brilliant, if controversial, maverick of English physics. (The four equations we now call Maxwell's equations were, in fact, first written down in their modern form by Heaviside, as discussed in Appendix III.) As one fellow Maxwellian put it:

> Maxwell, like every other pioneer who does not live to explore the country he opened out, had not had time to investigate the most direct means of access to the country, or the most systematic way of exploring it. This has been reserved for Oliver Heaviside to do. Maxwell's treatise is cumbered with the debris of his brilliant lines of assault... Oliver Heaviside has cleared those away, has opened up a direct route, has made a broad road, and has explored a considerable tract of country.
>
> George FitzGerald, 1893

It was not until 1888, some 24 years after Maxwell published his paper 'A dynamical theory of the electromagnetic field', and 15 years after the publication of his *Treatise on Electricity and Magnetism*, that Hertz's experiments on electromagnetic waves finally and irreversibly shifted the status of the Faraday–Maxwell theory of electromagnetism from speculative to definitive.

At around the same time, physicists began to appreciate the need for some mathematical apparatus that could handle these new and all-pervasive vector fields. The formal mathematical position at the time, as championed by the Scot, Peter Guthrie Tait, was that physicists should adopt Hamilton's quaternions for that purpose. However, it is no coincidence that we rarely teach quaternions to undergraduates: their algebraic structure is not at all intuitive. The vector calculus that we do teach, and that most physicists use in their research, was developed independently in the late 1880s by the US physicist Willard Gibbs and the British physicist and engineer Oliver Heaviside. However, just as some mathematicians had resisted the vector fields of Faraday and Maxwell, so some of the mathematicians of the time were resistant to the Gibbs–Heaviside system of vector calculus, Tait being the most vocal opponent. Both Tait and Heaviside relished a good scientific fight, and so they soon squared up to each other. It was clear that the gloves were already off when, in 1890, Tait referred to Gibbs in the following terms:

> Prof. Willard Gibbs must be ranked as one of the retarders of Quaternion progress, in virtue of his pamphlet on Vector Analysis; a sort of hermaphrodite monster.

Gibbs provided a dignified and measured rebuttal the following year in *Nature*. However, the colourful Heaviside felt no such inhibition when he hit back at Tait in 1893. Echoing Voltaire's lampooning of the fictitious philosopher Dr Pangloss in *Candide*, he wrote:

> But supposing, as is generally supposed, vector algebra is something 'awfully difficult', involving metaphysical considerations of an abstruse nature, only to be thoroughly understood by consummately profound metaphysicomathematicians, such as Prof. Tait, for example. Well, if so, there would not be the slightest chance for vector algebra and analysis to ever become generally useful.

Of course, such intemperate exchanges made it all the more difficult for competing views to be reconciled. So there followed a long 'struggle for existence' (Gibbs' words) for the new system of vector calculus, with a very public and often acrimonious exchange of views between Tait and Heaviside. However, by 1910 common sense had prevailed, with the

Gibbs–Heaviside system broadly accepted (see §1.6). It has ever since been used with little modification, and has become the workhorse of electrodynamics.

It is interesting that Heaviside led the charge in both of these skirmishes, paving the way for modern electrodynamics. Sadly (and there may be a moral here), Heaviside paid a price for taking on the scientific establishment. Self-educated, of humble origins and stubbornly independent, he relished provoking the mathematicians a little too much, and so made several enemies in the process. At the time of his death in 1925, the *Encyclopaedia Britannica* listed amongst his many achievements (i) the prediction in 1902, along with the American A. E. Kennelly, of a conducting layer in the upper atmosphere, then called the Heaviside (or Kennelly–Heaviside) layer, whose existence was confirmed experimentally in 1924, (ii) his work on transmission lines (starting with an audacious paper at age 26), which proved pivotal in the development of practical telegraphy, (iii) predicting, in 1889, the change in mass of charges moving at relativistic speeds, predating special relativity, and (iv) predicting, in 1902, the electromagnetic field of, and radiation from, accelerating charges moving at relativistic speeds. A century later, however, the Heaviside layer has been renamed the ionosphere, Heaviside's equations describing the propagation of signals along transmission lines are known simply as the telegrapher's equations, and undergraduate textbooks tend to associate the names of several well-known physicists with the fields of accelerating charges, but until the third edition of Jackson's classic monograph appeared in 1999, one rarely saw Heaviside's name listed amongst them. Indeed, virtually the only public acknowledgment Heaviside has been left with is the humble Heaviside step function. So perhaps the moral is, if you want history to treat you kindly, don't pick fights with the great and the good!

On that cautionary note, perhaps it is time to set aside Maxwell and the Maxwellians and return to those more conventional matters promised at the beginning of this preface. Regards the content of the book, this is intended as a *second-level* course for advanced undergraduates in physics, applied mathematics and engineering science. It is assumed that the reader has already taken an introductory course in quasi-static electric and magnetic fields and is familiar with circuit theory, and with the laws of Coulomb, Ampère and Faraday. It is also assumed that the reader is well versed in vector calculus, which is used ruthlessly throughout the text. Those readers who are a little rusty on this topic would do well to work through Chapter 1 before embarking on subsequent chapters.

Chapter 2 is somewhat unusual for texts on electrodynamics and perhaps needs a little explanation. There are many phenomena observed in electromagnetism that are *not* peculiar to the subject, but rather manifest themselves across a broad range of physical systems. These are often best introduced and understood in the context of a simpler mechanical system, such as acoustics. Examples include d'Alembert's solution of the wave equation, the idea of the dispersion of wave energy (and the associated concept of group velocity) in a wave-bearing system *not* governed by the simple wave equation, the inversion of Poisson's equation in the spirit of Gauss and Biot–Savart, and the use of retarded potentials to describe the propagation of waves from localised sources. These ubiquitous concepts are much more readily understood in the context of simple mechanical systems, such as acoustics, and so they have been grouped together and discussed in Chapter 2 under the banner of the properties of partial differential equations.

Chapter 3 may look a little precipitous, if not ambitious, since it goes straight to the heart of Maxwell's equations. It is intended to provide the reader with an overview that helps put the many specific results of subsequent chapters into perspective. In particular, what one learns in the context of quasi-static fields is often incorrect when time dependence is introduced, and it is important to understand this from the outset. So Chapter 3 is intended to help establish a roadmap for the rest of the book.

The remainder of the book follows a fairly conventional outline, starting with electrostatics and magnetostatics, working though Faraday's law and Maxwell's equations, and ending with waves, radiation and special relativity. The various applications, particularly MHD, are dealt with on route. The text is not meant to be encyclopaedic. Indeed, in an attempt to restrict the overall length, some topics are given only a cursory treatment (optical fibres and antenna arrays), whilst others have been omitted completely (diffraction and Cherenkov radiation). I hope the reader will forgive these omissions and take solace from the fact that these topics are covered in depth elsewhere.

The emphasis throughout is on physical ideas, rather than mathematical structure. Where mathematics helps expose an idea, say in radiation or the field of moving charges, it is used to full effect, but in those cases where it is in danger of obscuring an idea, such as optical fibres, it takes a back seat. The author makes no apology for this approach, though he has lived long enough to know that this will not be to everyone's taste.

Perhaps the matter of notation also requires a brief comment. In electromagnetism, more than any other subject, one rapidly runs out of symbols. In order to cope with this, some authors use a wide range of subscripts or invoke unusual fonts for particular quantities. Both strategies can become irksome if overused by an author determined never to use the same symbol for more than one quantity. We have adopted a more pragmatic approach. For example, $V$ is used both for volume, as in $dV$ in a volume integral, and for voltage, on the grounds that the reader is certain to know in a given context which is intended. Just in case there is any confusion, a list of commonly used symbols is provided after this preface.

I would like to thank Andrew Flewitt for several helpful suggestions, and the team at Oxford University Press, who were a delight to work with. Finally, I must confess that I am blessed with a long-suffering wife who patiently endured those long monastic silences that frequently accompanied the writing of this book.

P. A. Davidson, Cambridge, 2018

# CONTENTS

# COMMONLY USED SYMBOLS

## Geometry

| | | |
|---|---|---|
| position vector: $\mathbf{x}$ | displacement vector: $\mathbf{r}$ | unit vector: $\hat{\mathbf{e}}$ |
| element of length: $d\mathbf{r}$ | surface element: $d\mathbf{S}$ | volume element: $d\mathbf{x} = dx\,dy\,dz$ |

A prime is used to distinguish between the location of an observer at $\mathbf{x}$ and a source (charge or current) at $\mathbf{x}'$. The displacement vector $\mathbf{r} = \mathbf{x} - \mathbf{x}'$ then takes us from the source at $\mathbf{x}'$ to the observer at $\mathbf{x}$. A prime is also used to distinguish between the time $t$ at the observer and the retarded time $t'$ at the source, so that $t' = t - |\mathbf{r}|/c$.

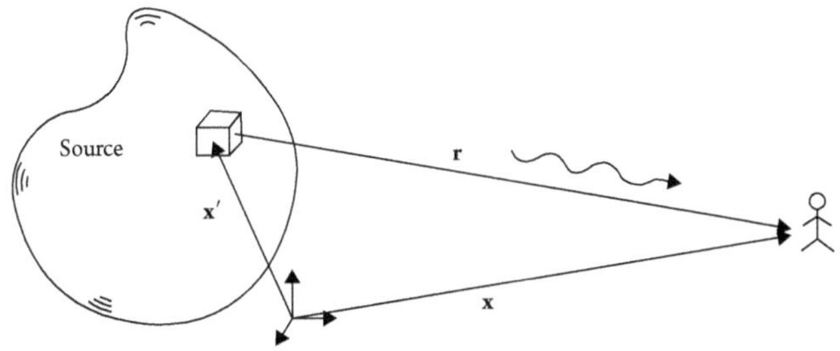


## Polar Coordinates

Cylindrical polar coordinates are $(s, \theta, z)$, where $\theta$ is the azimuthal angle. Spherical polar coordinates are $(r, \theta, \phi)$, where $\theta$ is the polar angle and $\phi$ the azimuthal angle.

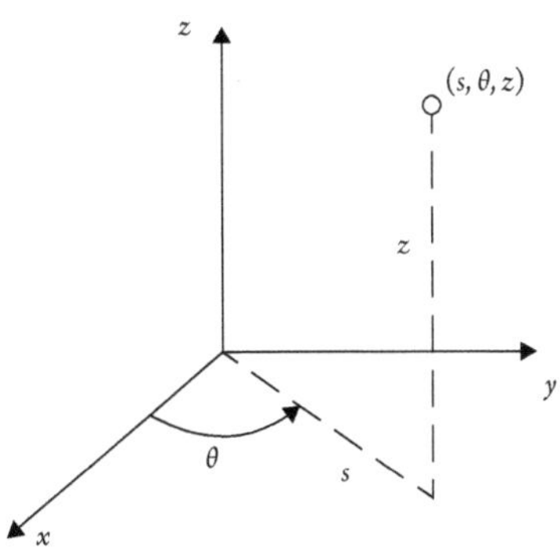


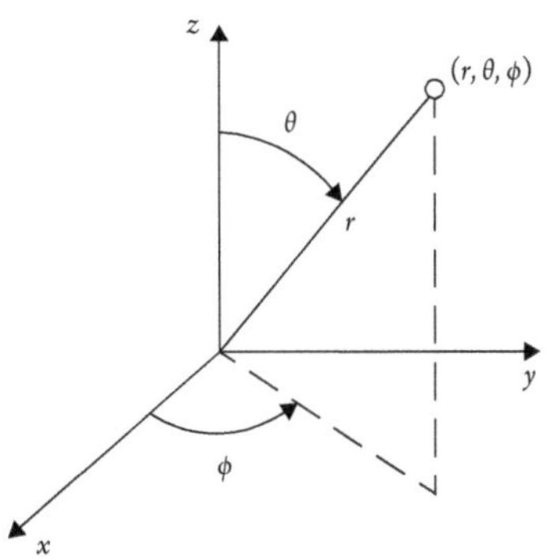

## Electrical Circuits

voltage: $V$
current: $I$
resistance: $R$
capacitance: $C$
inductance: $L$
impedance: $Z$

## Physical and Material Constants

permittivity: $\varepsilon$
permittivity of free space: $\varepsilon_0$
relative permittivity: $\varepsilon_r = \varepsilon/\varepsilon_0$
electric susceptibility: $\chi_e = \varepsilon_r - 1$
electrical conductivity: $\sigma$
charge relaxation time: $\tau_e = \varepsilon_0/\sigma$
permeability: $\mu$
permeability of free space: $\mu_0$
relative permeability: $\mu_r = \mu/\mu_0$
magnetic susceptibility: $\chi_m = \mu_r - 1$
mass density: $\rho$
magnetic diffusivity: $\lambda = 1/\sigma\mu_0$

## Fields

electric field: $\mathbf{E}$
electric polarisation: $\mathbf{P}$
electric displacement: $\mathbf{D}$
magnetic field: $\mathbf{B}$
magnetisation: $\mathbf{M}$
auxiliary magnetic field: $\mathbf{H}$
current density: $\mathbf{J}$
vector potential: $\mathbf{A}$
scalar potential: $\varphi$

## Charge and Current

point charge: $q$
total charge: $Q$
charge of an electron: $-q_e$
charge density: $\rho_e$
free charge density: $\rho_f$
bound charge density: $\rho_b$
surface charge density: $\sigma_e$
free surface charge: $\sigma_f$
bound surface charge: $\sigma_b$
free current density: $\mathbf{J}_f$
polarisation current: $\mathbf{J}_{\text{pol}}$
bound current density: $\mathbf{J}_{\text{mag}}$
surface current density: $\mathbf{K}$
free surface current: $\mathbf{K}_f$
bound surface current: $\mathbf{K}_{\text{mag}}$

## Energy and Force

potential energy: $U$
electric energy: $U_E$
magnetic energy: $U_B$
force on charge: $\mathbf{f}$
force per unit volume: $\mathbf{F}$
torque: $\mathbf{T}$

## Velocities and Waves

speed of sound: $a$
speed of light in vacuum: $c$
speed of light in dielectric: $\upsilon$
velocity of a medium: $\mathbf{u}$
velocity of a charge $q$: $\mathbf{u}_q$
velocity of prescribed curve: $\mathbf{v}$
phase velocity: $\mathbf{c}_p$
group velocity: $\mathbf{c}_g$
angular frequency: $\varpi$
wavenumber: $k = \varpi/c_p$
wave vector: $\mathbf{k}$
wavelength: $\lambda = 2\pi/k$

. . . and to those philosophers who pursue this inquiry zealously yet continuously, combining experiment with analogy, suspicious of their preconceived notions, paying more respect to the fact than a theory, not too hasty to generalise, and above all things, willing at every step to cross-examine their own opinions, both by reasoning and experiment, no branch of knowledge can afford so fine and ready a field for discovery as this.

Michael Faraday, 1837

# PART I

## Mathematical Preliminaries

# 1

# Vector Calculus

Prof. Willard Gibbs must be ranked as one of the retarders of Quaternion progress, in virtue of his pamphlet on Vector Analysis; a sort of hermaphrodite monster.

P. G. Tait, 1890

I came later to see that, so far as the vector analysis I required was concerned, the quaternion was not only not required, but was a positive evil of no inconsiderable magnitude.

O. Heaviside, 1893

Vector calculus, as we now know it, was largely developed in the 1880s by two physicists: Josiah Willard Gibbs in the USA and Oliver Heaviside in the UK. Initially, their work encountered considerable resistance from factions within the scientific community, most famously from the applied mathematician and physicist Peter Guthrie Tait, who preferred quaternions. However, by 1910 the Gibbs–Heaviside system was reasonably well established and today it dominates much of the physical sciences, being the natural mathematical language of vector fields. In this chapter we offer a brief overview of vector calculus, aimed at those who have only a passing acquaintance with the topic.

## 1.1 Vectors and Vector Algebra

Vector calculus is concerned with the analysis of vector fields, that is, vector quantities like magnetic or electric fields that are continuously spread across space, taking different values at different points. However, perhaps it is helpful to start with the simpler ideas of a point vector and of vector algebra, starting with some elementary definitions.

### *1.1.1 Elementary Definitions and Properties of Vectors*

A vector is a quantity, such as force or velocity, that has both magnitude and direction. It is usually indicated by bold type, such as $\mathbf{F}$, whereas its magnitude is written as either $|\mathbf{F}|$ or $F$. Note that the notion of direction contains within it two distinct concepts: that of a line of action and that of a sense along that line of action. For example, it is not enough to say that a force acts along a north–south axis. Rather, to uniquely determine the direction of the force, we need to specify if it points to the north or to the south. This may seem a trivially obvious point, verging on the pedantic, but it turns out to be important.

The prototype vector, whose properties are intuitively obvious, and which we want all other vectors to emulate, is the displacement vector $\mathbf{r}$. So the addition of two vectors $\mathbf{a}$ and $\mathbf{b}$ is defined in a way that is consistent with adding two displacement vectors. That is, the two vectors are positioned nose-to-tail and $\mathbf{a}+\mathbf{b}$ is defined as that vector which takes us from the start of the first vector to the end of the second. Vector addition is clearly commutative: $\mathbf{a}+\mathbf{b}=\mathbf{b}+\mathbf{a}$. We also define $-\mathbf{a}$ to have the same magnitude and line of action as $\mathbf{a}$, but the opposite sense along that line of action, i.e. it points in the opposite direction to $\mathbf{a}$. The act of subtracting $\mathbf{a}$ from $\mathbf{b}$ then follows.

The product of a positive scalar $\rho$ with a vector $\mathbf{a}$ is simply defined as a vector that has the same direction as $\mathbf{a}$, but whose amplitude is $\rho\,|\mathbf{a}|$. Two vectors, on the other hand, can be multiplied in one of two ways. The dot product (or scalar product) of $\mathbf{a}$ and $\mathbf{b}$, denoted by $\mathbf{a}\cdot\mathbf{b}$, is a scalar whose magnitude is

$$\mathbf{a}\cdot\mathbf{b}=ab\cos\theta, \tag{1.1}$$

where $\theta$ is the angle between the two vectors when positioned tail-to-tail. Clearly $\mathbf{a}\cdot\mathbf{a}=a^2$, while two mutually perpendicular vectors have a dot product of zero. Moreover, the scalar product is commutative, $\mathbf{a}\cdot\mathbf{b}=\mathbf{b}\cdot\mathbf{a}$, and it is readily shown to be distributive,

$$\mathbf{a}\cdot(\mathbf{b}+\mathbf{c})=\mathbf{a}\cdot\mathbf{b}+\mathbf{a}\cdot\mathbf{c}. \tag{1.2}$$

By way of contrast, the cross product (sometimes called the vector product) is denoted by $\mathbf{a}\times\mathbf{b}$ and is a vector of magnitude

$$|\mathbf{a}\times\mathbf{b}|=ab\,|\sin\theta|, \tag{1.3}$$

so that $\mathbf{a}\times\mathbf{a}=0$. The line of action of $\mathbf{a}\times\mathbf{b}$ is the line normal to the plane containing $\mathbf{a}$ and $\mathbf{b}$ and its sense along that line is, by convention, determined by the right-hand rule, that is, in the direction of advance of a right-hand screw rotated as $\mathbf{a}$ swings to $\mathbf{b}$ through the smallest possible angle. This is illustrated in Figure 1.1. The vector product is readily shown to be distributive,

$$\mathbf{a}\times(\mathbf{b}+\mathbf{c})=\mathbf{a}\times\mathbf{b}+\mathbf{a}\times\mathbf{c}, \tag{1.4}$$

but it is not commutative. Rather, our right-hand rule demands that $\mathbf{a}\times\mathbf{b}=-\mathbf{b}\times\mathbf{a}$.

Note that vectors defined through the cross product are, in some sense, different to conventional vectors, such as force $\mathbf{F}$ or displacement $\mathbf{r}$. Both $\mathbf{F}$ and $\mathbf{r}$ have an unambiguous magnitude and direction, whereas a vector like torque $\mathbf{T}=\mathbf{r}\times\mathbf{F}$ has a definite magnitude and line of action, but its sense along that line of action is determined by an arbitrary sign convention, the right-hand rule. Physically, this means that a quantity like $\mathbf{T}$ does not really have a unique direction, but rather it possess a line of action and we choose to attribute a definite direction to $\mathbf{T}$ by adopting the arbitrary convention of the right-hand rule. Other

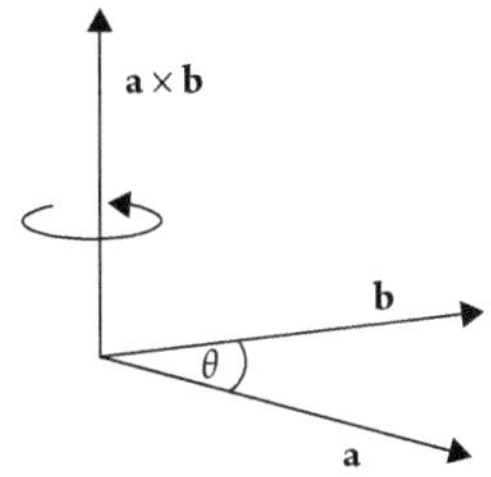


**Figure 1.1** The direction of $\mathbf{a} \times \mathbf{b}$ as determined by the right-hand rule.

quantities that behave like this are angular velocity and angular momentum, which are also defined through a cross product. So it seems that we have two distinct classes of vectors. One class, called *true vectors* (or *polar vectors*), are fully paid-up members of the vector family in that they have an unambiguous direction. The other class, called *axial vectors*, are defined through the cross product of two polar vectors and have a direction that is, in part, determined by an arbitrary sign convention. (Axial vectors are associated with planes, rather than directed lines.) It turns out that magnetic fields belong to this second class. So, if one day we all conspired to use the left-hand convention to define the cross product, all magnetic field lines would reverse direction!

The idea of an axial vector leads on to the related concept of a *pseudoscalar*. Suppose that we take the dot product of an axial vector, say $\mathbf{T}$, with a true vector, say velocity $\mathbf{u}$. Then $\mathbf{T}$ reverses direction on shifting from the right-hand to the left-hard rule, whereas $\mathbf{u}$ does not. It follows that the scalar product $\mathbf{u} \cdot \mathbf{T}$ changes sign upon changing convention. Scalars that have this disconcerting property are called pseudoscalars.

There is another subtlety that is worth mentioning at this point. Not every physical quantity that has magnitude and direction is a vector. Consider, for example, a rotation of 90° about a vertical axis. It has magnitude and a line of action, and we can give it a sense along its line of action using the right-hand rule. Suppose, therefore, that we label this rotation as $\mathbf{A}$ and treat it as a vector. Next consider a second rotation, this time of 90° about a horizontal axis pointing away from you, which we label as $\mathbf{B}$. We now perform a trivial experiment. Pick up a book and apply rotation $\mathbf{A}$ followed by $\mathbf{B}$, noting the orientation of the book after the composite action of $\mathbf{A}$ plus $\mathbf{B}$, which we write as $\mathbf{A} + \mathbf{B}$. Now return the book to its original orientation and apply the same two rotations in the reverse order, which we write as $\mathbf{B} + \mathbf{A}$. You will find that the book ends up in a different final orientation and we conclude that $\mathbf{A} + \mathbf{B}$ does not equal $\mathbf{B} + \mathbf{A}$, which contradicts the law of vector addition. Evidently finite rotations are not vectors. To get around this problem, we say a physical quantity is a vector if it has magnitude and direction *and* obeys the law of vector addition.

Having defined the scalar and cross products of two vectors, it is natural to ask how three vectors $\mathbf{a}$, $\mathbf{b}$ and $\mathbf{c}$ may be multiplied, which takes us to the so-called triple products. There are two types of triple product: the scalar triple product $\mathbf{a} \cdot (\mathbf{b} \times \mathbf{c})$ and the vector triple product $\mathbf{a} \times (\mathbf{b} \times \mathbf{c})$. These are readily shown to obey the following convenient rules:

$$\mathbf{a} \cdot (\mathbf{b} \times \mathbf{c}) = \mathbf{b} \cdot (\mathbf{c} \times \mathbf{a}) = \mathbf{c} \cdot (\mathbf{a} \times \mathbf{b}), \tag{1.5}$$

$$\mathbf{a} \times (\mathbf{b} \times \mathbf{c}) = (\mathbf{a} \cdot \mathbf{c})\mathbf{b} - (\mathbf{a} \cdot \mathbf{b})\mathbf{c}. \tag{1.6}$$

Note that the cyclic order of the vectors must be retained in (1.5). If it is not, a minus sign appears. Note also that (1.5) is a pseudoscalar if **a**, **b** and **c** are all true vectors.

We close this section by defining the ordinary derivative of a vector. Suppose the velocity **u** of a projectile is a known function of time, $\mathbf{u} = \mathbf{u}(t)$. Its time derivative at time $t$ is simply

$$\frac{d\mathbf{u}}{dt} = \lim_{\delta t \to 0} \frac{\mathbf{u}(t + \delta t) - \mathbf{u}(t)}{\delta t}, \tag{1.7}$$

where $\delta t$ is a short time increment. Alternatively, if **u** is a known function of position $s$ along its trajectory, then

$$\frac{d\mathbf{u}}{ds} = \lim_{\delta s \to 0} \frac{\mathbf{u}(s + \delta s) - \mathbf{u}(s)}{\delta s}. \tag{1.8}$$

If $f$, **a** and **b** are all known functions of a scalar $s$, then it may be shown that the derivatives of products of $f$, **a** and **b** follow the expected rules, such as

$$\frac{d}{ds}(f\mathbf{a}) = f\frac{d\mathbf{a}}{ds} + \frac{df}{ds}\mathbf{a}, \tag{1.9}$$

$$\frac{d}{ds}(\mathbf{a} \cdot \mathbf{b}) = \mathbf{a} \cdot \frac{d\mathbf{b}}{ds} + \frac{d\mathbf{a}}{ds} \cdot \mathbf{b}, \tag{1.10}$$

$$\frac{d}{ds}(\mathbf{a} \times \mathbf{b}) = \mathbf{a} \times \frac{d\mathbf{b}}{ds} + \frac{d\mathbf{a}}{ds} \times \mathbf{b}. \tag{1.11}$$

### *1.1.2 Vectors in Cartesian Coordinate Systems*

Note that, so far, none of our definitions have involved a coordinate system, and so they are known as *coordinate-free definitions*. This is important because, in a given frame of reference, the laws of nature, and by implication the physical quantities that appear in them, must be independent of the choice of coordinate system in that frame. So vectors have a life that is independent of any coordinate system we might choose to adopt. Nevertheless, it is often convenient to introduce a coordinate system and work with the components of a vector within that system.

The simplest such coordinate system is the Cartesian one in which we have three mutually orthogonal axes $x$, $y$ and $z$, and three unit vectors **i**, **j** and **k** that point along the $x$, $y$ and $z$ axes, respectively. Note that the convention is to always use a *right-handed coordinate system* in which $\mathbf{i} \times \mathbf{j} = \mathbf{k}$, $\mathbf{j} \times \mathbf{k} = \mathbf{i}$ and $\mathbf{k} \times \mathbf{i} = \mathbf{j}$, and it turns out to be important to conform to this convention, as we shall see. (By contrast, a left-handed coordinate system has $\mathbf{i} \times \mathbf{j} = -\mathbf{k}$ etc. under the right-hand rule, and retains $\mathbf{i} \times \mathbf{j} = \mathbf{k}$ *only if we switch to the left-hand rule for cross products*.)

Let $a_x$, $a_y$ and $a_z$ be the projections of a vector **a** along the $x$, $y$ and $z$ axes, respectively. Then the law of vector addition tells us that we can write **a** as

$$\mathbf{a} = a_x\mathbf{i} + a_y\mathbf{j} + a_z\mathbf{k},$$

where $a_x$, $a_y$ and $a_z$ are known as the components of **a** in this particular coordinate system. For example, the position vector **r** is the displacement vector that takes us from the origin to the point $(x, y, z)$, and is clearly given by

$$\mathbf{r} = x\mathbf{i} + y\mathbf{j} + z\mathbf{k}.$$

Often it is convenient to perform vector operations in terms of the components of a vector, rather than work with the vector itself. For example, suppose the vector **a** is a known function of time, $\mathbf{a} = \mathbf{a}(t)$, so that its components are also known functions of $t$. Since **i**, **j** and **k** are all constants, (1.9) tells us that the time derivative of **a** in component form is simply

$$\frac{d\mathbf{a}}{dt} = \frac{da_x}{dt}\mathbf{i} + \frac{da_y}{dt}\mathbf{j} + \frac{da_z}{dt}\mathbf{k}.$$

Component representation is particularly useful when it comes to evaluating products of vectors. For example, the scalar product of **a** and **b** is, using the distributive law (1.2),

$$\mathbf{a} \cdot \mathbf{b} = a_x b_x + a_y b_y + a_z b_z. \tag{1.12}$$

Moreover, the distributive law (1.4) allows us to write the vector product of **a** and **b** as

$$\mathbf{a} \times \mathbf{b} = (a_y b_z - a_z b_y)(\mathbf{j} \times \mathbf{k}) + (a_z b_x - a_x b_z)(\mathbf{k} \times \mathbf{i}) + (a_x b_y - a_y b_x)(\mathbf{i} \times \mathbf{j}), \tag{1.13}$$

and the fact that $\mathbf{i} \times \mathbf{j} = \mathbf{k}$, $\mathbf{j} \times \mathbf{k} = \mathbf{i}$ and $\mathbf{k} \times \mathbf{i} = \mathbf{j}$ in a right-handed coordinate system yields

$$\mathbf{a} \times \mathbf{b} = (a_y b_z - a_z b_y)\,\mathbf{i} + (a_z b_x - a_x b_z)\mathbf{j} + (a_x b_y - a_y b_x)\,\mathbf{k}. \tag{1.14}$$

This is often rewritten symbolically in determinant form as

$$\mathbf{a} \times \mathbf{b} = \begin{vmatrix} \mathbf{i} & \mathbf{j} & \mathbf{k} \\ a_x & a_y & a_z \\ b_x & b_y & b_z \end{vmatrix}, \tag{1.15}$$

where it is understood that one expands about the top row. Note, however, that (1.14) and (1.15) do not apply in a left-handed coordinate system unless we switch to a left-hand rule for cross products. For example, if we retain the right-hand rule to define cross products in left-handed coordinate systems, then we have $\mathbf{i} \times \mathbf{j} = -\mathbf{k}$, and so on. In such cases (1.13) demands that a minus sign appear in front of the determinant in (1.15).

It is disconcerting that the expression relating the components of a cross product, (1.14) or (1.15), might have to be modified to include a minus sign if we move from right-handed to left-handed coordinates. One way to avoid this difficulty is to agree that, whenever we find ourselves in a left-handed coordinate system, we switch from a right-hand to a left-hand rule in the definition of the cross product. In that way $\mathbf{i} \times \mathbf{j} = \mathbf{k}$ continues to hold in the left-handed coordinate system and so (1.13) tells us that (1.14) and (1.15) remain valid in both coordinate systems. The cross product $\mathbf{a} \times \mathbf{b}$ defined in this way is sometimes called a *pseudovector*, a subclass of the family of pseudotensors. (Indeed, strictly speaking, a cross product is not a vector at all, but rather an antisymmetric second-rank tensor.) Unfortunately, this strategy has the effect of reversing the physical direction of $\mathbf{a} \times \mathbf{b}$ whenever we change to left-handed coordinates, because $\mathbf{a} \times \mathbf{b}$ must now be evaluated using a left-hand rule. However, this need not overly concern us, because physically $\mathbf{a} \times \mathbf{b}$ really only has a line of action, and not a unique sense along that line of action. In adopting such a strategy we have conceded that $\mathbf{a} \times \mathbf{b}$ is not really a vector, whose physical direction is independent of the coordinate system. In any event, this convention is commonly used, so that $\mathbf{a} \times \mathbf{b}$, defined as a pseudovector, is deemed to have reversed direction on moving to a left-handed coordinate system. (Of course, an alternative and simpler strategy is to simply outlaw all left-handed coordinate systems!)

Now consider two *right-handed* Cartesian coordinate systems that share a common origin and have the same $z$ axis, but have different orientations of their $x$ and $y$ axes. Let system 1 have coordinates $(x, y, z)$ and unit vectors $\mathbf{i}$, $\mathbf{j}$ and $\mathbf{k}$, while system 2 locates a point using coordinates $(x', y', z)$ and unit vectors $\mathbf{i}'$, $\mathbf{j}'$ and $\mathbf{k}$. Suppose that system 2 is rotated relative to system 1 by an angle $\theta$, the right-hand rule being used to define the positive direction of $\theta$ relative to the $z$ axis. Evidently the unit vectors in the two systems are related by $\mathbf{i} \cdot \mathbf{i}' = \cos\theta$, $\mathbf{j} \cdot \mathbf{i}' = \sin\theta$, $\mathbf{i} \cdot \mathbf{j}' = -\sin\theta$ and $\mathbf{j} \cdot \mathbf{j}' = \cos\theta$. Now, by definition, any vector $\mathbf{a}$ must be the same in both coordinate systems, and so

$$\mathbf{a} = a_x\mathbf{i} + a_y\mathbf{j} + a_z\mathbf{k} = a_{x'}\mathbf{i}' + a_{y'}\mathbf{j}' + a_z\mathbf{k}. \tag{1.16}$$

Taking the dot product of (1.16) first with $\mathbf{i}'$ and then with $\mathbf{j}'$ shows that the components of $\mathbf{a}$ satisfy the geometrical transformation laws

$$a_{x'} = a_x \cos\theta + a_y \sin\theta,$$
$$a_{y'} = -a_x \sin\theta + a_y \cos\theta.$$

If they did not transform in this way, the vector would not be the same (would not have the same length and direction) in the two coordinate systems. So vectors can be identified by the way in which their components transform under a rotation of the coordinate system. This generalises in an obvious way to three-dimensional rotations, with

$$a_{x'} = a_x(\mathbf{i} \cdot \mathbf{i}') + a_y(\mathbf{j} \cdot \mathbf{i}') + a_z(\mathbf{k} \cdot \mathbf{i}'), \tag{1.17}$$

and so on for $a_{y'}$ and $a_{z'}$. Note that the dot products in (1.17) are the cosines of the angles between the various unit vectors in the two coordinate systems. Some authors regard this geometrical transformation law as so important that they like to use it to define a vector. That is to say, a vector is defined to be any set of three components that transforms according to (1.17), along with the equivalent expressions for $a_{y'}$ and $a_{z'}$. It is readily confirmed that the dot product is an invariant (independent of coordinate system) under this transformation law,

$$\mathbf{a} \cdot \mathbf{b} = a_x b_x + a_y b_y + a_z b_z = a_{x'} b_{x'} + a_{y'} b_{y'} + a_{z'} b_{z'}, \tag{1.18}$$

as it must be. A rather longer calculation is required to show that the cross product is also an invariant under this transformation rule,

$$\mathbf{a} \times \mathbf{b} = \begin{vmatrix} \mathbf{i} & \mathbf{j} & \mathbf{k} \\ a_x & a_y & a_z \\ b_x & b_y & b_z \end{vmatrix} = \begin{vmatrix} \mathbf{i}' & \mathbf{j}' & \mathbf{k}' \\ a_{x'} & a_{y'} & a_{z'} \\ b_{x'} & b_{y'} & b_{z'} \end{vmatrix}. \tag{1.19}$$

### *1.1.3 Mirror Reflections: Polar Versus Axial Vectors Revisited*

It is of interest to consider what happens when we view vectors, and the space that they inhabit, in a mirror. This notion of a mirror image brings us back to the idea that there are two distinct classes of vectors: *polar vectors* (or true vectors) and *axial vectors*.

Consider the displacement vector **r**, viewed in a mirror, as shown in Figure 1.2(a). Now compare this with the angular velocity of a disc, $\boldsymbol{\Omega}$, and its mirror image (Figure 1.2(b)). In the latter case we have used the right-hand rule to determine the direction of the angular velocity vector from the direction of rotation in both the original and the mirror images. Clearly the two vectors transform in different ways. The true vector **r** transforms in the way you would expect, reversing the direction of its component normal to the mirror, whilst the axial vector $\boldsymbol{\Omega}$ does not reverse its component normal to the plane of reflection. It reflects in quite a different way. We conclude that polar and axial vectors behave in different ways when reflected in a mirror, and this is yet another manifestation of the fact that, unlike true vectors, axial vectors do not really have an unambiguous direction along their line of action, but rather we attribute one to them through the arbitrary choice of the right-hand rule.

Some authors prefer to discuss the distinction between true and axial vectors by considering their transformation properties under a reflection of the coordinate system, rather than though their properties in a mirror image. For example, what happens if we reverse the $x$ axis? Such an approach is more in line with our discussion of rotated coordinate systems in §1.1.2, in which the vector remains fixed in space and we change the coordinate system in which we choose to view that vector. (Such a strategy is referred to as a *passive* approach to reflections.) In particular, it is informative to consider the transformation properties of the

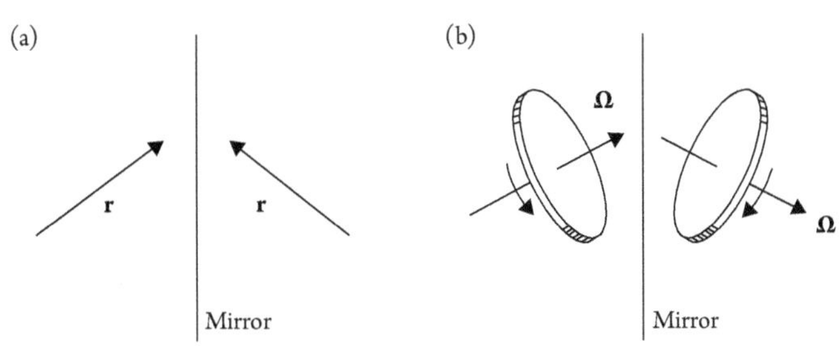


**Figure 1.2** (a) The displacement vector **r** and its mirror image. (b) A rotating disc and its mirror image, with the angular velocity $\boldsymbol{\Omega}$ determined by the right-hand rule. Note that, unlike **r**, the axial vector $\boldsymbol{\Omega}$ is not properly reflected in the mirror.

cross product under an *inversion* of the coordinates, in which we reverse the $x$, $y$ and $z$ axes and their corresponding unit vectors.

Note that such a transformation changes our right-handed coordinate system into a left-handed one. So, if we decide to adopt the convention discussed in §1.1.2, of using the left-hand rule to define the cross product in a left-handed coordinate system, then (1.19) should remain valid under an inversion of the coordinates, but all pseudovectors should reverse their physical direction. Let us see if this is so. Under an inversion of the coordinates, we have

$$\mathbf{i} \cdot \mathbf{i}' = \mathbf{j} \cdot \mathbf{j}' = \mathbf{k} \cdot \mathbf{k}' = -1, \tag{1.20}$$

and so (1.17) tells us that the components of a true vector transform according to $a_{x'} = -a_x$, and so on. (As we reverse the unit vectors, so we must multiply each component of **a** by $-1$ in order to keep the direction of the original vector fixed in space.) Clearly, under such a transformation, (1.18) remains valid and the dot product is an invariant. So what about the cross product? Let us see what happens when we use (1.19) to determine the cross product under an inversion of the coordinates. If **a** and **b** are true vectors, then $a_{x'} = -a_x$, $b_{x'} = -b_x$ and so on, and so the components of $\mathbf{c} = \mathbf{a} \times \mathbf{b}$ do *not* reverse sign according to (1.19). For example, (1.19) yields

$$c_{z'} = a_{x'} b_{y'} - a_{y'} b_{x'} = a_x b_y - a_y b_x = c_z. \tag{1.21}$$

Now the unit vectors change direction under our transformation and so (1.21) demands that the pseudovector **c** change its physical direction under an inversion of the coordinate system. This is exactly what we would expect from the discussion in §1.1.2: in order to ensure that (1.19) holds in both left- and right-handed coordinate systems, we agree to use the left-hand rule to define cross products in left-handed coordinates. It follows that **c** must reverse its physical direction under an inversion of the coordinates.

### *1.1.4 From Point Vectors to Vector Fields*

We close this section by moving from point vectors to vector fields. The classic example of a vector field is the velocity distribution within a moving fluid, $\mathbf{u} = \mathbf{u}(x, y, z)$. At each point in space **u** takes a particular value and in general this value varies continuously with position. In fact, **u** could also be a function of time, $\mathbf{u} = \mathbf{u}(\mathbf{r}, t)$, but for the present we shall neglect any time dependence.

Field lines are a way of visualising a vector field and are lines drawn in space such that at every point the line is tangent to the local vector field. In fluid mechanics these are called streamlines, while an example from electromagnetism is the magnetic field line. If the spatial distribution of a vector field is known, then the equations for the corresponding field lines may be obtained as follows. Let $d\mathbf{r}$ be the infinitesimal displacement vector that links two adjacent points on a field line, $\mathbf{r}$ and $\mathbf{r} + d\mathbf{r}$, drawn within the vector field $\mathbf{u} = \mathbf{u}(x, y, z)$. Then, by definition, $d\mathbf{r} = \lambda \mathbf{u}(\mathbf{r})$ for some $\lambda$ and it follows that

$$\frac{dx}{u_x} = \frac{dy}{u_y} = \frac{dz}{u_z}. \tag{1.22}$$

If the spatial dependences of $u_x$, $u_y$ and $u_z$ are known, then (1.22) may be integrated to find the instantaneous trajectory of the field lines, as discussed in Exercise 1.1.

One consequence of moving from point vectors to vector fields is that ordinary derivatives cease to be of much interest, and attention shifts to partial derivatives. However, this involves only a minor extension of the definition (1.8). For example, if we want the partial derivative of **u** with respect to $x$, keeping $y$ and $z$ constant, we simply adapt the definition (1.8) to read

$$\frac{\partial \mathbf{u}}{\partial x} = \lim_{\delta x \to 0} \frac{\mathbf{u}(x + \delta x, y, z) - \mathbf{u}(x, y, z)}{\delta x}. \tag{1.23}$$

The rules for differentiating products, such as (1.9)–(1.11), also extend in an obvious way to partial derivatives, as does the idea of differentiating a vector through differentiating its Cartesian components:

$$\frac{\partial \mathbf{u}}{\partial x} = \frac{\partial u_x}{\partial x}\mathbf{i} + \frac{\partial u_y}{\partial x}\mathbf{j} + \frac{\partial u_z}{\partial x}\mathbf{k}. \tag{1.24}$$

## 1.2 Line and Surface Integrals

We now turn from vector algebra to the calculus of vector fields, starting with line and surface integrals.

### *1.2.1 Line Integrals, Conservative Fields and the Scalar Potential*

Suppose we have a vector field $\mathbf{B}(x, y, z)$, and in the space in which this field lives we draw a curve C that runs from point P to point Q. Note that this curve is, in general, independent of **B** and in particular it need not follow a field line of **B** (Figure 1.3(a)). We define the *line integral* of **B** along the curve C as follows. We chop up C into lots of infinitesimal displacement vectors $d\mathbf{r}$, each of which links two adjacent points on the curve, **r** and $\mathbf{r} + d\mathbf{r}$, as shown in Figure 1.3(b). For each $d\mathbf{r}$ we form the dot product $\mathbf{B}(\mathbf{r}) \cdot d\mathbf{r}$. In other words, we take the component of **B** tangent to the curve and multiply it by the length $|d\mathbf{r}|$. We now add all such contribution along the curve from P to Q. In the limit as each $|d\mathbf{r}| \to 0$ the resulting summation is called the line integral of **B** along C from P to Q, denoted by

$$\Gamma = \int_{P,\,C}^{Q} \mathbf{B} \cdot d\mathbf{r}. \tag{1.25}$$

Note that the symbol $\Gamma$ is often used for line integrals. Note also that, in general, the value of $\Gamma$ depends not only on **B** and on the start and end points, but also on the curve C taken to get from P to Q (see Exercise 1.2).

When the vector field in question happens to be a force **F**, the line integral is clearly the work done by **F** on a particle in moving through the field along curve C from P to Q.

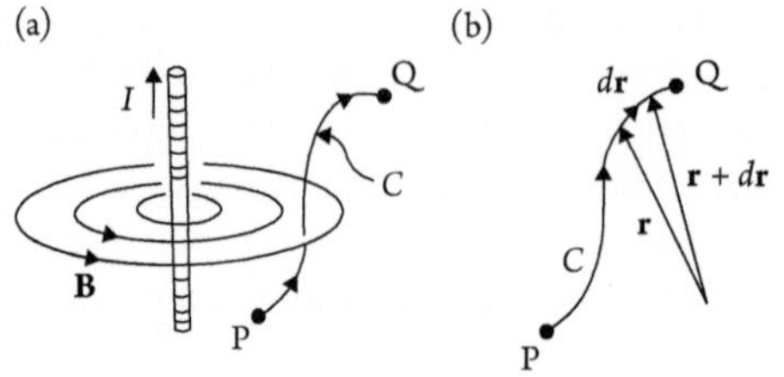


**Figure 1.3** (a) The path along which a line integral is to evaluated need not be a field line. (b) A line integral is constructed from many infinitesimal displacement vectors $d\mathbf{r}$.

However, line integrals need not always represent the work done by a force and indeed they frequently appear in electrodynamics with a different physical interpretation.

Often we wish to evaluate the line integral along a closed path, and such integrals are denoted by

$$\Gamma = \oint_C \mathbf{B} \cdot d\mathbf{r}. \tag{1.26}$$

We shall see that closed line integrals of this type appear frequently in the fundamental equations of electrodynamics.

Certain vector fields have the special property that all possible closed line integrals are zero,

$$\boxed{\oint_C \mathbf{B} \cdot d\mathbf{r} = 0 \qquad \text{(conservative fields)},} \tag{1.27}$$

and such fields are called *conservative* (by physicists) or *irrotational* (in mathematics). An example of a conservative field is a gravitational force field. Conservative fields have the useful property that integrals of the type (1.25) depend on P and Q, and of course on **B**, but not on the path taken to get from P to Q. This is easily seen by considering two distinct curves linking P to Q, say $C_1$ and $C_2$, and then constructing a closed curve C, obtained by traversing $C_1$ from P to Q and then $C_2$ from Q to P. Let $\Gamma_1$ and $\Gamma_2$ be the line integrals from P to Q along $C_1$ and $C_2$, respectively. Then, noting that $d\mathbf{r}$ reverses when we traverse a curve in the opposite direction, we have

$$\Gamma_1 - \Gamma_2 = \int_{P,C_1}^{Q} \mathbf{B} \cdot d\mathbf{r} - \int_{P,C_2}^{Q} \mathbf{B} \cdot d\mathbf{r} = \int_{P,C_1}^{Q} \mathbf{B} \cdot d\mathbf{r} + \int_{Q,C_2}^{P} \mathbf{B} \cdot d\mathbf{r} = \oint_C \mathbf{B} \cdot d\mathbf{r}. \tag{1.28}$$

Evidently, when **B** is conservative, (1.27) and (1.28) demand that $\Gamma_1 = \Gamma_2$, i.e. the line integral is path-independent, as claimed.

When a vector field is conservative (but only when it is conservative), it is meaningful to introduce a scalar field $\varphi(x, y, z)$, defined by

$$\varphi_P = -\int_{\text{datum}}^{P} \mathbf{B} \cdot d\mathbf{r}, \tag{1.29}$$

where the subscript P on $\varphi_P$ indicates that this is the value of $\varphi$ at point P. Note that (1.29) only makes sense if **B** is conservative, since otherwise the value of $\varphi_P$ would depend on the path taken to reach P and so $\varphi_P$ would not have a unique value. In cases where the vector field is a force, such as a gravitational force field, $\varphi(x, y, z)$ defined in this way is simply the potential energy of the field. More generally, $\varphi(x, y, z)$ defined by (1.29) is called the *scalar potential* of **B**.

Perhaps two comments are in order at this point. First, there is always an arbitrary constant associated with a scalar potential, which is a consequence of the fact that the datum point can be chosen at random. Second, in many mathematics texts the minus sign is dropped from the definition (1.29), simply as a matter of convenience. However, one then loses the direct link to potential energy in those cases where the vector is a force.

### 1.2.2 *Surface Integrals and Solenoidal Fields*

Surface integrals are usually used as a means of evaluating the *flux* of a vector field through any given surface.

Consider the situation shown in Figure 1.4 where an open surface $S$ sits in a region of space inhabited by the vector field $\mathbf{B}(x, y, z)$. We define the flux of **B** through $S$ as follows. Let us divide $S$ into many small elements of area $\delta S_i$, and for each $\delta S_i$ we introduce two vectors $\mathbf{r}_i$ and $\mathbf{n}_i$, which are the position vector that locates the centre of $\delta S_i$ and the unit normal to the surface at location $\mathbf{r}_i$. (The subscript $i$ means the $i$th element of area of the surface.) The flux of **B** through $\delta S_i$ is then defined to be $\delta S_i$ times the normal component of **B** evaluated at the centre of the element, $[\mathbf{B}(\mathbf{r}_i) \cdot \mathbf{n}_i]\delta S_i$. It is convenient to rewrite this flux

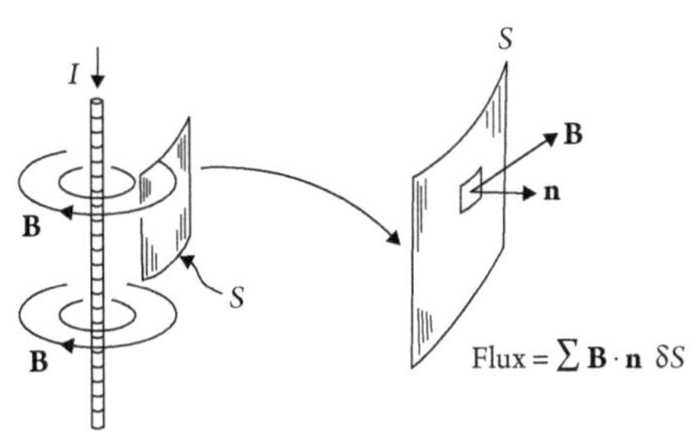


**Figure 1.4** The flux of **B** through the surface S.

as $\mathbf{B}(\mathbf{r}_i)\cdot(\delta S\mathbf{n})_i = \mathbf{B}(\mathbf{r}_i)\cdot\delta\,\mathbf{S}_i$, where $\delta\,\mathbf{S}_i = (\delta S\mathbf{n})_i$ is known as the vector area. Adding all such contributions across the surface and taking the limit as each $\delta S_i \to 0$ gives the net flux of **B** through the surface:

$$\Phi = \int_S \mathbf{B}\cdot d\mathbf{S}. \tag{1.30}$$

The integral on the right is an example of a surface integral.

Note that it is common to use the symbol $\Phi$ to represent a flux. Note also that there are two directions normal to the surface at any one point, so there is some ambiguity in the definition of **n**. However, for an open surface this ambiguity is readily handled on a case-by-case basis by specifying in which direction we are calculating the flux, for example from left to right as opposed to right to left.

Perhaps the most common example of a flux is the rate of flow of mass through a surface $S$. If $\rho(x, y, z)$ is the density field in the fluid and $\mathbf{u}(x, y, z)$ the velocity field, then the rate of flow of mass through an element of area $\delta S$ is $\rho(\mathbf{u}\cdot\mathbf{n})\delta S$. Note that only the normal component of **u** caries mass through the surface and so contributes to the mass flux. The net flow of mass through a finite surface $S$ is normally given the symbol $\dot{m}$. This is evidently

$$\dot{m} = \int_S (\rho\mathbf{u})\cdot d\mathbf{S}. \tag{1.31}$$

On occasions one needs to calculate the flux of a vector field through a closed surface and such a surface integral is denoted by

$$\Phi = \oint_S \mathbf{B}\cdot d\mathbf{S}. \tag{1.32}$$

An important sign convention now comes into play. It is universally agreed that the unit normal vector to the surface, **n**, always points outward, from the enclosed volume $V$ to the exterior world. With this sign convention, field lines entering the enclosed volume through $S$ make a negative contribution to (1.32), while those leaving $V$ make a positive contribution.

At this point an interesting possibility arises in which all the positive and negative contributions to $\Phi$ cancel, i.e. there is as much flux entering as leaving the enclosed volume. It turns out that many vector fields have this property, and we call them *solenoidal vector fields*. Thus a solenoidal field is one for which all integrals of the form (1.32) are zero:

$$\boxed{\oint_S \mathbf{B}\cdot d\mathbf{S} = 0 \qquad \text{(solenoidal fields).}} \tag{1.33}$$

The classic example of a solenoidal vector field is the velocity distribution in a fluid of uniform density. Here (1.31) applied to a closed surface *fixed in space* becomes

$$\dot{m}/\rho = \oint_S \mathbf{u} \cdot d\mathbf{S} = 0,$$

the net mass flux out through $S$ being zero because mass is conserved within the fixed surface $S$. Loosely speaking, we may think of solenoidal fields as being the 'what flows in must flow out' vector fields. Note that the subdivision between solenoidal and non-solenoidal fields is entirely independent of that between conservative and non-conservative fields. It turns out that all magnetic fields are solenoidal.

Solenoidal fields have the useful property that the flux through an open surface, say $S_1$, bounded by a curve $C$ is equal to the flux through any other surface that spans the same curve $C$. Of course, the various fluxes must all be evaluated in the same direction, say from left to right. In any event, the key point is that if $\mathbf{B}$ is solenoidal, then the fluxes of $\mathbf{B}$ through all surfaces spanning a given closed curve $C$ are equal. This is readily proved as follows. Consider two non-intersecting open surfaces $S_1$ and $S_2$ that span the same closed curve $C$. We choose a given direction to evaluate the fluxes of $\mathbf{B}$ through $S_1$ and $S_2$, say left to right, and let $\Phi_1$ and $\Phi_2$ be the resulting fluxes. Now consider the closed surface $S$ composed of $S_1$ and $S_2$, but with the unit normal vectors on one of the two open surfaces, say $S_1$, reversed so that $\mathbf{n}$ always points outward on the closed surface $S$. The net flux out of $S$ is then $\Phi = \Phi_2 - \Phi_1$, since reversing the direction of $\mathbf{n}$ on $S_1$ changes the sign of $\Phi_1$. So if $\mathbf{B}$ is solenoidal, and hence $\Phi = 0$, then $\Phi_2 = \Phi_1$, which establishes the result.

## 1.3 The Gradient, Divergence and Curl

We now turn from integral calculus to differential calculus, starting with the concepts of the *gradient*, the *divergence* and the *curl*.

### 1.3.1 *The Gradient and the Scalar Potential Revisited*

The operation of taking the gradient of a scalar field, say $\varphi(x, y, z)$, produces a vector field, written $\text{grad}\,\varphi$. The vector $\text{grad}\,\varphi$ is defined as that vector which, when dotted into the infinitesimal displacement vector $d\mathbf{r}$, gives the change in $\varphi$ across $d\mathbf{r}$:

$$\delta\varphi = \varphi(\mathbf{r} + d\mathbf{r}) - \varphi(\mathbf{r}) = (\text{grad}\,\varphi) \cdot d\mathbf{r}. \tag{1.34}$$

If $\mathbf{n}$ is a unit vector in the direction of $d\mathbf{r}$, and $ds$ is the length of $d\mathbf{r}$, then $d\mathbf{r} = (ds)\mathbf{n}$ and hence the rate of change of $\varphi$ in the direction of $\mathbf{n}$ is clearly

$$\frac{d\varphi}{ds} = (\text{grad}\,\varphi) \cdot \mathbf{n}. \tag{1.35}$$

This is called the *directional derivative* of $\varphi$ and its value depends on the orientation of $\mathbf{n}$, being greatest when $\mathbf{n}$ points normal to the surfaces of constant $\varphi$. Thus $\text{grad}\,\varphi$ is a vector that points in the direction of maximum rate of change of $\varphi$ and its magnitude is equal to the rate of change of $\varphi$ in that direction. In short, $\text{grad}\,\varphi$ is everywhere perpendicular to surfaces of constant $\varphi$, pointing 'uphill'.

An explicit expression for $\text{grad}\,\varphi$ in Cartesian coordinates is readily obtained from the definition (1.34). A truncated Taylor expansion gives

$$\delta\varphi = \frac{\partial\varphi}{\partial x}\delta x + \frac{\partial\varphi}{\partial y}\delta y + \frac{\partial\varphi}{\partial z}\delta z = \left(\frac{\partial\varphi}{\partial x}\mathbf{i} + \frac{\partial\varphi}{\partial y}\mathbf{j} + \frac{\partial\varphi}{\partial z}\mathbf{k}\right)\cdot d\mathbf{r}, \tag{1.36}$$

and we conclude that

$$\boxed{\text{grad}\,\varphi = \frac{\partial\varphi}{\partial x}\mathbf{i} + \frac{\partial\varphi}{\partial y}\mathbf{j} + \frac{\partial\varphi}{\partial z}\mathbf{k}.} \tag{1.37}$$

At this point it is convenient to introduce the *vector operator* $\nabla$, defined as

$$\boxed{\nabla = \mathbf{i}\frac{\partial}{\partial x} + \mathbf{j}\frac{\partial}{\partial y} + \mathbf{k}\frac{\partial}{\partial z},} \tag{1.38}$$

so that $\text{grad}\,\varphi$ can be written symbolically as $\nabla\varphi$. $\nabla$ is known as the *del operator* and constitutes a convenient shorthand. Although $\nabla$ looks like a vector, and indeed often behaves like a vector, it is *not* a vector. In particular, $\nabla$ is not a vector *multiplying* $\varphi$, but rather a vector operator *acting on* $\varphi$. To emphasise the difference, suppose that $T$ and $p$ are *independent* scalar fields, say temperature and pressure, so that $\nabla T$ and $\nabla p$ point at right angles to the surfaces of constant $T$ and $p$, respectively. Then, in general, we find $(\nabla T)\times(\nabla p)\neq 0$, despite the fact that if $\mathbf{A}$ is any vector field, then $(\mathbf{A}T)\times(\mathbf{A}p)=0$.

The gradient frequently crops up in diffusion laws. For example, Fourier's law of heat conduction relates the heat flux density vector (i.e. the heat flow per unit area) $\mathbf{q}$ to the local temperature gradient by

$$\mathbf{q} = -k\nabla T. \tag{1.39}$$

Here $k$ is the thermal conductivity, which is proportional to the thermal diffusivity, and the minus sign is required since heat flows 'downhill' to lower temperature. There are many other examples of *flux-gradient laws*, which generally take the form

$$\text{flux vector} = -\text{diffusivity}\;\nabla(\text{scalar}). \tag{1.40}$$

Let us now consider some of the mathematical properties of the gradient. It is readily confirmed that $\text{grad}\,\varphi$ satisfies the vector identities

$$\begin{aligned}\nabla(f+g) &= \nabla f + \nabla g,\\ \nabla(fg) &= f\nabla g + g\nabla f.\end{aligned} \tag{1.41}$$

Moreover, in some sense the gradient constitutes the inverse operation to a line integral, since the definition (1.34) yields

$$\int_{\mathrm{P}}^{\mathrm{Q}} (\nabla\varphi)\cdot d\mathbf{r} = \int_{\mathrm{P}}^{\mathrm{Q}} d\varphi = \varphi_{\mathrm{Q}} - \varphi_{\mathrm{P}}. \tag{1.42}$$

The gradient is therefore closely related to the concept of a conservative field and to the associated concept of a scalar potential, as we now discuss.

Consider the particular vector field **C** defined by $\mathbf{C} = -\nabla\varphi$ for some single-valued $\varphi$. Then (1.42) combined with the definition (1.27) tells us that **C** is a conservative field. Actually, we can make a stronger statement: not only does the definition $\mathbf{C} = -\nabla\varphi$ ensure that **C** is conservative, but also, if we are given a field **C** and told that it is conservative, then the combination of (1.29) and (1.42) tells us that we can always find a scalar field $\varphi$ such that $\mathbf{C} = -\nabla\varphi$. Moreover, (1.29) tells us that $\varphi$ is the scalar potential of **C**.

So far we have considered only Cartesian coordinates. Frequently it is more convenient to adopt polar coordinates, either cylindrical polar coordinates or spherical polar coordinates, whose properties are summarized in Appendix I. To obtain an explicit expression for grad$f$ in polar coordinates, one can either perform a simple coordinate transformation on (1.37) or else return to the coordinate-free definition (1.34) and apply this definition in the appropriate coordinate system. Either way, an expression for grad$f$ is readily found. For example, in cylindrical polar coordinates $(s, \theta, z)$ we have

$$\delta\mathbf{r} = (\delta s)\hat{\mathbf{e}}_s + (s\delta\theta)\hat{\mathbf{e}}_\theta + (\delta z)\hat{\mathbf{e}}_z, \tag{1.43}$$

where $s$ and $\theta$ are the two-dimensional polar coordinates in the $x$–$y$ plane and $\hat{\mathbf{e}}$ indicates a unit vector. It follows from (1.34) that

$$\nabla f = \frac{\partial f}{\partial s}\hat{\mathbf{e}}_s + \frac{1}{s}\frac{\partial f}{\partial \theta}\hat{\mathbf{e}}_\theta + \frac{\partial f}{\partial z}\hat{\mathbf{e}}_z. \tag{1.44}$$

### *1.3.2 The Divergence, Gauss' Theorem and the Laplacian*

The divergence of a vector field is a scalar field. In Cartesian coordinates the divergence of the vector field $\mathbf{B}(x, y, z)$ is defined as

$$\boxed{\operatorname{div}\mathbf{B} = \frac{\partial B_x}{\partial x} + \frac{\partial B_y}{\partial y} + \frac{\partial B_z}{\partial z},} \tag{1.45}$$

which can be written symbolically in terms of the del operator as

$$\operatorname{div}\mathbf{B} = \left(\mathbf{i}\frac{\partial}{\partial x} + \mathbf{j}\frac{\partial}{\partial y} + \mathbf{k}\frac{\partial}{\partial z}\right)\cdot(B_x\mathbf{i} + B_y\mathbf{j} + B_z\mathbf{k}) = \nabla\cdot\mathbf{B}. \tag{1.46}$$

It is readily confirmed that the definition (1.45) yields the vector identities

$$\nabla \cdot (\mathbf{A} + \mathbf{B}) = \nabla \cdot \mathbf{A} + \nabla \cdot \mathbf{B}, \tag{1.47}$$

$$\nabla \cdot (f\mathbf{A}) = f\nabla \cdot \mathbf{A} + \nabla f \cdot \mathbf{A}. \tag{1.48}$$

Note that $\mathbf{B}$ and $\nabla$ do not commute, in the sense the $\nabla \cdot \mathbf{B}$ and $\mathbf{B} \cdot \nabla$ are not the same thing, with

$$\mathbf{B} \cdot \nabla = B_x \frac{\partial}{\partial x} + B_y \frac{\partial}{\partial y} + B_z \frac{\partial}{\partial z}$$

being yet another differential operator. Once again, we see the danger of treating $\nabla$ as a vector, though it is frequently convenient to do exactly that.

The physical significance of the definition (1.45) may be clarified by considering the flux of $\mathbf{B}$ out through the surface of a small rectangular volume $\delta V = \delta x \delta y \delta z$. Consider, for example, the fluxes out through those faces whose normal is parallel to the $x$ axis. If the volume element runs from $x_0$ to $x_0 + \delta x$, then the net flux out of $\delta V$ through these two faces is

$$\Phi = B_x(x_0 + \delta x)(\delta y \delta z) - B_x(x_0)(\delta y \delta z) = \frac{\partial B_x}{\partial x} \delta x \delta y \delta z = \frac{\partial B_x}{\partial x} \delta V.$$

By symmetry, the contributions from the other two pairs of faces must have the same form, but with $y$ or $z$ substituted for $x$. We conclude, therefore, that the net flux out of our small rectangular volume is

$$\boxed{\oint \mathbf{B} \cdot d\mathbf{S} = (\nabla \cdot \mathbf{B}) \delta V.} \tag{1.49}$$

This is the key result that gives the concept of divergence some clear physical interpretation. It turns out that (1.49) is valid not just for a rectangular domain, but for any small volume enclosing the point in question. In fact (1.49) is regarded by some authors as sufficiently important that it is adopted as the *coordinate-free definition* of the divergence. That is to say, $\nabla \cdot \mathbf{B}$ at location $\mathbf{r}$ is defined as the flux of $\mathbf{B}$ out through a small, closed surface surrounding $\mathbf{r}$, divided by the volume enclosed by that surface. One can then work backwards from such a definition to show that (1.45) follows from (1.49).

We now have a simple physical interpretation of the formal mathematical definition of the divergence, (1.45). At locations where the field lines are, on average, diverging from a point, in the sense that $\oint \mathbf{B} \cdot d\mathbf{S}$ is positive for any small, closed surface surrounding that point, the divergence is positive. Conversely, at locations where the field lines are, on average, converging to a point, in the sense that $\oint \mathbf{B} \cdot d\mathbf{S}$ is negative for any surface surrounding that point, the divergence is negative.

As with the gradient, the form of the divergence in polar coordinates may be determined either by performing a coordinate transformation on the Cartesian definition or by returning to the coordinate-free definition. Either way, we find that, in cylindrical polar coordinates $(s, \theta, z)$,

$$\nabla \cdot \mathbf{B} = \frac{1}{s}\frac{\partial}{\partial s}(sB_s) + \frac{1}{s}\frac{\partial B_\theta}{\partial \theta} + \frac{\partial B_z}{\partial z}. \tag{1.50}$$

Note that (1.49) tells us that any solenoidal vector field, such as the velocity distribution in an incompressible fluid, must have zero divergence in accordance with the definitions (1.33) and (1.49). In fact, we shall see below that we can make a stronger statement: not only does any solenoidal field necessarily have zero divergence, but any vector field with zero divergence must be solenoidal, in the sense of (1.33).

One of the more celebrated theorems in vector calculus, called Gauss' theorem, is closely related to (1.49). It states that, if $S$ is any surface enclosing a *finite* volume $V$, then

$$\boxed{\oint_S \mathbf{B} \cdot d\mathbf{S} = \int_V (\nabla \cdot \mathbf{B}) dV.} \tag{1.51}$$

A proof of Gauss' theorem follows more or less directly from (1.49). Suppose we chop up $V$ into many small volume elements $\delta V_i$, each of which obeys (1.49). If $\delta S_i$ is the surface enclosing the $i$th volume element, then, on summing over $V$,

$$\int_V (\nabla \cdot \mathbf{B}) dV = \sum_i \oint_{\delta S_i} \mathbf{B} \cdot d\mathbf{S}. \tag{1.52}$$

The sum on the right of (1.52) involves many small, closed surfaces, most of which are interior to $V$. However, the fluxes in and out of all the interior surfaces within $V$ cancel, because each interior surface is shared by two adjacent volume elements, with $\mathbf{n}$ reversed at their common boundary. This leaves only the contributions to (1.52) from the bounding surface $S$, and Gauss' theorem then follows.

We close this section by noting that the compound operation $\nabla \cdot (\nabla\varphi)$ appears so frequently that we give it its own name and symbol. It is called the *Laplacian* and given the symbol $\nabla^2\varphi$. In Cartesian coordinates it is

$$\nabla^2\varphi = \frac{\partial^2\varphi}{\partial x^2} + \frac{\partial^2\varphi}{\partial y^2} + \frac{\partial^2\varphi}{\partial z^2}, \tag{1.53}$$

while for radially symmetric problems expressed in spherical polar coordinates it takes the form

$$\nabla^2\varphi = \frac{1}{r^2}\frac{\partial}{\partial r}\left(r^2\frac{\partial\varphi}{\partial r}\right).$$

### 1.3.3 The Curl

Let us now turn to the third operation, called *taking the curl*, which converts one vector field into another vector field. In Cartesian coordinates this is defined as

$$\operatorname{curl}\mathbf{V} = \left(\frac{\partial V_z}{\partial y} - \frac{\partial V_y}{\partial z}\right)\mathbf{i} + \left(\frac{\partial V_x}{\partial z} - \frac{\partial V_z}{\partial x}\right)\mathbf{j} + \left(\frac{\partial V_y}{\partial x} - \frac{\partial V_x}{\partial y}\right)\mathbf{k}, \tag{1.54}$$

which is more conveniently expressed in terms of the del operator as

$$\boxed{\operatorname{curl}\mathbf{V} = \nabla\times\mathbf{V} = \begin{vmatrix} \mathbf{i} & \mathbf{j} & \mathbf{k} \\ \partial/\partial x & \partial/\partial y & \partial/\partial z \\ V_x & V_y & V_z \end{vmatrix}.} \tag{1.55}$$

Once again, we have found it convenient for notational purposes to think of $\nabla$ as a vector. However, as always, we must be careful not to push the vector nature of $\nabla$ too far. For example, it is readily confirmed that in general $\nabla\times\mathbf{V}$ is not perpendicular to $\mathbf{V}$, as would be the case if $\nabla\times\mathbf{V}$ were the genuine cross product of two vectors. Perhaps the safest strategy is to think of (1.55) as shorthand for (1.54).

The physical significance of the curl may be clarified with the aid of the following example. Let us evaluate the closed line integral of $\mathbf{V}$ around the edge of a small rectangular element of area $\delta\,\mathbf{S} = (\delta x\,\delta y)\mathbf{k}$, using the right-hand rule to determine the direction of integration. Labelling three corners of the rectangle P, Q and R, as shown in Figure 1.5, we have

$$\oint \mathbf{V}\cdot d\mathbf{r} \approx V_{x,\mathrm{P}}\,\delta x + V_{y,\mathrm{Q}}\,\delta y - V_{x,\mathrm{R}}\,\delta x - V_{y,\mathrm{P}}\,\delta y = (V_{y,\mathrm{Q}} - V_{y,\mathrm{P}})\,\delta y - (V_{x,\mathrm{R}} - V_{x,\mathrm{P}})\,\delta x,$$

or

$$\oint \mathbf{V}\cdot d\mathbf{r} = \left(\frac{\partial V_y}{\partial x} - \frac{\partial V_x}{\partial y}\right)\delta x\,\delta y = (\nabla\times\mathbf{V})\cdot d\mathbf{S}. \tag{1.56}$$

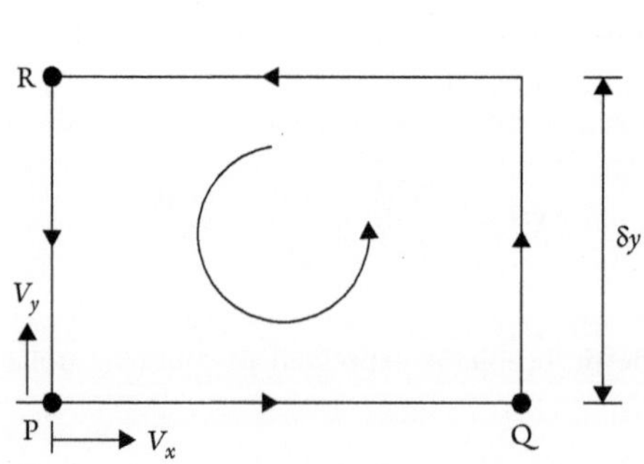


**Figure 1.5** The rectangular element of area used to derive (1.56).

It turns out that (1.56) generalises to any small, closed, planar curve $C$ encircling the point at which the curl is evaluated:

$$\boxed{\Gamma = \oint_C \mathbf{V} \cdot d\mathbf{r} = (\nabla \times \mathbf{V}) \cdot d\mathbf{S}.} \tag{1.57}$$

It follows immediately from (1.57) that any conservative field, as defined by (1.27), has zero curl. We shall return to this point shortly.

The expression (1.57) does for the curl what (1.49) does for the divergence. It is regarded by many authors as sufficiently important that it is adopted as the *coordinate-free definition* of the curl. That is to say, the various components of $\nabla \times \mathbf{V}$ are defined as the line integral of $\mathbf{V}$ around a small, closed, planar curve of appropriate orientation, divided by the area enclosed by that curve. One can then work backwards from such a definition to show that (1.54) follows from (1.57). By contrast, in cylindrical polar coordinates $(s, \theta, z)$, the definition (1.57) yields

$$\nabla \times \mathbf{V} = \frac{1}{s} \begin{vmatrix} \hat{\mathbf{e}}_s & s\hat{\mathbf{e}}_\theta & \hat{\mathbf{e}}_z \\ \partial/\partial s & \partial/\partial\theta & \partial/\partial z \\ V_s & sV_\theta & V_z \end{vmatrix}. \tag{1.58}$$

A number of useful vector identities come along with the curl. The two most important are

$$\boxed{\begin{aligned} &\nabla \cdot (\nabla \times \mathbf{A}) = 0, \\ &\nabla \times (\nabla\varphi) = 0, \end{aligned}} \tag{1.59}$$

for any $\mathbf{A}$ or $\varphi$. Three others are

$$\nabla \times (\varphi\mathbf{A}) = \nabla\varphi \times \mathbf{A} + \varphi\nabla \times \mathbf{A}, \tag{1.60}$$

$$\nabla \cdot (\mathbf{A} \times \mathbf{B}) = \mathbf{B} \cdot \nabla \times \mathbf{A} - \mathbf{A} \cdot \nabla \times \mathbf{B}, \tag{1.61}$$

$$\nabla \times \nabla \times \mathbf{A} = \nabla(\nabla \cdot \mathbf{A}) - \nabla^2\mathbf{A}, \tag{1.62}$$

where $\nabla^2\mathbf{A}$ is defined in Cartesian coordinates as

$$\nabla^2\mathbf{A} = \mathbf{i}\nabla^2 A_x + \mathbf{j}\nabla^2 A_y + \mathbf{k}\nabla^2 A_z.$$

Identities like these are most readily established in Cartesian coordinates by expanding all terms and showing that the right-hand side equals the left-hand side. More identities are given in Appendix I.

## 1.4 Stokes' Theorem and Conservative Fields Revisited

Stokes' theorem follows from the coordinate-free definition (1.57), in much the same way that Gauss' theorem follows from (1.49). It states that, if $S$ is a two-sided surface bounded by a closed curve $C$, then

$$\boxed{\Gamma = \oint_C \mathbf{V} \cdot d\mathbf{r} = \int_S (\nabla \times \mathbf{V}) \cdot d\mathbf{S},} \tag{1.63}$$

where $C$ is traversed in a direction consistent with the right-hand rule. A schematic proof proceeds as follows. The surface is broken up into many elements of area $\delta\, \mathbf{S}_i$, each with bounding curve $\delta C_i$. We can then apply (1.57) to each element of area and sum over the surface $S$. The result is

$$\int_S (\nabla \times \mathbf{V}) \cdot d\mathbf{S} = \sum_i \oint_{\delta C_i} \mathbf{V} \cdot d\mathbf{r}. \tag{1.64}$$

As with the proof of Gauss' theorem, there is now a great deal of cancellation on the right of (1.64). Those parts of $\delta C_i$ that lie 'inside' the surface, away from its edge $C$, are shared between two adjacent surface elements. Moreover, the right-hand rule demands that when the line integral of $\mathbf{V}$ is evaluated for two adjacent surface elements, the direction of $d\mathbf{r}$ on their common boundary is in opposite directions. It follows that all the internal contributions to the right of (1.64) cancel, and the only remaining contributions correspond to those line elements $d\mathbf{r}$ that lie on the edge of the surface, $C$. Stokes' theorem then follows.

Let us now return to the idea of a conservative field. In §1.2.1 we defined conservative fields through the properties of their line integrals,

$$\oint_C \mathbf{V} \cdot d\mathbf{r} = 0 \qquad \text{(definition of a conservative field).}$$

Subsequently, in §1.3.1, we showed that if $\mathbf{V} = -\nabla\varphi$ for some $\varphi$, then $\mathbf{V}$ is conservative. The converse is also true, if $\mathbf{V}$ is conservative, then we can always find a scalar field $\varphi$ such that $\mathbf{V} = -\nabla\varphi$. This follows from (1.29) with $\varphi$ being the scalar potential. Thus we have

$$\mathbf{V} = -\nabla\varphi \quad \Leftrightarrow \quad \oint_C \mathbf{V} \cdot d\mathbf{r} = 0\,. \tag{1.65}$$

We now see that there is a link between conservative fields and the curl. Specifically, the coordinate-free definition (1.57) tells us that all conservative fields have zero curl, which is consistent with the vector identity $\nabla \times (\nabla\varphi) = 0$. Stokes' theorem also tells us that the converse is true: if the curl of a vector field is zero, then it is conservative. Thus we also have

$$\nabla \times \mathbf{V} = 0 \quad \Leftrightarrow \quad \oint \mathbf{V} \cdot d\mathbf{r} = 0\,. \tag{1.66}$$

Combining all of these results yields

$$\boxed{\mathbf{V} = -\nabla\varphi \quad \Leftrightarrow \quad \oint_C \mathbf{V}\cdot d\mathbf{r} = 0 \quad \Leftrightarrow \quad \nabla\times\mathbf{V} = 0.} \tag{1.67}$$

(the three properties of conservative fields)

Note that some authors define conservative fields, not through the properties of their line integrals, but through one of the other two properties noted in (1.67). Of course, it is simply a matter of taste which is to be preferred.

## 1.5 Solenoidal Fields Revisited: The Vector Potential and Helmholtz's Decomposition

In §1.2.2 we defined solenoidal vector fields through the properties of their surface integrals,

$$\oint_S \mathbf{B}\cdot d\mathbf{S} = 0 \quad \text{(definition of a solenoidal field).} \tag{1.68}$$

Moreover, the coordinate-free definition of the divergence, (1.49), tells us that all solenoidal vector fields have zero divergence. Gauss' theorem yields the converse; if a vector field has zero divergence, it is solenoidal. In summary, we have

$$\oint \mathbf{B}\cdot d\mathbf{S} = 0 \quad \Leftrightarrow \quad \nabla\cdot\mathbf{B} = 0. \tag{1.69}$$

Now consider the vector identity $\nabla\cdot(\nabla\times\mathbf{A}) = 0$. Evidently, if $\mathbf{B}$ takes the form $\mathbf{B} = \nabla\times\mathbf{A}$, then it is solenoidal. We shall show below that the converse is also true: if $\mathbf{B}$ is solenoidal, then it is always possible to find a vector field $\mathbf{A}$ such that $\mathbf{B} = \nabla\times\mathbf{A}$. So we also have

$$\mathbf{B} = \nabla\times\mathbf{A} \quad \Leftrightarrow \quad \nabla\cdot\mathbf{B} = 0. \tag{1.70}$$

Combining all of these results yields

$$\boxed{\mathbf{B} = \nabla\times\mathbf{A} \quad \Leftrightarrow \quad \oint \mathbf{B}\cdot d\mathbf{S} = 0 \quad \Leftrightarrow \quad \nabla\cdot\mathbf{B} = 0} \tag{1.71}$$

(the three properties of solenoidal fields).

As with conservative fields, the preferred definition of a solenoidal field varies from author to author. Some authors define solenoidal fields, not through the properties of their surface integrals, but though one of the other two properties listed above. Of course, ultimately, this makes no difference.

The vector field **A** in (1.71) is called the *vector potential* for **B**. Often, but not always, it is convenient to take **A** to have zero divergence, in which case the vector potential is defined by

$$\nabla \times \mathbf{A} = \mathbf{B}, \quad \nabla \cdot \mathbf{A} = 0.$$

Given **B**, this fixes the distribution of **A**, since the requirements for a vector field to be uniquely determined are that its divergence and curl be specified and suitable boundary conditions be given.

We close with a useful decomposition of vector fields known as Helmholtz's decomposition. This states that it is always possible to decompose a vector field into a part that has zero divergence and a part that has zero curl. That is, given **B**, we can always find a vector potential **A** and a scalar potential $\varphi$ such that

$$\boxed{\mathbf{B} = \nabla \times \mathbf{A} - \nabla \varphi,} \tag{1.72}$$

The governing equation for the scalar potential is then found by taking the divergence of (1.72),

$$\nabla^2 \varphi = -\nabla \cdot \mathbf{B}, \tag{1.73}$$

which, as we shall see in Chapter 2, has a unique solution if suitable boundary conditions are given for $\varphi$. The equivalent governing equation for **A** depends on how we specify its divergence. As mentioned above, a common approach is to take $\nabla \cdot \mathbf{A} = 0$, so that the curl of (1.72), combined with the identity (1.62), yields

$$\nabla^2 \mathbf{A} = -\nabla \times \mathbf{B}. \tag{1.74}$$

Again, given **B**, this may be solved for **A** if suitable boundary conditions are specified. In an infinite domain these boundary conditions are usually taken to be that **A** and $\varphi$ vanish at infinity.

This completes our brief overview of vector calculus. Readers who feel the need for a more leisurely discussion could do worse than consult Arfken (1985). We close this chapter with a historical footnote.

## 1.6 An Aside: The Troubled History of Vector Calculus

Just as the full significance of Maxwell's *Treatise on Electricity and Magnetism* took a surprisingly long time to be appreciated by the wider physics community (see Appendix III), so the introduction of vector calculus, which is so well adapted to the study of electromagnetism, faced surprising resistance from factions of the nineteenth-century scientific community. In fact, as some of the quotations below suggest, at times the struggle to establish the modern system of vector calculus verged on the acrimonious. (The sources of these quotations may be found in the splendid historical survey by Crowe (1994).)

It is probably fair to say that modern vector calculus has its roots in the development of quaternions by the great Irish physicist and mathematician William Rowan Hamilton

(1805–1865). While directed quantities in two dimensions can be represented by complex numbers, Hamilton appreciated the need for a mathematical representation of directed quantities in three dimensions. This need became all the more pressing when the experimentalist Michael Faraday (1791–1867) insisted that electromagnetic phenomena could only be truly understood in the context of vector fields. As James Clerk Maxwell put it:

> The way in which Faraday made use of lines of force in co-ordinating the phenomena of magneto-electric induction shews him to have been in reality a mathematician of a very high order—one from whom the mathematicians of the future may derive valuable and fertile methods.

Lord Kelvin (1824–1907), or Sir William Thompson as he was known before he was ennobled, was more forthright about Faraday:

> One of the most brilliant steps made in philosophical exposition of which any instance existed in the history of science was that in which Faraday stated, in three or four words, intensely full of meaning, the law of magnetic attraction or repulsion.... Mathematicians were content to investigate the general expression of the resultant force experienced by a globe of soft iron in all such cases; but Faraday, without any mathematics, devined the result of the mathematical investigations. Indeed, the whole language of the magnetic field and 'lines of force' is Faraday's. It must be said for the mathematicians that they greedily accepted it, and have ever since been most zealous in using it to the best advantage.

The advantage of quaternions, as originally conceived by Hamilton and as subsequently championed and developed by the Scottish applied mathematician and physicist Peter Guthrie Tait (Figure 1.6), is the distinction made between scalar and vector quantities, as well as the introduction of the operator $\nabla$ by Tait. However, the somewhat contrived algebraic structure of quaternions is not so intuitive and failed to engage many of the prominent physicists of the time. Indeed, shortly after Tait's death in 1901, Kelvin wrote of his friend, and of the book that they had co-authored, *Treatise on Natural Philosophy*:

> We [Kelvin and Tait] have had a thirty-eight years' war over quaternions. He had been captivated by the originality and extraordinary beauty of Hamilton's genius in this respect, and had accepted, I believe, definitely from Hamilton to take charge of quaternions after his death, which he has most loyally executed. Times without number I offered to let quaternions into Thompson and Tait [their Treatise], if he could only show that in any case our work would be helped by their use. You will see that from the beginning to the end they were never introduced.

The English physicist and engineer Oliver Heaviside (Figure 1.7) was characteristically blunter. In the preface to Volume 1 of his 1893 book, *Electromagnetic Theory*, we find the following:

> Hamilton discovered or invented a remarkable system of mathematics, and since his death the quaternion mantle has adorned the shoulders of Prof. Tait, who has repeatedly advocated the claims of quaternions. Prof. Tait in particular emphasises its

**Figure 1.6** The Scottish mathematician and physicist P. G. Tait (1831–1901).

**Figure 1.7** The ingenious English physicist and engineer Oliver Heaviside (1850–1925).

> great power, simplicity, and perfect naturalness, on the one hand, and on the other tells the physicist that it is exactly what he wants for his physical purposes. It is also known that physicists, with great obstinacy, have been careful (generally speaking) to have nothing to do with quaternions... and so they go tinkering at it, trying to make it a little more intelligible, very much to the disgust of Prof. Tait, who would preserve the quaternion stream pure and undefiled.

Unhappy with quaternions, Heaviside set about devising a simpler system of vector calculus, writing in 1893:

> There was a time, indeed, when I, although recognising the appropriateness of vector analysis in electromagnetic theory (and in mathematical physics generally), did think it was harder to understand and to work with than the Cartesian analysis. But that was before I had thrown off the quaternion old-man-of-the-sea who fastened himself on my shoulders when reading the only accessible treatise on the subject—Prof. Tait's Quaternions. But I came later to see that, so far as the vector analysis I required was concerned, the quaternion was not only not required, but was a positive evil of no inconsiderable magnitude.

Perhaps the only dominant physicist of the time to flirt with quaternions was Maxwell. A close friend of both Tait and Kelvin, and greatly impressed by Faraday's vision of vector fields filling all space, Maxwell decided to discuss vectors in general, and the quaternion system in particular, in his pivotal 1887 *Treatise on Electricity and Magnetism*. In the opening chapter, Maxwell emphasises time and again the need to think of vectors as directed physical quantities in space, and not simply as some abstract ordered triplet associated with a coordinate system. Heaviside shared Maxwell's view that the essence of vectors is that they are physical quantities, independent of coordinate systems. Always happy to pick a fight with the British mathematicians of the time, in 1893 Heaviside teases the mathematicians with the following polemic:

> It is a noteworthy fact that ignorant men have long been in advance of the learned about vectors. Ignorant people, like Faraday, naturally think in vectors. They know nothing of their formal manipulation, but if they think about vectors, they think of them *as* vectors, that is, directed magnitudes. No ignorant man could or would think about the three components of a vector separately, and disconnected from one another. That is a devise of learned mathematicians, to enable them to evade vectors. The device is often useful, especially for calculating purposes, but for general purposes of reasoning the manipulation of the scalar components instead of the vector itself is entirely wrong.

Perhaps the most important consequence of Maxwell mentioning quaternions in his treatise was that it caught the eye of the American physicist Josiah Willard Gibbs (Figure 1.8). Like Heaviside, Gibbs found quaternions cumbersome and set about developing a simpler system of vector calculus. The fruits of his efforts were printed privately in the pamphlet *Elements of Vector Analysis*, the first half in 1881 and the second in 1884. Gibbs sent out 130 copies of this text to a variety of scientists and mathematicians, and when a copy landed

**Figure 1.8** The American physicist J. W. Gibbs (1839–1903).

on Tait's desk he was not impressed. In the preface to the third edition of his *Treatise on Quaternions* he writes in 1890:

> Prof. Willard Gibbs must be ranked as one of the retarders of Quaternion progress, in virtue of his pamphlet on Vector Analysis; a sort of hermaphrodite monster.

Tait was particularly disparaging on the matter of notation. Gibbs responded to Tait a year later in the journal *Nature*:

> ... if my offence had been solely in the matter of notation, it would have been less accurate to describe my production as a monstrosity, than to characterise its dress code as uncouth.

He went on to suggest that the differences between his system and Tait's were not merely notational, but rather conceptual. As Gibbs was developing vector calculus in the USA, so Heaviside was doing precisely the same in London, the two working independently. As Heaviside later noted:

> But on proceeding to apply quaternions to the development of electrical theory, I found it very inconvenient. Quaternions was in its vectorial aspects anti-physical and unnatural, and did not harmonise with common scalar mathematics. So I dropped out the quaternion altogether and kept to pure scalars and vectors, using a very simple vectorial algebra in my papers from 1883 onward.... Up to 1888 I imagined that I was the only one doing vectorial work on positive physical principles; but then I received a copy of Prof. Gibbs Vector Analysis (unpublished, 1881–4). This was a sort of condensed synopsis of a treatise. Though different in appearance, it was essentially

> the same vectorial algebra and analysis to which I had been led.... As time went on, and after a period during which the diffusion of pure vectorial analysis made much progress, in spite of the disparagement of the Edinburgh school of scorners (one of whom said my work was 'a disgrace to the Royal Society', to my great delight), it was most gratifying to find that Prof. Tait softened in his harsh judgements.

Whether Tait and Heaviside were ever truly reconciled is rather doubtful, since the Gibbs–Heaviside system of vector calculus was slow to gain acceptance. As Gibbs anticipated in 1888, 'I believe that a Kampf ums Dasein [struggle for existence] is just commencing between the different methods...', and indeed vigorous debate continued throughout the 1890s involving many prominent scientists. The tone of that debate is well captured by Lord Rayleigh's whimsical aside, 'Behold how these vectorists love one another.'

In any event, by 1910 the debate was all but over and modern vector analysis was firmly established. Readers who are intrigued by this footnote to the history of mathematical physics could do worse than consult the scholarly work by Crowe (1994).

## EXERCISES

1.1 Consider the two-dimensional vector field $\mathbf{u} = -y\mathbf{i} + x\mathbf{j}$. Use (1.22) to show that the field lines for this particular vector field are $x^2 + y^2 = C$, where $C$ is a constant of integration. Sketch the field lines, noting that changing the value of $C$ is equivalent to moving from one field line to another.

1.2 Consider the two-dimensional vector field $\mathbf{A} = x\mathbf{i} + 2xy\mathbf{j}$. Calculate the line integral of $\mathbf{A}$ from (0, 0) to (1, 1) along the paths $(0,0)\rightarrow(1,0)\rightarrow(1,1)$ and $(0,0)\rightarrow(0,1)\rightarrow(1,1)$ and confirm that the line integral is path-dependent.

1.3 Consider two scalar fields $p$ and $T$ that are functionally related, $p = p(T)$. Show that $\nabla p = (dp/dT)\nabla T$ and hence $\nabla p \times \nabla T = 0$. Now show that $\nabla p \times \nabla T = 0$ follows trivially from a consideration of surfaces of constant $p$ and $T$.

1.4 A velocity field $\mathbf{u}$ is solenoidal and restricted to the domain $V$, with zero normal component at the boundary. Show that $\mathbf{u} = (\mathbf{u} \cdot \nabla)\mathbf{x}$ and hence $\int_V \mathbf{u}\, dV = 0$.

1.5 Show that $\nabla \cdot (\nabla^2 \mathbf{A}) = \nabla^2(\nabla \cdot \mathbf{A})$.

1.6 Consider a hemispherical surface of radius $R$ centred on the origin and lying above the $x - y$ plane. Let $\mathbf{A} = -y\mathbf{i} + x\mathbf{j} + 2z\mathbf{k}$ and $\mathbf{B} = \nabla \times \mathbf{A}$. Calculate the flux of $\mathbf{B}$ out through the open surface $r = R$ by (i) using Stokes' theorem, (ii) using Gauss' theorem to transfer the flux calculation to the $x - y$ plane, and (iii) direct calculation.

## REFERENCES

Arfken, G., 1985, *Mathematical Methods for Physicists*, 3rd ed. Academic Press.

Crowe, M. J., 1994, *A History of Vector Analysis*. Dover.

# 2

# The Physical Signatures of Three Important Partial Differential Equations

His had been an intellectual decision founded on the conviction that if a little knowledge was a dangerous thing, a lot was lethal.

Tom Sharpe (*Porterhouse Blue*, 1974)

## 2.1 Three Important Second-Order Partial Differential Equations

### *2.1.1 The Classification of Linear Second-Order Partial Differential Equations*

Many, but certainly not all, of the partial differential equations (PDEs) encountered in physics and engineering are *second-order* and *linear*. There are plenty of important exceptions, though. For example, the equation governing the flexural vibrations of a thin, straight beam is

$$\frac{\partial^2 \eta}{\partial t^2} + K^2 \frac{\partial^4 \eta}{\partial x^4} = 0, \tag{2.1}$$

which is fourth-order in the spatial coordinate $x$. (Here $\eta(x,t)$ is the lateral displacement of the centreline of the beam and $K$ is a constant, as discussed in §2.3.3.) Another notable exception is the Navier–Stokes equation in fluid mechanics

$$\frac{\partial \mathbf{u}}{\partial t} + (\mathbf{u} \cdot \nabla)\mathbf{u} = -\nabla(p/\rho) + \nu \nabla^2 \mathbf{u}, \quad \nabla \cdot \mathbf{u} = 0, \tag{2.2}$$

where $\mathbf{u}(\mathbf{x},t)$ is the velocity field, $p(\mathbf{x},t)$ the pressure field, and $\rho$ and $\nu$ the density and kinematic viscosity of the fluid. This governs the motion of an incompressible, viscous fluid and it is non-linear because of the term $(\mathbf{u} \cdot \nabla)\mathbf{u}$. Indeed, it is precisely this non-linearity that keeps fluid dynamicists gainfully employed trying to understand chaotic processes such as turbulence and the weather.

However, in electrodynamics life is a little simpler and most of the PDEs one encounters are both second-order and linear. It turns out that, by and large, we are good at solving linear equations, in part because the principle of superposition applies. By way of contrast, we are terrible at solving non-linear equations. For example, we know of only a handful of exact non-trivial analytical solutions of the steady non-linear Navier–Stokes equation, perhaps no more than twenty or so in total.

Second-order linear PDEs come in one of three canonical forms. First, all diffusion processes are governed by the so-called *diffusion equation*. For example, the diffusion of heat through a thermally conducting solid is governed by

$$\boxed{\frac{\partial T}{\partial t} = \alpha \nabla^2 T,} \tag{2.3}$$

where $T(\mathbf{x}, t)$ is the temperature field and $\alpha$ the *thermal diffusivity*. Similarly, we shall see that a magnetic field $\mathbf{B}$ diffuses through an electrically conducting solid, such as a metal, according to

$$\frac{\partial \mathbf{B}}{\partial t} = \lambda \nabla^2 \mathbf{B}. \tag{2.4}$$

Here the *magnetic diffusivity* $\lambda$ is given by $\lambda = 1/\sigma\mu$, $\sigma$ being the electrical conductivity of the solid and $\mu$ its permeability.

On the other hand, the propagation of many (but certainly not all) types of waves is governed by the so-called *wave equation*. For example, small-amplitude acoustic waves propagating through still air are governed by

$$\boxed{\frac{\partial^2 p}{\partial t^2} = a^2 \nabla^2 p,} \tag{2.5}$$

where $p(\mathbf{x}, t)$ is the pressure field and $a$ the speed of sound. Similarly, we shall see that magnetic fields propagate through a vacuum in accordance with

$$\frac{\partial^2 \mathbf{B}}{\partial t^2} = c^2 \nabla^2 \mathbf{B}, \tag{2.6}$$

where $c$ is the speed of light. It turns out that $c^2 = 1/\varepsilon_0\mu_0$, where $\varepsilon_0$ and $\mu_0$ are the *permittivity* and *permeability* of free space, respectively.

The third canonical form of second-order PDE relates to steady fields. For example, from (2.3) we see that a steady temperature distribution is governed by

$$\boxed{\nabla^2 T = 0.} \tag{2.7}$$

Similarly, from (2.6) we see that in magnetostatics a free-space magnetic field $\mathbf{B}$ satisfies

$$\nabla^2 \mathbf{B} = 0. \tag{2.8}$$

This third equation is known as *Laplace's equation.* Often Laplace's equation appears with a *source term* on the right, such as

$$\boxed{\nabla^2 T = -s(\mathbf{x}),} \tag{2.9}$$

where in this particular case $s(\mathbf{x})$ represents sources of heat. In this *inhomogeneous* form, Laplace's equation is known as *Poisson's equation.* Sometimes, if the source on the right of Poisson's equation varies very slowly with time, we write the equation in a quasi-steady form, for example

$$\nabla^2 \mathbf{B} = -\mathbf{s}(\mathbf{x}, t), \tag{2.10}$$

where in this case $\mathbf{s}(\mathbf{x}, t)$ is related to those electrical currents that give rise to the magnetic field $\mathbf{B}$. (The precise form of $\mathbf{s}(\mathbf{x}, t)$ in (2.10) is discussed in §2.4.3.)

Note that these three canonical equations—the diffusion equation, the wave equation and Poisson's equation (or Laplace's equation)— are not only mathematically distinct, but represent physically different processes. In (2.3) and (2.4) information spreads diffusively, while in (2.5) and (2.6) it spreads at a constant speed—the wave speed. By contrast, information appears to propagate at infinite speed in (2.10), in the sense that any local change in the source $\mathbf{s}(\mathbf{x}, t)$ is instantaneously felt at all points in space. This is because in writing an equation like (2.10) we have removed, or 'filtered out', any transient processes associated with diffusion or wave propagation, those processes being deemed to have occurred very rapidly (in effect, instantaneously) by comparison with the timescale for changes in $\mathbf{s}(\mathbf{x}, t)$.

Sometimes linear second-order partial differential equations look rather different to the diffusion, wave or Laplace equations, say of the form,

$$A\frac{\partial^2 f}{\partial x_1^2} + 2B\frac{\partial^2 f}{\partial x_1 \partial x_2} + C\frac{\partial^2 f}{\partial x_2^2} = F\left(x_1, x_2, \frac{\partial f}{\partial x_1}, \frac{\partial f}{\partial x_2}\right), \tag{2.11}$$

where $A$, $B$ and $C$ are constants and $(x_1, x_2) \equiv (x, y)$, or else $(x_1, x_2) \equiv (x, t)$. In such cases a suitable change of variables converts (2.11) into one of our three canonical forms. In particular, if $B^2 = AC$ (a so-called parabolic equation), then (2.11) can be converted into a one-dimensional inhomogeneous diffusion-like equation. On the other hand, if $B^2 > AC$ (a hyperbolic equation), then (2.11) can be converted into a one-dimensional inhomogeneous wave-like equation. Finally, if $B^2 < AC$ (an elliptic equation), then (2.11) can be converted into a two-dimensional Poisson-like equation. This method of classification extends in an obvious way to higher dimensions. There is much incentive, therefore, to study individually the properties of the diffusion, wave and Poisson equations. Let us consider each in a little more detail.

### 2.1.2 *A Key Property of the Diffusion Equation: The Idea of Diffusion Length*

Solutions of the diffusion equation can often be conveniently represented in terms of a quantity known as the *diffusion length.* Indeed, the existence of a diffusion length is the hallmark of many diffusive processes and it is an extremely useful concept, as we shall see. Perhaps this is best illustrated by a specific example.

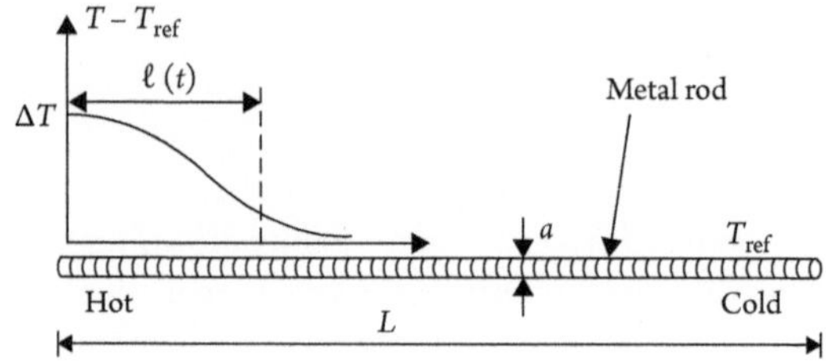


**Figure 2.1** The diffusion of heat down a long, thin rod.

Suppose that we have a long, thin metal rod running from $x = 0$ to $x = L$, and that the rod has a circular cross-section of radius $a$, with $a << L$, as shown in Figure 2.1. Suppose also that the rod is initially at the reference temperature $T_{\text{ref}}$ and that at time $t = 0$ we start to heat the end $x = 0$, maintaining it at the elevated temperature $T_{\text{ref}} + \Delta T$ for $t > 0$. Clearly it takes a finite time for heat to diffuse down the rod, and common sense tells us that after a time $t$ the heat introduced at $x = 0$ will have spread some finite distance, say $\ell(t)$, along the rod. Of course, the precise value of $\ell(t)$ will depend on how we choose to define this length; it could be defined as the point at which $T - T_{\text{ref}}$ drops to 20% of $\Delta T$, or perhaps 1% of $\Delta T$, but the precise definition need not concern us at this point.

Let us now restrict ourselves to times such that $a << \ell(t) << L$ so that, as far as the diffusion of heat is concerned, the rod may be considered infinitesimally thin and infinitely long. During this restricted time period the length $\ell(t)$ is known as the diffusion length for the particular problem in hand. This one-dimensional diffusion process is governed by (2.3) in the form

$$\frac{\partial T}{\partial t} = \alpha \frac{\partial^2 T}{\partial x^2}, \tag{2.12}$$

and so our diffusion length can depend on $\alpha$ and $t$, but not on $a$ and $L$ for as long as $a << \ell(t) << L$. We now deploy one of the most underrated tools in the analysis of physical systems: *dimensional analysis*. Since there are no parameters other than $\alpha$ and $t$ that can influence $\ell$ during our restricted time period, we have $\ell = \ell(\alpha, t)$. Moreover, all diffusivities have the dimensions of $\text{m}^2/\text{s}$, and so we have three parameters—namely $\ell$, $\alpha$ and $t$—which we label $P = 3$, and these have between them two dimensions—namely length and time—which we label $D = 2$. The Buckingham Pi theorem then tells us that we can construct only $P - D = 1$ independent dimensionless group from these three parameters, and by inspection this can be written as $\ell/\sqrt{\alpha t}$. Now dimensional analysis tells us that one dimensionless group can depend on only another dimensionless group, and since the linearity of (2.12) excludes any dependence of $\ell$ on $\Delta T/T_{\text{ref}}$ (see below), there is no other dimensionless group on which $\ell/\sqrt{\alpha t}$ can depend. We conclude that

$$\ell = \hat{c}\sqrt{\alpha t}, \tag{2.13}$$

where $\hat{c}$ is a dimensionless constant of order unity whose precise numerical value will depend on exactly how we choose to define $\ell$. Note that this expression confirms that the higher the thermal diffusivity, or the longer we wait, the further heat will diffuse. Equation (2.13) clearly represents excellent value for money! Despite the fact that we have yet to solve any equations, we already know that heat diffuses a distance $\ell \sim \sqrt{\alpha t}$ in a time $t$.

In fact we can use dimensional analysis to help find the precise form of the temperature distribution. First, we note that $T_{\text{ref}}$ is only a reference temperature and so cannot influence the diffusion process. Second, we note that the linearity of (2.12) means that increasing or decreasing $\Delta T$ will increase or decrease $T - T_{\text{ref}}$ proportionately, but not alter the spatial distribution of $(T - T_{\text{ref}})/\Delta T$. So we work with the dimensionless temperature scale $(T - T_{\text{ref}})/\Delta T$ and observe that, as long as $a << \ell(t) << L$,

$$\frac{T - T_{\text{ref}}}{\Delta T} = F(\alpha, x, t), \tag{2.14}$$

for some function $F$. However, since one dimensionless group can depend only on another dimensionless group, we are obliged to put the right-hand side of (2.14) into dimensionless form. As before, the Buckingham Pi theorem then tells us that this dimensionless form must be

$$\frac{T - T_{\text{ref}}}{\Delta T} = F\left(\frac{x}{\ell(t)}\right), \quad \ell = \sqrt{\alpha t}, \tag{2.15}$$

where we now *define* $\ell(t)$ as $\ell = \sqrt{\alpha t}$. This is an example of a *self-similar solution*, so-called because the solution always looks the same from one instant to the next as long as we keep normalising $x$ by the diffusion length $\ell(t)$. Thus time does not appear explicitly in (2.15), but rather implicitly via $\ell(t)$.

Finding the exact temperature distribution is now straightforward. We write $\eta = x/\sqrt{\alpha t}$ for the rescaled spatial coordinate and substitute (2.15) into the diffusion equation. After a little work we find $2F''(\eta) + \eta F'(\eta) = 0$, which may be integrated once to give $F'(\eta) \sim \exp(-\eta^2/4)$. A second integration, subject to the boundary conditions $F(0) = 1$ and $F(\infty) = 0$, yields

$$F(\eta) = 1 - \frac{1}{\sqrt{\pi}} \int_0^{\eta} \exp(-\eta^2/4) d\eta = 1 - \text{erf}(\eta/2). \tag{2.16}$$

However, perhaps the precise details of this solution are not so interesting—the more important results are (2.13) and (2.15). These assert that, for as long as $a << \ell(t) << L$, so that no geometric lengthscale enters into the problem, dimensional analysis tells us that heat diffuses a distance $\ell \sim \sqrt{\alpha t}$ in a time $t$ and that the temperature distribution is self-similar in accordance with (2.15). We shall return to these ideas in §2.2.1.

### 2.1.3 *A Key Property of the Wave Equation: d'Alembert's Solution*

Let us now turn to the wave equation. Waves that are governed by this equation (but *only* waves governed by this equation) have the very special property that pulses of arbitrary profile propagate without change of shape. This is known as d'Alembert's solution and it is a surprisingly general result.

To get across the essential ideas let us consider the particular case of waves on a stretched string (Figure 2.2). These are governed by

$$\frac{\partial^2 \eta}{\partial t^2} = c^2 \frac{\partial^2 \eta}{\partial x^2}, \qquad c = \sqrt{\frac{T_s}{\rho_s}}, \tag{2.17}$$

where $\eta(x,t)$ is the lateral displacement of the string, $c$ the wave speed, $T_s$ the tension in the string and $\rho_s$ its mass per unit length. The key point is that a general solution of (2.17) turns out to be

$$\eta(x,t) = F_f(x - ct) + F_b(x + ct), \tag{2.18}$$

where $F_f$ and $F_b$ are arbitrary functions of their arguments. The proof of (2.18) is straightforward. Let us write $\chi^{\pm} = x \pm ct$. Then from (2.18) we have

$$\frac{\partial^2 \eta}{\partial t^2} = c^2[F_f''(\chi^-) + F_b''(\chi^+)] = c^2 \frac{\partial^2 \eta}{\partial x^2}, \tag{2.19}$$

which evidently satisfies (2.17).

It is clear from Figure 2.3 that $F_f(x - ct)$ represents a *forward-travelling* pulse that propagates at the speed $c$ without any change of shape, while $F_b(x + ct)$ represents a *backward-travelling* pulse that also propagates at speed $c$, again without change of shape. The generality of the solution (2.18) is rather remarkable, and waves that can realise such a form, which turn out to be any waves governed by *the* wave equation, are called *non-dispersive waves.* However, it is worth repeating here that waves governed by a wave-like PDE, but not by *the* wave equation, do not in general admit d'Alembert's solution. Examples include waves on the surface of a pond and the flexural vibrations of a beam governed by (2.1). In such cases wave pulses typically spread out as they propagate, and these waves are known as *dispersive waves.* We shall pick up this story again in §2.3.4.

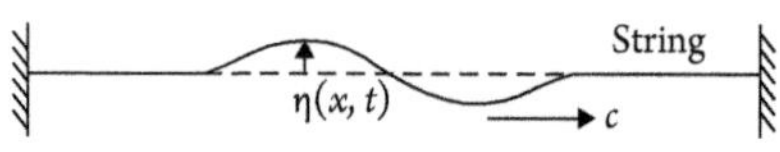


**Figure 2.2** Waves on a stretched string.

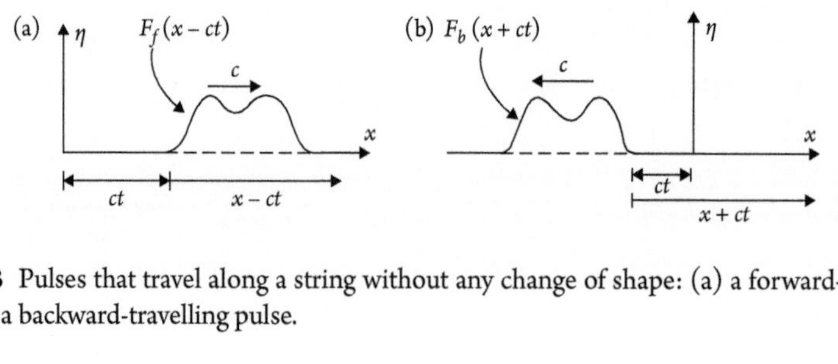


**Figure 2.3** Pulses that travel along a string without any change of shape: (a) a forward-travelling pulse; (b) a backward-travelling pulse.

### 2.1.4 A Key Property of Poisson's Equation: Green's Inversion Formula

We close this section by discussing solutions of the Poisson equation (2.9),

$$\nabla^2 T = -s(\mathbf{x}), \tag{2.20}$$

where the source $s$ is a known function of position. Here an important result is that, in an infinite domain, this may be inverted, or we might say solved, by the integral equation

$$T(\mathbf{x}) = \frac{1}{4\pi}\int \frac{s(\mathbf{x}')}{|\mathbf{x}-\mathbf{x}'|}d\mathbf{x}', \tag{2.21}$$

where $\mathbf{x}'$ is a dummy variable that samples the various points at which the source may be found and $d\mathbf{x}' = dx'\,dy'\,dz'$ is a volume element. This is sometimes referred to as the Green's function solution of Poisson's equation.

Some hint of where (2.21) comes from may be obtained by considering the particular case in which $s(\mathbf{x})$ is a three-dimensional delta function located at the origin, i.e. $s(\mathbf{x}) = \delta(\mathbf{x})$. In such cases $s(\mathbf{x})$ is a simply a blip concentrated at the origin that has the integral property

$$\int s(\mathbf{x})\,d\mathbf{x} = \int \delta(\mathbf{x})\,d\mathbf{x} = 1. \tag{2.22}$$

(Those unfamiliar with delta functions should consult Appendix I.) Clearly the solution for $T$ in such cases must be radially symmetric. Consider the particular trial solution

$$T = \frac{1}{4\pi\,|\mathbf{x}|} = \frac{1}{4\pi r}, \qquad \text{for } s(\mathbf{x}) = \delta(\mathbf{x}). \tag{2.23}$$

Evidently this is singular at the origin, but nevertheless it is well defined elsewhere and is clearly a solution of (2.20) away from the origin, since, in spherical polar coordinates,

$$\nabla^2 T = \frac{1}{r^2}\frac{\partial}{\partial r}\left[r^2\frac{\partial}{\partial r}\left(\frac{1}{4\pi r}\right)\right] = 0, \quad r \neq 0. \tag{2.24}$$

Of course, this is not enough to show that (2.23) is *the* solution of (2.20) for a unit delta function located at the origin (for example, $r^{-1}$ also satisfies (2.24)), and so we also need to consider what happens in the vicinity of the origin. We now encounter the problem that both $s(\mathbf{x})$ and the trial solution (2.23) diverge at $|\mathbf{x}| = 0$. However, the essential feature of a delta function is that, although its pointwise value is not well defined at the origin, it always integrates across the origin to give unity. So all we can or should ask of our trial solution in this respect is that its Laplacian also integrates across the origin to give unity. That is, integrating (2.20) with $s(\mathbf{x}) = \delta(\mathbf{x})$ over any volume that encloses the origin yields

$$\int_V \nabla^2 T\,d\mathbf{x} = -\int s(\mathbf{x})\,d\mathbf{x} = -\int \delta(\mathbf{x})\,d\mathbf{x} = -1, \tag{2.25}$$

and so all we should ask of our trial solution is that it does the same. So let us integrate the Laplacian of the trial solution (2.23) over a sphere of radius $r$ centred on the origin, invoking Gauss' theorem in order to side-step the singularity at $|\mathbf{x}| = 0$:

$$\int_V \nabla^2 T \, dV = \oint_S \nabla T \cdot d\mathbf{S} = -\oint_S \frac{1}{4\pi r^2} \, dS = -1.$$

Evidently our trial solution (2.23) satisfies (2.20) in an integral sense for any volume that encloses the origin, including one that shrinks onto the origin. So we conclude that (2.23) does indeed represent the solution to (2.20) for $s(\mathbf{x}) = \delta(\mathbf{x})$. In short,

$$\nabla^2 \left( \frac{1}{|\mathbf{x}|} \right) = -4\pi \delta(\mathbf{x}). \tag{2.26}$$

We can now go from the specific solution (2.23) to the more general Green's function solution (2.21) in two short steps. First we consider what happens if the delta function is displaced from the origin to location $\mathbf{x}'$, and has integral strength $s'$, rather than unity, i.e. $s(\mathbf{x}) = s'\delta(\mathbf{x} - \mathbf{x}')$. Noting that $|\mathbf{x}|$ in (2.23) has the meaning of 'distance from the source', it is evident that this expression now generalises to

$$T(\mathbf{x}) = \frac{s'}{4\pi |\mathbf{x} - \mathbf{x}'|}, \quad \text{for } s(\mathbf{x}) = s'\delta(\mathbf{x} - \mathbf{x}'). \tag{2.27}$$

Finally, we replace the continuous source distribution $s(\mathbf{x}')$ in (2.20) by an equivalent distribution of delta functions of the appropriate magnitude. Invoking superposition then takes us from (2.27) back to Green's solution (2.21).

The Green's function solution is a remarkable result that has many repercussions. For example, it leads directly to the famous Biot–Savart law for quasi-static magnetic fields. We shall consider some of its immediate consequences and potential generalisations in §2.4.

## 2.2 The Diffusion Equation Revisited

### 2.2.1 *More on Self-Similar Solutions*

Let us now return to the diffusion equation and explore some of its properties. We start by refining our ideas of diffusion length and of self-similar solutions of the one-dimensional diffusion equation. Let us return to the heated rod of §2.1.2 (Figure 2.1). The key finding here is that, as long as the geometric lengthscales $a$ and $L$ play no role in the diffusion process, dimensional analysis tells us that heat diffuses a distance $\ell \sim \sqrt{\alpha t}$ along the rod in a time $t$, and that the temperature distribution is self-similar in accordance with

$$\frac{T - T_{\text{ref}}}{\Delta T} = F\left( \frac{x}{\ell(t)} \right), \quad \ell = \sqrt{\alpha t}. \tag{2.28}$$

The caveat that $a$ and $L$ play no role, which applies whenever $a << \ell(t) << L$, turns out to be crucial. This can be seen in one of two ways. Formally, if the geometric lengthscales $a$ and $L$ are significant, then none of the dimensional arguments deployed in §2.1.2 remain valid, and so we cannot reach any concrete conclusions. Alternatively, it is clear on physical grounds that for very short time periods, $\ell(t) \sim a$, the temperature field must be treated as three-dimensional and so (2.12) does not apply, while at large times, $\ell(t) \sim L$, heat will start to pile up at the far end of the rod (if it is thermally insulated) and the self-similar solution will progressively be destroyed. So we must be cautious as to when we might expect such self-similar solutions to apply. Still, the idea that, in the absence of heat pilling up at boundaries, it diffuses a distance $\sim \sqrt{\alpha t}$ in a time $t$ is a useful one.

Of course, it is not just heat that diffuses through solids. A magnetic field trapped in an electrically conducting solid will diffuse in accordance with (2.4):

$$\frac{\partial \mathbf{B}}{\partial t} = \lambda \nabla^2 \mathbf{B}, \tag{2.29}$$

where $\lambda = 1/\sigma\mu$. Moreover, the momentum in a viscous fluid in which the pressure and non-linear inertial forces, $\nabla(p/\rho)$ and $(\mathbf{u} \cdot \nabla)\mathbf{u}$, are negligible (lava flows, lubricating oils, etc.) diffuses according to (2.2):

$$\frac{\partial \mathbf{u}}{\partial t} = \nu \nabla^2 \mathbf{u}. \tag{2.30}$$

Note that the appropriate diffusivity is now $\nu$, the kinematic viscosity of the fluid.

Let us consider an example of diffusion in a viscous fluid. Suppose a viscous fluid fills the half-space $y > 0$ and is in contact with a solid plate at $y = 0$. At $t = 0$ the plate is given a horizontal velocity $V$, dragging the fluid with it as shown in Figure 2.4. The question is, can we predict the horizontal velocity distribution in the fluid, $u(y, t)$?

The governing equation for $u(y, t)$ is evidently (2.30), in which we replace the vector $\mathbf{u}$ by its horizontal component $u$. The boundary condition on $u(y, t)$ comes in the form of the *no-slip condition* for viscous fluids, which says that fluid that touches a solid surface moves with that surface, which in this case yields $u(y = 0) = V$. Since there is no imposed lengthscale in this problem, we expect to find a self-similar velocity profile, and so we look for a self-similar solution of the form

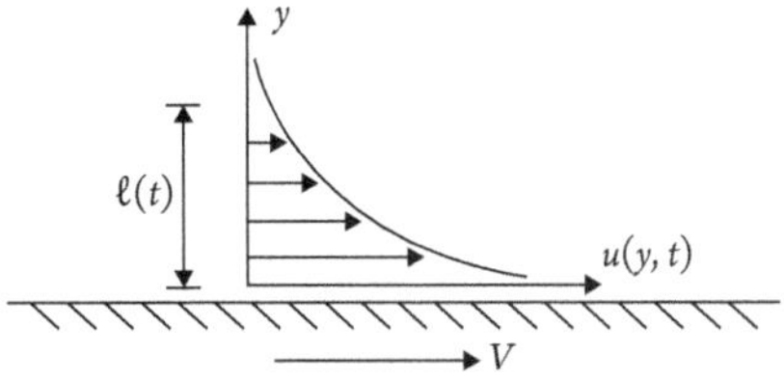


**Figure 2.4** A viscous fluid is in contact with a solid plate at $y = 0$. At $t = 0$ the plate is given a horizontal velocity $V$, and we wish to predict the fluid velocity $u(y, t)$.

$$\frac{u}{V} = F\left(\frac{y}{\ell(t)}\right) = F(\eta), \quad \ell = \sqrt{\nu t}. \tag{2.31}$$

Substituting this into (2.30) yields

$$2F''(\eta) + \eta F'(\eta) = 0, \tag{2.32}$$

exactly as for our thermal problem in §2.1.2. This may be integrated once to give $F'(\eta) \sim \exp(-\eta^2/4)$, and a second integration, subject to the boundary conditions $F(0) = 1$ and $F(\infty) = 0$, yields

$$\frac{u}{V} = F(\eta) = 1 - \frac{1}{\sqrt{\pi}} \int_0^\eta \exp(-\eta^2/4) d\eta = 1 - \text{erf}(\eta/2). \tag{2.33}$$

This is the same error-function profile we encountered in §2.1.2 for heat diffusion.

### 2.2.2 *The Diffusion of Oscillatory Fields*

Another restriction on self-similar solutions is that the boundary conditions must be steady. An important unsteady case is one in which the boundary conditions are oscillatory. To see how this changes things, let us redo the diffusion problem above, only now we oscillate the plate, rather than move it impulsively. The governing equation and boundary condition become

$$\frac{\partial u}{\partial t} = \nu \nabla^2 u, \quad u(y = 0) = V \cos \varpi t. \tag{2.34}$$

This may be solved by looking for solutions of the form $u(y, t) = \Re\{\hat{u}(y) \exp(j\varpi t)\}$, where $\hat{u}(y)$ is complex, $\Re$ indicates the real part of what follows, and $j = \sqrt{-1}$. Clearly $\hat{u}''(y) = (j\varpi/\nu)\hat{u}$ and hence

$$\hat{u}''(y) = \frac{(1+j)^2 \hat{u}}{\delta^2}, \qquad \delta = \sqrt{\frac{2\nu}{\varpi}}. \tag{2.35}$$

A little effort then yields

$$u(y, t) = V \exp\left(-\frac{y}{\delta}\right) \cos\left(\varpi t - \frac{y}{\delta}\right), \tag{2.36}$$

which is illustrated in Figure 2.5. Note that the oscillatory motion in the fluid is more or less restricted to a layer of thickness $\delta \sim \sqrt{2\nu/\varpi}$.

The analogous problem in electromagnetism is a magnetic field oscillating with constant amplitude adjacent to a semi-infinite conducting solid, say a metal block. Simply replacing (2.30) by (2.29), and $\nu$ by $\lambda = 1/\sigma\mu$, tells us that the magnetic field will penetrate only a distance

$$\delta = \sqrt{\frac{2\lambda}{\varpi}} = \sqrt{\frac{2}{\sigma\mu\varpi}} \tag{2.37}$$

into the metal. In electrical engineering $\delta = \sqrt{2/\sigma\mu\varpi}$ is referred to as the *skin depth*.

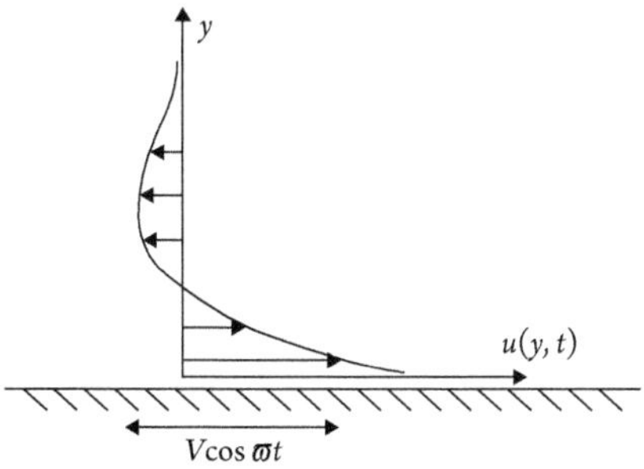


**Figure 2.5** A viscous fluid is in contact with a solid plate at $y = 0$. The plate oscillates with velocity $V \cos \varpi t$ and the oscillatory motion is transmitted to the fluid.

### 2.2.3 *Spherically Symmetric Diffusion, Kelvin and the Core of the Earth*

So far we have restricted attention to one-dimensional diffusion. However, one frequently encounters situations where the diffusion is radially symmetric, and so we now consider such cases. In spherical polar coordinates radially symmetric diffusion is governed by

$$\frac{\partial T}{\partial t} = \alpha \frac{1}{r^2} \frac{\partial}{\partial r} \left( r^2 \frac{\partial T}{\partial r} \right), \quad r = |\mathbf{x}|. \tag{2.38}$$

In such cases it is often useful to make the substitution $T(r,t) - T_0 = \Gamma(r,t)/r$, where $T_0$ is some reference temperature. Equation (2.38) then reverts to the familiar one-dimensional form

$$\frac{\partial \Gamma}{\partial t} = \alpha \frac{\partial^2 \Gamma}{\partial r^2}. \tag{2.39}$$

As an example of where this substitution proves useful, consider the following problem. A spherical cavity of radius $R$ is imbedded in a large solid mass and is filled with liquid. Initially the liquid and solid are all at temperature $T_0$ and at $t = 0$ the liquid temperature is raised to $T_0 + \Delta T$. The question is how heat diffuses into the solid through the spherical surface $r = R$. The obvious strategy is to look for a self-similar solution of the form

$$\frac{\Gamma(r,t)}{R \Delta T} = F\left( \frac{r - R}{\sqrt{\alpha t}} \right) = F(\eta), \quad F(0) = 1,\ F(\infty) = 0, \tag{2.40}$$

where $r$ is measured from the centre of the cavity and $\Gamma$ determined by (2.39). However, this is exactly the problem we solved when considering the one-dimensional flow of heat along a rod in §2.1.2, and also the viscous diffusion of momentum from an impulsively started plate in §2.2.1. So, without any work, we can lift the solution directly from either (2.16) or (2.33):

$$\frac{\Gamma(r,t)}{R \Delta T} = F(\eta) = 1 - \frac{1}{\sqrt{\pi}} \int_0^{\eta} \exp(-\eta^2/4)\, d\eta. \tag{2.41}$$

The temperature field is then

$$\frac{T(r,t)-T_0}{\Delta T} = \frac{R}{r}\left[1 - \frac{1}{\sqrt{\pi}}\int_0^{\eta} \exp(-\eta^2/4)\,d\eta\right], \tag{2.42}$$

with a corresponding temperature gradient at the surface $r = R$ of

$$\frac{\partial T}{\partial r} = -\frac{\Delta T}{\sqrt{\pi \alpha t}}\left(1 + \frac{\sqrt{\pi \alpha t}}{R}\right). \tag{2.43}$$

The second term on the right is a correction to the planar solution due to curvature and is negligible for times during which the diffusion length is much smaller than $R$.

In 1846 Kelvin used a calculation not unlike this to predict the age of the earth, with estimates of $\partial T/\partial r$ being determined from temperature gradients measured in deep mines. He concluded that the earth is around $10^8$ years old, an estimate that was later revised down to $2 \times 10^7$ years. This immediately triggered vigorous debate with both the geologists and the biologists, who were convinced the earth must be much older. The geologists, most notably Charles Lyell, had come to this conclusion through estimates of how long it would take for natural forces (earthquakes, volcanic eruptions, and erosion by wind and water) to cause the geological changes that had occurred, while Darwin required an older earth because a long time is required for natural selection to be effective. Indeed, in the first edition of his *Origin of the Species* Darwin writes:

> He who can read Sir Charles Lyell's grand work on the *Principles of Geology*, ... yet does not admit how incomprehensibly vast has been the past period of time, may at once close this volume.

So Darwin had nailed his colours to the mast, but it was left to his advocate Thomas Huxley to argue the case. In 1869, some nine years after his celebrated clash with Bishop Wilberforce, the tenacious biologist Huxley decided to take Kelvin to task. He did not doubt the correctness of Kelvin's mathematics, but questioned the assumptions that underpinned his analysis. Kelvin in turn was characteristically steadfast in the defence of his estimate, and decades of debate followed. When Huxley died in 1895 this debate was still unresolved, but the following year radioactivity was discovered and the arguments finally began to swing in favour of the Darwinists. Matters came to a head in 1904 when Rutherford, in a public lecture, noted that radioactive heating in the earth's core could mean that Kelvin's calculation significantly underestimated its age. We now know that the biologists were correct; the earth is around $5 \times 10^9$ years old.

Around the same time, a second important estimate of diffusion in the earth's core was underway, although this was much less well publicised. Towards the end of the nineteenth century there was great uncertainty as to the origin of the electrical currents deep within the earth's core, currents that are needed to support the geomagnetic field. One idea was that those currents, and their associated magnetic field, could have been trapped in the earth's iron core when it formed. The problem, though, is that a magnetic field trapped within a conducting solid will, unless supported by some external agency, decay as a result of Ohmic diffusion, a process that is governed by (2.29). The question, then, is how long can a primordial magnetic field trapped in the earth's core survive. One way to answer this is to use the idea of diffusion length to estimate the time required by a magnetic field to diffuse across the metallic core. If we infer from (2.43) the estimate $\ell \sim \sqrt{\pi\lambda t}$, where $\lambda$ is the magnetic diffusivity, and equate the diffusion length to the core radius $R_C$, we come up with a decay time of $t \sim R_C^2/\pi\lambda$. (Actually, an exact analysis gives the decay time for the slowest mode as $t \sim R_C^2/\pi^2\lambda$.) The English applied mathematician Horace Lamb undertook this calculation and published the result in 1883, concluding that a magnetic field trapped in a metal sphere the size of the earth would decay over a period of $\sim 10^6$ years. This is much too rapid to account for the present magnetic field, geological records showing the field to be $\sim 10^9$ years old. Clearly the origin of the geomagnetic field had to be sought elsewhere and many theories were proposed and subsequently found wanting, including the Hall effect, the rotation of negative electrostatic charges in the earth's crust, and thermoelectric currents. It was not until 1919 that a fluid dynamo, driven by convection, was finally suggested by the Irish physicist Joseph Larmor, an idea that would lead to a century of active research.

## 2.3 Waves and the Wave Equation Revisited

We now turn from the diffusion equation to the wave equation. We start by picking up on the discussion of §2.1.3 where we looked at transverse waves on a string.

### 2.3.1 *From Waves on a String to Alfvén Waves in a Plasma*

Faraday knew that magnetic field lines behave as if they are in tension. In particular, as we shall see in Chapter 8, a thin isolated flux tube experiences a tensile stress of $B^2/\mu$ along its axis, and if the flux tube has a cross-sectional area of $A$, then the tensile force resulting from this Faraday tension is $(B^2/\mu)A$. Now, a plucked violin string, with instantaneous radius of curvature of $R$ and carrying a tension $T_s$, experiences a transverse force per unit length of $T_s/R$, directed towards the centre of curvature of the string. Likewise, a magnetic flux tube carrying a tensile stress of $B^2/\mu$ experiences a transverse force per unit volume of $B^2/\mu R$ directed towards the centre of curvature of the tube.

It also turns out that, if the magnetic field lines sit in an electrically conducting fluid, such as a plasma, then they tend to 'stick' to the fluid, provided the conductivity of the medium is sufficiently high. That is to say, if the plasma moves, it drags the field lines along with it, as if the field and plasma were glued together. (This is discussed at length in Chapter 20.) Consider, then, what happens if we apply an impulsive force that pushes a jet of highly conducting fluid past a magnetic field. The field lines will be swept along with the flow and the resulting curvature of the lines will create a back reaction on the fluid (Figure 2.6). The greater the distortion of the field lines, the larger the back reaction, and so eventually the fluid will come to rest and the Faraday tension will then reverse the flow, pushing the fluid and field lines back towards their equilibrium position. However, the finite inertia of the

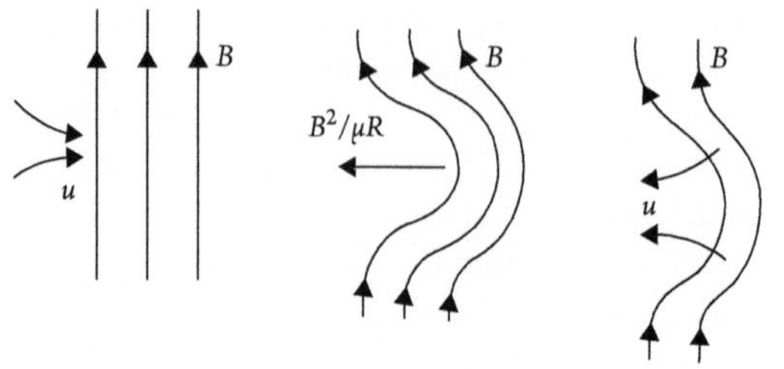


**Figure 2.6** In a highly conducting fluid the magnetic field lines behave like strings in tension, glued to the fluid. They oscillate if perturbed.

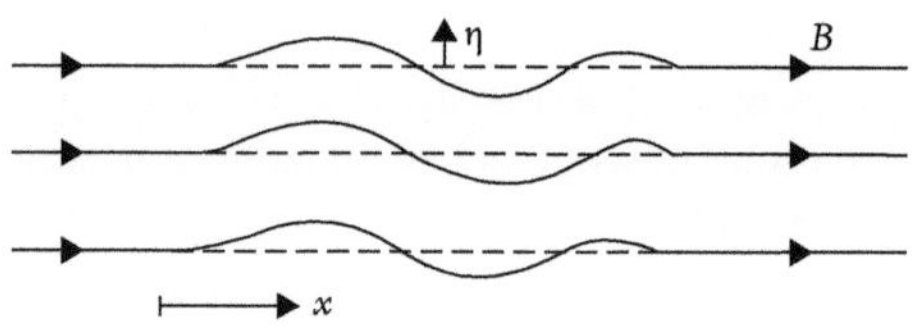


**Figure 2.7** Alfvén waves in a plasma.

fluid will carry it, and hence also the magnetic field lines, past the equilibrium position and so the entire process now begins in reverse. Evidently oscillations result and these are called Alfvén waves. Such waves are ubiquitous in astrophysics.

As noted in §2.1.3, a plucked violin string has a wave speed of $c = \sqrt{T_s/\rho_s}$, where $\rho_s$ is the mass per unit length of the string. Now, we have seen that a thin magnetic flux tube of cross-sectional area $A$ behaves as if it carries a tension of $T = (B^2/\mu)A$, and of course it has a mass per unit length of $\rho A$, where $\rho$ is the density of the plasma. This suggests that Alfvén waves have a wave speed of $c = \sqrt{B^2/\rho\mu}$, and we shall see in Chapter 21 that this is indeed the case. So Alfvén waves are transverse waves that travel along magnetic field lines, rather like waves on a violin string (Figure 2.7). Moreover, just like violin strings, they admit d'Alembert's solution, and so wave pulses of arbitrary profile can propagate along the field lines without change of shape. We shall return to the topic of Alfvén waves in Chapter 21, where we discuss their properties in considerable detail.

## 2.3.2 *Spherically Symmetric Solutions of the Wave Equation*

Let us now transfer attention from one-dimensional transverse waves to spherically symmetric waves, taking acoustics as a prototype. Such radially symmetric waves are governed by (2.5) in the form

$$\frac{\partial^2 p}{\partial t^2} = a^2 \frac{1}{r^2}\frac{\partial}{\partial r}\left(r^2 \frac{\partial p}{\partial r}\right), \tag{2.44}$$

where $a$ is the speed of sound. As with the diffusion equation in §2.2.3, it pays to introduce the auxiliary field $\widehat{p}\,(r,t) = rp(r,t)$, which is governed by the more familiar one-dimensional equation,

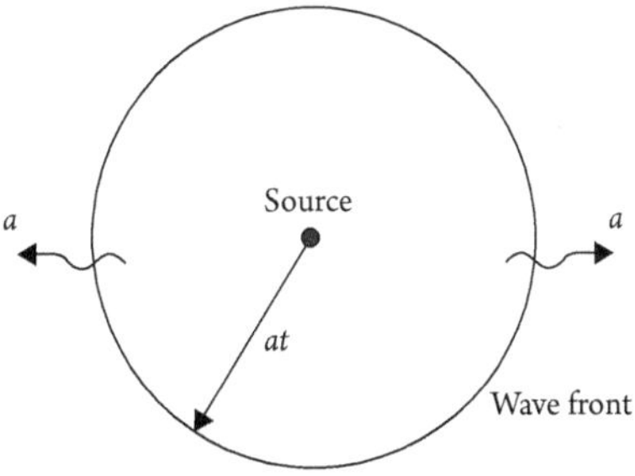


**Figure 2.8** A spherical wave front created by a pulse at the origin.

$$\frac{\partial^2 \widehat{p}}{\partial t^2} = a^2 \frac{\partial^2 \widehat{p}}{\partial r^2}. \tag{2.45}$$

It follows immediately from the discussion in §2.1.3 that $\widehat{p}\,(r,t)$ admits d'Alembert's solution (2.18) and so, by implication, the general solution of (2.44) is

$$p(r,t) = \frac{1}{r}[F_{\text{out}}(t - r/a) + F_{\text{in}}(t + r/a)]. \tag{2.46}$$

Here $F_{\text{out}}(t - r/a)$ represents an outward-travelling pulse of arbitrary profile that propagates at the speed $a$ without change of shape, while $F_{\text{in}}(t + r/a)$ represents an inward-travelling pulse that also propagates at speed $a$ without change of shape. Typically we are interested in waves that are initiated at the origin, rather than incoming waves, and in such cases (2.46) simplifies to

$$p(r,t) = \frac{F_{\text{out}}(t - r/a)}{r}. \tag{2.47}$$

Thus, for example, if we have a localised source of waves at the origin that takes the form of a blip at $t = 0$, or, more formally, a delta function in time, $\delta(t)$, we would expect a wave front to propagate outward as the spherical surface $r = at$ according to

$$p(r,t) \sim \frac{\delta(t - r/a)}{r}. \tag{2.48}$$

This is shown in Figure 2.8 and we shall see in §2.5.1 that (2.48) is substantially correct.

### 2.3.3 *Some Wave-Bearing Systems Not Governed by the Wave Equation*

One of the most fundamental ideas in the theory of wave propagation is that there is a sharp distinction between those waves that are governed by *the* wave equation (with a constant wave speed) and those that are governed by a similar, though different, wave-like PDE. The former are called *non-dispersive waves*, because wave packets travel without change of shape in accordance with d'Alembert's solution, while the latter are called *dispersive waves*, because wave packets typically spread out (disperse) as they propagate. This is a consequence of the fact that d'Alembert's solution works only for *the* wave equation, but not for its relatives. We shall explain why this distinction is so important in the next section, but first we need to introduce some examples of dispersive waves governed by wave-like PDEs. To keep the algebra simple, we shall restrict the discussion to one-dimensional wave systems. Let us start with the flexural vibrations of a beam.

Consider a thin, straight beam of mass per unit length $\rho_b$, Young's modulus $E$ and a cross-section whose second moment of inertia about its centreline is $I$ (Figure 2.9). Then, as is well known, transverse oscillations of the beam are governed by

$$\rho_b \frac{\partial^2 \eta}{\partial t^2} + IE \frac{\partial^4 \eta}{\partial x^4} = 0, \tag{2.49}$$

where $\eta(x,t)$ is the lateral displacement of the centreline of the beam and the coordinate $x$ runs along the centreline. If we look for travelling wave disturbances of the form $\eta(x,t) \sim \cos(kx - \varpi t)$, $k$ being the wavenumber of a disturbance of wavelength $\lambda = 2\pi/k$, then we obtain the *dispersion relationship*

$$\varpi^2 = (IE/\rho_b)k^4, \quad \text{or } \varpi = Kk^2, \tag{2.50}$$

where $K = \sqrt{IE/\rho_b}$. (The dispersion relationship in any wave-bearing system is the relationship between the angular frequency $\varpi$ and the wavenumber $k$ corresponding to a sinusoidal disturbance. This is usually written in the form $\varpi = \varpi(k)$.) The velocity of the wave crests, $c_p = \varpi/k$, is called the *phase velocity*. In the case of flexural vibrations of a beam this is given by $c_p = \varpi/k = Kk$.

Next consider gravity waves on the surface of a pond, idealised as a frictionless fluid of undisturbed depth $h$ (Figure 2.10). Let the change in elevation of the surface due to a disturbance be $\eta(x,t)$, and again we look for travelling waves of the form $\eta(x,t) \sim \cos(kx - \varpi t)$. In this case the dispersion relationship turns out to be

$$\varpi^2 = gk\tanh(kh), \tag{2.51}$$

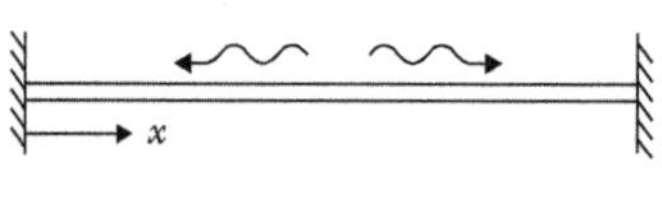


**Figure 2.9** Transverse oscillations of a beam.

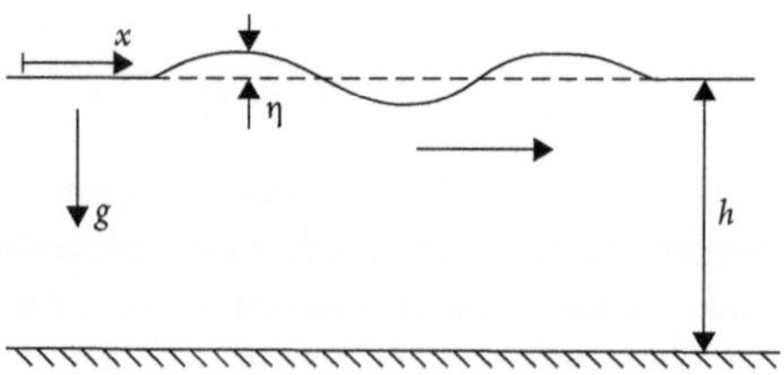


**Figure 2.10** Wave on the surface of a pond.

where $g$ is the acceleration due to gravity (see e.g. Acheson, 1990). For the case of deep-water waves (large $kh$), this simplifies to $\varpi = \sqrt{g|k|}$ and the phase velocity is evidently $c_p = \sqrt{g/|k|}$.

By way of comparison, consider one-dimensional acoustic waves, say waves in an organ pipe. Such waves are governed by the wave equation,

$$\frac{\partial^2 p}{\partial t^2} = a^2 \frac{\partial^2 p}{\partial x^2}, \tag{2.52}$$

and travelling waves of the form $p(x,t) \sim \cos(kx - \varpi t)$ have the dispersion relationship $\varpi = \pm ak$ and a phase velocity of $c_p = \varpi / k = \pm a$. Thus the wave crests move at the speed of sound, irrespective of the value of $k$. The key difference between acoustic waves and surface gravity waves, or waves on a beam, is that the phase velocity of acoustic waves is independent of $k$, whereas those systems governed by a wave-like PDE (but not *the* wave equation) have a phase velocity that depends on the wavenumber. In short, when waves are governed by the wave equation, the waves crests corresponding to different Fourier modes (different wavenumbers) all travel at the same speed—the wave speed—whereas in systems governed by a wave-like PDE, the waves crests corresponding to different wavenumbers travel at different speeds. This is the basis of the distinction between dispersive and non-dispersive waves, as we now discuss.

### 2.3.4 *Dispersive Versus Non-Dispersive Waves and d'Alembert Explained*

As indicated in the previous section, it is natural when investigating a one-dimensional wave-bearing system to look for travelling wave solutions of the form

$$\eta(x,t) \sim \exp[j(kx - \varpi t)] = \exp[jk(x - c_p t)], \tag{2.53}$$

where the real part is understood and $c_p = \varpi / k$, the phase speed, is the speed of the wave crests. In part, the idea behind using a trial solution like (2.53) is that we can always Fourier-decompose a more complicated initial condition into a series of such terms with varying wavenumber, and then take advantage of linearity to study each Fourier mode separately. Having determined how each Fourier mode in the initial condition propagates, we can then use superposition to reconstruct the full wave pattern at some later time. A common (and useful) convention when writing an expression like (2.53) is to take $\varpi$ as positive, but allow $k$ to take either sign. In such cases $k > 0$ corresponds to forward-travelling wave crests, $c_p > 0$, and $k < 0$ to backward-travelling wave crest, $c_p < 0$, as indicated in Figure 2.11.

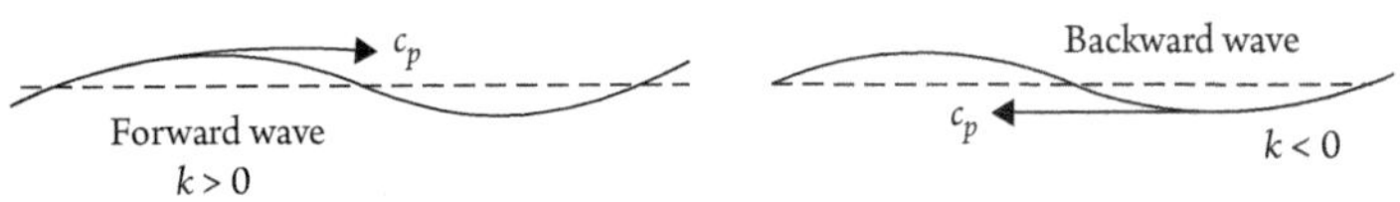


**Figure 2.11** Forward ($k > 0$) and backward ($k < 0$) travelling wave crests.

Having adopted (2.53) as a trial solution, the governing PDE for the system in hand then yields the dispersion relationship, $\varpi = \varpi(k)$. Those waves for which $\varpi(k)$ is linear in $k$, which are those systems governed by the wave equation (with a constant wave speed), are called non-dispersive, and all other types of waves are classified as dispersive. The essential point about non-dispersive waves is that the phase speed, and something we will meet in the next section called the group velocity, are independent of wavenumber and equal to the wave speed, say $c$. This, in turn, explains why the wave equation admits d'Alembert-like solutions in which wave pulses of arbitrary profile travel without change of shape. The idea is the following. Suppose our initial conditions consist of a localised pulse. To determine what happens to this pulse at later times, the natural strategy is to Fourier-decompose, or rather Fourier-transform, the initial pulse into a set of Fourier modes, each of given amplitude and wavenumber. As the system is linear, each mode then propagates independently of all the others. To determine the shape of the disturbance at some later time, we simply reconstruct the disturbance by adding together all the Fourier modes, which mathematically amounts to performing the inverse Fourier transform. However, for non-dispersive waves governed by the wave equation, the phase speed and group velocity are the same for all Fourier modes, and equal to $c$. So, after a given time $t$, all the Fourier modes will have moved the same distance, $ct$, and so when we reconstruct the pulse at time $t$, it has exactly the same shape as it had at time $t = 0$, just shifted by the distance $ct$.

The same argument explains why, in a dispersive system, a wave pulse changes shape as it propagates. After Fourier-decomposing the initial disturbance, each Fourier mode will travel at a different speed and so will propagate a different distance in a time $t$. When we reconstruct the disturbance at the later time $t$, by adding all the Fourier modes together, we get a shape that is different to that of the initial pulse because the various Fourier modes will have changed their relative positions. Typically, in dispersive systems, the pulse starts out as localised, because of the way it is generated, and then spreads out as it propagates. (Think of throwing a stone into a pond and the ripples that spread out from the point of impact.) Eventually, when the various Fourier modes are well separated in space, the initially localised pulse becomes dispersed, adopting the form of a *slowly modulated wave train*, sometimes called a *wave packet*. At any one location within such a wave train the disturbance looks like a single Fourier mode of given amplitude and wavenumber, but as we move along the wave train the amplitude and wavenumber slowly change. Typically, the amplitude of the modes of the various wavenumbers within a wave train, $a(k)$, is sharply peaked around the dominant wavenumber of the disturbance, $k_0$, as indicated in Figure 2.12. For example, in the case of a stone falling into a pond, $k_0 \sim 2\pi/(\text{stone size})$.

### 2.3.5 *The Concept of Group Velocity for Dispersive Waves*

Dispersive and non-dispersive waves differ in a second crucial respect: the speed at which a wave packet propagates as a whole in a dispersive system is *not* the speed at which the wave crests move. To focus our thoughts, let us once again consider flexural vibrations of a beam and surface gravity waves on a pond (Figure 2.13).

The speed at which a wave packet propagates as a whole in a dispersive system is called the *group velocity* of that wave packet, and it is not the same as the phase velocity of the

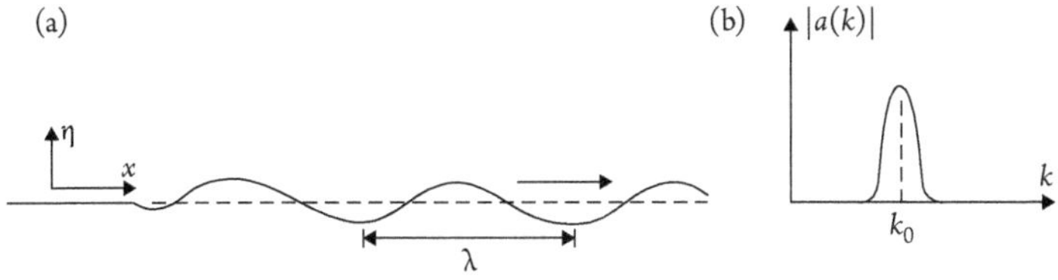


**Figure 2.12** (a) A slowly modulated wave train. (b) The variation of the amplitude of the Fourier modes as a function of wavenumber.

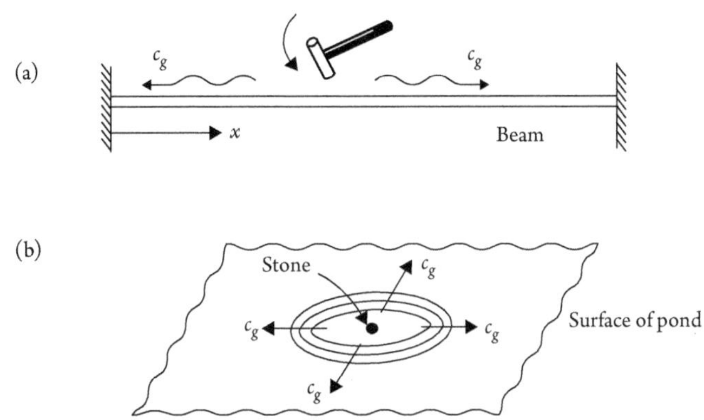


**Figure 2.13** Ripples generated in (a) a beam and (b) the surface of a pond.

dominant wavenumber within the wave packet. For example, if you strike a beam with a hammer, the resulting wave crests travel at half the speed of the wave packet within which they sit, or if you throw a stone into a deep pond, individual wave crests travel at twice the speed at which the overall patch of ripples disperses from the point of impact. This is one of the most counterintuitive aspects of wave dispersion and clearly needs to be explained. Let us start with the stone thrown into a pond.

Consider the following experiment. You throw a stone into a deep pond and watch the ripples spread out in the form of an annulus (Figure 2.14(a)). After some time the waves become reasonably dispersed and a cross section through the wave train looks something like that shown in Figure 2.14(b), where the dominant wavenumber will be of the order of $2\pi/(\text{stone size})$. If you focus attention on a particular wave crest within the wave packet, it will have an angular frequency $\varpi = \sqrt{g|k|}$ and travel at a phase speed $c_p = \sqrt{g/|k|}$, exactly as suggested in §2.3.3. (For simplicity, we shall ignore the fact that the dispersion is axisymmetric, rather than planar.) However, if you track the centre of the wave packet, then it travels at the slower speed $c_g = d\varpi/dk = \frac{1}{2}\sqrt{g/|k|}$, which we call the *group velocity* of the wave packet.

So, how can the fluid accommodate this difference in speed? On closer inspection you will find that ripples seem to appear out of nowhere at the rear of the wave packet, grow as they ripple through the wave packet at a speed $c_p = \sqrt{g/|k|}$, and then fade away as they approach

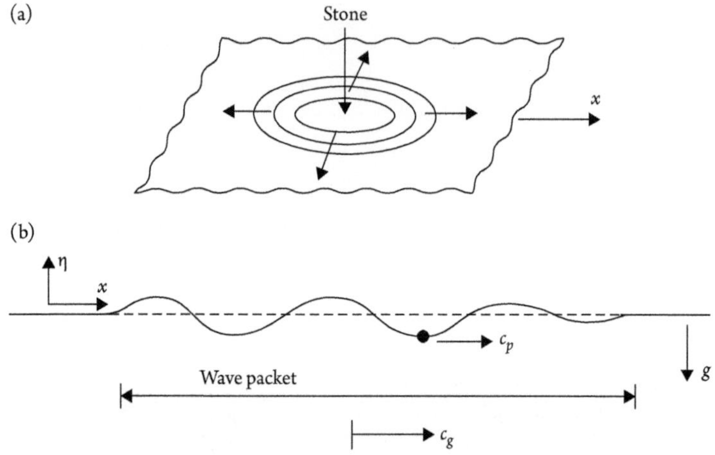


**Figure 2.14** (a) Waves on the surface of a pond. (b) Cross-section through a wave packet.

the far end of the wave train. If you did a similar experiment with a hammer and a beam, you would find the wave crests travelling at $c_p = \varpi/k = Kk$, in accordance with (2.50), but the overall wave packet that contains the waves moving at a group velocity $c_g = d\varpi/dk = 2Kk$, which is twice the speed of the wave crests.

The key question then is: why do wave packets travel at the speed $c_g = d\varpi/dk$, and not the phase speed $c_p = \varpi/k$? To understand this, we need to perform a little Fourier analysis (Kelvin's favourite mathematical toolkit!). Suppose that we have a slowly modulated wave train, as shown in Figure 2.12. Waves of wavenumber $k$ (wavelength $\lambda = 2\pi/k$) propagate as

$$\eta(x,t) \sim a(k)\exp[j(kx - \varpi t)], \quad \varpi = \varpi(k), \tag{2.54}$$

where $a(k)\,dk$ is the amplitude of the waves in the range $dk$. Superposition then tells us that

$$\eta(x,t) = \int_{-\infty}^{\infty} a(k)\exp[j(kx - \varpi t)]\,dk, \tag{2.55}$$

and we recognise that $a(k)$ is simply the Fourier transform of the initial disturbance $\eta(x,0)$. Now, if $a(k)$ is centred on $k_0$, as suggested in Figure 2.12, then we may write

$$\varpi(k) = \varpi(k_0) + (k - k_0)c_g(k_0) + O((k - k_0)^2), \tag{2.56}$$

where $c_g = d\varpi/dk$. We now substitute (2.56) into (2.55) and neglect the quadratic term in the expansion (2.56) on the assumption that $a(k)$ is small when $k$ is remote from $k = k_0$. This yields

$$\eta(x,t) = \int_{-\infty}^{\infty} a(k)\exp\{j[k_0 x - \varpi(k_0)t + (k-k_0)(x - c_g(k_0)t)]\}\,dk,$$

or equivalently,

$$\eta(x,t) = \exp[j(k_0 x - \varpi(k_0)t)] \int_{-\infty}^{\infty} a(k)\exp\{j[(k-k_0)(x - c_g(k_0)t)]\}\,dk. \tag{2.57}$$

Finally, treating $k - k_0$ as a dummy variable in (2.57), we conclude that (2.57) takes the general form

$$\eta(x,t) = F(x - c_g(k_0)t)\exp[j(k_0 x - \varpi(k_0)t)], \tag{2.58}$$

where the shape of the function $F$ depends on the amplitude distribution $a(k)$. Evidently this is a travelling wave of wavenumber $k_0$ and frequency $\varpi(k_0)$ which is modulated by an envelope (or spatially varying amplitude), as represented by $F(x - c_g(k_0)t)$. This is shown schematically in Figure 2.15 and it is clear that the overall wave packet, as represented by $F(x - c_g(k_0)t)$, propagates at the speed $c_g(k_0) = (d\varpi/dk)_0$.

So far, we have shown that a slowly modulated wave train, or wave packet, propagates at the speed $c_g(k_0) = (d\varpi/dk)_0$, which is the group velocity corresponding to the dominant wavenumber in the wave packet. It is possible to refine this analysis somewhat and establish a slightly stronger result. Specifically, it turns out that $c_g(k) = d\varpi/dk$ is the speed at which you must travel to continually observe waves of wavelength $\lambda = 2\pi/k$. By implication, $c_g(k)$ is the speed at which waves of wavenumber $k$ transport energy across the medium.

Although we have restricted the discussion to one-dimensional wave packets, the entire analysis generalises in an obvious way to three dimensions. For example, solutions of *the* wave equation (with a constant wave speed) are non-dispersive, such as the pressure waves described by (2.47). When considering three-dimensional dispersive waves, one first looks for travelling waves of the form

$$\eta(x,t) \sim \exp[j(\mathbf{k}\cdot\mathbf{x} - \varpi t)], \tag{2.59}$$

where $\mathbf{k}$ is now a *wave vector*, which has a direction as well as a magnitude. Substituting this trial solution into the governing PDE for the particular system at hand yields the

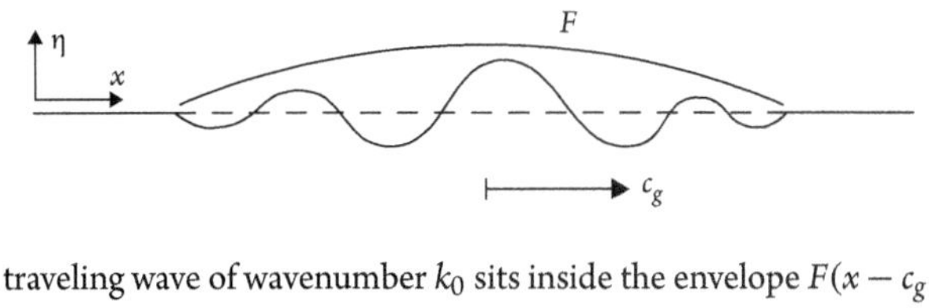


**Figure 2.15** A traveling wave of wavenumber $k_0$ sits inside the envelope $F(x - c_g(k_0)t)$.

dispersion relationship $\varpi = \varpi(\mathbf{k})$. The group velocity is then defined as the vector $\mathbf{c}_g$ whose components are

$$(\mathbf{c}_g)_i = \frac{\partial \varpi}{\partial k_i}. \tag{2.60}$$

As in one dimension, $\mathbf{c}_g$ has the properties that (i) $\mathbf{c}_g(\mathbf{k}_0)$ is the velocity at which a wave packet dominated by the wave vector $\mathbf{k}_0$ propagates away from the source of the disturbance, (ii) $\mathbf{c}_g(\mathbf{k})$ is the velocity at which you must travel in order to keep seeing waves of wave vector $\mathbf{k}$ and (iii) $\mathbf{c}_g(\mathbf{k})$ is the velocity at which the energy of waves of wave vector $\mathbf{k}$ propagates through the medium.

In three dimensions the group velocity and phase velocity $\mathbf{c}_p(\mathbf{k}) = (\varpi/k^2)\mathbf{k}$ of dispersive waves usually have different magnitudes *and* directions. For example, internal gravity waves in a stratified fluid have $\mathbf{c}_g$ perpendicular to $\mathbf{c}_p$, so that the wave energy propagates *parallel* to the wave crests! (See Exercise 2.2 at the end of this chapter.)

## 2.4 More on Inverting Poisson's Equation

Let us now transfer attention from the wave equation to Poisson's equation $\nabla^2 V = -s(\mathbf{x})$. Our primary interest here is to explore in a little more detail Green's inversion integral

$$\boxed{V(\mathbf{x}) = \frac{1}{4\pi}\int \frac{s(\mathbf{x}')}{|\mathbf{x}-\mathbf{x}'|}\,d\mathbf{x}'.} \tag{2.61}$$

### 2.4.1 *An Example of the Use of Green's Function*

Let us consider a specific example designed to bring out the utility of (2.61). We have already seen that the motion of a viscous incompressible fluid is governed by

$$\frac{\partial \mathbf{u}}{\partial t} + (\mathbf{u}\cdot\nabla)\mathbf{u} = -\nabla\left(\frac{p}{\rho}\right) + \nu\nabla^2\mathbf{u}, \quad \nabla\cdot\mathbf{u} = 0, \tag{2.62}$$

where $\mathbf{u}(\mathbf{x},t)$ is the velocity field, $p(\mathbf{x},t)$ the pressure field, and $\rho$ and $\nu$ the density and viscosity of the fluid, respectively. Using the identity

$$\nabla(u^2/2) = (\mathbf{u}\cdot\nabla)\mathbf{u} + \mathbf{u}\times(\nabla\times\mathbf{u}),$$

this is often rewritten as

$$\frac{\partial \mathbf{u}}{\partial t} = -\nabla\left(\frac{\hat{p}}{\rho}\right) + \mathbf{u}\times(\nabla\times\mathbf{u}) + \nu\nabla^2\mathbf{u}, \tag{2.63}$$

where $\hat{p} = p + \frac{1}{2}\rho u^2$ is sometimes called the stagnation pressure. Taking the divergence of (2.63), and recalling that the velocity field of an incompressible fluid is solenoidal, we find

$$\nabla^2\left(\frac{\hat{p}}{\rho}\right) = -\nabla\cdot[(\nabla\times\mathbf{u})\times\mathbf{u}] = -s(\mathbf{x}). \tag{2.64}$$

Now consider the situation where $\nabla\times\mathbf{u}$ is localised in space, say the velocity field in the wake of a submarine's propeller. If we wish to determine the resulting pressure field, then we may invert (or solve) this Poisson equation using (2.61):

$$\hat{p}(\mathbf{x}) = \frac{\rho}{4\pi}\int\frac{s(\mathbf{x}')}{|\mathbf{x}-\mathbf{x}'|}\,d\mathbf{x}'. \tag{2.65}$$

Such a relationship is particularly useful if you want to locate a submarine by listening for its far-field pressure signature!

Two comments are in order at this point. First, usually when we encounter Poisson's equation $\nabla^2 V = -s(\mathbf{x})$, the source $s(\mathbf{x})$ is localised in space, say some given distribution of charges or currents, yet the thing we are solving for, $V(\mathbf{x})$, pervades all space. For example, it is possible to show (though it is a tedious calculation) that the pressure field given by (2.65) falls off as $|\mathbf{x}|^{-3}$ at large distances from the source.

Second, if the velocity field in the wake of our submarine varies with time, and we acknowledge this explicitly by rewriting (2.65) as

$$\hat{p}(\mathbf{x},t) = \frac{\rho}{4\pi}\int\frac{s(\mathbf{x}',t)}{|\mathbf{x}-\mathbf{x}'|}\,d\mathbf{x}', \tag{2.66}$$

we see immediately that something strange is going on. According to (2.66), a fluctuation in $\mathbf{u}$ behind our submarine is *instantaneously* felt in the far-field pressure distribution. Of course, this is impossible because changes in $\mathbf{u}$ are transmitted to the far field through acoustic waves, and these carry information at the speed of sound, $a$. So what has gone wrong? The answer is that the notion of an incompressible fluid is an idealisation (all fluids are to some degree compressible), and in this idealised limit the speed of sound is infinitely fast. If we incorporated a small but finite compressibility into our equations, there would appear a time lag between a fluctuation in $\mathbf{u}$ at $\mathbf{x}'$ and its appearance in the far-field pressure at $\mathbf{x}$. We shall return to this point in §2.4.4.

### 2.4.2 *A Glimpse at Gauss' Law of Electrostatics*

There is a close link between the Green's function solution (2.61) and Gauss' law of electrostatics, as we now show. Recall that we derived (2.61) by first considering the case where the source is a delta function located at the origin, $s(\mathbf{x}) = \delta(\mathbf{x})$. In particular, we showed that the solution to $\nabla^2 V = -\delta(\mathbf{x})$ is

$$V = \frac{1}{4\pi|\mathbf{x}|} = \frac{1}{4\pi r}, \tag{2.67}$$

and that (2.61) then follows from the principle of superposition.

Let us now turn to electrostatics, pre-empting some of the things we shall say in Chapter 4. In electrostatics the electric field **E** is irrotational, while its divergence is fixed by the distribution of charges:

$$\nabla\cdot\mathbf{E}=\frac{\rho_e}{\varepsilon_0},\quad \nabla\times\mathbf{E}=0. \tag{2.68}$$

Here $\rho_e$ is the total charge density, free charges plus bound charges, $\varepsilon_0$ is the permittivity of free space, and the expression $\nabla\cdot\mathbf{E}=\rho_e/\varepsilon_0$ is known as Gauss' law in differential form. By contrast, Gauss' law in integral form is clearly

$$\boxed{\oint_S \mathbf{E}\cdot d\mathbf{S}=\frac{1}{\varepsilon_0}\int_V \rho_e dV=\frac{Q}{\varepsilon_0},} \tag{2.69}$$

where $Q$ is the total charge enclosed by the surface $S$. Since $\nabla\times\mathbf{E}=0$, it is conventional to introduce a scalar potential for **E**, defined by $\mathbf{E}=-\nabla\varphi$. The governing equation for $\varphi$ is evidently the Poisson equation

$$\nabla^2\varphi=-\frac{\rho_e}{\varepsilon_0}. \tag{2.70}$$

Now consider the case of a point charge $q$ sitting at the origin (Figure 2.16). The integral version of Gauss' law applied to a spherical surface or radius $r$ centred on the origin tells us that the radially symmetric electric field surrounding $q$ must satisfy $|\mathbf{E}|\,(4\pi r^2)=q/\varepsilon_0$. It follows that

$$\mathbf{E}=-\frac{\partial\varphi}{\partial r}\hat{\mathbf{e}}_r=\frac{q}{4\pi r^2\varepsilon_0}\hat{\mathbf{e}}_r, \tag{2.71}$$

where $\hat{\mathbf{e}}_r$ is a unit radial vector. Evidently the electrostatic potential associated with our point charge is $\varphi=q/4\pi r\varepsilon_0$. However (2.70) for a point charge at the origin is

$$\nabla^2\varphi=-\frac{q}{\varepsilon_0}\delta(\mathbf{x}), \tag{2.72}$$

and so we conclude that the solution of $\nabla^2\varphi=-\delta(\mathbf{x})$ is $\varphi=1/4\pi r$. We have arrived back at (2.67), and by implication (2.61), through a consideration of Gauss' law alone.

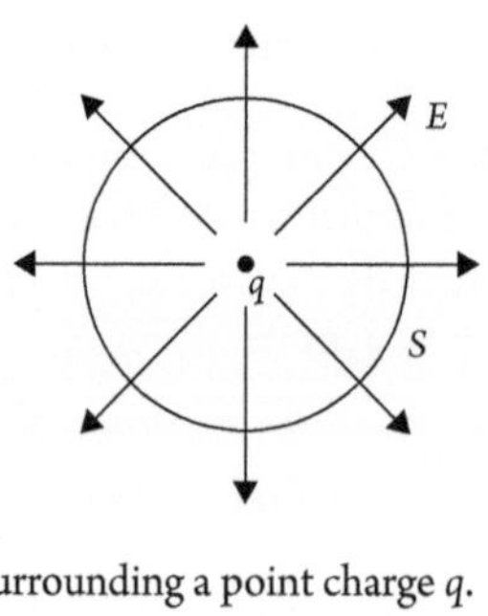


**Figure 2.16** The electric field **E** surrounding a point charge $q$.

### 2.4.3 A Glimpse at the Biot–Savart Law of Magnetostatics

There is also a close link between the Green's function solution (2.61) and the Biot–Savart law of magnetostatics, as we now show. In Chapter 3 we shall see that all magnetic fields are solenoidal, while the curl of a steady magnetic field $\mathbf{B}$ is determined by Ampère's law:

$$\nabla \cdot \mathbf{B} = 0, \qquad \nabla \times \mathbf{B} = \mu \mathbf{J}. \tag{2.73}$$

Here $\mu$ is the permeability of the medium and $\mathbf{J}$ is the current density vector, a vector that points in the direction of maximum current flow and has the units of amperes per square metre.

Since $\mathbf{B}$ is solenoidal, we have the identity $\nabla \times \nabla \times \mathbf{B} = -\nabla^2 \mathbf{B}$, and so the curl of (2.73) yields

$$\nabla^2 \mathbf{B} = -\mu \nabla \times \mathbf{J}. \tag{2.74}$$

The individual components of (2.74) may now be inverted using (2.61), which yields

$$\mathbf{B}(\mathbf{x}) = \frac{\mu}{4\pi} \int \frac{\nabla' \times \mathbf{J}'}{|\mathbf{r}|} d\mathbf{x}', \qquad \mathbf{r} = \mathbf{x} - \mathbf{x}', \tag{2.75}$$

where $\mathbf{J}' = \mathbf{J}(\mathbf{x}')$ and $\nabla'$ operates on $\mathbf{x}'$ while keeping $\mathbf{x}$ constant. Note that we use $\mathbf{r}$ for the displacement vector that takes us from the source at $\mathbf{x}'$ to the point where $\mathbf{B}$ is measured, as shown in Figure 2.17.

It is conventional to rewrite the integrand in (2.75) using the expression

$$\frac{\nabla' \times \mathbf{J}'}{|\mathbf{r}|} = \nabla' \times \left(\frac{\mathbf{J}'}{|\mathbf{r}|}\right) + \mathbf{J}' \times \nabla' \left(\frac{1}{|\mathbf{r}|}\right),$$

where the first term on the right can be converted from a volume integral into a surface integral, and so integrates to zero whenever $\mathbf{J}$ is localised in space. Moreover, it is readily confirmed that

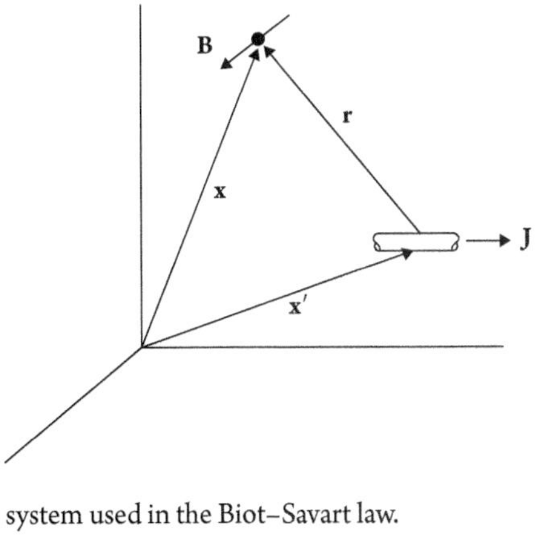


**Figure 2.17** Coordinate system used in the Biot–Savart law.

$$\nabla'\left(\frac{1}{|\mathbf{r}|}\right) = -\nabla\left(\frac{1}{|\mathbf{r}|}\right) = \frac{\mathbf{r}}{|\mathbf{r}|^3}, \tag{2.76}$$

where $\nabla$ operates on $\mathbf{x}$, keeping $\mathbf{x}'$ constant. So, provided the currents are localised in space, (2.75) can be rewritten as

$$\boxed{\mathbf{B}(\mathbf{x}) = \frac{\mu}{4\pi}\int \frac{\mathbf{J}(\mathbf{x}') \times \mathbf{r}}{|\mathbf{r}|^3}\, d\mathbf{x}', \quad \mathbf{r} = \mathbf{x} - \mathbf{x}'.} \tag{2.77}$$

This is the famous Biot–Savart law. Its utility is clear: when the current density is a known localised function of position, the corresponding steady magnetic field may be calculated directly from (2.77). In effect, this states that a small volume element $d\mathbf{x}'$ located at $\mathbf{x}'$ and carrying a current density of $\mathbf{J}(\mathbf{x}')$ induces a magnetic field at point $\mathbf{x}$ that is given by

$$d\mathbf{B}(\mathbf{x}) = \frac{\mu}{4\pi}\frac{\mathbf{J}(\mathbf{x}') \times \mathbf{r}}{|\mathbf{r}|^3}\, d\mathbf{x}'. \tag{2.78}$$

### 2.4.4 *Information Propagation and Hidden Physics*

We conclude our discussion of inverting the Poisson equation by picking up on a theme introduced at the end of §2.4.1. Recall that we noted something odd about (2.66). In particular, it suggests that a fluctuation in $\mathbf{u}$ behind our submarine is felt *instantaneously* in the far-field pressure distribution. In reality, of course, this is impossible because changes in $\mathbf{u}$ are transmitted to the far field through acoustic waves, and these propagate at the speed of sound, $a$. Of course, there is nothing wrong with Green's inversion of the Poisson equation (2.64). Rather, the Poisson equation itself is missing some key physics. In particular, by assuming the water is incompressible, we limited ourselves to an idealised fluid in which the speed of sound is infinite.

Something similar happens with the Biot–Savart law. Let us restrict ourselves to free-space fields and to situations where the currents are changing relatively slowly. In this case the Biot–Savart law is often written in the quasi-steady form

$$\mathbf{B}(\mathbf{x},t) = \frac{\mu_0}{4\pi}\int \frac{\mathbf{J}(\mathbf{x}',t) \times \mathbf{r}}{|\mathbf{r}|^3}\, d\mathbf{x}', \quad \mathbf{r} = \mathbf{x} - \mathbf{x}', \tag{2.79}$$

where $\mu_0$ is the permeability of free space. This suggests that a fluctuation in $\mathbf{J}$ at location $\mathbf{x}'$ *instantaneously* changes the magnetic field at some remote point $\mathbf{x}$. Again, this is impossible because changes in $\mathbf{J}$ are transmitted to remote locations through the propagation of electromagnetic waves, and these waves travel at the speed of light, $c$. As before, there is nothing wrong with Green's inversion of the Poisson equation (2.74), but rather the Poisson equation itself is missing some essential physics. In particular, Ampère's law is valid only in strictly steady situations. When the fields are time-dependent, it must be replaced by something called the Ampère–Maxwell equation. (This equation is introduced in Chapter 3, but the precise details do not matter right now.) As we shall see, this has the effect of changing our Poisson equation (2.74) into an inhomogeneous wave equation,

$$\nabla^2 \mathbf{B} - \frac{1}{c^2}\frac{\partial^2 \mathbf{B}}{\partial t^2} = -\mu \nabla \times \mathbf{J}. \tag{2.80}$$

So, by using Ampère's law to write (2.79), we were, in effect, considering an idealised case in which $c \to \infty$.

This is typical of the Poisson equation and its inversion. The absence of time in the solution seems to imply that when the source term changes, the influence of this change is felt instantaneously at all points, and this cannot be true in practice. The problem invariably lies in the derivation of the Poisson equation itself, where transient processes, such as wave propagation or diffusion, are suppressed (some would say *filtered out*) on the assumption that these act very rapidly on the time scales of interest.

## 2.5 Inverting the Inhomogeneous Wave Equation: Retarded Potentials

Rather remarkably, it is possible to invert (or solve) an inhomogeneous wave equation such as (2.80), in much the same way as (2.61) inverts the Poisson equation. To focus our thoughts, let us consider the particular case of acoustic waves.

### 2.5.1 *Retarded Potentials in Acoustics*

The governing equation for acoustic waves can be written in the form

$$\nabla^2 p - \frac{1}{a^2}\frac{\partial^2 p}{\partial t^2} = -s(\mathbf{x}, t), \tag{2.81}$$

where $a$ is the speed of sound and the source $s(\mathbf{x}, t)$ may, for example, be that in (2.64), although other acoustic sources are possible. For $a \to \infty$ Green's inversion integral tells us that

$$p(\mathbf{x}, t) = \frac{1}{4\pi}\int \frac{s(\mathbf{x}', t)}{|\mathbf{r}|} d\mathbf{x}', \quad \mathbf{r} = \mathbf{x} - \mathbf{x}', \tag{2.82}$$

and so the question is, what happens when $a$ is finite? The answer is simple, beautiful and intuitive. It is

$$\boxed{p(\mathbf{x}, t) = \frac{1}{4\pi}\int \frac{s(\mathbf{x}', t')}{|\mathbf{r}|} d\mathbf{x}', \quad t' = t - |\mathbf{r}|/a,} \tag{2.83}$$

where $\mathbf{r} = \mathbf{x} - \mathbf{x}'$ and $t'$ is called the *retarded time* (see e.g. Morse & Ingard, 1987). Evidently, all we need to do is replace $t$ on the right of (2.82) by $t'$! For reasons that will make more sense when we apply this in electrodynamics, the solution (2.83) is known as the *retarded potential.*

The physical interpretation of (2.83) is clear: it takes a time $|\mathbf{r}|/a$ for an event at $\mathbf{x}'$, where $s$ is evaluated, to reach $\mathbf{x}$, where $p$ is measured. The proof of (2.83) is a little more tricky, although it follows more or less the same line of reasoning as the proof of Green's inversion integral in §2.1.4. We start by considering a delta function at the origin whose strength varies with time: $s(\mathbf{x},t) = S(t)\delta(\mathbf{x})$. The resulting pressure field $p$ is radially symmetric and from (2.47) it must take the form

$$p(r,t) = \frac{F(t-r/a)}{r}, \tag{2.84}$$

for some $F$. However, for $a \to \infty$ our wave equation reverts to a Poisson equation and so the solution (2.84) must revert to the Green's function solution

$$p(r,t) = \frac{S(t)}{4\pi r}.$$

This tells us that $F$ is $S/4\pi$ and hence

$$p(r,t) = \frac{S(t-r/a)}{4\pi r}. \tag{2.85}$$

Now suppose the source were located at the point $\mathbf{x}'$, rather than at the origin, i.e. $s(\mathbf{x},t) = S'(t)\delta(\mathbf{x}-\mathbf{x}')$. Then evidently (2.85) generalises to

$$p(\mathbf{x},t) = \frac{S'(t-|\mathbf{r}|/a)}{4\pi\,|\mathbf{r}|}, \qquad \mathbf{r} = \mathbf{x}-\mathbf{x}'. \tag{2.86}$$

The final step is to replace the continuous source distribution $s(\mathbf{x}')$ in (2.81) by an equivalent distribution of delta functions of the appropriate magnitude. Invoking superposition then takes us from (2.86) back to the retarded potential (2.83) and the proof is complete. Retarded potentials play a pivotal role in the theory of acoustic radiation, and they are equally important in electrodynamics, where they may be used to invert wave equations like (2.80).

### 2.5.2 *Retarded Potentials and Moving Sources*

We close this chapter with a rather subtle point that frequently crops up in the theory of acoustic radiation, as well as in electromagnetic radiation. This relates to the generation of waves by moving sources. As above, we shall use the particular case of acoustic waves to get across the general idea, restricting the discussion to sources moving at subsonic speeds. So let us return to (2.81),

$$\nabla^2 p - \frac{1}{a^2}\frac{\partial^2 p}{\partial t^2} = -s(\mathbf{x},t), \tag{2.87}$$

where $s(\mathbf{x},t)$ is now taken to be a small, compact source of fixed shape and fixed overall strength $S$ that moves along a known trajectory $\mathbf{x}_0(t)$. (Here $\mathbf{x}_0(t)$ locates the *centre* of the moving source, and by the phrase *compact* we mean of a size smaller than the wavelength of

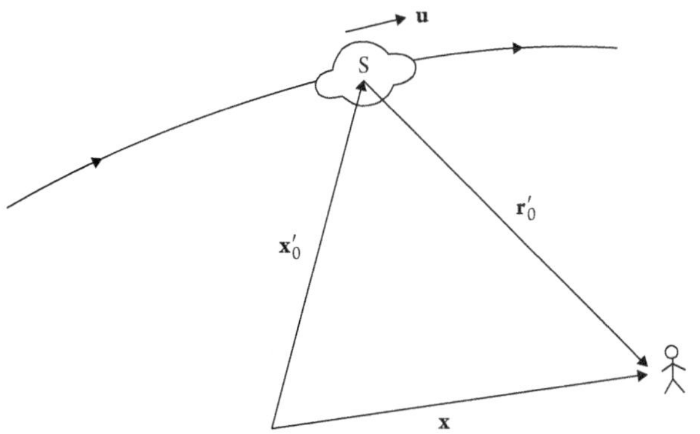


**Figure 2.18** A compact source of magnitude $S$ moves along a known trajectory $\mathbf{x}_0(t)$.

the waves.) The situation is as shown in Figure 2.18, where, as usual, we take $\mathbf{r} = \mathbf{x} - \mathbf{x}'$ to be the displacement vector that takes us from the source to the observer at $\mathbf{x}$.

Now if the source is small enough, one might have expected the pressure field $p(\mathbf{x}, t)$ to be given by (2.86) in the form

$$p(\mathbf{x},t) = \frac{S}{4\pi \left|\mathbf{r}_0'\right|}, \quad \mathbf{r}_0' = \mathbf{x} - \mathbf{x}_0', \tag{2.88}$$

where $\mathbf{r}_0'$ is evaluated at the retarded time $t' = t - \left|\mathbf{r}_0'\right|/a$. However, rather surprisingly, this is *not* the case. In fact, it turns out that the correct solution is (see e.g. Morse & Ingard, 1987),

$$p(\mathbf{x},t) = \frac{S}{4\pi\left(\left|\mathbf{r}_0'\right| - \mathbf{u}' \cdot \mathbf{r}_0'/a\right)}, \quad \mathbf{u}' = \frac{d\mathbf{x}_0'}{dt}, \tag{2.89}$$

where $\mathbf{u}'$ is the velocity of the source at the retarded time $t' = t - \left|\mathbf{r}_0'\right|/a$. Clearly the difference between (2.88) and (2.89) is crucial whenever the Mach number $\left|\mathbf{u}'\right|/a$ is not small.

To understand where (2.89) comes from, we must look more carefully at the retarded potential

$$p(\mathbf{x},t) = \frac{1}{4\pi}\int \frac{s(\mathbf{x}',t')}{|\mathbf{r}|}\, d\mathbf{x}', \quad t' = t - |\mathbf{r}|/a, \tag{2.90}$$

where $\mathbf{r} = \mathbf{x} - \mathbf{x}'$. Let us consider the very special case where the source is cylindrical in shape, of length $\ell$ and cross-sectional area $A$, has a uniform strength per unit volume of $s = S/A\ell$, and moves directly towards the observer at speed $u'$. This is shown in Figure 2.19. Crucially we recognise that, although the source is compact, it still has a finite size. It follows that, for a given time $t$ at the observer, we must evaluate contributions to the integral in (2.90) from the rear and front of the source at different retarded times $t' = t - |\mathbf{r}|/a$, and we label this difference in retarded times $\Delta t'$. Moreover, the source will have moved a finite

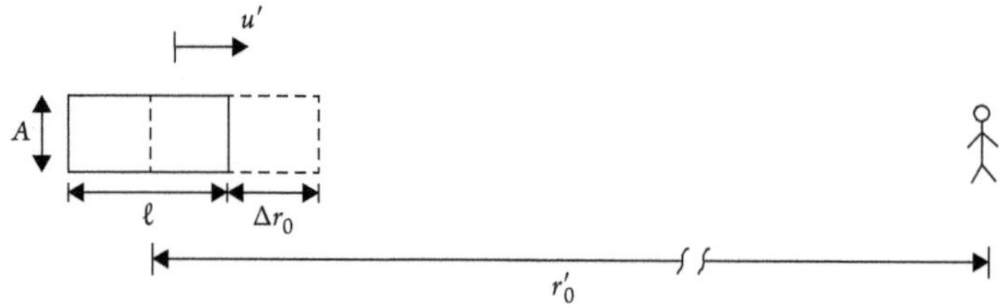


**Figure 2.19** A cylindrical source moves at speed $u'$ towards the observer.

distance $\Delta r_0$ during this time difference, with $\Delta r_0 = u'\,\Delta t'$. We can relate $\Delta r_0$ to $\ell$ and $u'$ as follows. First we note that the difference in position between the back of the source at the earlier time $t'$ and the front of the source at the later time $t' + \Delta t'$ is simply $\ell + u'\,\Delta t'$. Next we apply the expression $t' = t - |\mathbf{r}|/a$ first to the back of the source at time $t'$ and then to the front of the source at the later time $t' + \Delta t'$. This gives

$$\Delta t' = \frac{\ell + u'\,\Delta t'}{a},$$

or equivalently $\Delta t' = \ell/(a - u')$. It follows that $\Delta r_0$ is related to $\ell$ and $u'$ according to

$$\Delta r_0 = u'\,\Delta t' = \ell\frac{u'}{a - u'}. \tag{2.91}$$

To summarise, then, the cylindrical source moves a distance $\Delta r_0$ in the time interval $\Delta t'$, which is the difference between the time at which we first start evaluating contributions to the integral (2.90) from the rear of the source to the time when we finish adding contributions to the integral from the front of the source. Thus the source sweeps through, or perhaps we should say *occupies*, a length $\ell + \Delta r_0 = \ell/(1 - u'/a)$ during the period over which the integral in (2.90) is evaluated. It follows that, if we approximate $|\mathbf{r}|$ by $|\mathbf{r}'_0|$ in the denominator of (2.90) (on the assumption that $\ell << |\mathbf{r}'_0|$), then the integral takes the value

$$p(\mathbf{x},t) = \frac{1}{4\pi}\int \frac{s(\mathbf{x}',t')}{|\mathbf{r}|}\,d\mathbf{x}' = \frac{sA(\ell + \Delta r_0)}{4\pi\,|\mathbf{r}'_0|} = \frac{S}{4\pi\,|\mathbf{r}'_0|\,(1 - u'/a)}. \tag{2.92}$$

This should be compared with (2.89) for the case in which $\mathbf{u}'$ and $\mathbf{r}'_0$ are parallel.

Although we have taken a particular shape for the source (a cylinder), and restricted the case to a source moving directly towards the observer, it turns out that the precise shape of the compact source does not matter, and the components of $\mathbf{u}'$ perpendicular to $\mathbf{r}'_0$ play no role (see e.g. Morse & Ingard, 1987). Moreover, the size of the source does not enter into the final result when expressed in terms of $S = sA\ell$. In short, (2.92) generalises directly to (2.89), where the source may be taken as small as we wish.

Evidently the far-field pressure ahead of a source is increased by a factor of $(1 - u'/a)^{-1}$ by virtue of the movement of that source. The same factor of $(1 - u'/a)^{-1}$ occurs in the Doppler shift in frequency, although the two phenomena are rather different.

* * *

This concludes our brief overview of linear second-order partial differential equations. Our discussion has necessarily been brief, and strongly biased towards the physical phenomena exhibited by these equations, rather than their formal mathematical properties or techniques for solving them. However, the interested reader will find a more detailed mathematical discussion in Jeffrey (2002), while their applications in acoustics and fluid mechanics are discussed in Morse & Ingard (1987) and Acheson (1990), respectively.

## EXERCISES

2.1 A spherical cavity of radius $R$ is embedded in an otherwise infinite, thermally conducting solid. The temperature on the surface of the cavity $|\mathbf{x}| = R$ oscillates as $T = T_0 + \Delta T \cos \omega t$, where $T_0$ is the temperature of the solid for $|\mathbf{x}| >> R$. Find $T(r,t)$ for $r = |\mathbf{x}| > R$ using the same method that led to (2.42).

2.2 Small-amplitude internal gravity waves in a stratified fluid are governed by

$$\frac{\partial^2}{\partial t^2}(\nabla^2 u_z) + N^2 \left( \frac{\partial^2}{\partial x^2} + \frac{\partial^2}{\partial y^2} \right) u_z = 0, \quad N^2 = -\frac{g}{\rho_0}\frac{d\rho_0}{dz},$$

where $\rho_0$ is the unperturbed density, $u_z$ the vertical velocity, and $\mathbf{g} = -g\hat{\mathbf{e}}_z$ the gravitational acceleration. Consider the case of $N =$ constant and show that the corresponding dispersion relationship is

$$\varpi = \frac{Nk_\perp}{|\mathbf{k}|}, \quad k_\perp = \sqrt{k_x^2 + k_y^2}.$$

Hence show that the group velocity is given by

$$\mathbf{c}_g = \frac{N}{|\mathbf{k}|^3 k_\perp}[\mathbf{k} \times (\mathbf{k} \times \mathbf{k}_\parallel)] = \frac{N}{|\mathbf{k}|^3 k_\perp}(k_\parallel^2 \mathbf{k}_\perp - k_\perp^2 \mathbf{k}_\parallel),$$

where $\mathbf{k}_\parallel = k_z \hat{\mathbf{e}}_z$. Finally show that $\mathbf{c}_g$ is perpendicular to the phase velocity $\mathbf{c}_p$.

2.3 There is another way of getting to the Biot–Savart law. The vector potential $\mathbf{A}$ for a quasi-steady magnetic field $\mathbf{B} = \nabla \times \mathbf{A}$ is solenoidal and related to the current density $\mathbf{J}$ by Ampère's law:

$$\nabla^2 \mathbf{A} = -\nabla \times \mathbf{B} = -\mu_0 \mathbf{J}.$$

The Green's function solution of this Poisson equation is evidently

$$\mathbf{A}(\mathbf{x}) = \frac{\mu_0}{4\pi} \int \frac{\mathbf{J}(\mathbf{x}')}{|\mathbf{x} - \mathbf{x}'|} dV'.$$

Show that this differentiates with respect to $\mathbf{x}$ (keeping $\mathbf{x}'$ constant) to give the Biot–Savart law,

$$\mathbf{B}(\mathbf{x}) = -\frac{\mu_0}{4\pi}\int \frac{\mathbf{r}\times\mathbf{J}(\mathbf{x}')}{r^3}\,dV', \quad \mathbf{r} = \mathbf{x} - \mathbf{x}'.$$

*Hint:* Use (2.76) along with the identity $\nabla \times (\mathbf{J}'/|\mathbf{r}|) = \nabla(1/|\mathbf{r}|) \times \mathbf{J}'$.

## REFERENCES

Acheson, D. J., 1990, *Elementary Fluid Dynamics*. Oxford University Press.
Jeffrey, A., 2002, *Applied Partial Differential Equations*. Academic Press.
Morse, P. M. & Ingard, K. U., 1987, *Theoretical Acoustics*. Princeton University Press.

# PART II
## The Fundamentals of Electrodynamics

# 3

# A Bird's Eye View of Electromagnetism

> Experiment is the sole source of truth. It alone can teach us something new; it alone can give us certainty.
>
> Henri Poincaré (*Science and Hypothesis*, 1902)

> When it was proposed to establish laboratories at Cambridge, Todhunter, the mathematician, objected that it was unnecessary for students to see experiments performed, since the results could be vouched for by their teachers, all of them of the highest character, and many of them clergymen of the Church of England.
>
> Bertrand Russell (*The Scientific Outlook*, 1931)

In this chapter we shall, for simplicity, leave aside dielectric and magnetic materials and focus instead on electromagnetic fields in vacuum or in good electrical conductors, such as metals or certain plasmas. Our aim is to provide an overall theoretical framework, a sort of skeleton, onto which the subsequent chapters can add detail and nuance. To that end, we shall borrow freely from the chapters that follow, anticipating many of the results to come. The topics which concern us are the Lorentz force, Gauss' law, charge conservation, Ampère's law, the Biot–Savart law, Faraday's law and finally Maxwell's generalisation of Ampère's law. This culminates in what are now known as Maxwell's equations of electrodynamics, one of the crowning achievements of nineteenth-century science.

## 3.1 The Lorentz Force, the Electric Field and Charge Conservation

In general, a charged particle moving with velocity $\mathbf{u}_q$ and carrying a charge $q$ experiences three electromagnetic forces. These are

$$\mathbf{f} = q(\mathbf{E}_s + \mathbf{E}_i) + q\mathbf{u}_q \times \mathbf{B}. \tag{3.1}$$

The first term on the right is the electrostatic force $q\mathbf{E}_s$, $\mathbf{E}_s$ being the electrostatic field. This is known as the *Coulomb force* and it arises from the mutual attraction or repulsion of charges. The second contribution, $q\mathbf{E}_i$, is the force exerted on the charge by an electric field $\mathbf{E}_i$ that has been *induced* by a changing magnetic field $\mathbf{B}(t)$. The third term is the *Lorentz force*, which arises whenever a charge moves through a magnetic field $\mathbf{B}$. Let us consider each of these forces in turn.

*Coulomb's law*, which we shall discuss in detail in Chapter 4, describes the force of attraction or repulsion between static charges. Amongst other things, it tells us that $\mathbf{E}_s(\mathbf{x})$, the electrostatic force on a unit test charge at location $\mathbf{x}$, is irrotational. Moreover *Gauss' law of electrostatics*, which we introduced in §2.4.2, says that the divergence of $\mathbf{E}_s$ is fixed by the distribution of charges. (In Figure 2.16, for example, the electric field lines diverge from the point charge.) Taken together, these laws tell us that an electrostatic field satisfies

$$\nabla \times \mathbf{E}_s = \mathbf{0}, \qquad \nabla \cdot \mathbf{E}_s = \frac{\rho_e}{\varepsilon_0}, \tag{3.2}$$

where $\varepsilon_0$ is *the permittivity of free space* and $\rho_e(\mathbf{x})$ is the charge density in coulombs/m$^3$. Given the distribution of charges $\rho_e$, this yields the corresponding distribution of $\mathbf{E}_s$, since the requirements for a vector field to be uniquely determined are that its divergence and curl be specified and suitable boundary conditions given.

Since $\mathbf{E}_s$ is irrotational, we may introduce the *electrostatic potential* $\varphi_s$, defined by $\mathbf{E}_s = -\nabla\varphi_s$. It follows that electrostatic fields are governed by the Poisson equation

$$\nabla^2\phi_s = -\frac{\rho_e}{\varepsilon_0}. \tag{3.3}$$

In §2.4.2 we showed that, in infinite domains, this may be inverted, or we might say solved, using the integral relationship (the Green's function solution)

$$\varphi_s(\mathbf{x}) = \frac{1}{4\pi\varepsilon_0}\int \frac{\rho_e(\mathbf{x}')}{|\mathbf{r}|}\,d\mathbf{x}', \qquad \mathbf{r} = \mathbf{x} - \mathbf{x}'. \tag{3.4}$$

(Here $\mathbf{x}'$ is a dummy variable that samples the charge density at different points and $d\mathbf{x}' = dx'\,dy'\,dz'$ is the usual volume element.)

On the other hand, the induced electric field $\mathbf{E}_i$ has zero divergence, while its curl is governed by the differential form of *Faraday's law*, $\nabla \times \mathbf{E} = -\partial\mathbf{B}/\partial t$. Together these tell us that

$$\nabla \cdot \mathbf{E}_i = 0, \qquad \nabla \times \mathbf{E}_i = -\frac{\partial\mathbf{B}}{\partial t}. \tag{3.5}$$

Given $\partial\mathbf{B}/\partial t$, and suitable boundary conditions, (3.5) is sufficient to uniquely determined the distribution of, $\mathbf{E}_i$.

It is conventional to define the total electric field as $\mathbf{E} = \mathbf{E}_s + \mathbf{E}_i$, which is evidently governed by

$$\boxed{\begin{array}{cc} \nabla \cdot \mathbf{E} = \dfrac{\rho_e}{\varepsilon_0}, & \nabla \times \mathbf{E} = -\dfrac{\partial \mathbf{B}}{\partial t}. \\ \text{(Gauss' law)} & \text{(Faraday's law)} \end{array}} \tag{3.6}$$

(We shall use the term Gauss' *law* for (3.2) and Gauss' *theorem* for the divergence theorem in vector calculus.) The total electromagnetic force on a moving charge $q$ is then

$$\boxed{\mathbf{f} = q\mathbf{E} + q\mathbf{u}_q \times \mathbf{B},} \tag{3.7}$$

and indeed it is customary to use (3.7) to *define* the electric and magnetic fields $\mathbf{E}$ and $\mathbf{B}$. Thus the electric field $\mathbf{E}$ is defined as the force on a unit test charge at rest in the observer's frame of reference. The additional force associated with motion of the test charge is then used to define $\mathbf{B}$.

It is important to note the frame of reference used to define $\mathbf{E}$ in any given situation, since the value of $\mathbf{E}$ changes if we change the frame of reference. Suppose, for example, that in the laboratory frame there is an electric field $\mathbf{E}$ and a magnetic field $\mathbf{B}$. The electric field is defined by the force on a unit test charge at rest in the laboratory frame, and if the test charge then moves with velocity $\mathbf{u}_q$, the additional force $q\mathbf{u}_q \times \mathbf{B}$ is used to define $\mathbf{B}$. Now consider a second frame of reference that moves with velocity $\mathbf{u}_q$ relative to the laboratory frame and in which the charge is at rest. Then the force on the charge must now be interpreted as due to an electric field only, say $\mathbf{E}_{\text{rel}}$, where the subscript 'rel' indicates relative to the moving frame. Relativistic corrections apart, we now have $\mathbf{f} = q(\mathbf{E} + \mathbf{u}_q \times \mathbf{B}) = q\mathbf{E}_{\text{rel}}$. Evidently the electric field in the frame of reference moving with velocity $\mathbf{u}_q$ is

$$\mathbf{E}_{\text{rel}} = \mathbf{E} + \mathbf{u}_q \times \mathbf{B}. \tag{3.8}$$

It turns out that, if $\mathbf{u}_q$ is much less than the speed of light, then the magnetic fields $\mathbf{B}$ and $\mathbf{B}_{\text{rel}}$ are equal, as discussed in Chapter 23.

Let us now consider the influence of an electric field $\mathbf{E}$ on an electrical conductor, starting with the current induced by an electric field. The current density in a conductor, $\mathbf{J}$, is a vector that points in the direction of maximum current flow and its component in any given direction is equal to the current per unit area in that direction. Its numerical value is proportional to the force experienced by the free charges in the conductor, and so in a stationary conductor we have $\mathbf{J}$ proportional to $\mathbf{E}$. This is reflected in the usual form of Ohm's law, $\mathbf{J} = \sigma \mathbf{E}$, where $\sigma$ is the *electrical conductivity* and $\mathbf{J}$ is called the *conduction current*. In a moving conductor, however, Ohm's law is a little more complicated. It turns out that the drift velocity of the free charge carriers relative to the conductor is usually very small and so the free charges more or less move with the conductor. It follows that we must use the electric field measured in a frame of reference moving with the local velocity of the conductor to evaluate the forces on the free charges and hence to evaluate the conduction current,

**J**. Ohm's law then generalises to

$$\mathbf{J} = \sigma \mathbf{E}_{\text{rel}} = \sigma(\mathbf{E} + \mathbf{u} \times \mathbf{B}), \tag{3.9}$$

where **u** is the local velocity of the conductor (which may vary from place to place).

Next we consider the electromagnetic force exerted on a unit volume of a conductor as a result of the Coulomb and Lorentz forces in (3.7). We start by noting that the forces exerted on the free charges are ultimately transmitted to the conductor. Moreover, if (3.7) is summed over a unit volume of the conductor, then $\sum q$ becomes the charge density $\rho_e$ and $\sum q\mathbf{u}_q$ ($\mathbf{u}_q$ being the velocity of the charges) becomes the current density **J**. So, the electromagnetic force per unit volume is then

$$\boxed{\mathbf{F} = \rho_e \mathbf{E} + \mathbf{J} \times \mathbf{B}.} \tag{3.10}$$

Note that $\rho_e$ and **J** are not, in general, independent fields. Rather, they are related by the principle of conservation of charge. This requires that the net current leaving a given volume through a closed surface $S$ must equal the rate of depletion of charge within that volume:

$$\oint_S \mathbf{J} \cdot d\mathbf{S} = -\frac{\partial}{\partial t} \int_V \rho_e \, dV.$$

Using Gauss' theorem, this can be rewritten as

$$\int_V \nabla \cdot \mathbf{J} \, dV = -\int_V \frac{\partial \rho_e}{\partial t} \, dV.$$

Since this is true for any and all volumes $V$, we may equate the integrands above to give the *charge conservation equation*

$$\boxed{\nabla \cdot \mathbf{J} = -\frac{\partial \rho_e}{\partial t}.} \tag{3.11}$$

However, in *good* conductors, say metals, the charge density is tiny and so the right-hand side of (3.11) is negligible, giving $\nabla \cdot \mathbf{J} = 0$. We may demonstrate this for the particular case of a stationary conductor as follows. Charge conservation, Ohm's law $\mathbf{J} = \sigma \mathbf{E}$ and Gauss' law $\nabla \cdot \mathbf{E} = \rho_e / \varepsilon_0$ combine to yield

$$\frac{\partial \rho_e}{\partial t} = -\nabla \cdot \mathbf{J} = -\nabla \cdot (\sigma \mathbf{E}) = -\frac{\sigma}{\varepsilon_0} \rho_e.$$

This can be rewritten as

$$\frac{\partial \rho_e}{\partial t} + \frac{\rho_e}{\tau_e} = 0, \quad \tau_e = \frac{\varepsilon_0}{\sigma}, \tag{3.12}$$

where the quantity $\tau_e$ is called the *charge relaxation time*. For a typical metallic conductor $\tau_e$ has a value of around $10^{-14}$ seconds—extremely small! Of course the solution to (3.12) is $\rho_e = \rho_e(0) \exp(-t/\tau_e)$, and so any excess charge density that, at $t = 0$, lies in the interior of a conductor will move extremely rapidly to the surface under the action of the electrostatic repulsion forces. It follows that $\rho_e$ is almost always zero in stationary conductors, and so $\nabla \cdot \mathbf{J} = 0$. It also follows that $\rho_e \mathbf{E}$ in (3.10) is zero within the interior of such a conductor, though there may be forces arising from surface charges.

If the various parts of the conductor move through a magnetic field with a velocity $\mathbf{u}(\mathbf{x}, t)$, then a little more work is required to show that the right-hand side of (3.11) is negligible. Ohm's law now becomes $\mathbf{J} = \sigma(\mathbf{E} + \mathbf{u} \times \mathbf{B})$ and (3.12) generalises to

$$\frac{\partial \rho_e}{\partial t} + \frac{\rho_e}{\tau_e} = -\sigma \nabla \cdot (\mathbf{u} \times \mathbf{B}). \tag{3.13}$$

In this case a small but finite charge can build up within the interior of the conductor. However, we are interested in events that take place on a timescale much longer than $\tau_e$ (we exclude events like batteries being turned on), and so we may drop $\partial \rho_e / \partial t$ from (3.13), to give $\rho_e = -\sigma \tau_e \nabla \cdot (\mathbf{u} \times \mathbf{B})$. Crucially, dropping $\partial \rho_e / \partial t$ in (3.13) is equivalent to ignoring $\partial \rho_e / \partial t$ in (3.11), and so once again we conclude that $\nabla \cdot \mathbf{J} = 0$. Moreover, in a good conductor $\rho_e$ is too small for $\rho_e \mathbf{E}$ to make any significant contribution to (3.10) within the interior of the conductor. That is, if $\ell$ is a typical lengthscale of the motion, then $\rho_e \mathbf{E}$ is of order $(\sigma \tau_e u B / \ell)(J/\sigma) \sim (\tau_e u / \ell)(JB) << JB$, and so the ratio of $\rho_e \mathbf{E}$ to $\mathbf{J} \times \mathbf{B}$ is tiny. Within good conductors, then, (3.10) and (3.11) simplify to $\mathbf{F} = \mathbf{J} \times \mathbf{B}$ and $\nabla \cdot \mathbf{J} = 0$.

## 3.2 Ampère's Law, the Biot–Savart Law and the Ampère–Maxwell Equation

Since $\nabla \cdot [\nabla \times (\cdot)] = 0$, Faraday's law (3.6) ensures that $\partial \mathbf{B} / \partial t$ is solenoidal. In fact, it turns out that we can make the stronger statement that $\mathbf{B}$ is itself solenoidal, $\nabla \cdot \mathbf{B} = 0$. Moreover, a given distribution of $\mathbf{B}$ is clearly controlled by the distribution of currents that generates it. In magnetostatics this relationship takes the form of *Ampère's law*, which states that the magnetic field and current density vector $\mathbf{J}$ are related by $\nabla \times \mathbf{B} = \mu_0 \mathbf{J}$, where $\mu_0$ is the *permeability of free space*. Thus, in magnetostatics $\mathbf{B}$ is governed by

$$\boxed{\nabla \cdot \mathbf{B} = 0, \qquad \nabla \times \mathbf{B} = \mu_0 \mathbf{J}.} \tag{3.14}$$

Given $\mathbf{J}$, this is sufficient to uniquely determine $\mathbf{B}$, since both its curl and divergence are then specified.

In §2.4.3 we showed that, in an infinite domain, (3.14) may be inverted (we might say solved) using the *Biot–Savart law*. That is, when the current density is a known, localised function of position, the magnetic field may be calculated directly from

$$\boxed{\mathbf{B}(\mathbf{x}) = \frac{\mu_0}{4\pi} \int \frac{\mathbf{J}(\mathbf{x}') \times \mathbf{r}}{|\mathbf{r}|^3} d\mathbf{x}', \qquad \mathbf{r} = \mathbf{x} - \mathbf{x}',} \tag{3.15}$$

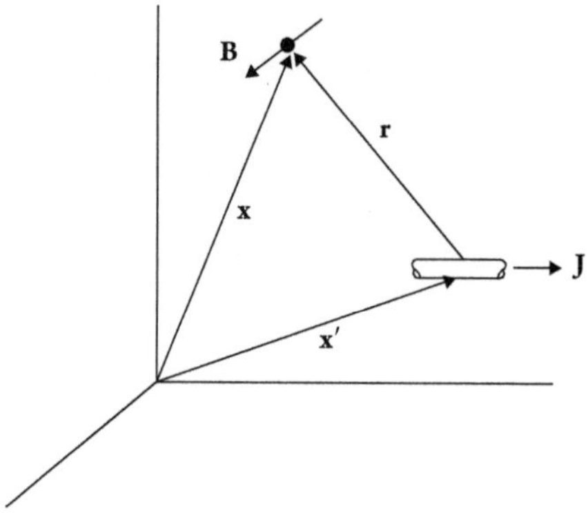


**Figure 3.1** Coordinate system used in the Biot–Savart law

where $\mathbf{x}'$ is a dummy variable that samples the current density at different points, as indicated in Figure 3.1. This reveals the true nature of Ampère's law, which tells us about the magnetic field produced by a given distribution of currents. We shall provide an alternative derivation of (3.15) in §3.4. (Note that the Biot–Savart law enforces not only Ampère's law, but also $\nabla \cdot \mathbf{B} = 0$, as discussed in Exercise 3.4.)

It is important to note that Ampère's law, while often applied to slowly varying currents and magnetic fields, strictly only applies to steady situations, or else in cases where the charge density is everywhere negligible. This is evident from the fact that the vector identity $\nabla \cdot [\nabla \times (\cdot)] = 0$, combined with (3.14), requires that $\nabla \cdot \mathbf{J} = 0$, which is inconsistent with (3.11) whenever $\partial \rho_e / \partial t$ is significant. This discrepancy was rectified by Maxwell, who modified Ampère's law to read,

$$\boxed{\nabla \times \mathbf{B} = \mu_0 \left( \mathbf{J} + \varepsilon_0 \frac{\partial \mathbf{E}}{\partial t} \right).} \tag{3.16}$$

This is the *Ampère–Maxwell equation* and the additional term $\varepsilon_0 \partial \mathbf{E}/\partial t$ is known as the *displacement current*. The necessity of the last term in (3.16) can be seen if we take the divergence of (3.16). Noting that $\nabla \cdot [\nabla \times (\cdot)] = 0$, and using Gauss' law (3.6), this yields

$$\nabla \cdot \mathbf{J} = -\varepsilon_0 \frac{\partial}{\partial t} \nabla \cdot \mathbf{E} = -\frac{\partial \rho_e}{\partial t}, \tag{3.17}$$

which is just the charge conservation equation.

The additional term in the Ampère–Maxwell equation (the displacement current) plays a crucial role whenever electromagnetic waves are important. However, it can often be neglected when considering slowly varying fields, or else fields in good electrical conductors, where (3.12) and (3.13) tell us that the charge density $\rho_e$ is tiny and so charge conservation reduces to the statement that $\nabla \cdot \mathbf{J} = 0$.

## 3.3 Faraday's Law in Differential and Integral Form

We have already noted that Faraday's law in *differential form* tells us about the electric field induced by a time-dependent magnetic field,

$$\nabla \times \mathbf{E} = -\frac{\partial \mathbf{B}}{\partial t}. \tag{3.18}$$

The integral version of Faraday's law, on the other hand, is a rather more subtle statement. It is couched in terms of magnetic fluxes and the so-called *electromotive force* (EMF for short). So let us start by explaining what we mean by an EMF. The EMF around a closed curve $C$ is defined as the force exerted on a unit test charge that *moves with the loop*, the force being integrated once around that loop. If parts of the loop move relative to the laboratory frame, then that force is the electric field measured in the local frame of reference moving with $C$. That is to say,

$$\text{EMF} = \oint_C \mathbf{E}_{\text{rel}} \cdot d\mathbf{r}, \tag{3.19}$$

where $C$ is a closed curve that deforms in space with a prescribed velocity $\mathbf{v}(\mathbf{x}, t)$, and $\mathbf{E}_{\text{rel}}$ for the line element $d\mathbf{r}$ is the electric field in a frame of reference that moves with $d\mathbf{r}$, $\mathbf{E}_{\text{rel}} = \mathbf{E} + \mathbf{v} \times \mathbf{B}$. Note that, in general, $\mathbf{v}$ may vary as we move around the loop. Note also that, when considering the case of fields in conductors, the loop $C$ need *not* move with the conductor, but rather may execute some entirely different motion. To emphasise that point, we shall use $\mathbf{u}(\mathbf{x}, t)$ to indicate the motion of a conductor (and by implication the free charges within it) and $\mathbf{v}(\mathbf{x}, t)$ for the prescribed velocity of the curve $C$.

The integral version of Faraday's law relates the EMF around such a closed loop to the rate of change of magnetic flux through that loop. It is

$$\boxed{\text{EMF} = \oint_C \mathbf{E}_{\text{rel}} \cdot d\mathbf{r} = -\frac{d}{dt}\int_S \mathbf{B} \cdot d\mathbf{S},} \tag{3.20}$$

where $S$ is any surface that spans the curve $C$. It is important to note that the magnetic flux in (3.20) may change because $\mathbf{B}$ is changing with time (with $C$ fixed), or because the loop $C$ is moving uniformly across an inhomogeneous magnetic field, or because $C$ is changing shape, or indeed any combinations of these processes. In all possible situations the *flux rule* (3.20) says the induced EMF is (minus) the rate of change of flux through $C$.

If $C$ is fixed and $\mathbf{B}$ varies in time, the induced EMF is called a *transformer* EMF, and if $\mathbf{B}$ is constant and part of $C$ is moving, it is called a *motional* EMF. These two extreme cases are shown in Figure 3.2, where the curve $C$ in Figure 3.2(b) moves with the conductor. It is natural to interpret events in Figure 3.2(a), the transformer EMF, in terms of an induced electric field $\mathbf{E}_i$ and the differential form of Faraday's law, $\nabla \times \mathbf{E} = -\partial \mathbf{B}/\partial t$, which generates that induced electric field. On the other hand, events in Figure 3.2(b) (the motional EMF) are naturally interpreted in terms of the Lorentz force acting on the charges in the

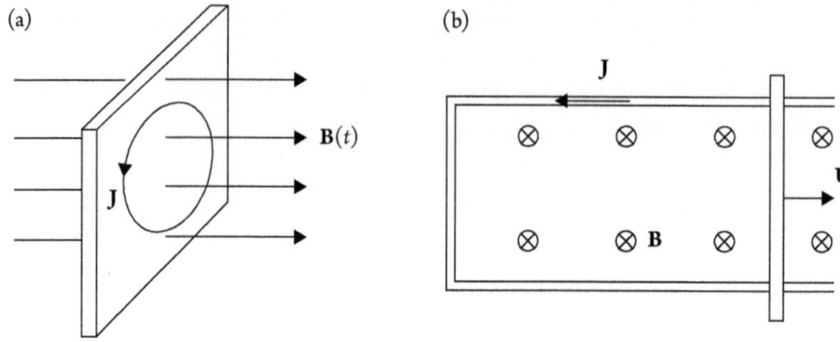


**Figure 3.2** An EMF can be generated by either (a) the variation of the magnetic field with time or else (b) movement of part of the circuit C through the magnetic field.

moving conductor. The flux rule tells us that the origin of the EMF is immaterial; whatever its physical source, the induced EMF always equals (minus) the rate of change of magnetic flux through *C*.

It is easy to show that the differential form of Faraday's law corresponds directly to the special case in which the curve *C* in (3.20) is at rest in a laboratory frame and the EMF arises from a magnetic field that is time-dependent (Figure 3.2a). That is, in such cases Faraday's integral equation simplifies to

$$\oint_C \mathbf{E} \cdot d\mathbf{r} = \int_S (\nabla \times \mathbf{E}) \cdot d\mathbf{S} = -\int_S \frac{\partial \mathbf{B}}{\partial t} \cdot d\mathbf{S}, \tag{3.21}$$

and since this is true for *any* surface *S*, we may equate the integrands in the two surface integrals. This brings us back to the differential form of Faraday's law, $\nabla \times \mathbf{E} = -\partial \mathbf{B}/\partial t$.

Actually, we can also derive the more general form of Faraday's integral equation (in which the loop *C* is allowed to move and deform) starting from $\nabla \times \mathbf{E} = -\partial \mathbf{B}/\partial t$, although this requires considerably more effort. We first need an important kinematic result. Suppose that $\mathbf{G}$ is any solenoidal vector field, $\nabla \cdot \mathbf{G} = 0$, *C* is a closed curve that deforms in space with a prescribed velocity $\mathbf{v}(\mathbf{x}, t)$ and *S* is any surface that spans the closed curve *C* (Figure 3.3). Then the key kinematic result is

$$\frac{d}{dt}\int_S \mathbf{G} \cdot d\mathbf{S} = \int_S \frac{\partial \mathbf{G}}{\partial t} \cdot d\mathbf{S} - \oint_C (\mathbf{v} \times \mathbf{G}) \cdot d\mathbf{r}. \tag{3.22}$$

We shall prove (3.22) in Chapter 10, but the idea behind it is the following. The flux of $\mathbf{G}$ through *S* changes for one of two reasons. First, even if *S* were fixed in space, there would be a change in flux whenever $\mathbf{G}$ was time-dependent. This is the surface integral on the right of (3.22). Second, if the curve *C* moves, it may expand at points to include additional flux, or perhaps contract at other points to exclude flux. Moreover, the surface area swept out by a line element $d\mathbf{r}$ of the curve *C* in a time $\delta t$ is $d\mathbf{S} = (\mathbf{v}\delta t) \times d\mathbf{r}$. Note that the relative orientation of $d\mathbf{S}$ and $d\mathbf{r}$ is fixed in accordance with the right-hand rule,

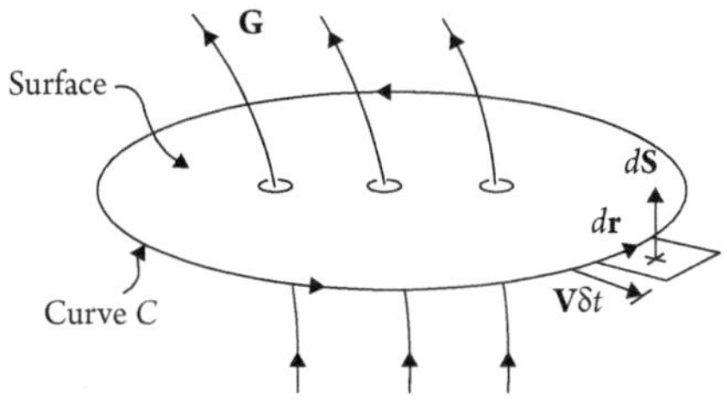


**Figure 3.3** The surface swept out by a line element $d\mathbf{r}$ in a time $\delta t$ is $d\mathbf{S} = (\mathbf{v}\delta t) \times d\mathbf{r}$.

as shown in Figure 3.3. We conclude that the change in the flux of $\mathbf{G}$ through $S$ by virtue of the movement of the boundary element $d\mathbf{r}$ is $\mathbf{G} \cdot d\mathbf{S} = \mathbf{G} \cdot (\mathbf{v} \times d\mathbf{r})\, \delta t$. Rearranging the scalar triple product gives $\mathbf{G} \cdot d\mathbf{S} = -[(\mathbf{v} \times \mathbf{G}) \cdot d\mathbf{r}]\, \delta t$, which explains the line integral on the right of (3.22).

Let us now return to the differential form of Faraday's law,

$$\nabla \times \mathbf{E} = -\frac{\partial \mathbf{B}}{\partial t}, \tag{3.23}$$

where $\mathbf{E}$ is measured in the laboratory frame of reference. Integrating this over an open surface $S$ and invoking (3.22) yields

$$\oint_C \mathbf{E} \cdot d\mathbf{r} = -\int_S \frac{\partial \mathbf{B}}{\partial t} \cdot d\mathbf{S} = -\frac{d}{dt}\int_S \mathbf{B} \cdot d\mathbf{S} - \oint_C (\mathbf{v} \times \mathbf{B}) \cdot d\mathbf{r}, \tag{3.24}$$

where $\mathbf{v}(\mathbf{x}, t)$ is the prescribed velocity of the bounding curve $C$. Combining the two line integrals in (3.24) then gives

$$\oint_C (\mathbf{E} + \mathbf{v} \times \mathbf{B}) \cdot d\mathbf{r} = -\frac{d}{dt}\int_S \mathbf{B} \cdot d\mathbf{S}. \tag{3.25}$$

Finally we invoke the idea of the Lorentz force $\mathbf{f} = q\mathbf{v} \times \mathbf{B}$ acting on a unit test charge that moves with the curve $C$. This tells us that, in a frame of reference moving with velocity $\mathbf{v}$, the electric field is $\mathbf{E}_{\text{rel}} = \mathbf{E} + \mathbf{v} \times \mathbf{B}$. We have arrived back at the integral version of Faraday's law:

$$\text{EMF} = \oint_C \mathbf{E}_{\text{rel}} \cdot d\mathbf{r} = -\frac{d}{dt}\int_S \mathbf{B} \cdot d\mathbf{S}. \tag{3.26}$$

In short, the integral version of Faraday's law can be derived from the differential form, provided that we use the Lorentz force to tell us how $\mathbf{E}$ transforms from one frame of reference to another.

For the particular case where the curve $C$ lies within a conductor and *moves with that conductor*, so we have $\mathbf{v}(\mathbf{x}, t) = \mathbf{u}(\mathbf{x}, t)$, Ohm's law tells us that $\mathbf{E}_{\text{rel}}$ in (3.26) may be written in terms of current density: $\mathbf{E}_{\text{rel}} = \mathbf{E} + \mathbf{u} \times \mathbf{B} = \mathbf{J}/\sigma$. In such cases Faraday's law adopts the particularly useful form

$$\frac{1}{\sigma}\oint_C \mathbf{J} \cdot d\mathbf{r} = -\frac{d}{dt}\int_S \mathbf{B} \cdot d\mathbf{S}. \tag{3.27}$$

$$(C \text{ moves with the conductor})$$

Although we have derived Faraday's integral law using the differential form (3.23), with the Lorentz force slipped in at the last minute, it is important to recall that the induced EMFs in the two cases shown in Figure 3.2 have different physical origins. In Figure 3.2(a) the EMF is caused by an unsteady magnetic field inducing an electric field through $\nabla \times \mathbf{E} = -\partial \mathbf{B}/\partial t$. In Figure 3.2(b), on the other hand, the EMF is generated by the Lorentz force $\mathbf{f} = q\mathbf{u} \times \mathbf{B}$ acting directly on the moving charges in the conductor. Apparently, it just so happens that the induced EMF in both cases can be described by the same flux rule. This odd coincidence set Einstein thinking, as we shall see in Chapter 22.

## 3.4 The Vector Potential

Since $\mathbf{B}$ is solenoidal, $\nabla \cdot \mathbf{B} = 0$, then we may introduce a *vector potential* for $\mathbf{B}$ defined in the usual way through

$$\boxed{\nabla \times \mathbf{A} = \mathbf{B}.} \tag{3.28}$$

This automatically ensures that $\mathbf{B}$ is solenoidal, since $\nabla \cdot \nabla \times \mathbf{A} = 0$. However, as discussed in §1.5, a vector potential is not uniquely determined by an expression like $\nabla \times \mathbf{A} = \mathbf{B}$. Rather, to unambiguously define $\mathbf{A}$, we must specify both its curl and its divergence. So we need to decide on the value of $\nabla \cdot \mathbf{A}$. This is known as setting the *gauge* for $\mathbf{A}$. It turns out that the most convenient choice for $\nabla \cdot \mathbf{A}$ depends upon the circumstances.

For cases where we may ignore displacement currents, such as quasi-steady electric fields, or else cases where the charge density is everywhere negligible, the Ampère–Maxwell equation reverts to Ampère's law. In such situations the most convenient choice of gauge is

$$\nabla \cdot \mathbf{A} = 0. \tag{3.29}$$

This is known as the *Coulomb gauge*. However, for those cases where we cannot ignore the displacement current in the Ampère–Maxwell equation, such as in the radiation of waves, it turns out to be preferable to choose

$$\nabla \cdot \mathbf{A} = -\frac{1}{c^2}\frac{\partial \varphi}{\partial t}, \tag{3.30}$$

where $c = 1/\sqrt{\mu_0 \varepsilon_0}$ is the speed of light and the scalar function $\varphi$ is defined below in (3.34). This is called the *Lorenz gauge* for $\mathbf{A}$.

We shall return to (3.30) in §3.6, where we shall explain why it is so convenient when considering electromagnetic waves. In the interim, we consider the quasi-steady case in which displacement currents may be ignored and Ampère's law applies. Then the Coulomb gauge, combined with Ampère's law, yields a Poisson equation for **A**:

$$\nabla^2 \mathbf{A} = -\nabla \times \mathbf{B} = -\mu_0 \mathbf{J}. \tag{3.31}$$

Given the Green's function solution (3.4) to the Poisson equation (3.3), it is clear that (3.31) inverts to yield

$$\mathbf{A}(\mathbf{x}) = \frac{\mu_0}{4\pi}\int \frac{\mathbf{J}(\mathbf{x}')}{|\mathbf{r}|}\,d\mathbf{x}', \qquad \mathbf{r} = \mathbf{x} - \mathbf{x}'. \tag{3.32}$$

We now take the curl of (3.32) with respect to **x**, while keeping $\mathbf{x}'$ constant. Noting that $\nabla \times (\mathbf{J}'/|\mathbf{r}|) = \nabla(1/|\mathbf{r}|) \times \mathbf{J}'$, $\mathbf{J}'$ being independent of **x**, and that $\nabla(1/|\mathbf{r}|) = -\mathbf{r}/|\mathbf{r}|^3$, this takes us back to the Biot–Savart law (3.15). (See also Exercise 2.3 at the end of Chapter 2.)

Finally, let us now substitute **A** for **B** in Faraday's equation. This yields

$$\nabla \times \mathbf{E} = -\nabla \times \left(\frac{\partial \mathbf{A}}{\partial t}\right), \tag{3.33}$$

which uncurls to give

$$\boxed{\mathbf{E} = -\frac{\partial \mathbf{A}}{\partial t} - \nabla\varphi,} \tag{3.34}$$

where $\varphi$ is an arbitrary scalar potential. Given (3.2) and (3.5), it is natural to assume that $\mathbf{E}_i = -\partial \mathbf{A}/\partial t$ and $\mathbf{E}_s = -\nabla\varphi$, so that $\varphi$ can now be interpreted as the electrostatic potential $\varphi_s$. For the particular case where we set the gauge for **A** as $\nabla \cdot \mathbf{A} = 0$, so that $\mathbf{E}_i = -\partial \mathbf{A}/\partial t$ is consistent with $\nabla \cdot \mathbf{E}_i = 0$, this is indeed the correct interpretation. When the Lorenz gauge is used, however, the scalar potential $\varphi$ in (3.34) in not the electrostatic potential, as we shall see.

## 3.5 Maxwell's Equations, Electromagnetic Energy and the Poynting Vector

Perhaps we should now summarise the fundamental equations of electrodynamics. We have

| | |
|---|---|
| $\mathbf{f} = q(\mathbf{E} + \mathbf{u} \times \mathbf{B})$, | $\nabla \cdot \mathbf{J} = -\dfrac{\partial \rho_e}{\partial t}$. |
| (force law) | (charge conservation) |

(3.35, 3.36)

Moreover, for materials that are neither dielectric nor magnetic, we have seen that

| | |
|---|---|
| $\nabla \cdot \mathbf{E} = \dfrac{\rho_e}{\varepsilon_0}$, | $\nabla \times \mathbf{E} = -\dfrac{\partial \mathbf{B}}{\partial t}$, |
| (Gauss' law) | (Faraday's law in differential form) |

(3.37, 3.38)

$$\nabla \cdot \mathbf{B} = 0, \qquad \nabla \times \mathbf{B} = \mu_0 \left( \mathbf{J} + \varepsilon_0 \frac{\partial \mathbf{E}}{\partial t} \right). \tag{3.39, 3.40}$$

(solenoidal nature of **B**) (Ampère–Maxwell equation)

These last four equation are known collectively as *Maxwell's equations*. (Figure 3.4 shows Maxwell.) Note that these laws are not all independent. For example, (3.39) follows from the divergence of Faraday's law, while the divergence of the Ampère–Maxwell equation, combined with Gauss' law, yields charge conservation.

We are now in a position to derive an energy equation for electromagnetic fields. The dot product of **B** with Faraday's law, combined with the dot product of **E** with the Ampère–Maxwell equation, yields

$$\frac{\partial}{\partial t}\left( \frac{\mathbf{B}^2}{2\mu_0} + \frac{\varepsilon_0 \mathbf{E}^2}{2} \right) + \mathbf{J} \cdot \mathbf{E} = -\frac{1}{\mu_0}(\mathbf{B} \cdot \nabla \times \mathbf{E} - \mathbf{E} \cdot \nabla \times \mathbf{B}),$$

from which

$$\frac{\partial}{\partial t}\left( \frac{\mathbf{B}^2}{2\mu_0} + \frac{\varepsilon_0 \mathbf{E}^2}{2} \right) = -\mathbf{J} \cdot \mathbf{E} - \frac{1}{\mu_0}\nabla \cdot (\mathbf{E} \times \mathbf{B}). \tag{3.41}$$

**Figure 3.4** James Clerk Maxwell. (Published courtesy of King's College London Archives.)

As we shall see, $\mathbf{B}^2/2\mu_0$ is the magnetic energy per unit volume, while $\varepsilon_0\mathbf{E}^2/2$ is the energy per unit volume of the electric field. Integrating over a volume $V$ we find

$$\boxed{\frac{d}{dt}\int_V \left(\frac{\mathbf{B}^2}{2\mu_0}+\frac{\varepsilon_0\mathbf{E}^2}{2}\right)dV = -\int_V \mathbf{J}\cdot\mathbf{E}\,dV - \oint_S \frac{\mathbf{E}\times\mathbf{B}}{\mu_0}\cdot d\mathbf{S},} \tag{3.42}$$

where $S$ is the surface that encloses the volume $V$. In a vacuum, where there are no currents, this simplifies to

$$\frac{d}{dt}\int_V \left(\frac{\mathbf{B}^2}{2\mu_0}+\frac{\varepsilon_0\mathbf{E}^2}{2}\right)dV = -\oint_S \frac{\mathbf{E}\times\mathbf{B}}{\mu_0}\cdot d\mathbf{S}. \tag{3.43}$$

In a conductor, on the other hand, the generalised form of Ohm's law, $\mathbf{J}=\sigma(\mathbf{E}+\mathbf{u}\times\mathbf{B})$, requires

$$\boxed{\frac{d}{dt}\int_V \left(\frac{\mathbf{B}^2}{2\mu_0}+\frac{\varepsilon_0\mathbf{E}^2}{2}\right)dV = -\frac{1}{\sigma}\int_V \mathbf{J}^2\,dV - \int_V \mathbf{u}\cdot(\mathbf{J}\times\mathbf{B})dV - \oint_S \frac{\mathbf{E}\times\mathbf{B}}{\mu_0}\cdot d\mathbf{S}.} \tag{3.44}$$

Equations (3.43) and (3.44) are examples of energy budgets and the vector $\mathbf{E}\times\mathbf{B}/\mu_0$ is called the *Poynting vector*. In (3.43) conservation of energy tells us that the integral on the right must represent the rate at which electromagnetic energy flows out of $V$ through the surface $S$. It follows that the Poynting vector represents the flux of electromagnetic energy per unit area leaving through the surface $S$. In (3.44) there are two additional terms on the right. The loss of electromagnetic energy within $V$ represented by the first integral is called the Joule (or Ohmic) dissipation and it represents the rate at which electromagnetic energy is irreversibly lost to heat as a result of the flow of currents in a conductor of finite resistivity. The second integral on the right is the rate of working of the Lorentz force on the conductor and it may be positive or negative. This represents an exchange of electromagnetic and mechanical energy. These various energy relationships are discussed in, for example, Feynman, Leighton & Sands (1964).

## 3.6 Electromagnetic Waves, Gauges and Retarded Potentials Revisited

Maxwell's displacement current is central to nearly all of electrodynamics, in large part because it leads to the entirely new phenomenon of electromagnetic waves, a phenomenon that is absent in the pre-Maxwell equations. This is clear from the curl of the Ampère–Maxwell equation,

$$\nabla^2\mathbf{B} = -\mu_0\nabla\times\mathbf{J} - \frac{1}{c^2}\frac{\partial}{\partial t}\nabla\times\mathbf{E}, \quad c=\frac{1}{\sqrt{\mu_0\varepsilon_0}}, \tag{3.45}$$

which, combined with Faraday's law, yields the inhomogeneous wave equation

$$\nabla^2 \mathbf{B} - \frac{1}{c^2}\frac{\partial^2 \mathbf{B}}{\partial t^2} = -\mu_0 \nabla \times \mathbf{J}. \tag{3.46}$$

Compare this with the curl of Ampère's law, which excludes displacement currents,

$$\nabla^2 \mathbf{B} = -\mu_0 \nabla \times \mathbf{J}. \tag{3.47}$$

Clearly Maxwell's displacement current has transformed a Poisson equation into an inhomogeneous wave equation, with the source term $-\mu_0 \nabla \times \mathbf{J}$. In a vacuum, where the source term is absent, we have

$$\nabla^2 \mathbf{B} - \frac{1}{c^2}\frac{\partial^2 \mathbf{B}}{\partial t^2} = 0, \tag{3.48}$$

which describes the propagation of electromagnetic waves travelling at the speed of light, $c = 1/\sqrt{\mu_0 \varepsilon_0}$.

Electromagnetic waves are often best expressed in terms of the vector potential **A**, rather than **B**. Before doing this, however, we need to return to the issue of how to choose the optimal gauge (divergence) for the vector potential. As noted in §3.4, the simplest choice in magnetostatics, where we have $\nabla \cdot \mathbf{J} = 0$, is the Coulomb gauge, $\nabla \cdot \mathbf{A} = 0$. However, when **J** is not solenoidal, and so charge conservation demands that displacement currents be retained, it is preferable to choose an alternative definition of $\nabla \cdot \mathbf{A}$. In particular, it is preferable to choose the *Lorenz gauge*,

$$\nabla \cdot \mathbf{A} = -\frac{1}{c^2}\frac{\partial \varphi}{\partial t}, \tag{3.49}$$

where $\varphi$ is the scalar potential defined by (3.34).

To see why this is so we must return to the 'uncurled' version of Faraday's law (3.34):

$$\mathbf{E} = -\frac{\partial \mathbf{A}}{\partial t} - \nabla \varphi. \tag{3.50}$$

Let us use (3.50) to substitute for **E** in the Ampère–Maxwell equation, while replacing **B** by $\nabla \times \mathbf{A}$:

$$\nabla \times \nabla \times \mathbf{A} = \mu_0 \mathbf{J} - \frac{1}{c^2}\frac{\partial}{\partial t}\left(\frac{\partial \mathbf{A}}{\partial t} + \nabla \varphi\right). \tag{3.51}$$

We now convert $\nabla \times \nabla \times \mathbf{A}$ into a Laplacian using $\nabla^2 \mathbf{A} = \nabla(\nabla \cdot \mathbf{A}) - \nabla \times \nabla \times \mathbf{A}$, while making no assumption about the form of $\nabla \cdot \mathbf{A}$. This yields a cumbersome equation involving both the vector potential **A** and the scalar potential $\varphi$:

$$\nabla^2 \mathbf{A} - \nabla(\nabla \cdot \mathbf{A}) - \frac{1}{c^2}\frac{\partial^2 \mathbf{A}}{\partial t^2} = -\mu_0 \mathbf{J} + \frac{1}{c^2}\frac{\partial}{\partial t}\nabla \varphi. \tag{3.52}$$

Next, we substitute for **E** in Gauss' law, which gives us a second equation relating the scalar and vector potentials:

$$\nabla^2\varphi + \frac{\partial}{\partial t}\nabla\cdot\mathbf{A} = -\frac{\rho_e}{\varepsilon_0}. \tag{3.53}$$

Comparing (3.52) and (3.53) it appears that our two potentials are coupled, being governed by a pair of rather complicated equations. However we can tidy things up if we recall that we have some freedom in the choice of $\nabla\cdot\mathbf{A}$. For example, adopting the Coulomb gauge $\nabla\cdot\mathbf{A}=0$ helps tidy up (3.53), converting it into the standard Poisson equation for an electrostatic potential, but unfortunately this still leaves us with both potentials in (3.52). By contrast, if we adopt the Lorenz gauge (3.49), we obtain an enormous simplification:

$$\boxed{\nabla^2\varphi - \frac{1}{c^2}\frac{\partial^2\varphi}{\partial t^2} = -\frac{\rho_e}{\varepsilon_0},} \tag{3.54}$$

$$\boxed{\nabla^2\mathbf{A} - \frac{1}{c^2}\frac{\partial^2\mathbf{A}}{\partial t^2} = -\mu_0\mathbf{J}.} \tag{3.55}$$

The vector and scalar potentials are now decoupled, each potential being governed by a wave equation, with $\rho_e/\varepsilon_0$ or $\mu_0\mathbf{J}$ acting as a source term. These simple, elegant equations makes it immediately clear that information propagates away from the sources $\rho_e/\varepsilon_0$ and $\mu_0\mathbf{J}$ at the speed of light, carried off by electromagnetic waves.

We might compare these equations with the equivalent quasi-steady results, in which we used the Coulomb gauge $\nabla\cdot\mathbf{A}=0$. These are the Poisson equations (3.3) and (3.31):

$$\nabla^2\varphi_s = -\frac{\rho_e}{\varepsilon_0}, \quad \nabla^2\mathbf{A} = -\mu_0\mathbf{J}. \tag{3.56}$$

Here there is apparently no time delay between changes in the sources and the corresponding changes in the potentials, and indeed we might think of (3.56) as being the wave equations (3.54) and (3.55) in the limit of $c\to\infty$.

Now, we have seen that these quasi-steady Poisson equations may be inverted using the Green's function solution to give (3.4) and (3.32):

$$\varphi_s(\mathbf{x},t) = \frac{1}{4\pi\varepsilon_0}\int\frac{\rho_e(\mathbf{x}',t)}{|\mathbf{r}|}d\mathbf{x}', \tag{3.57}$$

$$\mathbf{A}(\mathbf{x},t) = \frac{\mu_0}{4\pi}\int\frac{\mathbf{J}(\mathbf{x}',t)}{|\mathbf{r}|}d\mathbf{x}', \tag{3.58}$$

where $\mathbf{r}=\mathbf{x}-\mathbf{x}'$. (By including time in these expressions, we now explicitly note that $\varphi_s$ and $\rho_e$ are evaluated at the same instant, as are **A** and **J**.) However, we also learned how to invert the inhomogeneous wave equation in §2.5.1, and so (3.54) and (3.55) may be similarly inverted (i.e. solved). In particular, we may borrow the inversion integral (2.83) for acoustic waves, which immediately tells us that the solutions to (3.54) and (3.55) are

$$\boxed{\varphi(\mathbf{x},t) = \frac{1}{4\pi\varepsilon_0}\int \frac{\rho_e(\mathbf{x}',t')}{|\mathbf{r}|}\,d\mathbf{x}',} \tag{3.59}$$

$$\boxed{\mathbf{A}(\mathbf{x},t) = \frac{\mu_0}{4\pi}\int \frac{\mathbf{J}(\mathbf{x}',t')}{|\mathbf{r}|}\,d\mathbf{x}',} \tag{3.60}$$

where $t' = t - r/c$ is called the *retarded time*. Evidently, all we need do is replace $t$ by $t'$ in (3.57) and (3.58)! Again we see that information can be transmitted no faster than it can be carried off in the form of waves, and so a fluctuation in $\rho_e$ or $\mathbf{J}$ at location $\mathbf{x}'$ is not felt instantaneously at the remote position $\mathbf{x}$, but rather a time $|\mathbf{r}|/c$ later. The potentials governed by (3.59) and (3.60) are known as the *retarded potentials*, and they will play a pivotal role in our discussion of electromagnetic radiation.

* * *

This completes our brief overview of electromagnetism. In our attempt to build a minimalist theoretical framework, we have had to leave much unsaid, glossing over many of the more delicate theoretical issues, sidestepping the conceptual links between electrodynamics and special relativity, turning a blind eye to the fact that many materials are either dielectric or magnetic, and completely ignoring the many interesting applications in physics and engineering. In the rest of this book we shall build on this skeletal framework, filling in the conceptual gaps, teasing out the links to special relativity, embracing the complex nature of real materials, and providing hints as to the many fascinating applications in nature and technology.

## EXERCISES

3.1 A circular coil carrying a steady current $I$ sits in a uniform, externally imposed magnetic field and is free to rotate. Show that it tends to orientate itself in such a way that the total magnetic field linking the coil is a maximum.

3.2 A liquid metal flows through a uniform magnetic field with velocity $\mathbf{u}(\mathbf{x}, t)$. The imposed magnetic field is negligibly perturbed by the induced currents. Show that the condition for there to be no net charge distribution within the fluid is that $\mathbf{B} \cdot (\nabla \times \mathbf{u}) = 0$.

3.3 Consider a perfectly conducting medium, in which $\sigma \to \infty$. Use (3.27) to show that the magnetic flux linking any closed curve $C$ that moves with the conductor is conserved for all time.

3.4 Use (3.32) to show that the Biot–Savart law may be rewritten in the form

$$\mathbf{B}(\mathbf{x}) = \frac{\mu_0}{4\pi}\int \nabla\left(\frac{1}{|\mathbf{r}|}\right) \times \mathbf{J}'\, d\mathbf{x}', \quad \mathbf{r} = \mathbf{x} - \mathbf{x}'.$$

Now use the identities $\nabla \cdot (\mathbf{c} \times \mathbf{d}) = \mathbf{d} \cdot \nabla \times \mathbf{c} - \mathbf{c} \cdot \nabla \times \mathbf{d}$ and $\nabla \times (\nabla(\cdot)) = 0$ to confirm that

$$\nabla \cdot [\nabla(|\mathbf{r}|^{-1}) \times \mathbf{J}'] = \mathbf{J}' \cdot [\nabla \times \nabla(|\mathbf{r}|^{-1})] - \nabla(|\mathbf{r}|^{-1}) \cdot (\nabla \times \mathbf{J}') = 0,$$

where $\nabla$ operates on $\mathbf{x}$, keeping $\mathbf{x}'$ constant. Hence show that the Biot–Savart law yields $\nabla \cdot \mathbf{B} = 0$.

3.5 The *magnetic helicity* of a magnetic field is defined as the volume integral

$$h_m = \int \mathbf{A} \cdot \mathbf{B}\, dV.$$

Consider a magnetic field composed exclusively of two thin magnetic flux tubes that are interlinked like two adjacent links in a chain. Let the tubes have volumes $V_1$ and $V_2$, centrelines $C_1$ and $C_2$, and magnetic fluxes $\Phi_1$ and $\Phi_2$. Show that the magnetic helicity for this simple geometry is

$$h_m = \int_{V_1} \mathbf{A} \cdot \mathbf{B}\, dV + \int_{V_2} \mathbf{A} \cdot \mathbf{B}\, dV = \Phi_1 \oint_{C_1} \mathbf{A} \cdot d\mathbf{l} + \Phi_2 \oint_{C_2} \mathbf{A} \cdot d\mathbf{l}.$$

Hence use Stokes' theorem to show that $h_m = \pm 2\Phi_1\Phi_2$, where the positive sign is applicable when the two flux tubes are linked in a right-handed fashion, and the negative sign when the connection is left-handed.

## REFERENCE

Feynman R. P., Leighton R. B. & Sands M., 1964, *The Feynman Lectures on Physics*, Volume II. Addison-Wesley.

# 4

# The Foundations of Electrostatics

On an ordinary day over flat desert country, or over the sea, as one goes upward from the surface of the ground the electric potential increase by about 100 volts per meter. . . . This means that outdoors the potential at the height of your nose is 200 volts higher than the potential at your feet! You might ask: 'Why don't we just stick a pair of electrodes out in the air one metre apart and use the 100 volts to power our electric lights?' Or you might wonder: 'If there is *really* a potential difference of 200 volts between my nose and my feet, why is it I don't get a shock when I go out into the street?'

Richard Feynman (*Lectures on Physics*, Volume II, 1964)

By the end of this chapter you will have the answer to Feynman's whimsical questions.

## 4.1 Coulomb's Law, the Electrostatic Field and Poisson's Potential

### *4.1.1 Coulomb's Law and the Electric Field*

In 1785 the military engineer Charles-Augustin de Coulomb published the results of his now celebrated series of experiments on electrostatic forces. He investigated the force of attraction or repulsion between two small charged bodies and showed that this force obeys an inverse square law (something Henry Cavendish had already demonstrated some fourteen years earlier), is proportional to the charges on the two bodies, and lies along the line connecting the bodies. These results are embodied in *Coulomb's law* for stationary point charges, which states that the force exerted by a charge $q'$ on a second charge $q$ is, in SI units,

$$\boxed{\mathbf{f} = \frac{1}{4\pi\varepsilon_0}\frac{qq'}{|\mathbf{r}|^2}\hat{\mathbf{e}}.} \tag{4.1}$$

Here $\mathbf{r}$ is the displacement vector that takes us from $q'$ to $q$ and $\hat{\mathbf{e}} = \mathbf{r}/|\mathbf{r}|$ is a unit vector in the direction of the displacement vector, as shown in Figure 4.1. (We shall use $\mathbf{f}$ to denote

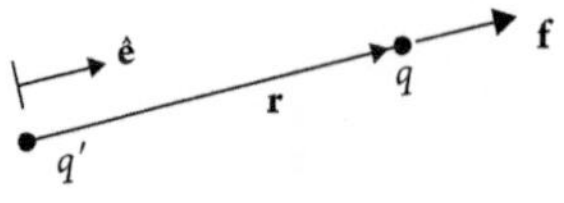


**Figure 4.1** The force exerted by a point charge $q'$ on a second charge $q$.

the net force on individual charges and reserve **F** for the force per unit volume exerted on matter.)

Note that the force in (4.1) is repulsive if the charges are like-signed and attractive if they have opposite signs. The constant $\varepsilon_0$ is called the *permittivity of free space* and, since $q$ is measured in coulombs (C), it has units of $\mathrm{C^2\,N^{-1}\,m^{-2}}$. The value of the pre-factor in (4.1) is $(4\pi\varepsilon_0)^{-1} = 8.99 \times 10^9 \mathrm{N\,m^2\,C^{-2}}$. This might be compared with the equivalent pre-factor in Newton's universal law of gravity, $G = 6.67 \times 10^{-11}\mathrm{N\,m^2\,kg^{-2}}$, which is an inverse square law just like (4.1), but with mass replacing charge. Evidently, *if* we are allowed to swap kilograms for coulombs, electrostatic forces are much stronger that gravitational forces.

If there are many point charges $q_i'$, all exerting electrostatic forces on the charge $q$, then superposition applies and the net force acting on $q$ is evidently

$$\mathbf{f} = \frac{1}{4\pi\varepsilon_0}\sum_i \frac{q_i' q}{|\mathbf{r}_i|^2}\hat{\mathbf{e}}_i. \tag{4.2}$$

The electric field **E** at a point **x** is *defined* as the force per unit charge, $\mathbf{f}(\mathbf{x})/q$, exerted on a small stationary test charge $q$ placed at **x**. If the charges that establish the field **E** are mobile, then the test charge placed at **x** could, in principle, influence the location of the surrounding charges, and hence alter the very field it is intended to measure. To avoid this complication, we might imagine that the test charge $q$ has an infinitesimal amount of charge and that **E** is defined through

$$\mathbf{E}(\mathbf{x}) = \lim_{q\to 0}\frac{\mathbf{f}(\mathbf{x})}{q},$$

although the quantization of charge makes this argument a little tricky in practice.

The electric field $\mathbf{E}(\mathbf{x})$ established by a single stationary point charge located at $\mathbf{x}'$ is evidently

$$\mathbf{E}(\mathbf{x}) = \frac{1}{4\pi\varepsilon_0}\frac{q(\mathbf{x}')}{|\mathbf{r}|^2}\hat{\mathbf{e}} = \frac{1}{4\pi\varepsilon_0}\frac{q(\mathbf{x}')\mathbf{r}}{|\mathbf{r}|^3}, \qquad \mathbf{r} = \mathbf{x} - \mathbf{x}', \tag{4.3}$$

where **r** takes us from the location of the charge to the point where **E** is being measured (Figure 4.2). If there many such static charges, and they are sufficiently numerous and small that the charge per unit volume can be approximated by a smooth function of position, $\rho_e(\mathbf{x})$, then superposition yields

$$\boxed{\mathbf{E}(\mathbf{x}) = \frac{1}{4\pi\varepsilon_0}\int \frac{\rho_e(\mathbf{x}')\mathbf{r}}{|\mathbf{r}|^3}d\mathbf{x}', \quad \mathbf{r} = \mathbf{x} - \mathbf{x}',} \tag{4.4}$$

where $\rho_e(\mathbf{x})$ is the charge density and $d\mathbf{x}' = dx'\,dy'\,dz'$ is the usual volume element.

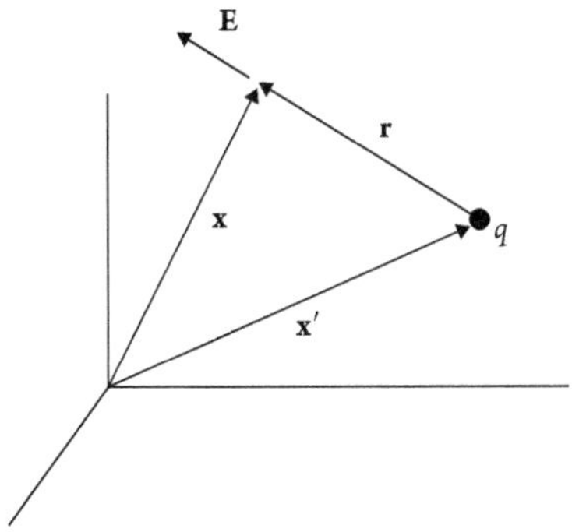


**Figure 4.2** The definition of **r** used in (4.4) and (4.8).

Now, if $\hat{\mathbf{e}}_r$ is a unit vector in the radial direction and $r = |\mathbf{x}|$ the distance from the origin, then it is readily confirmed using spherical polar coordinates that $\nabla \times (\hat{\mathbf{e}}_r/r^2) = 0$. Similarly, through a shift in origin, we have $\nabla \times (\mathbf{r}/|\mathbf{r}|^3) = 0$, where $\mathbf{r} = \mathbf{x} - \mathbf{x}'$ and $\nabla\times$ operates on $\mathbf{x}$ while holding $\mathbf{x}'$ constant. It is therefore clear from (4.3) that $\nabla \times \mathbf{E} = 0$ for a stationary point charge. More generally (4.4) yields

$$\nabla \times \mathbf{E} = \frac{1}{4\pi\varepsilon_0}\int \rho_e(\mathbf{x}')\nabla \times \left(\frac{\mathbf{r}}{|\mathbf{r}|^3}\right) d\mathbf{x}' = 0, \tag{4.5}$$

since the operator $\nabla\times$ may be taken inside the integral sign and $\rho_e(\mathbf{x}')$ is independent of $\mathbf{x}$. Evidently an electrostatic field is conservative (mathematicians would say irrotational), and this follows from the fact that the Coulomb force is a central force, like the gravitational force in Newton's universal law of gravity. From (1.67) we now have

$$\mathbf{E} = -\nabla\varphi \qquad \oint_C \mathbf{E}\cdot d\mathbf{r} = 0, \tag{4.6}$$

for some scalar function $\varphi$ and for any closed curve C. Here $\varphi$ is the scalar potential for the vector field **E**, which in this particular context is called the *electrostatic potential*. We shall add the subscript *s* to $\varphi$ to remind us that we are dealing here with electrostatics. The units of $\varphi_s$ are volts (V) and so **E** is measured in V/m.

For a point charge $q$ at the origin the electric field is evidently radial and spherically symmetric (Figure 4.3) and so in spherical polar coordinates we have

$$E_r(r) = -\frac{\partial\varphi_s}{\partial r} = \frac{q}{4\pi\varepsilon_0 r^2},$$

from which

$$\varphi_s = \frac{q}{4\pi\varepsilon_0 r}, \tag{4.7}$$

where we have taken the constant of integration to be zero. For a continuous distribution of charge this generalises to

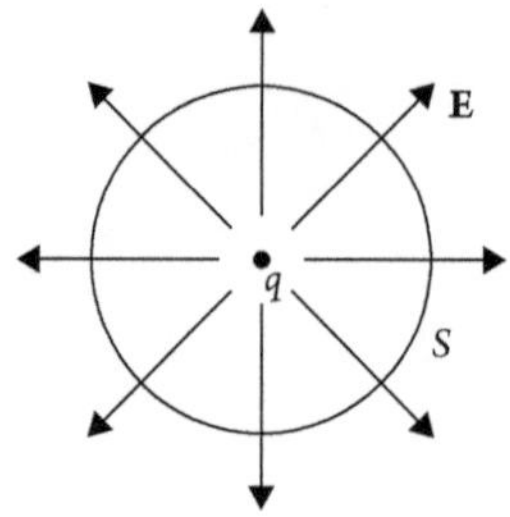


**Figure 4.3** The electric field **E** surrounding a point charge $q$.

$$\boxed{\varphi_s(\mathbf{x}) = \frac{1}{4\pi\varepsilon_0}\int \frac{\rho_e(\mathbf{x}')}{|\mathbf{r}|}\, d\mathbf{x}', \quad \mathbf{r} = \mathbf{x} - \mathbf{x}',} \tag{4.8}$$

where $\mathbf{x}'$ is a dummy variable that samples the charge density distribution, $\rho_e(\mathbf{x}')\,d\mathbf{x}'$ is the charge in the volume element $d\mathbf{x}'$and, as always, $\mathbf{r}$ takes us from the location of the charge to the point of measurement (Figure 4.2). Noting that

$$\nabla\left(\frac{1}{|\mathbf{r}|}\right) = -\frac{\mathbf{r}}{|\mathbf{r}|^3}, \tag{4.9}$$

where $\nabla$ operates on $\mathbf{x}$ while holding $\mathbf{x}'$ constant, it is readily confirmed that the gradient of (4.8) takes us back to (4.4):

$$\mathbf{E}(\mathbf{x}) = -\nabla\varphi_s(\mathbf{x}) = -\frac{1}{4\pi\varepsilon_0}\int \rho_e(\mathbf{x}')\nabla\left(\frac{1}{|\mathbf{r}|}\right) d\mathbf{x}' = \frac{1}{4\pi\varepsilon_0}\int \frac{\rho_e(\mathbf{x}')\mathbf{r}}{|\mathbf{r}|^3}\, d\mathbf{x}'.$$

(Once again, we have used the fact that the operator $\nabla$ may be taken inside the integral sign and that $\rho_e(\mathbf{x}')$ is independent of $\mathbf{x}$.)

### *4.1.2 The Physical Interpretation of the Electrostatic Potential*

From the definition $\mathbf{E} = -\nabla\varphi_s$, the electrostatic force on a small test charge $q$ is evidently $\mathbf{f} = -q\nabla\varphi_s$. If we wish to move this test charge at constant velocity through the electrostatic field, then we must apply an external balancing force of $\mathbf{f}_{\text{ext}} = q\nabla\varphi_s$ to the charge. It follows that the work done *by the external force* against the electric field in moving the charge from $\mathbf{x}_1$ to $\mathbf{x}_2$ is

$$W = q\int_{\mathbf{x}_1}^{\mathbf{x}_2} \nabla\varphi_s \cdot d\mathbf{r} = q[\varphi_s(\mathbf{x}_2) - \varphi_s(\mathbf{x}_1)]. \tag{4.10}$$

Note that this is independent of the path taken from $\mathbf{x}_1$ to $\mathbf{x}_2$, which is a direct result of the property $\oint \mathbf{E}\cdot d\mathbf{r} = 0$. (Think of two distinct paths linking $\mathbf{x}_1$ and $\mathbf{x}_2$. If the line integral of $\mathbf{E}$ along those two paths had different values, then the closed line integral defined by traversing one path from $\mathbf{x}_1$ to $\mathbf{x}_2$ and then the second path from $\mathbf{x}_2$ back to $\mathbf{x}_1$ would demand that $\oint \mathbf{E} \cdot d\mathbf{r} \neq 0$, yielding a contradiction.) Evidently $q\varphi_s(\mathbf{x})$ is the *potential energy* of the charge $q$ by virtue of the prescribed force field $q\mathbf{E}(\mathbf{x})$.

As with all potentials, we are free to choose where we define the datum point (or zero point) for $\varphi_s$, and it is often convenient to take $\varphi_s$ to be zero at infinity, as we did in (4.7) and (4.8). In such cases

$$W = q \int_{\infty}^{\mathbf{x}} \nabla \varphi_s \cdot d\mathbf{r} = q\varphi_s(\mathbf{x}), \tag{4.11}$$

and so $q\varphi_s(\mathbf{x})$ is the work that needs to be done *against* the electric field to bring the charge $q$ from infinity to the point $\mathbf{x}$. For the particular case where $\mathbf{E}$ is associated with a point charge $q'$ at the origin, (4.7) gives $q\varphi_s = qq'/(4\pi\varepsilon_0 r)$.

*Equipotential surfaces* are surfaces defined by a constant value of $\varphi_s$, the simplest example being the concentric spherical equipotential surfaces surrounding a point charge. On the other hand, *lines of force* (often called *field lines*) are lines drawn in space that are everywhere parallel to $\mathbf{E}$ and normal to the equipotential surfaces. In the case of a point charge these point radially outward from the charge (Figure 4.3).

A more intricate example is shown in Figure 4.4. This shows the two-dimensional electric field near the edge of two large parallel metal plates carrying equal and opposite charges. In effect, this is the 'fringing' electric field at the edge of a parallel-plate capacitor. Between the parallel plates the field lines are perpendicular to the two plates, while the equipotential

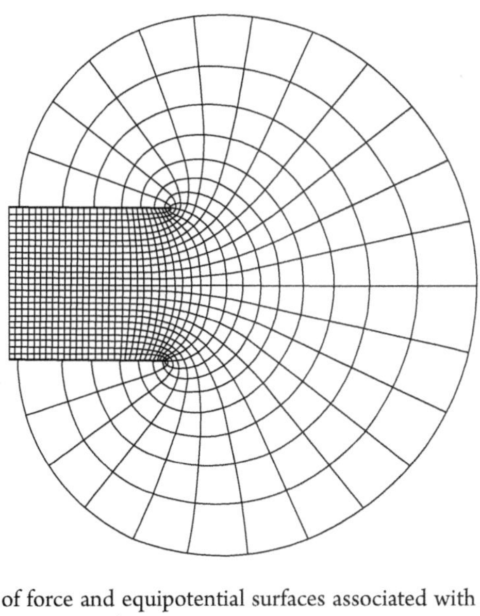

**Figure 4.4** The lines of force and equipotential surfaces associated with two parallel plates carrying equal and opposite charge. (Reproduced from J. C. Maxwell. *A Treatise on Electricity and Magnetism*, 3rd ed. Clarendon Press, 1891.)

surfaces initially run parallel to the plates and then radiate outward from the edge of the capacitor.

## 4.2 Gauss' Law in Differential and Integral Form

We first met Gauss' law of electrostatics in §2.4.2. Let us start with the electric field of a point charge $q$ located at the origin. In spherical polar coordinates this takes the form (Figure 4.3)

$$\mathbf{E} = \frac{q}{4\pi\varepsilon_0}\frac{\hat{\mathbf{e}}_r}{r^2}, \tag{4.12}$$

which may be shown to have zero divergence at all points other than the origin (where the divergence is ill defined). It also satisfies

$$\oint_{S(r)} \mathbf{E}\cdot d\mathbf{S} = \frac{q}{\varepsilon_0},$$

for any spherical surface $S(r)$ centred on the origin. Using Gauss' theorem of vector calculus we can rewrite these statements as

$$\nabla\cdot\mathbf{E} = 0 \quad (\text{for } \mathbf{x} \neq 0), \qquad \int_{V(r)} \nabla\cdot\mathbf{E}\, dV = \frac{q}{\varepsilon_0}, \tag{4.13}$$

where $V(r)$ is the spherical volume enclosed by $S(r)$. Thus $\nabla\cdot\mathbf{E}$ is zero everywhere except at the origin, where it has a finite integral value spread over an infinitesimally small volume. In terms of the three-dimensional delta function we recognise that

$$\nabla\cdot\mathbf{E} = \frac{q}{\varepsilon_0}\delta(\mathbf{x}),$$

since, by definition, $\int\delta(\mathbf{x})\, dV = 1$ for any volume that encloses the origin (see Appendix I). Combining the two statements in (4.13) we have

$$\int_V \nabla\cdot\mathbf{E}\, dV = \frac{q}{\varepsilon_0},$$

for *any* volume $V$ that encloses the charge, the volume integral being zero if it does not enclose $q$. Superposition now tells us that, if we have a continuous distribution of charge, of density $\rho_e(\mathbf{x})$, then

$$\int_V \nabla\cdot\mathbf{E}\, dV = \frac{1}{\varepsilon_0}\int_V \rho_e\, dV = \frac{Q}{\varepsilon_0}, \tag{4.14}$$

where $V$ is any volume and $Q$ is the total charge within $V$. Since this is true for any and all volumes $V$, we may equate the integrands in (4.14) and so

$$\boxed{\nabla \cdot \mathbf{E} = \frac{\rho_e}{\varepsilon_0}.} \tag{4.15}$$

This is *Gauss' law* in differential form, which is one of the four Maxwell equations. When combined with the observation that electrostatic fields are conservative, we have

$$\nabla \times \mathbf{E} = 0, \qquad \nabla \cdot \mathbf{E} = \frac{\rho_e}{\varepsilon_0}. \tag{4.16}$$

This unambiguously defines **E**, since any vector field is uniquely determined if its divergence and curl are known (and suitable boundary conditions are given). In integral form Gauss' law states that

$$\boxed{\oint_S \mathbf{E} \cdot d\mathbf{S} = \frac{Q}{\varepsilon_0},} \tag{4.17}$$

where $Q$ is the total charge enclosed by the surface $S$. We shall see in the next section that, when combined with some form of geometrical symmetry, (4.17) often enables the electric fields associated with simple charge distributions to be determined.

Finally, following Poisson in 1812 (who was inspired by the use of potentials in gravitational theory), we may combine Gauss' law with (4.6) to yield a Poisson equation for the potential:

$$\nabla^2 \varphi_s = -\frac{\rho_e}{\varepsilon_0}. \tag{4.18}$$

This equation may be inverted (perhaps we should say solved) using (4.8) to give

$$\varphi_s(\mathbf{x}) = \frac{1}{4\pi\varepsilon_0} \int \frac{\rho_e(\mathbf{x}')}{|\mathbf{r}|} d\mathbf{x}', \quad \mathbf{r} = \mathbf{x} - \mathbf{x}', \tag{4.19}$$

which is a particular example of a Green's function solution, as discussed in §2.1.4.

## 4.3 Electric Fields Associated with Particularly Simple Charge Distributions

In order to illustrate the utility of (4.17), we now consider some particularly simple charge distributions. Let us begin with a uniformly charged sheet of infinite extent that has a charge per unit area of $\sigma_e$. If there are no charges other than those on the sheet, then symmetry tells us that **E** must be normal to the sheet and mirror-symmetric about the sheet. The field is as shown in Figure 4.5, where **E** diverges from the sheet. To determine $|\mathbf{E}|$ either side of the sheet, we apply (4.17) to a rectangular control volume that straddles the sheet, as shown in

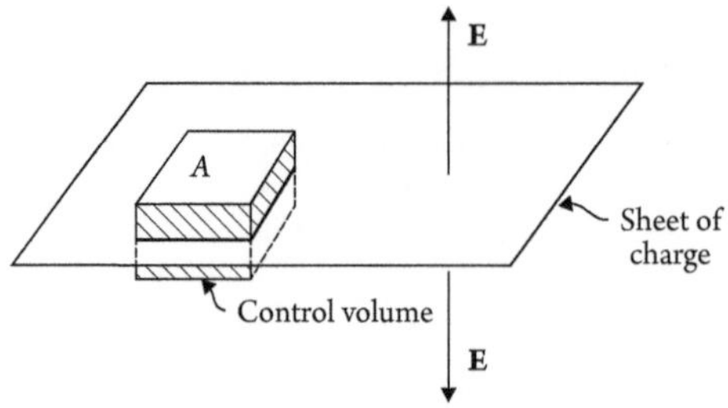


**Figure 4.5** The electric field either side of a uniformly charged sheet.

Figure 4.5. If $A$ is the area of the control volume parallel to the sheet and $Q$ the enclosed surface charge, then we have

$$2\,|\mathbf{E}|\,A = \frac{Q}{\varepsilon_0} = \frac{\sigma_e A}{\varepsilon_0},$$

from which

$$|\mathbf{E}| = \frac{\sigma_e}{2\varepsilon_0}. \tag{4.20}$$

Note that the jump in $\mathbf{E}$ across the sheet, $\Delta\mathbf{E}$, can be written in the form

$$|\Delta\mathbf{E}|_n = \frac{|\sigma_e|}{\varepsilon_0}, \tag{4.21}$$

where the subscript $n$ indicates the jump in the component of $\mathbf{E}$ normal to the surface. We shall return to this expression shortly.

Let us now consider two uniformly charged sheets separated by a distance $d$, with the top sheet carrying a positive surface charge per unit area of $\sigma_e$ and the bottom one carrying an equal but opposite surface charge per unit area of $-\sigma_e$. This is shown in Figure 4.6, where again the sheets are assumed to be infinite in extent. By superimposing two solutions of the form shown in Figure 4.5, we conclude that the electric field is zero outside the two sheets and equal to $|\mathbf{E}| = \sigma_e/\varepsilon_0$ between the sheets. Note that the jump in $\mathbf{E}$ across either sheet is again given by (4.21), which is an inevitable consequence of applying Gauss' law (4.17) to a rectangular control volume that straddles either sheet.

Let us now take each sheet to be of a large but finite area $A$. Then we have a simple model of a parallel-plate capacitor. Since $\mathbf{E}$ is uniform between the plates, except near the lateral edges, the difference in potential between the two plates is simply $\Delta\varphi_s = |\mathbf{E}|d = \sigma_e d/\varepsilon_0$. If we use $\Delta V$ to denote this difference in potential and $Q$ for the net charge on either plate, $Q = \sigma_e A$, then the potential difference (or voltage drop) across the parallel-plate capacitor is

$$\Delta V = \frac{Q}{\varepsilon_0 A/d} = \frac{Q}{C},$$

where $C = \varepsilon_0 A/d$ is the *capacitance* of the parallel-plate capacitor.

As our final example, consider a sphere of radius $R$ centred on the origin and containing a uniform charge density $\rho_e$, as shown in Figure 4.7. In this case symmetry tells us that

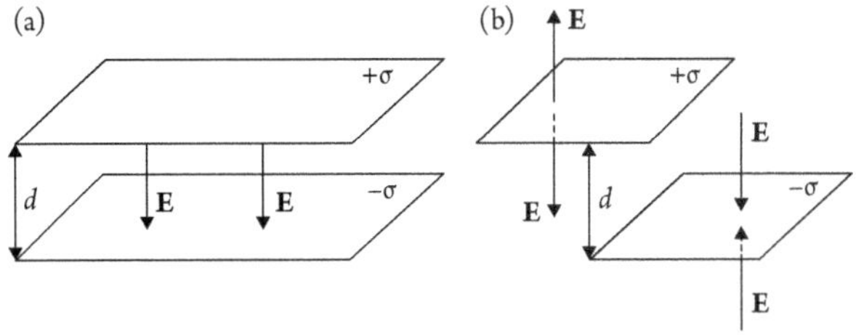


**Figure 4.6** (a) The electric field associated with a pair of uniformly charged sheets carrying equal and opposite surface charges. (b) The same problem viewed as the superposition of two individual sheets.

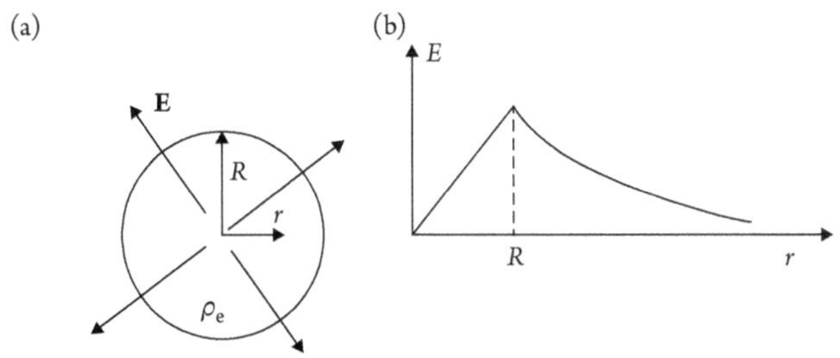


**Figure 4.7** (a) A uniformly charged sphere. (b) The radial variation of the electric field.

**E** is radially symmetric and (4.17) applied to a spherical control volume of radius $r < R$ yields $|\mathbf{E}|\,4\pi r^2 = \frac{4}{3}\pi r^3(\rho_e/\varepsilon_0)$, from which $|\mathbf{E}| = \rho_e r/(3\varepsilon_0)$. Similarly, (4.17) applied to a spherical control volume of radius $r > R$ yields $|\mathbf{E}| = \rho_e R^3/(3\varepsilon_0 r^2)$. Thus the electric field grows linearly up to a radius of $r = R$ and thereafter falls off as $r^{-2}$ with an intensity that is exactly the same as if all the charge were located at the origin. In summary, the electric field associated with a sphere of uniform charge is

$$\mathbf{E}(\mathbf{r}) = \begin{cases} \dfrac{\rho_e \mathbf{r}}{3\varepsilon_0}, & |\mathbf{r}| < R, \\ \dfrac{R^3}{|\mathbf{r}|^3}\dfrac{\rho_e \mathbf{r}}{3\varepsilon_0}, & |\mathbf{r}| > R. \end{cases} \tag{4.22}$$

## 4.4 The Electric Dipole

As another example of a simple charge distribution (but one where symmetry will not help us much), we consider the field of an electric dipole centred at the origin. Suppose we have a positive charge $q$ sitting on the $z$ axis at location $z = d/2$ and just below it a negative charge $-q$ at $z = -d/2$, as shown in Figure 4.8(a). From (4.7) or (4.8) it is clear that the potential for this pair of point charges is given by

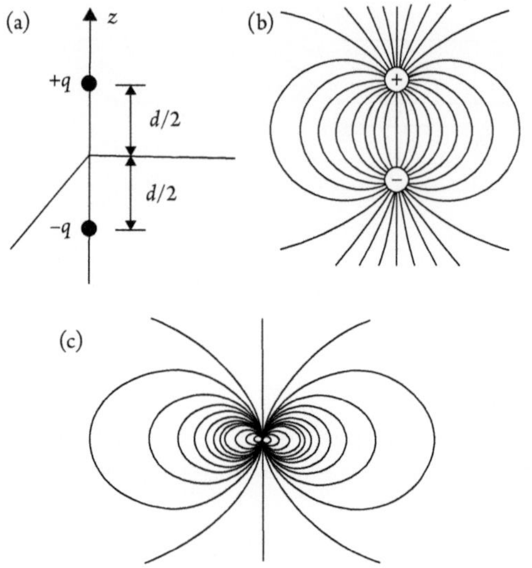


**Figure 4.8** (a) An electric dipole consisting of a pair of point charges of opposite sign. (b) The electric field close to the charge. (c) The far-field dipole corresponding to (4.25).

$$\varphi_s = \frac{q}{4\pi\varepsilon_0}\left[\frac{1}{\sqrt{x^2+y^2+(z-d/2)^2}} - \frac{1}{\sqrt{x^2+y^2+(z+d/2)^2}}\right]. \tag{4.23}$$

Now we are interested in distances from the dipole that are large by comparison with $d$ (known as the *far field*) and so we may use the expansion

$$[x^2+y^2+(z\pm d/2)^2]^{-1/2} \approx (r^2 \pm zd)^{-1/2} \approx r^{-1}\left(1 \mp \frac{zd}{2r^2}\right)$$

to simplify (4.23):

$$\varphi_s = \frac{(qd)z}{4\pi\varepsilon_0 r^3}, \qquad r >> d.$$

Introducing the *electric dipole moment* $\mathbf{p} = q\mathbf{d}$, where $\mathbf{d} = d\hat{\mathbf{e}}_z$ is the displacement vector that takes us from $-q$ to $q$, this potential may be put into the coordinate-free form

$$\varphi_s = \frac{\mathbf{p}\cdot\mathbf{r}}{4\pi\varepsilon_0 r^3} = -\frac{1}{4\pi\varepsilon_0}\mathbf{p}\cdot\nabla\left(\frac{1}{r}\right), \tag{4.24}$$

where we have used (4.9) to obtain the expression on the far right. The electric field now follows from $\mathbf{E} = -\nabla\varphi_s$:

$$\mathbf{E} = \frac{1}{4\pi\varepsilon_0}\nabla\left[\mathbf{p}\cdot\nabla\left(\frac{1}{r}\right)\right]. \tag{4.25}$$

Returning to Cartesian coordinates, the components of $\mathbf{E}$ corresponding to (4.25) with $\mathbf{p} = p\hat{\mathbf{e}}_z$ are readily shown to be

$$\mathbf{E} = \frac{p}{4\pi\varepsilon_0 r^5}(3xz,\ 3yz,\ 3z^2 - r^2),$$

and the resulting field-lines are shown in Figure 4.8(c).

Many vector fields, including magnetic fields produced by a localised system of currents or a velocity field associated with some localised vortices, have exactly the same far-field structure as (4.25), only $\mathbf{p}$ is replaced by the appropriate dipole source. Such a structure is known generically as the *far-field dipole.*

## 4.5 The Elimination of Electrostatic Fields within the Interior of Conductors

So far we have focused on the fields generated by point charges, or by a *prescribed* distribution of charge, such as a uniformly charged sheet or a uniformly charged sphere. In this section and the next we consider fields in the presence of conductors, where the charges are free to move in response to the forces of mutual attraction or repulsion, or perhaps under the influence of an imposed electric field.

### 4.5.1 *A Macroscopic Approach to Charges and Electric Fields in Matter*

When considering conductors, or indeed other types of matter, it is natural to ask if a macroscopic approach to electric fields and charge distributions, in which $\mathbf{E}(\mathbf{x})$ and $\rho_e(\mathbf{x})$ vary smoothly on the geometric lengthscale of the body, is a legitimate point of view. After all, the microscopic reality is one of a sea of negatively charged electrons immersed in a lattice of positively charged nuclei. Fortunately, it turns out that a macroscopic approach *is* appropriate for most purposes, in which the discrete charges of the electrons and protons can be replaced by a smooth continuous distribution of charge, of density $\rho_e(\mathbf{x})$.

The reason is that nuclei are almost invariably extremely small and closely packed, much smaller than the macroscopic dimensions of the body, $L$. Thus one can usually identify a spatial scale over which the charges can be meaningfully averaged, say $\ell_{\text{ave}}$, which is simultaneously much larger than the microscopic scale of the nuclei, say $\ell_{\text{nuclei}}$, yet much smaller than the macroscopic scale of the body, $L$. The requirements for such an approach to work are that (i) $\ell_{\text{ave}}/\ell_{\text{nuclei}}$ must be sufficiently large for the spatial and temporal averaging of the charge to be statistically meaningful and (ii) $\ell_{\text{ave}}/L$ is sufficiently small to justify the use of integral calculus at the macroscopic level. We shall assume these conditions are met in the chapters that follow.

### 4.5.2 *There is No Electrostatic Field Inside a Conductor*

It is easy to show that there can be no macroscopic electrostatic field inside a conductor. In a sense, this follows directly from Ohm's law, $\mathbf{J} = \sigma\mathbf{E}$, where $\mathbf{J}$ is the current density vector introduced in §3.1 and $\sigma$ the electrical conductivity. This tells us that a non-zero electric field would drive a current, which is inconsistent with our assumption of a static charge distribution. However, this does not explain how **E** is eliminated from the interior of a conductor. To understand what is going on, we must temporarily set aside electrostatics and return to the principle of charge conservation, as set out in §3.1.

Consider a control volume $V$ embedded within a conductor and enclosed by a surface $S$. Charge conservation requires that the net rate of flow of charge out through $S$ is equal to the rate of depletion of charge within $V$:

$$\oint_S \mathbf{J} \cdot d\mathbf{S} = -\frac{\partial}{\partial t} \int_V \rho_e \, dV. \tag{4.26}$$

Gauss' theorem then tells us that

$$\int_V \nabla \cdot \mathbf{J} \, dV = -\int_V \frac{\partial \rho_e}{\partial t} dV, \tag{4.27}$$

and since this applies to any volume $V$, we may equate the integrands in (4.27) to give the *charge conservation equation*:

$$\nabla \cdot \mathbf{J} = -\frac{\partial \rho_e}{\partial t}. \tag{4.28}$$

However, Ohm's law, $\mathbf{J} = \sigma \mathbf{E}$, combined with Gauss' law, $\nabla \cdot \mathbf{E} = \rho_e/\varepsilon_0$, requires the divergence of **J** within a conductor to satisfy $\nabla \cdot \mathbf{J} = \sigma \rho_e/\varepsilon_0$. Substituting for $\nabla \cdot \mathbf{J}$ in (4.28) then yields

$$\frac{\partial \rho_e}{\partial t} + \frac{\rho_e}{\tau_e} = 0, \quad \tau_e = \frac{\varepsilon_0}{\sigma}, \tag{4.29}$$

whose solution is $\rho_e = \rho_e(0) \exp(-t/\tau_e)$, where $\tau_e$ is the charge relaxation time.

As noted in §3.1, $\tau_e$ is usually extremely small, being around $10^{-18}$ s in a metal. (Actually, in practice, non-ideal effects tend to increase this to around $10^{-14}$ s, but this is still extremely fast!) Thus any excess charge that exists within the conductor at $t = 0$ rapidly migrates to the boundaries, where it accumulates to form a surface charge density $\sigma_e$. It is the powerful electrostatic repulsion forces that drive any excess charge to the boundary on such a short timescale. In brief, excess free charges of like sign want to get as far away from each other as possible and they do this by moving to the boundary of the conductor.

It now follows from Gauss' law that, after some extremely rapid transient, the electric field within a conductor satisfies $\nabla \cdot \mathbf{E} = 0$. However, once the charges settle down to their equilibrium configuration we also have the electrostatic constraint of $\nabla \times \mathbf{E} = 0$, and so **E** in the conductor is governed by $\nabla \cdot \mathbf{E} = 0$ and $\nabla \times \mathbf{E} = 0$. This tells us that **E** *could be,* though *need not be,* zero. The final step in the argument is to introduce Ohm's law, $\mathbf{J} = \sigma \mathbf{E}$, which tells us that a non-zero **E** would drive a current, which is inconsistent with a static charge distribution. We have arrived at the following important conclusion:

*The charge density and electrostatic field are everywhere zero within a conducting body, all excess charge being located at the surface.*

Moreover, given that $\mathbf{E} = -\nabla\varphi_s = 0$, the entire body is an equipotential. It turns out that distributing the surface charges in such a way that the conducting body is an equipotential minimises the electrostatic energy of those charges (see Exercise 4.5).

At this point it is natural to ask what happens if the conductor is of finite extent and sits in an externally imposed electric field. What stops the external field getting into the conductor? We shall now show that it is blocked (some people say *screened*) by the surface charges.

### 4.5.3 *Surface Charges and Boundary Conditions*

Let us now consider the outer surface of such a conductor. It is an equipotential surface and supports a surface charge density $\sigma_e$ that, in general, varies around the perimeter of the conductor. If we apply Gauss' theorem (4.17) to a small, thin, cylindrical control volume that straddles the boundary, like that shown in Figure 4.9(a), we conclude that the jump in $E_n$ across the surface, $\Delta E_n = [E_n]_{\text{outside}} - [E_n]_{\text{inside}}$, satisfies (4.21) in the form $\Delta E_n = \sigma_e/\varepsilon_0$, where the subscript $n$ indicates the component of $\mathbf{E}$ in the direction of the unit normal vector $\hat{\mathbf{n}}$ (Figure 4.9(a)). Moreover, in order to satisfy $\oint \mathbf{E} \cdot d\mathbf{r} = 0$ applied to the loop shown in Figure 4.9(b), the jump in the *tangential* component of $\mathbf{E}$ across the surface must satisfy $\Delta\mathbf{E}_t = 0$. We conclude that the jump conditions at the surface of the conductor are

$$\Delta E_n = \frac{\sigma_e}{\varepsilon_0}, \quad \Delta\mathbf{E}_t = 0. \tag{4.30}$$

However, we also know that $\mathbf{E} = 0$ within the interior of the conductor, and so these jump conditions yield, immediately outside the conductor,

$$\mathbf{E}_n = \frac{\sigma_e}{\varepsilon_0}\hat{\mathbf{n}}, \qquad \mathbf{E}_t = 0, \tag{4.31}$$

where $\hat{\mathbf{n}}$ is a unit normal vector pointing outward from the surface. These boundary conditions must be satisfied at all points just outside the conductor.

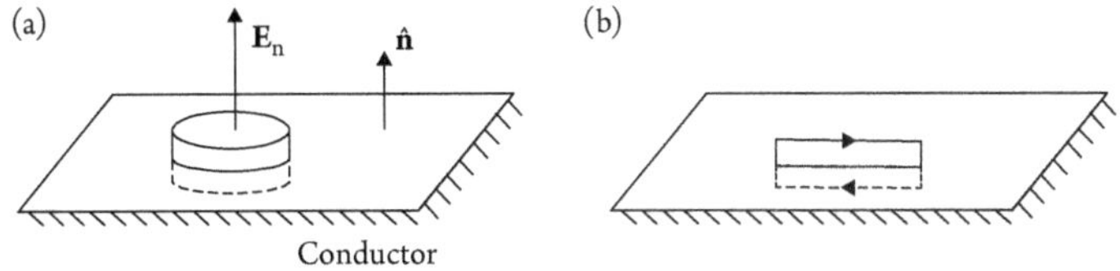


**Figure 4.9** (a) Gauss' law applied to this cylindrical volume yields the jump condition (4.30) for $E_n$. (b) (4.6) applied to the loop shown gives the jump condition $\Delta\mathbf{E}_t = 0$.

Note that, provided $\sigma_e \neq 0$, the electric field is normal to the surface of the conductor. When $\sigma_e$ is positive, the normal field at the surface, $\mathbf{E}_n$, points outwards, from the interior of the body to the external world, whereas a negative $\sigma_e$ corresponds to an inward-pointing $\mathbf{E}_n$. It is this jump in $\mathbf{E}_n$ at the boundary, caused by the surface charges, that shields the interior of a conductor from an externally applied electric field.

## 4.6 Simple Examples of Electric Fields in the Presence of Conductors

### *4.6.1 Electrostatic Shielding*

Let us now consider some simple examples of electric fields in the presence of conductors. Suppose a conductor sits in an external electric field, as shown in Figure 4.10(a). We know there can be no field within the conductor, and so we conclude that the external field must be excluded from the conductor by surface charges. These charges migrate from the interior to the surface immediately after the external field is applied. They arrange themselves in such a way as to set up an additional electric field that exactly cancels the externally applied field at all points within the conductor.

Now suppose that the conductor is hollow, as shown in Figure 4.10(b). Can an electric field develop within the cavity owing to the accumulation of surface charges on the interior surface of the conductor? The short answer is no, provided that there are no charges within the cavity itself. Let us see why this is so. The first point to note is that, because $\mathbf{E} = 0$ within the conductor, Gauss' law (4.17) applied to a surface $S$ lying entirely within the conductor and enclosing the cavity requires

$$\oint_S \mathbf{E} \cdot d\mathbf{S} = 0 = \frac{Q}{\varepsilon_0},$$

where $Q$ is the net charge on the interior surface. So, if surface charges *do* appear on the interior surface, there must be equal amounts of positive and negative charge. However, if such

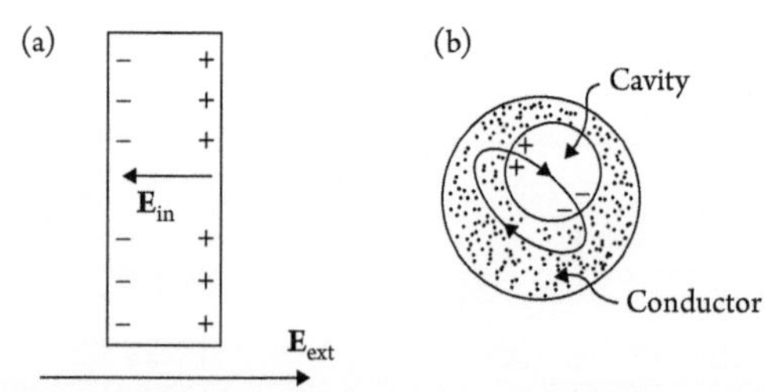


**Figure 4.10** (a) A conductor excludes an external electric field by arranging for charges to accumulate at its surface. These set up a second internal field that is equal and opposite to the externally applied field. (b) If the conductor is hollow, integration of **E** along the curve shown confirms that the cavity is also shielded from any external field.

a charge does develop on the inside surface, we can evaluate $\oint \mathbf{E} \cdot d\mathbf{r}$ around the closed loop shown in Figure 4.10(b). This starts at a positive charge on the interior surface, traverses the cavity to a point where there is some negative surface charge and then completes the circuit within the conductor. That part of the integral which takes us from the positive to the negative charge within the cavity must be non-zero and positive, while the contribution to the line integral from the conductor is zero (since $\mathbf{E} = 0$ in the conductor). This then requires $\oint \mathbf{E} \cdot d\mathbf{r} \neq 0$, which contravenes (4.6). So we conclude that there can be no surface charges on the inner boundary and no electric field within the cavity. Indeed, if one tried to establish regions of equal and opposite positive and negative surface charge, the charges would simply slide along the surface and cancel.

These arguments apply whether or not there is an external electric field, and so we conclude that a conductor not only shields itself from an external electric field, but also any cavity within it. Thus a closed conducting wall isolates its interior contents from any external electrostatic field, which is the basis of the shielding of sensitive electrical equipment.

### 4.6.2 *Sharp Conductors and Lightning Rods*

We close this section with a discussion of electric fields in and around sharp, slender conductors, such as needles or narrow cones. The motivation is to explain why, at very high voltages, air will ionise at the tip of such conductors, leading to an electrical discharge. As we shall see, the key point is that the electric field in a tapered conductor, say a narrow cone, tends to increase roughly as the inverse of the local diameter of the cone, so that near the tip extremely intense electric fields are produced, and these can be strong enough to ionise the surrounding air.

We shall provide a formal analysis of this problem in §5.2.3. Here we merely offer a heuristic discussion of the phenomenon. Let us start with a seemingly unrelated geometry, suggested as a steppingstone to sharp conductors in a number of texts. Consider a sequence of charged spheres of varying radii $a_i$, all charged and connected to each other by wires that ensure they are all at the same potential. The centres of the spheres sit on a straight line and are positioned so that the edges of the spheres are all tangent to an imaginary cone of half-angle $\beta$ (Figure 4.11a). The surface charge on each sphere will preferentially sit near its intersection with of the imaginary cone, rather than near the centreline of the cone, the charges on any one sphere being repelled by the charges in the adjacent spheres. Nevertheless, if the spheres are well separated, we can get some hint as to what happens if we assume that the surface charge on each sphere is roughly uniform.

For the $i$th sphere, then, the surface electric field $E_i$ and electrostatic potential $\varphi_0$ are related to the net charge on that sphere, $Q_i$, by

$$E_i \approx \frac{Q_i}{4\pi\varepsilon_0 a_i^2}, \qquad \varphi_0 \approx \frac{Q_i}{4\pi\varepsilon_0 a_i}, \tag{4.32}$$

where $\varphi_0$ is common to all the spheres. This, in turn, yields

$$E_i \approx \frac{\varphi_0}{a_i} \approx \frac{\varphi_0}{\beta r_i}, \tag{4.33}$$

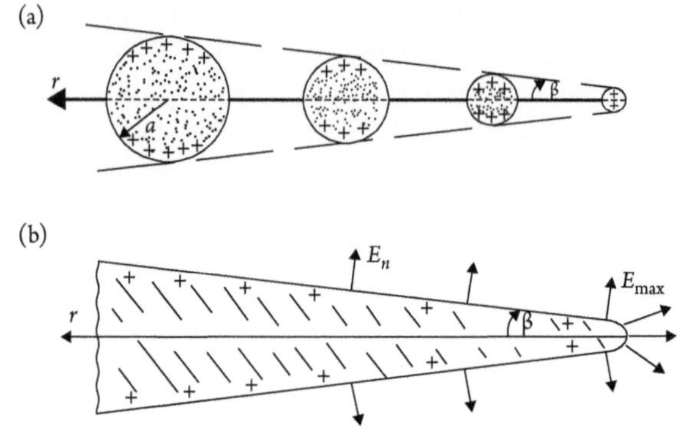


**Figure 4.11** (a) A sequence of charged metal spheres sit inside an imaginary cone of half-angle $\beta$ and are connected by a central wire. (b) A charged metal cone of half-angle $\beta$ and tip radius $a$.

where the radial coordinate $r$ runs along the axis of the imaginary cone (Figure 4.11(a)) and $r_i$ marks the distance from the apex of the cone to the centre of the $i$th sphere.

The key point about (4.32) and (4.33) is that, although the larger spheres carry more charge than the smaller ones, the intensity of the electric field increases as we move to the smaller spheres. The suggestion is that, perhaps, the surface electric field of a slender conical conductor, $E_n$, scales with the local curvature of the surface. It should be said immediately, however, that this is a dangerous argument, and indeed for a body of arbitrary shape there is *no* simple relationship between the magnitude of $E_n$ and the local surface curvature (Price & Crowley, 1985). Nevertheless, it *is* the case that a sharp point produces a singular electric field (see §5.1.3) and the simple model problem shown in Figure 4.11(a) happens to work reasonably well as an analogue for the particular case of a slender conical conductor. In particular, our simple model suggests that the surface field of a narrow cone of half-angle $\beta$ and tip radius $a$ might behave something like

$$E_n \sim \frac{\varphi_0}{\beta r}, \qquad E_{\text{max}} \sim \frac{\varphi_0}{a}. \tag{4.34}$$

The exact solution to the problem of a charged cone, framed in terms of Legendre functions, is given in Jackson (1999) and we shall discuss this in detail in §5.2.3. For small cone angles the surface field does indeed take the form $E_n \sim 1/(\beta r)$, although there are logarithmic corrections. Specifically, for $\beta << 1$, an asymptotic expansion of the Legendre functions yields

$$E_n \sim \frac{\nu\varphi_0}{\beta r^{1-\nu}}, \qquad \nu = \frac{1}{2\ln(2/\beta)} << 1. \tag{4.35}$$

Examples of field lines and equipotential surfaces in the vicinity of sharp conductors are shown in Figure 4.12. Notice the bunching of the equipotential surfaces around the tip of the conductor in Figure 4.12(b). This is the primary reason why large fields develop at the edges of slender conductors.

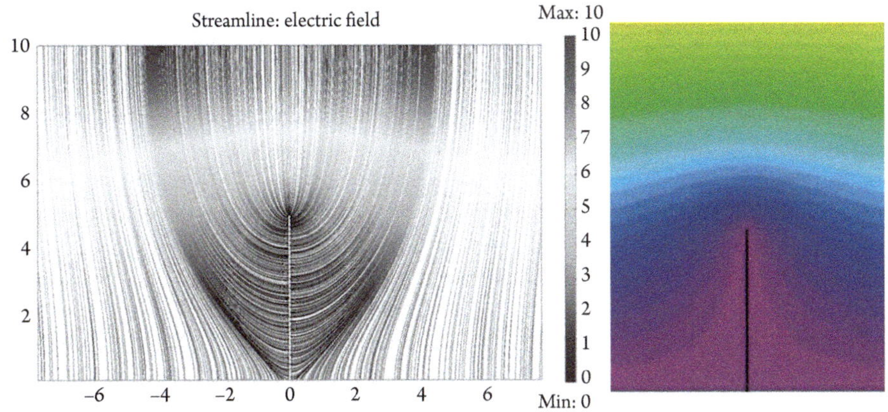


**Figure 4.12** The electric field-lines (left) and equipotential surfaces (right) surrounding a needle. (Courtesy of Tim Wilkinson and John Robertson, University of Cambridge.)

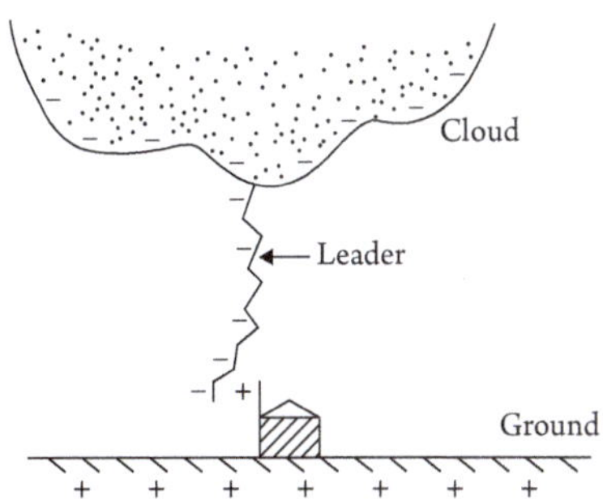


**Figure 4.13** Leaders, filled with negative charge, migrate downwards from a thundercloud towards the ground. The large electric field near the tip of a lightning rod ionises the surrounding air, so that the leader discharges through the rod, rather than the house.

Evidently very large electric fields can build up at the tip of a needle-like conductor, and when high voltages are applied this can lead to an electrical discharge. Typically what happens is that the strong field surrounding the tip accelerates nearby free charges in the air to extremely high speeds, and these then collide with, and ionise, the surrounding air molecules. This, in turn, releases yet more charges and the rate of ionisation is increased until electrical breakdown occurs, resulting in a spark. For air at normal temperatures and pressures the electric field required for ionisation is around $10^6$–$10^7$ V/m.

This is, of course, how lightning rods work. In thunderstorms a vast potential difference builds up between the ground and the thundercloud (many millions of volts), the cloud being at a negative potential relative to the ground. Long, thin, jagged strings of negative charge and ionised air molecules, called *leaders*, migrate downwards from the cloud towards the ground (Figure 4.13). Within a leader the air is conducting, and when such a leader touches the ground (or any 'grounded' object, such as a building) the electrical circuit is closed, in the sense that there is a continuous conducting path all the way up to the cloud, filled with negative charge. The negative charge in the leader then discharges to the ground, starting with the lowest part of the leader and working its way up to the cloud. This is the lightning strike, and it allows the thundercloud to discharge some of its negative charge, perhaps around 20 C, with an associated current of ~10 kA. This, in turn, heats the air in the leader to temperatures as high as 50 000 K. Note that lighting is initiated at or near the ground and then flashes upwards in a so-called *return stroke*. In any event, if a sharp, grounded conductor (a lightning rod) projects above the surrounding buildings, the large voltage difference between it and an adjacent leader causes breakdown to occur preferentially near the tip of the conductor. The thundercloud can then discharge through the lightning rod, rather than through the building to which it is attached.

* * *

This completes our brief introduction to the laws of electrostatics. In Chapter 5 we shall explore some of the mathematical consequences of these laws. In the meantime, readers hungry for more details could do worse than consult Feynman, Leighton & Sands (1964) or Purcell (2011), both of which have excellent introductions to the subject.

---

## EXERCISES

4.1 Show that, if there exists a local minimum in an electrostatic potential, then, on a small surface enclosing that minimum,

$$\oint \mathbf{E}\cdot d\mathbf{S} = \int \nabla\cdot\mathbf{E}\,dV = -\oint \nabla\varphi_s\cdot d\mathbf{S} < 0.$$

Now use Gauss' law to show that no such local minimum (or maximum) can exist in a charge-free region of space. Hence show that a charged particle in free space cannot be held in *stable* equilibrium by an electrostatic field. This is known as *Earnshaw's theorem*.

4.2 A uniformly charged rod of length $L$ and carrying a charge per unit length of $\lambda$ sits on the $x$ axis and is centred at the origin. Show that the field a distance $z$ above the midpoint of the rod is given by

$$\mathbf{E} = \frac{\lambda}{4\pi\varepsilon_0 z}\frac{L\hat{\mathbf{e}}_z}{\sqrt{z^2+(L/2)^2}}.$$

4.3 Show that the dipole moment introduced by (4.24), $\mathbf{p} = qd\hat{\mathbf{e}}_z$, can be rewritten as

$$\mathbf{p} = \int \mathbf{x}'\rho_e(\mathbf{x}')\,d\mathbf{x}',$$

which provides a convenient generalised definition of $\mathbf{p}$ for charge distributions of zero net charge. Now use (4.25) to confirm that the resulting far-field form of $\mathbf{E}$ can be written as

$$\mathbf{E} = \frac{1}{4\pi\varepsilon_0}(\mathbf{p}\cdot\nabla)\nabla\left(\frac{1}{r}\right),$$

and hence confirm that in cylindrical polar coordinates $(s, \theta, z)$ the far field is

$$\mathbf{E} = \frac{p}{4\pi\varepsilon_0 r^3}\left(\frac{3sz}{r^2}\hat{\mathbf{e}}_s + \frac{3z^2 - r^2}{r^2}\hat{\mathbf{e}}_z\right), \quad r^2 = s^2 + z^2.$$

Find the equivalent expression in spherical polar coordinates.

4.4 An *electric quadrupole* consists of three charges sitting on the $z$ axis: a positive charge of $q$ at $z = d/2$, a negative charge of $-2q$ at $z = 0$ and another positive charge of $q$ at $-d/2$. Show that the far-field electrostatic potential is, to leading order in $d/r$,

$$\varphi_s = \frac{qd^2}{16\pi\varepsilon_0 r^3}\left(\frac{3z^2}{r^2} - 1\right), \qquad r >> d.$$

Note that this is much weaker than the dipole potential (4.24).

4.5 Given that the energy density of an electric field is $\varepsilon_0\mathbf{E}^2/2$, show that the total electrostatic energy associated with a given distribution of charges can be written as

$$U_E = \frac{1}{2}\int \varphi_s\rho_e\, dV.$$

Consider two metal spheres of radii $R_1$ and $R_2$ that, between them, must share the total charge $Q$. The spheres are well separated so that the charge distribution on each sphere is negligibly perturbed by the other sphere, and hence the surface charge on each sphere may be treated as uniform. Show that the total energy of the charges is minimised if the charge is divided up according to $Q_1/R_1 = Q_2/R_2$, where $Q = Q_1 + Q_2$. Hence show that this minimum-energy state is automatically realised if we connect the two spheres by a wire to form a single equipotential body. This is an example of *Thomson's theorem*, which states: if we have one or more bodies, each carrying a fixed amount of *surface charge*, then the total electrostatic energy is minimised when the surface charges distribute so as to make each body an equipotential (see e.g. Jackson, 1999). Thus an electrostatic field in the presence of conductors represents a minimum-energy state.

## REFERENCES

Feynman, R. P., Leighton, R. B. & Sands, M., 1964, *The Feynman Lectures on Physics*, Volume II. Addison-Wesley.

Jackson, J. D., 1999, *Classical Electrodynamics*, 3rd ed. Wiley.

Price, R. H. & Crowley, R. J., 1985, The lightning-rod fallacy. *Am. J. Phys.* **53**(9), 843.

Purcell, E. M., 2011, *Electricity and Magnetism*, 2nd ed. Cambridge University Press.

# 5

# Solving for Electrostatic Fields, the Multipole Expansion and Electrostatic Energy

Electricity is of two kinds, positive and negative. The difference is, I presume, that one comes a little more expensive, but is more durable; the other is a cheaper thing, but the moths get into it.

Stephen Leacock (*Literary Lapses*, 1911)

In Chapter 4 we laid the foundations of electrostatics, illustrating the basic theory with particularly simple examples of charge distributions and their associated electric fields. In this chapter we build on these foundations and consider somewhat more intricate (though still idealised) arrangements of charges and conductors. We also introduce the concepts of electrostatic energy and the electric stress tensor.

## 5.1 Methods for Finding Electrostatic Fields

### *5.1.1 Equations, Boundary Conditions and a Uniqueness Theorem*

We shall consider electrostatic fields that are established in part by a prescribed distribution of charge and in part by mobile charges on the surfaces of conducting bodies. In regions of space that are free of charge, Gauss' law demands that $\mathbf{E}$ be solenoidal, and since electrostatic fields are also irrotational, the electric field in such a region is governed by

$$\nabla \times \mathbf{E} = 0, \qquad \nabla \cdot \mathbf{E} = 0. \tag{5.1}$$

These two equations allow us to introduce the auxiliary fields $\varphi_s$ and $\mathbf{C}$, defined through

$$\mathbf{E} = -\nabla \varphi_s, \quad \mathbf{E} = \nabla \times \mathbf{C}, \quad \nabla \cdot \mathbf{C} = 0. \tag{5.2}$$

Of course, these are the scalar and vector potentials introduced in §§1.4 and 1.5 and it is clear from (5.1) and (5.2) that both $\varphi_s$ and $\mathbf{C}$ satisfy Laplace's equation,

$$\nabla^2\varphi_s = 0, \quad \nabla^2\mathbf{C} = 0. \tag{5.3}$$

For the (somewhat exceptional) case of a two-dimensional electric field $\mathbf{E}(x,y) = (E_x, E_y, 0)$ the vector potential takes the form $\mathbf{C} = \psi(x,y)\hat{\mathbf{e}}_z$, and so we have

$$E_x = -\frac{\partial\varphi_s}{\partial x} = \frac{\partial\psi}{\partial y}, \qquad E_y = -\frac{\partial\varphi_s}{\partial y} = -\frac{\partial\psi}{\partial x}. \tag{5.4}$$

(two-dimensional fields only)

Clearly $\nabla\varphi_s \cdot \nabla\psi = 0$, and so lines of constant $\psi$ are perpendicular to the equipotential surfaces and are therefore the field lines for $\mathbf{E}$ (i.e. lines of force). A function analogous to $\psi$ appears in magnetostatics, where it is sometimes called the *flux function* of the magnetic field, and in two-dimensional fluid mechanics, where is it known as the streamfunction of the velocity field. (The equivalent function in axisymmetric problems, known as the *Stokes flux function*, or Stokes streamfunction, is discussed in Exercise 5.1.)

Returning to three dimensions, within a conductor we have $\mathbf{E} = 0$ and so each conductor constitutes an equipotential. Moreover, in §4.5.3 we showed that the electric field at the boundary of a conductor has zero tangential component while the normal component is related to the surface charge density $\sigma_e$ by

$$\mathbf{E}_n = \frac{\sigma_e}{\varepsilon_0}\hat{\mathbf{n}} \tag{5.5}$$

(here $\hat{\mathbf{n}}$ is a unit normal vector that points outwards from the surface of the conductor). Thus, provided $\sigma_e \neq 0$, the field lines at the surface of a conductor are perpendicular to that surface.

On the other hand, if there is a prescribed distribution of charge on a surface, then the jump in the normal component of $\mathbf{E}$ across that surface is given by (4.30), $\Delta E_n = \sigma_e/\varepsilon_0$. There are therefore two common types of boundary conditions in electrostatics. A specified distribution of $\varphi_s$ on all the boundaries constitutes a so-called *Dirichlet problem*, and it corresponds physically to fixing the potential on all bounding surfaces (such as on the surfaces of conducting bodies). On the other hand, if the surface charge distribution $\sigma_e$ is specified on some boundary, then $\partial\varphi_s/\partial n$ is usually known. This constitutes an example of a *Neumann boundary condition*.

Since the governing equations are linear, it is intuitively plausible that any given arrangement of conductors and charges has an associated electric field that is unique. This may be confirmed formally using 'proof by contradiction'. Suppose there exist two distinct potentials $\varphi_s = \varphi_1$ and $\varphi_s = \varphi_2$, both of which satisfy Laplace's equation as well as the boundary conditions, i.e. a specified variation of $\varphi_s$ on the boundary or else a given distribution of $\partial\varphi_s/\partial n$. Now consider the difference function $\hat{\varphi} = \varphi_2 - \varphi_1$. This also satisfies $\nabla^2\hat{\varphi} = 0$

and its boundary conditions are either $\hat{\varphi} = 0$ or $\partial\hat{\varphi}/\partial n = 0$. On integrating $\hat{\varphi}\nabla\hat{\varphi}$ over the bounding surface of the domain, Gauss' theorem yields

$$\oint \hat{\varphi}\nabla\hat{\varphi} \cdot d\mathbf{S} = \int \nabla \cdot (\hat{\varphi}\nabla\hat{\varphi})\, dV = \int [\hat{\varphi}\nabla^2\hat{\varphi} + (\nabla\hat{\varphi})^2]\, dV = \int (\nabla\hat{\varphi})^2 dV. \tag{5.6}$$

However, the surface integral on the left must be zero by virtue of the boundary conditions on $\hat{\varphi}$, and so we conclude that

$$\int (\nabla\hat{\varphi})^2\, dV = 0,$$

which is possible only if $\nabla\hat{\varphi}$ is everywhere zero. This, in turn, tells us that $\varphi_1$ and $\varphi_2$ differ by, at most, a constant, and so their electric fields are identical.

### 5.1.2 *The Method of Images: A Simple Example*

There are a number of well-developed analytical methods available for determining the electric field in a given situation, which we now discuss. Let us start with the simplest method, leaving the more complicated cases for §§5.1.3 and 5.2.

Sometimes we encounter a situation where the field $\mathbf{E}$ associated with a given distribution of charge is known for the particular case of an infinite domain, and we wish to find the corresponding field when $\mathbf{E}$ is restricted to some subdomain, perhaps due to the confining influence of a conducting boundary. In such cases the *method of images* is often helpful. As a trivial example, suppose we wish to find the electric field associated with a point charge $q$ located a distance $a$ above a plane conducting surface that is grounded (i.e. at zero potential) and of infinite extent. To fix our thoughts, let the conducting surface be the $x$–$y$ plane and let the charge sit on the $z$ axis at $z = a$, as shown in Figure 5.1(a). The presence of the conducting surface ensures that the field lines on the $x$–$y$ plane must be normal to that plane, and this is achieved through the development of surface charges that shield the region below the $x$–$y$ plane from the electric field created by $q$.

It is clear from symmetry that the solution to our problem of a point charge in a semi-infinite domain is identical (at least for $z > 0$) to that of two point charges in an infinite domain of equal and opposite sign and located on the $z$ axis at $z = a$ and $z = -a$. That is, $\mathbf{E}$ on the $x$–$y$ plane in the two-charge problem satisfies the boundary condition of zero tangential field, exactly as required in the original problem. So we can mimic the presence of the conducting plane by replacing the conducting surface by an *image charge* of $-q$ at the mirror point below the $x$–$y$ plane. We have arrived back at the dipole problem of §4.4, as shown in Figure 5.1(b).

We can now use the analysis of §4.4 to determine the electric field in the region $z > 0$. Note, in particular, that the field at the $x$–$y$ plane is

$$\mathbf{E} = \frac{-qa}{2\pi\varepsilon_0(a^2 + s^2)^{3/2}}\hat{\mathbf{e}}_z, \quad z = 0, \tag{5.7}$$

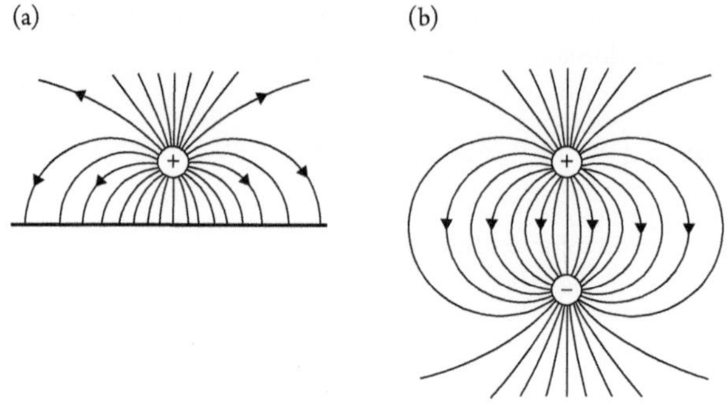


**Figure 5.1** (a) A point charge $q$ sits above a plane conducting surface. (b) The same field in $z > 0$ is generated by equal and opposite charges above and below the $x$–$y$ plane.

where $s$ is the distance from the $z$ axis in the $x$-$y$ plane. The surface charge distribution is then given by (5.5) as

$$\sigma_e = \frac{-qa}{2\pi(a^2 + s^2)^{3/2}}. \tag{5.8}$$

Note also that this charge distribution integrates over the conducting surface to give

$$\int \sigma_e \, dA = -q, \tag{5.9}$$

as it must (but make sure you understand why it must!). Finally, note that the force on the point charge $q$ is the same in the original and image problems, i.e. $\mathbf{f} = q\mathbf{E}$, and so is simply

$$\mathbf{f} = -\frac{q^2}{4\pi\varepsilon_0(2a)^2}\hat{\mathbf{e}}_z. \tag{5.10}$$

### 5.1.3 *The Method of Complex Potentials for Two-Dimensional Fields*

Let us now turn to the rather special case of two-dimensional fields $\mathbf{E}(x,y) = (E_x, E_y, 0)$. Admittedly, this is rarely a good approximation in practice, but it is worthy of consideration since there is a substantial mathematical apparatus available for calculating such fields. The apparatus in question is called the *theory of complex potentials*. We start by returning to (5.4):

$$\frac{\partial \psi}{\partial x} = \frac{\partial \varphi_s}{\partial y}, \quad \frac{\partial \psi}{\partial y} = -\frac{\partial \varphi_s}{\partial x}, \tag{5.11}$$

where both $\psi$ and $\varphi_s$ satisfy Laplace's equation. Such relationships are well known in the theory of complex variables, where they are called the *Cauchy–Riemann equations*. The point is this: consider the complex function

$$F(z) = \phi(x,y) + j\chi(x,y), \tag{5.12}$$

where $z = x + jy$ and all partial derivatives of $\phi$ and $\chi$ with respect to $x$ and $y$ are assumed continuous. Now suppose that $F$ is an analytic function of $z$, i.e. $dF/dz$ at the point $z$ has a unique value that is independent of the direction of $dz$ in the complex plane. Then

$$\frac{\partial F}{\partial x} = \frac{\partial \phi}{\partial x} + j\frac{\partial \chi}{\partial x} = \frac{dF}{dz}\frac{\partial z}{\partial x} = \frac{dF}{dz}$$

and

$$\frac{\partial F}{\partial y} = \frac{\partial \phi}{\partial y} + j\frac{\partial \chi}{\partial y} = \frac{dF}{dz}\frac{\partial z}{\partial y} = j\frac{dF}{dz},$$

from which

$$\frac{\partial \phi}{\partial x} + j\frac{\partial \chi}{\partial x} = \frac{\partial \chi}{\partial y} - j\frac{\partial \phi}{\partial y}.$$

Evidently $\phi$ and $\chi$ obey

$$\frac{\partial \phi}{\partial x} = \frac{\partial \chi}{\partial y}, \quad \frac{\partial \phi}{\partial y} = -\frac{\partial \chi}{\partial x},$$

and we conclude that the real and imaginary parts of $F(z)$ automatically satisfy the Cauchy–Riemann equations and so are individually harmonic. It follows that *any* analytic function

$$F(z) = \psi + j\varphi_s \tag{5.13}$$

will represent a solution of the two-dimensional electrostatic equations. Such a function is known as a *complex potential*. Moreover, since

$$\frac{dF}{dz} = \frac{\partial \psi}{\partial x} + j\frac{\partial \varphi_s}{\partial x} = -E_y - jE_x, \tag{5.14}$$

we have $|dF/dz| = |\mathbf{E}|$, which provides a simple method for evaluating $|\mathbf{E}|$ at all points.

Of course, exactly what geometrical configuration any given function $F(z)$ represents can only be determined after the event by dissecting the details of the solution. In this sense complex potentials constitute an indirect method, and experienced practitioners of this method tend to fall back on a collection of cunning tricks they have acquired, or patterns they have spotted, and then cultivated, over years of experience. On the positive side, applied mathematicians and theoretical physicists have studiously catalogued the properties of such complex potentials, mostly in the latter half of the nineteenth century and often in the context of inviscid fluid dynamics. (Helmholtz was one of the pioneers of this method in fluid mechanics.) So there is a positive cornucopia of solutions from which to choose! (See, for example, Lamb (1932) for examples of two-dimensional irrotational flows that have their counterparts in electrostatics.) We illustrate the power of this method with two rather simple examples, followed by one less straightforward case.

Perhaps the simplest mathematical form for the complex potential is the power law $F(z) = jV_0 + Cz^n$, where $V_0$ and $C$ are constants. Evidently, in terms of two-dimensional polar coordinates $(r,\theta)$, this yields

$$\psi = Cr^n \cos n\theta, \quad \varphi_s = V_0 + Cr^n \sin n\theta. \tag{5.15}$$

Let us now rewrite $n$ as $n = \pi/\alpha$, thus defining an angle $\alpha$. Then the potential $\varphi_s$ is equal to $V_0$ on the radial surfaces $\theta = 0$ and $\theta = \alpha$, while it is symmetric about the bisector $\alpha/2$. We conclude that, for $\alpha < \pi$, (5.15) represents the two-dimensional electric field in the corner region defined by two intersecting conducting surfaces located at $\theta = 0$ and $\theta = \alpha$, the surfaces being held at potential $V_0$. This is illustrated in Figure 5.2(a), which shows the equipotential surfaces. Also, for $\alpha > \pi$, we have the solution to the electric field in and around a two-dimensional wedge composed of two conducting surfaces separated by the half-angle $\beta = \pi - \alpha/2$. This is shown in Figure 5.2(b).

The latter case is of some interest since it represents the two-dimensional analogue of the sharp conical conductor discussed in §4.6.2, and so we might anticipate a singularity in $|\mathbf{E}|$ at the tip of the wedge. This is readily confirmed. The magnitude of $\mathbf{E}$, written in terms of the half-angle of the wedge, $\beta$, is

$$|\mathbf{E}| = \left|\frac{dF}{dz}\right| = nC\,|z|^{n-1} = nCr^{-p},\ p = \frac{\pi/2 - \beta}{\pi - \beta},$$

and for small values of $\beta$ we have $|\mathbf{E}| \sim r^{-1/2}$. This might be contrasted with the axisymmetric case of §4.6.2, in which $|\mathbf{E}| \sim r^{-1}$. We shall return to the problem of a charged cone in §5.2.3.

Next we consider the complex potential $F(z) = -jE_0\sqrt{z^2 - a^2}$. What does this represent? Well, for $|z| >> a$ we have $F(z) = -jE_0 z$, and, given $F'(z) = -E_y - jE_x$, we conclude that $\mathbf{E} = E_0\hat{\mathbf{e}}_x$ in the far field. Conversely we shall see that, for $y = 0$ and $|x| < a$, we have $\varphi_s = 0$. We might anticipate, therefore, that this complex potential corresponds to a grounded conducting plate located on the $x$ axis at $-a < x < a$ and immersed in the externally imposed field $\mathbf{E} = E_0\hat{\mathbf{e}}_x$. It turns out that this is indeed the case and that the field is as shown in Figure 5.3.

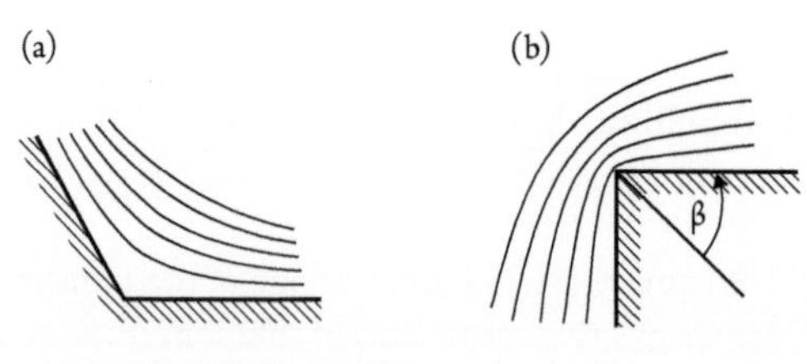


**Figure 5.2** (a) For $\alpha < \pi$ (5.15) gives the field in the corner bounded by two intersecting conducting surfaces. The equipotential surfaces for $n = 3/2$ are shown. (b) For $\alpha > \pi$ this represents the field near a conducting wedge. Equipotential surfaces for $n = 2/3$ are shown.

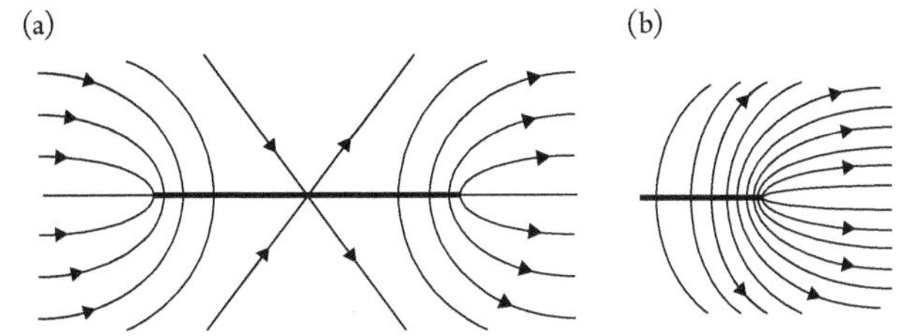


**Figure 5.3** A uniform field $\mathbf{E} = E_0\hat{\mathbf{e}}_x$ is perturbed by a conducting plate located at $-a < x < a$. (a) Overall view. (b) Field near the right-hand edge of the plate.

The field lines and equipotential surfaces are most readily obtained from $F^2 = E_0^2(a^2 - z^2)$, which yields

$$\psi^2 - \varphi_s^2 = E_0^2(a^2 + y^2 - x^2), \quad \psi\varphi_s = -E_0^2 xy.$$

These are indeed consistent with the picture of a uniform field $\mathbf{E} = E_0\hat{\mathbf{e}}_x$ perturbed by a thin conducting plate located at $-a < x < a$. In particular, on the $x$ axis we find

$$\varphi_s = 0, \quad \psi^2 = E_0^2(a^2 - x^2) \quad \text{for } |x| < a,$$
$$\psi = 0, \quad \varphi_s^2 = E_0^2(x^2 - a^2) \quad \text{for } |x| > a,$$

and so $\varphi_s = 0$ for $|x| < a$, as claimed above. Moreover, the components of $\mathbf{E}$ at the surface of the plate are

$$E_x = -\frac{\partial\varphi_s}{\partial x} = 0, \quad E_y = -\frac{\partial\psi}{\partial x} = \pm E_0\frac{x}{\sqrt{a^2 - x^2}}. \tag{5.16}$$

A consideration of the field-line geometry (the field lines bend *into* the plate for $x < 0$ and *out* of the plate for $x > 0$) tells us that the upper sign in (15.16) applies to the top surface of the plate and the lower sign to the bottom surface. Note that, as with the wedge problem above, we have an $|\mathbf{E}| \sim r^{-1/2}$ singularity at the edges of the plate, where in this case $r = a - |x|$.

Given the surface distribution of $E_y$, the surface charge density on either side of the plate can be determined from (5.5). It is

$$\sigma_e = \varepsilon_0 E_0 \frac{x}{\sqrt{a^2 - x^2}}$$

on both surfaces, being negative where the field lines converge on the plate and positive where they diverge. Note that the field lines in Figure 5.3(a) are everywhere normal to the plate, other than at $x = 0$, where we have $\sigma_e = 0$.

Let us now consider a slightly more intricate example. Figure 5.4 shows the two-dimensional electric field near the edge of a parallel-plate capacitor. The complex potential

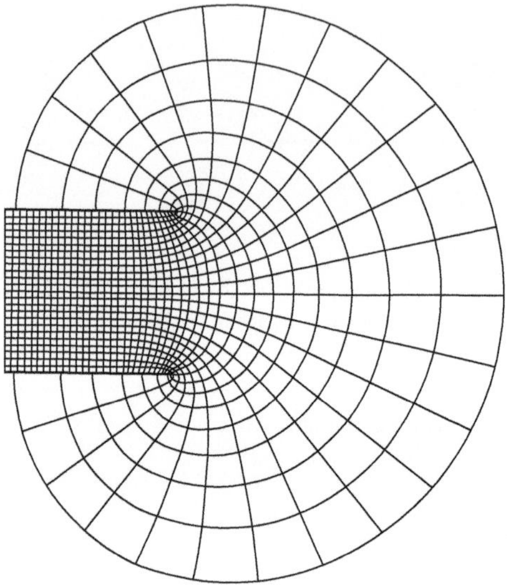

**Figure 5.4** The lines of force and equipotential surfaces at the edge of a parallel-plate capacitor. The equipotential lines radiate outward from the edge of the capacitor, while the lines of force run from plate to plate. (Reproduced from J. C. Maxwell. *A Treatise on Electricity and Magnetism*, 3rd ed. Clarendon Press, 1891.)

for this problem started off life as the inviscid irrotational flow from a large reservoir into a two-dimensional channel, as formulated by Helmholtz. It was subsequently reworked by Maxwell as the 'fringing field' of a parallel-plate capacitor.

It turns out that the complex potential for this problem takes the form of the inverse function $2\pi z/d = F + \exp(F)$. It is a little tricky unpicking the structure of **E** from this inverse function, although the details are spelt out in Lamb (1932) for the equivalent inviscid fluid problem. (See also the discussion in Exercise 5.8 for the electrostatic case.) In summary, the two plates of the capacitor lie at $y = \pm d/2$ and are held at the potentials $\varphi_s = \pm\pi$. The plates are semi-infinite in extent, running from $x = -\infty$ to $x = -d/2\pi$, and the view shown in Figure 5.4 is focused on the edges of the plates. The equipotential lines radiate outward, while the field lines run from plate to plate, mostly inside the capacitor, but some externally. The solution is singular at the edges of the plates, as might be expected from (5.16).

## 5.2 Examples of Axisymmetric Fields

We now consider non-planar fields. In order to keep the algebra manageable, we shall restrict ourselves to axisymmetric geometries.

### *5.2.1 A Point Charge Near a Grounded Conducting Sphere*

We start with a problem that was solved in its most elegant form by a young Lord Kelvin (then William Thomson) in 1848. Consider the case of a point charge located a distance $d$ from the centre of a conducting sphere of radius $a$ (with $d > a$), as shown in Figure 5.5(a). The sphere is assumed to be grounded and so is at zero potential.

Evidently the charge $q$ will attract negative charges to the surface of the sphere closest to $q$, and push positive charges to those parts of the surface that are furthest from $q$. This arrangement of positive and negative surface charge must be such as to exactly cancel the electric field due to $q$ within the conducting sphere. As we shall see, if we are interested *only* in the electric field external to the sphere, it turns out to be permissible to replace all of these surface charges by a single point charge of carefully chosen magnitude located within the sphere. This is, of course, a great simplification, reminiscent of the method of images shown in Figure 5.1. However, understanding how this works requires some rather delicate geometrical considerations.

As emphasised in Maxwell (1891), the key to this problem is to spot the following geometrical trick. A sphere is the locus of all points for which the ratio of the distances from two fixed points is a constant. Moreover, if those points are separated by a distance $b$, and the ratio of the distances from the two points is $e = r_1/r_2$, where $e > 1$ by construction, then the radius of the resulting sphere is

$$a = \frac{eb}{e^2 - 1}. \tag{5.17}$$

(The enthusiastic or sceptical reader might wish to try proving this for themselves.) Now consider two point charges $q$ and $q'$ separated by a distance $b$ and with $q > 0$ and $q' < 0$. If the distance from $q$ is $r_1$ and that from $q'$ is $r_2$, then the net potential associated with these two charges is

$$\varphi_s = \frac{q}{4\pi\varepsilon_0 r_1} - \frac{|q'|}{4\pi\varepsilon_0 r_2}. \tag{5.18}$$

We now consider the equipotential surface defined by $\varphi_s = 0$. This satisfies

$$e = \frac{r_1}{r_2} = \frac{q}{|q'|} = \text{constant}, \tag{5.19}$$

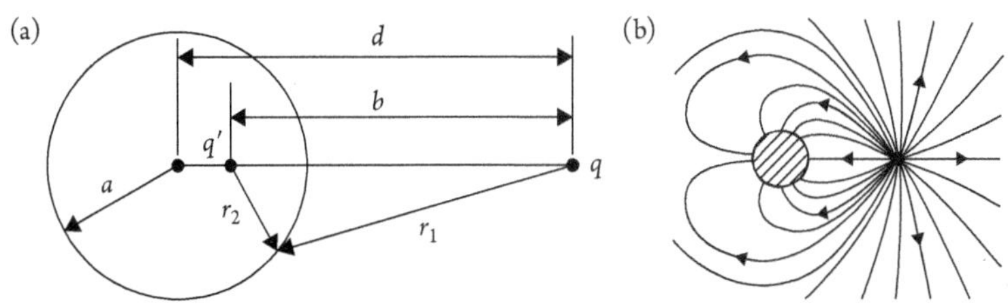


**Figure 5.5** (a) A point charge $q$ adjacent to a conducting sphere may be replaced by two charges $q$ and $q'$, with $q'$ placed at the 'image point' shown. (b) Lines of force for $d = 4a$.

and from the geometrical considerations above we see that this equipotential surface is a sphere whose radius is given by (5.17). It follows immediately that, as far as the field *outside* our conducting sphere is concerned, we may replace the combination of a sphere and a point charge $q$ by two point charges $q$ and $q'$. The location of $q'$, which clearly lies on the line connecting $q$ with the centre of the sphere, is then chosen to satisfy (5.17). In particular, if $q'$ is a distance $c$ from the centre of the sphere, so that $b = d - c$ and $a - c$ is the distance from $q'$ to the closest point on the spherical surface, then (5.17) requires

$$a = \frac{e(d-c)}{e^2 - 1}, \quad \text{where} \quad e = \frac{r_1}{r_2} = \frac{d-a}{a-c}. \tag{5.20}$$

This yields $c = a^2/d$. Finally, the requirement that $e = (d-a)/(a-c)$ gives us the ratio of the magnitudes of the two point charges:

$$e = -\frac{q}{q'} = \frac{d}{a}. \tag{5.21}$$

Note that, as $d$ tends to $a$, so that $q$ becomes close to the surface of the sphere, $q$ and $q'$ increasingly become equal and opposite, which corresponds to the image system shown in Figure 5.1 for a plane surface.

In summary, then, we can replace the surface charges on the conducting sphere by an image charge $q' = -(a/d)q$ located a distance $c = a^2/d$ from the centre of the sphere. The potential outside the sphere is then

$$\varphi_s = \frac{q}{4\pi\varepsilon_0 r_1} - \frac{aq/d}{4\pi\varepsilon_0 r_2}, \tag{5.22}$$

from which $\mathbf{E}$ may be calculated. The more patient reader may wish to confirm for themselves that (5.22) does indeed give $\varphi_s = 0$ on the surface of a sphere of radius $a$.

The underlying lesson here is that it is often possible to replace an arrangement of spherical conductors and point charges by a simpler system of point charges. However, the required arrangement of equivalent point charges is often far from obvious and requires a familiarity with the appropriate literature. Jackson (1999) and Lorrain & Corson (1970) are both good starting points in this respect.

### 5.2.2 *A Conducting Sphere in a Uniform Electric Field*

We now employ a similar strategy to evaluate the electric field arising from a conducting sphere sitting in a uniform externally imposed field $\mathbf{E}_0$. The sphere has radius $a$ and the field lines are shown in Figure 5.6. In order to exclude the external field from the interior of the conductor, charges must accumulate on the surface of the sphere, and symmetry tells us that this distribution of surface charge must be antisymmetric about the plane through the centre of the sphere and normal to $\mathbf{E}_0$. One consequence of this is that there must be as much negative as positive surface charge on the sphere, which in turn suggests that we try replacing the conducting sphere not by a point charge, as in the last example, but by a

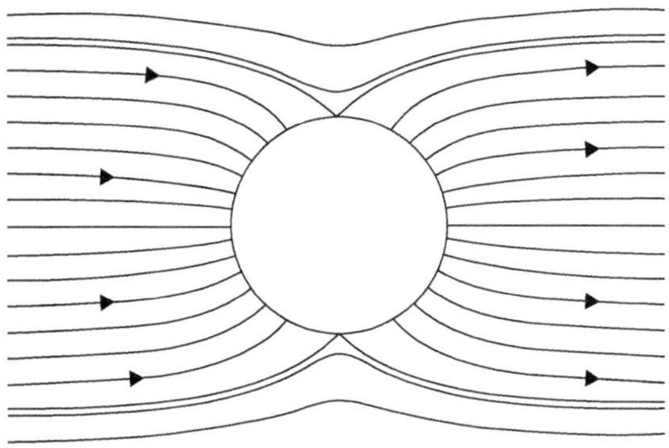

**Figure 5.6** The electric field arising from a conducting sphere sitting in an externally imposed uniform field $\mathbf{E}_0$. (Reprinted from Lord Kelvin (William Thomson). *Reprint of Papers on Electrostatics and Magnetism.* Macmillan, 1872, p. 489.)

dipole of the type introduced in §4.4. So the 'trick' in this case is to replace (or at least *try* to replace—it's not yet a done deal!) the conducting sphere by a *point dipole* located at the centre of the sphere. Such a dipole has a potential given by (4.24),

$$\varphi_s = \frac{\mathbf{p} \cdot \mathbf{r}}{4\pi\varepsilon_0 r^3}, \tag{5.23}$$

where $\mathbf{r}$ is measured from the centre of the point dipole.

In order to pursue this, let us adopt spherical polar coordinates centred on the sphere, with the $z$ axis aligned with $\mathbf{E}_0$ and $\theta$ the polar angle measured from $\mathbf{E}_0$. Symmetry demands that $\mathbf{p}$ be aligned with $\mathbf{E}_0$ and so in this coordinate system we have $\mathbf{p} \cdot \mathbf{r} = pr\cos\theta = pz$. This suggests that we look for a combination of the potential $\varphi_s = -E_0 z = -E_0 r\cos\theta$, which represents the uniform external field $\mathbf{E}_0 = E_0\hat{\mathbf{e}}_z$ in the absence of the conducting sphere, with the potential (5.23) in the form

$$\varphi_s = \frac{pz}{4\pi\varepsilon_0 r^3} = \frac{pr\cos\theta}{4\pi\varepsilon_0 r^3}. \tag{5.24}$$

The combined potential of this 'replacement system' is then

$$\varphi_s = -E_0 r\cos\theta\left(1 - \frac{p}{4\pi\varepsilon_0 r^3 E_0}\right) = -E_0 z\left(1 - \frac{p}{4\pi\varepsilon_0 r^3 E_0}\right). \tag{5.25}$$

This will be equivalent to our original combination of a conducting sphere in a uniform field provided that we can arrange for $\varphi_s$ to be zero on the surface of the sphere, $r = a$. But this is readily arranged by choosing our dipole moment according to $p = (4\pi\varepsilon_0)E_0 a^3$, so that (5.25) simplifies to

$$\varphi_s = -E_0 r\cos\theta\left(1 - \frac{a^3}{r^3}\right). \tag{5.26}$$

For $r > a$ this is indeed the potential associated with a conducting sphere embedded in the uniform field $\mathbf{E}_0$, as can be seen from the fact that it satisfies Laplace's equation for $r > a$, ensures $\varphi_s = 0$ on the spherical surface $r = a$ and recovers $\mathbf{E} = \mathbf{E}_0$ for large $r$. (Actually, we can deduce (5.26) directly from (5.22) by taking $\mathbf{E}_0$ to be the result of two equal and opposite point charges placed either side of the sphere at asymptotically large distances.)

The electric field on the surface of the sphere is now readily calculated and after a little algebra we find that

$$\mathbf{E}(r = a) = 3E_0 \cos\theta\, \hat{\mathbf{e}}_r, \tag{5.27}$$

which, in turn, requires a surface charge density of

$$\sigma_e = 3\varepsilon_0 E_0 \cos\theta. \tag{5.28}$$

This charge is positive at the front of the sphere and negative at the back. Note that the field lines in Figure 5.6 are normal to the sphere everywhere other than at $\theta = \pm\pi/2$, where $\sigma_e = 0$.

### 5.2.3 *Charged Cones and Sharp Conductors Revisited*

As our third and final example of an axisymmetric electric field, we return to the problem of a charged conical conductor, first introduced in §4.6.2. Consider a conducting cone of half-angle $\beta$ whose apex sits at the origin and that extends downward from $\mathbf{x} = 0$ into the region $z < 0$. Adopting spherical polar coordinates, with $\theta$ the polar angle measured from the $z$ axis, the surface of the cone is $\theta = \pi - \beta = \alpha$ (Figure 5.7(a)). The cone is charged and we wish to determine the electric field in the vicinity of its tip.

The solution to this problem lies in the use of Legendre functions, which are discussed in Appendix I. For the present purposes the main point is this. In any axisymmetric geometry (i.e. one in which there is azimuthal symmetry) a particularly important solution of Laplace's equation $\nabla^2\varphi_s = 0$ is

$$\varphi_s = Ar^\nu P_\nu(\cos\theta) = Ar^\nu P_\nu(z/r), \tag{5.29}$$

where $P_\nu$ is the *Legendre function of the first kind of order* $\nu$. The function $P_\nu$ is governed by Legendre's equation

$$(1 - x^2)\frac{d^2P}{dx^2} - 2x\frac{dP}{dx} + \nu(\nu + 1)P = 0, \tag{5.30}$$

and its properties are documented in, for example, Abramowitz & Stegun (1965). It is regular in the range $-1 < \cos\theta \le 1$ but singular at $\cos\theta = -1$, unless $\nu$ is an integer. In order that the potential remain finite near the origin, we shall restrict ourselves to cases in which $\nu > 0$.

The conical surface in Figure 5.7(a) must be an equipotential, which we take to be $\varphi_s = 0$, and so (5.29) is a legitimate solution to the problem at hand provided that we can arrange for $P_\nu(\cos\alpha) = 0$. This constitutes an eigenvalue problem for $\nu$. For a given $\alpha$ there are

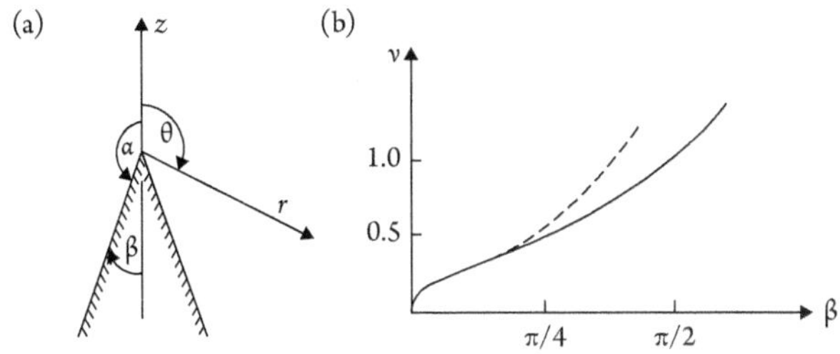


**Figure 5.7** (a) A charged conical conductor has its tip at the origin and sits below the $x-y$ plane. (b) The order $\nu$ corresponding to the smallest root of $P_\nu(\cos\alpha)=0$ as a function of $\beta$. The dashed line is the approximation (5.32).

multiple values of $\nu$ that satisfy $P_\nu(\cos\alpha)=0$, but the value that dominates (5.29) near the tip of the cone is the lowest value of $\nu$, since $\varphi_s \sim r^\nu$. Thus we seek the smallest root of $P_\nu(\cos\alpha)=0$. When expressed in terms of $\beta$, rather than $\alpha$, this root turns out to be a monotonically increasing function of $\beta$, taking the value $\nu=0$ at $\beta=0$ and rising to $\nu=1$ for $\beta=\pi/2$ (Jackson, 1999). This variation in $\nu(\beta)$ is shown in Figure 5.7(b).

Since we are interested in narrow cone angles $\beta << 1$, we must focus on small values of $\nu$. For small $\nu$ the Legendre function takes the simple form (Abramowitz & Stegun, 1965)

$$P_\nu(\cos\theta)=1+2\nu\ln[\cos(\theta/2)], \tag{5.31}$$

and the requirement that $P_\nu(\cos\alpha)=0$ now translates to $2\nu\ln[\sin(\beta/2)]=-1$. For small $\beta$ this simplifies to the expression

$$\nu=\frac{1}{2\ln(2/\beta)}. \tag{5.32}$$

From Figure 5.7(b) we see that this is a good approximation up to cone half-angles of $\beta\approx\pi/4$. The electric field on the surface of the cone can now be obtained by differentiating (5.29), and for small $\nu$ this takes the form given in §4.6.2:

$$E_n=\frac{A}{r^{1-\nu}}\frac{1}{\beta\ln(2/\beta)}, \quad \nu=\frac{1}{2\ln(2/\beta)}. \tag{5.33}$$

For small cone angles this leads to the singular field $|\mathbf{E}|\sim r^{-(1-\nu)}\to r^{-1}$, which is responsible for the electrical breakdown observed near the tips of sharp conductors and explains the operation of lightning conductors.

A second application of the solution (5.29) occurs in fluid dynamics. Consider a vertical capillary tube filled with a conducting liquid and with the lower end open to the atmosphere. If the fluid column is not too high, then it can be held in equilibrium against gravity by the action of surface tension, with a smoothly curved surface protruding just below the end of the capillary tube. Suppose now that an electric field is applied to the liquid by placing its surface at an elevated potential $V_0$. Then electrostatic forces act on the liquid,

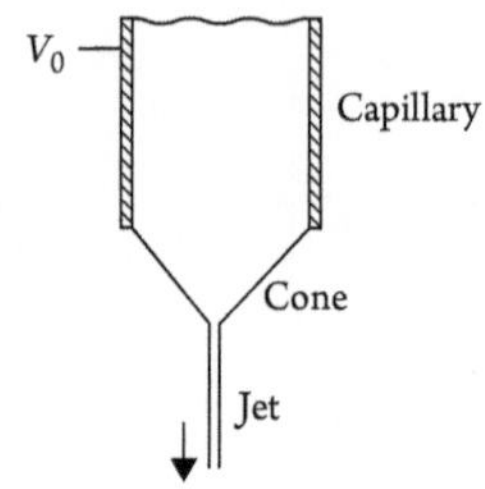


**Figure 5.8** A Taylor cone.

taking the form of electric surface stresses. (We shall discuss such electrostatic stresses in §5.5.2.) As the surface potential is increased, so these stresses rise and increasingly the force balance is between surface tension and electrostatic forces, rather than between surface tension and gravity. Moreover, as the importance of gravity diminishes, so the liquid surface is observed to change shape, shifting progressively to a conical profile, except close to the tip, which is inevitably rounded (Figure 5.8). (The tendency for liquid drops to deform into a cone when subject to an electric field was noted by William Gilbert in 1600.) This cone is known as a *Taylor cone* in honour of the Cambridge physicist G. I. Taylor, who established many of the key properties of this process. In particular, Taylor showed that the conical shape arises naturally from a balance between electrostatic stresses and surface tension, and that the same force balance requires that (5.29) take the specific form

$$\varphi_s = V_0 + Ar^{1/2}P_{1/2}(\cos\theta).$$

The cone half-angle can now be predicted from the requirement that $P_{1/2}(\cos\alpha) = 0$, and this yields a cone half-angle of 49.3°, consistent with experiment.

However, this is not the end of the story. At some critical potential the equilibrium of the fluid around the tip of the cone is lost. A continuous stream of liquid is then ejected from the tip in the form of an extremely narrow jet. The jet in turn becomes unstable and disintegrates into charged droplets of a size comparable to the diameter of the jet. This phenomenon forms the basis of many important technological processes, such as inkjet printing.

## 5.3 The Average Electric Field in a Spherical Control Volume

We now turn to a problem that, at first glance, seems somewhat academic: that of the average electric field within a spherical control volume due to an arbitrary distribution of charge. The charge may lie inside the spherical control volume, outside the control volume or both. The calculation itself is a little dull, especially if done by brute force integration. However, the end result is surprisingly simple and so has considerable generality. It also turns out to be rather useful when we come to discuss dielectrics (insulators) in Chapter 6.

### 5.3.1 A Statement of the Key Result

Suppose we have an arbitrary field of charge $\rho_e(\mathbf{x})$ in an infinite domain. Within this field of charge we identify a spherical control volume of radius $|\mathbf{x}| = R$, which may or may not include some of the charge. If we now integrate $\mathbf{E}(\mathbf{x})$ over this volume, which we denote by $V_R$, then it is possible to establish two related results. First, if all the charge lies *outside* $V_R$, then

$$\int_{V_R} \mathbf{E}(\mathbf{x})\, d\mathbf{x} = V_R \mathbf{E}(0), \tag{5.34}$$

where $\mathbf{E}(0)$ is the field at the centre of the spherical volume. This is a kind of 'mean-value theorem', in that $\mathbf{E}(0) = V_R^{-1} \int \mathbf{E}(\mathbf{x})\, d\mathbf{x}$. Second, if all the charge lies *inside* $V_R$, then

$$\int_{V_R} \mathbf{E}(\mathbf{x})\, d\mathbf{x} = -\frac{1}{3\varepsilon_0} \int_{V_R} \mathbf{x}' \rho_e(\mathbf{x}')\, d\mathbf{x}', \tag{5.35}$$

where $\mathbf{x}' = 0$ is the centre of $V_R$. Of course, if some of the charge lies inside and some outside, superposition yields

$$\int_{V_R} \mathbf{E}(\mathbf{x})\, d\mathbf{x} = V_R \mathbf{E}_{q\mathrm{out}}(0) - \frac{1}{3\varepsilon_0} \int_{V_R} \mathbf{x}' \rho_e(\mathbf{x}')\, d\mathbf{x}', \tag{5.36}$$

where the subscript on $\mathbf{E}_{q\mathrm{out}}(0)$ is to remind us that this is the field at the centre of $V_R$ due to the charge lying *outside* $V_R$ only.

Let us digress for a moment to discuss the integral on the right of (5.36). We have met this integral before. In Exercise 4.3 we defined it to be the *electric dipole moment* of a localised distribution of charge, usually given the symbol $\mathbf{p}$,

$$\boxed{\mathbf{p} = \int \mathbf{x}' \rho_e(\mathbf{x}')\, d\mathbf{x}'.} \tag{5.37}$$

A special case of (5.37) is the physical example introduced in §4.4, where the dipole consists of two point charges of equal and opposite sign, $q$ and $-q$, separated by a short displacement $\mathbf{d}$ and with $\mathbf{p}$ defined as $\mathbf{p} = q\mathbf{d}$. (You might confirm that substituting the two point charges $q$ and $-q$ into (5.37) returns a dipole moment $\mathbf{p} = q\mathbf{d}$.) Indeed, we shall see in §5.4.1 that a localised distribution of charge that is neutral overall, $Q = \int \rho_e\, d\mathbf{x}' = 0$, and has a dipole moment defined by (5.37) has the far-field potential

$$\varphi_s = -\frac{1}{4\pi\varepsilon_0} \mathbf{p} \cdot \nabla \left( \frac{1}{|\mathbf{x}|} \right). \tag{5.38}$$

This is exactly the same far-field potential that we found for the special case of the two adjacent point charges $q$ and $-q$. So the definition (5.37) is a natural generalisation of the dipole moment introduced in §4.4.

Returning to the problem at hand, we may rewrite (5.36) in terms of $\mathbf{p}$ as

$$\boxed{\int_{V_R} \mathbf{E}(\mathbf{x})\, d\mathbf{x} = V_R \mathbf{E}_{q\mathrm{out}}(0) - \frac{\mathbf{p}_{q\mathrm{in}}}{3\varepsilon_0},} \tag{5.39}$$

where the subscript on $\mathbf{p}_{q\mathrm{in}}$ is to remind us that the dipole moment is that due to the charge lying *inside* $V_R$ only. Note the minus sign in front of $\mathbf{p}_{q\mathrm{in}}$. This can be understood with reference to Figure 5.1(b), which shows that the field between the charges in a dipole is, on average, antiparallel to the dipole moment.

We shall now provide a proof of (5.39), first by brute force integration and then using an elegant trick that removes much of the tedious algebra.

### 5.3.2 *A Direct Proof by Brute Force Integration*

We start by writing $\mathbf{E}(\mathbf{x})$ in terms of the integral of the charge density using (4.4):

$$\mathbf{E}(\mathbf{x}) = \frac{1}{4\pi\varepsilon_0}\int \rho_e(\mathbf{x}')\frac{\mathbf{x}-\mathbf{x}'}{|\mathbf{x}-\mathbf{x}'|^3}\, d\mathbf{x}' = -\frac{1}{4\pi\varepsilon_0}\int \rho_e(\mathbf{x}')\nabla\left(\frac{1}{|\mathbf{x}-\mathbf{x}'|}\right) d\mathbf{x}',$$

where $\nabla$ operates on $\mathbf{x}$ while holding $\mathbf{x}'$ constant. It follows that the integral of $\mathbf{E}$ over $V_R$ can be written as

$$\int_{V_R} \mathbf{E}(\mathbf{x})\, d\mathbf{x} = -\frac{1}{4\pi\varepsilon_0}\int \rho_e(\mathbf{x}')\left[\int_{V_R} \nabla\left(\frac{1}{|\mathbf{x}-\mathbf{x}'|}\right) d\mathbf{x}\right] d\mathbf{x}', \tag{5.40}$$

where we have swapped the order of integration. However, we also have the integral theorem

$$\int \nabla f\, dV = \oint f\, d\mathbf{S},$$

for any scalar function $f$, and so we can rewrite (5.40) in the particularly convenient form

$$\int_{V_R} \mathbf{E}(\mathbf{x})\, d\mathbf{x} = -\frac{1}{4\pi\varepsilon_0}\int \rho_e(\mathbf{x}')\left(\oint_{S_R} \frac{d\mathbf{S}}{|\mathbf{x}-\mathbf{x}'|}\right) d\mathbf{x}', \tag{5.41}$$

where $S_R$ is the spherical surface $|\mathbf{x}| = R$.

Our task now is to evaluate the surface integral in (5.41). To this end, we adopt spherical polar coordinates aligned with $\mathbf{x}'$, so that $\theta$ is the polar angle that takes us from $\mathbf{x}'$ to $\mathbf{x}$, as shown in Figure 5.9. Note that all of the contributions to the surface integral, other than those aligned with $\mathbf{x}'$, integrate to zero by virtue of symmetry. Then

$$|\mathbf{x}-\mathbf{x}'|^2 = (R\cos\theta - |\mathbf{x}'|)^2 + (R\sin\theta)^2$$

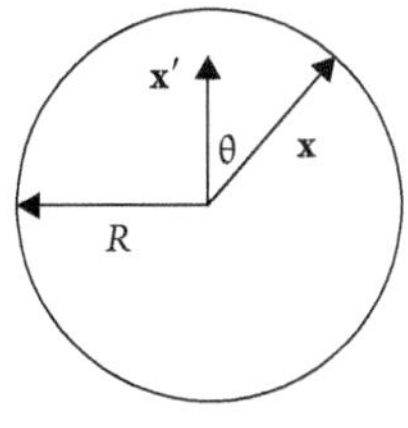


**Figure 5.9** Geometry used to evaluate the surface integral in (5.41).

and hence

$$\oint_{S_R} \frac{d\mathbf{S}}{|\mathbf{x}-\mathbf{x}'|} = \int_0^{\pi} \frac{(2\pi R \sin\theta)(\cos\theta\,\mathbf{x}'/|\mathbf{x}'|)(R\,d\theta)}{\sqrt{(R\cos\theta - |\mathbf{x}'|)^2 + (R\sin\theta)^2}}$$

$$= 2\pi R^2 \frac{\mathbf{x}'}{|\mathbf{x}'|} \int_{-1}^{1} \frac{\chi\, d\chi}{\sqrt{R^2 + |\mathbf{x}'|^2 - 2R|\mathbf{x}'|\,\chi}}, \tag{5.42}$$

where $\chi = \cos\theta$. The integral on the second line of (5.42) may now be evaluated using the indefinite integral

$$\int \frac{\chi\, d\chi}{\sqrt{a - b\chi}} = \frac{2}{b^2}\left(\frac{a - b\chi}{3} - a\right)\sqrt{a - b\chi},$$

and after a little work we obtain

$$\oint_{S_R} \frac{d\mathbf{S}}{|\mathbf{x}-\mathbf{x}'|} = \begin{cases} \frac{4\pi}{3}\mathbf{x}', & |\mathbf{x}'| < R, \quad (5.43a) \\ \frac{4\pi}{3}\frac{R^3}{|\mathbf{x}'|^3}\mathbf{x}', & |\mathbf{x}'| > R. \quad (5.43b) \end{cases}$$

Thus our integral of $\mathbf{E}$ over $V_R$ becomes

$$\int_{V_R} \mathbf{E}(\mathbf{x})\, d\mathbf{x} = \frac{V_R}{4\pi\varepsilon_0} \int_{|\mathbf{x}'|>R} \rho_e(\mathbf{x}') \frac{(-\mathbf{x}')}{|\mathbf{x}'|^3}\, d\mathbf{x}' - \frac{1}{3\varepsilon_0} \int_{V_R} \mathbf{x}' \rho_e(\mathbf{x}')\, d\mathbf{x}', \tag{5.44}$$

or equivalently

$$\int_{V_R} \mathbf{E}(\mathbf{x})\, d\mathbf{x} = V_R \mathbf{E}_{q\mathrm{out}}(0) - \frac{\mathbf{p}_{q\mathrm{in}}}{3\varepsilon_0}. \tag{5.45}$$

We have arrived back at (5.39), albeit after a great deal of integration. Although this particular derivation of (5.39) is rather long-winded, some of the results we have deduced on route, such as (5.43a, b), will prove useful when it comes to our discussion of magnetostatics.

### 5.3.3 A Simpler Proof Using an Elegant Trick

There is a faster way of getting to the same result that sidesteps some of this tedious integration. The approach is to rewrite (5.40) in its more natural form

$$\int_{V_R} \mathbf{E}(\mathbf{x})\, d\mathbf{x} = \int \rho_e(\mathbf{x}') \left( \frac{1}{4\pi\varepsilon_0} \int_{V_R} \frac{\mathbf{x}-\mathbf{x}'}{|\mathbf{x}-\mathbf{x}'|^3}\, d\mathbf{x} \right) d\mathbf{x}'. \tag{5.46}$$

Next, it will be convenient for what follows to swap the roles of the dummy variables $\mathbf{x}$ and $\mathbf{x}'$ in (5.46), so that

$$\int_{V_R} \mathbf{E}(\mathbf{x}')\, d\mathbf{x}' = -\int \left( \frac{1}{4\pi\varepsilon_0} \int_{V_R} \rho_e(\mathbf{x}) \frac{\mathbf{x}-\mathbf{x}'}{|\mathbf{x}-\mathbf{x}'|^3}\, d\mathbf{x}' \right) d\mathbf{x}, \tag{5.47}$$

since $\rho_e(\mathbf{x})$ is independent of $\mathbf{x}'$. We now evaluate the inner integral in (5.47) using a trick. In §4.3 we noted that a charged sphere $V_R$ centred on the origin and of *uniform charge density* $\rho_e$ creates an electric field

$$\mathbf{E}(\mathbf{x}) = \begin{cases} \frac{\rho_e \mathbf{x}}{3\varepsilon_0}, & |\mathbf{x}| < R, \\ \frac{R^3}{|\mathbf{x}|^3} \frac{\rho_e \mathbf{x}}{3\varepsilon_0}, & |\mathbf{x}| > R. \end{cases}$$

So, for our uniformly charged sphere, we have

$$\mathbf{E}(\mathbf{x}) = \begin{cases} \frac{1}{4\pi\varepsilon_0} \int_{V_R} \rho_e \frac{\mathbf{x}-\mathbf{x}'}{|\mathbf{x}-\mathbf{x}'|^3}\, d\mathbf{x}' = \frac{\rho_e \mathbf{x}}{3\varepsilon_0}, & |\mathbf{x}| < R, \quad (5.48) \\ \frac{1}{4\pi\varepsilon_0} \int_{V_R} \rho_e \frac{\mathbf{x}-\mathbf{x}'}{|\mathbf{x}-\mathbf{x}'|^3}\, d\mathbf{x}' = \frac{V_R}{4\pi\varepsilon_0} \frac{\rho_e \mathbf{x}}{|\mathbf{x}|^3}, & |\mathbf{x}| > R. \quad (5.49) \end{cases}$$

This allows us to evaluate the inner integral in (5.47), $\rho_e(\mathbf{x})$ being constant in that integral. This gives

$$\int_{V_R} \mathbf{E}(\mathbf{x}')\, d\mathbf{x}' = -\int_{|\mathbf{x}|<R} \frac{\mathbf{x}\rho_e(\mathbf{x})}{3\varepsilon_0}\, d\mathbf{x} - \frac{V_R}{4\pi\varepsilon_0} \int_{|\mathbf{x}|>R} \frac{\mathbf{x}\rho_e(\mathbf{x})}{|\mathbf{x}|^3}\, d\mathbf{x}$$

and swapping the roles of $\mathbf{x}$ and $\mathbf{x}'$ once again yields

$$\int_{V_R} \mathbf{E}(\mathbf{x})\, d\mathbf{x} = \frac{V_R}{4\pi\varepsilon_0} \int_{|\mathbf{x}'|>R} \rho_e(\mathbf{x}') \frac{(-\mathbf{x}')}{|\mathbf{x}'|^3}\, d\mathbf{x}' - \frac{1}{3\varepsilon_0} \int_{V_R} \mathbf{x}' \rho_e(\mathbf{x}')\, d\mathbf{x}'.$$

We have arrived back at (5.45):

$$\int_{V_R} \mathbf{E}(\mathbf{x})\, d\mathbf{x} = V_R \mathbf{E}_{qout}(0) - \frac{\mathbf{p}_{qin}}{3\varepsilon_0}. \tag{5.50}$$

We shall have reason to return to this result in this chapter and the next.

## 5.4 The Electrostatic Field Far From its Source

In many branches of physics or applied mathematics it is natural to examine a field, say a magnetic field or a velocity field, at large distances from its source, and electrostatics is no exception. In the absence of waves, such fields usually fall off as an inverse power law of the distance from the source, as for example in the electric dipole field introduced in §4.4. This leads one to the realisation that 'events' are rarely localised in space, but rather cast a long shadow, and this is true as much in electromagnetism and classical fluid mechanics as it is in the archetypal example of Newton's universal law of gravitation. In this section we examine in a systematic way the far field of a localised distribution of static charges. In Chapter 8 we shall perform a similar analysis for quasi-steady magnetic fields.

### *5.4.1 The Far Field as a Multipole Expansion: Dipoles and Quadrupoles*

We start with (4.8) for the electrostatic potential arising from a given distribution of static charges $\rho_e(\mathbf{x})$:

$$\varphi_s(\mathbf{x}) = \frac{1}{4\pi\varepsilon_0}\int \frac{\rho_e(\mathbf{x}')}{|\mathbf{r}|}\,d\mathbf{x}', \quad \mathbf{r} = \mathbf{x} - \mathbf{x}'. \tag{5.51}$$

Here we have in mind the situation where $\rho_e(\mathbf{x})$ is localised in space, say exponentially small outside a sphere of radius $a$ surrounding the origin. Then for $|\mathbf{x}| >> a$ we may expand $|\mathbf{r}|^{-1}$ in (5.51) as a power series in $\mathbf{x}'$. For example, the Taylor expansion

$$\frac{1}{|\mathbf{x}+\delta\mathbf{x}|} = \frac{1}{|\mathbf{x}|} + \frac{\partial}{\partial x_i}\left(\frac{1}{|\mathbf{x}|}\right)\delta x_i + \frac{1}{2}\frac{\partial^2}{\partial x_i \partial x_j}\left(\frac{1}{|\mathbf{x}|}\right)\delta x_i\,\delta x_j + \cdots$$

yields

$$\frac{1}{|\mathbf{x}-\mathbf{x}'|} = \frac{1}{|\mathbf{x}|} - \frac{\partial}{\partial x_i}\left(\frac{1}{|\mathbf{x}|}\right)x_i' + \frac{1}{2}\frac{\partial^2}{\partial x_i \partial x_j}\left(\frac{1}{|\mathbf{x}|}\right)x_i' x_j' + \cdots, \tag{5.52}$$

where

$$\frac{\partial^2}{\partial x_i \partial x_j}\left(\frac{1}{|\mathbf{x}|}\right) = \frac{3x_i x_j - \mathbf{x}^2\delta_{ij}}{|\mathbf{x}|^5}. \tag{5.53}$$

Alternatively, if such a Taylor expansion is not to your taste, then precisely the same expression may be obtained by applying the binomial expansion

$$\frac{1}{(1+\varepsilon)^{1/2}} = 1 - \frac{1}{2}\varepsilon + \frac{3}{8}\varepsilon^2 + \cdots$$

to

$$\frac{1}{|\mathbf{x}-\mathbf{x}'|} = \frac{1}{|\mathbf{x}|}\left(1 - \frac{2\mathbf{x}\cdot\mathbf{x}'}{\mathbf{x}^2} + \frac{\mathbf{x}'^2}{\mathbf{x}^2}\right)^{-1/2},$$

and gathering terms. Either way, substituting (5.52) and (5.53) into (5.51) yields

$$\varphi_s(\mathbf{x}) = \frac{1}{4\pi\varepsilon_0}\left[\int \rho_e(\mathbf{x}')\,d\mathbf{x}'\frac{1}{|\mathbf{x}|} - \int \mathbf{x}'\rho_e(\mathbf{x}')d\mathbf{x}'\cdot\nabla\left(\frac{1}{|\mathbf{x}|}\right)\right.$$
$$\left.+\int \frac{x_i'x_j'}{2}\rho_e(\mathbf{x}')\,d\mathbf{x}'\frac{3x_ix_j - \mathbf{x}^2\delta_{ij}}{|\mathbf{x}|^5} + \cdots\right].$$

This is usually rewritten as

$$\boxed{\varphi_s(\mathbf{x}) = \frac{1}{4\pi\varepsilon_0}\left[\frac{Q}{|\mathbf{x}|} - \mathbf{p}\cdot\nabla\left(\frac{1}{|\mathbf{x}|}\right) + Q_{ij}\frac{3x_ix_j - \mathbf{x}^2\delta_{ij}}{|\mathbf{x}|^5} + \cdots\right],} \tag{5.54}$$

where $Q = \int \rho_e\,d\mathbf{x}'$ is the net charge, $\mathbf{p}$ is the *electric dipole moment* of the charge distribution,

$$\mathbf{p} = \int \mathbf{x}'\rho_e(\mathbf{x}')\,d\mathbf{x}', \tag{5.55}$$

and $Q_{ij}$ is the *electric quadrupole tensor*, defined as

$$Q_{ij} = \frac{1}{2}\int x_i'x_j'\rho_e(\mathbf{x}')\,d\mathbf{x}'. \tag{5.56}$$

It is clear that the rate at which the far-field potential falls off depends crucially on whether or not $Q$ and $\mathbf{p}$ are finite. With the exception of Debye–Hückel screening, which we shall discuss shortly, it is rather unusual for a local charge distribution not to have a finite dipole or quadrupole moment, and so it is common to terminate expansions like (5.54) at the quadrupole term.

Perhaps some comments are in order at this point. First, the dipole moment $\mathbf{p}$, as defined by (5.55), is independent of the choice of origin *only* if there is no net charge, i.e. $Q = 0$. Second, some authors prefer to define the electric quadrupole tensor in the traceless form

$$\hat{Q}_{ij} = \frac{1}{2}\int (3x_i'x_j' - \mathbf{x}'^2\delta_{ij})\rho_e(\mathbf{x}')\,d\mathbf{x}' = 3Q_{ij} - Q_{kk}\delta_{ij}. \tag{5.57}$$

Noting that $\mathbf{x}^2\delta_{ij}x_i'x_j' = \mathbf{x}^2\mathbf{x}'^2 = x_ix_j\delta_{ij}\mathbf{x}'^2$, the quadrupole contribution to (5.54) can be rewritten in terms of $\hat{Q}_{ij}$ as

$$\frac{\hat{Q}_{ij}}{4\pi\varepsilon_0}\frac{x_ix_j}{|\mathbf{x}|^5}. \tag{5.58}$$

Third, consider the charge distribution introduced in Exercise 4.4, consisting of three point charges $q$, $-2q$ and $q$, located on the $z$ axis at $z = d/2$, $z = 0$ and $z = -d/2$. Then (5.55) tells us that $Q = \mathbf{p} = 0$, and so the far-field potential must be dominated by the quadrupole term in (5.54). Moreover, it is readily confirmed that, for this particular charge distribution,

$$\frac{Q_{ij}}{4\pi\varepsilon_0}\frac{3x_ix_j - \mathbf{x}^2\delta_{ij}}{|\mathbf{x}|^5} = \frac{qd^2}{16\pi\varepsilon_0}\frac{3z^2 - \mathbf{x}^2}{|\mathbf{x}|^5}, \tag{5.59}$$

which is consistent with the results of Exercise 4.4.

### 5.4.2 *A Second Look at the Dipole Moment and its Field*

Let us now consider the case of a localised charge distribution that is neutral overall, $Q = 0$, but has finite dipole and quadrupole moments. Then, to leading order, the far-field potential is

$$\varphi_s = \frac{\mathbf{p}\cdot\mathbf{x}}{4\pi\varepsilon_0\,|\mathbf{x}|^3} = -\frac{1}{4\pi\varepsilon_0}\mathbf{p}\cdot\nabla\left(\frac{1}{|\mathbf{x}|}\right), \tag{5.60}$$

with a corresponding electric field (see Exercise 4.3),

$$\mathbf{E}(\mathbf{x}) = -\nabla\varphi_s = \frac{1}{4\pi\varepsilon_0}\nabla\left[\mathbf{p}\cdot\nabla\left(\frac{1}{|\mathbf{x}|}\right)\right] = \frac{1}{4\pi\varepsilon_0}(\mathbf{p}\cdot\nabla)\nabla\left(\frac{1}{|\mathbf{x}|}\right). \tag{5.61}$$

Expanding (5.61), we find that the electric field can be written in the form

$$\mathbf{E}(\mathbf{x}) = \frac{1}{4\pi\varepsilon_0}\frac{3\mathbf{x}(\mathbf{p}\cdot\mathbf{x}) - \mathbf{x}^2\mathbf{p}}{|\mathbf{x}|^5}, \tag{5.62}$$

which is commonly used in place of (5.61).

It is important to remember, however, that (5.62) is only the leading-order term in a power-law expansion of $|\mathbf{x}|^{-1}$. It holds only asymptotically for distances that are very large relative to the scale of the dipole, say $d$. Since it is only the ratio of $|\mathbf{x}|$ to $d$ that matters here, rather than think about (5.62) as corresponding to $|\mathbf{x}| \to \infty$ (with fixed $d$), we might consider (5.62) as being the field corresponding to $|\mathbf{x}|$ fixed and $d \to 0$. This particular point of view leads naturally to the idealisation of the *point electric dipole*, and (5.62) is sometimes referred to as the electric field of a point dipole, though of course it does not apply very close to the origin, i.e. where $r \sim d \to 0$.

Now the azimuthal symmetry of (5.62) means that it integrates to zero over any spherical volume centred on the origin, say $V_R$. However, (5.39) demands that

$$\int_{V_R} \mathbf{E}(\mathbf{x})\,d\mathbf{x} = -\frac{\mathbf{p}}{3\varepsilon_0}. \tag{5.63}$$

Thus the failure of (5.62) to represent the field at the origin has led to the incorrect integral value of $\mathbf{E}$. However, this can be 'patched up' if we add to (5.62) the field

$$\mathbf{E} = -\frac{\mathbf{p}}{3\varepsilon_0}\delta(\mathbf{x}), \tag{5.64}$$

where $\delta(\mathbf{x})$ is the three-dimensional delta function. Of course, this additional term does not change $\mathbf{E}$ for $r > d$ (as $d \to 0$) and so does not influence (5.62) in the region where it applies. Rather, the role of (5.64) is simply to ensure that the integral constraint (5.63) is satisfied. So the electric field of a point dipole located at position $\mathbf{x}_o$ is often written as

$$\mathbf{E}(\mathbf{x}) = \frac{1}{4\pi\varepsilon_0}\frac{3\mathbf{r}(\mathbf{p}\cdot\mathbf{r}) - \mathbf{r}^2\mathbf{p}}{|\mathbf{r}|^5} - \frac{\mathbf{p}}{3\varepsilon_0}\delta(\mathbf{r}), \tag{5.65}$$

where $\mathbf{r} = \mathbf{x} - \mathbf{x}_0$. Since any real dipole has a finite size, we need to be a little careful in how we interpret (5.65). The key point is that (5.65) cannot be applied on scales $r \sim d$, but, as the size of the dipole shrinks, so a larger and larger region of space is encompassed by (5.65). By adding the delta function to (5.65), we ensure that our solution not only applies at all points away from $\mathbf{x}_0$, but also integrates across $\mathbf{x}_0$ to give the correct value.

### 5.4.3 *An Aside: Long-Range Interactions Versus Debye–Hückel Screening*

We suggested earlier that it is a little unusual for a localised distribution of charge to have zero dipole and quadrupole moments. The implication is that there are usually strong long-range electrostatic fields (and corresponding long-range forces) that fall off slowly as power laws from their sources. A similar situation arises with quasi-steady magnetic fields and with velocity fields in an incompressible fluid. Here again there are usually long-range forces that fall off as power laws, typically as the inverse cube of the distance. These long-range interactions ensure that every part of the dynamical system communicates with every other part, and this can make it difficult to characterise the overall behaviour. (For example, certain integrals may diverge, or be only conditionally convergent.) It is of interest, therefore, to ask if there is a situation in which these long-range interactions are partially, if not wholly, suppressed. In electrostatics it turns out that there is, and the elimination of the expected long-range interactions is called Debye–Hückel screening, as we now discuss.

Consider a situation in which the charges are mobile, but we may still apply the laws of electrostatics, say in an electrolyte. If we distinguish between the positive charges $q^+$ and the negative ones $q^-$, and let $n^+(\mathbf{x})$ and $n^-(\mathbf{x})$ be the density of positive and negative charges, then we can write Poisson's equation as

$$\nabla^2\varphi_s = -\frac{\rho_e}{\varepsilon_0} = -\frac{q^+n^+ - q^-n^-}{\varepsilon_0}. \tag{5.66}$$

Now consider some local region within the electrolyte, say a small subvolume $V_R$. Then the long-range forces exerted by the charges within $V_R$ on the rest of the electrolyte will depend on the values of $Q$, $\mathbf{p}$, $Q_{ij}$ etc. for those charges instantaneously contained within $V_R$. One might expect, for example, that it is unlikely that $Q$ will be zero if the charges are non-uniformly distributed. If so, then the charges within $V_R$ will behave like a monopole, with a long-range electric field that falls off with the inverse square of the distance from $V_R$. However, there is a natural mechanism that tends to ensure electrical neutrality within $V_R$: if initially there is some excess positive charge, for example, this will attract surrounding negative charges into $V_R$, and this will continue until there is zero net charge. So there *is* the possibility that certain regions within the electrolyte might be partially, if not wholly, electrically decoupled from their surroundings. Let us see how this works for the case of a colloidal suspension of charged particles dispersed in an electrolyte (Figure 5.10).

Suppose that the colloidal particles, which are much larger than the ions in the electrolyte, carry a net positive charge and that the electrolyte itself consists of a sea of positive and negative ions dissolved in water. Then the colloidal particles will attract the surrounding negative ions and repel the positive ones, and so we might anticipate that each colloidal particle will surround itself with a shell of negative ions, creating a region of overall charge neutrality. However, the situation is a little more complicated than this because each ion in the electrolyte moves under the influence of the fields of all the other ions, and not just because of an adjacent colloidal particle. Moreover, there will random thermal agitation of the ions (collisions) that could offset, at least to some degree, the systematic tendency for neutral clusters of charge to form. Given these complications, we wish to determine how efficiently each colloidal particle can electrically isolate itself from its surroundings by building up a shell of negative charge.

To simplify the problem, we shall assume that the positive and negative ions both carry the same magnitude of charge, say $q_e$, and so in the region immediately surrounding one of the colloidal particles (5.66) becomes

$$\nabla^2 \varphi_s = -\frac{q_e n^+ - q_e n^-}{\varepsilon_0}. \tag{5.67}$$

Moreover, we shall assume that the colloidal particle and its surrounding charge distribution are both spherically symmetric, and so

$$\frac{1}{r^2}\frac{d}{dr}\left(r^2 \frac{d\varphi_s}{dr}\right) = -\frac{q_e n^+ - q_e n^-}{\varepsilon_0},$$

where $r$ is the distance from the centre of the colloidal particle. We rewrite this in the more convenient form

$$\frac{1}{r}\frac{d^2}{dr^2}(r\varphi_s) = -\frac{q_e n^+ - q_e n^-}{\varepsilon_0}. \tag{5.68}$$

If the ion density distributions $n^+(\mathbf{x})$ and $n^-(\mathbf{x})$ were known, then this would determine the distribution of $\varphi_s$, and hence give us the electric field and forces acting on the ions. However, the ions, in turn, respond to these electrostatic forces and move, so there is a second half to this problem that needs to be considered.

Let us first consider how a field of ions position themselves in a prescribed electrostatic field. The Boltzmann law of statistical mechanics tells us that particles in thermal equilibrium with a conservative force field arrange themselves so that their mean number density behaves as

$$n(x) = n_0 \exp\left[-\frac{U(x)}{k_B T}\right], \tag{5.69}$$

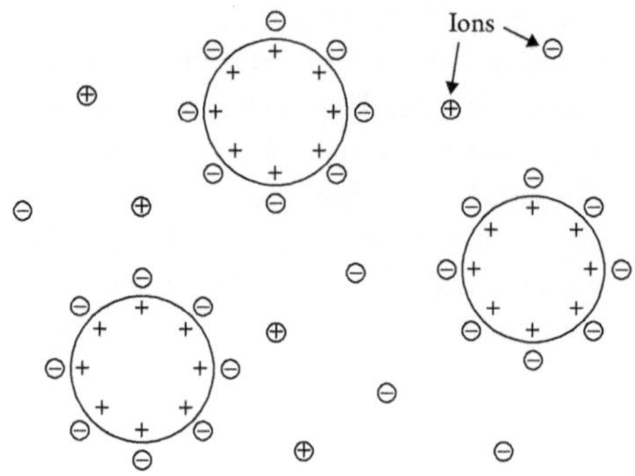


**Figure 5.10** Positively charged colloidal particles surrounded by shells of negative ions.

where $k_B$ is Boltzmann's constant, $T$ the absolute temperature and $U(x)$ the potential energy of a given particle by virtue of the force acting on that particle. We saw in §4.1.2 that, in the case of an electrostatic force field acting on a charge $q$, the potential energy of that charge is $q\varphi_s$ and so, according to the Boltzmann distribution, our ions will distribute themselves according to

$$n^{\pm}(r) = n_0 \exp\left(\mp\frac{q_e\varphi_S}{k_B T}\right),$$

for some constant $n_0$ that is indicative of the overall ion concentration. Substituting these number densities into (5.68), we obtain a single closed equation for $\varphi_S$:

$$\frac{1}{r}\frac{d^2}{dr^2}(r\varphi_s) = \frac{2q_e n_0}{\varepsilon_0}\sinh\left(\frac{q_e\varphi_S}{k_B T}\right). \tag{5.70}$$

This is most readily solved for high temperature or low potential, because in either case the sinh function in (5.70) may be replaced by $q_e\varphi_S/k_B T$, which then simplifies to

$$\frac{d^2}{dr^2}(r\varphi_s) = \frac{2q_e^2 n_0}{\varepsilon_0 k_B T} r\varphi_s.$$

Since the growing exponential solution of this equation is physically unacceptable, we are left with

$$\varphi_s = \frac{R}{r}V_0 \exp\left(-\frac{r-R}{D}\right), \quad r > R, \tag{5.71}$$

where $V_0$ is the potential at the surface of the colloidal particle, located at $r = R$, and

$$D^2 = \frac{\varepsilon_0 k_B T}{2q_e^2 n_0}. \tag{5.72}$$

The lengthscale $D$ is called the Debye length (after the physical chemist Peter Debye) and is a measure of the thickness of the shell of negative charge that surrounds each positive colloidal particle. Notice that $D$ increases if the thermal energy of the ions rises, but decreases if the ion concentration $n_0$ is increased.

This is in many ways a remarkable result. It says that the electric field of each colloidal particle is blocked (some say screened) by a shell of negative ions that are attracted to that particle. Moreover, this screening is very efficient in the sense that the electric field generated by each particle decays exponentially fast, so there are none of those long-range algebraic power laws we normally expect to see in electrostatics. This will happen to all of the colloidal particles, and so they are effectively electrically isolated from one another, unless they get so close that their ion shells start to overlap, in which case they repel each other.

There are many important consequences of Debye screening. For example, in order to maintain a stable colloidal suspension, it is necessary to avoid aggregation of the suspended particles, since coagulation will cause the particles to precipitate out of solution. The tendency for particles to aggregate is driven, in part, by the interparticle van der Waals forces that attract adjacent particles to each other. On the other hand, if two particles get close enough for their ion shells to overlap, then they will be repelled. It turns out that coagulation is suppressed provided the Debye length is sufficiently large to stop the van der Waals forces taking hold. However, the Debye length decreases if the ion concentration is increased, and so, if there are enough ions in the solution, electrostatic repulsion ceases to be an efficient means of preventing aggregation. Therefore, adding salt to a colloidal suspension can cause it to precipitate out. Debye layers play an important role in many heterogeneous fluid systems, such as milk, paint and ink, to name but a few.

## 5.5 Electrostatic Energy and Forces

### *5.5.1 The Potential Energy of a Set of Charges and Energy Density of an Electric Field*

So far we have said very little about the energy of an electrostatic field, other than to note that the potential energy of a point charge $q$ in a *prescribed* electric field $\mathbf{E}$ (with potential $\varphi_s$) is $q\varphi_s$. We now rectify this omission.

Suppose we have a set of $N$ point charges $q_i$ located at positions $\mathbf{x}_i$ in an infinite domain and with each charge held in place by a force of constraint $\mathbf{f}_i = -q_i\mathbf{E}(\mathbf{x}_i)$. Then the overall potential for this system of charges is evidently

$$\varphi_s(\mathbf{x}) = \frac{1}{4\pi\varepsilon_0}\sum_{i=1}^{N}\frac{q_i}{|\mathbf{x}-\mathbf{x}_i|}. \tag{5.73}$$

Since each charge sits in the electric field created by all the other charges, it possesses a potential energy $q_i\varphi_s(\mathbf{x}_i)$, and, on adding all such contributions, we may associate a potential energy with the system of charges. Crucially, however, the net potential energy is *not* the naïve estimate $\sum_{i=1}^{N} q_i\varphi_s(\mathbf{x}_i)$, but rather $\frac{1}{2}\sum_{i=1}^{N} q_i\varphi_s(\mathbf{x}_i)$. Let us see where the factor of $\frac{1}{2}$ comes from.

The global potential energy is most readily evaluated by calculating how much energy is released if, one by one, we let the charges $q_i$ slowly recede to infinity, the electrostatic forces being in balance with the forces of constraint at all times. The calculation proceeds as follows. The potential at the location of charge $q_1$ is

$$\varphi_s(\mathbf{x}_1) = \frac{1}{4\pi\varepsilon_0}\sum_{i=2}^{N}\frac{q_i}{|\mathbf{x}_1 - \mathbf{x}_i|},$$

and so, as it recedes to infinity, with all other charges fixed in place, it liberates a potential energy

$$U_1 = \frac{q_1}{4\pi\varepsilon_0}\sum_{i=2}^{N}\frac{q_i}{|\mathbf{x}_1 - \mathbf{x}_i|}.$$

We now let charge $q_2$ slowly recede to infinity, and in doing so it releases the additional potential energy

$$U_2 = \frac{q_2}{4\pi\varepsilon_0}\sum_{i=3}^{N}\frac{q_i}{|\mathbf{x}_2 - \mathbf{x}_i|},$$

and so it goes on, with the $j$th charge releasing the energy

$$U_j = \frac{q_j}{4\pi\varepsilon_0}\sum_{i=j+1}^{N}\frac{q_i}{|\mathbf{x}_j - \mathbf{x}_i|}. \tag{5.74}$$

Evidently, the total potential energy released as the various charges are slowly withdrawn to infinity is

$$\begin{aligned} U_{\mathrm{E}} = {} & \frac{q_1}{4\pi\varepsilon_0}\left(0 + \frac{q_2}{r_{12}} + \frac{q_3}{r_{13}} + \frac{q_4}{r_{14}} + \frac{q_5}{r_{15}} + \cdots + \frac{q_N}{r_{1N}}\right) \\ & + \frac{q_2}{4\pi\varepsilon_0}\left(0 + 0 + \frac{q_3}{r_{23}} + \frac{q_4}{r_{24}} + \frac{q_5}{r_{25}} + \cdots + \frac{q_N}{r_{2N}}\right) \\ & + \frac{q_3}{4\pi\varepsilon_0}\left(0 + 0 + 0 + \frac{q_4}{r_{34}} + \frac{q_5}{r_{35}} + \cdots + \frac{q_N}{r_{3N}}\right) \\ & + \cdots, \end{aligned}$$

where $r_{ij} = \left|\mathbf{x}_i - \mathbf{x}_j\right|$. This can be rewritten in the more symmetric form

$$\begin{aligned} 2U_{\mathrm{E}} = {} & \frac{q_1}{4\pi\varepsilon_0}\left(0 + \frac{q_2}{r_{12}} + \frac{q_3}{r_{13}} + \frac{q_4}{r_{14}} + \frac{q_5}{r_{15}} + \cdots + \frac{q_N}{r_{1N}}\right) \\ & + \frac{q_2}{4\pi\varepsilon_0}\left(\frac{q_1}{r_{21}} + 0 + \frac{q_3}{r_{23}} + \frac{q_4}{r_{24}} + \frac{q_5}{r_{25}} + \cdots + \frac{q_N}{r_{2N}}\right) \\ & + \frac{q_3}{4\pi\varepsilon_0}\left(\frac{q_1}{r_{31}} + \frac{q_2}{r_{32}} + 0 + \frac{q_4}{r_{34}} + \frac{q_5}{r_{35}} + \cdots + \frac{q_N}{r_{3N}}\right) \\ & + \cdots, \end{aligned}$$

which is equivalent to

$$U_E = \frac{1}{2}\sum_{i=1}^{N} q_i \varphi_s(\mathbf{x}_i), \tag{5.75}$$

where $\varphi_s(\mathbf{x}_i)$ is the potential associated with all the charges of the original system, except $q_i$. The equivalent result for a continuous field of charge is evidently

$$U_E = \frac{1}{2}\int \rho_e \varphi_s \, d\mathbf{x}. \tag{5.76}$$

Although we have interpreted $U_E$ as the energy released as the charges are removed to infinity, it is equally valid to think of it as the energy required to assemble the initial charge distribution, by moving charges inward from infinity against the repulsive forces of the other charges already in place. Either way, the expression for $U$ given by (5.76) is the potential energy of the system of charges.

Let us now rewrite this in terms of the electric field $\mathbf{E}$. The product of $\varphi_s$ with Poisson's equation

$$\nabla^2 \varphi_s = -\frac{\rho_e}{\varepsilon_0}$$

yields

$$\varphi_s \nabla^2 \varphi_s = \nabla \cdot (\varphi_s \nabla \varphi_s) - (\nabla \varphi_s)^2 = -\frac{\rho_e \varphi_s}{\varepsilon_0},$$

or equivalently

$$\rho_e \varphi_s = \varepsilon_0 \mathbf{E}^2 - \nabla \cdot (\varepsilon_0 \varphi_s \nabla \varphi_s).$$

It follows that

$$U_E = \frac{1}{2}\int \rho_e \varphi_s \, d\mathbf{x} = \frac{\varepsilon_0}{2}\int \mathbf{E}^2 \, d\mathbf{x} - \frac{\varepsilon_0}{2}\oint \varphi_s \nabla \varphi_s \cdot d\mathbf{S},$$

and with $\varphi_s$ zero at infinity, this simplifies to

$$\boxed{U_E = \frac{1}{2}\int \rho_e \varphi_s d\mathbf{x} = \frac{\varepsilon_0}{2}\int \mathbf{E}^2 \, d\mathbf{x}.} \tag{5.77}$$

As noted in §3.5, the quantity $\varepsilon_0 \mathbf{E}^2/2$ is the *energy density of the electric field*. This raises the question: should we regard this energy as a form of potential energy associated with the charges, or the intrinsic energy of the electric field, which is distributed throughout all space? At this simple level, either interpretation is legitimate, but as we move from electrostatics to electrodynamics, we shall find the latter interpretation to be more compelling.

### 5.5.2 *The Electric Stress Tensor*

In this subsection we consider the electrostatic stress system associated with the Coulomb force $\mathbf{f} = q\mathbf{E}$. We start, however, by setting aside electrostatics and offering instead a brief discussion of conventional mechanical stresses, if only to introduce some notation.

Consider a small cube of material, $\delta V = \delta x \delta y \delta z$, aligned with coordinates $(x, y, z)$, centred on location $(x_0, y_0, z_0)$ and part of a larger material sample that is subject to stress. Each surface element of our small cube will experience a force that is exerted on it by the adjacent material and the magnitude of that force will be proportional to its area. For example, the force $\delta\mathbf{F}$ exerted on the surface element $\delta A = \delta x \delta y$ shown in Figure 5.11, whose normal is $\hat{\mathbf{e}}_z$, is conventionally written as the force per unit area:

$$\frac{\delta\mathbf{F}^{(\text{face}\,z)}}{\delta x \delta y} = \tau_{xz}\hat{\mathbf{e}}_x + \tau_{yz}\hat{\mathbf{e}}_y + \tau_{zz}\hat{\mathbf{e}}_z. \tag{5.78}$$

Here $\tau_{xz}$ etc. are the *mechanical stresses* acting on the element of area $\delta x \delta y$, evaluated at the centre of that face, $(x_0, y_0, z_0 + \delta z/2)$. The rule is that the first subscript on $\tau_{ij}$ tells us about the direction of the associated component of force, while the second subscript identifies the normal to the surface. The force per unit area on the companion face below, whose normal is $-\hat{\mathbf{e}}_z$, is by convention written as

$$\frac{\delta\mathbf{F}^{(\text{face}\,-z)}}{\delta x \delta y} = -\tau_{xz}\hat{\mathbf{e}}_x - \tau_{yz}\hat{\mathbf{e}}_y - \tau_{zz}\hat{\mathbf{e}}_z, \tag{5.79}$$

the negative sign being used because in equilibrium these forces will balance those on the face above. By extension, the net force acting on the surface element $\delta y \delta z$, whose normal is $\hat{\mathbf{e}}_x$, is written as

$$\frac{\delta\mathbf{F}^{(\text{face}\,x)}}{\delta y \delta z} = \tau_{xx}\hat{\mathbf{e}}_x + \tau_{yx}\hat{\mathbf{e}}_y + \tau_{zx}\hat{\mathbf{e}}_z. \tag{5.80}$$

A torque balance on the cube shown in Figure 5.11 then requires that $\tau_{xz} = \tau_{zx}$ etc., so we may swap the order of the indices on $\tau_{ij}$ with impunity.

If these stresses are now allowed to vary continuously in space, then small differences in stress between companion faces can create a net force on the volume element $\delta V = \delta x \delta y \delta z$. For example, small differences in the stresses on the top and bottom of the cube produce a force

$$\frac{\delta\mathbf{F}^{(\text{faces}\,z,-z)}}{\delta x \delta y \delta z} = \frac{\partial \tau_{xz}}{\partial z}\hat{\mathbf{e}}_x + \frac{\partial \tau_{yz}}{\partial z}\hat{\mathbf{e}}_y + \frac{\partial \tau_{zz}}{\partial z}\hat{\mathbf{e}}_z.$$

A little algebra then shows that the net force arising from all six faces is $\delta F_i = (\partial \tau_{ij}/\partial x_j)\,\delta V$. Note that this spatial variation in stress does not alter the requirement that $\tau_{ij} = \tau_{ji}$ in order to maintain angular equilibrium.

Let us now return to electrostatics and to the stress system associated with the Coulomb force $\mathbf{f} = q\mathbf{E}$. When the charges $q$ are located within matter, the force $\mathbf{f} = q\mathbf{E}$ is transmitted

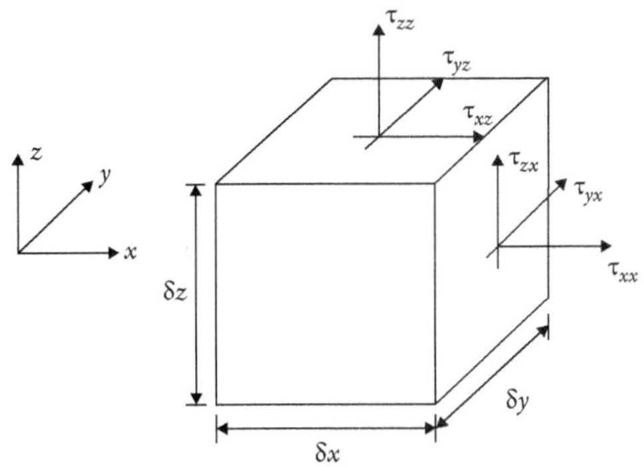


**Figure 5.11** Mechanical stresses acting on a small cube $\delta V = \delta x \delta y \delta z$.

directly to that material, taking the form $\mathbf{F} = \rho_e \mathbf{E}$ per unit volume. Invoking Gauss' law, the Coulomb force per unit volume may be rewritten as

$$\mathbf{F} = \rho_e \mathbf{E} = (\varepsilon_0 \nabla \cdot \mathbf{E})\mathbf{E}. \tag{5.81}$$

This, in turn, can be manipulated into the form

$$F_i = \varepsilon_0 \frac{\partial E_j}{\partial x_j} E_i = \varepsilon_0 \left[ \frac{\partial}{\partial x_j}(E_i E_j) - E_j \frac{\partial E_i}{\partial x_j} \right] = \varepsilon_0 \left[ \frac{\partial}{\partial x_j}(E_i E_j) - E_j \frac{\partial E_j}{\partial x_i} \right],$$

where we have used the fact that $\nabla \times \mathbf{E} = 0$ to substitute $\partial E_j / \partial x_i$ for $\partial E_i / \partial x_j$. We conclude that that the Coulomb force per unit volume can be written as

$$F_i = \rho_e E_i = \frac{\partial}{\partial x_j}(\varepsilon_0 E_i E_j) - \frac{\partial}{\partial x_i}\left( \frac{1}{2} \varepsilon_0 \mathbf{E}^2 \right), \tag{5.82}$$

or equivalently

$$\boxed{F_i = \rho_e E_i = \frac{\partial \tau_{ij}^E}{\partial x_j}, \qquad \tau_{ij}^E = \varepsilon_0 \left( E_i E_j - \frac{1}{2} \mathbf{E}^2 \delta_{ij} \right).} \tag{5.83}$$

The point about (5.83) is that, as far as calculating the Coulomb force is concerned, the integral of the body force $\mathbf{F} = \rho_e \mathbf{E}$ over any given volume $V$ is entirely equivalent to the effect of an imaginary system of stresses $\tau_{ij}^E$ acting over the surface of that volume,

$$\int_V \rho_e E_i \, dV = \oint_S \tau_{ij}^E \, dS_j. \tag{5.84}$$

The quantity $\tau_{ij}^E$ is called the *electric stress tensor*, or the electric contribution to *Maxwell's stress tensor*, and it satisfies $\tau_{ij}^E = \tau_{ji}^E$, as it must. Its utility lies in the fact that it is frequently more convenient to think of the electrostatic forces as arising from a set of surface stresses, rather than a distributed body force. If the material is in equilibrium under a combination of mechanical stresses $\tau_{ij}$ and electrostatic forces, then the equation of equilibrium is simply

$$\frac{\partial}{\partial x_j}(\tau_{ij} + \tau_{ij}^E) = 0.$$

### 5.5.3 *Force, Torque and Energy of an Electric Dipole in an External Field*

We close this chapter by returning to the concept of the electric dipole, evaluating the net force and torque exerted on such a dipole by an *externally imposed*, slowly varying field $\mathbf{E}_0(\mathbf{x})$. Our motivation is to set the scene for the topic of the next chapter—dielectrics.

Consider a localised distribution of charge that is centred at the origin and is electrically neutral, $Q = \int \rho_e \, dV = 0$, and which has a dipole moment, defined by (5.55), of

$$\mathbf{p} = \int \mathbf{x}\rho_e(\mathbf{x})\, d\mathbf{x}. \tag{5.85}$$

The net torque (evaluated about the origin) experienced by this dipole is evidently

$$\mathbf{T}_d = \int \mathbf{x} \times [\rho_e(\mathbf{E}_\rho + \mathbf{E}_0)]\, d\mathbf{x},$$

where $\mathbf{E}_\rho$ is the field associated with the local charge distribution and $\mathbf{E}_0$ is the externally imposed field. However, a closed system of charges cannot exert any net force or any net torque on itself (see Exercise 5.5), and so

$$\mathbf{T}_d = \int \mathbf{x} \times (\rho_e \mathbf{E}_0)\, d\mathbf{x}. \tag{5.86}$$

If $\mathbf{E}_0$ varies slowly in space on the scale of the dipole, then we may approximate its value in (5.86) by $\mathbf{E}_0(0)$, which then simplifies to

$$\boxed{\mathbf{T}_d = \left(\int \mathbf{x}\rho_e\, d\mathbf{x}\right) \times \mathbf{E}_0(0) = \mathbf{p} \times \mathbf{E}_0(0).} \tag{5.87}$$

Note that $\mathbf{T}_d$ tends to align $\mathbf{p}$ with $\mathbf{E}_0(0)$, which turns out to be important for certain dielectrics.

There is, of course, an additional contribution to $\mathbf{T}_d$ arising from the spatial variation of $\mathbf{E}_0$, and it is readily shown that this is related to the quadrupole component, (5.56), of the charge distribution, as discussed in Exercise 5.3. However, this second contribution is smaller than (5.87) when $\mathbf{E}_0(\mathbf{x})$ is a slowly varying function of position.

Let us now calculate the force on the dipole, which turns out to be a little more subtle. First we note that the net forces associated with $\mathbf{E}_\rho$ and $\mathbf{E}_0(0)$ are both zero,

$$\mathbf{F}_d = \int \rho_e[\mathbf{E}_\rho + \mathbf{E}_0(0)]\, d\mathbf{x} = \int \rho_e \mathbf{E}_0(0)\, d\mathbf{x} = Q\mathbf{E}_0(0) = 0, \tag{5.88}$$

since $Q = 0$ and a closed system of charges cannot exert any net force on itself. Thus the dipole can experience a force *only* if $\mathbf{E}_0$ is spatially varying. Since we are assuming that $\mathbf{E}_0(\mathbf{x})$ varies slowly with position, at least on the scale of the dipole, we may approximate $\mathbf{E}_0(\mathbf{x})$ by the Taylor expansion

$$E_{0i}(\mathbf{x}) = E_{0i}(0) + \left[\frac{\partial E_{0i}}{\partial x_j}\right]_0 x_j. \tag{5.89}$$

We conclude that the net force on the dipole is

$$\mathbf{F}_d = \int \rho_e \mathbf{E}_0(\mathbf{x})\, d\mathbf{x} = \int \rho_e x_j\, d\mathbf{x} \left[\frac{\partial E_{0i}}{\partial x_j}\right]_0 = (\mathbf{p}\cdot\nabla)\,\mathbf{E}_0(\mathbf{x}). \tag{5.90}$$

Noting that $\nabla \times \mathbf{E}_0 = 0$, we may rewrite this in the more convenient form

$$\boxed{\mathbf{F}_d = (\mathbf{p}\cdot\nabla)\,\mathbf{E}_0(\mathbf{x}) = \nabla[\mathbf{p}\cdot\mathbf{E}_0(\mathbf{x})],} \tag{5.91}$$

as becomes clear on expanding the gradient on the far right.

Finally we note that the dipole possesses some potential energy by virtue of the presence of the external field. This potential energy exists over and above the self-energy of the charge distribution, (5.76), and arises from the fact that the potential energy of a point charge $q$ in a *prescribed* electric field $\mathbf{E}_0$ (with potential $\varphi_0$) is $q\varphi_0$. Once again, we assume that $\mathbf{E}_0(\mathbf{x})$ varies slowly with position, and so we may use a local approximation to $\varphi_0$, say $\varphi_0 = A - \mathbf{E}_0(0)\cdot\mathbf{x}$, for some constant $A$. The additional potential energy arising from the external field $\mathbf{E}_0(\mathbf{x})$ is then

$$U_d = \int \rho_e \varphi_0\, dV = AQ - \int [\rho_e \mathbf{E}_0(0)\cdot\mathbf{x}]\, dV = -\left(\int \rho_e \mathbf{x}\, dV\right)\cdot\mathbf{E}_0(0), \tag{5.92}$$

which we can write as

$$\boxed{U_d = -\mathbf{p}\cdot\mathbf{E}_0(0).} \tag{5.93}$$

Note that $U_d$ is a minimum when $\mathbf{p}$ is aligned with $\mathbf{E}_0(0)$, zero if the dipole is orthogonal to $\mathbf{E}_0(0)$ and maximal if $\mathbf{p}$ is antiparallel to $\mathbf{E}_0(0)$. We shall see in Chapter 8 that an analogous expression exists for magnetic dipoles (current loops) siting in an imposed magnetic field. Equations (5.87) and (5.93) tell us that there is a natural tendency for dipoles to align with the prevailing electric field, thus reducing their potential energy, as shown in Figure 5.12. For example, the maximum-energy state in Figure 5.12 is unstable to small rotations, while

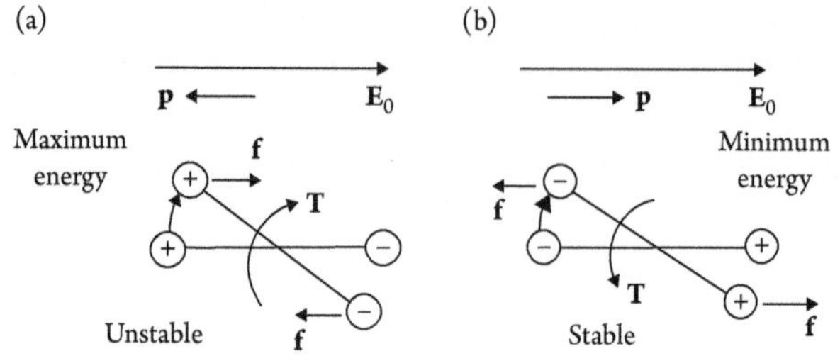


**Figure 5.12** The potential energy of a dipole is (a) maximal when **p** is antiparallel to $\mathbf{E}_0$ and (b) minimal when **p** is aligned with $\mathbf{E}_0$. For arbitrary orientations the torque **T** tends to align the dipole with the ambient electric field, $\mathbf{E}_0$.

the minimum-energy state is stable, as indicated by the direction of the torque **T**. This is important in a gas of polar molecules, where the molecular dipoles tend to align with an imposed field, a phenomenon known as *orientation polarization.*

* * *

This completes our overview of electrostatics. Readers interested in more details might wish to consult Griffiths (2013), Purcell (2011) or Lorrain & Corson (1970), all of which have thoughtful introductions to the subject, or else Jackson (1999) who provides an authoritative and complete account.

## EXERCISES

5.1 In §5.1.1 we saw that electric field lines in two dimensions are represented by $\psi =$ constant, where the function $\psi(x,y)$ is defined by $\mathbf{E} = \nabla \times \mathbf{C} = \nabla \times [\psi(x,y)\hat{\mathbf{e}}_z]$. The equivalent field lines in two-dimensional axisymmetric problems, $\mathbf{E}(s,z) = (E_s, 0, E_z)$ in $(s,\theta,z)$ coordinates, are defined by the lines $\Psi =$ constant, where the function $\Psi(s,z)$ is defined by $\mathbf{E} = \nabla \times \mathbf{C} = \nabla \times \{[\Psi(s,z)/s]\hat{\mathbf{e}}_\theta\}$. In the equivalent problem in magnetostatics $\Psi(s,z)$ is known as the *Stokes flux function,* while in fluid mechanics it is known as the *Stokes streamfunction.* Confirm that $\nabla\varphi_s \cdot \nabla\Psi = 0$, and so lines of constant $\Psi$ are perpendicular to the equipotential surfaces.

5.2 Consider a localised field of charge in an infinite domain that sits within a spherical control volume of radius $R$, which we denote by $V_R$. Let the root mean square charge density and the mean electric field within $V_R$ be defined as

$$\rho_{\text{rms}} = \left( V_R^{-1} \int_{V_R} \rho_e^2 \, d\mathbf{x} \right)^{1/2}, \quad \bar{\mathbf{E}} = V_R^{-1} \int_{V_R} \mathbf{E}(\mathbf{x}) \, d\mathbf{x}.$$

Use the Schwarz inequality to show that

$$|\bar{\mathbf{E}}| \leq \frac{R\rho_{\text{rms}}}{\sqrt{15}\varepsilon_0}.$$

5.3 We wish to calculate the net torque $\mathbf{T}$ exerted by an externally imposed electric field $\mathbf{E}_0(\mathbf{x})$ on a distribution on charge $\rho_e(\mathbf{x})$ that is localised near the origin and electrically neutral, $Q = 0$. The torque is evaluated about the origin. When $\mathbf{E}_0$ is uniform the torque is given by (5.87), $\mathbf{T} = \mathbf{p} \times \mathbf{E}_0(0)$. Now consider the case where $\mathbf{E}_0(\mathbf{x})$ varies slowly in space (on the scale of the charge distribution) and can be approximated by

$$E_{0i}(\mathbf{x}) = \mathrm{E}_{0i}(0) + \left[\frac{\partial E_{0i}}{\partial x_j}\right]_0 x_j.$$

Show that an additional torque now arises, given by

$$\mathbf{T} = 2\left(Q_{ij}\frac{\partial}{\partial x_j}\right) \times \mathbf{E}_0(\mathbf{x}),$$

where $Q_{ij}$ is the quadrupole contribution to the charge distribution, (5.56). Confirm that, in general, this is smaller in magnitude than (5.87).

5.4 Consider the problem of a point charge adjacent to a grounded conducting sphere, as shown in Figure 5.5. Consider now the case where the sphere is not grounded but rather is charged and maintained at a fixed potential $V_0$. Show that the method of solution used for the grounded sphere may again be used, only now an additional point charge must be placed at the centre of the sphere. What is the magnitude of the additional point charge?

5.5 Consider a localised distribution on charge $\rho_e(\mathbf{x})$. Use (5.83),

$$\rho_e E_i = \frac{\partial \tau_{ij}^E}{\partial x_j}, \quad \tau_{ij}^E = \varepsilon_0\left(E_i E_j - \frac{1}{2}\mathbf{E}^2\delta_{ij}\right),$$

along with the fact that $|\mathbf{E}|_\infty = |\mathbf{E}|(\mathbf{x} \to \infty) = O(r^{-2})$, to show that the net Coulomb force exerted on the charge distribution by itself is zero. Of course, this is a direct consequence of momentum conservation in an isolated system. Now show that the same is true for the net electrostatic torque exerted on the distribution by itself.

5.6 Consider the dipole shown in Figure 5.1(b), where we take the $x$–$y$ plane to correspond to the midplane of the dipole. Consider the electric stress tensor acting across the plane $z = 0$ on the lower half of the dipole, i.e. the stress exerted by the upper charge on the lower charge. Show that this consists of an upward stress of $(\varepsilon_0\mathbf{E}^2/2)\hat{\mathbf{e}}_z$ distributed across the plane $z = 0$, consistent with the two charges pulling on each other.

5.7 Consider the complex potential

$$F(z) = \psi + j\varphi_s = -\frac{j\lambda}{2\pi\varepsilon_0}\ln\frac{z-a}{z+a}.$$

Show that this represents a two-dimensional dipole-like field corresponding to a line of charge of strength $\lambda$ located on the $x$-axis at $x = a$ combined with an equal and opposite line of charge at $x = -a$. The field lines and equipotential lines are shown in Figure 5.13.

5.8 Consider the two-dimensional field at the edge of a parallel-plate capacitor as shown in Figure 5.4. It is described by the complex potential $F(z) = \psi + j\varphi_s$, where $F$ is defined through the inverse function $z = F + \exp(F)$. Show that

$$x = \psi + e^{\psi}\cos\varphi_s, \quad y = \varphi_s + e^{\psi}\sin\varphi_s,$$

and that consequently the midplane $y = 0$ is at zero potential. Also confirm that when $\varphi_s = \pm\pi$ we have $y = \pm\pi$ and $x = \psi - \exp\psi$. If we hold $y$ constant at $y = \pm\pi$, and vary $\psi$ from $\psi = -\infty$ to $\psi = 0$ and then from $\psi = 0$ to $\psi = \infty$, the value of $x$ varies from $x = -\infty$ to $x = -1$ and then back to $x = -\infty$. Hence show that the two plates of the capacitor lie at $y = \pm\pi$, are held at the potentials $\varphi_s = \pm\pi$, and extend from $x = -\infty$ to $x = -1$.

5.9 Show that the energy required to assemble a set of point dipoles in an infinite domain, otherwise free of charge, is

$$U_d = -\frac{1}{2}\sum_{i=1}^{N}\mathbf{p}_i\cdot\mathbf{E}(\mathbf{x}_i),$$

where $\mathbf{E}(\mathbf{x}_i)$ is the field associated with all the dipoles, except $\mathbf{p}_i$. This is the potential energy of a system of (electrically neutral) dipoles.

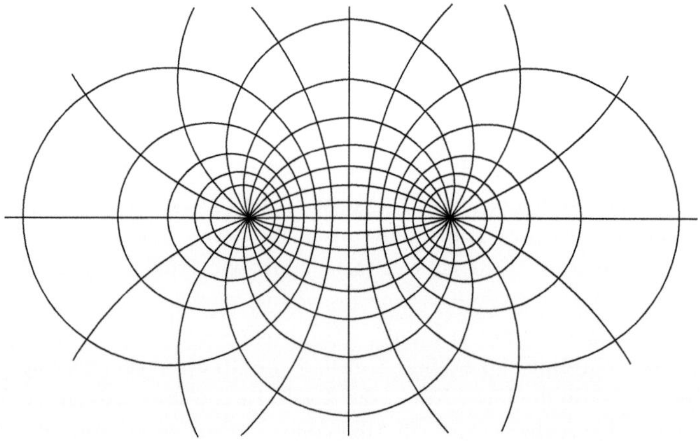

**Figure 5.13** Field lines and equipotential surfaces for equal and opposite line charges at $x = \pm a$. (Exercise 5.7).

## REFERENCES

Abramowitz, M. & Stegun, I. A., 1965, *Handbook of Mathematical Functions*. Dover.
Griffiths, D. J., 2013, *Introduction to Electrodynamics*, 4th ed. Pearson.
Jackson, J. D., 1999, *Classical Electrodynamics*, 3rd ed. Wiley.
Lamb, H., 1932, *Hydrodynamics*, 6th ed. Cambridge University Press.
Lorrain, P. & Corson, D., 1970, *Electromagnetic Fields and Waves*, 2nd ed. W. H. Freeman.
Maxwell, J. C., 1891, *A Treatise on Electricity and Magnetism*, 3rd ed., Clarendon Press.
Purcell, E., 2011, *Electricity and Magnetism*, 2nd ed. Cambridge University Press.

# 6

# Dielectrics

Faraday, in his mind's eye, saw lines of force traversing space where the mathematicians saw centres of force attracting at a distance: Faraday sought the seat of the phenomena in real actions going on in the medium, they were satisfied that they had found it in a power of action at a distance... When I translated what I considered to be Faraday's ideas into mathematical form, I found that in general the results of the two methods coincided, but that Faraday's methods resembled those in which we begin with the whole and arrive at the parts by analysis, while the ordinary mathematical methods were founded on the principle of beginning with the parts and building up the whole by synthesis.

Maxwell (*A Treatise on Electricity and Magnetism*, 1873)

Dielectrics are electrical insulators. In such a material there are no free charges, but rather the electrical charges are bound to individual atoms or molecules. At first sight it might seem that such a material would have negligible influence on an ambient electric field, since there can be no bulk movement of charge in response to an imposed field. However, as Faraday showed in 1837, even insulators can have a profound influence on an ambient electric field.

Faraday was deeply distrustful of the notion of action at a distance, and consequently unhappy with the contemporary mathematical interpretation of Coulomb's law, in which remote charges attracted or repelled each other with nothing of significance happening in the medium between the charges. Rather, Faraday was convinced that, in both electricity and magnetism, the medium between two charged bodies or two current-carrying wires must play a pivotal role, and of course this would eventually lead him to introduce the notion of electric and magnetic fields pervading all space. In order to challenge the view that the medium played no role in electrostatics, he decided to investigate the effect of introducing insulating material into the gap between the two charged surfaces of a capacitor. In Faraday's words:

The question may be stated thus: suppose $A$ an electrified plate of metal suspended in air, and $B$ and $C$ two exactly similar plates, placed parallel to and on each side of $A$ at equal distances and insulated; $A$ will then induce equally towards $B$ and $C$. If in this position of the plates some other dielectric than air, as shell-lac, be introduce between $A$ and $C$, will the induction between them remain the same?

Faraday quickly found (as Cavendish had before him) that the introduction of an insulator increased the capacitance $C = Q/V$ of a capacitor by an amount that is purely a property of the insulating material introduced. (Here $Q$ is the charge on either of the capacitor plates and $V$ the potential difference, or voltage drop, between the two surfaces.) He called this property of the insulator its 'specific inductive capacity'. Today we refer to it as the dielectric constant of the material.

## 6.1 The Polarisation of Dielectrics

### 6.1.1 *An Experiment with Capacitors*

Consider a parallel-plate capacitor in which there is a vacuum between the plates. Its capacitance is defined as $C = Q/V$ and, as discussed in §4.3, Gauss' law yields $C = \varepsilon_0 A/d$, where $A$ is the area of each plate and $d$ the gap between the plates. Now suppose we place a strip of plastic or glass between the plates of the capacitor, more or less filling the gap, as shown in Figure 6.1. Then, as Faraday discovered, the capacitance $C = Q_p/V$ increases, $Q_p$ being the charge on the *upper plate* of the capacitor. The capacitance is still proportional to $A/d$, but the constant of proportionality is larger. Let us write this as

$$C = \varepsilon_r \frac{\varepsilon_0 A}{d}, \quad \varepsilon_r > 1, \tag{6.1}$$

where $\varepsilon_r$ is the *dielectric constant* of the insulator, often called the *relative permittivity* of the dielectric. Typical values of $\varepsilon_r$ are listed in Table 6.1 for a number of common dielectrics at room temperature. The values for the gases are noticeably lower than those for the liquids and solids, but this simply because of the large spaces between the gas molecules.

Let us try to understand why $C$ increases. Suppose we keep the voltage across the capacitor, and hence the electric field in the dielectric, $|\mathbf{E}|$, constant during the introduction of the dielectric strip. Since $C = Q_p/V$, the charge on the plate must increase as we insert the insulator. We now apply Gauss' law to the control volume shown in Figure 6.1, which includes the upper surface of the insulator and the lower edge of the upper plate. This tells us that $|\mathbf{E}| = (Q_p + Q_d)/A\varepsilon_0$, where $Q_d$ is any net charge that has appeared on the

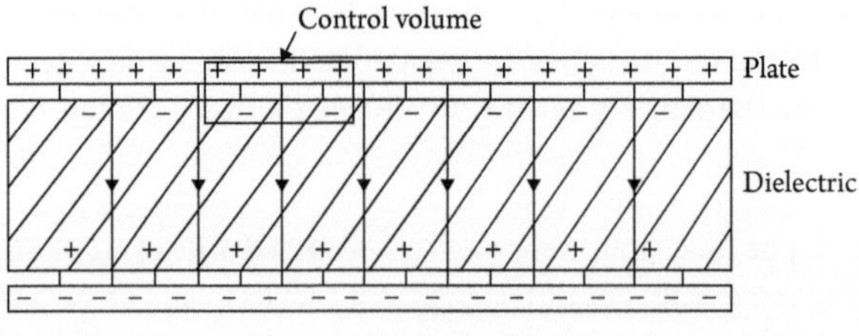


**Figure 6.1** When an insulator fills the gap of a parallel-plate capacitor, the capacitance of the latter increases. The implication is that charges must appear on the surfaces of the insulator.

**Table 6.1** Dielectric constants for some insulators at room temperature and 1 kHz

| **Material** | **Dielectric constant $\varepsilon_r$** |
|---|---|
| Air | 1.0005 |
| Methane | 1.0009 |
| Water | 80 |
| Benzene | 2.3 |
| Paper | 3.5 |
| Polyethylene | 2.3 |
| Pyrex glass | 4.2 |

upper surface of the dielectric. Since $|\mathbf{E}|$ is held constant as $Q_p$ rises, we are forced to the conclusion that negative charges appear on the upper surface of the insulator, charges that exactly counter the increase in positive charge on the upper capacitor plate.

It is not difficult to understand where this negative charge comes from. Although the charges in the insulator are bound to individual molecules, each atom consists of positive and negative charges that can be displaced relative to each other when an external electric field is applied. In Figure 6.1 the capacitor sets up an electric field that is directed downwards, and so in each atom within the dielectric the electrons are pulled slightly upwards while the positive nuclei are pushed slightly downwards. These movements of charge are tiny, yet nevertheless they leave each atom with a small but finite dipole moment. The dielectric is said to be *polarised* by the external electric field. Of course the insulator must maintain overall electrical neutrality, but nevertheless these tiny movements of charge allow negative charge to accumulate on the upper surface of the dielectric and positive charge on the lower surface.

The surface charge on the dielectric is called *bound charge* and its density is given the symbol $\sigma_b$, whereas the surface charge density on the capacitor plates, which is due to free charges, is denoted by $\sigma_f$. The sum of the two gives the net surface charge at the interface of the dielectric and the capacitor plates, $\sigma_e = \sigma_f + \sigma_b$. Gauss' law applied to the control volume shown in Figure 6.1 then gives $|\mathbf{E}| = (\sigma_f - |\sigma_b|)/\varepsilon_0$. Since we are holding $|\mathbf{E}|$ constant in our thought experiment, the appearance of surface charges $|\sigma_b|$ on the upper surface of the dielectric must be compensated for by a corresponding increase in $\sigma_f$ in the upper capacitor plate, and it is this that causes the increase in capacitance.

Let us now estimate the dielectric constant $\varepsilon_r$ in terms of the induced bound charge. If $(\sigma_f)_0 = \varepsilon_0 |\mathbf{E}|$ is the initial charge density on the upper plate, before the dielectric is inserted, then, by definition, $\varepsilon_r = \sigma_f/(\sigma_f)_0$, i.e. the relative permittivity is the ratio of $Q_p$ after and before the dielectric is inserted into the gap. It follows from Gauss' law that $\varepsilon_0 |\mathbf{E}| = (\sigma_f)_0 = \sigma_f - |\sigma_b|$ and hence

$$\varepsilon_r = 1 + \frac{|\sigma_b|}{\varepsilon_0 |\mathbf{E}|}. \tag{6.2}$$

Thus $\varepsilon_r - 1$ is a measure of the amount of bound charge induced by a given electric field. We shall return to this expression in §6.4.

### *6.1.2 Mechanisms of Polarisation: Electronic Versus Orientation Polarisation*

Notice that water has a much higher dielectric constant than the other insulators in Table 6.1. This suggests that there may be more than one way in which an insulator becomes polarised. If that is indeed the case, then the nature of the resulting molecular dipoles may vary from material to material. Two of the more common situations are called electronic and orientation polarisation, and these are particularly prevalent in gases and liquids. So let us start with dielectrics which are fluids, leaving the more complicated solid dielectrics until later.

*Electronic polarisation* occurs in atoms or molecules that have no intrinsic dipole moment. In such cases molecular dipoles are induced because an external electric field displaces the positive nucleus and negative electrons of an atom in opposite directions. Such displacements are miniscule (a tiny fraction of the size of the atom) because the restoring forces are large. However, they do occur and, provided the external electric field is not too large, they tend to be proportional to the magnitude of the applied field. This suggests that the strengths of the induced molecular dipole moments are, in some statistical sense, proportional to the local electric field acting on the molecules. If $\mathbf{p}$ is the mean molecular dipole moment at some point in the dielectric, then we might write

$$\mathbf{p} = \alpha\varepsilon_0\mathbf{E}_{\text{loc}}, \tag{6.3}$$

where $\mathbf{E}_{\text{loc}}$ is some suitable local measure of the electric field (which we shall define later). The constant $\alpha$ is sometimes called the *atomic* or *molecular polarisability* and has the dimensions of volume.

For very simple atoms crude estimates of $\alpha$ are available. The simplest model is that of a point nucleus of charge $q$ immersed in a spherical electron cloud of uniform charge density and overall charge $-q$. Suppose that, as the result of an ambient electric field $\mathbf{E}_{\text{loc}}$, the nucleus is displaced relative to the centre of the electron cloud by a distance $\mathbf{d}$ while the cloud itself remains approximately spherical. Then equilibrium is established when the electric field arising from the electron cloud and acting on the displaced nucleus is equal and opposite to $\mathbf{E}_{\text{loc}}$. If the cloud has volume $V_{\text{atom}}$, then, according to (4.22), equilibrium requires

$$\mathbf{E}_{\text{loc}} \approx \frac{q\mathbf{d}}{3\varepsilon_0 V_{\text{atom}}} = \frac{\mathbf{p}}{3\varepsilon_0 V_{\text{atom}}},$$

and we conclude that $\alpha \approx 3V_{\text{atom}}$. More sophisticated models also give $\alpha \sim V_{\text{atom}}$, as discussed in Feynman, Leighton & Sands (1964) and Purcell (2011).

In other cases the molecules carry a permanent dipole moment and are free to rotate, water or water vapour being the most obvious examples. These are called polar molecules. In the absence of an external field the random orientation of the polar molecules ensures that the spatially averaged molecular dipole moment is zero. However, if an external field is applied, then each molecular dipole experiences a torque that tends to align it with the local electric field, as discussed in §5.5.3. This systematic orientation of the molecular dipoles is, of course, to some degree offset by random thermal collisions in the fluid, and so some sort of statistical equilibrium inevitably results in which there is a mean alignment of the molecular dipole moments with the local electric field. This is called *orientation polarisation*. As with electronic polarisation, the resulting dipole moment per unit volume tends to be proportional to the local electric field.

A simple model of orientation polarisation can be constructed by assuming that the likely orientation of the polar molecules relative to the ambient electric field $\mathbf{E}_{\rm loc}$ follows a Boltzmann distribution, such as that in (5.69). If $\theta$ is the angle between $\mathbf{E}_{\rm loc}$ and any one dipole, and $n(\theta)$ is the relative number of molecules with orientation $\theta$, then the Boltzmann distribution gives

$$n(\theta) = n_0 \exp\left[-\frac{U(\theta)}{k_B T}\right].$$

Here $T$ is temperature, $k_B$ is the Boltzmann constant, $U(\theta) = -\mathbf{p}_0 \cdot \mathbf{E}_{\rm loc} = -p_0 E_{\rm loc} \cos\theta$ is the potential energy of a polar molecule by virtue of $\mathbf{E}_{\rm loc}$ (see (5.93)), and $p_0$ is the magnitude of the intrinsic dipole moment on each polar molecule. Thus we have

$$n(\theta) = n_0 \exp\left(\frac{p_0 E_{\rm loc} \cos\theta}{k_B T}\right) \approx n_0 \left(1 + \frac{p_0 E_{\rm loc} \cos\theta}{k_B T}\right),$$

provided that $\mathbf{E}_{\rm loc}$ is not too large. The mean molecular dipole moment averaged over $\theta$, $\mathbf{p}$, is then parallel to $\mathbf{E}_{\rm loc}$ (by symmetry) and its magnitude is proportional to the integral of $n(\theta) p_0 \cos\theta$ over all possible angles. The integration is a little complicated because it involves solid angles (see e.g. Lorrain & Corson, 1970), but after a little algebra one finds

$$\mathbf{p} = \frac{p_0^2}{3k_B T}\mathbf{E}_{\rm loc}. \tag{6.4}$$

The $1/T$ dependence of $\mathbf{p}$ on temperature is known as Curie's law.

So far we have focused on fluid dielectrics. While the polarisation of gases and liquids is sometimes (but not always) easy to picture, that of solids is frequently much more complicated and depends on the internal structure of the material. Some solids have a built-in polarisation that is there even in the absence of an external field, rather like a permanent magnet. Examples include certain crystalline substances or materials like wax that have long polar molecules that can be preferentially orientated. Changes in the polarisation of such materials can often be detected when they are heated (called pyroelectricity) or subject to strain (e.g. piezoelectric crystals). Some crystalline materials contain rotatable dipoles and so are prone to a form of orientation polarisation. Other solids have no intrinsic polarisation but are prone to electronic polarisation of the atoms, just like in fluids, while some ionic crystals, such as NaCl, can be polarised through *ionic polarisation*, in which the negative and positive ions are pulled in different directions by an external field. One problem with such crystals, however, is that it is not always clear which charges should be classified as bound and which as free, as emphasised by Purcell (2011). In short, the polarisation of solid dielectrics is rarely simple.

## 6.2 The Polarisation Vector and Bound Charge Distributions

Let us now try to build a macroscopic model of dielectrics based on the idea that an externally applied field will induce a sea of tiny molecular dipoles that are aligned, at least in some statistical sense, with the local electric field. Of course, the electric field associated with these dipoles augments the externally applied field. We also expect the formation of molecular dipoles to be associated with the accumulation of bound charges, as in the simple capacitor discussed above. So we seek a general theory of dielectrics that relates the mean molecular dipole moment per unit volume to the accumulation of bound charges within the insulator and to the macroscopic electric field in and around the dielectric that arises directly from the molecular dipoles. Such macroscopic models go back to the start of the twentieth century, and in particular to the work of Lorentz.

### 6.2.1 *The Polarisation Vector, Bound Charges and the Exterior Electric Field*

Suppose that the mean molecular dipole moment is $\mathbf{p}$ and that there are $N$ such dipoles per unit volume. Then the dipole moment per unit volume is $\mathbf{P} = N\mathbf{p}$, which is known as the *electric polarisation vector*. This is a function of position and also of the applied electric field.

Consider a point *exterior* to the dielectric. The molecular dipoles are so small that we may use the far-field approximation to calculate the external field established by these dipoles. For example, the potential at an external point $\mathbf{x}$ induced by the polarisation of the volume element $dV'$ is

$$d\varphi_s(\mathbf{x}) = \frac{\mathbf{P}(\mathbf{x}')\cdot\mathbf{r}}{4\pi\varepsilon_0 r^3}\,dV' = -\frac{1}{4\pi\varepsilon_0}\mathbf{P}(\mathbf{x}')\cdot\nabla\left(\frac{1}{r}\right)dV', \quad \mathbf{r} = \mathbf{x} - \mathbf{x}',$$

where, as usual, $\nabla$ operates on $\mathbf{x}$ while holding $\mathbf{x}'$ constant. If $V_d$ is the volume of the dielectric, then, given that $\nabla'(1/r) = -\nabla(1/r)$, we have

$$\varphi_s(\mathbf{x}) = -\frac{1}{4\pi\varepsilon_0}\int_{V_d}\mathbf{P}(\mathbf{x}')\cdot\nabla\left(\frac{1}{r}\right)dV' = \frac{1}{4\pi\varepsilon_0}\int_{V_d}\mathbf{P}(\mathbf{x}')\cdot\nabla'\left(\frac{1}{r}\right)dV'. \qquad (6.5)$$

Finally we note that,

$$\mathbf{P}(\mathbf{x}')\cdot\nabla'\left(\frac{1}{r}\right) = \nabla'\cdot\left[\frac{\mathbf{P}(\mathbf{x}')}{r}\right] - \frac{1}{r}\nabla'\cdot\mathbf{P}(\mathbf{x}'),$$

and so Gauss' theorem yields

$$\boxed{\varphi_s(\mathbf{x}) = \frac{1}{4\pi\varepsilon_0}\oint_{S_d}\frac{\mathbf{P}(\mathbf{x}')}{r}\cdot d\mathbf{S}' - \frac{1}{4\pi\varepsilon_0}\int_{V_d}\frac{\nabla'\cdot\mathbf{P}(\mathbf{x}')}{r}\,dV'.} \tag{6.6}$$

Of course, we must add to this the potential associated with the externally applied electric field.

Rather remarkably, (6.6) is *exactly* the potential that would be established by the volumetric and surface charge distributions

$$\boxed{\rho_b = -\nabla\cdot\mathbf{P}, \quad \sigma_b = \mathbf{P}\cdot\mathbf{n},} \tag{6.7}$$

where $\mathbf{n}$ is the (outward-pointing) unit normal to the surface $S_d$. As we shall see, these are not imaginary charges, but rather they represent the accumulation of bound charges within the dielectric, exactly as occurred with the simple case of a parallel-plate capacitor discussed above. In terms of these bound charges our external field is

$$\varphi_s(\mathbf{x}) = \frac{1}{4\pi\varepsilon_0}\oint_{S_d}\frac{\sigma_b(\mathbf{x}')}{r}\,dS' + \frac{1}{4\pi\varepsilon_0}\int_{V_d}\frac{\rho_b(\mathbf{x}')}{r}\,dV'. \tag{6.8}$$

So, as far as the external electric field is concerned, a sea of induced molecular dipoles may be replaced by the bound charge distributions $\rho_b = -\nabla\cdot\mathbf{P}$ and $\sigma_b = \mathbf{P}\cdot\mathbf{n}$.

Let us try to understand how the induced molecular dipoles cause bound charges to accumulate within the dielectric. To that end, we suppose that each dipole is associated with the displacement of a positive charge $q$ by the distance $\mathbf{d}$, with $\mathbf{p} = q\mathbf{d}$, as in §4.4. (In reality, positive and negative charges are displaced relative to each other, but this simple cartoon in which only the positive charges move is adequate for our present purposes.) Then the net charge that passes across a surface element $d\mathbf{S}$ located within the dielectric is

$$dQ = Nq(\mathbf{d}\cdot d\mathbf{S}) = N\mathbf{p}\cdot d\mathbf{S} = \mathbf{P}\cdot d\mathbf{S}. \tag{6.9}$$

If $d\mathbf{S}$ lies at the surface of the dielectric, then the net bound charge that passes through that surface element as a result of polarisation is evidently $(\mathbf{P}\cdot\mathbf{n})\,dS = \sigma_b\,dS$, from which we obtain $\sigma_b = \mathbf{P}\cdot\mathbf{n}$, in accordance with (6.7). On the other hand, the net bound charge that passes through a small closed surface *within* the dielectric is, from (6.9),

$$Q = \oint\mathbf{P}\cdot d\mathbf{S} = \int\nabla\cdot\mathbf{P}\,dV \approx \nabla\cdot\mathbf{P}\,\delta V, \tag{6.10}$$

where $\delta V$ is the enclosed volume. So we conclude that the net loss of bound charge from $\delta V$ as a result of polarisation is $\nabla\cdot\mathbf{P}\,\delta V$. The corresponding *accumulation* of bound charge per unit volume is then $\rho_b = -\nabla\cdot\mathbf{P}$, exactly as given in (6.7).

In short, the volumetric and surface charge distributions itemised in (6.7) correspond to *real* charge distributions associated with the accumulation of bound charges within the dielectric. Note that these bound charge distributions leave the dielectric electrically neutral overall, as they must. That is, the net bound charge is

$$Q_b = \oint_{S_d} \sigma_b \, dS' + \int_{V_d} \rho_b \, dV' = \oint_{S_d} \mathbf{P}' \cdot d\mathbf{S}' - \int_{V_d} \nabla' \cdot \mathbf{P}' \, dV' = 0. \tag{6.11}$$

Note also that the laws of Gauss and Coulomb apply to any net accumulation of charge, *irrespective of whether that charge is free charge or bound charge.* Thus, for example, Gauss' law is

$$\boxed{\nabla \cdot \mathbf{E} = \frac{\rho_e}{\varepsilon_0} = \frac{\rho_f + \rho_b}{\varepsilon_0},} \tag{6.12}$$

where $\rho_f$ is that part of the net charge density $\rho_e$ which is associated with the accumulation of excess free charges and a given point.

### 6.2.2 *The Electric Field Induced by a Uniformly Polarised Sphere*

So far we have studiously avoided the thorny problem of how we determine $\mathbf{P}$ from the externally applied electric field, other than to hint that in many cases $\mathbf{P}$ is proportional to the local electric field. For example, in the case of electronic polarisation we have suggested that the molecular dipole moment can be modelled as $\mathbf{p} = \alpha\varepsilon_0\mathbf{E}_{\text{loc}}$. The complication here is that it is not so obvious what a given molecule interprets as the ambient, or local, field $\mathbf{E}_{\text{loc}}$. We shall postpone any discussion of this somewhat delicate question until §6.4. In the meantime we shall assume that, somehow, $\mathbf{P}$ is a known function of position and ask what the corresponding electric field is. (Of course, we must add to this the externally applied field that created the polarisation in the first place.)

Let us take as a simple example a spherical volume of dielectric that has radius $R$ and volume $V_R$ and in which the polarisation vector $\mathbf{P}$ is uniform. We adopt spherical polar coordinates in which $z$ is aligned with $\mathbf{P}$ and so $\theta$ is the polar angle measured from $\mathbf{P}$ (see Figure 6.2(a)). From (6.7) we have $\rho_b = 0$ (i.e. no bound charges within the sphere) and a surface charge density of $\sigma_b = \mathbf{P} \cdot \mathbf{n} = P\cos\theta$. However, we have already met this surface charge distribution in a somewhat different problem: according to (5.28), $\sigma_f = (3\varepsilon_0 E_0)\cos\theta$ is the surface charge that is induced in a conducting sphere when subject to a uniform external electric field $\mathbf{E}_0$, where $\theta$ is the polar angle measured from $\mathbf{E}_0$. This is shown in Figure 6.2(b). We conclude that the electric fields outside the dielectric and conducting spheres, induced by their respective surface charges, are identical, provided of course that $\mathbf{P} = 3\varepsilon_0\mathbf{E}_0$.

Now, in §5.2.2 we showed that the *external* electric field in the case of a conducting sphere is the superposition of the uniform imposed field $\mathbf{E}_0$ and the field due to a dipole moment of strength $\mathbf{p} = (3\varepsilon_0\mathbf{E}_0)V_R$ located at the centre of the sphere. Since the imposed field $\mathbf{E}_0$ is a consequence of remote charges, we conclude that the external electric field associated

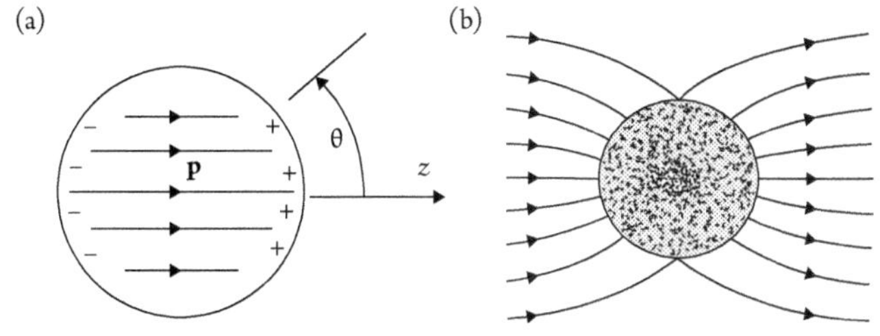


**Figure 6.2** (a) The surface bound charge associated with a uniformly polarized sphere. (b) The distortion of a uniform electric field by a conducting sphere.

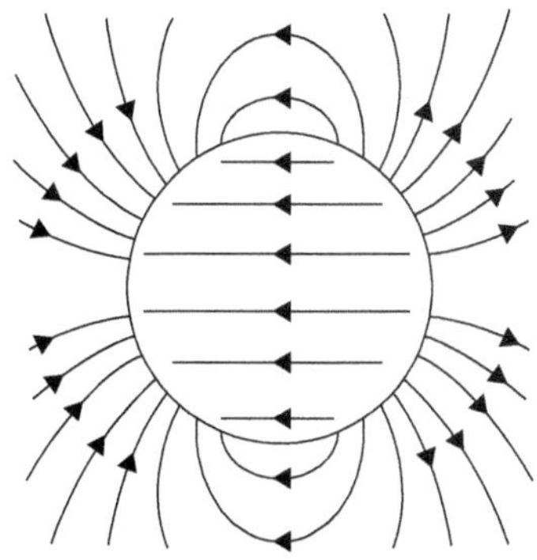

**Figure 6.3** The electric field for a dielectric sphere of uniform polarization **P**.

exclusively with the surface charge distribution $\sigma_f = (3\varepsilon_0 E_0)\cos\theta$ is that of the point dipole $\mathbf{p} = (3\varepsilon_0 \mathbf{E}_0)V_R$. By implication, the external electric field generated by the bound charge $\sigma_b = P\cos\theta$ on the spherical dielectric is that of a point dipole located at the centre of $V_R$ and of strength $\mathbf{p} = \mathbf{P}V_R$. In short, as far as the *external* electric field is concerned, the uniform distribution of molecular dipoles within $V_R$ may be replaced by one equivalent dipole located at the centre of the dielectric sphere and of strength $\mathbf{p} = \mathbf{P}V_R$.

There is one other interesting thing to note about this problem. In the case of the conducting sphere the surface charge distribution $\sigma_f = (3\varepsilon_0 E_0)\cos\theta$ creates an electric field within the sphere that is equal and opposite to the externally imposed field. (Remember that the net electric field in a conductor is zero.) Thus the charge distribution $\sigma_f = (3\varepsilon_0 E_0)\cos\theta$ creates the field $-\mathbf{E}_0$ within $V_R$. It follows that the bound charge distribution $\sigma_b = P\cos\theta$ on the dielectric sphere is associated with a uniform electric field of

$$\mathbf{E} = -\frac{\mathbf{P}}{3\varepsilon_0} \tag{6.13}$$

within $V_R$, as shown in Figure 6.3. Note that **E** within $V_R$ is antiparallel to **P**. We shall return to (6.13) in §6.2.4.

### 6.2.3 *The Polarisation Current Density*

Although we are interested here in electrostatics, when it comes to dynamics it will prove useful to consider what happens when the bound charge distribution varies with time. In particular, we shall see that an associated current density is required to redistribute the bound charges in space. To understand why this must be, so let us return to the notion of charge conservation.

In §4.5.2 we saw that charge conservation relates the current density to changes in the charge distribution by

$$\nabla \cdot \mathbf{J} = -\frac{\partial \rho_e}{\partial t}. \tag{6.14}$$

When writing this, we had in mind conductors in which $\rho_e$ represents free charges, $\rho_e = \rho_f$, and $\mathbf{J}$ is the associated conduction current, which we might write as $\mathbf{J}_f$. However, in dielectrics we now see that there is a second source of charge, $\rho_e = \rho_b$, and so charge conservation applied to the bound charge tells us that there must be a second source of current density:

$$\nabla \cdot \mathbf{J}_{\rm pol} = -\frac{\partial \rho_b}{\partial t}. \tag{6.15}$$

The current $\mathbf{J}_{\rm pol}$ is known as the *polarisation current density*, and since $\rho_b = -\nabla \cdot \mathbf{P}$ we may rewrite (6.15) as

$$\nabla \cdot \left( \mathbf{J}_{\rm pol} - \frac{\partial \mathbf{P}}{\partial t} \right) = 0. \tag{6.16}$$

Of course *one* particular solution of (6.16) is

$$\mathbf{J}_{\rm pol} = \frac{\partial \mathbf{P}}{\partial t}, \tag{6.17}$$

although it is important to keep in mind that this is not the most general solution. We shall return to this expression when we come to discuss the dynamics of dielectrics.

### 6.2.4 *The Macroscopic Electric Field at an Interior Point in a Dielectric*

Let us now consider the electric field within the dielectric itself, whose calculation turns out to be a little more delicate than that of the exterior field induced by polarisation. Here we must restrict ourselves to the macroscopic field in the dielectric, i.e. the electric field that has been averaged over the scale of the molecules. Also, once again, we are concerned only with the electric field established by the polarisation, and not with the externally applied field that induced that polarisation in the first place.

We wish to calculate the macroscopic electric field at a point $\mathbf{x}_0$ within the dielectric that arises from all the molecular dipoles within the insulator. For those dipoles that are remote from $\mathbf{x}_0$ (i.e. many molecular lengthscales away from $\mathbf{x}_0$) it will be convenient to use the far-field approximation for the dipoles in the same way as we did to calculate the exterior electric field. To that end, we consider a spherical control volume $V_R$ centred on $\mathbf{x}_0$ whose radius $R$ is much larger than the molecular scale but much smaller than the characteristic lengthscale for variations in the macroscopic fields **E** and **P**. We now divide all the molecular dipoles into those that lie inside $V_R$ and those that lie outside. By virtue of our choice of $R$ we may (i) treat **P** and **E** as locally uniform within $V_R$ and (ii) use the far-field approximation to determine the contribution to $\mathbf{E}(\mathbf{x}_0)$ that comes from those dipoles that lie outside $V_R$.

Consider first the contribution to $\mathbf{E}(\mathbf{x}_0)$ that comes from the dipoles that lie outside $V_R$. This is

$$\mathbf{E}(\mathbf{x}_0) = -\frac{1}{4\pi\varepsilon_0}\nabla\left[\oint_{S_d}\frac{\sigma_b(\mathbf{x}')}{r}dS' + \int_{V_d-V_R}\frac{\rho_b(\mathbf{x}')}{r}dV' + \oint_{S_R}\frac{\sigma_b(\mathbf{x}')}{r}dS'\right], \tag{6.18}$$

where $\mathbf{r} = \mathbf{x}_0 - \mathbf{x}'$, $V_d$ is the volume of the dielectric and $S_d$ its outer surface. Note that the three terms on the right of (6.18) have been calculated as if $\mathbf{x}_0$ were an exterior point, using the far-field estimate for the potential established by the molecular dipoles. In particular, we have applied (6.8) to the volume $V_d - V_R$, which is bounded by the two surfaces $S_d$ and $S_R$.

Now in §5.3 we saw that the electric field associated with the all charges inside $V_R$ must satisfy

$$\int_{V_R}\mathbf{E}(\mathbf{x})\,d\mathbf{x} = -\frac{\mathbf{p}_{qin}}{3\varepsilon_0}, \tag{6.19}$$

where $\mathbf{p}_{qin} = \mathbf{P}(\mathbf{x}_0)V_R$ is the net dipole moment associated with the charges within $V_R$. So the total field at $\mathbf{x}_0$ is

$$\mathbf{E}(\mathbf{x}_0) = -\frac{1}{4\pi\varepsilon_0}\nabla\left[\oint_{S_d}\frac{\sigma_b(\mathbf{x}')}{r}dS' + \int_{V_d-V_R}\frac{\rho_b(\mathbf{x}')}{r}dV' + \oint_{S_R}\frac{\sigma_b(\mathbf{x}')}{r}dS'\right] - \frac{\mathbf{P}(\mathbf{x}_0)}{3\varepsilon_0}. \tag{6.20}$$

However, the last two terms on the right of (6.20) exactly cancel. To see why, let us adopt spherical polar coordinates in which $z$ is aligned with $\mathbf{P}(\mathbf{x}_0)$ and so $\theta$ is the polar angle measured from $\mathbf{P}(\mathbf{x}_0)$. The bound charge on the surface $S_R$ on the right of (6.20) is that due to the polarisation **P** that lies *outside* $V_R$, and so we have $\sigma_b = \mathbf{P}(\mathbf{x}_0)\cdot\hat{\mathbf{n}}$, where the unit vector $\hat{\mathbf{n}}$ points *inward* from the surface of $V_R$ towards $\mathbf{x}_0$. Equivalently, we may write $\sigma_b = -\mathbf{P}(\mathbf{x}_0)\cdot\mathbf{n} = -P(\mathbf{x}_0)\cos\theta$, where **n** is the unit vector that points radially *outward* from the surface of $V_R$. However, (6.13) tells us that the bound charge distribution $\sigma_b = P(\mathbf{x}_0)\cos\theta$ creates a uniform electric field within $V_R$ of $\mathbf{E} = -\mathbf{P}(\mathbf{x}_0)/3\varepsilon_0$, and so the surface charge $\sigma_b = -\mathbf{P}(\mathbf{x}_0)\cdot\mathbf{n}$ must induce a uniform electric field within $V_R$ of $\mathbf{E} = \mathbf{P}(\mathbf{x}_0)/3\varepsilon_0$. Rather remarkably, this exactly cancels the last contribution on the right of (6.20). So we conclude that, for any point inside the dielectric,

$$\mathbf{E}(\mathbf{x}_0) = -\frac{1}{4\pi\varepsilon_0}\nabla\left[\oint_{S_d}\frac{\sigma_b(\mathbf{x}')}{r}\,dS' + \int_{V_d}\frac{\rho_b(\mathbf{x}')}{r}\,dV' - \int_{V_R}\frac{\rho_b(\mathbf{x}')}{r}\,dV'\right]. \tag{6.21}$$

If the bound charge is taken to be uniform within $V_R$, $\rho_b(\mathbf{x}') = \rho_b(\mathbf{x}_0)$, then the third contribution to (6.21) is the electric field at the centre of a uniformly charged sphere, which is zero by symmetry. So we obtain the surprisingly simple result that

$$\mathbf{E}(\mathbf{x}_0) = -\frac{1}{4\pi\varepsilon_0}\nabla\left[\oint_{S_d}\frac{\sigma_b(\mathbf{x}')}{r}\,dS' + \int_{V_d}\frac{\rho_b(\mathbf{x}')}{r}\,dV'\right]. \tag{6.22}$$

Comparing this with (6.8) shows that it is exactly the same expression as for an exterior point. The fact that (6.22) applies equally to interior and exterior points is a manifestation of the fact that the bound charges are in fact real charges, representing the accumulation of charge within the dielectric. Of course, we must add to (6.22) any externally applied electric field, such as that which induced the polarisation in the first place.

## 6.3 The Combined Electric Field, Gauss' Law and the Electric Displacement

Let us now add the externally applied electric field to that generated by the molecular dipoles. We start by returning to Gauss' law. As already mentioned, it is important to appreciate that bound charges are not some fictitious mathematical intermediary, but rather represent the real accumulation of charges within the dielectric. Thus the total charge density is $\rho_e = \rho_f + \rho_b$, and the *total* electric field is governed by

$$\boxed{\begin{aligned}\nabla\cdot(\varepsilon_0\mathbf{E}) &= \rho_f + \rho_b,\\ \nabla\times\mathbf{E} &= 0.\end{aligned}} \tag{6.23}$$

(If you are worried about keeping track of the surface charges, you might consider them to be included in $\rho_f$ and $\rho_b$ by demanding that the surface charge occupy a thin surface layer of finite volume.)

Since the bound charge density is given by $\rho_b = -\nabla\cdot\mathbf{P}$, Gauss' law may be rewritten as

$$\nabla\cdot(\varepsilon_0\mathbf{E} + \mathbf{P}) = \rho_f. \tag{6.24}$$

For largely historical reasons, it has become conventional at this point to introduce an auxiliary field $\mathbf{D}$, called the *electric displacement vector* and defined as

$$\boxed{\mathbf{D} = \varepsilon_0\mathbf{E} + \mathbf{P}.} \tag{6.25}$$

Evidently Gauss' law rewritten in terms of **D** takes a particularly simple form. In particular, (6.23) becomes

$$\boxed{\begin{aligned}\nabla \cdot \mathbf{D} &= \rho_f, \\ \nabla \times \mathbf{E} &= 0.\end{aligned}} \tag{6.26}$$

Given a specified distribution of $\rho_f$, we now have the option of working with **D** and **E**, rather than with **E** and $\rho_b$ (or equivalently **E** and **P**). Note that the integral form of (6.26) is

$$\oint \mathbf{D} \cdot d\mathbf{S} = \int \rho_f \, dV = Q_f,$$

which is often helpful in determining **D**, as discussed in §6.4.3.

So why introduce the auxiliary field **D**? Historically it was not known that **P**, and its associated electric field, owed their existence to bound charges. Rather, it was thought that the only real charges were the free charges. If that were true, then (6.26) would be a natural way of formulating the observed phenomena. So the historical logic behind writing (6.26) is that only the free charges now appear as a source term in Gauss' law.

Although we now know that bound charge exists, and that (6.23) governs the resulting electric field, many authors still find it more convenient to work with **D** rather than **E**. In part, this is because $\rho_f$ is often known in advance, whereas $\rho_b$ is not, and in part because the spatial structure of **D** is often easier to picture than that of **E**. However, it is important to keep in mind that any apparent advance in writing (6.26) is largely illusory. The problem with (6.26) is that **D**, rather than **E**, becomes the object of Gauss' law. This is particularly problematic when we recall that a vector is only uniquely determined once both its divergence and its curl are specified. When working with **E** this is fine, since its divergence and curl are given by (6.23), although we still need to determine the bound charge, or equivalently $\nabla \cdot \mathbf{P}$, to find **E**. However, if we choose to work with **D** and with (6.26), the problem is not closed unless we somehow specify $\nabla \times \mathbf{D}$, or equivalently $\nabla \times \mathbf{P}$, which in general is non-zero (see Exercise 6.5). But what is $\nabla \times \mathbf{D}$? Admittedly, for particularly simple geometries (those with planar, spherical or cylindrical symmetry) the symmetry in the problem often automatically ensures that $\nabla \times \mathbf{D} = 0$ (see Exercise 6.1). Working with (6.26) is then profitable. However, in general the lack of knowledge of $\nabla \times \mathbf{D}$ is a fundamental stumbling block to working with **D** alone.

In any event, whatever field we work with, **E** or **D**, we still need some means of pinning down the polarisation, **P**. Only then do we have a closed system of equations.

## 6.4 Relating the Polarisation Vector to the Local Electric Field

It seems that we can no longer put off the question of how the polarisation **P** is related to the externally imposed electric field. We have already suggested that the molecular dipole moments are proportional to the local macroscopic field experienced by the molecules, $\mathbf{E}_{\text{loc}}$.

For example, in the case of electronic polarisation, (6.3) gives the molecular dipole moment for fluids as

$$\mathbf{p} = \alpha\varepsilon_0 \mathbf{E}_{\rm loc}, \tag{6.27}$$

where, for simple atoms, the molecular polarisability $\alpha$ is of the order of the volume of the atoms. In the case of orientation polarisation, on the other hand, (6.4) gives the polarisation as

$$\mathbf{P} = \frac{Np_0^2}{3k_BT}\mathbf{E}_{\rm loc}. \tag{6.28}$$

In either case the polarisation is proportional to $\mathbf{E}_{\rm loc}$. So the question is: what is $\mathbf{E}_{\rm loc}$?

### 6.4.1 *The Concept of a Local Field*

It is clear that the mean macroscopic electric field experienced by any given molecule is not simply the externally applied electric field, but rather the sum of the externally applied field and the macroscopic field induced by all the surrounding molecular dipoles. So the local mean field experienced by a given molecule sitting at location $\mathbf{x}_0$ is the macroscopic field $\mathbf{E}(\mathbf{x}_0)$ minus the contribution to $\mathbf{E}(\mathbf{x}_0)$ that comes from the molecule itself. In order to isolate the molecule, let us return to the idea introduced in §6.2.4 of a small spherical volume $V_R$ centred on $\mathbf{x}_0$. If we shrink the size of $V_R$ until it includes only the molecule in question, then, according to the results of §5.3, the electric field induced by the molecule, $\mathbf{E}_{\rm mol}$, satisfies

$$\int_{V_R} \mathbf{E}_{\rm mol}(\mathbf{x})\, d\mathbf{x} = -\frac{\mathbf{p}_{\rm mol}}{3\varepsilon_0}, \tag{6.29}$$

where $\mathbf{p}_{\rm mol}$ is the dipole moment of the molecule. Since $\mathbf{p}_{\rm mol} \approx \mathbf{P}(\mathbf{x}_0)V_R$, we conclude that the contribution to $\mathbf{E}(\mathbf{x}_0)$ from the molecule itself is $\mathbf{E}_{\rm mol} \approx -\mathbf{P}(\mathbf{x}_0)/3\varepsilon_0$. The local mean field experienced by the molecule is then

$$\boxed{\mathbf{E}_{\rm loc} \approx \mathbf{E} + \frac{\mathbf{P}}{3\varepsilon_0}.} \tag{6.30}$$

Of course, this is all a little rough and ready. For example, the molecule is unlikely to be particularly spherical and so it may not be very meaningful to think about isolating a single molecule using a spherical volume like $V_R$. For example, if the molecule and control volume are disc-like, with an axis parallel to $\mathbf{E}$, then it is not hard to show that (6.30) becomes $\mathbf{E}_{\rm loc} = \mathbf{E} + \mathbf{P}/\varepsilon_0$. Still, let us use (6.30) as a guide to the local field and see where this leads when combined with the hypothesis that the polarisation is proportional to $\mathbf{E}_{\rm loc}$.

### 6.4.2 *Linear Dielectrics and the Relative Permittivity Revisited*

From the definition of the molecular polarisability $\alpha$ we have $\mathbf{P} = N\alpha\varepsilon_0\mathbf{E}_{\text{loc}}$. Combining this with our estimate of $\mathbf{E}_{\text{loc}}$ yields

$$\mathbf{P} = N\alpha\varepsilon_0\left(\mathbf{E} + \frac{\mathbf{P}}{3\varepsilon_0}\right),$$

or equivalently

$$\mathbf{P} = \frac{N\alpha}{1-(N\alpha/3)}\varepsilon_0\mathbf{E}. \tag{6.31}$$

The equivalent result for orientation polarisation is equally simple to obtain. The key point about (6.31), or its equivalent for polar dielectrics, is that we have a *linear relationship* between $\mathbf{P}$ and $\varepsilon_0\mathbf{E}$. Actually, not all dielectrics exhibit such a simple linear relationship; however, many do and so we shall restrict ourselves to these. Given that $\mathbf{P}$ is (usually) proportional to $\varepsilon_0\mathbf{E}$, the question then arises as to how we might frame that linear relationship between $\mathbf{P}$ and $\varepsilon_0\mathbf{E}$ in a convenient way.

To that end, let us return to our discussion of capacitors in §6.1. This led us to the idea of the *dielectric constant*, or *relative permittivity*, of a dielectric, denoted by $\varepsilon_r$ and determined for a capacitor by (6.2):

$$\varepsilon_r = 1 + \frac{|\sigma_b|}{\varepsilon_0\,|\mathbf{E}|}.$$

Since the polarisation in a parallel-plate capacitor is uniform, we have $\rho_b = 0$ and a surface charge density of $\sigma_b = \mathbf{P}\cdot\mathbf{n}$. So, for a capacitor, our dielectric constant satisfies

$$\varepsilon_r - 1 = \frac{|\mathbf{P}|}{\varepsilon_0\,|\mathbf{E}|}. \tag{6.32}$$

For any linear dielectric this clearly generalises to

$$\boxed{\mathbf{P} = (\varepsilon_r - 1)\varepsilon_0\mathbf{E}.} \tag{6.33}$$

This provides a convenient means of characterising the polarisation induced in a linear dielectric by an electric field. In particular, the relative permittivity has the advantage that it can be readily measured in an experiment. Moreover, $\varepsilon_r$ may be estimated for a given class of dielectric through expressions like (6.31). So the widespread convention is to use the constitutive equation (6.33) to close the system of equations in §6.3. For example, (6.33) allows $\rho_b = -\nabla\cdot\mathbf{P}$ to be evaluated in terms of $\mathbf{E}$, which when added to (6.23) yields a closed system of equations. In terms of $\varepsilon_r$, the constitutive relationship between $\mathbf{D}$ and $\mathbf{E}$ for a linear dielectric becomes

$$\boxed{\mathbf{D} = \varepsilon_r\varepsilon_0\mathbf{E}.} \tag{6.34}$$

Although $\varepsilon_r$ already has two names—the dielectric constant and the relative permittivity—it is traditional to introduce yet more notation surrounding the constitutive equation (6.33). The quantity $\chi_e = \varepsilon_r - 1$ is often called the *electric susceptibility*, while $\varepsilon = \varepsilon_r\varepsilon_0$ is known simply as the *permittivity* of the dielectric. In the interests of economy, we shall restrict ourselves to $\varepsilon_r$ when characterising a linear dielectric.

We close this section by returning to the case of electronic polarisation in a non-polar fluid dielectric. In such cases the relative permittivity for liquids and gases can be estimated from (6.31) as

$$\varepsilon_r = 1 + \frac{N\alpha}{1 - (N\alpha/3)}. \tag{6.35}$$

Of course, this rests on the somewhat rough and ready estimate of $\mathbf{E}_{\text{loc}}$ given by (6.30). Expression (6.35) is known as the Clausius–Mossotti equation and is commonly used to estimate the relative permittivity of liquid dielectrics. The validity of (6.35) is put to the test for a number of liquids in, for example, Feynman, Leighton & Sands (1964) and, somewhat surprisingly, it turns out to give reasonable estimates. This lends tentative support to the estimate of $\mathbf{E}_{\text{loc}}$ given by (6.30). For gases (6.35) reduces to the simpler expression $\varepsilon_r = 1 + N\alpha$, simply because the number density is small and so $N\alpha << 1$.

### 6.4.3 The Parallel-Plate Capacitor Revisited

We close by returning to Faraday's experiments with capacitors. How should we interpret his findings in terms of the discussion in §6.4.2? If we choose to work in terms of **D**, then $\nabla \cdot \mathbf{D} = \rho_f$ tells us that the **D**-lines are continuous in the absence of free charges and so they begin and end on the free charges. In the case of a capacitor the **D**-lines begin and end on the plates and are continuous in between. This is shown in Figure 6.4(a). In integral form we have

$$\oint \mathbf{D} \cdot d\mathbf{S} = \int \rho_f \, dV = Q_f, \tag{6.36}$$

and so a disc-like control volume at the edge of the positively charged plate tells us that $|\mathbf{D}|$ is equal to the surface charge density on the plate, $D = \sigma_f$. In the narrow air gap between the plate and the dielectric we have $\mathbf{D} = \varepsilon_0\mathbf{E}$ and so $E_{\text{gap}} = \sigma_f/\varepsilon_0$, which we could have deduced directly from Gauss' law applied to the disc-like control volume. (Here we take the dielectric constant of air to be unity.) Within the dielectric, on the other hand, we have $\mathbf{D} = \varepsilon_r\varepsilon_0\mathbf{E}$, and so the electric field there takes the lower value of $E_d = \sigma_f/\varepsilon_r\varepsilon_0$.

If we choose to work with **E**, however, the picture is that shown in Figure 6.4(b). The **E**-lines begin and end on the real charges, and this includes the bound charges. There is therefore a discontinuity in **E** across the surface of the dielectric. In particular, **E** drops by an amount $E_{\text{gap}} - E_d = |\sigma_b|/\varepsilon_0$ as we pass into the dielectric from the left. This is due to the negative bound charges $\sigma_b = -|\mathbf{P}|$ on the surface of the dielectric. Since (6.33) requires $P = (\varepsilon_r - 1)\varepsilon_0 E_d$, we have

$$\varepsilon_0(E_{\text{gap}} - E_d) = |\sigma_b| = P = (\varepsilon_r - 1)\varepsilon_0 E_d,$$

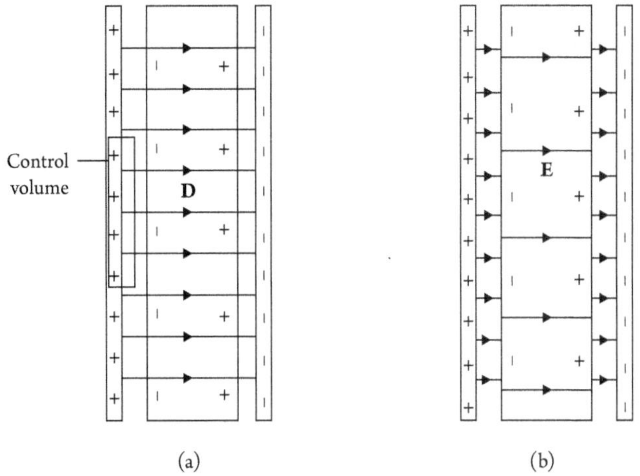


**Figure 6.4** A capacitor with a dielectric core: (a) the electric displacement field **D**; (b) the electric field **E**. Notice that **D** is continuous but **E** is not.

from which $E_d = E_{\text{gap}}/\varepsilon_r = \sigma_f/\varepsilon_r\varepsilon_0$. Of course, this is exactly what we found when working with **D**, rather than **E**.

Notice that, although **E** is the fundamental field in this problem, and **D** is somewhat of an artificial construct, it is easier to start with **D** and then deduce **E** using the constitutive equation (6.34). This is because, in this simple geometry, planar symmetry automatically ensures that $\nabla \times \mathbf{D} = 0$ and so **D** is determined simply by Gauss' law. In more complicated geometries, where $\nabla \times \mathbf{D} \neq 0$, the determination of **D** is more difficult.

## EXERCISES

6.1 A long, thin, straight rod carries a uniform charge per unit length of $\lambda$. Show that the electric field is radial and of magnitude $\lambda/2\pi\varepsilon_0 s$, where $s$ is the radial distance from the rod. The rod is now wrapped in a cylindrical insulator of radius $R$. Use (6.36) to find an expression for **D** as a function of $s$. Confirm that, for $s > R$, we recover the electric field for the uninsulated rod. (Notice that symmetry here automatically ensures $\nabla \times \mathbf{D} = 0$, so that knowing the divergence of **D** is sufficient to uniquely determine **D**.)

6.2 A spherical annulus of solid dielectric is centred on the origin and has a built-in polarisation **P**, which points in the radial direction and is a function of distance from the origin, $r$, only. The inner and outer radii of the dielectric shell are $R_1$ and $R_2$ and the radial variation of **P** is given by $\mathbf{P} = (c/r^2)\hat{\mathbf{e}}_r$. There are no free charges. Find **D** and hence deduce $\mathbf{E}(r)$.

6.3 Repeat Exercise 6.2, only this time first find the bound charges and then find $\mathbf{E}(r)$ using Gauss' law.

6.4 A thin circular disc of dielectric material has a uniform polarisation **P** parallel to the axis of the disc. It has radius $R$ and thickness $t$. Find the electrostatic potential outside the disc on its axis as a function of axial distance from the disc.

6.5 Show that, even for a linear dielectric, $\nabla \times \mathbf{D}$ is in general non-zero at the edges of the dielectric. (*Hint:* expand $\nabla \times (\varepsilon_r \varepsilon_0 \mathbf{E})$, treating $\varepsilon_r$ as a smoothly varying function and then consider the limit as $\varepsilon_r$ becomes a step function.)

6.6 Show that, if a small piece of dielectric is placed adjacent to a charged sphere, it is attracted to that sphere. (A sketch showing the bound charge is all that is needed.)

6.7 By definition, (5.37) gives the net dipole moment of a dielectric as

$$\mathbf{p} = \int \rho_e \mathbf{x}\, dV = \int \rho_b \mathbf{x}\, dV + \oint \sigma_b \mathbf{x}\, dS.$$

Substituting for the bound charge using (6.7) gives us

$$p_i = -\int x_i \nabla \cdot \mathbf{P}\, dV + \oint x_i \mathbf{P} \cdot d\mathbf{S}.$$

Confirm that this yields

$$\mathbf{p} = \int \mathbf{P}\, dV,$$

as it must.

## REFERENCES

Feynman, R. P., Leighton, R. B. & Sands, M., 1964, *The Feynman Lectures on Physics*, Volume II. Addison-Wesley.

Lorrain, P. & Corson, D., 1970, *Electromagnetic Fields and Waves*, 2nd ed. W. H. Freeman.

Purcell, E., 2011, *Electricity and Magnetism*, 2nd ed. Cambridge University Press.

# 7

# Magnetostatics I

## *The Magnetic Field, Ampère's Law and the Biot–Savart Law*

Monsieur Ampère is so clumsy that when his apparatus does not move, he reportedly pushes to shift it.

Pierre-Simon Laplace, *circa* 1821

I am naturally sceptical in the matter of theories and therefore you must not be angry with me for not admitting the one you have advanced immediately. Its ingenuity and applications are astonishing and exact but I cannot comprehend how the currents are produced and particularly if they be supposed to exist round each particle and I wait for further proofs of their existence before I finally admit them.

Letter from Faraday to Ampère, 1822

Ampère's pioneering work on magnetic fields initially met with some scepticism, not least from his compatriot and rival Biot. However, his ideas soon dominated the subject.

So far, our primary concern has been the mutual interaction of charges at rest. When charges are in motion they experience an additional force, over and above that of Coulomb. This is called the *Lorentz force* and it is used to define magnetic fields. In this chapter and the next we restrict ourselves to *static magnetic fields* in vacuum and in the presence of conductors. We focus in particular on the relationship between magnetic fields and the currents that create them, the resulting spatial structure of magnetic fields, and the forces they exert on conducting material. We leave ferromagnetism for Chapter 9 and the inductive action of an unsteady magnetic field for Chapter 10. Here we focus on Ampère's law and the Biot–Savart law, both of which tell us (in rather different ways) about the spatial structure of a magnetic field generated by a given distribution of current. Let us start, however, with some definitions.

## 7.1 The Lorentz Force, the Magnetic Field and the Vector Potential

A charged particle moving with velocity $\mathbf{u}_q$ and carrying a charge $q$ experiences two electromagnetic forces:

$$\mathbf{f} = q\mathbf{E} + q\mathbf{u}_q \times \mathbf{B}. \tag{7.1}$$

The first term is the Coulomb force and it is used to define the electric field **E**, while the second term is the Lorentz force and this is used to define the magnetic field **B**. That is to say, the electric field **E** is defined as the force on a unit test charge at rest in the observer's frame of reference, while the additional force associated with motion of the test charge is then used to define **B**. Note that the Lorentz force does no work on a charged particle, since it is perpendicular to $\mathbf{u}_q$. In SI units magnetic fields are measured in tesla (T) or in webers per square meter ($\text{Wb/m}^2$), although in practice many authors still prefer the older unit of gauss (G), where 1 tesla = $10^4$ gauss. (The surface magnetic field of the earth is around half a gauss.)

Two subtleties immediately arise from the definitions of **E** and **B**, one relating to the vector nature of **B** and the other to the frame of reference used to defined **E**. Let us start with the former. Since force and velocity are both true vectors (polar vectors), the appearance of the cross product in (7.1) tells us that **B** is a *pseudovector* (an axial vector), which changes direction if we move from a right-hand convention for cross products to a left-hand convention. (The definition of a pseudovector is given in §1.1.3.) Thus magnetic fields have a definite magnitude and line of action, but their direction along that line of action is fixed by a somewhat arbitrary sign convention.

The second subtlety relates to the frame of reference used to define **E** and **B**. Suppose that there is an electric field **E** and a magnetic field **B** in a laboratory frame. The electric field is defined by the force on a test charge at rest in the laboratory frame. If the test charge now moves with velocity $\mathbf{u}_q$ in that frame, the additional force $q\mathbf{u}_q \times \mathbf{B}$ is used to define **B**. However, suppose we decide to repeat our force measurement in a second frame of reference that moves with velocity $\mathbf{u}_q$ relative to the laboratory frame, and in which the charge is at rest. (In order to avoid any relativistic considerations, we shall take $\mathbf{u}_q$ to be much smaller than the speed of light.) Then, as noted in Chapter 3, the force on the charge is unchanged but it must now be interpreted as due to an electric field only. So we have

$$\mathbf{f} = q(\mathbf{E} + \mathbf{u}_q \times \mathbf{B}) = q\mathbf{E}_{\text{rel}}, \tag{7.2}$$

where **E** is measured in the laboratory frame and the subscript 'rel' indicates 'relative to the moving frame'. Evidently the electric field in the frame of reference moving with velocity $\mathbf{u}_q$ is

$$\mathbf{E}_{\text{rel}} = \mathbf{E} + \mathbf{u}_q \times \mathbf{B}. \tag{7.3}$$

It turns out that, when $\mathbf{u}_q$ is much smaller than the speed of light, the magnetic fields **B** and $\mathbf{B}_{\text{rel}}$ are equal (see Exercise 7.10).

Let us now turn to the divergence of **B**. Faraday's law of induction

$$\nabla \times \mathbf{E} = -\frac{\partial \mathbf{B}}{\partial t}, \tag{7.4}$$

which is one of Maxwell's four fundamental equations, ensures that $\partial \mathbf{B}/\partial t$ is solenoidal, since $\nabla \cdot [\nabla \times (\cdot)] = 0$ for any vector field. In fact, as noted in Chapter 3, it turns out that all of the experimental evidence points to the fact that **B** is itself solenoidal,

$$\nabla \cdot \mathbf{B} = 0, \tag{7.5}$$

which is another of Maxwell's four equations. If we compare this with Gauss' law, $\nabla \cdot \mathbf{E} = \rho_e/\varepsilon_0$, we see that there is no magnetic analogue of electrical charges on which the magnetic field lines may begin and end. Rather, the **B**-lines are continuous in space, forming closed loops or chaotic tangles, or else receding to infinity. Some people characterise this by saying there are *no magnetic charges*, or *no magnetic monopoles*.

Since $\nabla \cdot \mathbf{B} = 0$, it is often useful to picture a magnetic field in terms of *magnetic flux tubes* (Figure 7.1), which are imaginary tubes, usually thin, whose boundary is composed of magnetic field lines. The flux of the magnetic field, $\Phi = \int \mathbf{B} \cdot d\mathbf{S}$, is clearly constant along such a flux tube.

It also follows from $\nabla \cdot \mathbf{B} = 0$ that we may introduce a vector potential for **B**, defined (in part) by $\mathbf{B} = \nabla \times \mathbf{A}$. However, a vector field is only unambiguously defined once both its divergence and curl are specified (and suitable boundary conditions given), and so we must also specify the divergence of **A**. The choice of divergence is largely a matter of convenience and the simplest option is to demand $\nabla \cdot \mathbf{A} = 0$, which is known as taking the *Coulomb gauge* for **A**. This is the convention universally adopted in magnetostatics. (We will find it convenient to modify our convention for $\nabla \cdot \mathbf{A}$ when it comes to electrodynamics, where we shall adopt the so-called *Lorenz gauge* for **A**.) Of course, we also need appropriate boundary

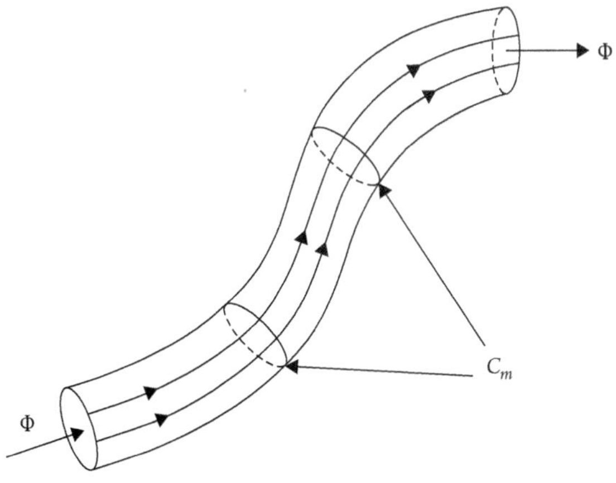


**Figure 7.1** A magnetic flux tube.

conditions for **A**, and in an infinite domain it is natural to demand that **A** is zero at infinity. So, in magnetostatics, the vector potential **A** is usually defined by

$$\nabla \times \mathbf{A} = \mathbf{B}, \qquad \nabla \cdot \mathbf{A} = 0, \qquad \mathbf{A}(\mathbf{x} \to \infty) = 0. \tag{7.6}$$

Note that, from Stokes' theorem, the magnetic flux within a flux tube can be expressed in terms of the line integral of **A** along any closed curve $C$ that encircles the flux tube,

$$\Phi = \oint_C \mathbf{A} \cdot d\mathbf{r}.$$

We shall see that, while it is natural to work with **B** rather than **A** in elementary applications of electromagnetism, the vector potential turns out to be particularly convenient when it comes to more advanced aspects of electrodynamics.

We close this section with an aside about knotted magnetic field lines. As noted in Exercise 3.5, the *magnetic helicity* of some part of a magnetic field is defined as

$$h_m = \int_V \mathbf{A} \cdot \mathbf{B}\, dV, \tag{7.7}$$

where the volume $V$ is chosen such that $\mathbf{B} \cdot d\mathbf{S} = 0$ at all points on its surface. To gain some appreciation of the significance of $h_m$, consider a magnetic field composed exclusively of two thin magnetic flux tubes that are linked as shown in Figure 7.2. Let the tubes have volumes $V_1$ and $V_2$, centrelines $C_1$ and $C_2$, and magnetic fluxes $\Phi_1$ and $\Phi_2$. If $dV$ is a short portion of one of the flux tubes, of length $d\mathbf{l}$, then $\mathbf{B}\, dV = \Phi\, d\mathbf{l}$, and so the global magnetic helicity for this simple geometry is

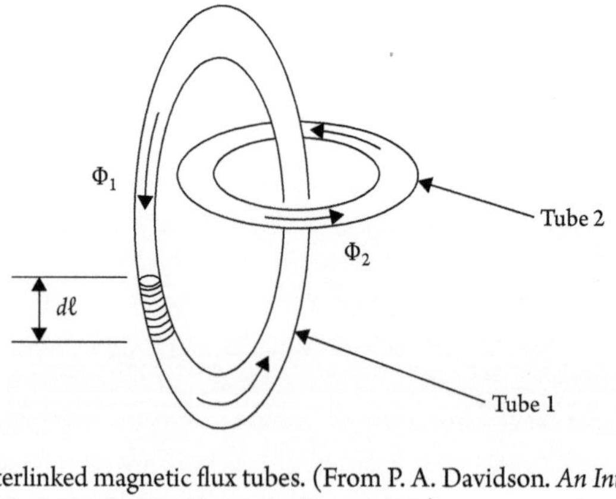


**Figure 7.2** Two interlinked magnetic flux tubes. (From P. A. Davidson. *An Introduction to Magnetohydrodynamics*, 2nd ed. Cambridge University Press, 2017.)

$$h_m = \int_{V_1} \mathbf{A} \cdot \mathbf{B}\, dV + \int_{V_2} \mathbf{A} \cdot \mathbf{B}\, dV = \Phi_1 \oint_{C_1} \mathbf{A} \cdot d\mathbf{l} + \Phi_2 \oint_{C_2} \mathbf{A} \cdot d\mathbf{l}.$$

However, from Stokes' theorem,

$$\oint_{C_1} \mathbf{A} \cdot d\mathbf{l} = \pm\Phi_2, \qquad \oint_{C_2} \mathbf{A} \cdot d\mathbf{l} = \pm\Phi_1,$$

where the positive sign is applicable when the two flux tubes are linked in a right-handed fashion and the negative sign when the connection is left-handed. Evidently, for this idealised geometry, $h_m = \pm 2\Phi_1\Phi_2$. On the other hand, if the flux tubes were not linked, the magnetic helicity would be zero, whereas if the tubes were wound around each other $n$ times, then $h_m = \pm 2n\Phi_1\Phi_2$. (The integer $n$ is often called the winding number.) Even a single flux tube may possess helicity if it is knotted.

Magnetic helicity is often used as a topological measure of the degree of entanglement of magnetic flux tubes, and it plays an important role in certain plasmas, such coronal flux loops. We shall return to this topic in Chapters 8 and 20.

## 7.2 The Flow of Current: Ohm's Law, Ohmic Heating and Charge Conservation

As a prelude to discussing Ampère's law, which links the spatial structure of $\mathbf{B}$ to that of the current, it is convenient to recall the definition of the current density vector $\mathbf{J}$. It will also be helpful to consider some of the consequences of the flow of current, such as Ohmic heating, electromechanical energy conversion and charge redistribution.

### *7.2.1 Ohm's Law in Stationary and Moving Conductors*

We introduced the current density vector in §3.1. This is a vector that points in the direction of maximum current flow and its component in any one direction is the current per unit area flowing in that direction. In a conductor $\mathbf{J}$ is composed of free charges moving under the influence of force (7.1) and it is known as the *conduction current*. Let us denote this current density by $\mathbf{J}_f$, the subscript standing for *free*.

The vector $\mathbf{J}_f$ is proportional to the force experienced by the free charges, and the magnitude of the constant of proportionality may be estimated for metals (and many plasmas) as follows. Consider a stationary conductor, or else a moving conductor in which we adopt a frame of reference that moves with the local velocity of the conductor. Let the free charge carriers, of which there are $N$ per unit volume, move with a velocity $\mathbf{u}_d$ relative to the conductor, which is known as the *drift velocity*. Then the force on each charge is

$$\mathbf{f} = q\mathbf{E} + q\mathbf{u}_d \times \mathbf{B},$$

where $\mathbf{E}$ is the electric field measured in a frame of reference in which the conductor is locally stationary. To determine the mean drift velocity at any one location requires the use of quantum mechanics. However, there is a simple non-quantum cartoon, known as the Drude model, that is not too misleading and is adequate for our purposes. Suppose that the mean drift velocity arises from a balance between the combined Coulomb and Lorentz forces, which accelerate the charge carriers, and a mean drag caused by multiple inelastic collisions of the free charges within the bulk of the conductor. Then we might write this balance as

$$q\mathbf{E} + q\mathbf{u}_d \times \mathbf{B} = \frac{m\mathbf{u}_d}{\tau_c},$$

where $\tau_c$ is some measure of the average time between collisions and $m$ the mass of the charge carriers. If we now sum this force balance over all the free charges in a unit volume of the conductor, then $\sum q$ becomes $Nq$ and $\sum q\mathbf{u}_d$ becomes the conduction current $\mathbf{J}_f$:

$$Nq\mathbf{E} + \mathbf{J}_f \times \mathbf{B} = \sum \frac{m\mathbf{u}_d}{\tau_c},$$

or equivalently

$$\mathbf{E} + (\mathbf{J}_f \times \mathbf{B})/Nq = \frac{m}{Nq^2\tau_c} \sum q\mathbf{u}_d.$$

Experiments show that the right-hand side of this equation is proportional to the conduction current, as we would expect, and so

$$\mathbf{E} + (\mathbf{J}_f \times \mathbf{B})/Nq = \frac{\mathbf{J}_f}{\sigma},$$

where the constant of proportionality, $\sigma = Nq^2\tau_c/m$, is the *electrical conductivity*.

The second term on the left of this force balance is known as the Hall effect and it is negligible in nearly all conductors, such as metals and collision-dominated plasmas. So, in nearly all conductors we end up with Ohm's law in its simple, familiar form:

$$\mathbf{J}_f = \sigma \mathbf{E}. \tag{7.8}$$

This is all that is required for stationary conductors. However, from the discussion above it is clear that, in moving conductors, the electric field in (7.8) must be interpreted as that measured in a frame of reference moving with the local velocity of the conductor. Ohm's law then generalises to

$$\mathbf{J}_f = \sigma \mathbf{E}_{\text{rel}} = \sigma(\mathbf{E}_{\text{lab}} + \mathbf{u} \times \mathbf{B}),$$

where $\mathbf{u}$ is the local velocity of the conductor and $\mathbf{E}_{\text{lab}}$ is the field measured in the laboratory frame of reference. Now, in a moving conductor the *total current density* $\mathbf{J}$ comprises the conduction current $\mathbf{J}_f$ and the so-called *convection current* $\rho_e\mathbf{u}$ due to bulk movement of the conductor. However, we saw in §3.1 that in good conductors $\rho_e$, and hence the convection current, is tiny and hence $\mathbf{J}$ is completely dominated by the conduction current: $\mathbf{J} = \mathbf{J}_f$. So the generalised version of Ohm's law is normally written

$$\boxed{\mathbf{J} = \sigma(\mathbf{E} + \mathbf{u} \times \mathbf{B}),} \tag{7.9}$$

where it is understood that $\mathbf{J} = \mathbf{J}_f$ and that $\mathbf{E}$ is measured in the laboratory frame.

### *7.2.2 The Macroscopic Lorentz Force, Ohmic Heating and Electromechanical Energy Conversion*

Let us now evaluate the macroscopic force transmitted to, and energy generated within, a conducting material as a result of the Coulomb and Lorentz forces acting on the free charges. We start with the force. If (7.1) is summed over a unit volume of the conductor (including both free and bound charges), then $\sum q$ becomes the charge density $\rho_e$ and $\sum q\mathbf{u}_q$ becomes the total current density $\mathbf{J}$. Since the forces exerted on the charges are ultimately transmitted to the conductor, the macroscopic electromagnetic force per unit volume of a conductor is evidently

$$\mathbf{F} = \rho_e \mathbf{E} + \mathbf{J} \times \mathbf{B}. \tag{7.10}$$

However, in nearly all conductors, stationary or moving, the charge density is tiny, so much so that (7.10) almost invariably simplifies to $\mathbf{F} = \mathbf{J} \times \mathbf{B}$, as discussed in §3.1.

Let us now consider energy. The flow of current though a conductor is associated with an irreversible loss of electromagnetic (or mechanical) energy, as is evident from (3.44). Although this loss emerges naturally from the macroscopic equations, it is helpful to understand its origins. The underlying idea is the following. The electromagnetic forces acting on the free charge carriers not only impart momentum to the free charges, which subsequently manifests itself as a force acting on the conductor, they also impart kinetic energy to the free charges. This energy is subsequently lost by the charge carriers as they undergo inelastic collisions, and it reappears as energy in the bulk conductor, typically as heat or as bulk kinetic energy. Since the Lorentz force $q\mathbf{u}_q \times \mathbf{B}$ does no work on the charge carriers, the rate of working of the electromagnetic force is associated purely with the electric field, being $\mathbf{u}_q \cdot (q\mathbf{E})$ per charge. Summing over a unit volume of the conductor, $\sum q\mathbf{u}_q$ becomes the current density $\mathbf{J}$, which is completely dominated by the conduction current. So the rate of working of the electric field on the conductor is $\mathbf{J} \cdot \mathbf{E}$ per unit volume:

$$\boxed{\text{rate of working of } \mathbf{E} \text{ on conductor} = \mathbf{J} \cdot \mathbf{E}}$$

In an electrical circuit this energy is supplied by an external source, such as a battery, but in the absence of such an external source this energy must be supplied by the electromagnetic field itself. This is why the term $\mathbf{J} \cdot \mathbf{E}$ appears as a drain on electromagnetic energy on the right-hand side of the unsteady energy equation (3.42).

In a stationary conductor the energy transferred to the free charges, $\mathbf{J} \cdot \mathbf{E}$, is lost by the charge carriers through inelastic collisions and simply ends up as heat in the conductor. Ohm's law in a stationary conductor, $\mathbf{J} = \sigma\mathbf{E}$, then gives $\mathbf{J}^2/\sigma$ as the rate of generation of heat per unit volume. This is called *Joule heating* or *Ohmic heating* and it is evidently a drain on the electromagnetic energy in (3.44). In a moving conductor, however, there is the possibility that some of the electrical power $\mathbf{J}\cdot\mathbf{E}$ manifests itself as mechanical energy in the conductor, say as kinetic energy. This is apparent from the generalised form of Ohm's law (7.9), which yields $\mathbf{J}\cdot\mathbf{E}=\mathbf{J}^2/\sigma-\mathbf{J}\cdot(\mathbf{u}\times\mathbf{B})$. Rearranging the triple product gives

$$\boxed{\mathbf{J}\cdot\mathbf{E}=\frac{\mathbf{J}^2}{\sigma}+\mathbf{u}\cdot(\mathbf{J}\times\mathbf{B}),} \tag{7.11}$$

(rate of working of $\mathbf{E}$ = Joule heating + rate of working of $\mathbf{J}\times\mathbf{B}$ on conductor)

where $\mathbf{u}$ is the local velocity of the conductor. The electrical power $\mathbf{J}\cdot\mathbf{E}$ now manifests itself partly as irreversible Ohmic heating and partly as the rate of working of the macroscopic Lorentz force acting on the conductor, $\mathbf{u}\cdot(\mathbf{J}\times\mathbf{B})$. This second term may be positive or negative and allows for an exchange of electromagnetic and mechanical energy (see Exercises 7.1–7.3). It is pivotal in electrical engineering and in nearly all applications of magnetohydrodynamics. For example, it is the mechanism by which buoyant motion in the core of the earth maintains the geomagnetic field.

### 7.2.3 *Charge Conservation Revisited*

We close this section by recalling that $\rho_e$ and $\mathbf{J}$ are not independent fields, but are related by the conservation of charge. That is, the net current leaving a given volume $V$ through a closed surface $S$ is equal to the rate of depletion of charge within that volume:

$$\oint_S \mathbf{J}\cdot d\mathbf{S}=-\frac{\partial}{\partial t}\int_V \rho_e\,dV.$$

Invoking Gauss' theorem gives

$$\int_V \nabla\cdot\mathbf{J}\,dV=-\int_V \frac{\partial\rho_e}{\partial t}\,dV,$$

and equating the integrands in the volume integrals yields the local charge conservation equation,

$$\boxed{\nabla\cdot\mathbf{J}=-\frac{\partial\rho_e}{\partial t}.} \tag{7.12}$$

Evidently, in static situations $\mathbf{J}$ is solenoidal. Moreover, in §3.1 we showed that in good conductors (metals and hot plasmas) the charge density is usually tiny and so the right-hand side of (7.12) is negligible. (We exclude situations where we deliberately place charges on the surface of the conductor, as in a capacitor.) The charge density in good conductors is also so small that the Coulomb force may be neglected in the macroscopic force (7.10). So, in good electrical conductors, the charge conservation equation and the macroscopic Lorentz force simplify to

$$\boxed{\nabla \cdot \mathbf{J} = 0, \qquad \mathbf{F} = \mathbf{J} \times \mathbf{B}.} \tag{7.13}$$

(charge conservation and Lorentz force in good conductors)

## 7.3 Ampère's Law and the Biot–Savart Law

### 7.3.1 *Ampère's Law*

In magnetostatics the curl of a magnetic field is given by *Ampère's law*. This states that the magnetic field and current density are related by $\nabla \times \mathbf{B} = \mu_0 \mathbf{J}$, where $\mu_0$ is the *permeability of free space*. Thus in magnetostatics $\mathbf{B}$ is governed by the two equations

$$\boxed{\nabla \cdot \mathbf{B} = 0, \qquad \nabla \times \mathbf{B} = \mu_0 \mathbf{J}.} \tag{7.14}$$

Note that Ampère's law demands that $\nabla \cdot \mathbf{J} = 0$, since $\nabla \cdot [\nabla \times (\cdot)] = 0$ for any vector field. This is consistent with the static equation (7.13), as it must be, but not in general with the dynamic equation (7.12). This warns us that Ampère's law will need to be modified whenever $\partial \rho_e / \partial t$ makes a significant contribution to the charge conservation equation. We shall discuss the form of that modification in Chapter 13.

Given the distribution of $\mathbf{J}$, (7.14) is sufficient to uniquely determine $\mathbf{B}$, since both its curl and divergence are then specified. In particularly simple geometries symmetry often automatically enforces $\nabla \cdot \mathbf{B} = 0$, in which case $\mathbf{B}$ can often be determined simply by applying Ampère's law in integral form. This integral version of Ampère's law is known as *Ampère's circuital law* and takes the form

$$\oint_C \mathbf{B} \cdot d\mathbf{r} = \mu_0 \int_S \mathbf{J} \cdot d\mathbf{S} = \mu_0 I, \tag{7.15}$$

where $S$ is any open surface that spans the curve $C$ and $I$ is the net current passing though the surface $S$.

As a simple example, consider the case of a long straight wire carrying a current $I$, as shown in Figure 7.3. Symmetry tells us that $\mathbf{B}$ points in the azimuthal direction and has a magnitude that depends only on the distance from the wire, $r$. Such a field automatically satisfies $\nabla \cdot \mathbf{B} = 0$. Applying (7.15) to a curve that coincides with a magnetic field line then yields the field distribution

$$\mathbf{B} = \frac{\mu_0 I}{2\pi r} \hat{\mathbf{e}}_\theta. \tag{7.16}$$

We note in passing that two such parallel wires carrying currents in the same direction will, according to (7.13), attract each other with a force per unit length of $F = \mu_0 I_1 I_2 / 2\pi d$,

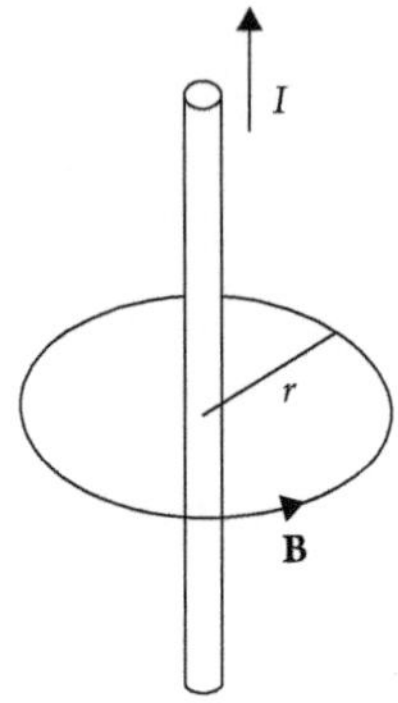


**Figure 7.3** The integral version of Ampère's law applied to a wire.

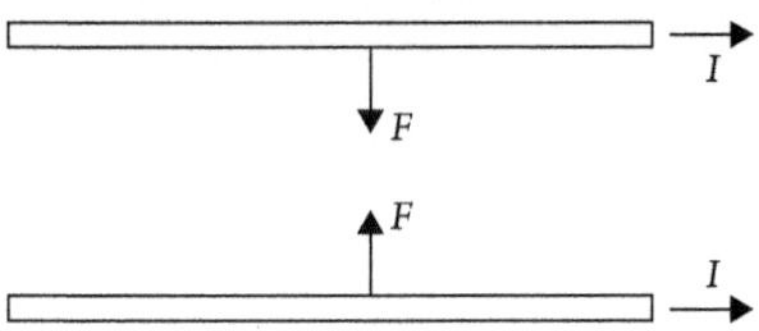


**Figure 7.4** Two parallel wires carrying currents in the same direction attract each other.

where $d$ is the distance between the wires (Figure 7.4). If the two wires carry current in opposite directions, then they repel each other.

### *7.3.2 The Biot–Savart Law*

Despite the ease with which Ampère's circuital law yields (7.16), symmetry only guarantees that $\nabla \cdot \mathbf{B} = 0$ in very simple geometries. In general we need to invoke both parts of (7.14) in order to determine **B**. In infinite domains this is relatively straight forward because (7.14) may be inverted, or solved, using the *Biot–Savart law*.

Suppose that **J** is a known localised function of position. Then the Biot–Savart law tells us that the corresponding magnetic field may be calculated directly from

$$\boxed{\mathbf{B}(\mathbf{x}) = \frac{\mu_0}{4\pi} \int \frac{\mathbf{J}(\mathbf{x}') \times \mathbf{r}}{|\mathbf{r}|^3} d\mathbf{x}', \qquad \mathbf{r} = \mathbf{x} - \mathbf{x}',} \tag{7.17}$$

where $\mathbf{x}'$ is a dummy variable that samples the current density at different points, as indicated in Figure 7.5. The Biot–Savart law may also be written in terms of the vector potential **A**, as we saw in §3.4. This takes the particularly simple form

$$\boxed{\mathbf{A}(\mathbf{x}) = \frac{\mu_0}{4\pi} \int \frac{\mathbf{J}(\mathbf{x}')}{|\mathbf{r}|} d\mathbf{x}', \qquad \mathbf{r} = \mathbf{x} - \mathbf{x}'.} \tag{7.18}$$

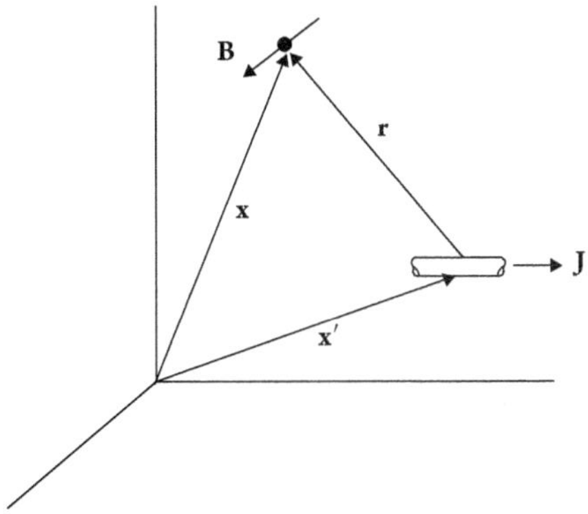


**Figure 7.5** Coordinate system used in the Biot–Savart law.

The link between (7.18) and the Biot–Savart law is straightforward and rests on the identity

$$\nabla \times \left(\frac{\mathbf{J}'}{|\mathbf{r}|}\right) = \nabla\left(\frac{1}{|\mathbf{r}|}\right) \times \mathbf{J}',$$

where $\mathbf{J}' = \mathbf{J}(\mathbf{x}')$ and $\nabla$ operates on $\mathbf{x}$ while treating $\mathbf{x}'$ as constant. Since $\nabla(1/|\mathbf{r}|) = -\mathbf{r}/|\mathbf{r}|^3$, this identity can be written as

$$\nabla \times \left(\frac{\mathbf{J}'}{|\mathbf{r}|}\right) = \mathbf{J}' \times \frac{\mathbf{r}}{|\mathbf{r}|^3}, \tag{7.19}$$

from which

$$\mathbf{B}(\mathbf{x}) = \nabla \times \mathbf{A}(\mathbf{x}) = \frac{\mu_0}{4\pi}\int \nabla \times \left(\frac{\mathbf{J}'}{|\mathbf{r}|}\right) d\mathbf{x}' = \frac{\mu_0}{4\pi}\int \frac{\mathbf{J}' \times \mathbf{r}}{|\mathbf{r}|^3}\, d\mathbf{x}'. \tag{7.20}$$

We have arrived back at the Biot–Savart law (7.17).

We shall return to the Biot–Savart law in §7.8, where we discuss its mathematical origins. There are two approaches to showing that (7.17) is equivalent to the pair of equations in (7.14). One is to show by direct calculation that taking the divergence and curl of the Biot–Savart law yields $\nabla \cdot \mathbf{B} = 0$ and $\nabla \times \mathbf{B} = \mu_0 \mathbf{J}$. The second is to use Ampère's law plus $\nabla \cdot \mathbf{B} = 0$ to deduce Poisson equations for both **B** and **A**, equations in which the source term is **J** or else its curl. Inverting those Poisson equations then leads us directly to the Biot–Savart law (7.17) and its companion equation for **A**, (7.18). We shall outline both approaches in §7.8. First, however, in order to demonstrate the utility of the Biot–Savart law, we consider some simple examples of magnetostatic fields.

# 7.4 Some Simple Examples of Static Magnetic Fields

## 7.4.1 *Current Sheets, Current Rings and Solenoids*

We start with the somewhat idealised problem of a uniform sheet of current which is confined to the $x$–$y$ plane and flows in the $x$ direction (Figure 7.6). If the magnitude of the surface current density is $K$ (A/m), then the surface current density vector is $\mathbf{K} = K\hat{\mathbf{e}}_x$. Perhaps the first question is: what is the direction of $\mathbf{B}$ induced by this current sheet? It is clear from the Biot–Savart law that there can be no component in the $x$ direction, since $\mathbf{B}$ is at right angles to $\mathbf{K}$. Nor can there be any component in the direction of $z$, because of symmetry. To see why, use the Biot–Savart law to find the field on the $z$ axis. Any contribution to $B_z$ that comes from the current located at $y$ is exactly cancelled by a contribution from the current at $-y$. So we conclude that $\mathbf{B} = B_y(z)\hat{\mathbf{e}}_y$, i.e. $\mathbf{B}$ is everywhere parallel to the current sheet, at right angles to the direction of current flow, and independent of $x$ and $y$ because of homogeneity in $x$ and $y$. Moreover, since $\nabla \times \mathbf{B} = \mu_0\mathbf{J}$, $B_y$ must be uniform above and below the sheet, where $\mathbf{J} = 0$, but possibly discontinuous across the $x$–$y$ plane. Finally, it is clear from the Biot–Savart law that $\mathbf{B}$ points in the negative $y$ direction above the current sheet and in the positive $y$ direction below. In short, we have $\mathbf{B} = \mp B_0\hat{\mathbf{e}}_y$, where $B_0$ is a positive constant.

Given this information, we can now find $\mathbf{B}$ by applying Ampère's circuital law to the dashed rectangular curve $C$ shown in Figure 7.6. If the rectangle has length $\ell$ in the $y$ direction, then

$$\oint_C \mathbf{B} \cdot d\mathbf{r} = \mu_0 K\ell, \tag{7.21}$$

from which we conclude that $\mathbf{B} = \mp(\mu_0 K/2)\hat{\mathbf{e}}_y$, where the negative sign applies above the current sheet and the positive sign below. Note that there is a jump in magnetic field across the current sheet of $\Delta\mathbf{B} = \mu_0\mathbf{K} \times \mathbf{n}$, where $\mathbf{n}$ is the unit normal to the surface pointing in the direction of the jump.

Consider now the field induced by a circular wire of radius $R$ and carrying a current $I$. It is natural to adopt cylindrical polar coordinates $(s, \theta, z)$ whose origin lies at the centre of the current loop. Clearly $\mathbf{B}$ is axisymmetric, sits in the $s$–$z$ plane, and has the shape

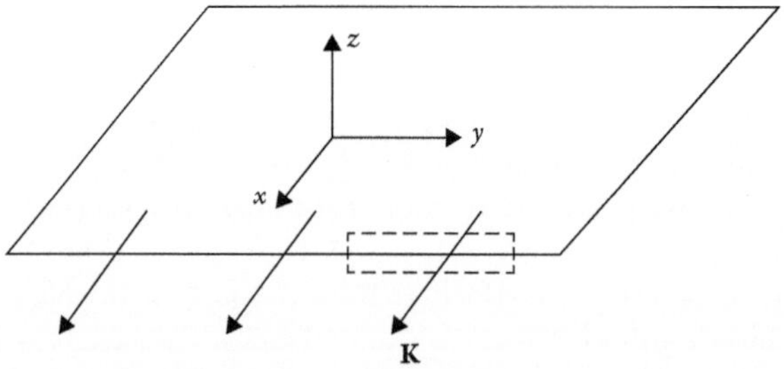


**Figure 7.6** A uniform current sheet $\mathbf{K} = K\hat{\mathbf{e}}_x$ confined to the $x$–$y$ plane induces a magnetic field in the $y$ direction that is antisymmetric about the $x$–$y$ plane.

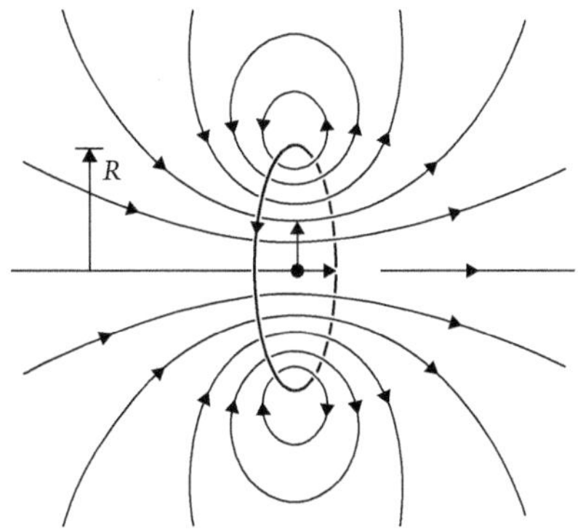


**Figure 7.7** The magnetic field induced by a current loop of radius $R$.

shown in Figure 7.7. We note in passing that, since parallel currents attract and antiparallel currents repel (Figure 7.4), is it clear that the Lorentz force acting on the current loop points radially outwards, placing the current ring in tension. Now if $I$ flows in the positive azimuthal direction, then the flux of **B** through the loop is in the positive $z$ direction. The field at the centre of the ring is then, from the Biot–Savart law,

$$\mathbf{B}(\mathbf{x}=0) = \frac{\mu_0}{4\pi}\int_0^{2\pi} \frac{(I\hat{\mathbf{e}}'_\theta)\times(-R\hat{\mathbf{e}}'_s)}{R^3} R\,d\theta' = \frac{\mu_0 I}{2R}\hat{\mathbf{e}}_z. \tag{7.22}$$

More generally, it is easy to show (see Exercise 7.4) that anywhere on the axis of the loop the magnetic field is given by

$$\mathbf{B}(z) = \frac{\mu_0 I}{2R}\frac{1}{(1+z^2/R^2)^{3/2}}\hat{\mathbf{e}}_z. \tag{7.23}$$

Note that $|\mathbf{B}|$ falls off as $|\mathbf{x}|^{-3}$ at large distances from the loop.

For reasons that will become apparent, it is conventional to write such expressions in a slightly different way. We define the *magnetic dipole moment* **m** of any localised distribution of current to be

$$\boxed{\mathbf{m} = \frac{1}{2}\int \mathbf{x}\times\mathbf{J}\,dV.} \tag{7.24}$$

For our simple planar current loop, then, we have $\mathbf{m} = I(\pi R^2)\hat{\mathbf{e}}_z$. In terms of **m**, the field at the centre of the ring is evidently

$$\mathbf{B}(\mathbf{x}=0) = \frac{\mu_0\mathbf{m}}{2\pi R^3}. \tag{7.25}$$

It is of interest to consider the field induced by a number of such current loops, arranged sequentially along the $z$ axis. Figure 7.8 shows the cases of two and four current rings and

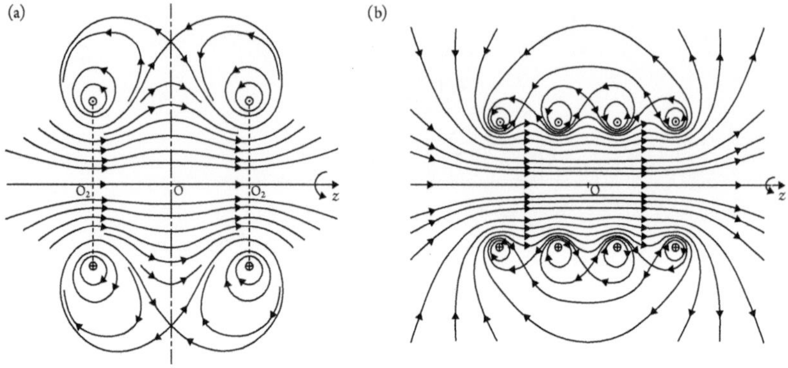


**Figure 7.8** The magnetic field induced by (a) two and (b) four current loops of equal radius and current. (From E. Durand. *Magnétostatique*. Masson, 1968.)

it is clear that, as the number of current loops increases, the magnetic field encircled by the rings becomes increasingly uniform and axial. In the limit of a large number of closely packed current rings, we obtain a good approximation to a solenoid.

We close this subsection by considering the magnetic field of a cylindrical solenoid. This consists of a tightly wound helical coil carrying a current $I$ and of overall axial extent $0 < z < L$, where $L$ is larger than the radius of the solenoid, $R$. If we ignore the fact that the current has a slight axial component by virtue of the helical winding, then we may model this as a cylindrical current sheet of radius $R$ in which the current is purely azimuthal. If there are $N$ turns per unit length, then, in cylindrical polar coordinates $(s, \theta, z)$ aligned with the axis of the solenoid, the surface current density is $\mathbf{K} = IN\hat{\mathbf{e}}_\theta$.

The magnetic field in the case of a relatively short solenoid, of aspect ratio $L = 4R$, is shown in Figure 7.9 and the field inside the solenoid is evidently much larger than that outside. In the limit of $L >> R$ it is clear from $\nabla \cdot \mathbf{B} = 0$ that the field outside the solenoid (and away from its ends) will be negligible by comparison with that inside, and that the internal field will be almost purely axial (except near the ends). In such a limit Ampère's circuital law applied to a rectangular curve $C$ that straddles the cylindrical current sheet gives the internal field as

$$\mathbf{B}_{\text{in}} = \mu_0 K\hat{\mathbf{e}}_z = \mu_0 NI\hat{\mathbf{e}}_z. \tag{7.26}$$

Note that, as with the planar current sheet, we have $\Delta\mathbf{B} = \mu_0\mathbf{K} \times \mathbf{n}$, where $\mathbf{n}$ is now the unit normal to the surface pointing towards the axis of the solenoid and $\Delta\mathbf{B} = \mathbf{B}_{\text{in}}$. Note also that, as with a single current loop, the Lorentz force acting on the solenoid points radially outwards, placing the helical coil in tension.

### 7.4.2 *Free-Space Magnetic Fields and Potential Theory Revisited*

In the absence of local currents, steady magnetic fields in free space, like their electrostatic counterparts, have zero divergence and curl:

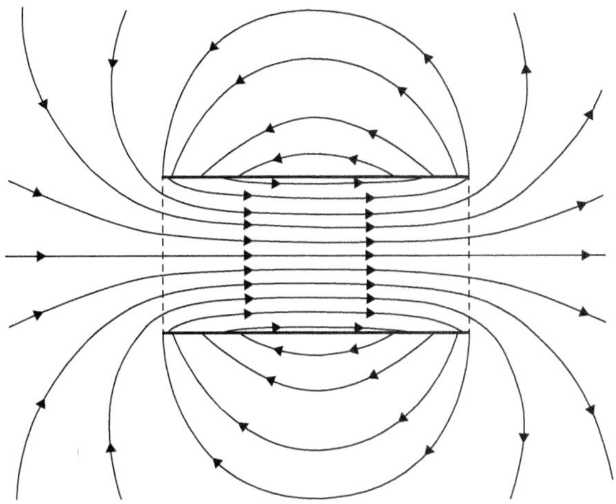

**Figure 7.9** The magnetic field of a relatively short solenoid. (From E. Durand. *Magnétostatique*. Masson, 1968.)

$$\nabla \times \mathbf{B} = 0, \qquad \nabla \cdot \mathbf{B} = 0. \tag{7.27}$$

The first of these two equations allow us to introduce the *magnetic scalar potential* $\chi$, defined through $\mathbf{B} = -\nabla\chi$, and so we find that

$$\mathbf{B} = -\nabla\chi, \quad \mathbf{B} = \nabla \times \mathbf{A}, \quad \nabla \cdot \mathbf{A} = 0 \tag{7.28}$$

are the magnetostatic counterparts of (5.2). Of course, in free space $\chi$ and $\mathbf{A}$ both satisfy Laplace's equation:

$$\nabla^2\chi = 0, \quad \nabla^2\mathbf{A} = 0. \tag{7.29}$$

For the exceptional case of a two-dimensional magnetic field $\mathbf{B}(x, y) = (B_x, B_y, 0)$, the vector potential takes the form $\mathbf{A} = \psi(x, y)\hat{\mathbf{e}}_z$, and we obtain

$$B_x = -\frac{\partial\chi}{\partial x} = \frac{\partial\psi}{\partial y}, \qquad B_y = -\frac{\partial\chi}{\partial y} = -\frac{\partial\psi}{\partial x}. \tag{7.30}$$

Since $\mathbf{B} \cdot \nabla\psi = 0$, the lines of constant $\psi$ are parallel to $\mathbf{B}$ and so represent the field lines of $\mathbf{B}$. The function $\psi$ is called the *flux function* of the magnetic field. Because (7.30) is identical to (5.4), we may, once again, use the theory of complex potentials to calculate planar magnetic fields. Indeed, we can lift many of the solutions from the electrostatic (or irrotational flow) literature and reinterpret them as steady two-dimensional magnetic fields. For example, just as Maxwell reinterpreted Helmholtz's irrational flow into a two-dimensional channel as the fringing electric field at the edge of a capacitor (Figure 5.4), so we, in turn, may reinterpret this solution as the magnetic field entering or leaving a magnetic pole piece. The equipotential surfaces of the electrostatic problem then become magnetic field lines, and we obtain the magnetic field shown in Figure 7.10. However, it is relatively rare that one may treat a magnetic field as two-dimensional and so we shall not dwell on such solutions.

### 7.4.3 *An Aside: Isolated Flux Tubes, Sunspots and Solar Flares*

The case of a long solenoid, discussed in §7.4.1, may be thought of as an extreme example of a concentrated (or isolated) flux tube in which the field lines a bunched up in a narrow tube and then allowed to close outside the tube in a region of weaker, more dispersed field (Figure 7.11). Such isolated flux tubes occur sometimes in astrophysical plasmas, the most striking case being the surface of the sun, where they are commonly observed. As with a solenoid, currents must encircle such flux tubes, maintaining the intense magnetic field within the tube. Since opposite currents repel, the Lorentz forces within such a flux tube point radially outward, so that the plasma pressure is smaller inside the flux tube than outside.

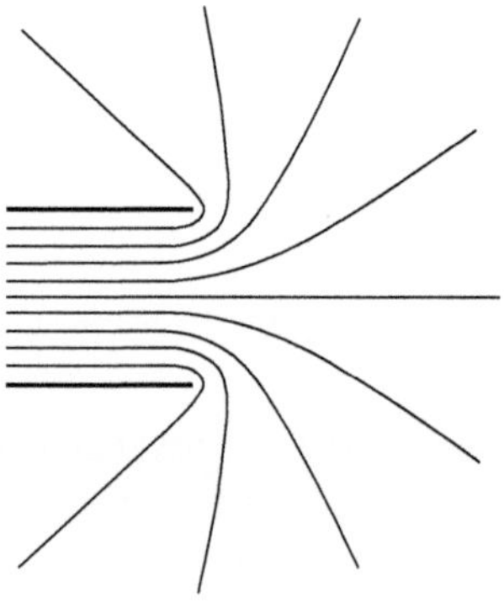

**Figure 7.10** The potential solution of Figure 5.4 (the fringing field at the edge of a capacitor) may be reinterpreted as a magnetic field entering or leaving a magnetic pole.

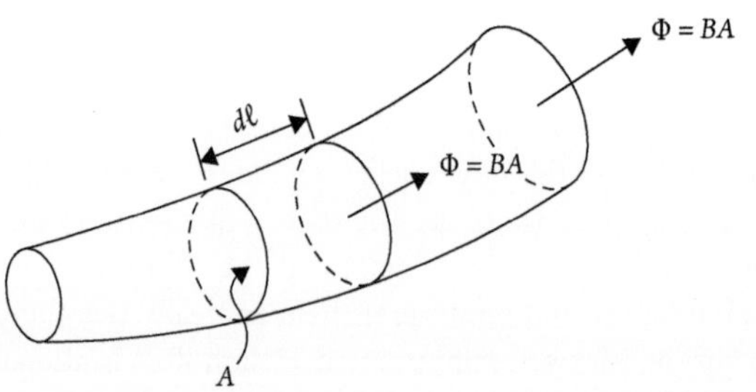


**Figure 7.11** An isolated flux tube. (From P. A. Davidson. *An Introduction to Magnetohydrodynamics*, 2nd ed. Cambridge University Press, 2017.)

Particularly intense flux tubes are generated within the interior of the sun near the base of the so-called convection zone. This is caused by strong differential rotation at the base of the convection zone, a motion that is well adapted to stretching and intensifying the local magnetic field in the east–west direction. This east–west magnetic field can be extremely intense, in excess of $10^4$ G, and this means that the plasma pressure, and hence density, within isolated flux tubes is significantly lower than that of the surrounding plasma. The flux tubes are therefore buoyant and parts of these tubes tend to rise up towards the surface of the sun, where they occasionally erupt though the photosphere (the surface of the sun) and emerge into the corona, as indicated in Figure 7.12. At those points where the flux tubes intersect the photosphere the locally intense magnetic field supresses the convective motions in the plasma and the surface temperature falls. These are sunspots, which appear relatively dark in images of the photosphere.

Some of the flux tubes that float up into the corona form prominences, which are arch-like tubular structures around $10^5$ km in length and filled with a billion tons of plasma. Quiescent prominences are stable, long-lived structures that survive for many weeks, while other prominences erupt when the structure of their magnetic field becomes unstable.

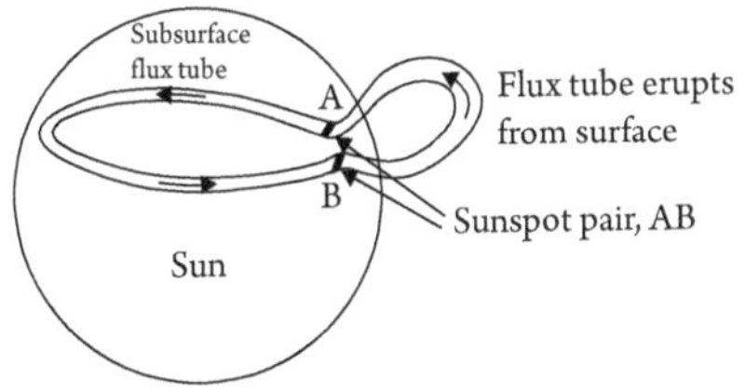


**Figure 7.12** Schematic of sunspot formation by a buoyant flux tube. (From P. A. Davidson. *An Introduction to Magnetohydrodynamics*, 2nd ed. Cambridge University Press, 2017.)

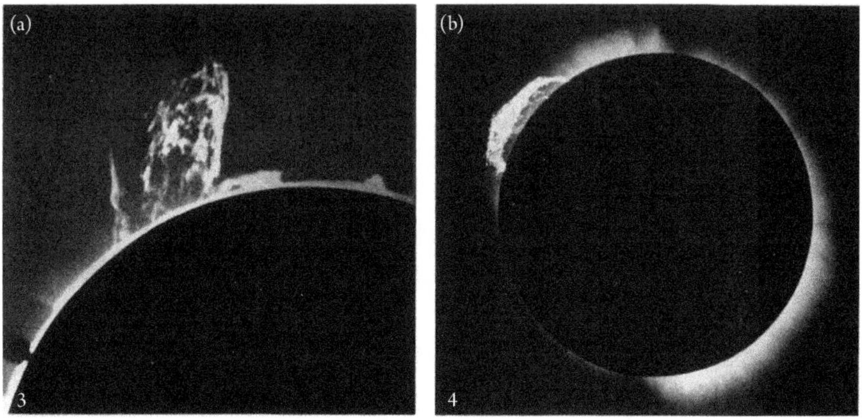


**Figure 7.13** Magnetic activity in the solar atmosphere: (a) an eruptive prominence; (b) a quiescent prominence.

Eruptive prominences are associated with spectacular events like solar flares and coronal mass ejections (Figure 7.13). We shall discus flux tubes, sunspots and solar flares in Chapter 21.

## 7.5 The Kinematics of Axisymmetric Magnetic Fields

When dealing with axisymmetric fields in cylindrical polar coordinates $(s,\ \theta,\ z)$, it is natural to divide a magnetic field into *poloidal*, $\mathbf{B}_p(s,z) = (B_s,\ 0,\ B_z)$, and *azimuthal*, $\mathbf{B}_\theta(s,z) = (0,\ B_\theta,\ 0)$, components. For example, the field induced by the current in a long straight wire is azimuthal, while the dipole-like magnetic field of a circular current loop, or the external magnetic field of Saturn, are examples of poloidal fields. (The external field of Saturn is very close to axisymmetric, while that of the earth is not entirely.) Note that axial symmetry ensures that $\nabla \cdot \mathbf{B}_\theta = 0$, so that both $\mathbf{B}_\theta$ and $\mathbf{B}_p$ are separately solenoidal and we may define a vector potential for each. Note also that the term toroidal is sometimes used in place of azimuthal, especially in the fusion plasma community, and sometimes meridional is substituted for poloidal. We shall not use the terms toroidal or meridional here.

Axial symmetry demands that the curl of an azimuthal field be poloidal, while that of a poloidal field is azimuthal. (You might try checking this for yourself.) So Ampère's law tells us that the current associated with a poloidal magnetic field is azimuthal, while the current that supports an azimuthal field is poloidal:

$$\nabla \times \mathbf{B}_p = \mu_0 \mathbf{J}_\theta, \qquad \nabla \times \mathbf{B}_\theta = \mu_0 \mathbf{J}_p. \tag{7.31}$$

This is illustrated in Figure 7.14 for the particular case where the current distribution is restricted to a sphere, such as in the core of a planet.

Let us focus now on the poloidal component of a magnetic field. In §7.4.2 we saw that magnetic field lines in two-dimensional fields are represented by the curves $\psi = \text{constant}$,

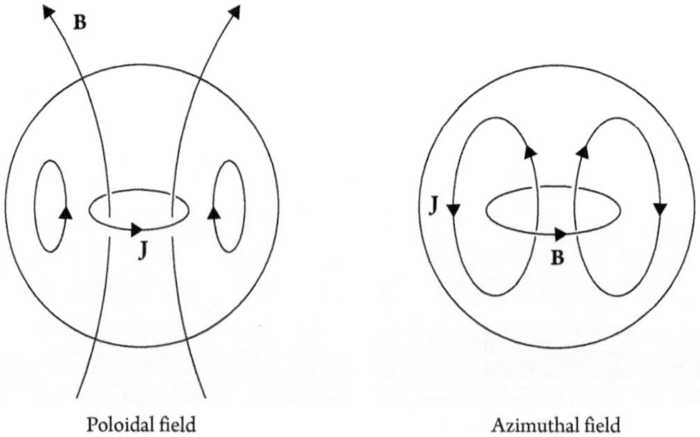


**Figure 7.14** Poloidal and azimuthal magnetic fields and the currents that support them.

where the *flux function* $\psi(x, y)$ is defined by $\mathbf{B} = \nabla \times \mathbf{A} = \nabla \times [\psi(x, y)\hat{\mathbf{e}}_z]$. The equivalent field lines in an axisymmetric poloidal field $\mathbf{B}_p(s, z) = (B_s, 0, B_z)$ are given by the surfaces $\Psi = \text{constant}$, where the *Stokes flux function* $\Psi(s, z)$ is defined by

$$\mathbf{B}_p = \nabla \times \mathbf{A}_\theta = \nabla \times \left[\frac{\Psi(s, z)}{s}\hat{\mathbf{e}}_\theta\right]. \tag{7.32}$$

In component form we have

$$\mathbf{B}_p = \left(-\frac{1}{s}\frac{\partial \Psi}{\partial z}, 0, \frac{1}{s}\frac{\partial \Psi}{\partial s}\right), \tag{7.33}$$

from which we see that $\mathbf{B}_p \cdot \nabla \Psi = 0$, so that surfaces of constant $\Psi$ are parallel to the magnetic field lines, as they must be.

The usual physical interpretation of $\Psi$ is that the magnetic flux $\Phi$ between two poloidal field lines $\Psi_1$ and $\Psi_2$ (strictly, these are two magnetic surfaces $\Psi_1$ and $\Psi_2$) is given by $2\pi(\Psi_2 - \Psi_1)$. This can be seen by integrating the axial magnetic field across the annular gap that separates $\Psi_1$ and $\Psi_2$:

$$\Phi = \int_{s_1}^{s_2} 2\pi s B_z \, ds = 2\pi \int_{s_1}^{s_2} \frac{\partial \Psi}{\partial s} \, ds = 2\pi(\Psi_2 - \Psi_1). \tag{7.34}$$

Combining (7.31) and (7.32), we see that the azimuthal current required to support a poloidal field is related to the Stokes flux function by

$$\mu_0 \mathbf{J}_\theta = \nabla \times \mathbf{B}_p = -\nabla^2 \mathbf{A}_\theta = -\nabla^2 \left[\frac{\Psi(s, z)}{s}\hat{\mathbf{e}}_\theta\right]. \tag{7.35}$$

Since the unit vector $\hat{\mathbf{e}}_\theta$ lying inside the Laplacian in (7.35) is not a constant, it is convenient to expand the term on the far right using $\nabla^2(\cdot) = -\nabla \times \nabla \times (\cdot)$. After a little effort we can extract $\hat{\mathbf{e}}_\theta$ from the Laplacian and we find

$$\mu_0 \mathbf{J}_\theta = \nabla \times \mathbf{B}_p = -s^{-1}(\nabla_*^2 \Psi)\, \hat{\mathbf{e}}_\theta, \tag{7.36}$$

where $\nabla_*^2$ is the Laplacian-like operator,

$$\nabla_*^2 \equiv \frac{\partial^2}{\partial z^2} + s\frac{\partial}{\partial s}\left(\frac{1}{s}\frac{\partial}{\partial s}\right). \tag{7.37}$$

The operator $\nabla_*^2$ is sometimes referred to as the *Stokes operator.*

Returning to the more general case in which $\mathbf{B} = \mathbf{B}_p + \mathbf{B}_\theta$, we see that the most efficient way to represent an axisymmetric magnetic field is to use the two scalar functions $\Psi(s, z)$ and $B_\theta(s, z)$. The fields $\mathbf{B}$ and $\mathbf{J}$ are then given by

$$\mathbf{B} = \nabla \times \left(\frac{\Psi}{s}\hat{\mathbf{e}}_\theta\right) + B_\theta \hat{\mathbf{e}}_\theta, \tag{7.38}$$

$$\mu_0 \mathbf{J} = \nabla \times (B_\theta \hat{\mathbf{e}}_\theta) - s^{-1}(\nabla_*^2 \Psi)\, \hat{\mathbf{e}}_\theta. \tag{7.39}$$

We shall find it convenient to adopt this decomposition when dealing with axisymmetric magnetic fields, as in the case below.

# 7.6 The Magnetic Dipole

## 7.6.1 *The Current Ring Revisited*

Let us now return to the planar current loop of §7.4.1. Our aim is to use the Biot–Savart law to find the structure of the magnetic field, both in the near field and in the far field. As before, we take the loop to have radius $R$ and carry a current $I$, and we adopt cylindrical polar coordinates $(s, \theta, z)$ centred on the loop (Figure 7.15). From (7.24) the dipole moment of the loop is

$$\mathbf{m} = \frac{1}{2}\int \mathbf{x} \times \mathbf{J}\, dV = I(\pi R^2)\hat{\mathbf{e}}_\mathbf{z}. \tag{7.40}$$

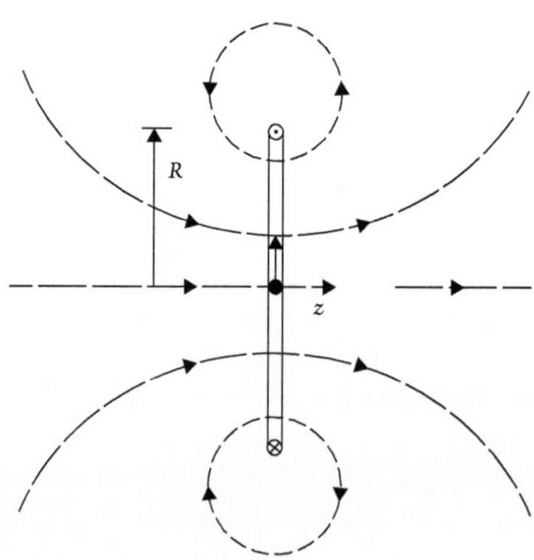


**Figure 7.15** A current loop of radius $R$ and carrying a current $I$.

The Biot–Savart law now yields

$$\mathbf{A}(\mathbf{x}) = \frac{\mu_0}{4\pi}\int_0^{2\pi} \frac{I\hat{\mathbf{e}}'_\theta}{|\mathbf{r}|} R\,d\theta' , \qquad \mathbf{r} = \mathbf{x} - R\hat{\mathbf{e}}'_s,$$

and with a little effort it is possible to show that

$$|\mathbf{r}|^2 = |\mathbf{x}|^2 + R^2 - 2sR\cos\tilde{\theta}, \qquad \tilde{\theta} = \theta - \theta'.$$

Substituting for $|\mathbf{r}|$ and $\hat{\mathbf{e}}'_\theta$ in the Biot–Savart integral, we obtain

$$\mathbf{A}(\mathbf{x}) = \frac{\mu_0 IR}{4\pi}\int_0^{2\pi} \frac{\cos\tilde{\theta}}{\sqrt{|\mathbf{x}|^2 + R^2 - 2sR\cos\tilde{\theta}}}\, d\tilde{\theta}\ \hat{\mathbf{e}}_\theta,$$

and so we conclude that the Stokes flux function for our current loop is given by the rather cumbersome integral

$$\Psi(s,z) = \frac{\mu_0 IRs}{4\pi}\int_0^{2\pi} \frac{\cos\tilde{\theta}}{\sqrt{|\mathbf{x}|^2 + R^2 - 2sR\cos\tilde{\theta}}}\, d\tilde{\theta}\ .$$

This is most conveniently tidied up through the use of elliptic integrals. In particular, introducing

$$k^2 = \frac{4Rs}{z^2 + (s+R)^2},$$

we find (see Exercise 7.8)

$$\Psi(s,z) = \frac{\mu_0 I\sqrt{Rs}}{2\pi}\left[\left(\frac{2}{k} - k\right)K(k) - \frac{2}{k}E(k)\right], \tag{7.41}$$

where $K(k)$ and $E(k)$ are the complete elliptic integrals of the first and second kind, whose properties are well documented (e.g. in Abramowitz and Stegun, 1965). (Those readers too impatient to attempt Exercise 7.8 will find a derivation of (7.41) in Jackson (1999), although he uses spherical polar, rather than cylindrical polar, coordinates. Our discussion follows Batchelor (1967), who derives (7.41) in the context of vortex rings.)

The expression (7.41) yields the magnetic field at all points in space, but of particular interest are (i) the asymptotic form at large distances from the current loop and (ii) the field on the axis of the loop. Both of these cases correspond to $k \ll 1$. The low-$k$ expansions of $K(k)$ and $E(k)$ are

$$K(k) = \frac{\pi}{2}\left(1 + \frac{k^2}{4} + \frac{9}{64}k^4 + \cdots\right),$$

$$E(k) = \frac{\pi}{2}\left(1 - \frac{k^2}{4} - \frac{3}{64}k^4 + \cdots\right),$$

and so for small $k$ we find

$$\Psi = \frac{\mu_0\,|\mathbf{m}|}{4\pi}\frac{s^2}{[z^2+(s+R)^2]^{3/2}}.$$

On the axis of the loop this yields

$$B_z = \frac{1}{s}\frac{\partial\Psi}{\partial s} = \frac{\mu_0\,|\mathbf{m}|}{2\pi}\frac{1}{(z^2+R^2)^{3/2}},$$

which is (7.23), while for large distances $|\mathbf{x}| >> R$ and for any $s$, the vector potential is

$$\mathbf{A} = \frac{\Psi}{s}\hat{\mathbf{e}}_\theta = \frac{\mu_0}{4\pi}\frac{|\mathbf{m}|\,s}{|\mathbf{x}|^3}\hat{\mathbf{e}}_\theta.$$

This is normally rewritten in the coordinate-free form

$$\boxed{\mathbf{A} = \frac{\mu_0}{4\pi}\frac{\mathbf{m}\times\mathbf{x}}{|\mathbf{x}|^3} = -\frac{\mu_0}{4\pi}\mathbf{m}\times\nabla\left(\frac{1}{|\mathbf{x}|}\right),} \tag{7.42}$$

and the corresponding magnetic field for $|\mathbf{x}| >> R$ is evidently

$$\boxed{\mathbf{B} = \nabla\times\mathbf{A} = \frac{\mu_0}{4\pi}(\mathbf{m}\cdot\nabla)\nabla\left(\frac{1}{|\mathbf{x}|}\right),} \tag{7.43}$$

or in component form

$$\mathbf{B} = \frac{\mu_0 m}{4\pi\,|\mathbf{x}|^3}\left(\frac{3sz}{|\mathbf{x}|^2}\hat{\mathbf{e}}_s + \frac{3z^2-|\mathbf{x}|^2}{|\mathbf{x}|^2}\hat{\mathbf{e}}_z\right).$$

The structure of this dipolar field is shown in Figure 7.16.

The key result here is (7.42) and (7.43). We shall see in Chapter 8 that *any* steady localised current distribution, not just a circular current loop, has a far-field vector potential given by (7.42), the only restriction being that the dipole moment of the current distribution, defined as

$$\mathbf{m} = \frac{1}{2}\int \mathbf{x}\times\mathbf{J}\,dV,$$

is finite. Comparing (7.43) with Exercise 4.3, we see that electric and magnetic dipoles have exactly the same far-field structure. It is necessary only to replace $\mathbf{p}/\varepsilon_0$ by $\mu_0\mathbf{m}$. By contrast,

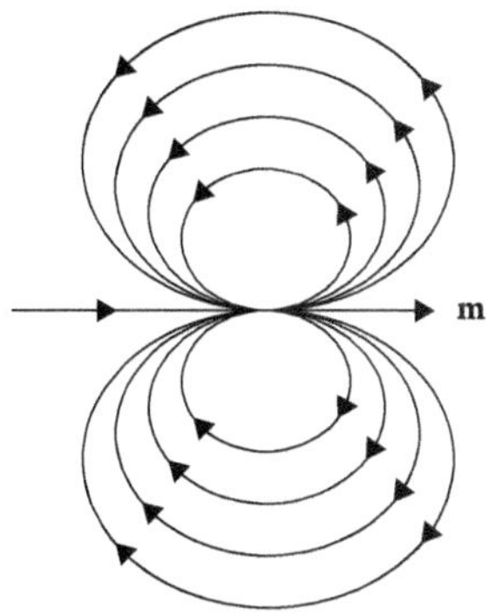


**Figure 7.16** The dipole magnetic field at large distances from a circular current loop.

the near fields are very different, and in any event depend on the particular arrangement of charges or currents used to construct the electrostatic or magnetic dipole.

### *7.6.2 The Dipole Moment as a Local Measure of the Mean Magnetic Field*

In §5.3.1 we saw that, when charges are confined to a spherical volume of radius $R$, say $V_R$, the integral of the electric field over that volume is proportional to the electric dipole moment of the charge distribution, $\mathbf{p}$:

$$\int_{V_R} \mathbf{E}(\mathbf{x})\, d\mathbf{x} = -\frac{\mathbf{p}}{3\varepsilon_0}, \quad \mathbf{p} = \int_{V_R} \mathbf{x}\rho_e(\mathbf{x})\, d\mathbf{x}. \tag{7.44}$$

This result is quite general in the sense that it holds for any distribution of charge, the only requirement being that all of the charge is confined to $V_R$. We shall see in Chapter 8 that there is an equivalent result for currents confined to a sphere, where it is necessary only to replace $\mathbf{E}$ by $\mathbf{B}$ and $-\mathbf{p}/\varepsilon_0$ by $2\mu_0\mathbf{m}$:

$$\int_{V_R} \mathbf{B}(\mathbf{x})\, d\mathbf{x} = \frac{2}{3}\mu_0\mathbf{m}, \quad \mathbf{m} = \frac{1}{2}\int_{V_R} \mathbf{x} \times \mathbf{J}\, d\mathbf{x}. \tag{7.45}$$

Again this result is entirely general, valid for any localised distribution of current, the only restriction being that $\mathbf{J}$ must be confined to the sphere $V_R$ (Figure 7.17).

At first sight the minus sign in the exchange $\mathbf{p}/\varepsilon_0 \leftrightarrow -2\mu_0\mathbf{m}$ is surprising, given that the far-field structures of electric and magnetic dipoles are equivalent under the exchange $\mathbf{p}/\varepsilon_0 \leftrightarrow \mu_0\mathbf{m}$. However, the volume integrals on the left of (7.44) and (7.45) are dominated by the near field and not by the far field. (This is evident from the fact that the radius of $V_R$ may be varied arbitrarily in (7.44) and (7.45) and this makes no difference to the value of

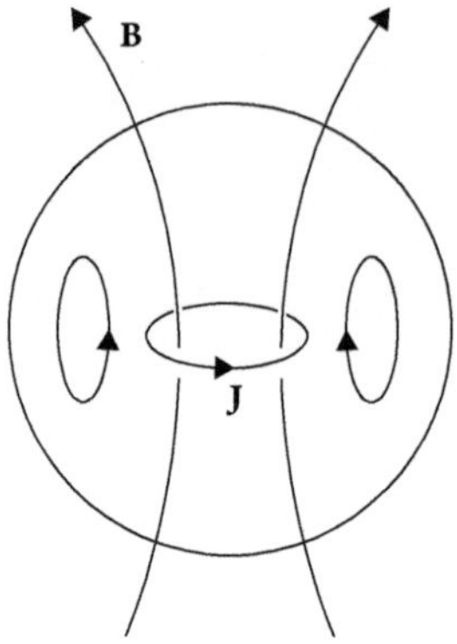


**Figure 7.17** The magnetic field generated by currents confined to a sphere of radius $R$.

the integrals of **E** or **B**, as long as the volume encloses all of the charge or current.) Moreover, it is clear from a comparison of Figures 7.15 and 5.1(b) that while **B** and **m** are parallel in the near field, **E** and **p** are antiparallel. So the minus sign is to be expected.

In any event, (7.45) gives us a convenient measure of the mean magnetic field within $V_R$:

$$\bar{\mathbf{B}} = \frac{1}{V_R}\int_{V_R} \mathbf{B}(\mathbf{x})\, d\mathbf{x} = \frac{2}{3}\frac{\mu_0 \mathbf{m}}{V_R}.$$

This relationship between $\bar{\mathbf{B}}$ and the magnetic dipole moment proves particularly useful when it comes to estimating the mean magnetic field in the conducting cores of planets.

## 7.7 Another Aside: Planetary Magnetic Fields and the Dipole-Rotation Scaling Law

The magnetic field of the earth is not the result of permanent magnetism, since the temperature in the earth's interior is well above the Curie point at which ferromagnetic material loses such magnetism. Indeed, the temperature at the centre of the earth is similar to that at the surface of the sun. Nor is the earth's magnetic field simply the relic of some primordial field trapped within the interior during its formation, since that would have decayed long ago through Ohmic dissipation. (As discussed in §2.2.3, such a field decays on a timescale of $t \sim R_C^2/\pi^2\lambda \sim 10^4$ years, where $R_C$ is the radius of the conducting core and $\lambda = 1/\sigma\mu_0$ the magnetic diffusivity.) Rather, planetary magnetic fields are maintained, despite Joule dissipation, through the convective action of their conducting fluid interiors, as indicated in Figure 7.18.

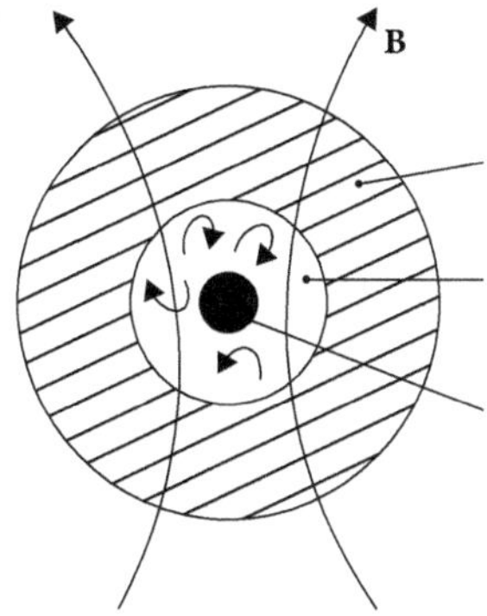


**Figure 7.18** Motion in the earth's core maintains the terrestrial magnetic field.

In brief, in the core of the earth there is a self-excited fluid dynamo that converts mechanical energy into magnetic energy, with motion of the conducing fluid across the magnetic field lines generating currents that in turn maintain the original magnetic field. In terms of energetics, (3.44) tells us that the energy of the earth's magnetic field may be maintained by convection provided that

$$\int_{V_C} \mathbf{u}\cdot(\mathbf{J}\times\mathbf{B})\,dV = -\frac{1}{\sigma}\int_{V_C} \mathbf{J}^2 dV,$$

where $V_C$ is the volume of the conducting core. That is, the rate of working of the Lorentz force on the conductor must be negative, so that the conductor does work on the magnetic field, and at a rate that exactly compensates for the Joule dissipation. The precise manner by which this occurs is still not entirely understood, although plausible cartoons have been put forward, supported by numerical simulations (see Chapter 21).

In fact, it is not just the earth that maintains a magnetic field through dynamo action in its core. Mercury, Ganymede (Jupiter's largest moon), Jupiter, Saturn, Uranus and Neptune all have magnetic fields that probably owe their existence to fluid dynamos within their cores. (Mars exhibits some remnant magnetism suggesting that it too once had a fluid dynamo, now extinct, while Venus has yet to develop a magnetic field.) Interestingly, the external magnetic fields of the terrestrial planets (Mercury and the earth) and the gas giants (Jupiter and Saturn) are all remarkably similar, being dipolar and with a magnetic axis more or less aligned with the rotation axis. (By contrast, the observable magnetic fields of the ice giants, Uranus and Neptune, are more complicated, with strongly tilted magnetic axes.) Even the intensity of the magnetic fields within these planets, when suitably normalised, are very similar, as discussed below. This similarity is, perhaps, surprising since the sizes and structures of these planets are quite different, as shown in Figure 7.19 and Table 7.1. For example, the terrestrial planets have iron cores and rocky mantles, while the gas giants are composed of hydrogen and helium, with the high pressure at their centres creating metallic hydrogen, which is conducting.

There is much that we do not know about the interior of the planets, but the three things that can be measured with some degree of confidence are the size of the planet,

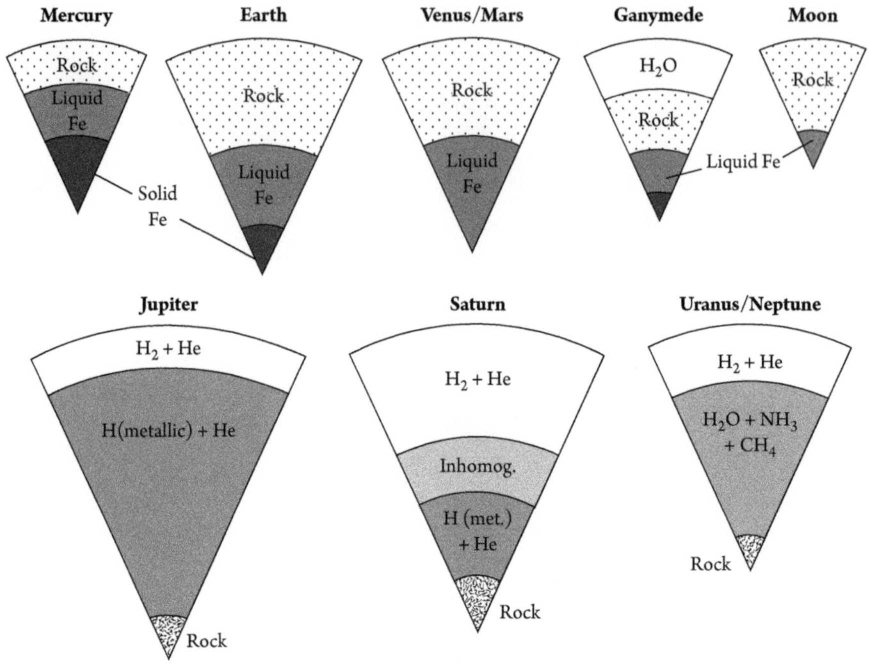


**Figure 7.19** Approximate structure of the planets. (Figure courtesy of Uli Christensen.)

from which we can infer the size of the conducting core, $R_C$, how fast the planet rotates, $\Omega$, and the dipole moment of the magnetic field, $\mathbf{m}$. From the dipole moment we can calculate the mean axial field in the conducting core, $\bar{B}_z$, using (7.45). It order to compare the magnetic fields across the various planets, we must agree on a suitably normalised version of $\bar{B}_z$, and, given the measured values of $R_C$ and $\Omega$, it is natural to normalise $\bar{B}_z$ in the *dimensionless* form

$$\Pi_B = \frac{\bar{B}_z/\sqrt{\rho\mu_0}}{\Omega R_C},$$

where $\rho$ is the density of the fluid in the conducting core. Estimates of $\Pi_B$ for the planets in our solar system are given in the final column of Table 7.1, where we have included for comparison a fully convective low-mass M-dwarf star, which is thought to have a dynamo similar to that of Jupiter. It is extraordinary how uniform $\Pi_B$ is across all of these astrophysical objects (Davidson, 2016), despite the wide range of sizes, rotation rates and interior structures. The reasons for this uniformity are still poorly understood and we shall return to planetary dynamos in Chapter 21.

**Table 7.1** Properties of the planets. The mean axial magnetic fields in the conducting cores, $\bar{B}_z$, are obtained using (7.45)

| **Planet/Star** | **Rotation period (days)** | **Radius of planet/star ($10^3$ km)** | **Core radius $R_C$ ($10^3$ km)** | **Dipole moment ($10^{22}$ A m$^2$)** | **Mean $B_z$ in core (G)** | **Angle of dipole** | $\frac{\bar{B}_z/\sqrt{\rho\mu_0}}{\Omega R_C} \times 10^6$ |
|---|---|---|---|---|---|---|---|
| Mercury | 58.6 | 2.44 | 1.8 | 0.004 | 0.014 | 5° | 5.5 |
| Earth | 1 | 6.37 | 3.48 | 7.9 | 3.7 | 11° | 13 |
| Jupiter | 0.413 | 69.9 | 55 | 150 000 | 18 | 9.6° | 5.2 |
| Saturn | 0.440 | 58.2 | 29 | 4600 | 3.8 | <0.1° | 2.3 |
| Uranus | 0.718 | 25.3 | ~18 | 390 | ~1.3 | 59° | ~1.0 |
| Neptune | 0.671 | 24.6 | ~20 | 200 | ~0.5 | 47° | ~0.3 |
| V374 Pegasi | 0.44 | 200 | — | — | ~800 | — | ~17 |

Source: P. A. Davidson. *An Introduction to Magnetohydrodynamics*, 2nd ed. Cambridge University Press, 2017.

## 7.8 The Mathematical Origins of the Biot–Savart Law

We have yet to show that the Biot–Savart law is equivalent to Ampère's law (plus $\nabla \cdot \mathbf{B} = 0$). We remedy that omission now.

### 7.8.1 *The Biot–Savart Law as a Solution of Poisson's Equation*

Although the Biot–Savart law was originally determined experimentally, it is now usually thought of as the Green's function solution of Poisson's equation, as discussed in §2.4.3. For example, taking the curl of Ampère's law and invoking the solenoidal nature of **B** yields the Poisson equation

$$\nabla^2 \mathbf{B} = -\mu_0 \nabla \times \mathbf{J}. \tag{7.46}$$

Compare this with Gauss' law written in terms of the electrostatic potential,

$$\nabla^2 \varphi_s = -\frac{\rho_e}{\varepsilon_0}, \tag{7.47}$$

whose solution in an infinite domain is (4.19):

$$\varphi_s(\mathbf{x}) = \frac{1}{4\pi\varepsilon_0} \int \frac{\rho_e(\mathbf{x}')}{|\mathbf{r}|} \, d\mathbf{x}', \quad \mathbf{r} = \mathbf{x} - \mathbf{x}'. \tag{7.48}$$

Evidently the solution of (7.46) in an infinite domain is

$$\mathbf{B}(\mathbf{x}) = \frac{\mu_0}{4\pi} \int \frac{\nabla' \times \mathbf{J}(\mathbf{x}')}{|\mathbf{r}|} \, d\mathbf{x}', \qquad \mathbf{r} = \mathbf{x} - \mathbf{x}'. \tag{7.49}$$

Both (7.48) and (7.49) are examples of the Green's function solution of Poisson's equation, as discussed in §2.4. To link back to the Biot–Savart law, we first note that

$$\frac{\nabla' \times \mathbf{J}'}{|\mathbf{r}|} = \nabla' \times \left(\frac{\mathbf{J}'}{|\mathbf{r}|}\right) - \nabla'\left(\frac{1}{|\mathbf{r}|}\right) \times \mathbf{J}',$$

where $\nabla'$ operates on $\mathbf{x}'$ while treating $\mathbf{x}$ as constant. Moreover,

$$\nabla\left(\frac{1}{|\mathbf{r}|}\right) = -\frac{\mathbf{r}}{|\mathbf{r}|^3} = -\nabla'\left(\frac{1}{|\mathbf{r}|}\right), \tag{7.50}$$

from which,

$$\frac{\nabla' \times \mathbf{J}'}{|\mathbf{r}|} = \nabla' \times \left(\frac{\mathbf{J}'}{|\mathbf{r}|}\right) + \frac{\mathbf{J}' \times \mathbf{r}}{|\mathbf{r}|^3}. \tag{7.51}$$

We now use (7.51) to substitute for $\nabla' \times \mathbf{J}'$ in the Green's function solution (7.49). The first term on the right of (7.51) converts to a surface integral, which is zero since **J** is assumed to be localised in space, while the second term converts (7.49) into the Biot–Savart law (7.17).

An alternative, though ultimately equivalent, derivation of the Biot–Savart law uses the vector potential. We start by using $\nabla \cdot \mathbf{B} = 0$ to invoke a vector potential $\mathbf{A}$ and then substitute $\mathbf{A}$ for $\mathbf{B}$ in Ampère's law. Adopting the Coulomb gauge for $\mathbf{A}$ (i.e. $\nabla \cdot \mathbf{A} = 0$) then yields the Poisson equation

$$\nabla^2 \mathbf{A} = -\mu_0 \mathbf{J}. \tag{7.52}$$

From (7.48) we know that the solution of (7.52) in an infinite domain is

$$\mathbf{A}(\mathbf{x}) = \frac{\mu_0}{4\pi} \int \frac{\mathbf{J}(\mathbf{x}')}{|\mathbf{r}|} d\mathbf{x}', \qquad \mathbf{r} = \mathbf{x} - \mathbf{x}', \tag{7.53}$$

and we have already shown in §7.3.2 that (7.53) is equivalent to the Biot–Savart law.

So we have two ways of getting from the Green's function solution (7.48) to the Biot–Savart law. Of the two, the route involving the vector potential will prove more useful when we come to generalise the Biot–Savart law to time-dependent problems.

### 7.8.2 *Confirmation that the Biot–Savart Law Yields the Correct Values of $\nabla \cdot \mathbf{B}$ and $\nabla \times \mathbf{B}$*

We may confirm retrospectively that the Biot–Savart returns the correct values of both $\nabla \cdot \mathbf{B}$ and $\nabla \times \mathbf{B}$, i.e. it enforces $\nabla \cdot \mathbf{B} = 0$ and $\nabla \times \mathbf{B} = \mu_0 \mathbf{J}$. For example, using

$$\nabla \left( \frac{1}{|\mathbf{r}|} \right) = -\frac{\mathbf{r}}{|\mathbf{r}|^3},$$

the Biot–Savart law may be rewritten in the form

$$\mathbf{B}(\mathbf{x}) = \frac{\mu_0}{4\pi} \int \nabla \left( \frac{1}{|\mathbf{r}|} \right) \times \mathbf{J}' \, d\mathbf{x}', \qquad \mathbf{r} = \mathbf{x} - \mathbf{x}', \tag{7.54}$$

from which

$$\nabla \cdot \mathbf{B} = \frac{\mu_0}{4\pi} \int \nabla \cdot \left[ \nabla \left( \frac{1}{|\mathbf{r}|} \right) \times \mathbf{J}' \right] d\mathbf{x}'. \tag{7.55}$$

However the two identities,

$$\nabla \cdot (\mathbf{c} \times \mathbf{d}) = \mathbf{d} \cdot \nabla \times \mathbf{c} - \mathbf{c} \cdot \nabla \times \mathbf{d}$$

and $\nabla \times [\nabla(\cdot)] = 0$, yield

$$\nabla \cdot \left[ \nabla \left( \frac{1}{|\mathbf{r}|} \right) \times \mathbf{J}' \right] = \mathbf{J}' \cdot \left[ \nabla \times \nabla \left( \frac{1}{|\mathbf{r}|} \right) \right] - \nabla \left( \frac{1}{|\mathbf{r}|} \right) \cdot (\nabla \times \mathbf{J}') = 0,$$

thus ensuring that $\nabla \cdot \mathbf{B} = 0$. (Note that $\nabla \times \mathbf{J}' = 0$ here because $\nabla$ operates on $\mathbf{x}$ while keeping $\mathbf{x}'$ constant.)

The related proof that the curl of the Biot–Savart law ensures $\nabla \times \mathbf{B} = \mu_0 \mathbf{J}$ proceeds along similar lines, although it is a little more intricate and requires the use of (2.26),

$$\nabla^2 \left( \frac{1}{|\mathbf{r}|} \right) = -4\pi\,\delta(\mathbf{x} - \mathbf{x}'), \tag{7.56}$$

where $\delta(\mathbf{x} - \mathbf{x}')$ is the three-dimensional Dirac delta function. (The derivation of (7.56) is discussed in §2.1.4.) The proof starts by noting that the curl of (7.54) yields

$$\nabla \times \mathbf{B} = \frac{\mu_0}{4\pi} \int \nabla \times \left[ \nabla \left( \frac{1}{|\mathbf{r}|} \right) \times \mathbf{J}' \right] d\mathbf{x}'. \tag{7.57}$$

Next, using standard vector identities, the integrand may be expanded to give

$$\begin{aligned} \nabla \times \left[ \nabla \left( \frac{1}{|\mathbf{r}|} \right) \times \mathbf{J}' \right] &= (\mathbf{J}' \cdot \nabla) \nabla \left( \frac{1}{|\mathbf{r}|} \right) - \mathbf{J}' \nabla^2 \left( \frac{1}{|\mathbf{r}|} \right) \\ &= -(\mathbf{J}' \cdot \nabla') \nabla \left( \frac{1}{|\mathbf{r}|} \right) + 4\pi\,\delta(\mathbf{x} - \mathbf{x}') \mathbf{J}', \end{aligned}$$

where we have used (7.56) and $(\mathbf{J}' \cdot \nabla)\nabla(1/|\mathbf{r}|) = -(\mathbf{J}' \cdot \nabla')\nabla(1/|\mathbf{r}|)$. Let us rewrite this as

$$\nabla \times \left[ \nabla \left( \frac{1}{|\mathbf{r}|} \right) \times \mathbf{J}' \right] = -\nabla' \cdot \left[ \nabla \left( \frac{1}{|\mathbf{r}|} \right) \mathbf{J}' \right] + \nabla \left( \frac{\nabla' \cdot \mathbf{J}'}{|\mathbf{r}|} \right) + 4\pi\,\delta(\mathbf{x} - \mathbf{x}') \mathbf{J}'.$$

The divergence on the right integrates to zero using Gauss' theorem combined with the fact that $\mathbf{J}$ is localised in space. Moreover, noting that $\int \mathbf{J}'\,\delta(\mathbf{x} - \mathbf{x}')\,d\mathbf{x}' = \mathbf{J}(\mathbf{x})$, the remaining terms integrate to give

$$\nabla \times \mathbf{B} = \mu_0 \mathbf{J}(\mathbf{x}) + \frac{\mu_0}{4\pi} \nabla \left( \int \frac{\nabla' \cdot \mathbf{J}'}{|\mathbf{r}|} d\mathbf{x}' \right). \tag{7.58}$$

Finally, since $\nabla \cdot \mathbf{J} = 0$ in magnetostatics, we recover $\nabla \times \mathbf{B} = \mu_0 \mathbf{J}$.

## 7.9 From Statics to Dynamics: A Second Glimpse at Retarded Potentials

We have seen that, given a steady localised distribution of $\rho_e$ or of $\mathbf{J}$, the electrostatic scalar potential $\varphi_s$ or magnetostatic vector potential $\mathbf{A}$ is given by

$$\varphi_s(\mathbf{x}, t) = \frac{1}{4\pi\varepsilon_0} \int \frac{\rho_e(\mathbf{x}', t)}{|\mathbf{r}|} d\mathbf{x}', \tag{7.59}$$

or

$$\mathbf{A}(\mathbf{x},t) = \frac{\mu_0}{4\pi}\int \frac{\mathbf{J}(\mathbf{x}',t)}{|\mathbf{r}|}\,d\mathbf{x}', \tag{7.60}$$

where $\mathbf{r} = \mathbf{x} - \mathbf{x}'$. As discussed in §7.8.1, these may be interpreted as the Green's function solutions of the static Poisson equations (7.47) and (7.52):

$$\nabla^2\varphi_s = -\frac{\rho_e}{\varepsilon_0}, \qquad \nabla^2\mathbf{A} = -\mu_0\mathbf{J}. \tag{7.61}$$

Note that we have explicitly included time in (7.59) and (7.60) to remind the reader that, for quasi-steady sources, $\varphi_s$ and $\rho_e$ are evaluated at the same instant, as are $\mathbf{A}$ and $\mathbf{J}$.

Of course, it is clear that these expressions cannot hold when the sources $\rho_e$ and $\mathbf{J}$ are time-dependent, because they imply that changes in $\rho_e$ and $\mathbf{J}$ are immediately communicated to all points in space, resulting in instantaneous changes in $\varphi_s$ and $\mathbf{A}$. In reality, information can propagate no faster than the speed of light, so there must be a time lag between fluctuations in $\rho_e$ and $\mathbf{J}$ at location $\mathbf{x}'$ and the resulting changes in $\varphi_s$ and $\mathbf{A}$ at position $\mathbf{x}$. Indeed, we noted that trouble was brewing at the beginning of §7.3.1, when we observed that Ampère's law does not, in general, hold in time-dependent problems because it demands $\nabla\cdot\mathbf{J} = 0$, which is inconsistent with the charge conservation equation

$$\nabla\cdot\mathbf{J} = -\frac{\partial\rho_e}{\partial t} \tag{7.62}$$

in those cases where $\partial\rho_e/\partial t$ plays any significant role. Since we have interpreted (7.60) as the Green's function solution of Ampère's law (plus $\nabla\cdot\mathbf{B} = 0$), it follows that a modification to Ampère's law must result in a corresponding change to (7.60). So the question now arises: how should the expressions (7.59) and (7.60) be modified when the sources are time-dependent?

We discussed this briefly in §3.6 and will offer a more detailed discussion in Chapter 13, but perhaps it is worth recalling the key ideas now. There are three changes that occur as we move away from statics. The first is that Ampère's law is generalised to the Ampère–Maxwell equation

$$\nabla\times\mathbf{B} = \mu_0\left(\mathbf{J} + \varepsilon_0\frac{\partial\mathbf{E}}{\partial t}\right). \tag{7.63}$$

Invoking Gauss' law, $\nabla\cdot\mathbf{E} = \rho_e/\varepsilon_0$, it is evident that the divergence of (7.63) is consistent with the unsteady charge conservation equation (7.62), unlike Ampère's law. Second, Faraday's law,

$$\nabla\times\mathbf{E} = -\frac{\partial\mathbf{B}}{\partial t},$$

may be uncurled to give

$$\mathbf{E} = -\frac{\partial\mathbf{A}}{\partial t} - \nabla\varphi, \tag{7.64}$$

for some function $\varphi$. This is then used to define an electric potential $\varphi$, which is a generalisation of the electrostatic potential $\varphi_s$. The third, and final, step is to abandon the Coulomb gauge for the vector potential, $\nabla \cdot \mathbf{A} = 0$, and replace it by the *Lorenz gauge*,

$$\nabla \cdot \mathbf{A} = -\frac{1}{c^2}\frac{\partial \varphi}{\partial t}, \tag{7.65}$$

where $c = 1/\sqrt{\varepsilon_0 \mu_0}$ is the speed of light. Notice that our new expressions for $\nabla \times \mathbf{B}$, $\nabla\varphi$ and $\nabla \cdot \mathbf{A}$ in (7.63)–(7.65) reduce to their static counterparts when all time derivatives are set to zero.

While (7.63) and (7.64) are, perhaps, natural extensions of their static counterparts, the Lorenz gauge, (7.65), seems to have appeared like a rabbit from a hat! Where did it come from? We discussed this briefly in §3.6. In principle, we can choose the gauge for $\mathbf{A}$ any way we see fit, although if we change the gauge for $\mathbf{A}$, then the form of $\varphi$ in (7.64) must alter in such a way that $\mathbf{E}$ remains unchanged, as discussed in Exercise 7.9. However, the Lorenz gauge is particularly convenient since it is the *only* choice that yields *decoupled* wave equations for the vector and scalar potentials, as shown below. So one way to think about (7.65) is that it provides a convenient simplification in the governing equations for the scalar and vector potentials.

Let us see how this works out. Taking the divergence of (7.64) and substituting for $\nabla \cdot \mathbf{A}$ and $\nabla \cdot \mathbf{E}$ using the Lorenz gauge and Gauss' law yields a *wave equation* for $\varphi$:

$$\boxed{\nabla^2 \varphi - \frac{1}{c^2}\frac{\partial^2 \varphi}{\partial t^2} = -\frac{\rho_e}{\varepsilon_0}.} \tag{7.66}$$

(The Coulomb gauge, by contrast, takes us back to the Poisson equation (7.61), so that $\varphi$ then becomes the electrostatic potential $\varphi_s$.) Moreover, substituting $\mathbf{A}$ for $\mathbf{B}$ in (7.63), rewriting $\nabla \times \nabla \times \mathbf{A}$ in terms of $\nabla^2 \mathbf{A}$ using the identity (1.62), and evaluating $\partial \mathbf{E}/\partial t$ using (7.64) yields

$$\nabla^2 \mathbf{A} - \frac{1}{c^2}\frac{\partial^2 \mathbf{A}}{\partial t^2} = -\mu_0 \mathbf{J} + \nabla\left(\nabla \cdot \mathbf{A} + \frac{1}{c^2}\frac{\partial \phi}{\partial t}\right).$$

The Lorenz gauge then gives us

$$\boxed{\nabla^2 \mathbf{A} - \frac{1}{c^2}\frac{\partial^2 \mathbf{A}}{\partial t^2} = -\mu_0 \mathbf{J}.} \tag{7.67}$$

So the judicious choice of the Lorenz gauge has yielded *decoupled wave equations* for $\varphi$ and $\mathbf{A}$. These two wave equations might be compared with their static analogues (7.61). Evidently, we have gone from Poisson equations to inhomogeneous wave equations in which information propagates at the *finite speed c*. Indeed, we might think of (7.61) as being the wave equations (7.66) and (7.67) in the limit of $c \to \infty$.

These wave equations make it immediately clear that information propagates away from the sources $\rho_e$ and $\mathbf{J}$ at the speed of light, carried away by electromagnetic waves. Moreover, we learned how to invert such inhomogeneous wave equations in §2.5.1, albeit in the context of acoustic waves. Equations (7.66) and (7.67) may be similarly inverted, and this tells us that their solutions in an infinite domain are

$$\boxed{\varphi(\mathbf{x},t) = \frac{1}{4\pi\varepsilon_0}\int \frac{\rho_e(\mathbf{x}',t')}{|\mathbf{r}|}\,d\mathbf{x}',} \tag{7.68}$$

$$\boxed{\mathbf{A}(\mathbf{x},t) = \frac{\mu_0}{4\pi}\int \frac{\mathbf{J}(\mathbf{x}',t')}{|\mathbf{r}|}\,d\mathbf{x}',} \tag{7.69}$$

where $t' = t - r/c$ is called the *retarded time,* and $\varphi$ and $\mathbf{A}$ are known as the *retarded potentials*. Comparing (7.59) and (7.60) with (7.68) and (7.69), we see that all we need do is replace $t$ by $t'$ at the source, which is a remarkably simple and convenient result! Of course, this reflects the fact that information can be transmitted no faster than it can be carried off in the form of waves, and so a fluctuation in $\rho_e$ or $\mathbf{J}$ at location $\mathbf{x}'$ is not felt instantaneously at the remote position $\mathbf{x}$, but rather a time $|\mathbf{r}|/c$ later. We shall return to the idea of retarded potentials time and again in subsequent chapters, where we shall see that they constitute a powerful approach to electromagnetic wave propagation.

## EXERCISES

7.1 A frictionless metal slide moves with velocity $\mathbf{u} = u\hat{\mathbf{e}}_x$ along two conducting rails in the presence of an imposed uniform magnetic field $\mathbf{B} = B_0\hat{\mathbf{e}}_z$, as shown in Figure 7.20. The slide and rails are made of the same material and have density $\rho$, conductivity $\sigma$ and a common thickness $T$. The slide and rails have unit depth into the page, the rails are separated by a distance $W$, and the instantaneous position of the slide is a distance $L(t)$ from the end of the rails.

The motion of the slide through the magnetic field induces a force on the free charges, $q\mathbf{u}_q \times \mathbf{B}$, which in turn drives a current through the slide of $\mathbf{J} = -|\mathbf{J}|\,\hat{\mathbf{e}}_y$. Since the slide and rails have a common thickness, the current density has the same

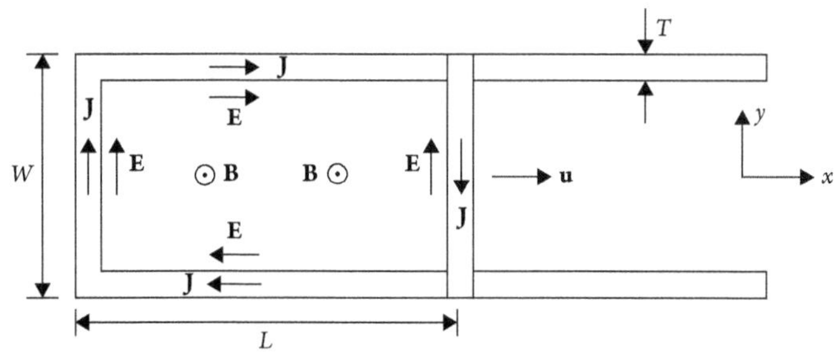


**Figure 7.20** A frictionless metal slide moves along two conducting rails in the presence of a uniform magnetic field.

magnitude in both. The magnitude of the current is sufficiently low that the imposed field **B** is negligibly perturbed, and since **B** is (almost) constant, Faraday's law gives $\nabla \times \mathbf{E} \approx 0$.

Confirm that the direction of the electric field in the rails must be as shown, and use $\nabla \times \mathbf{E} = 0$ to show that the direction of the electric field in the slide is such as to oppose the flow of current. The generalised version of Ohm's law, (7.9), gives in this case

$$\oint \mathbf{J} \cdot d\mathbf{r} = \sigma \oint \mathbf{u} \times \mathbf{B} \cdot d\mathbf{r}.$$

Use this to confirm that the current density in, and Lorentz force acting on, the slide are given by

$$\mathbf{J} = -\frac{\sigma\, u B_0 W}{2(L+W)}\hat{\mathbf{e}}_y, \quad \mathbf{J} \times \mathbf{B} = -\frac{\sigma B_0^2 W}{2(L+W)}\mathbf{u}.$$

If the slide has an initial velocity $u_0$, and changes in $L$ are small, $L \approx L_0$, confirm that the slide decelerates according to

$$u = u_0 e^{-t/\tau}, \quad \frac{1}{\tau} = \frac{\sigma B_0^2}{\rho} \frac{W}{2(L_0+W)}.$$

Conservation of energy tells us that this loss in kinetic energy must reappear as heat in the slide and rails.

7.2 Show that, for the slide and rails in Exercise 7.1, the Joule dissipation rate and the rate of working of the Lorentz force are given by

$$\int \frac{\mathbf{J}^2}{\sigma} dV = \frac{(\rho W T)u^2}{\tau} = -\frac{d}{dt}\frac{(\rho W T)u^2}{2}, \quad \int \mathbf{u} \cdot (\mathbf{J} \times \mathbf{B})\, dV = -\frac{(\rho W T)u^2}{\tau}.$$

The first of these expressions confirms that the lost kinetic energy reappears as heat. Now compare these two expressions with the energy equation (7.11):

$$\mathbf{J} \cdot \mathbf{E} = \frac{\mathbf{J}^2}{\sigma} + \mathbf{u} \cdot (\mathbf{J} \times \mathbf{B}).$$

Evidently, the rate of working of the electric field on the slide and rails integrates to zero, $\int \mathbf{J} \cdot \mathbf{E}\, dV = 0$. Show that this is a consequence of $\nabla \times \mathbf{E} = 0$ and of continuity of current:

$$\int \mathbf{J} \cdot \mathbf{E}\, dV = I \oint \mathbf{E} \cdot d\mathbf{r} = 0.$$

7.3 Now apply the energy equation (7.11) to the rails and slide separately. Show that, in the rails, the rate of working of the electric field on the free charges, $\mathbf{J} \cdot \mathbf{E}$, is positive and equal to the Joule dissipation rate, while, in the slide, the rate of working of the electric field is negative. Show that, in the slide, $|\mathbf{u} \cdot (\mathbf{J} \times \mathbf{B})|$ is equal to the sum of $|\mathbf{J} \cdot \mathbf{E}|$ and the Joule dissipation.

7.4 Consider the current ring of radius $R$ and current $I$ shown in Figure 7.7. Use the Biot–Savart law with cylindrical polar coordinates $(s,\ \theta,\ z)$ centred on the loop to show that, on the axis of the ring,

$$\mathbf{B}(s=0) = \frac{\mu_0}{4\pi}\int_0^{2\pi} \frac{(I\hat{\mathbf{e}}'_\theta)\times(z\hat{\mathbf{e}}_z - R\hat{\mathbf{e}}'_s)}{(R^2+z^2)^{3/2}} R\,d\theta' = \frac{\mu_0 I R^2}{2(R^2+z^2)^{3/2}}\hat{\mathbf{e}}_z.$$

7.5 Consider a solenoid of length $L$, radius $R$, $N$ turns per unit length, and carrying a current $I$. Use the result from Exercise 7.4 to show that the centreline magnetic field at the midpoint of the solenoid is

$$B_z = \frac{\mu_0 N I}{2}\frac{L}{\sqrt{R^2+(L/2)^2}}$$

and that at the ends is

$$B_z = \frac{\mu_0 N I}{2}\frac{L}{\sqrt{R^2+L^2}}.$$

7.6 Consider a long straight metal rod of radius $R$ and carrying a uniform axial current $I$. Show that in cylindrical polar coordinates the magnetic field and Lorentz force vary with distance from the centre of the rod according to

$$B_\theta = \frac{\mu_0 I s}{2\pi R^2}, \qquad \mathbf{J}\times\mathbf{B} = -\frac{2B_\theta^2}{\mu_0 s}\hat{\mathbf{e}}_s; \quad s \le R.$$

Find an expression for the compressive stress at the centre of the rod in terms of the magnetic field at the edge of the rod, $B_\theta(s=R)$.

7.7 Consider a toroidal coil consisting of a long thin solenoid of radius $a$ and current $I$ whose axis forms a circle of radius $R$, with $R \gg a$. Cylindrical polar coordinates $(s,\ \theta,\ z)$ are centred on the torus such that its major axis is $s = R$. Show that the magnetic field within the interior of the toroidal coil is $\mu_0 N I/2\pi s$, where $N$ is the total number of turns.

7.8 The derivation of the Stokes flux function (7.41) for current rings starts with the Biot–Savart integral

$$\Psi = \frac{\mu_0 I}{4\pi} R s \int_0^{2\pi} \frac{\cos\theta}{\sqrt{z^2+s^2+R^2-2sR\cos\theta}}\,d\theta = \frac{\mu_0 I}{4\pi}\Sigma.$$

Show that the denominator in the integrand can be rewritten as

$$\sqrt{z^2+s^2+R^2-2sR\cos\theta} = \frac{2\sqrt{Rs}}{k}\sqrt{1-k^2\cos^2(\theta/2)}, \qquad k^2 = \frac{4Rs}{z^2+(s+R)^2}.$$

Hence confirm that

$$\Sigma = \sqrt{Rs}\int_0^{\pi}\left[\left(\frac{2}{k}-k\right)\frac{1}{\sqrt{1-k^2\cos^2(\theta/2)}}-\frac{2}{k}\sqrt{1-k^2\cos^2(\theta/2)}\right]d\theta,$$

or equivalently

$$\Sigma = 2\sqrt{Rs}\int_0^{\pi/2}\left[\left(\frac{2}{k}-k\right)\frac{1}{\sqrt{1-k^2\sin^2\gamma}}-\frac{2}{k}\sqrt{1-k^2\sin^2\gamma}\right]d\gamma,\quad 2\gamma=\pi-\theta.$$

Finally, deduce (7.41) using the definitions of the complete elliptic integrals $K$ and $E$.

7.9 We are free to choose the gauge of $\mathbf{A}$, $\nabla\cdot\mathbf{A}$, in any way that is convenient, the only restriction on $\mathbf{A}$ being that $\nabla\times\mathbf{A}=\mathbf{B}$. However, we have *defined* the electric potential through (7.64), so if we change the gauge for $\mathbf{A}$, then the form of $\varphi$ must alter in such a way that $\mathbf{E}$ remains unchanged. (Both $\mathbf{B}$ and $\mathbf{E}$ must be independent of the choice of gauge.) Show that we may alter the gauge of $\mathbf{A}$, leaving $\mathbf{B}$ unchanged, through the transformation $\mathbf{A}'=\mathbf{A}+\nabla\Lambda$, for some function $\Lambda$, but that the electric potential must then change according to $\varphi'=\varphi-\partial\Lambda/\partial t$. The function $\Lambda$ is known as the *gauge function*. Evidently the Lorenz gauge is a particularly convenient choice for unsteady problems, since it decouples the governing equations for $\mathbf{A}$ and $\varphi$, leading to (7.66) and (7.67). However, the Coulomb gauge remains popular in certain branches of physics.

7.10 Suppose we have fields $\mathbf{E}$ and $\mathbf{B}$ measured in a laboratory frame of reference and their counterparts $\mathbf{E}_{rel}$ and $\mathbf{B}_{rel}$ measured in an inertial frame that moves with velocity $\mathbf{u}$ relative to the laboratory frame. In Chapter 23 we shall see that, when $u$ is much less than the speed of light $c$, so that certain relativistic corrections may be ignored, $\mathbf{E}$ and $\mathbf{B}$ transform according to

$$\mathbf{E}_{rel}=\mathbf{E}+\mathbf{u}\times\mathbf{B},\qquad \mathbf{B}_{rel}=\mathbf{B}-\mathbf{u}\times\mathbf{E}/c^2. \tag{a}$$

Since $u\ll c$, it is natural to simplify the second of these to $\mathbf{B}_{rel}=\mathbf{B}$. However, some care must be exercised because $\mathbf{E}$ and $\mathbf{B}$ do not have the same dimensions in SI units. Suppose we adopt a different set of units in which $\mathbf{E}$ is rescaled as $\mathbf{E}'=\mathbf{E}/c$, so that $\mathbf{E}'$ and $\mathbf{B}$ now have the same dimensions. Show that, in these new units, the fields transform in the more symmetric way

$$\mathbf{E}'_{rel}=\mathbf{E}'+\mathbf{u}\times\mathbf{B}/c,\qquad \mathbf{B}_{rel}=\mathbf{B}-\mathbf{u}\times\mathbf{E}'/c. \tag{b}$$

(You may continue to assume that $u\ll c$.) Now use Faraday's law to show that $E\sim(L/T)B$ and $E'\sim(L/cT)B$, where $T$ and $L$ are characteristic time- and length-scales. In topics such as electrical circuits and magnetohydrodynamics we usually have $L/cT\ll 1$ and $E'\ll B$, while for radiation in a vacuum we find $L/cT\sim 1$ and $E'\sim B$. Show that, provided $u \ll c$ and $E'\le O(B)$, the transformation (b) also reduces to $\mathbf{B}_{rel}=\mathbf{B}$.

7.11 Show that, in a current-free region of space,

$$\oint \nabla(\mathbf{B}^2/2) \cdot d\mathbf{S} = \int (\nabla B_i)^2 dV \geq 0.$$

Hence show that there can be no local maximum of $\mathbf{B}^2$ in such a region.

## REFERENCES


Abramowitz, M & Stegun, I. A., 1965, *Handbook of Mathematical Functions*. Dover.

Batchelor, G. K., 1967, *An Introduction to Fluid Dynamics*. Cambridge University Press.

Davidson, P. A., 2016, Dynamos driven by helical waves: scaling laws for numerical dynamos and for the planets, *Geophys. J. Int.* **207**(2) 680.

Jackson, J. D., 1999, *Classical Electrodynamics*, 3rd ed. Wiley.

# 8

# Magnetostatics II

## *Dipoles, Force Distributions and Energy*

> One of the most brilliant steps made in philosophical exposition of which any instance existed in the history of science was that in which Faraday stated, in three or four words, intensely full of meaning, the law of magnetic attraction or repulsion.... Mathematicians were content to investigate the general expression of the resultant force experienced by a globe of soft iron in all such cases; but Faraday, without any mathematics, divined the result of the mathematical investigations. Indeed, the whole language of the magnetic field and 'lines of force' is Faraday's. It must be said for the mathematicians that they greedily accepted it, and have ever since been most zealous in using it to the best advantage.
>
> Lord Kelvin, 1872

In this chapter we continue our discussion of static magnetic fields, exploring some of the consequences of the Lorentz force and introducing the idea of magnetic energy. The topics that concern us are the forces and torques experienced by local current distributions, the energy of a static distribution of current, and various representations of the Lorentz force, such as Faraday's tension and Maxwell's stresses. We start, however, by returning to the humble dipole, in part because this dominates the far field of a localised distribution of current and in part because it provides a convenient starting point for discussing forces and energy.

## 8.1 The Dipole Revisited: Localised Current Distributions and their Far Field

In §7.4.1 we introduced the magnetic dipole moment $\mathbf{m}$ in the context of a circular current loop of radius $R$ and current $I$, as shown in Figure 8.1. We defined $\mathbf{m}$ to be a vector aligned with the axis of the loop, say in the $\hat{\mathbf{e}}_z$ direction, and of magnitude $|\mathbf{m}| = I(\pi R^2)$. However, at the time we suggested that a suitably generalised definition of $\mathbf{m}$, appropriate for any localised distribution of current, is

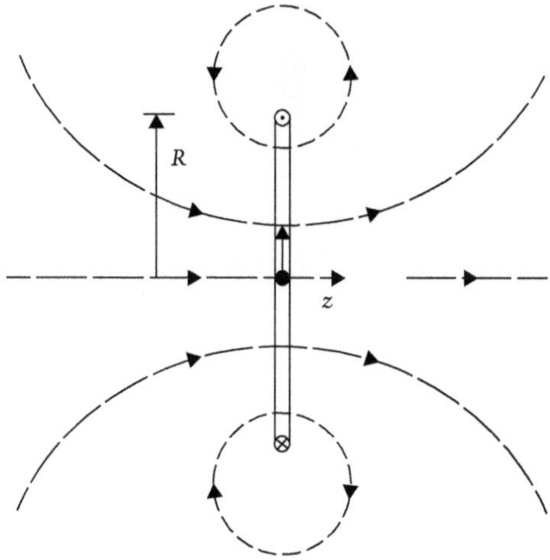


**Figure 8.1** A current loop of radius $R$ and carrying a current $I$.

$$\mathbf{m} = \frac{1}{2}\int \mathbf{x} \times \mathbf{J}\, dV, \tag{8.1}$$

which reduces to $\mathbf{m} = I(\pi R^2)\hat{\mathbf{e}}_z$ for the specific case of a circular current loop. Moreover, in §7.6.1 we showed that the magnetic field at large distances from such a current loop is

$$\mathbf{B}(\mathbf{x}) = \frac{\mu_0}{4\pi}(\mathbf{m}\cdot\nabla)\nabla\left(\frac{1}{|\mathbf{x}|}\right), \tag{8.2}$$

and we asserted that *any* localised distribution of *steady* current has the same far-field, provided that $\mathbf{m}$ is defined as in (8.1). We shall now show why the definition (8.1) is indeed a natural generalisation of $\mathbf{m} = I(\pi R^2)\hat{\mathbf{e}}_z$ and that (8.2) is the corresponding leading-order term in a far-field expansion of $\mathbf{B}(\mathbf{x})$.

Consider a localised distribution of current centred on the origin (Figure 8.2) and the Biot–Savart integral for its vector potential,

$$\mathbf{A}(\mathbf{x}) = \frac{\mu_0}{4\pi}\int \frac{\mathbf{J}(\mathbf{x}')}{|\mathbf{x}-\mathbf{x}'|}\, d\mathbf{x}'. \tag{8.3}$$

We are interested in the form of $\mathbf{A}$ at large distances from the current. In the limit of large $\mathbf{x}$, the expansion (5.52) tells us that

$$\frac{1}{|\mathbf{x}-\mathbf{x}'|} = \frac{1}{|\mathbf{x}|} - \frac{\partial}{\partial x_i}\left(\frac{1}{|\mathbf{x}|}\right)x_i' + C_{ij}x_i'x_j' + \cdots, \qquad C_{ij}(\mathbf{x}) = \frac{1}{2}\frac{\partial^2}{\partial x_i \partial x_j}\left(\frac{1}{|\mathbf{x}|}\right), \tag{8.4}$$

and so we have

$$\mathbf{A}(\mathbf{x}) = \frac{\mu_0}{4\pi|\mathbf{x}|}\int \mathbf{J}(\mathbf{x}')\, d\mathbf{x}' - \frac{\mu_0}{4\pi}\frac{\partial}{\partial x_i}\left(\frac{1}{|\mathbf{x}|}\right)\int x_i'\mathbf{J}(\mathbf{x}')\, d\mathbf{x}' + \frac{\mu_0}{4\pi}C_{ij}\int x_i'x_j'\mathbf{J}(\mathbf{x}')\, d\mathbf{x}' + \cdots. \tag{8.5}$$

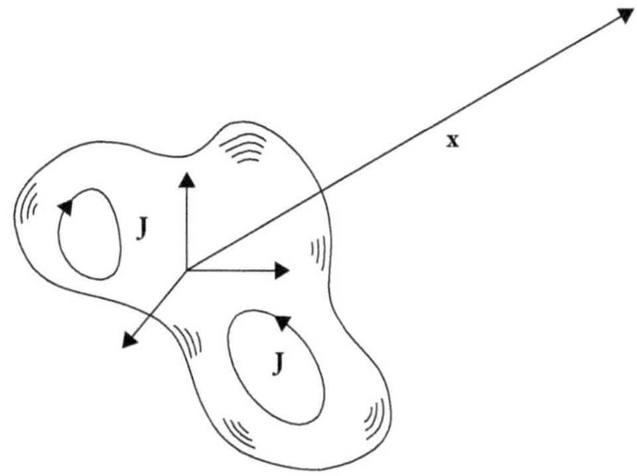


**Figure 8.2** A localised distribution of current.

The first integral in this expansion is zero because $\nabla \cdot \mathbf{J} = 0$ yields $J_i = \mathbf{J} \cdot \nabla x_i = \nabla \cdot (x_i \mathbf{J})$, and Gauss' theorem then tells us that $\mathbf{J}$ integrates to zero for any localised distribution of current. Turning to the second integral, this may be rearranged by noting that $\nabla \cdot \mathbf{J} = 0$ gives $x_i J_j + x_j J_i = \mathbf{J} \cdot \nabla(x_i x_j) = \nabla \cdot (x_i x_j \mathbf{J})$, and so $x_i J_j + x_j J_i$ also integrates to zero. So the second term on the right can be rewritten as

$$\frac{\partial}{\partial x_i}\left(\frac{1}{|\mathbf{x}|}\right)\int x'_i J'_j \, d\mathbf{x}' = \frac{1}{2}\frac{\partial}{\partial x_i}\left(\frac{1}{|\mathbf{x}|}\right)\int (x'_i J'_j - x'_j J'_i)\, d\mathbf{x}' = -\nabla\left(\frac{1}{|\mathbf{x}|}\right) \times \mathbf{m},$$

and (8.5) becomes

$$\boxed{\mathbf{A}(\mathbf{x}) = \frac{\mu_0}{4\pi}\nabla\left(\frac{1}{|\mathbf{x}|}\right) \times \mathbf{m} + \frac{\mu_0}{4\pi} C_{ij}\int x'_i x'_j \mathbf{J}(\mathbf{x}')\, d\mathbf{x}' + \cdots,} \tag{8.6}$$

where $\mathbf{m}$ is defined by (8.1). The curl of (8.6) then gives the *multipole expansion* for $\mathbf{B}$ as

$$\boxed{\mathbf{B}(\mathbf{x}) = \frac{\mu_0}{4\pi}(\mathbf{m} \cdot \nabla)\nabla\left(\frac{1}{|\mathbf{x}|}\right) + \frac{\mu_0}{4\pi}(\nabla C_{ij}) \times \int x'_i x'_j \mathbf{J}(\mathbf{x}')\, d\mathbf{x}' + \cdots,} \tag{8.7}$$

where the first term is the dipole contribution to $\mathbf{B}$, the second is the quadruple contribution, and so on. Note that, unlike in electrostatics, there is no monopole term.

So we see that the definition (8.1) appears quite naturally in a multipole expansion of a localised current distribution and that (8.2) is indeed the leading-order term in that expansion. Note also that, since $\nabla \cdot \mathbf{J} = 0$ ensures that $\mathbf{J}$ integrates to zero, the choice of origin of coordinates makes no difference to the value of $\mathbf{m}$ in (8.1).

A common situation is where $\mathbf{J}$ takes the form of a filamentary current loop $C$. In such cases $\mathbf{x} \times \mathbf{J}\, dV = I(\mathbf{x} \times d\mathbf{r})$, where $I$ is the current in the loop and $d\mathbf{r}$ is part of the axis of the loop. Moreover, applying Stokes' theorem to the vector field $\mathbf{c} \times \mathbf{x}$, where $\mathbf{c}$ is any constant vector, yields $2\int_S d\mathbf{S} = \oint_C \mathbf{x} \times d\mathbf{r}$, where $S$ is any surface that spans the curve $C$ (see Exercise 8.8). Equation (8.1) now simplifies to

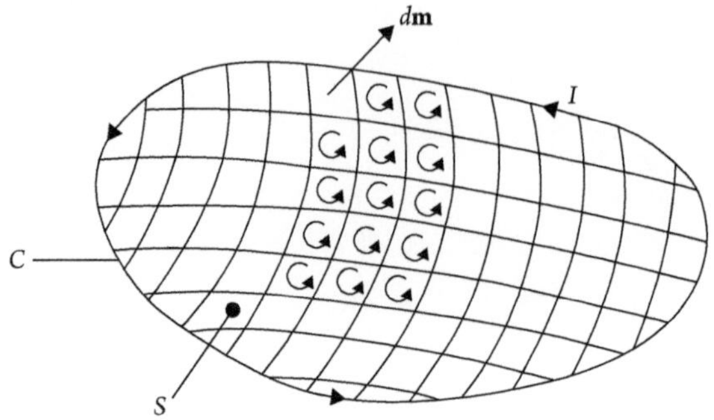


**Figure 8.3** The dipole moment of a large non-planar current loop can be considered as the sum of the many infinitesimal current loops surrounding the elements of area $d\mathbf{S}$.

$$\boxed{\mathbf{m} = \frac{1}{2} I \oint_C \mathbf{x} \times d\mathbf{r} = I \int_S d\mathbf{S}.} \tag{8.8}$$

A geometrical interpretation of (8.8) is straightforward. Suppose, for example, that the loop is planar and consists of a single turn lying in the $x$–$y$ plane, spanned by the flat surface $S$. Then the triangular contribution to $S$ that is swept out by rotating $\mathbf{x}$ in the $x$–$y$ plane by a small angle is readily shown to be $d\mathbf{S} = \mathbf{x} \times d\mathbf{r}/2$, and so we have $\mathbf{m} = I \int d\mathbf{S}$, as in (8.8). This is consistent with $\mathbf{m} = I(\pi R^2)\hat{\mathbf{e}}_z$ for a circular current loop.

If the current loop $C$ is non-planar, then we proceed as follows. Any surface that spans $C$ may be broken up into many elements of area $d\mathbf{S}$, each of which is approximately planar. We now imagine that the current $I$ circulates around the perimeter of each surface element $d\mathbf{S}$ in accordance with the right-hand rule, as shown in Figure 8.3. When all such infinitesimal currents loops are summed, the end result is simply the current $I$ circulating around the bounding curve $C$, because all other current elements $I\, d\mathbf{r}$ cancel. So the dipole moment for the current loop $C$ is the sum of the dipole moments of all the imaginary infinitesimal current loops surrounding the planar surface elements $d\mathbf{S}$, each of which is given by $d\mathbf{m} = I\, d\mathbf{S}$. Once again, we find that $\mathbf{m} = I \int d\mathbf{S}$.

## 8.2 The Torque, Force and Energy of a Dipole Sitting in an External Magnetic Field

### *8.2.1 The Magnetic Torque Acting on a Dipole in an External Magnetic Field*

We now turn from kinematics to dynamics. Consider a filamentary current loop in the form of a rectangle of sides $a$ and $b$, as shown in Figure 8.4. Suppose that this current loop is immersed in an externally imposed magnetic field $\mathbf{B}_0(\mathbf{x})$, which varies slowly on the scale of the loop, so that $\mathbf{B}_0(\mathbf{x})$ may be considered to be approximately uniform in the vicinity of the

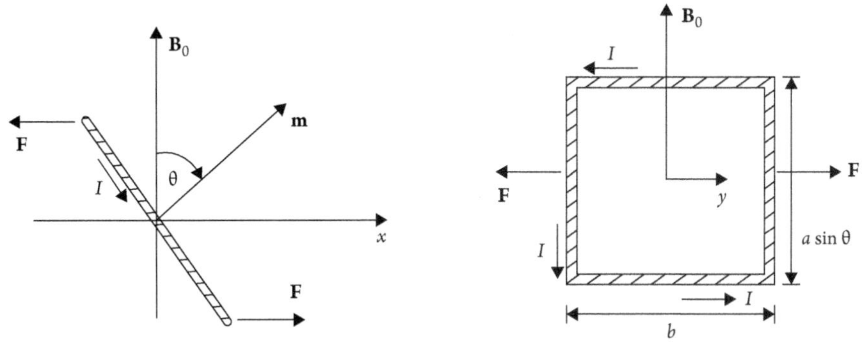


**Figure 8.4** A rectangular current loop sits in an externally imposed magnetic field $\mathbf{B}_0$.

loop. The total magnetic field is then $\mathbf{B} = \mathbf{B}_0 + \mathbf{B}_{\text{loop}}(\mathbf{x})$, where $\mathbf{B}_{\text{loop}}$ is the local magnetic field generated by the current $I$ in the loop. We take Cartesian coordinates centred on the loop, with $\mathbf{B}_0(0)$ aligned with the $z$ axis, sides $b$ parallel to the $y$ axis and sides $a$ lying parallel to the $x$–$z$ plane. The dipole moment $\mathbf{m}$, which is normal to the loop, makes an angle $\theta$ to $\mathbf{B}_0$.

Let us now calculate the Lorentz torque acting on the loop. The first thing to note is that an *isolated system of currents cannot exert any net force or torque on itself*, and so the net torque associated with $I$ and $\mathbf{B}_{\text{loop}}$ is zero. So let us restrict ourselves to the field $\mathbf{B}_0$. The Lorentz forces acting on the two sides of length $b$ are in the directions shown on the left of Figure 8.4 and of magnitude $|\mathbf{F}| = IB_0 b$. Since there is no torque arising from the current in the sides of length $a$, the net torque on the loop is

$$\mathbf{T} = -a\sin\theta\,(IB_0 b)\hat{\mathbf{e}}_y = -mB_0\sin\theta\,\hat{\mathbf{e}}_y = \mathbf{m}\times\mathbf{B}_0. \tag{8.9}$$

The expression $\mathbf{T} = \mathbf{m}\times\mathbf{B}_0$ might be compared with $\mathbf{T} = \mathbf{p}\times\mathbf{E}_0$ for an electric dipole $\mathbf{p}$ (see §5.5.3). Note that this torque tends to align the dipole moment with $\mathbf{B}_0$, just as the equivalent electrostatic torque acts to align $\mathbf{p}$ with $\mathbf{E}_0$.

The expression $\mathbf{T} = \mathbf{m}\times\mathbf{B}_0$ turns out to be remarkably general. Consider a filamentary current loop $C$ of arbitrary shape and carrying a current $I$. Then $\mathbf{J}\,dV = I\,d\mathbf{r}$ where $d\mathbf{r}$ is part of the axis of the loop, and so

$$\mathbf{T} = \int \mathbf{x}\times(\mathbf{J}\times\mathbf{B}_0)\,dV = I\oint_C \mathbf{x}\times(d\mathbf{r}\times\mathbf{B}_0). \tag{8.10}$$

For a *uniform* $\mathbf{B}_0$, the identity

$$2\mathbf{x}\times(d\mathbf{r}\times\mathbf{B}_0) = (\mathbf{x}\times d\mathbf{r})\times\mathbf{B}_0 + d\mathbf{r}\cdot\nabla[\mathbf{x}\times(\mathbf{x}\times\mathbf{B}_0)]$$

then gives

$$\mathbf{T} = \frac{1}{2}I\left(\oint_C \mathbf{x}\times d\mathbf{r}\right)\times\mathbf{B}_0 + I\oint_C \nabla(\sim)\cdot d\mathbf{r} = \frac{1}{2}I\left(\oint_C \mathbf{x}\times d\mathbf{r}\right)\times\mathbf{B}_0,$$

since $\nabla(\sim)\cdot d\mathbf{r}$ integrates to zero around a closed loop. Using (8.8) to evaluate the remaining line integral tells us that

$$\mathbf{T} = \frac{1}{2}I\left(\oint_C \mathbf{x}\times d\mathbf{r}\right)\times\mathbf{B}_0 = \mathbf{m}\times\mathbf{B}_0, \tag{8.11}$$

exactly as we found for the rectangular current loop.

More generally, let us consider any localised distribution of current. The net torque associated with the locally induced magnetic field is necessarily zero, while the torque associated with the externally applied field $\mathbf{B}_0$ is evidently $\mathbf{T} = \int \mathbf{x}\times(\mathbf{J}\times\mathbf{B}_0)\,dV$. For a uniform field $\mathbf{B}_0$, this may be rewritten using the identity

$$2\mathbf{x}\times(\mathbf{J}\times\mathbf{B}_0) = (\mathbf{x}\times\mathbf{J})\times\mathbf{B}_0 + \mathbf{J}\cdot\nabla[\mathbf{x}\times(\mathbf{x}\times\mathbf{B}_0)]. \tag{8.12}$$

Since $\nabla\cdot\mathbf{J} = 0$, the second term on the right of (8.12) may be written as a divergence, and this yields

$$\mathbf{T} = \int \mathbf{x}\times(\mathbf{J}\times\mathbf{B}_0)\,dV = \frac{1}{2}\int \mathbf{x}\times\mathbf{J}\,dV\times\mathbf{B}_0 + \oint(\sim)\mathbf{J}\cdot d\mathbf{S}. \tag{8.13}$$

Since the current distribution is localised, the surface integral in (8.13) is zero and hence

$$\boxed{\mathbf{T} = \int \mathbf{x}\times(\mathbf{J}\times\mathbf{B}_0)\,dV = \mathbf{m}\times\mathbf{B}_0.} \tag{8.14}$$

Note that, since the choice of origin of coordinates makes no difference to the value of $\mathbf{m}$, the torque $\mathbf{T}$ in (8.14) is independent of the origin of coordinates.

Evidently the result $\mathbf{T} = \mathbf{m}\times\mathbf{B}_0$ is quite ubiquitous. This may be understood from Figure 8.3. The net torque associated with a filamentary current loop $C$ can be considered to be the sum of the torques associated with many infinitesimal current loops surrounding the surface elements $d\mathbf{S}$ sitting on a surface $S$ that spans $C$. Each of these infinitesimal torques is given by $d\mathbf{T} = d\mathbf{m}\times\mathbf{B}_0$ and so summing over these elementary torques gives $\mathbf{T} = (\sum d\mathbf{m})\times\mathbf{B}_0$. Moreover, we have already seen that the dipole moment of the filamentary current loop $C$ is given by $\mathbf{m} = \sum d\mathbf{m}$, and so we conclude that $\mathbf{T} = \mathbf{m}\times\mathbf{B}_0$. This lies behind the result (8.11). In a similar way, any three-dimensional current distribution may considered to be the sum of many filamentary current loops, and so summing over all of these current loops again gives the net torque as $\mathbf{T} = \mathbf{m}\times\mathbf{B}_0$, which is (8.14).

### *8.2.2 The Net Lorentz Force Acting on a Dipole in an External Magnetic Field*

Let us now turn to the Lorentz force acting on a localised distribution of current. As noted above, the net force associated with the locally induced magnetic field $\mathbf{B}_{\text{loop}}$ is zero, and so the only net force is that generated by the externally imposed field. However, we must now allow for the fact that $\mathbf{B}_0(\mathbf{x})$ varies (slowly) on the scale of the current distribution, otherwise the net force arising from $\mathbf{B}_0$ is also zero. That is, if we take $\mathbf{B}_0$ to be strictly uniform, then

$$\mathbf{F} = \int (\mathbf{J} \times \mathbf{B}_0)\, dV = \int \mathbf{J}\, dV \times \mathbf{B}_0 = 0, \tag{8.15}$$

because $\nabla \cdot \mathbf{J} = 0$ yields $\nabla \cdot (x_i \mathbf{J}) = \mathbf{J} \cdot \nabla x_i = J_i$, thus ensuring that $\mathbf{J}$ integrates to zero. So let us make the *local* approximation,

$$\mathbf{B}_0(\mathbf{x}) = \mathbf{B}_0(0) + \left[\frac{\partial \mathbf{B}_0}{\partial x_i}\right]_0 x_i,$$

from which

$$\mathbf{F} = \int (\mathbf{J} \times \mathbf{B}_0)\, dV = \int x_i \mathbf{J}\, dV \times \left[\frac{\partial \mathbf{B}_0}{\partial x_i}\right]_0. \tag{8.16}$$

Next we recall that $\nabla \cdot \mathbf{J} = 0$ gives $\nabla \cdot (x_i x_j \mathbf{J}) = \mathbf{J} \cdot \nabla (x_i x_j) = x_i J_j + x_j J_i$, and so Gauss' theorem tells us that $x_i J_j + x_j J_i$ integrates to zero for any localised distribution of $\mathbf{J}$. So we may rewrite (8.16) as

$$F_i = \varepsilon_{ijk} \int x_n J_j\, dV \left[\frac{\partial B_{0k}}{\partial x_n}\right]_0 = \varepsilon_{ijk} \frac{1}{2} \int (x_n J_j - x_j J_n)\, dV \left[\frac{\partial B_{0k}}{\partial x_n}\right]_0, \tag{8.17}$$

where $\varepsilon_{ijk}$ is the Levi-Civita symbol, which allows us to write cross products in suffix notation. The integral on the right is $\mathbf{m}$, and so $\mathbf{F}$ depends on only $\mathbf{m}$ and $[\partial \mathbf{B}_0 / \partial x_i]_0$. Invoking $\nabla \cdot \mathbf{B} = 0$, (8.17) simplifies to $F_i = \mathbf{m} \cdot [\partial \mathbf{B}_0 / \partial x_i]_0$. (You might try proving this for yourself. If you are unfamiliar with suffix notation, simply focus on the $x$ component of (8.17) and expand everything in Cartesian coordinates.) However, $\mathbf{m}$ is a constant, and so we find that the net force is simply

$$\boxed{\mathbf{F} = \nabla (\mathbf{m} \cdot \mathbf{B}_0).} \tag{8.18}$$

Here the gradient in $\mathbf{B}_0$ is evaluated at the centre of the dipole, $\mathbf{x} = 0$. Once again we find that the integral effect of the Lorentz forces acting on a localised current distribution may be expressed very simply in terms of $\mathbf{m}$ and $\mathbf{B}_0$.

### 8.2.3 *The Potential Energy of a Dipole in an External Magnetic Field*

Note that (8.18) is a conservative force, and so we may associate a form of potential energy with a dipole immersed in an external magnetic field. The usual argument proceeds as follows. Suppose that $\mathbf{B}_0(\mathbf{x})$ is fixed and varies slowly on the scale of the dipole, but nevertheless falls to zero at infinity. Now consider a dipole consisting of a current loop in which the *current is held constant* at all times, say with the help of a battery. The loop starts off located at $\mathbf{x} = \infty$, where $\mathbf{B}_0 = 0$, and is slowly pulled towards the location $\mathbf{x} = 0$ with the help of an external mechanical force $\mathbf{F}_{\text{mech}} = -\nabla (\mathbf{m} \cdot \mathbf{B}_0)$, which exactly balances the

Lorentz force at all times. The work done by $\mathbf{F}_{\text{mech}}$ in pulling the current loop in from infinity is evidently

$$W_{\text{mech}} = \int \mathbf{F}_{\text{mech}} \cdot d\mathbf{r} = \int \nabla(-\mathbf{m} \cdot \mathbf{B}_0) \cdot d\mathbf{r} = -\mathbf{m} \cdot \mathbf{B}_0(0).$$

Now, as long as the current $I$ is maintained constant in the loop, this work is recoverable by letting the loop recede back to infinity. So we may associated a potential energy with the dipole, equal to the work done in pulling the dipole in from the far field:

$$\boxed{U_{\text{mech}} = -\mathbf{m} \cdot \mathbf{B}_0.} \tag{8.19}$$

Alternatively, we might think of the dipole being held fixed at the origin by mechanical forces, but allowed to rotate. As before, we insist that $I$ is held constant during movement of the loop, a caveat that will turn out to be important. We may slowly rotate the dipole by applying a mechanical torque $\mathbf{T}_{\text{mech}} = -\mathbf{m} \times \mathbf{B}_0$ that just balances the Lorentz torque. Suppose, for example, that the dipole in Figure 8.4 has $\mathbf{m}$ initially parallel to $\mathbf{B}_0$, so that $\theta = 0$, and is then rotated to the orientation shown in the figure, in which $\mathbf{m}$ holds an angle of $\theta$ relative to $\mathbf{B}_0$. Then, from (8.9), at any intermediate angle $\theta_i$,

$$\mathbf{T}_{\text{mech}} = -\mathbf{m} \times \mathbf{B}_0 = mB_0 \sin\theta_i \, \hat{\mathbf{e}}_y,$$

and the work done in rotating the dipole away from $\theta = 0$ is $W_{\text{mech}} = mB_0(1 - \cos\theta)$. As before, we may associate a potential energy with this work, provide the current is held fixed in the loop. Moreover, the constant in our expression for the mechanical work is an artefact of the initial orientation of the loop, which was an arbitrary choice. So it may be dropped in the expression for potential energy, giving $U_{\text{mech}} = -mB_0 \cos\theta$. (The datum for potential energy is then $\theta = \pm\pi/2$.) Once again, we conclude that $U_{\text{mech}} = -\mathbf{m} \cdot \mathbf{B}_0$.

Perhaps a few comments are in order at this point. First, note that we have demanded that both $I$ and $\mathbf{B}_0(\mathbf{x})$ are held fixed as the loop slides or rotates. This will turn out to be important in §8.4. Second, (8.19) is analogous to the potential energy of an electrostatic dipole in an external electric field, as given by (5.93): $U_d = -\mathbf{p} \cdot \mathbf{E}_0$. In each case the torque tends to align the dipole moment with the external field, $\mathbf{E}_0$ or $\mathbf{B}_0$, and so the potential energy of the dipole is a minimum when the dipole is aligned with the external field. Finally, note that we have *not* labelled $U = -\mathbf{m} \cdot \mathbf{B}_0$ as magnetic energy. In fact we shall see shortly that the magnetic energy is something rather different.

We shall return to the topic of energy in §8.4, where we shall explore the relationship between $U = -\mathbf{m} \cdot \mathbf{B}_0$ and magnetic energy. First, however, we need another kinematic result, analogous the electrostatic findings of §5.3.

## 8.3 The Average Magnetic Field in a Spherical Control Volume

Suppose we have an arbitrary distribution of current in an infinite domain and consider a spherical control volume of radius $R$ embedded within the resulting magnetic field. We are interested in the integral of **B** over the control volume $V_R$, which we take to be centred on the origin. We start by writing that integral of **B** in terms of the vector potential,

$$\int_{V_R} \mathbf{B}(\mathbf{x})\,d\mathbf{x} = \int_{V_R} \nabla \times \mathbf{A}\,d\mathbf{x} = -\oint_{S_R} \mathbf{A} \times d\mathbf{S}. \tag{8.20}$$

The Biot–Savart integral for the vector potential then gives

$$\int_{V_R} \mathbf{B}(\mathbf{x})\,d\mathbf{x} = -\frac{\mu_0}{4\pi}\oint_{S_R}\left[\int_{V_\infty} \frac{\mathbf{J}(\mathbf{x}')}{|\mathbf{x}-\mathbf{x}'|}\,d\mathbf{x}'\right] \times d\mathbf{S}, \tag{8.21}$$

where $V_\infty$ indicates an integral over all space. Switching the order of integrations yields

$$\int_{V_R} \mathbf{B}(\mathbf{x})\,d\mathbf{x} = \frac{\mu_0}{4\pi}\int_{V_\infty}\left(\oint_{S_R} \frac{d\mathbf{S}}{|\mathbf{x}-\mathbf{x}'|}\right) \times \mathbf{J}(\mathbf{x}')\,d\mathbf{x}'. \tag{8.22}$$

However, we evaluated the surface integral in (8.22) when performing the equivalent integration of **E** in Chapter 5. In particular, (5.43a, b) tell us that

$$\oint_{S_R} \frac{d\mathbf{S}}{|\mathbf{x}-\mathbf{x}'|} = \begin{cases} \frac{4\pi}{3}\mathbf{x}', & |\mathbf{x}'| < R, \qquad (8.23)\\ V_R\frac{\mathbf{x}'}{|\mathbf{x}'|^3}, & |\mathbf{x}'| > R. \qquad (8.24)\end{cases}$$

So we have two important special cases. When all the current is exterior to the control volume, we find

$$\frac{1}{V_R}\int_{V_R} \mathbf{B}(\mathbf{x})d\mathbf{x} = \frac{\mu_0}{4\pi}\int_{V_\infty} \frac{\mathbf{J}(\mathbf{x}') \times (-\mathbf{x}')}{|\mathbf{x}'|^3}\,d\mathbf{x}' = \mathbf{B}(0), \tag{8.25}$$

and so the average magnetic field within $V_R$ is equal to the field at the centre of the control volume. (We obtained essentially the same result for electrostatic fields in Chapter 5.) If all the current lies inside the control volume, on the other hand, we obtain

$$\boxed{\int_{V_R} \mathbf{B}(\mathbf{x})\,d\mathbf{x} = \frac{\mu_0}{3}\int_{V_R} \mathbf{x}' \times \mathbf{J}'\,d\mathbf{x}' = \frac{2}{3}\mu_0\mathbf{m},} \tag{8.26}$$

so the average magnetic field within $V_R$ is proportional to the dipole moment of the current distribution. We introduced this result in §7.6.2, where we noted its utility in calculating the average magnetic field in the core of a planet. In §5.3 we showed that the equivalent result for an electrostatic field is

$$\int_{V_R} \mathbf{E}(\mathbf{x})\, d\mathbf{x} = -\frac{\mathbf{p}}{3\varepsilon_0}, \tag{8.27}$$

where $\mathbf{p}$ is the electric dipole moment. We shall have reason to return to (8.26) on several occasions.

We close this section with a comment on (8.26). Suppose the current is restricted to the inner region, $|\mathbf{x}| < \hat{R}$, and that we let $R/\hat{R} \to \infty$ in (8.26). It is tempting to conclude that the integral over all space of $\mathbf{B}$ is then $2\mu_0\mathbf{m}/3$. But it is not! The point is that the integrand on the left of (8.26) falls off as $|\mathbf{x}|^{-3}$ at large distances from the current, and so the integral of $\mathbf{B}$ over all space is convergent, but only conditionally convergent. It turns out that, no matter how large we make $R$, there is always an order-one contribution to $\int_{V_\infty} \mathbf{B}(\mathbf{x})\, d\mathbf{x}$ that comes from outside $V_R$. In the limit $R/\hat{R} \to \infty$ this contribution turns out to be $\mu_0\mathbf{m}/3$. (See, for example, the discussion of the equivalent fluid problem in Batchelor (1967, §7.2).) So we conclude that, for a localised current distribution,

$$\boxed{\int_{V_\infty} \mathbf{B}(\mathbf{x})\, d\mathbf{x} = \mu_0\mathbf{m}.} \tag{8.28}$$

## 8.4 The Potential Energy of a Steady Distribution of Currents

### 8.4.1 *The Potential Energy of a Dipole Revisited*

Let us now return to the problem of a magnetic dipole (a current loop) immersed in an external magnetic field $\mathbf{B}_0(\mathbf{x})$ that varies slowly on the scale of the loop, but nevertheless falls to zero at infinity. As before, we let $\mathbf{B}_{\text{loop}}$ be the local magnetic field generated by the current in the loop, $I_{\text{loop}}$, so that the total magnetic field is $\mathbf{B} = \mathbf{B}_0(\mathbf{x}) + \mathbf{B}_{\text{loop}}(\mathbf{x})$. In §8.2.3 we saw that, provided $I_{\text{loop}}$ and $\mathbf{B}_0(\mathbf{x})$ are both held constant as the loop slides or rotates, the dipole possess a potential energy of $U_{\text{mech}} = -\mathbf{m} \cdot \mathbf{B}_0$, where $\mathbf{B}_0$ is evaluated at the centre of the loop. This potential energy is the mechanical work required to pull the dipole into the field $\mathbf{B}_0(\mathbf{x})$ from an initial location at $\mathbf{x} = \infty$ (where $\mathbf{B}_0 = 0$), work that is recoverable provided $I_{\text{loop}}$ and $\mathbf{B}_0(\mathbf{x})$ are both held fixed.

Physically, the situation is as shown in Figure 8.5. Both $\mathbf{B}_0$ and $\mathbf{B}_{\text{loop}}$ are maintained by current flowing in conductors, say a solenoid in the case of $\mathbf{B}_0$ and a single current loop in the case of $\mathbf{B}_{\text{loop}}$. (The radius of the solenoid is taken to be much larger than that of the dipole, so that $\mathbf{B}_0$ varies slowly on the scale of the dipole.) In the derivation of $U_{\text{mech}} = -\mathbf{m} \cdot \mathbf{B}_0$ we had to assume that both $I_{\text{loop}}$ and $\mathbf{B}_0(\mathbf{x})$ were held constant as the dipole moves, and so batteries (or some other source) were required to keep $I_{\text{loop}}$ and $I_{\text{sol}}$ constant at all times. We now seek the relationship between $U_{\text{mech}}$ and the overall magnetic energy of the system.

In §3.5 we noted that the magnetic energy density is $\mathbf{B}^2/2\mu_0$, something we shall confirm in Chapter 11. So the total magnetic energy of our system can be written as

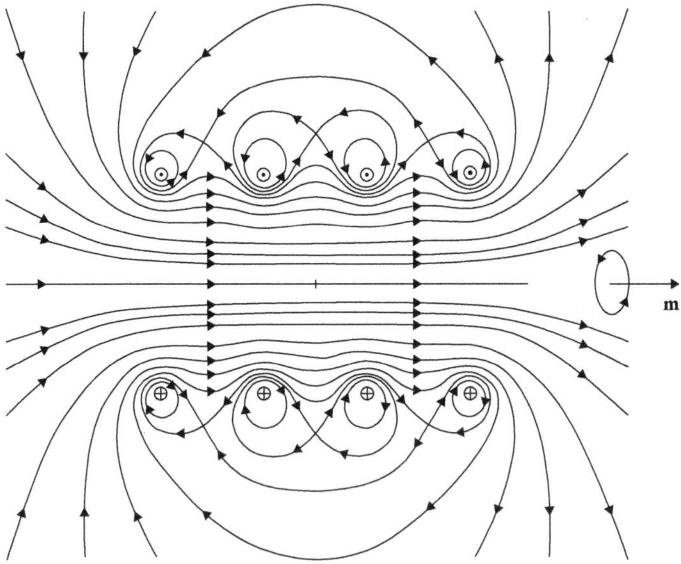


**Figure 8.5** A small dipole (a current loop) is immersed in a large-scale magnetic field $\mathbf{B}_0(\mathbf{x})$. Both $I_{\text{loop}}$ and $\mathbf{B}_0(\mathbf{x})$ are held constant as the dipole moves through $\mathbf{B}_0(\mathbf{x})$. (Adapted from E. Durand. *Magnétostatique*. Masson, 1968.)

$$U_B = \frac{1}{2\mu_0}\int_{V_\infty}(\mathbf{B}_0^2 + \mathbf{B}_{\text{loop}}^2)\,dV + \frac{1}{\mu_0}\int_{V_\infty}\mathbf{B}_0\cdot\mathbf{B}_{\text{loop}}\,dV = \text{self energy} + \text{mutual energy}. \tag{8.29}$$

If $\mathbf{B}_0$ is approximately uniform on the scale of the dipole, we may approximate this as

$$U_B \approx \frac{1}{2\mu_0}\int_{V_\infty}(\mathbf{B}_0^2 + \mathbf{B}_{\text{loop}}^2)\,dV + \frac{\mathbf{B}_0}{\mu_0}\cdot\int_{V_\infty}\mathbf{B}_{\text{loop}}\,dV = \frac{1}{2\mu_0}\int_{V_\infty}(\mathbf{B}_0^2 + \mathbf{B}_{\text{loop}}^2)\,dV + \mathbf{m}\cdot\mathbf{B}_0, \tag{8.30}$$

where we have used (8.28) to evaluate the mutual magnetic energy of the two fields. So we conclude that the magnetic energy that arises from the spatial overlap of $\mathbf{B}_0$ and $\mathbf{B}_{\text{loop}}$ is $U_{B\,\text{mutual}} = \mathbf{m}\cdot\mathbf{B}_0 = -U_{\text{mech}}$. At first sight this seems odd, since $U_{\text{mech}}$ is the mechanical work done by the force $\mathbf{F}_{\text{mech}} = -\nabla(\mathbf{m}\cdot\mathbf{B}_0)$ in pulling the dipole into the field $\mathbf{B}_0(\mathbf{x})$, and so we might have expected that conservation of energy demands $U_{B\,\text{mutual}} = U_{\text{mech}}$. Something is clearly amiss! The key to resolving this apparent paradox is to remember that power sources, such as batteries, are required to hold $I_{\text{loop}}$ and $\mathbf{B}_0(\mathbf{x})$ constant as the dipole is pulled into the field $\mathbf{B}_0$, and these sources can and do perform electrical work over and above that required to match the Ohmic dissipation in the system. Let us explore this in a little more detail.

As the dipole is pulled from the far field into $\mathbf{B}_0(\mathbf{x})$, the magnetic flux through the current loop changes, and Faraday's law tells us that this induces an EMF around the closed curve $C$ consisting of the wire loop and its battery. Specifically, (3.20) gives us

$$\oint_C \mathbf{E}_{\rm rel} \cdot d\mathbf{r} = -\frac{d}{dt} \int_S \mathbf{B} \cdot d\mathbf{S}, \tag{8.31}$$

where $\mathbf{E}_{\rm rel} = \mathbf{E} + \mathbf{u} \times \mathbf{B}$, $\mathbf{u}$ being the velocity of the loop. At those times when $\mathbf{E}_{\rm rel}$ opposes the flow of current, then the battery must do additional electrical work to maintain $I_{\rm loop}$, whereas when $\mathbf{E}_{\rm rel}$ tends to aid the flow of current, the battery does less work. (We shall leave to one side the practical issue of how a source like a battery can maintain a constant current.)

So let us find the work done by the battery as the loop is pulled in from the far field. The rate at which the electric field does work on the charge carriers in the conducting loop is given by (7.11). Integrating over the wire loop, we find

$$\int_{\rm conductor} \mathbf{J} \cdot \mathbf{E}\, dV = \int \frac{\mathbf{J}^2}{\sigma} dV + \mathbf{u} \cdot \int (\mathbf{J} \times \mathbf{B}_0)\, dV,$$

or, using $\mathbf{J}\, dV = I\, d\mathbf{r}$,

$$I \int_{\rm conductor} \mathbf{E} \cdot d\mathbf{r} = \int \frac{\mathbf{J}^2}{\sigma} dV - \mathbf{u} \cdot \mathbf{F}_{\rm mech}. \tag{8.32}$$

Moreover, since $\partial \mathbf{B}_0 / \partial t = 0$, we have $\nabla \times \mathbf{E} = 0$ and so $\oint \mathbf{E} \cdot d\mathbf{r} = 0$. Now the battery does work by pushing charge *against* $\mathbf{E}$ (like a pump pushing water uphill), and so, if the closed path in the line integral of $\mathbf{E}$ consists of the conducting loop *and* the battery, we have

$$\text{rate of working of battery} = -I \int_{\rm battery} \mathbf{E} \cdot d\mathbf{r} = I \int_{\rm conductor} \mathbf{E} \cdot d\mathbf{r}.$$

This combines with (8.32) to give

$$\text{rate of working of battery} = \int \frac{\mathbf{J}^2}{\sigma} dV - \mathbf{u} \cdot \mathbf{F}_{\rm mech}. \tag{8.33}$$

So we have the plausible statement that the rate of working of the battery plus the rate of working of $\mathbf{F}_{\rm mech}$ equals the Ohmic dissipation in the loop. We label the rate of working of the battery, *over and above* that needed to compensate for the Ohmic dissipation, the *reversible electrical power*. If we now integrate (8.33) over time, we find that the sum of the reversible electrical work $U^{\rm loop}_{\rm elect}$ and mechanical work done on the wire loop is zero:

$$U^{\rm loop}_{\rm elect} + U_{\rm mech} = 0. \tag{8.34}$$

(See Feynman, Leighton & Sands (1964) for an alternative derivation of (8.34).)

However, this is not the end of the story. As the loop is pulled through $\mathbf{B}_0(\mathbf{x})$, the movement of its magnetic field $\mathbf{B}_{\text{loop}}$ changes the magnetic flux through the coils that support $\mathbf{B}_0$. Once again an electric field is induced and the rate of working of the battery in the coils deviates from that required to match the Ohmic dissipation. We can determine the reversible electrical work done by this battery by noting that it does not matter whether the current loop is pulled towards the solenoid or the solenoid is pulled towards the loop: either way, the electrical work done by the battery in the solenoid is the same. All that matters is that the relative configuration of the current loop and solenoid is the same at the end of the process. Moreover, Newton's third law tells us that equal and opposite Lorentz forces act on the loop and solenoid, so that the same mechanical work is done irrespective of whether the loop moves towards the solenoid or the solenoid moves towards the loop. So let us now imagine that the loop is held fixed in space and the solenoid moves, bringing the two into the same relative configuration. We may now apply (8.34) to the solenoid and this yields

$$U_{\text{elect}}^{\text{sol}} + U_{\text{mech}} = 0. \tag{8.35}$$

Returning to the original case in which the loop moves and the solenoid is stationary, we conclude that

$$U_{\text{elect}}^{\text{loop}} = U_{\text{elect}}^{\text{sol}} = -U_{\text{mech}}. \tag{8.36}$$

Let us now do an energy balance on the overall system. The total amount of mechanical and reversible electrical work done as the two magnetic fields are pulled into a mutually overlapping configuration is

$$U_{\text{total}} = U_{\text{elect}}^{\text{loop}} + U_{\text{elect}}^{\text{sol}} + U_{\text{mech}} = -U_{\text{mech}}. \tag{8.37}$$

Moreover, in §8.2.3 we showed that $U_{\text{mech}} = -\mathbf{m} \cdot \mathbf{B}_0$ and so we conclude that the total reversible work done is $U_{\text{total}} = \mathbf{m} \cdot \mathbf{B}_0$. Conservation of energy then suggests that this should appear as the mutual magnetic energy of the two overlapping magnetic fields, $U_{B\,\text{mutual}} = \mathbf{m} \cdot \mathbf{B}_0$, which is exactly (8.30).

### *8.4.2 The Mutual Potential Energy of Steady Currents*

Let us explore in a little more detail the implications of the expression $U_{B\,\text{mutual}} = \mathbf{m} \cdot \mathbf{B}_0$. As is stands, this applies only to a small current loop in a slowly varying external field $\mathbf{B}_0(\mathbf{x})$ and it is natural to ask if this may be generalised to give the mutual potential energy of an arbitrary arrangement of current loops.

Consider a current loop $C$ sitting in an external magnetic field $\mathbf{B}_0(\mathbf{x})$ and carrying a current $I$. We remove the restriction that the characteristic lengthscale for $\mathbf{B}_0(\mathbf{x})$ is large compared with that of $C$, so the current loop can have any shape or size. Let $S$ be any surface that spans $C$ and let $S$ be broken up into many elements of area, $d\mathbf{S}$. We now imagine that the current $I$ circulates around the perimeter of each surface element $d\mathbf{S}$ in accordance with the right-hand rule, as shown in Figure 8.3. When all such infinitesimal currents loops are summed, the end result is simply the current $I$ circulating around $C$, because all other currents cancel. It follows that $\mathbf{B}_{\text{loop}}$, and hence $U_{B\text{ mutual}}$, associated with the current loop $C$ is the sum of the contributions from the surface elements $d\mathbf{S}$. Now, for each $d\mathbf{S}$ we may combine $dU_{B\text{ mutual}} = d\mathbf{m}\cdot\mathbf{B}_0$ with $\mathbf{m} = I\int d\mathbf{S}$ to give $dU_{B\text{ mutual}} = I\,d\mathbf{S}\cdot\mathbf{B}_0$, and summing all such contributions over the surface $S$ gives us

$$U_{B\text{ mutual}} = I\int_S \mathbf{B}_0\cdot d\mathbf{S} = I\oint_C \mathbf{A}_0\cdot d\mathbf{r}. \tag{8.38}$$

Now suppose we have two current loops, each sitting in the magnetic field of the other. Let the loops have currents $I_1$ and $I_2$, centrelines $C_1$ and $C_2$, and fields $\mathbf{B}_1(\mathbf{x})$ and $\mathbf{B}_2(\mathbf{x})$. Then the mutual potential energy of the two current loops can be written in one of two ways,

$$U_{B\text{ mutual}} = I_1\oint_{C_1}\mathbf{A}_2\cdot d\mathbf{r} = I_2\oint_{C_2}\mathbf{A}_1\cdot d\mathbf{r}, \tag{8.39}$$

depending on which loop we integrate around. This is most conveniently written in the symmetric form

$$U_{B\text{ mutual}} = \frac{1}{2}\left(I_1\oint_{C_1}\mathbf{A}_2\cdot d\mathbf{r} + I_2\oint_{C_2}\mathbf{A}_1\cdot d\mathbf{r}\right), \tag{8.40}$$

or, since $\mathbf{J}\,dV = I\,d\mathbf{r}$,

$$U_{B\text{ mutual}} = \frac{1}{2}\left(\int_{V_1}\mathbf{J}\cdot\mathbf{A}_2 dV + \int_{V_2}\mathbf{J}\cdot\mathbf{A}_1 dV\right). \tag{8.41}$$

This generalises in an obvious way to give the mutual potential energy of multiple current loops.

### 8.4.3 *The Total Magnetic Energy of Steady Currents*

Let us now see if we can get an expression for the total magnetic energy of any steady distribution of current. We shall see in Chapter 11 that, perhaps not surprisingly, the magnetic energy of an isolated current loop, say $C_1$, is

$$U_{B\text{ self}} = \frac{1}{2}I_1\oint_{C_1}\mathbf{A}_1\cdot d\mathbf{r} = \frac{1}{2}\int_{V_1}\mathbf{J}\cdot\mathbf{A}_1\,dV. \tag{8.42}$$

Combining this with (8.41) tells us that the total magnetic energy (self-energy plus mutual energy) of two current loops whose fields overlap is

$$U_B = \frac{1}{2}\left(\int_{V_1}\mathbf{J}\cdot\mathbf{A}\,dV + \int_{V_2}\mathbf{J}\cdot\mathbf{A}\,dV\right) = \frac{1}{2}\int_{V_1+V_2}\mathbf{J}\cdot\mathbf{A}\,dV. \tag{8.43}$$

Evidently the total magnetic energy for any steady localised distribution of currents in an infinite domain is simply

$$\boxed{U_B = \frac{1}{2}\int_{V_\infty} \mathbf{J}\cdot\mathbf{A}\,dV.} \tag{8.44}$$

It is readily shown that, for *steady currents*, this is equivalent to a magnetic energy density of $\mathbf{B}^2/2\mu_0$. First we note that Ampère's law allows us to rewrite (8.44) as

$$U_B = \frac{1}{2\mu_0}\int_{V_\infty} \mathbf{A}\cdot\nabla\times\mathbf{B}\,dV.$$

The identity $\mathbf{A}\cdot\nabla\times\mathbf{B} = \mathbf{B}^2 - \nabla\cdot(\mathbf{A}\times\mathbf{B})$ then yields

$$\boxed{U_B = \frac{1}{2\mu_0}\int_{V_\infty} \mathbf{B}^2\,dV,} \tag{8.45}$$

since the integral of $\nabla\cdot(\mathbf{A}\times\mathbf{B})$ converts to a surface integral whose integrand is zero. However, it is worth remembering that Ampère's law does not always apply in unsteady situations, and in such cases (8.44) and (8.45) are *not* equivalent. It turns out that $\mathbf{B}^2/2\mu_0$ is always the magnetic energy density, so that (8.45) is always correct, but that (8.44) is only guaranteed to give the correct magnetic energy for the case of steady currents.

## 8.5 How to Create the Perfect Dipole: The Rotating Charged Sphere

By way of an aside, we now describe how to generate a magnetic field that is a pure dipole. Consider a sphere of radius $R$ that rotates with angular velocity $\boldsymbol{\Omega}$ and carries a uniform surface charge density $\sigma_e$ (Figure 8.6). The rotating charges constitute a surface convection current $\mathbf{K}(\mathbf{x}') = \sigma_e\boldsymbol{\Omega}\times\mathbf{x}'$, where $\mathbf{x}'$ locates a point on the surface. The vector potential is evidently

$$\mathbf{A}(\mathbf{x}) = \frac{\mu_0}{4\pi}\oint_{|\mathbf{x}'|=R} \frac{\mathbf{K}(\mathbf{x}')}{|\mathbf{x}-\mathbf{x}'|}dS' = \frac{\mu_0\sigma_e R}{4\pi}\boldsymbol{\Omega}\times\oint_{|\mathbf{x}'|=R}\frac{d\mathbf{S}'}{|\mathbf{x}-\mathbf{x}'|}, \tag{8.46}$$

where the surface integral may be evaluated using (8.23) and (8.24). This yields the surprisingly simple solution

$$\mathbf{A}_{\text{in}}(\mathbf{x}) = \frac{1}{3}(\mu_0\sigma_e R)\,\boldsymbol{\Omega}\times\mathbf{x}, \qquad |\mathbf{x}| < R, \tag{8.47}$$

$$\mathbf{A}_{\text{out}}(\mathbf{x}) = \frac{\mu_0\sigma_e R V_R}{4\pi}\nabla\left(\frac{1}{|\mathbf{x}|}\right)\times\boldsymbol{\Omega}, \quad |\mathbf{x}| > R, \tag{8.48}$$

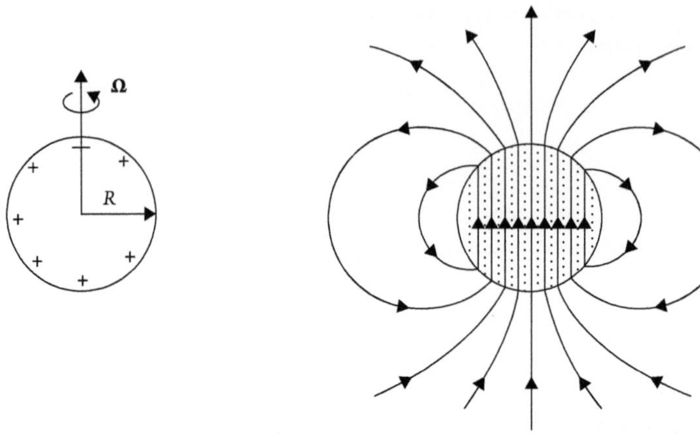


**Figure 8.6** The magnetic field generated by a rotating sphere carrying a surface charge.

where $V_R$ is the volume of the sphere. The internal magnetic field is evidently uniform and aligned with $\boldsymbol{\Omega}$,

$$\mathbf{B}_{\text{in}} = \nabla \times \mathbf{A}_{\text{in}} = \frac{2}{3}\mu_0\sigma_e R\,\boldsymbol{\Omega}, \quad |\mathbf{x}| < R, \tag{8.49}$$

while (8.6) tells us that the external magnetic field is that of a pure dipole in which the dipole moment is $\mathbf{m} = \sigma_e R V_R\,\boldsymbol{\Omega}$. Note that $\mathbf{B}_{\text{in}} V_R = 2\mu_0\mathbf{m}/3$, as demanded by (8.26).

This simple, elegant solution will prove useful when we discuss magnetic materials in Chapter 9, and it once formed the basis for a speculative explanation of the geomagnetic field (see Exercise 8.2).

## 8.6 Maxwell's Stresses and Faraday's Tension

We now turn to force distributions. There are two ways of rewriting the Lorentz force that prove to be particularly useful when dealing with quasi-static magnetic fields, particularly magnetic fields in conducting fluids. One is due to Faraday and another was introduced by Maxwell, and both play a prominent role in the description of astrophysical and fusion plasmas. Although Faraday's representation came first, and in many ways is the more useful of the two, we shall start with Maxwell's reworking of the Lorentz force.

We may rewrite the Lorentz force in terms of $\mathbf{B}$ alone using Ampère's law. Starting with the vector identity

$$(\nabla \times \mathbf{B}) \times \mathbf{B} = (\mathbf{B} \cdot \nabla)\mathbf{B} - \nabla(\mathbf{B}^2/2),$$

$\nabla \times \mathbf{B} = \mu_0\mathbf{J}$ gives us

$$\mathbf{J} \times \mathbf{B} = (\mathbf{B} \cdot \nabla)\left(\frac{\mathbf{B}}{\mu_0}\right) - \nabla\left(\frac{\mathbf{B}^2}{2\mu_0}\right). \tag{8.50}$$

It is often convenient to treat the two terms on the right of (8.50) separately. The force $-\nabla(\mathbf{B}^2/2\mu_0)$ is clearly irrotational and points from regions of high field to low field. In the case of a long solenoid, for example, it is the only contribution to $\mathbf{J} \times \mathbf{B}$ and it is responsible for the radial bursting force exerted on the solenoid coils. In a conducting fluid, such as a plasma, $\mathbf{B}^2/2\mu_0$ is called the *magnetic pressure,* usually denoted by $p_m$. Provided there are no free fluid surfaces, the primary role of $-\nabla p_m$ is simply to change the pressure distribution in the plasma. This can be seen in the following way. The net pressure force per unit volume acting on a small fluid parcel is $-\nabla p$, where $p$ is the fluid pressure (see §20.2.3), and so when $(\mathbf{B} \cdot \nabla)(\mathbf{B}/\mu_0)$ is zero, the equation of motion for a fluid parcel (other forces apart) can be written as

$$\rho \cdot (\text{acceleration}) = \mathbf{J} \times \mathbf{B} - \nabla p = -\nabla(p + p_m), \tag{8.51}$$

where $\rho$ is the fluid density. Evidently a static equilibrium is one possibility here, in which $-\nabla p$ balances $-\nabla p_m$. Alternatively, if there *is* motion of the plasma, perhaps driven by some other force, and there are no free surfaces, then it may be shown (see Chapter 20) that $-\nabla p_m$ is still balanced by part of the pressure gradient. Either way, the magnetic pressure changes the fluid pressure but has no influence on the motion. By way of contrast, the first term on the right of (8.50) is usually, though not always, rotational and so cannot in general be balanced by the pressure forces in a fluid. Consequently, this force usually drives, or at least shapes, the velocity distribution in a conducting fluid.

Whether in a fluid or a solid conductor, we can use $\nabla \cdot \mathbf{B} = 0$ to rewrite the $i$th component of the Lorentz force as

$$(\mathbf{J} \times \mathbf{B})_i = \mathbf{B} \cdot \nabla\left(\frac{B_i}{\mu_0}\right) - \frac{\partial}{\partial x_i}\left(\frac{\mathbf{B}^2}{2\mu_0}\right) = \frac{\partial}{\partial x_j}\left(\frac{B_i B_j}{\mu_0} - \frac{\mathbf{B}^2}{2\mu_0}\delta_{ij}\right),$$

where there is an implied summation over $j$. Thus the Lorentz force is mechanically equivalent to the imaginary set of stresses

$$\boxed{\tau_{ij}^M = \frac{B_i B_j}{\mu_0} - \frac{\mathbf{B}^2}{2\mu_0}\delta_{ij},} \tag{8.52}$$

in the sense that we can write $(\mathbf{J} \times \mathbf{B})_i = \partial \tau_{ij}^M / \partial x_j$. (See §5.5.2 for a discussion of the equivalence of stress systems and body forces.) The quantity $\tau_{ij}^M$ is called the *magnetic stress tensor,* or the magnetic contribution to *Maxwell's stress tensor.*

These expressions should be compared with their counterparts in electrostatics, in which the Coulomb force is expressed in terms of the *electric stress tensor.* The primary utility of (8.52) lies in the fact that it is often more useful to think of the Lorentz force as arising from a set of surface stresses, rather than as a distributed body force. For example, integrating (8.50) over a volume $V$, invoking $\nabla \cdot \mathbf{B} = 0$ and Gauss' theorem, shows that we can represent the integrated effect of the Lorentz force by surface stresses alone,

$$\int_V \mathbf{J} \times \mathbf{B}\, dV = \oint \frac{\mathbf{B}}{\mu_0}\mathbf{B} \cdot d\mathbf{S} - \oint \frac{\mathbf{B}^2}{2\mu_0}\, d\mathbf{S}. \tag{8.53}$$

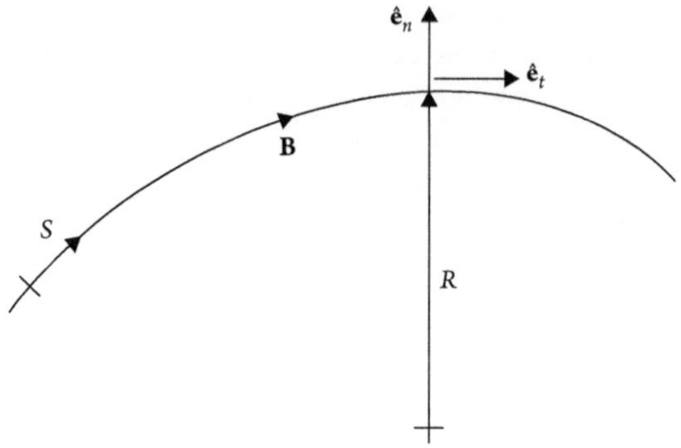


**Figure 8.7** The intrinsic coordinates used in (8.55).

The surface integrals on the right of (8.53) represent the cumulative effect of the distributed stress system $\tau_{ij}^M = (B_iB_j/\mu_0) - (\mathbf{B}^2/2\mu_0)\,\delta_{ij}$ acting over the surface of $V$.

We now turn to the second representation of $\mathbf{J}\times\mathbf{B}$. We start by writing $(\mathbf{B}\cdot\nabla)\mathbf{B}$ in terms of intrinsic coordinates:

$$(\mathbf{B}\cdot\nabla)\mathbf{B} = B\frac{\partial}{\partial s}(B\hat{\mathbf{e}}_t) = B\frac{\partial B}{\partial s}\hat{\mathbf{e}}_t - \frac{B^2}{R}\hat{\mathbf{e}}_n. \tag{8.54}$$

Here $\mathbf{B} = B\hat{\mathbf{e}}_t$, $\hat{\mathbf{e}}_t$ is a unit vector tangential to a magnetic field line, $\hat{\mathbf{e}}_n$ is the principal unit normal vector directed *away* from the local centre of curvature, $s$ is a coordinate measured along the field line and $R$ is the radius of curvature of the field line, as shown in Figure 8.7. ($R$ is defined in the usual way by $d\hat{\mathbf{e}}_t/ds = -\hat{\mathbf{e}}_n/R$). It follows that the Lorentz force can be written as

$$\mathbf{J}\times\mathbf{B} = \left[\frac{\partial}{\partial s}\left(\frac{B^2}{2\mu_0}\right)\hat{\mathbf{e}}_t - \frac{B^2}{\mu_0 R}\hat{\mathbf{e}}_n\right] - \nabla(B^2/2\mu_0). \tag{8.55}$$

In problems such as the magnetic confinement of laboratory plasmas, the last term on the right of (8.55) is of great importance, since it can act to pressurize the plasma. However, for conducting fluids in which there are no free surfaces, say the interior of a planet, the rotational part of $\mathbf{J}\times\mathbf{B}$ is usually of most interest, since it can drive motion in the fluid. In such cases the force $-\nabla p_m$ is usually absorbed into the pressure gradient and attention then focuses on

$$(\mathbf{B}\cdot\nabla)\left(\frac{\mathbf{B}}{\mu_0}\right) = \frac{\partial}{\partial s}\left(\frac{B^2}{2\mu_0}\right)\hat{\mathbf{e}}_t - \frac{B^2}{\mu_0 R}\hat{\mathbf{e}}_n. \tag{8.56}$$

Crucially, there is a Lorentz force transverse to the magnetic field lines that is associated with the *curvature* of those lines.

To get a feel for the physical origin of (8.56), consider a thin isolated flux tube of cross-sectional area $A$ and carrying a magnetic flux $\Phi = B(s)A(s)$ (Figure 8.8). In terms of Maxwell stresses, the force $\mu_0^{-1}(\mathbf{B}\cdot\nabla)\mathbf{B}$ can be represented by the action of the stress system $B_iB_j/\mu_0$, and for our isolated flux tube this reduces to a tensile stress $B^2/\mu_0$ acting along the axis of the flux tube. This tensile stress is known as *Faraday's tension* and it leads to a tensile force in the flux tube of $T(s) = (B^2/\mu_0)A$. Now we know from elementary mechanics that a string carrying a tensile force $T_s$ experiences a force per unit length of $\delta\mathbf{F} = (dT_s/ds)\hat{\mathbf{e}}_t - (T_s/R)\hat{\mathbf{e}}_n$, where $R$ the local radius of curvature of the string. If we substitute $T = (B^2/\mu_0)A = \Phi B/\mu_0$ into this expression, we see that our flux tube experiences a force

$$\delta\mathbf{F} = \frac{dT}{ds}\hat{\mathbf{e}}_t - \frac{T}{R}\hat{\mathbf{e}}_n = \frac{1}{\mu_0}\frac{d(\Phi B)}{ds}\hat{\mathbf{e}}_t - \frac{B^2 A}{\mu_0 R}\hat{\mathbf{e}}_n. \tag{8.57}$$

Noting that $\Phi$ is independent of $s$ and dividing throughout by $A$ takes us back to (8.56).

In summary, then, the normal and tangential components of $\mu_0^{-1}(\mathbf{B}\cdot\nabla)\mathbf{B}$ in (8.56) can be thought of as arising from the action of Faraday's tension $B^2/\mu_0$ acting along the axis of the flux tube, as indicated in Figure 8.8. This was Faraday's great insight. In many cases Faraday's tension provides a powerful qualitative means for picturing the effect of the Lorentz force. We may think of magnetic field lines as being in tension, carrying a tensile stress $B^2/\mu_0$, and exerting a force on the conductor, rather like stretched elastic bands.

It follows that, just as the sideways force on a plucked string, $\delta\mathbf{F}_n = -(T_s/R)\hat{\mathbf{e}}_n$, leads to waves on the string with a wave speed $c = \sqrt{T_s/\rho_s}$ (see §2.1.3), so the sideways force on a 'plucked' magnetic field line sitting in a plasma leads to oscillations of that plasma with a wave speed $c = \sqrt{(B^2/\mu_0)/\rho}$. (Here $\rho_s$ is the mass per unit length of the string and $\rho$ is the density of the plasma.) These are called Alfvén waves (Figure 8.9) and they are

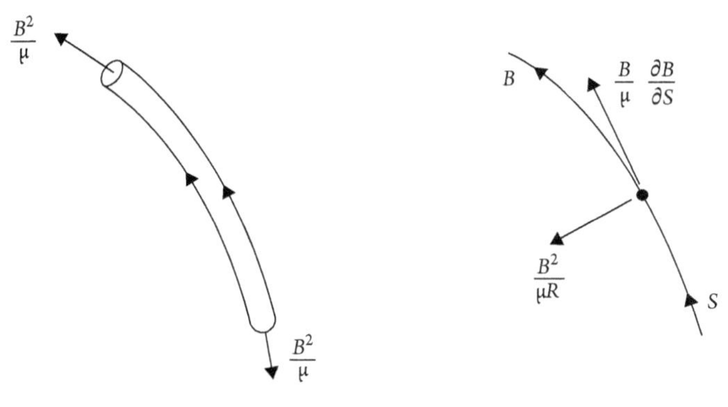


**Figure 8.8** The components of $\mu_0^{-1}(\mathbf{B}\cdot\nabla)\mathbf{B}$ acting on a thin flux tube can be interpreted as a consequence of the Faraday tension $B^2/\mu_0$ acting along the axis of that tube.

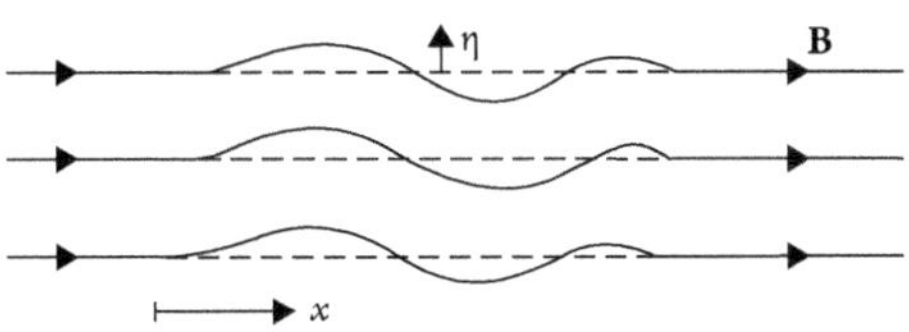


**Figure 8.9** Alfvén waves in a plasma.

ubiquitous in astrophysics. Alfvén waves were introduce in §2.3.1 and are discussed at length in Chapter 21.

Of course, in any given situation we must remember to add the magnetic pressure gradient, $-\nabla(\mathbf{B}^2/2\mu_0)$, to $\mu_0^{-1}(\mathbf{B}\cdot\nabla)\mathbf{B}$ to get the total Lorentz force, and indeed in certain very simple geometries it is not particularly helpful to distinguish between the two contributions, as discussed in Exercise 8.6. The magnetic pressure is, of course, particularly important in the magnetic confinement of fusion plasmas.

## 8.7 An Aside: The Difficulty of Confining Fusion Plasmas with Magnetic Fields

For commercial nuclear fusion to be achievable, it is essential that the hot plasma be kept away from the reactor walls and confined in a stable configuration for long periods of time. In tokamaks magnetic fields are used for this purpose, although to date confinement times have been restricted to less than a second, with the plasma plagued by a zoo of instabilities, as we now discuss.

A plasma in equilibrium is governed by (8.51) in the form $\mathbf{J}\times\mathbf{B}=\nabla p$. This requires that $\mathbf{J}\cdot\nabla p=\mathbf{B}\cdot\nabla p=0$, which in turn demands that the $\mathbf{J}$ and $\mathbf{B}$ fields lie on constant pressure surfaces. If the plasma is to be confined away from walls, then these surfaces must close within the plasma. Moreover, the surfaces cannot be simply connected, because any closed contour for which $\oint\mathbf{B}\cdot d\mathbf{r}$ is non-zero would then require current flowing across such a surface, which contravenes $\mathbf{J}\cdot\nabla p=0$. So spherical surfaces are not possible and the simplest option is a set of nested tori, as shown in Figure 8.10. In such confinement schemes one must arrange for the Lorentz force to point inwards, towards the so-called magnetic axis. In order to allow for a simple analysis, it is sometimes convenient to imagine the torus to have a relatively small cross section, so that the geometry may be may be locally 'unwrapped' to form a plasma cylinder.

Perhaps the simplest confinement system is that shown in Figure 8.11. An axial current is induced in the surface of the plasma and the resulting azimuthal field creates a radial Lorentz force that points inwards. (Remember, like-signed currents attract.) The form of

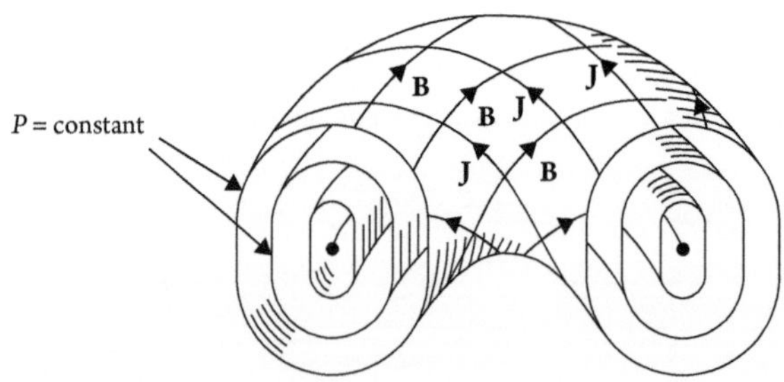


**Figure 8.10** In plasma confinement the **B** and **J** lines sit on constant-pressure surfaces. (Reprinted from P. A. Davidson. *An Introduction to Magnetohydrodynamics*, 2nd ed. Cambridge University Press, 2017.)

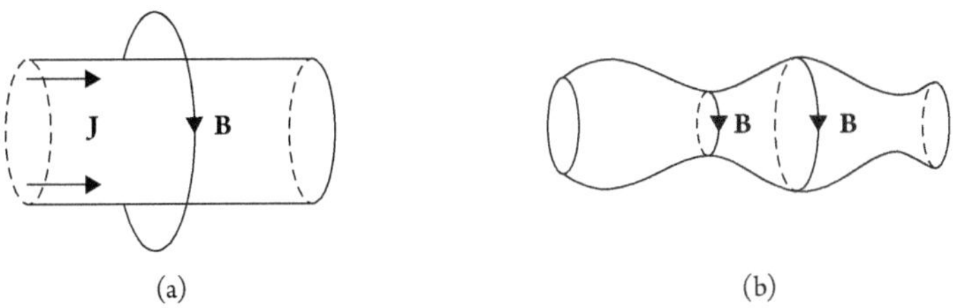


**Figure 8.11** The linear pinch: (a) the confinement principle; (b) instability of the pinch. (Reprinted from P. A. Davidson. *An Introduction to Magnetohydrodynamics*, 2nd ed. Cambridge University Press, 2017.)

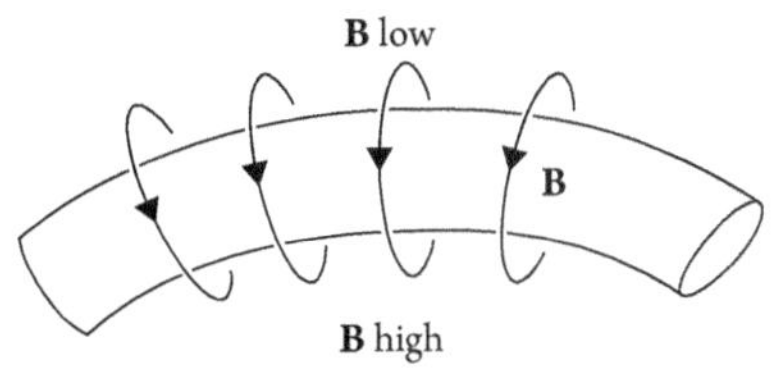


**Figure 8.12** The kink instability. (Reprinted from P. A. Davidson. *An Introduction to Magnetohydrodynamics*, 2nd ed. Cambridge University Press, 2017.)

the Lorentz force in this geometry is discussed in Exercise 8.6, where it is shown that the only significant term is the radial gradient in magnetic pressure, $\mathbf{J} \times \mathbf{B} = -\nabla p_m$. This configuration is known as the *linear pinch* and it is infamously unstable.

It is easy to see how the instability arises. Suppose $I$ is the total current passing along the cylinder. Then, from Ampère's circuital law, the surface field just outside the plasma is $B_\theta = \mu_0 I / 2\pi R$, where $R$ is the radius of the cylinder. If $R$ is locally perturbed, from $R$ to $R + \delta R$, then $B_\theta$ locally reduces by an amount $\delta B_\theta = -B_\theta \, \delta R / R$, causing a corresponding drop in magnetic pressure, $\delta p_m = B_\theta \, \delta B_\theta / \mu_0$. An instability then develops because there is a fall in magnetic pressure at those points where the radius increases and (by a similar line of reasoning) a rise in magnetic pressure where the radius decreases. This is sometimes called the *sausage-mode instability*. It can be stabilised by trapping an axial magnetic field $B_z$ within the plasma. The idea is that the sausage-mode instability will bend the axial field lines, compressing them where the plasma pinches inward and causing them to diverge at places where the plasma expands outward. However, we know from the discussion in §8.6 that the Faraday tension in the axial field lines will resist this bending and hence push back against the perturbation. In fact, it is readily shown that the plasma column is stable to axisymmetric disturbances provided that $B_z > B_\theta / \sqrt{2}$. Unfortunately, such a configuration is then unstable to non-axisymmetric disturbances in the form of the *kink instability*. Suppose that the plasma column bends, as shown in Figure 8.12. Then the external field lines are pressed together on the inside of the bend, and spread out on the outer side. Consequently the external magnetic pressure is increased on the concave side of the column and reduced on the opposite side. This produces a sideways force that amplifies the initial disturbance.

The sausage-mode and kink deformations are the simplest members of a complex zoo of magnetic instabilities that plague plasma confinement schemes. In fact, the stable confinement of plasmas by magnetic fields remains a largely unsolved problem. We take up this story again in Chapter 21.

## 8.8 Force-Free Fields and Magnetic Helicity Revisited

In certain astrophysical plasmas, such as parts of the solar atmosphere, the fluid pressure $p$ is too feeble to maintain the usual static balance, $\mathbf{J}\times\mathbf{B}=\nabla p$. In those cases equilibrium of the plasma demands that $\mathbf{J}\times\mathbf{B}=0$. Such magnetic fields are known as *force-free fields* and they are governed by $\nabla\times\mathbf{B}=\alpha\mathbf{B}$, for some $\alpha$. Vector fields that satisfy this equation are called *Beltrami fields*.

The divergence of $\nabla\times\mathbf{B}=\alpha\mathbf{B}$, combined with $\nabla\cdot\mathbf{B}=0$, tells us that $\mathbf{B}\cdot\nabla\alpha=0$, and so $\alpha$ is constant along a magnetic field line. If the field lines lie on closed $\alpha$-surfaces then those surfaces cannot be simply connected, and for precisely the same reason that constant-pressure surfaces in a fusion plasma cannot be simply connected. The simplest case is therefore that in which the magnetic field lines sit on nested tori, either closing on themselves after a finite number of loops (so-called rational surfaces), or else covering the surfaces ergodically. Alternatively, the field lines may fill a finite volume in which $\alpha$ is uniform, in which case the field is known as a constant-$\alpha$ force-free field. The curl of $\nabla\times\mathbf{B}=\alpha\mathbf{B}$, along with $\nabla\cdot\mathbf{B}=0$, tells us that constant-$\alpha$ fields satisfy the Helmholtz equation,

$$\nabla^2\mathbf{B}+\alpha^2\mathbf{B}=0. \tag{8.58}$$

Note that constant-$\alpha$ fields are necessarily localised or confined in some sense. This is because standard uniqueness theorems for elliptic PDEs (see §5.1.1) tell us that the only solution of (8.58) in an infinite domain that satisfies $\mathbf{B}=O(|\mathbf{x}|^{-3})$ at infinity is $\mathbf{B}=0$.

There is a close relationship between force-free fields and *magnetic helicity* $h_m$, defined by

$$h_m=\int_V \mathbf{A}\cdot\mathbf{B}\,dV, \tag{8.59}$$

where $V$ is any volume that satisfies $\mathbf{B}\cdot d\mathbf{S}=0$ on its bounding surface. (We introduced magnetic helicity in §7.1.) In particular, minimising magnetic energy subject to the conservation of magnetic helicity, and to a fixed normal component of $\mathbf{B}$ on the boundary, leads to a force-free magnetic field, as we now show.

Consider the ratio

$$\alpha=\int\mathbf{B}^2dV\Big/\int\mathbf{A}\cdot\mathbf{B}\,dV,$$

which is to be minimised. Infinitesimal changes to $\mathbf{B}$ about $\alpha=\alpha_0$ yield, to leading order,

$$2\int\mathbf{B}\cdot\delta\mathbf{B}\,dV=(\delta\alpha)\int\mathbf{A}\cdot\mathbf{B}\,dV+\alpha_0\int(\mathbf{A}\cdot\delta\mathbf{B}+\mathbf{B}\cdot\delta\mathbf{A})\,dV.$$

Evidently, stationary values of $\alpha$ correspond to

$$2\int \mathbf{B}\cdot\delta\mathbf{B}\,dV = \alpha_0\int(\mathbf{A}\cdot\delta\mathbf{B}+\mathbf{B}\cdot\delta\mathbf{A})\,dV = 2\alpha_0\int \mathbf{A}\cdot\delta\mathbf{B}\,dV + \alpha_0\oint(\mathbf{A}\times\delta\mathbf{A})\cdot d\mathbf{S},$$

where we have used $\nabla\cdot(\mathbf{A}\times\delta\mathbf{A}) = \mathbf{B}\cdot\delta\mathbf{A} - \mathbf{A}\cdot\delta\mathbf{B}$. Rearranging the scalar triple product $(\mathbf{A}\times\delta\mathbf{A})\cdot d\mathbf{S}$ and taking $\delta\mathbf{A}\times d\mathbf{S} = 0$ on the bounding surface eliminates the surface integral and we conclude that

$$\int(\mathbf{B}-\alpha_0\mathbf{A})\cdot\delta\mathbf{B}\,dV = 0. \tag{8.60}$$

(Note that taking the tangential component of **A** fixed on the boundary is equivalent to fixing the normal component of **B** on the boundary.) Since this holds for all possible choices of $\delta\mathbf{B}$, stationary values of $\alpha$ must correspond to $\mathbf{B}-\alpha_0\mathbf{A}=0$, from which we obtain $\nabla\times\mathbf{B}=\alpha_0\mathbf{B}$. Evidently, stationary values of magnetic energy correspond to a force-free magnetic field, and with a little more effort it is possible to show that this stationary state is in fact a minimum in energy.

In turns out that the dissipative relaxation of a plasma to a force-free magnetic field is a common feature of plasmas that have undergone the transient influence of decaying turbulence. The reason is as follows. Although it is far from obvious that it should be the case, turbulence in a highly conducting plasma usually (though not always) dissipates energy much faster than it changes the magnetic helicity (see e.g. Davidson, 2017). When this is indeed the case, a reasonable approximation to the end state of the exhausted turbulence is one in which the magnetic energy has been minimised subject to the conservation of magnetic helicity.

When there is no helicity present in the magnetic field, the story is a little different. It is readily confirmed that minimising the magnetic energy in a given domain, subject to $\delta\mathbf{B}\cdot d\mathbf{S}=0$ on its bounding surface, leads to a current-free (i.e. a potential) magnetic field within $V$. This is discussed in Exercise 8.7.

## EXERCISES

8.1 Show that the ratio of internal magnetic energy to kinetic energy for the rotating sphere of §8.5 is

$$\frac{\text{magnetic energy}}{\text{kinetic energy}} = \frac{5}{2}\left(\frac{\mathbf{B}_{\text{in}}/\sqrt{\rho\mu_0}}{\Omega R}\right)^2 = \frac{10}{9}\frac{\mu_0\sigma_e^2}{\rho},$$

where $\rho$ is the density of the sphere and $\sigma_e$ is the surface charge density.

8.2 Show that, for the rotating sphere of §8.5, the internal magnetic field is related to the external electric field by $|\mathbf{B}_{\text{in}}| = 2\Omega R\,|\mathbf{E}_{\text{out}}|/3c^2$, where $c = 1/\sqrt{\mu_0\varepsilon_0}$ is the speed of light. There is a strong radial electric field in the earth's atmosphere, of the order of $|\mathbf{E}_{\text{out}}| \sim 100\,\text{V/m}$. An early theory of the geomagnetic field attributed the earth's dipole magnetic field to rotating charges on the surface of the earth. For the earth $\mathbf{B} \sim 0.6$ G and $R \sim 6370$ km. Show that this hypothesis fails on an order-of-magnitude basis.

8.3 Consider the magnetic scalar potential defined by $\mathbf{B} = -\nabla\chi$. Show that, for the pure dipole of (8.6), $4\pi\chi = -\mu_0 \mathbf{m} \cdot \nabla(1/|\mathbf{x}|)$.

8.4 Show that, if the currents supporting $\mathbf{m}$ and $\mathbf{B}_0$ in (8.18) are non-overlapping, then the force on the dipole may be rewritten as $\mathbf{F} = (\mathbf{m} \cdot \nabla)\mathbf{B}_0$. In diamagnetic materials molecular dipoles are induced by $\mathbf{B}_0$ that are antiparallel to $\mathbf{B}_0$, whereas in paramagnetic materials molecular dipoles are induced that are parallel to $\mathbf{B}_0$. Show that diamagnetic materials are repelled by regions of strong magnetic field, whereas paramagnetic materials are attracted by regions of strong magnetic field.

8.5 Show that the potential energy of a static distribution of current may be written as

$$U_B = \frac{\mu_0}{8\pi} \iint \frac{\mathbf{J}(\mathbf{x}) \cdot \mathbf{J}(\mathbf{x}')}{|\mathbf{x} - \mathbf{x}'|} d\mathbf{x}' d\mathbf{x}.$$

8.6 Current flows along a cylindrical conductor and is a function of radius only. The associated magnetic field consists of circular field lines and so in two-dimensional polar coordinates we have $\mathbf{J} = J_z(r)\hat{\mathbf{e}}_z$ and $\mathbf{B} = B_\theta(r)\hat{\mathbf{e}}_\theta$. Use Ampère's law to find the relationship between $J_z$ and $B_\theta$ and hence show that the Lorentz force can be written as

$$\mathbf{J} \times \mathbf{B} = -\left[\frac{B_\theta^2}{\mu_0 r} + \frac{\partial}{\partial r}\left(\frac{B_\theta^2}{2\mu_0}\right)\right]\hat{\mathbf{e}}_r.$$

Confirm that the first term on the right arises from Faraday's tension in the field lines, while the second is the gradient in magnetic pressure. Also confirm that the net Lorentz force is irrotational. Note that in this particularly simple geometry there is little value in distinguishing between Faraday's tension and the magnetic pressure, since both contributions are irrotational. Finally, confirm that, when the current is concentrated in a thin surface layer at the outer radius of the conductor, the magnetic pressure term dominates.

8.7 Consider a potential (current-free) magnetic field $\mathbf{B}_0$ in a domain $V$ and with a scalar potential defined by $\mathbf{B}_0 = -\nabla\chi$. Let $\hat{\mathbf{B}}$ be any other magnetic field that can exist in the domain with the same normal component on the bounding surface of $V$. Finally, we introduce the solenoidal field $\Delta\mathbf{B}$ defined by $\hat{\mathbf{B}} = \mathbf{B}_0 + \Delta\mathbf{B}$ and $\Delta\mathbf{B} \cdot d\mathbf{S} = 0$ on the bounding surface. Show that

$$\int \hat{\mathbf{B}}^2 dV = \int \mathbf{B}_0^2 dV + \int (\Delta\mathbf{B})^2 dV - 2\oint \chi \Delta\mathbf{B} \cdot d\mathbf{S},$$

and hence deduce that the potential field represents a minimum energy state.

8.8 Consider the vector field $\mathbf{F} = \mathbf{c} \times \mathbf{x}$, where $\mathbf{c}$ is any constant vector. Apply Stokes' theorem to $\mathbf{F}$ to give

$$\int_S (2\mathbf{c}) \cdot d\mathbf{S} = \oint_C (\mathbf{c} \times \mathbf{x}) \cdot d\mathbf{r},$$

where $S$ is any surface that spans the closed curve $C$. Hence deduce that $2\int_S d\mathbf{S} = \oint_C \mathbf{x} \times d\mathbf{r}$.

8.9 Consider a circular current loop of radius $R$ that carries a current $I_1$. It is coplanar with a long straight wire that carries a current $I_2$. The distance from the wire to the centre of the loop is $d$, with $d >> R$, and the current in that part of the loop which is closest to the wire is in the same direction as that in the wire. Use symmetry to determine the direction of the net Lorentz force acting on the loop. Evaluate the Lorentz force $\mathbf{J} \times \mathbf{B}$ as a function of angular position in the loop and integrate this to show that the magnitude of the net Lorentz force is

$$F = \frac{\mu_0 I_1 I_2 R^2}{2d^2}.$$

Now derive the same result directly from the dipole force equation (8.18), $\mathbf{F} = \nabla(\mathbf{m} \cdot \mathbf{B}_0)$.

8.10 A thin circular disc of radius $R$ has a uniform surface charge density $\sigma_e$ and rotates about an axis perpendicular to its plane at a rate $\Omega$. Find the resulting dipole moment and use the results of Exercise 7.4 to show that the magnetic field on the axis of the disc, a distance $z$ from the disc, is

$$\mathbf{B} = \frac{2\mu_0 \mathbf{m}}{\pi R^4}\left(\frac{R^2 + 2z^2}{\sqrt{R^2 + z^2}} - 2z\right).$$

Find $\mathbf{B}$ for large $z/R$ and show that it is consistent with the far-field expansion (8.7).

## REFERENCES

Batchelor, G. K., 1967, *An Introduction to Fluid Dynamics*. Cambridge University Press.

Davidson, P. A., 2017, *An Introduction to Magnetohydrodynamics*, 2nd ed. Cambridge University Press.

Feynman R. P., Leighton R. B. & Sands M., 1964, *The Feynman Lectures on Physics*, Volume II. Addison-Wesley.

# 9

# Magnetic Fields in Matter

[Lodestone] not only attracts iron rings, but also imparts to them a similar power of attracting other rings; and sometimes you may see a number of pieces of iron and rings suspended from one another so as to form quite a long chain.

(Socrates, as recounted in one of Plato's dialogues, describing the antics of workers at the iron mines on the island of Samothrace)

I've seen those Samothracian iron rings leap up and iron filings in the brazen bowls seethe furiously, when underneath was set the magnet stone.

(Lucretius, *circa* 60 BC)

The magnetic properties of lodestone (i.e. magnetite, $Fe_3O_4$) were known to the ancient Chinese and Greeks, with the philosopher Thales of Miletus often being credited with the first observations in the sixth century BC. The Roman philosopher and poet Lucretius later informs us that the name *magnes* for Lodestone is derived from Magnesia, a district in Asia Minor that was rich in the ore. While Thales thought the observations provided evidence that lodestone had a soul, Lucretius attributed the phenomena to the exhalation of fine particles that stream off the lodestone and 'smites by blows the air asunder'. Of course, it was not until the advent of quantum mechanics that the observations could be adequately explained.

## 9.1 Three Types of Magnetism in Matter

### *9.1.1 Diamagnetism, Paramagnetism and Ferromagnetism*

The electrons orbiting around nuclei possess a microscopic dipole moment, and, like conventional dipoles, these can react to an ambient magnetic field. In the absence of an external field these molecular dipoles often have no macroscopic effect, because of their random orientation, but this randomness is lost in the presence of a mean magnetic field. For example, in some materials the molecular dipoles try to align with the local macroscopic field, albeit subject to other randomising effects, such as thermal agitation. The collective response of a sea of molecular dipoles to an external magnetic field can create a large-scale macroscopic field that adds to, or perturbs, the external one. When this happens, the material is said to be *magnetised*, or *magnetically polarized*.

There are three distinct ways in which atoms and molecules can react to an external magnetic field, and this leads to materials being classified as either *diamagnetic*, *paramagnetic* or *ferromagnetic*. Although these phenomena are all deeply rooted in quantum mechanics, simple cartoons based on classical physics can get across some of the ideas, at least in a qualitative sense. The first point to note is that electrons not only orbit nuclei, they also *spin* about an axis through their centres (Figure 9.1), and it turns out that both the orbit and the spin are associated with a dipole moment. In the case of the orbital motion, the origin of the dipole moment is clear, but perhaps it is less obvious for the electron spin. (Readers may wish to think of the spin dipole moment as like the rotating charged sphere of §8.5.) In diamagnetism it is the orbital dipole moment that is affected by the external magnetic field, while in paramagnetism and ferromagnetism it is the spin dipole moment that reacts to the field.

In diamagnetism the introduction of a mean magnetic field speeds up or slows down the orbital motion of an electron, and this changes the magnitude (but not the orientation) of its dipole moment $\mathbf{m}$. As we shall see in the next subsection, the net effect is to induce a systematic component of $\mathbf{m}$ that is *antiparallel* to the mean field $\mathbf{B}$. As discussed in Exercise 8.4, this means that diamagnetic materials are repelled by regions of strong magnetic field. Diamagnetism is a universal phenomenon that affects all materials. However, it is *extremely* weak, typically perturbing the external magnetic field by a tiny fraction of a percent.

In paramagnetism the torque $\mathbf{m} \times \mathbf{B}$ associated with the external magnetic field tends to *tilt* the axis of the *spin dipoles*, in much the same way as a current loop that is free to rotate tends to orientate itself with the local mean field. There is clearly an analogy here to orientation polarization in a dielectric, and Exercise 8.4 tells us that paramagnetic materials are attracted to regions of intense magnetic field. Paramagnetism tends to be stronger than diamagnetism, when it manifests itself, but it is still weak in the sense that it only perturbs the external magnetic field by a fraction of a percent. Moreover, it is restricted to certain

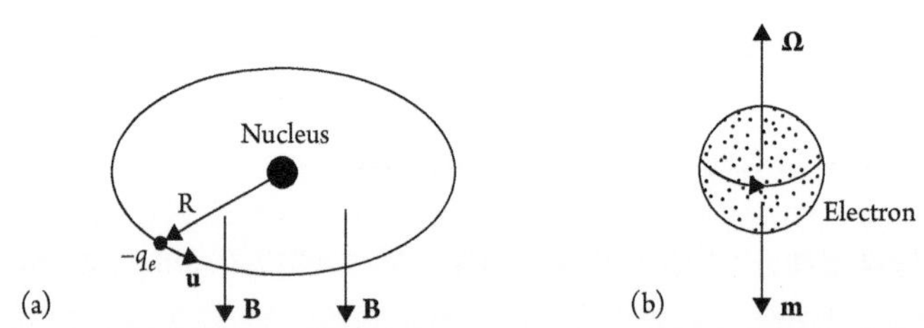


**Figure 9.1** (a) The orbit of an electron about a nucleus. (b) The spin of an electron about an axis through its centre. Both the orbit and the spin are associated with dipole moments.

materials only. The reason is that, as a result of the Pauli exclusion principle, most electrons come in pairs, in which the spin axes of the two electrons are constrained to point in opposite directions, resulting in no net dipole moment. Paramagnetism cannot operate in such cases. In atoms with an odd number of electrons, the unpaired electron usually occupies the outer orbit, and in crystals or compounds their spin is often paired off with an electron in an adjacent atom. However, a few elements have unpaired electrons whose spin remains uncoupled, and these paramagnetic materials include some of the rare earth and transition elements. Moreover, a few molecules, such as nitric oxide (NO), have unpaired electrons and these also exhibit paramagnetism.

Finally, ferromagnetism is found in compounds or alloys containing iron, nickel or cobalt (and occasionally in more exotic materials). It also involves the tilting of the spin axis of an unpaired electron, but it is orders of magnitude stronger than paramagnetism. It differs from diamagnetism and paramagnetism in that some residual magnetisation usually remains after the external magnetic field is removed, so that hysteresis (history) effects are important and permanent magnets common. Another key difference is that individual molecular dipoles react not just to an external magnetic field, but also to the field of neighbouring dipoles. In particular, small *domains* less than 1 mm across spontaneously form, within which all the molecular dipoles are aligned, and this occurs without the aid of an external magnetic field. However, in an unmagnetised ferromagnetic material these small domains are randomly orientated and so there is no overall magnetisation. The role of an external magnetic field is then to force those domains whose dipoles are aligned with the external field to grow at the expense of the domains that are misaligned, thus producing a mean magnetisation in the material.

### *9.1.2 A Non-Quantum Cartoon on Diamagnetism*

Somewhat surprisingly, there is a simple, non-quantum cartoon of diamagnetism that is not too misleading and does a reasonable job of describing the phenomenon. Consider an electron of charge $-q_e$ and mass $m_e$ that orbits a nucleus at a radius of $R$ and with a speed $u$ (Figure 9.2). If we consider a time-averaged picture, then the rotating charge constitutes a current of $I = -q_e/(2\pi R/u)$, and this in turn is associated a magnetic dipole of $\mathbf{m} = -(q_e u R/2)\hat{\mathbf{e}}$, where the unit vector $\hat{\mathbf{e}}$ points along the orbital axis.

In the absence of a mean magnetic field the radially inward acceleration of the electron is maintained by the attractive Coulomb force, according to

$$\frac{1}{4\pi\varepsilon_0}\frac{q_e^2}{R^2} = m_e\frac{u_0^2}{R}. \tag{9.1}$$

Now suppose that a mean magnetic field $\mathbf{B}$ is introduced that, for simplicity, is parallel or antiparallel with the orbital axis, $\mathbf{B} = B\hat{\mathbf{e}}$. This causes an additional force to act on the electron of $-q_e\mathbf{u} \times \mathbf{B}$, which is also directed radially inward, and so the radial force balance is adjusted to read

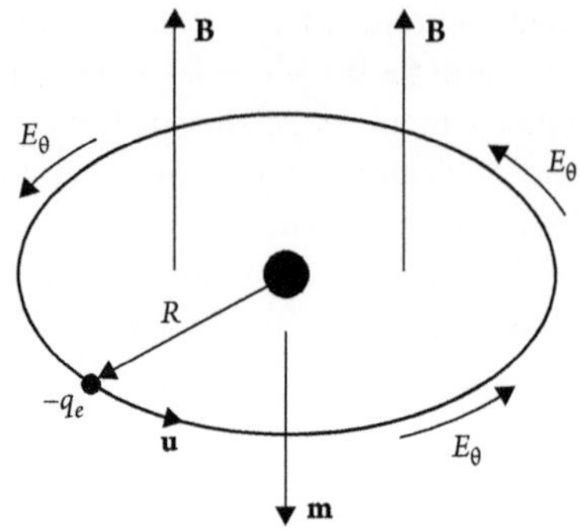


**Figure 9.2** The influence of an external field on the orbital motion of an electron.

$$\frac{1}{4\pi\varepsilon_0}\frac{q_e^2}{R^2} + q_e u_f B = m_e \frac{u_f^2}{R}, \tag{9.2}$$

where $u_f$ is the final orbital speed after the application of **B**. (We are assuming here that the introduction of **B** does not change $R$. This is a somewhat arbitrary assumption, but let us see where it leads.) Subtracting (9.1) from (9.2) now yields

$$q_e u_f B = m_e \frac{u_f^2 - u_0^2}{R} = m_e \frac{(u_f - u_0)(u_f + u_0)}{R} \approx \frac{2m_e u_f (u_f - u_0)}{R}, \tag{9.3}$$

where we have assumed that the change in speed is small. It follows that the perturbation in orbital speed is $\delta u = q_e RB/2m_e$, with a corresponding change in dipole moment of

$$\delta \mathbf{m} = -\frac{q_e \delta u R}{2}\hat{\mathbf{e}} = -\frac{q_e^2 R^2}{4m_e}\mathbf{B}. \tag{9.4}$$

Evidently, the change in dipole moment is antiparallel to **B**. If the molecular dipoles are randomly orientated before the introduction of **B**, they cease to be so afterwards because each dipole picks up a small bias in the opposite direction to **B**. This is diamagnetism.

Note that the acceleration or deceleration of the electron during the implementation of **B** cannot be caused by the Lorentz force itself, since this force does no work. Rather, during the imposition of **B** a transient azimuthal electric field $E_\theta$ is generated in accordance with Faraday's law, and it is this that accelerates or decelerates the electron. Consider the situation shown in Figure 9.2. Applying Faraday's law in integral form to the orbital path during the transient application of **B** allows us to calculate $E_\theta$:

$$\oint \mathbf{E}\cdot d\mathbf{r} = 2\pi R E_\theta = -\frac{d}{dt}\int \mathbf{B}\cdot d\mathbf{S} = -(\pi R^2)\frac{\partial B}{\partial t}.$$

This, in turn, yields the work done by the azimuthal Coulomb force $f_\theta$ during the application of the external field:

$$\int f_\theta \, d\ell = \int f_\theta \, u \, dt = \int (-q_e E_\theta) u \, dt = \int \frac{q_e R}{2} \frac{\partial B}{\partial t} u \, dt \approx \frac{1}{2} q_e R B u_f, \tag{9.5}$$

where the change in $u$ is assumed small. Equating this to the change in the kinetic energy of the electron gives us

$$\frac{1}{2} m_e (u_f^2 - u_0^2) \approx \frac{1}{2} q_e R B u_f, \tag{9.6}$$

which brings us back to (9.3). The fact that this second analysis yields the same result as the first lends support to the assumption that the introduction of **B** does not change $R$.

## 9.2 The Magnetisation Vector and Bound Current Distributions

We now seek to develop a model that allows us to predict the macroscopic magnetic field created by the collective organisation of the molecular dipoles. In many ways this mimics the analysis of §6.2, where we found the macroscopic electric field associated with molecular electric dipoles. The approach rests on the fact that the molecular dipoles are very small, so that the magnetic field generated by such a dipole can be approximated by the far-field contribution to **B** of a localised current distribution, as given by (8.7).

We start by introducing the *magnetisation vector* **M**, which is the net dipole moment per unit volume defined by $\mathbf{M}\, dV = \sum \mathbf{m}_{\text{mol}}$, where $\mathbf{m}_{\text{mol}}$ are the molecular dipoles in the volume element $dV$. Note that **M** is a macroscopic (locally averaged) field. Note also that it may be a consequence of diamagnetism, paramagnetism or ferromagnetism; it does not matter which. As in §6.2, the philosophy is to assume that $\mathbf{M}(\mathbf{x})$ is somehow given, and then explore the consequences of a prescribed distribution of **M**. This postpones the delicate problem of relating **M** to the externally applied field.

Starting with (8.6), replacing **m** by $\mathbf{M}\, dV$ and then integrating over the magnetised body, we find that the macroscopic vector potential either inside or outside the body is

$$\mathbf{A}(\mathbf{x}) = \frac{\mu_0}{4\pi} \int \nabla \left( \frac{1}{|\mathbf{r}|} \right) \times \mathbf{M}(\mathbf{x}') \, d\mathbf{x}', \qquad \mathbf{r} = \mathbf{x} - \mathbf{x}'. \tag{9.7}$$

Here, as usual, $\mathbf{x}$ locates the point where we measure the field and $\mathbf{x}'$ is a dummy variable that samples the source at different locations. Note that $\nabla$ operates on $\mathbf{x}$ while treating $\mathbf{x}'$ as a constant, and that (7.50) gives us

$$\nabla \left( \frac{1}{|\mathbf{r}|} \right) = -\frac{\mathbf{r}}{|\mathbf{r}|^3} = -\nabla' \left( \frac{1}{|\mathbf{r}|} \right). \tag{9.8}$$

Our task now is to manipulate the integral (9.7) into a more useful form. First we express the gradient in terms of $\nabla'$ using (9.8). Next, invoking the identity

$$\nabla' \times \left(\frac{\mathbf{M}'}{|\mathbf{r}|}\right) = \nabla'\left(\frac{1}{|\mathbf{r}|}\right) \times \mathbf{M}' + \frac{\nabla' \times \mathbf{M}'}{|\mathbf{r}|} \tag{9.9}$$

yields

$$\mathbf{A}(\mathbf{x}) = -\frac{\mu_0}{4\pi}\int \nabla'\left(\frac{1}{|\mathbf{r}|}\right) \times \mathbf{M}(\mathbf{x}')\, d\mathbf{x}' = \frac{\mu_0}{4\pi}\int \frac{\nabla' \times \mathbf{M}'}{|\mathbf{r}|}\, d\mathbf{x}' - \frac{\mu_0}{4\pi}\int \nabla' \times \left(\frac{\mathbf{M}'}{|\mathbf{r}|}\right) d\mathbf{x}'. \tag{9.10}$$

Finally we rewrite the integral on the far right as a surface integral, from which we obtain

$$\boxed{\mathbf{A}(\mathbf{x}) = \frac{\mu_0}{4\pi}\int \frac{\nabla' \times \mathbf{M}'}{|\mathbf{r}|}\, d\mathbf{x}' + \frac{\mu_0}{4\pi}\oint \frac{\mathbf{M}'}{|\mathbf{r}|} \times d\mathbf{S}'.} \tag{9.11}$$

Compare (9.11) with the Biot–Savart integral

$$\mathbf{A}(\mathbf{x}) = \frac{\mu_0}{4\pi}\int \frac{\mathbf{J}(\mathbf{x}')}{|\mathbf{r}|}\, d\mathbf{x}'. \tag{9.12}$$

The macroscopic field induced by magnetic polarization is evidently equivalent to a set of volumetric and surface current densities given by

$$\boxed{\mathbf{J}_{\text{mag}} = \nabla \times \mathbf{M}, \qquad \mathbf{K}_{\text{mag}} = \mathbf{M} \times \mathbf{n},} \tag{9.13}$$

where **n** is an (outward-pointing) unit normal vector to the closed surface. These are known as the *bound currents* associated with the magnetisation and, like **M** and **A**, they represent a macroscopic field. In terms of these bound currents, we have

$$\boxed{\mathbf{A}(\mathbf{x}) = \frac{\mu_0}{4\pi}\int \frac{\mathbf{J}'_{\text{mag}}}{|\mathbf{r}|}\, d\mathbf{x}' + \frac{\mu_0}{4\pi}\oint \frac{\mathbf{K}'_{\text{mag}}}{|\mathbf{r}|}\, dS',} \tag{9.14}$$

from which the magnetic field may be calculated. This is the most useful form of the magnetisation integral. Given a distribution of **M**, we can calculate the bound currents from (9.13) and then find the macroscopic magnetic field using the Biot–Savart law, treating the bound currents as if they were real currents. (Note that we have not addressed the delicate question that arises within the magnetised material of how to handle the near field of those dipoles that lie very close to **x**. However, this is discussed at length in Lorrain & Corson (1970) and the end result is that (9.14) requires no modification.)

There is, of course, a direct analogy to the use of bound charges in dielectrics to find the macroscopic electric field (see §6.2.1). Moreover, just as the bound charges in a dielectric represent the accumulation of real charge within the interior and on the surface of the dielectric, so it can be useful to think of the bound currents of (9.13) as being a consequence of real macroscopic currents, averaged (in some sense) over the molecules.

The usual argument goes like this. Suppose, for simplicity, that the local value of **M** points in the $z$ direction and consider a cubic volume of material $dV = dx\,dy\,dz$ that is very small from the macroscopic point of view but very large when compared with the size of the molecules. Then **M** is well defined and given by $\mathbf{M}\,dV = \sum \mathbf{m}_{\text{mol}}$. Now $\sum \mathbf{m}_{\text{mol}}$ is associated with some complicated three-dimensional arrangement of spinning and orbiting electrons whose description must be quantum-mechanical. However, as far as the macroscopic magnetic field is concerned, all that matters is $\mathbf{M}\,dV$, and so the same macroscopic field is obtained if we replace the three-dimensional sea of spinning electrons within $dV$ by a current $I_{\text{mag}}$ circulating around our cubic volume in the $x$–$y$ plane, whose magnitude is defined by $I_{\text{mag}}\,dx\,dy = M_z\,dV$ and whose direction is given by the right-hand rule. This is shown in Figure 9.3. We shall now show that the bound currents defined by (9.13) are just the sum of many such macroscopic current loops.

Consider first the case where **M** is uniform and points, say, in the $z$ direction. It is helpful to consider a slice of the material parallel to the $x$–$y$ plane that spans the magnetised body, is of thickness $t$ and has a perimeter $C$. If we divide up the slice into lots of rectangular volume elements $dV = t\,dx\,dy$, then, for each volume element, we have $I_{\text{mag}}\,dx\,dy = M_z\,dV$, or equivalently $I_{\text{mag}} = M_z t$ (Figure 9.4). But $M_z$ is uniform and so the current $I_{\text{mag}}$ is the same in each rectangular element, and when we sum over all such current loops the internal currents cancel in the usual way and we are just left with the current $I_{\text{mag}}$ circulating around the perimeter $C$. This is associated with a surface current density of magnitude $K_{\text{mag}} = I_{\text{mag}}/t = M_z$, and since the vector $\mathbf{K}_{\text{mag}}$ is perpendicular to **M** and to the unit normal vector **n**, we have $\mathbf{K}_{\text{mag}} = \mathbf{M} \times \mathbf{n}$, exactly as in (9.13). (The order of **M** and **n** in the cross product is determined by the right-hand rule applied to **K**.) Note, however, that **K** is a strange kind of current, in that no electron actually makes it all the way around the perimeter $C$, or even a short part of the way around $C$. Rather, all the orbiting electrons are bound to their parent nuclei: hence the name, *bound current.*

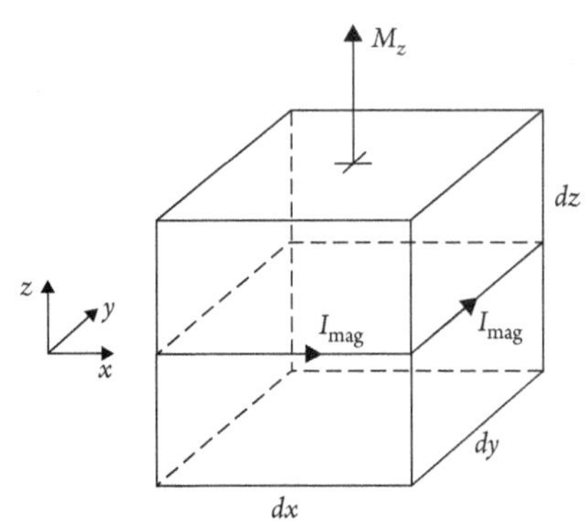


**Figure 9.3** The current $I_{\text{mag}}$ defined by $I_{\text{mag}}\,dx\,dy = M_z\,dV$ generates the same macroscopic magnetic field as the sea of spinning electrons contained within $dV$.

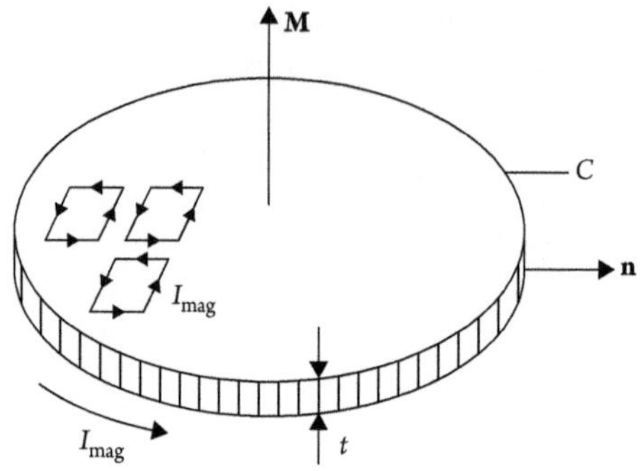


**Figure 9.4** When **M** is uniform, the current $I_{mag}$ that circulates around each loop is the same. Adding all the current loops gives a current $I_{mag}$ circulating around the perimeter.

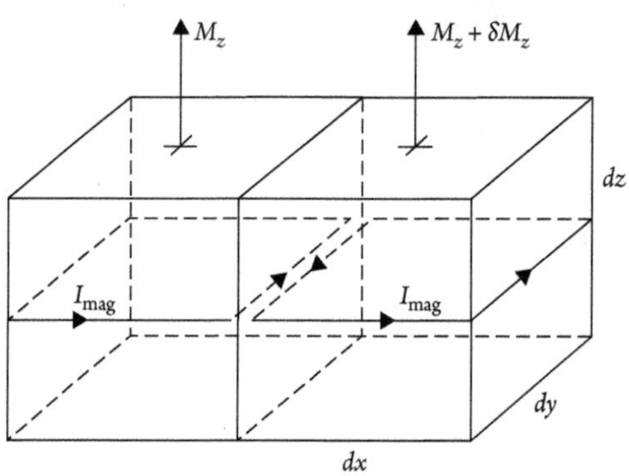


**Figure 9.5** The current $I_{mag}$, defined by $I_{mag}\, dx\, dy = M_z\, dV$, varies between adjacent volume elements when $M_z$ is non-uniform.

We now turn to the volumetric bound current $\mathbf{J}_{mag}$ that arises when **M** is non-uniform. For simplicity, we shall again consider the case where **M** points in the $z$ direction and focus attention on a slice of material lying parallel to the $x$–$y$ plane and of thickness $dz$. Consider, in particular, two adjacent volume elements $dV = dx\, dy\, dz$ whose centres are separated by $dx$, as shown in Figure 9.5. Around each volume element in the $x$–$y$ plane there circulates a current, given by $I_{mag} = M_z\, dz$, exactly as discussed above. However, this time the current is not the same in adjacent elements, and so we no longer get a perfect cancelation of currents at the interface. Rather, for the two cubes shown in Figure 9.5, there is a net current at the interface of $\delta I_y = -\delta M_z\, dz = -(\partial M_z/\partial x)\, dx\, dz$, where the minus sign comes from the right-hand rule. (From now on we shall drop the subscript 'mag' on the current $I$ for simplicity.) We may associate this interfacial current with a volumetric current density $J_y = \delta I_y/(dx\, dz)$, from which we find $J_y = -\partial M_z/\partial x$. If we now repeat the analysis for two volume elements separated in $y$, rather then $x$, then we find interfacial currents flowing in the $x$ direction, and these give rise to a current density $J_x = \partial M_z/\partial y$. Combining these two components of $\mathbf{J}_{\text{mag}}$ gives $\mathbf{J}_{\text{mag}} = \nabla \times \mathbf{M}$, exactly as in (9.13). All of this extends in an obvious way to three dimensions, although we shall not pause to confirm this.

## 9.3 The Uniformly Magnetised Sphere

In order to illustrate the utility of (9.13) and (9.14), let us consider the problem of a uniformly magnetised sphere of radius $R$, as shown in Figure 9.6. We shall take $\mathbf{M}$ to be uniform within the sphere and to point in the $z$ direction, so that in spherical polar coordinates (9.13) gives us $\mathbf{J}_{\text{mag}} = 0$ and $\mathbf{K}_{\text{mag}} = \mathbf{M} \times \hat{\mathbf{e}}_r = M \sin\theta\, \hat{\mathbf{e}}_\varphi$. So the surface bound current circulates in the azimuthal direction, is symmetric about the midplane and falls to zero at the top and bottom of the sphere. But we have already met this current distribution in §8.5, where we considered the rotating charged sphere. In that case the surface current was $\mathbf{K} = \sigma_e \boldsymbol{\Omega} \times \mathbf{x} = \sigma_e \Omega R \sin\theta\, \hat{\mathbf{e}}_\varphi$, where $\sigma_e$ is the surface charge density and $\boldsymbol{\Omega}$ the angular velocity of the sphere. So we may immediately lift the solution for our magnetised sphere from (8.47)–(8.49), but with $\mathbf{M}$ replacing $\sigma_e R \boldsymbol{\Omega}$. The vector potential outside the sphere is then given by (8.48) in the form

$$\mathbf{A}_{\text{out}}(\mathbf{x}) = \frac{\mu_0}{4\pi} \nabla \left( \frac{1}{|\mathbf{x}|} \right) \times (\mathbf{M} V_R), \quad |\mathbf{x}| > R,$$

which is exactly the vector potential for a point dipole located at the origin and of strength $\mathbf{m} = \mathbf{M} V_R$. Moreover, (8.49) tells us that the magnetic field inside the sphere is uniform, aligned with $\mathbf{M}$ and of magnitude

$$\mathbf{B}_{\text{in}} = \frac{2}{3} \mu_0 \mathbf{M}, \quad |\mathbf{x}| < R. \tag{9.15}$$

Note that $\mathbf{B}_{\text{in}} V_R = 2\mu_0 \mathbf{m}/3$, as demanded by (8.26).

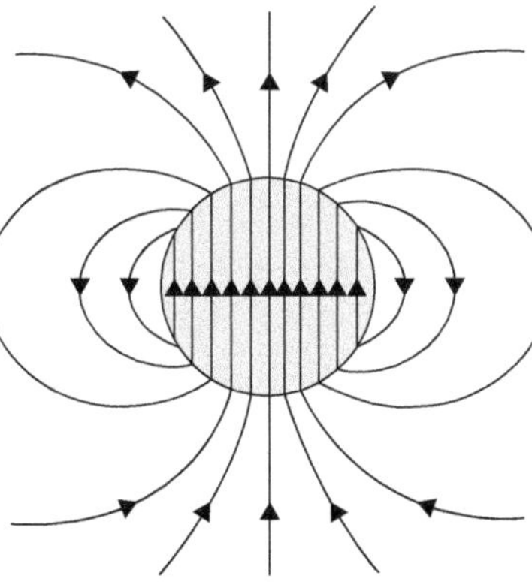

**Figure 9.6** The magnetic field of a uniformly magnetised sphere. (Adapted from H. C. Ohanian, *Classical Electrodynamics*. Allyn and Bacon, 1988.)

## 9.4 Ampère's Law Applied to Magnetic Materials: The H Field

We now turn from the Biot–Savart law to Ampère's law. Since we are treating the bound current as if it were a real current, at least as far as calculating the magnetic field is concerned, it must be included on the right-hand side of Ampère's law along with the conduction current. (We shall not worry about other forms of **J**, such as polarization current, at this stage.) Ampère's law then becomes, with the help of (9.13),

$$\nabla \times \mathbf{B} = \mu_0(\mathbf{J}_f + \mathbf{J}_{\text{mag}}) = \mu_0(\mathbf{J}_f + \nabla \times \mathbf{M}),$$

where $\mathbf{J}_f$ is the conduction current associated with the free changes. At this point it is conventional, and indeed convenient, to introduce an auxiliary field, denoted by **H** and defined by

$$\boxed{\mathbf{H} = \frac{\mathbf{B}}{\mu_0} - \mathbf{M}.} \tag{9.16}$$

Evidently, the units of **H** are the same as those of **M**, namely A/m. When recast in terms of **H**, Ampère's law take the particularly simple form

$$\boxed{\nabla \times \mathbf{H} = \mathbf{J}_f.} \tag{9.17}$$

This is convenient in the sense that we usually know, or control, the conduction current. For example, in the case of an electromagnet the conduction current is known in the coils that encircle the ferromagnetic material. So the philosophy behind (9.17) is to try and determine **H** from Ampère's circuital law in the form

$$\oint_C \mathbf{H} \cdot d\mathbf{r} = \int_S \mathbf{J}_f \cdot d\mathbf{S} = I_f \tag{9.18}$$

and then find **B** from (9.16), assuming **M** is known.

For example, suppose we have a long solenoid consisting of a soft iron core inside a tightly wound helical coil carrying a current $I$ and with $N$ turns per unit length (Figure 9.7). Then we may apply (9.18) in exactly the same way as we applied Ampère's circuital law to a non-magnetic solenoid in Chapter 7. That is, we take the curve $C$ to straddle the coil, lying parallel to the axis of the solenoid just inside and outside the coil. This tells us that **H** inside the solenoid is uniform and axial and of magnitude $H = NI$. If, given **H**, we can find **M** in the magnetised core (and we shall see how to do this in the next section), then we can get **B** from (9.16) and we have all the fields. However, the calculation of **H** is rarely this simple, in large part because **H** is usually not solenoidal. Let us explore this point in a little more detail.

In terms of **B**, (9.16) reads $\mathbf{B} = \mu_0(\mathbf{H} + \mathbf{M})$, and since $\nabla \cdot \mathbf{B} = 0$ we have $\nabla \cdot (\mathbf{H} + \mathbf{M}) = 0$. Now **M** is usually not solenoidal. For example, it is discontinuous at the ends of a

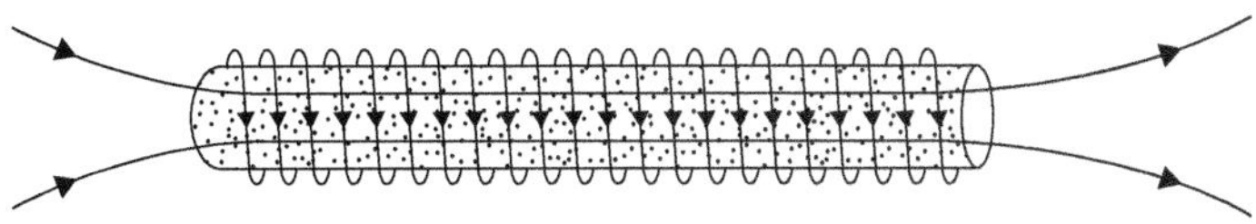

**Figure 9.7** A long solenoid with a soft iron core.

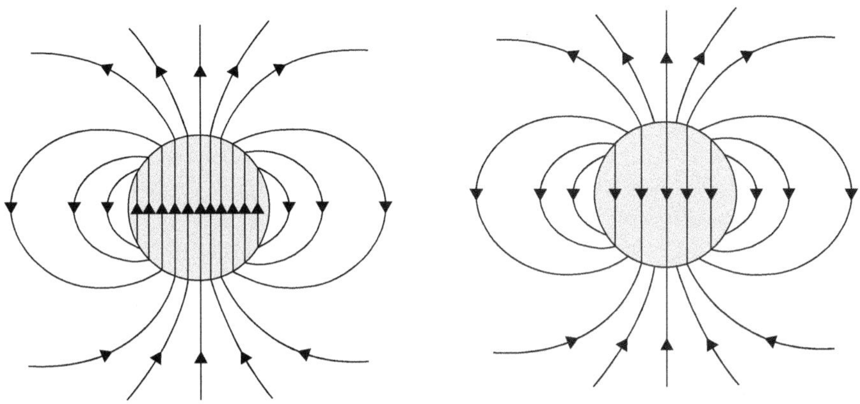

**Figure 9.8** A uniformly magnetised sphere. **M** points upwards. The **B** field is on the left and **H** on the right. Note that the **H** lines are discontinuous, with $\mathbf{H}_{\text{in}} = -\mathbf{B}/2\mu_0 = -\mathbf{M}/3$. (Adapted from H. C. Ohanian, *Classical Electrodynamics*. Allyn and Bacon, 1988.)

simple bar magnet, and at the surface of the magnetised sphere discussed in §9.3. It follows that, in general, $\nabla \cdot \mathbf{H} \neq 0$ and so the **H**-lines need *not* be continuous in the way the **B**-lines are. This is illustrated in Figure 9.8, which shows the magnetic field associated with a uniformly magnetised sphere. The **B**-lines on the left are continuous, while the **H**-lines on the right are discontinuous at the surface of the sphere, corresponding to a discontinuity in **M**. In fact, combining (9.15) with (9.16) tells us that, inside the sphere,

$$\mathbf{H}_{\text{in}} = -\frac{\mathbf{B}}{2\mu_0} = -\frac{\mathbf{M}}{3},$$

so that the **H**-lines *reverse direction* at the surface of the sphere.

Perhaps some comments are in order at this point. First, although **H** can be a useful intermediary for calculation purposes (when magnetic materials are present), it is important to keep in mind that **B** is the fundamental field and **H** merely a convenient construct. This point is argued forcefully in a number of texts, perhaps most notably in Purcell (2011). Second, there is no consensus in the literature as to what to call **H**, so we shall refrain from giving it a name. Some authors call it the magnetic field or magnetic field intensity, but this simply creates endless confusion between **H** and **B**. Third, the allure of (9.17) or (9.18) is somewhat illusory. The point is that a vector field is only uniquely determined once its curl and divergence are both specified. When working with **B**, Ampère's law is enough to fix the magnetic field (if $\mathbf{J} = \mathbf{J}_f + \mathbf{J}_{\text{mag}}$ is known), because we have $\nabla \cdot \mathbf{B} = 0$. However, in general

$\nabla \cdot \mathbf{H} \neq 0$, so that (9.17) does *not* uniquely determine **H**. Admittedly, for some particularly simple geometries (like our long solenoid) symmetry ensures that $\nabla \cdot \mathbf{H} = 0$, in which case (9.17) is sufficient to determine **H**. However, such situations are rather rare. As we shall see when we discuss magnetic circuits, one usually has to find a way to combine (9.18) with the fact that $\nabla \cdot \mathbf{B} = 0$.

## 9.5 The Magnetic Susceptibility and Relative Permeability

So far we have studiously avoided the problem of determining **M**, simply assuming that it is (somehow) known. We now introduce a widely used constitutive law that closes the loop and relates **M** back to **B**. This relationship is analogous to the one in which we related the electric displacement vector to the electric field in a dielectric material.

In diamagnetic and paramagnetic materials the induced magnetisation **M** is observed to be proportional to the externally imposed magnetic field, as suggested by (9.4) in the case of diamagnetism, and by the balance between the alignment torque $\mathbf{T} = \mathbf{m} \times \mathbf{B}$ and random thermal agitation in paramagnetism. (See (6.28) for the analogous behaviour in polar dielectrics.) So, for these materials we have a linear relationship between **M** and **B**, or equivalently between **M** and **H**. Traditionally this has been written in the form $\mathbf{M} = \chi_m \mathbf{H}$, where $\chi_m$ is a dimensionless material constant called the *magnetic susceptibility*. It is positive for paramagnetic materials and negative for diamagnetism. Typical values of $\chi_m$ are given in Table 9.1 and they are all very small.

In ferromagnetism the relationship is more complicated. In particular, hysteresis and the associated phenomenon of permanent (i.e. residual) magnetism normally plays an important role. However, for certain ferromagnetic materials, such as annealed *soft iron*, or silicon steel, hysteresis is relatively weak, especially if the imposed field is not too large. These *soft magnetic materials* exhibit only mild residual magnetism and are widely used in applications where hysteresis is undesirable, such as in transformer cores and electromagnets. To the extent that hysteresis and microstructural anisotropy may be neglected in soft magnetic materials, which is somewhat of an idealisation, we may think of **M** as being a *single-valued* function of **H**, with **M** uniquely determined by a *nonlinear* curve $\mathbf{M} = F(\mathbf{H})$. Such a relationship is known as a *magnetisation curve* and an example is shown in Figure 9.9(a). So, for soft iron subject to modest fields, it is often convenient to approximate the relationship

**Table 9.1** Magnetic susceptibilities of some common materials at room temperature

| Diamagnetic material | Magnetic susceptibility $\chi_m$ | Paramagnetic material | Magnetic susceptibility $\chi_m$ |
|---|---|---|---|
| Water | $-9 \times 10^{-6}$ | Aluminium | $2 \times 10^{-5}$ |
| Copper | $-9 \times 10^{-6}$ | Tungsten | $7 \times 10^{-5}$ |
| Silver | $-2 \times 10^{-5}$ | Platinum | $3 \times 10^{-4}$ |

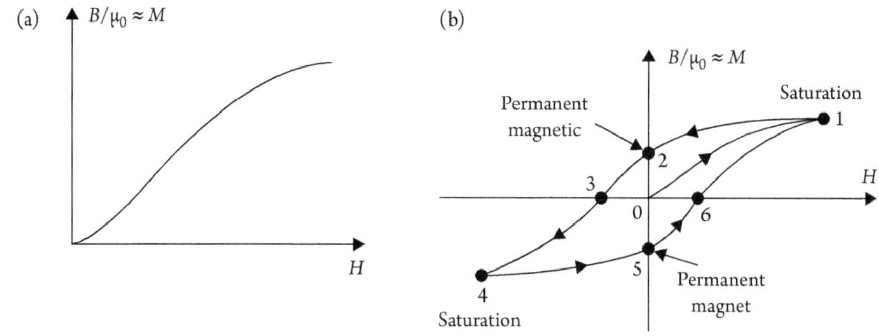


**Figure 9.9** *B–H* magnetisation curves for (a) an 'ideal' soft iron with negligible hysteresis or anisotropy and (b) a ferromagnetic material exhibiting significant hysteresis.

between **M** and **H** by the usual constitutive law $\mathbf{M} = \chi_m \mathbf{H}$, where $\chi_m$ is now a function of $|\mathbf{H}|$. Values of $\chi_m$ for soft iron are of the order of $10^3$ or $10^4$, i.e. vast by comparison with paramagnetic materials.

Actually, for historical reasons, magnetisation curves for ferromagnetic materials are usually plotted in the form $\mathbf{B} = F(\mathbf{H})$, rather than $\mathbf{M} = F(\mathbf{H})$. However, for soft magnetic materials **M** completely dominates the right-hand side of $\mathbf{B}/\mu_0 = \mathbf{H} + \mathbf{M}$ and so plotting **M** versus **H** is more or less equivalent to plotting $\mathbf{B}/\mu_0$ against **H**. Schematic examples of such plots are shown in Figure 9.9 for an 'ideal' soft iron (i.e. no hysteresis and no anisotropy) and for a ferromagnetic material exhibiting significant hysteresis—a so-called *magnetically hard* material. Particularly hard magnetic materials are used for making permanent magnets. We shall discuss hard magnetic materials and permanent magnets in §9.8, but in the meantime we shall focus on magnetically soft materials.

In summary, then, for diamagnetic and paramagnetic materials, and for soft ferromagnetic materials in which the imposed field is not too large and hysteresis may be (tentatively) ignored, we may adopt the constitutive law $\mathbf{M} = \chi_m \mathbf{H}$. This then yields

$$\boxed{\mathbf{B} = \mu_0(\mathbf{H} + \mathbf{M}) = \mu_0(1 + \chi_m)\mathbf{H} = \mu \mathbf{H},} \tag{9.19}$$

where $\mu = \mu_0(1 + \chi_m)$ is called the *permeability* of the material. Some authors write $\mathbf{B} = \mu_r \mu_0 \mathbf{H}$, where $\mu_r = 1 + \chi_m$ is known as the *relative permeability*. If $\chi_m$ is taken to be a constant in (9.19), then the material is said to be a *linear* medium. This is a good model for diamagnetic and paramagnetic materials, and a rough approximation to soft ferromagnetic materials subject to modest external fields. For soft ferromagnetic materials subject to larger fields, a better approximation is to allow $\chi_m$ to vary with $|\mathbf{H}|$, in which case the material is classified as *nonlinear*. Note, however, that (9.19) makes no sense for ferromagnetic materials in which permanent magnetism plays an important role.

It is now clear how to find **B** within the solenoid shown in Figure 9.7, which has a soft iron core. Ampère's circuital law in the form (9.18) tells us that $H = NI$, and then (9.19) gives us $B = \mu NI$. However, as already emphasised, this geometry is unusual in that symmetry ensures $\nabla \cdot \mathbf{H} = 0$, so that **H** follows directly from (9.18). In more complicated geometries (9.18) and (9.19) have to be combined with $\nabla \cdot \mathbf{B} = 0$ in order to find **B** or **H**.

## 9.6 Magnetic Circuits and Electromagnets

An important application of ferromagnetic materials is the core of an electromagnet. A simple example is shown in Figure 9.10. The iron core, which is typically soft iron or some other ferromagnetic material in which hysteresis is low, is energised by a coil with multiple turns, say $N_t$ in total, and a current $I$. Typically there is an airgap in the iron core (or magnetic yoke). The coil induces a magnetic field **B**, which, in turn, induces a strong magnetisation **M** in the yoke. At first glance it might be thought that this magnetisation is restricted to that part of the magnetic yoke which lies close to the coil, but in fact this is not the case. What is observed is that the entire magnetic core develops a strong magnetisation, with the magnetic field being very intense at all points inside the yoke and relatively weak outside. One way to understand this is to think of the **B**-field of the coil locally magnetising the yoke, which then creates a large local **B**-field in the ferromagnetic material, which in turn magnetises adjacent parts of the yoke, and so on.

So the hallmark of these kinds of *magnetic circuits* is that the magnetic field more or less runs parallel to the axis of the yoke, being very intense inside the core (and in the airgap) and weak outside. The magnetic field that lies outside the magnetic circuit is referred to as *leakage flux* and a well-designed electromagnet will minimise this. This forms the basis of an *approximate*, if reasonably effective, means of calculating the properties of electromagnets. The leakage flux is ignored and it is assumed that **B** runs parallel to the axis of the magnetic circuit and is uniform at any one cross-section. Also, for simplicity, hysteresis and anisotropy in the soft iron is assumed to be negligible, so that the permeability in the expression $\mathbf{B} = \mu\mathbf{H}$ is taken to be a single-valued scalar function of **H**, thus allowing for phenomena like saturation but not for history effects.

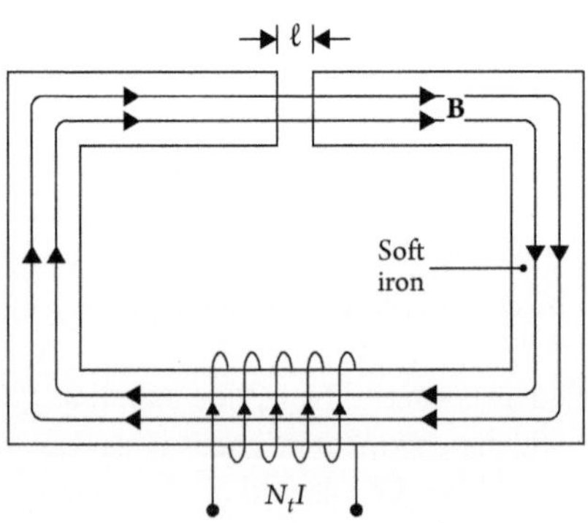


**Figure 9.10** An electromagnet with a single coil and an air gap.

One then treats the magnetic circuit rather like an electrical circuit, with the magnetic flux $\Phi$ being the analogue of current. The magnetic circuit is divided up into idealised segments, which in the case of Figure 9.10 are the yoke and the airgap. Each segment has associated with it a length measured along the centreline of the circuit, $\ell_i$, a cross-sectional area $S_i$ and a permeability $\mu_i$, which is $\mu_0$ in the case of the airgap. From these, each segment is assigned a so-called *reluctance* defined as $\ell_i/\mu_i S_i$, which is the analogue of electrical resistance. We now combine the fact that $\Phi$ is constant around the loop, which follows from $\nabla \cdot \mathbf{B} = 0$, with Ampère's circuital law in the form of (9.18). This gives us

$$\oint_C \mathbf{H} \cdot d\mathbf{r} = \sum_i H_i \ell_i = \sum_i \frac{B_i}{\mu_i} \ell_i = \Phi \sum_i \frac{\ell_i}{\mu_i S_i} = N_t I, \tag{9.20}$$

from which we can determine the magnetic flux in the circuit,

$$\Phi = \frac{N_t I}{\sum_i (\ell_i/\mu_i S_i)} = \frac{N_t I}{\sum \text{reluctances}}. \tag{9.21}$$

This is the analogue of $I = \text{EMF}/\sum$ resistances in a 'series' electrical circuit, with $N_t I$ playing the role of an EMF. In fact, the analogy to electrical circuits can be readily formalised in a quite general way, with the EMF, $I$ and resistance playing the roles of $N_t I$, $\Phi$ and reluctance. This can then be used to great effect to analyse quite complicated magnetic circuits, such as those with parallel branches or multiple coils. We shall not pause to explore this topic, despite its importance in electrical engineering, but readers will find a thoughtful discussion in, for example, Lorrain & Corson (1970).

## 9.7 Magnetic Shielding

The observation that magnetic flux tends to concentrate within high-permeability material leads to the phenomenon of *magnetic shielding*. The general idea is the following. If a closed shell constructed from soft iron (or some other soft ferromagnetic material) is placed in a magnetic field, it will tend to exclude the magnetic flux from the interior of the shell. The flux that would otherwise have sat within the shell is instead channelled through the soft iron wall of the shell. This is illustrated in Figure 9.11 for the particular case of a spherical shell of high-permeability material sitting in a uniform magnetic field. For finite values of $\chi_m$ the shielding is only ever partial, although the relative permeability of soft iron is so large that in practice the shielding is highly effective.

Some idea of the effectiveness of shielding comes from applying (9.18) to a closed curve constructed from the field line that passes through the centre of the spherical cavity and one of the adjacent field lines that enters and leaves the shell wall near the centreline. Let the shell wall have a thickness $t$, the cavity a radius $R$, and the soft iron a permeability $\mu = \mu_r \mu_0$. We shall assume that the relative permeability of the shell, $\mu_r$, is large in the sense that

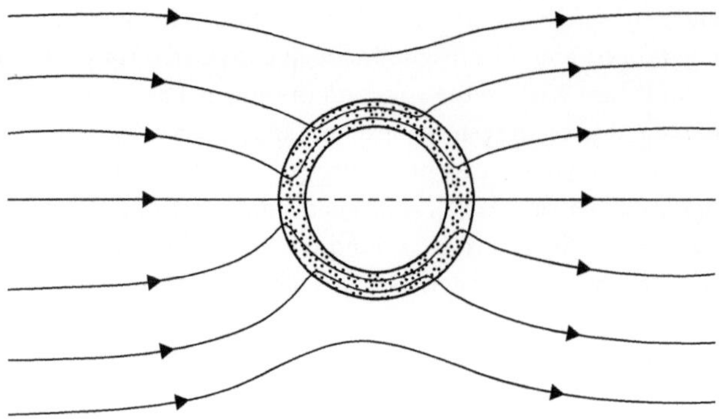

**Figure 9.11** A spherical shell constructed from soft iron sits in a uniform magnetic field.

$\mu_r >> R/t >> 1$. We traverse the closed curve by following the field line in the shell wall and then returning back along the centreline of the cavity. Equation (9.18) then yields

$$\oint_C \mathbf{H} \cdot d\mathbf{r} \sim \frac{RB_{\text{shell}}}{\mu} - \frac{RB_{\text{cavity}}}{\mu_0} (= 0).$$

However, the external field remote from the shell, $\mathbf{B}_0$, is related to the magnetic field within the soft iron by $B_0 R \sim B_{\text{shell}} t$ and so

$$B_{\text{cavity}} \sim \frac{R}{t\mu_r} B_0. \tag{9.22}$$

A similar estimate applies to two-dimensional shielding by cylindrical shells. An exact analysis of shielding shows that, in either case, the field inside the cavity is uniform and in the limit of $\mu_r >> R/t >> 1$ takes the form

$$\mathbf{B}_{\text{cavity}} = C \frac{R}{t\mu_r} \mathbf{B}_0, \tag{9.23}$$

where $C$ is a constant. The calculation is somewhat lengthy, involving Legendre functions in the spherical case, but it turns out that $C = 2$ for cylindrical shells (Zangwill, 2013) while $C = \frac{3}{2}$ for spherical shells.

## 9.8 Magnetisation Curves, Hysteresis and Permanent Magnets

So far we have said very little about hysteresis or about the associated phenomenon of permanent magnetism. So let us return to the magnetisation curve in Figure 9.9. This is reproduced in Figure 9.12 in a slightly different format, with $B$ plotted against $\mu_0 H$. To

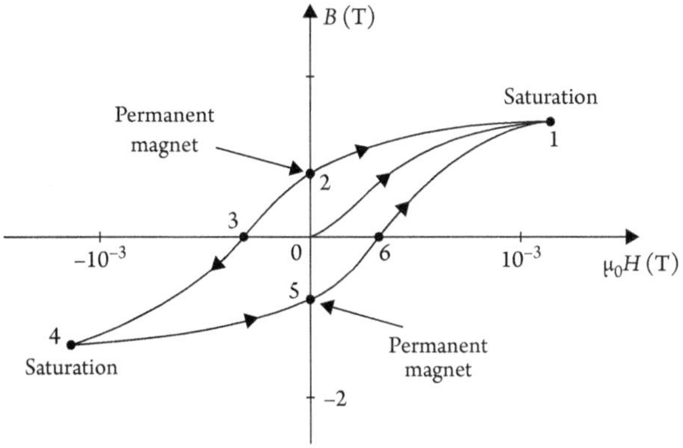


**Figure 9.12** *B–H* magnetisation curve for a ferromagnetic material exhibiting hysteresis.

understand what is going on in this figure, it is perhaps easiest to think of a cylindrical iron specimen sitting in a long solenoid (Figure 9.7). Then $H = NI$ and so $H$ is effectively the current in the solenoid. Recalling that $\mathbf{M} \approx \mathbf{B}/\mu_0$ in a ferromagnetic material, the figure then represents changes in the magnetisation $M$, as represented by $\mathbf{B}/\mu_0$, which result from changes in the current $I$, as represented by $H$.

Suppose we start out with an unmagnetised sample. Because of the interaction between adjacent molecular dipoles in the material (a quantum-mechanical effect), the sample is peppered with tiny *domains* within each of which all the molecular dipoles are aligned. These domains spontaneously form in any ferromagnetic material without the aid of a mean magnetic field. However, in an unmagnetised sample the domains are randomly orientated and so there is no overall magnetisation. We now turn up the current in the solenoid and subject the sample to a mean field **H**. The role of this imposed field is to shift the *domain boundaries* in such a way that those domains whose dipoles are more or less aligned with the external field grow at the expense of the domains which are strongly misaligned. As the current is increased, more misaligned domains give way to aligned domains and the overall magnetisation grows. This is curve $0 \to 1$ in Figure 9.12. Eventually all of the domains are approximately aligned with **H** and any further rise in current can only increase $M$ by rotating the molecular dipoles *within* the domains into a more perfect alignment with **H**. However, this is much harder than shifting domain boundaries and so the curve soon flattens off and the iron is said to become *saturated*. For soft iron the saturation field strength is around 2 T.

If the current is now reduced back to zero, we do not retrace the original curve, but rather follow curve $1 \to 2$. In particular, when $I$ reaches zero, we find that there is some residual magnetisation, called *remanence*. To eliminate this, a current must be applied in the opposite direction and we trace out curve $2 \to 3$. Further increases in negative current cause the magnetisation to grow in the reverse direction until saturation is again reached (curve $3 \to 4$). If the current is reduced back to zero, we follow curve $4 \to 5$ and again we find some residual magnetisation at zero current, only this time it is negative. A positive current is now required to eliminate this residual magnetisation (curve $5 \to 6$). Note that at points 2 and 5, where the current is zero but **M** finite, we have a permanent magnet.

A measure of the severity of the hysteresis is the value of $H$ at points 3 and 6, and this is called the *coercive force*. This may vary enormously from one ferromagnetic material to another, ranging from 1 to $10^4$ A/m. Materials with a very small coercive force are classified as being *magnetically soft* and those with a particularly large coercive force as *magnetically hard*. The former are preferred in applications involving alternating currents and the latter are required for producing permanent magnets.

The natural way to analyse permanent magnets is to take the magnetisation $\mathbf{M}$ as *prescribed*. The simplest (if somewhat idealised) permanent magnet to understand is a cylindrical magnet of radius $R$, length $L$ and a uniform magnetisation aligned with the cylinder axis. From (9.13) there are no internal bound currents and no surface currents at the ends of the magnet. So the only bound current circulates in the azimuthal direction on the cylindrical surface and takes the form $\mathbf{K}_{\text{mag}} = \mathbf{M} \times \mathbf{n} = M\hat{\mathbf{e}}_\theta$ in cylindrical polar coordinates aligned with $\mathbf{M}$. The magnetic field is therefore *identical* to that generated by a solenoid of the same radius and length and with a current per unit length of $NI = M$. The case of $L = 4R$ is shown in Figure 9.13 (a), where $\mathbf{B}$ is more or less aligned with $\mathbf{M}$.

The question now arises as to the direction of $\mathbf{H}$ inside the magnet. We might anticipate from the uniformly magnetised sphere shown in Figure 9.8 that $\mathbf{H}$ will be antiparallel to $\mathbf{B}$, and this turns out to be correct. To see that this is so, first note that (9.18) demands $\oint \mathbf{H} \cdot d\mathbf{r} = 0$, there being no free currents. Now consider the case of a very long magnet, $L >> R$. Then, from what we know about infinitely long solenoids, we have an internal magnetic field of $\mathbf{B}_\infty/\mu_0\,(= NI\hat{\mathbf{e}}_z$ in the solenoid$) = \mathbf{M}$, which requires $\mathbf{H}_\infty = 0$. (The subscript $\infty$ indicates that $L/R \to \infty$.) So, in this idealised case $\mathbf{H}$ is zero within the magnet, except near its ends. Next consider a magnet of finite length but with the same magnetisation $\mathbf{M}$, like the one shown in Figure 9.13. Then, from (9.16), the internal $\mathbf{H}$ and $\mathbf{B}$ fields are related by

$$\mathbf{H}_{\text{in}} = \frac{\mathbf{B}_{\text{in}}}{\mu_0} - \mathbf{M} = -\frac{\mathbf{B}_\infty - \mathbf{B}_{\text{in}}}{\mu_0}. \tag{9.24}$$

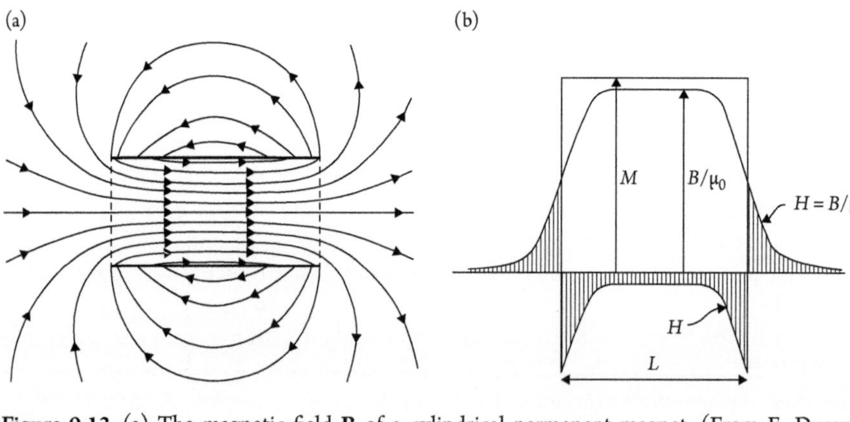


**Figure 9.13** (a) The magnetic field $\mathbf{B}$ of a cylindrical permanent magnet. (From E. Durand. *Magnétostatique*. Masson, 1968.) (b) The $\mathbf{H}$ field on the axis. Within the magnet, $\mathbf{H}$ is in the opposite direction to $\mathbf{B}$.

However, we know that for a given current per unit length, $NI$, the magnetic field in a short solenoid is weaker than that in an infinite solenoid (see Exercise 7.5), and so $|\mathbf{B}_{\text{in}}| < |\mathbf{B}_{\infty}|$ in (9.24). We conclude that $\mathbf{H}_{\text{in}}$ is antiparallel to both $\mathbf{M}$ and $\mathbf{B}_{\text{in}}$ in a cylindrical permanent magnet, as shown in Figure 9.13(b).

In fact, we could have guessed that this must be the case based purely on $\oint \mathbf{H} \cdot d\mathbf{r} = 0$. The point is that $\mathbf{H}$ is parallel to $\mathbf{B}$ outside the permanent magnet, so to ensure that $\mathbf{H}$ integrates to zero it must be antiparallel to $\mathbf{B}$ inside the magnet.

* * *

This completes our introduction to magnetic materials. Our discussion has been brief and therefore partial. We have omitted entire subject areas, such as the fascinating topic of magnetic fluids, in which colloidal suspensions of magnetic particles are subject to magnetic fields, resulting in many exotic phenomena. Readers searching for a more leisurely introduction will find excellent introductory chapters in Purcell (2011) and Lorrain & Corson (1970), with the former emphasising the physical ideas and the latter the mathematical development. Zangwill (2013) offers a more advanced treatment.

## EXERCISES

9.1 A large flat plate of thickness $t$ is magnetised with a uniform magnetisation $\mathbf{M}$ whose direction is parallel to the faces of the plate. Find $\mathbf{B}$ inside and outside the plate.

9.2 A thin circular disk of thickness $t$ and radius $R$, with $R >> t$, is magnetised with a uniform magnetisation $\mathbf{M}$ whose direction is normal to the faces of the disc. Show that the resulting bound current flows around the rim of the disc in the azimuthal direction. Find the resulting magnetic field on the axis of the disc using the results of Exercise 7.4.

9.3 A soft iron sphere of high relative permeability $\mu_r >> 1$ and negligible hysteresis sits in a uniform external magnetic field $\mathbf{B}_0 = \mu_0 \mathbf{H}_0$. Use (9.15) and Fig. 9.8 to show that the field within the sphere, $\mathbf{B}_{\text{in}}$, is

$$\mathbf{B}_{\text{in}} = \mathbf{B}_0 + \frac{2}{3}\mu_0 \mathbf{M} = \mu \mathbf{H}_{\text{in}} = \mu\left(\frac{\mathbf{B}_0}{\mu_0} - \frac{\mathbf{M}}{3}\right) .$$

Hence show that the induced magnetisation and magnetic field within the sphere are given by

$$\mathbf{M} = 3\frac{\mu_r - 1}{\mu_r + 2}\frac{\mathbf{B}_0}{\mu_0} \approx 3\frac{\mathbf{B}_0}{\mu_0}, \qquad \mathbf{B}_{\text{in}} = 3\frac{\mu_r}{\mu_r + 2}\mathbf{B}_0 \approx 3\mathbf{B}_0.$$

9.4 In §9.8 we showed that $\mathbf{H}$ in a cylindrical permanent magnet of uniform magnetisation is, on average, antiparallel to $\mathbf{B}$. Consider the spherical control surface that just circumscribes the magnet shown in Figure 9.13. Use (8.26) to show that, on integrating over this volume,

$$\int \mathbf{B}\, dV = \frac{2}{3}\mu_0 \mathbf{M} V_{\text{mag}} = \mu_0 \left( \left( \int \mathbf{H}\, dV + \mathbf{M} V_{\text{mag}} \right) \right),$$

where $V_{\text{mag}}$ is the volume of the magnet. Hence show that

$$\mu_0 \int \mathbf{H}\, dV = -\frac{1}{2} \int \mathbf{B}\, dV.$$

9.5 Find **H** on the centreline of a cylindrical permanent magnet of radius $R$, length $L$ and uniform magnetisation **M** aligned with the axis of the cylinder. Focus first on the midpoint and ends of the magnet using the results of Exercise 7.5 and then derive a general expression for any point on the axis.

9.6 A two-dimensional dipole is formed from two long straight wires carrying opposing currents. The wires sit on the axis of a soft-iron tubular shield of radius $R$, wall thickness $t$ and relative permeability $\mu_r$, with $\mu_r >> R/t >> 1$. Show, with the aid of a sketch and using (9.18), that the shielding reduces the external field by a factor of order $R/t\mu_r$.

## REFERENCES

Lorrain, P. & Corson, D., 1970, *Electromagnetic Fields and Waves*, 2nd ed. W. H. Freeman.
Purcell, E., 2011, *Electricity and Magnetism*, 2nd ed. Cambridge University Press.
Zangwill, A., 2013, *Modern Electrodynamics*. Cambridge University Press.

# 10

# Faraday's Law of Electromagnetic Induction

> What led me more or less directly to the special theory of relativity was the conviction that the electromotive force acting on a body in motion in a magnetic field was nothing else but an electric field.
>
> Albert Einstein, 1952

We shall explore the significance of this comment, and its connection to Faraday's law, below. Widely regarded as the prince of experimentalists, Faraday (Figure 10.1) not only discovered electromagnetic induction, paramagnetism, diamagnetism and dielectrics, he is also credited with a myriad of discoveries in chemistry and material science. But perhaps his greatest contribution to physics was the concept of the field. Distrustful of action at a distance, he insisted that the space between charges and currents played a central role in electromagnetic phenomena. Mathematicians were initially suspicious of Faraday's magnetic field and 'lines of force', but when Maxwell championed Faraday's modes of thought, translating them into mathematical form, a new era began. As Einstein put it:

> During the second half of the 19th century, in conjunction with the researches of Faraday and Maxwell, it became more and more clear that the description of electromagnetic processes in terms of fields was vastly superior to a treatment on the basis of the mechanical concepts of material points.... One psychological effect of this immense success was that the field concept, as opposed to the mechanistic framework of classical physics, gradually won greater independence.
>
> Einstein (1954)

Readers of an historical bent may wish to consult the brief sketch of Faraday in Appendix III. In this chapter we focus on Faraday's law of electromagnetic induction.

We introduced Faraday's law in §3.3. In differential form it says $\nabla \times \mathbf{E} = -\partial \mathbf{B}/\partial t$, which tells us about the electric field induced by a time-dependent magnetic field. We shall now fill in some of the gaps, explore the consequences of the law and illustrate the theory with a few examples. The discussion is non-relativistic, with $u << c$. We start by returning to Faraday's original discovery of electromagnetic induction in the autumn of 1831.

**Figure 10.1** Michael Faraday.

## 10.1 Electromagnetic Induction: One Physical Law or Two?

As Faraday discovered, when a wire is pulled through a magnetic field $\mathbf{B}$ with a velocity $\mathbf{u}$, a potential difference is generated in that wire, and if the wire is part of a conducting circuit, that potential difference drives a current through the circuit. This is illustrated in Figure 10.2(a), where a rectangular wire loop is pulled with a constant speed $u$ through the airgap in an electromagnet. (We shall assume, for simplicity, that the magnetic field between the two poles in the electromagnet is uniform and negligibly perturbed by the current in the wire.) The modern interpretation of the current observed in Figure 10.2(a) is that the Lorentz force $\mathbf{f} = q\mathbf{u} \times \mathbf{B}$ acts on the free charges in the wire to drive a current. It is customary to express this in terms of an *electromotive force*, or EMF, defined as the force acting on a unit test charge integrated around a closed curve, in this case the wire loop. For the particular geometry shown in Figure 10.2(a) we have

$$\text{EMF} = \oint \mathbf{u} \times \mathbf{B} \cdot d\mathbf{r} = -uBw, \tag{10.1}$$

where $w$ is the width of the wire loop into the page and the direction of integration is set by the right-hand rule relative to the vector $\mathbf{B}$. This is an example of what is called a *motional* EMF, i.e. an EMF generated by movement of a conductor through a magnetic field. Now it so happens that we may also write (10.1) in the form

$$\text{EMF} = -\frac{d\Phi}{dt}, \tag{10.2}$$

where $\Phi$ is the magnetic flux through the loop. This is usually called *Faraday's flux rule*, or *flux law*. Note that, although we have deduced (10.2) for only one very simple geometry, it turns out that the flux rule has great generality, as we shall see. Note also that any electric field that may be present cannot contribute to the EMF in (10.1), because the magnetic field is steady, and so $\nabla \times \mathbf{E} = 0$ and hence $\oint q\mathbf{E} \cdot d\mathbf{r} = 0$.

Now suppose that the loop is held stationary and the magnet is pushed from left to right with the same constant speed, $u$, as shown in Figure 10.2(b). Evidently, the same current will flow in the wire, since all that matters is the relative movement of the magnet and the wire loop. To be more precise, whether we assert that it is the magnet or the wire that is stationary simply depends on our frame of reference, and since both frames are inertial the same laws of physics must apply. Now if the current in the wire is unchanged by the change of frame of reference, we would expect that the EMF that drives that current to also be unchanged. However, in Figure 10.2(b) the loop is stationary and so there is no Lorentz force acting on the free charges, $\mathbf{f} = q\mathbf{u} \times \mathbf{B} = 0$. Evidently, the EMF must arise from an electric field that acts on the charges through the force $\mathbf{f} = q\mathbf{E}$. In short, in Figure 10.2(b) the changing magnetic field must *induce* an electric field $\mathbf{E}$. Equation (10.2) now reads

$$\text{EMF} = \oint \mathbf{E} \cdot d\mathbf{r} = -\frac{d\Phi}{dt} = -\frac{d}{dt}\int_S \mathbf{B} \cdot d\mathbf{S}, \tag{10.3}$$

where $S$ is any surface that spans the wire loop. Since the loop is stationary, we may take the time derivative inside the surface integral. Invoking Stokes' theorem, we now have

$$\int_S \nabla \times \mathbf{E} \cdot d\mathbf{S} = -\int_S \frac{\partial \mathbf{B}}{\partial t} \cdot d\mathbf{S}. \tag{10.4}$$

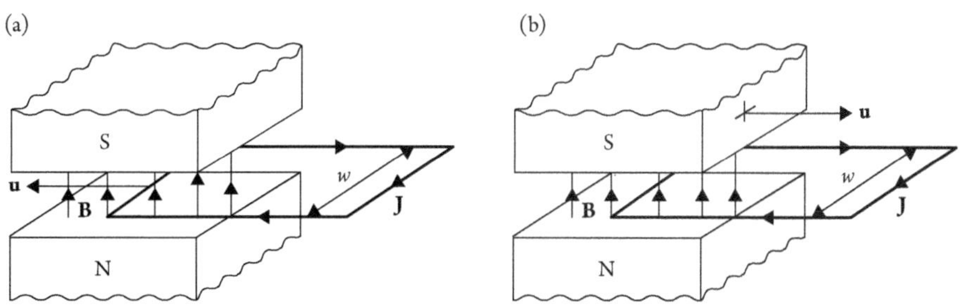


**Figure 10.2** Current is induced in a rectangular wire loop through the relative motion of the loop and a magnetic field. The induced current is assumed small so that the imposed magnetic field is negligibly perturbed. (a) The magnet is stationary and the wire loop moves. (b) The wire loop is stationary and the magnet moves.

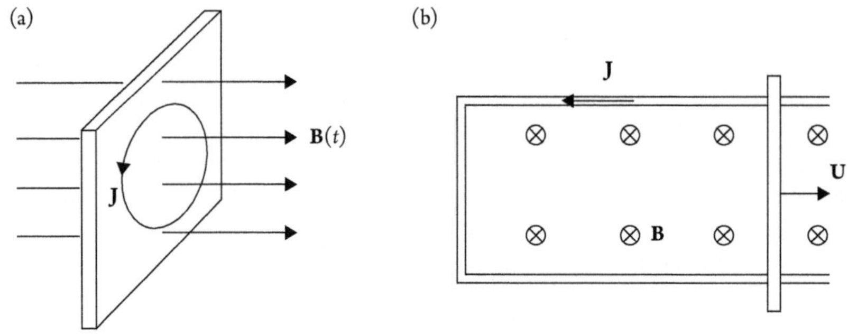


**Figure 10.3** (a) An EMF generated by the variation of the magnetic field with time is called a *transformer* EMF. (b) An EMF generated by the movement of part of the circuit C through the magnetic field, in this case a metal side, is called a *motional* EMF.

We interpret this to be a direct consequence of the differential form of *Faraday's law*,

$$\boxed{\nabla \times \mathbf{E} = -\frac{\partial \mathbf{B}}{\partial t},} \tag{10.5}$$

which, as noted in Chapter 3, tells us about the electric field induced by a time-dependent magnetic field $\mathbf{B}(t)$. Everything seems to be in harmony.

However, on reflection, there something odd about the discussion above. We normally think of Faraday's law (10.5) and the Lorentz force $\mathbf{f} = q\mathbf{u} \times \mathbf{B}$ as being two independent physical laws. Yet it seems that both are captured by Faraday's flux rule, at least in the simple examples discussed above. This extraordinary coincidence set Einstein thinking, as noted at the start of this chapter. In fact, Faraday demonstrated that the flux rule (10.2) applies not only to the two cases shown in Figure 10.2, it also applies to situations in which both the loop and the magnet are stationary and $\mathbf{B}$ changes with time, say by varying the current in the electromagnet. So the flux rule is a remarkably convenient and surprisingly general result. Cases in which an EMF is generated by a time-dependent magnetic field, such as those shown in Figures 10.2(b) and 10.3(a), are known as *transformer* EMFs.

Although we have considered a specific (and very simple) geometry in Figure 10.2, this example highlights a number of intriguing results, which we shall now explore in more detail. Perhaps the most important suggestion is that Faraday's flux rule is, in some sense, a more general statement than the differential form of Faraday's law, (10.5). So let us now derive the flux rule in its most general form, exploring the relationship between it, the Lorentz force and Faraday's differential law (10.5). We start by returning to the definition of an EMF.

## 10.2 The Definition of the Electromotive Force

Suppose that, in the laboratory frame of reference, we have an electric field $\mathbf{E}$ and a magnetic field $\mathbf{B}$. At each point in space the force acting on a test charge $q$ moving with velocity $\mathbf{u}$ is

$\mathbf{f} = q(\mathbf{E} + \mathbf{u} \times \mathbf{B})$. Now suppose that $C$ is a closed curve drawn in the same space that **E** and **B** inhabit. This curve may be stationary or it may move, it may deform or it may remain rigid, and it may coincide with a conducting loop or it may sit partially or wholly outside any conductor. It short, $C$ is any closed curve we care to consider. Moreover, let **v** be the velocity of each part of the curve, which in general will vary around the curve if it deforms in space. Then the EMF around the curve $C$ is defined in terms of the force on a unit test charge that moves with velocity **v**, i.e. moves with a line element $d\mathbf{r}$. Specifically, it is the closed line integral of this force around the curve $C$:

$$\boxed{\text{EMF} = \oint_C \frac{\mathbf{f}}{q} \cdot d\mathbf{r} = \oint_C (\mathbf{E} + \mathbf{v} \times \mathbf{B}) \cdot d\mathbf{r}.} \tag{10.6}$$

Perhaps some comments are in order at this point. First, it is important to emphasise that $C$ need not be embedded within a conductor, and, even if it is, it need not move with the conductor. Rather, it may lie outside any conducting material, or perhaps sit within a conducting medium but execute some motion entirely independent of that of the conducting material. For that reason, we use **v** for the velocity of the curve $C$ and **u** for the velocity field of any conducting material. (By implication, **u** is, more or less, also the velocity of the free charges within a conductor, since the drift velocity of the free charges relative to a conductor is usually very small.)

Second, as discussed in §3.1, $\mathbf{E} + \mathbf{v} \times \mathbf{B}$ is the electric field measured in a frame of reference moving with velocity **v** relative to the laboratory frame, $\mathbf{E}_{\text{rel}} = \mathbf{E} + \mathbf{v} \times \mathbf{B}$. (Recall that, in this chapter, we take $|\mathbf{v}| << c$, so that relativistic corrections may be ignored.) Thus we may rewrite (10.6) as

$$\text{EMF} = \oint_C \mathbf{E}_{\text{rel}} \cdot d\mathbf{r}. \tag{10.7}$$

Of course, if **v** varies around the curve $C$, i.e. the curve deforms in space, then the frame or reference in which $\mathbf{E}_{\text{rel}}$ is measured will also vary around the curve $C$.

Third, suppose that $C$ is indeed embedded within a conductor and moves with that conductor. (We refer to such a curve as a *material curve.*) Then we have $\mathbf{v} = \mathbf{u}$, where **u** is the local velocity of the conductor, which may vary from place to place. Since the current density in the conductor is given by Ohm's law

$$\mathbf{J} = \sigma \mathbf{E}_{\text{rel}} = \sigma(\mathbf{E} + \mathbf{u} \times \mathbf{B}), \tag{10.8}$$

the EMF in such cases is evidently related to the current density by

$$\text{EMF} = \frac{1}{\sigma} \oint_C \mathbf{J} \cdot d\mathbf{r} \quad \text{(material curve only)}. \tag{10.9}$$

Finally, if additional sources of current, such as batteries, are present, then the net force acting on a test charge $q$ moving with velocity $\mathbf{u}$ must be generalised to read

$$\mathbf{f} = \mathbf{f}_s + q(\mathbf{E} + \mathbf{u} \times \mathbf{B}), \tag{10.10}$$

where the subscript $s$ stands for *source*. Here $\mathbf{f}_s$ may be a chemical force, if we are dealing with batteries, or perhaps it may be proportional to a temperature gradient in the case of a thermocouple. In such cases the EMF is usually defined in the same way, as the line integral of the force on a unit test charge moving with the local velocity of the curve $C$. So (10.6) is then generalised to read

$$\text{EMF} = \oint_C (\mathbf{f}_s/q) \cdot d\mathbf{r} + \oint_C (\mathbf{E} + \mathbf{v} \times \mathbf{B}) \cdot d\mathbf{r}. \tag{10.11}$$

However, we shall not use the generalisation (10.11) in this chapter, although it will crop up later when we consider electrical circuits. In fact, when discussing electromagnetic induction, the EMF we have in mind is *always* (10.6), rather than (10.11).

## 10.3 From Faraday's Law in Differential Form to Faraday's Flux Rule

### *10.3.1 Faraday's Differential Law Versus Faraday's Flux Rule*

Faraday's law in differential form,

$$\nabla \times \mathbf{E} = -\frac{\partial \mathbf{B}}{\partial t}, \tag{10.12}$$

tells us how to calculate the electric field induced by a time-dependent magnetic field. It is one of Maxwell's four equations. Faraday's flux rule, by contrast, states that

$$\boxed{\text{EMF} = \oint_C (\mathbf{E} + \mathbf{v} \times \mathbf{B}) \cdot d\mathbf{r} = \oint_C \mathbf{E}_{\text{rel}} \cdot d\mathbf{r} = -\frac{d}{dt} \int_S \mathbf{B} \cdot d\mathbf{S},} \tag{10.13}$$

where $S$ is any open surface that spans the curve $C$.

As already emphasised, the magnetic flux in (10.13) may change because $\mathbf{B}$ is changing with time while $C$ is fixed in space (Figures 10.2(b) and 10.3(a)), or because the loop $C$ is moving uniformly across an inhomogeneous magnetic field (Figure 10.2(a)), or because $C$ is deforming within a prescribed magnetic field (Figure 10.3(b)). In all such cases the flux rule asserts that the induced EMF, as defined by (10.6), is minus the rate of change of flux through the curve $C$. In short, the flux rule (10.13) covers both transformer and motional EMFs, or indeed any combination of the two. This flux rule is sometimes known as the integral version of Faraday's law.

For the special case in which the curve $C$ in (10.13) is at rest in a laboratory frame, and the EMF arises from $\partial \mathbf{B}/\partial t$ alone, it is clear that (10.12) and (10.13) are entirely equivalent. That is, when the curve $C$ is at rest, Faraday's flux rule simplifies to

$$\oint_C \mathbf{E} \cdot d\mathbf{r} = \int_S (\nabla \times \mathbf{E}) \cdot d\mathbf{S} = -\int_S \frac{\partial \mathbf{B}}{\partial t} \cdot d\mathbf{S}. \tag{10.14}$$

Since this is true for any surface $S$, we may equate the integrands in the two surface integrals, which brings us back to the differential form of Faraday's law, (10.12).

Let us now see if we can *derive* the most general version of Faraday's flux rule, in which the loop $C$ is allowed to move and deform, starting from Faraday's law in differential form, $\nabla \times \mathbf{E} = -\partial \mathbf{B}/\partial t$. To that end, we need an important kinematic result, first introduced in Chapter 3.

### 10.3.2 *A Useful Kinematic Result*

We shall first state the result and then prove it. Suppose that $C$ is a closed curve that deforms in space with a prescribed velocity $\mathbf{v}(\mathbf{x}, t)$, $S$ is any open surface that spans the closed curve $C$ and $\mathbf{G}(\mathbf{x}, t)$ is any *solenoidal* vector field that links the surface $S$. Then the kinematic result we require is

$$\boxed{\frac{d}{dt}\int_S \mathbf{G} \cdot d\mathbf{S} = \int_S \frac{\partial \mathbf{G}}{\partial t} \cdot d\mathbf{S} - \oint_C (\mathbf{v} \times \mathbf{G}) \cdot d\mathbf{r}.} \tag{10.15}$$

The geometrical idea behind (10.15) is the following. The flux of $\mathbf{G}$ through $S$ may change for one of two reasons. First, even if $C$ and $S$ are both fixed in space, there may be a change in flux through $S$ whenever $\mathbf{G}$ is time-dependent. This is the surface integral on the right of (10.15). Second, if the curve $C$ moves, it may expand at points to include additional flux, or perhaps contract at other points to exclude flux. This is the origin of the line integral in (10.15).

The proof of (10.15) proceeds as follows. Consider first the boundary term. The surface element $d\mathbf{S}$ that is swept out by a line element $d\mathbf{r}$ of the curve $C$ in a time $\delta t$ has sides $d\mathbf{r}$ and $\mathbf{v}\delta t$, as shown in Figure 10.4. This vector area, which is normal to both $d\mathbf{r}$ and $\mathbf{v}\delta t$, can be written as $d\mathbf{S} = (\mathbf{v}\delta t) \times d\mathbf{r}$. (Of course, $d\mathbf{S} = d\mathbf{r} \times (\mathbf{v}\delta t)$ is also a possibility, but let's stick to the definition $d\mathbf{S} = (\mathbf{v}\delta t) \times d\mathbf{r}$ and see where it leads.) Thus the change in the flux of $\mathbf{G}$ through $S$ by virtue of movement of the boundary element $d\mathbf{r}$ is $\mathbf{G} \cdot d\mathbf{S} = \mathbf{G} \cdot (\mathbf{v} \times d\mathbf{r})\delta t$. We now need to check to see if this change in flux is positive or negative. By convention, the relative orientation of $\mathbf{G}$ and $d\mathbf{r}$ is fixed in accordance with the right-hand rule, as shown in Figure 10.4. So, if the loop in the vicinity of $d\mathbf{r}$ is expanding to accommodate more flux, then the definition $d\mathbf{S} = (\mathbf{v}\delta t) \times d\mathbf{r}$, as opposed to $d\mathbf{S} = d\mathbf{r} \times (\mathbf{v}\delta t)$, has $\mathbf{G} \cdot d\mathbf{S} > 0$. So we conclude that $\mathbf{G} \cdot d\mathbf{S} = \mathbf{G} \cdot (\mathbf{v} \times d\mathbf{r})\delta t$ represents an increase in the flux through $S$, provided of course that the curve $C$ is traversed in accordance with the right-hand rule.

Next, we rearrange the scalar triple product to give $\mathbf{G} \cdot d\mathbf{S} = -[(\mathbf{v} \times \mathbf{G}) \cdot d\mathbf{r}]\delta t$ and sum all such contributions around the closed curve $C$. If $\mathbf{G}$ were steady, so that movement of the

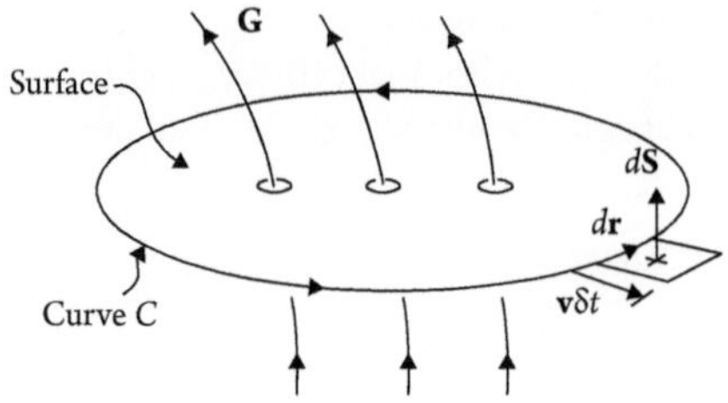


**Figure 10.4** The surface swept out by a line element $d\mathbf{r}$ in a time $\delta t$ is $d\mathbf{S} = (\mathbf{v}\delta t) \times d\mathbf{r}$.

boundary were the only source of a change in flux, then this procedure would give

$$\delta \int_S \mathbf{G} \cdot d\mathbf{S} = -\delta t \oint_C (\mathbf{v} \times \mathbf{G}) \cdot d\mathbf{r},$$

from which

$$\frac{d}{dt} \int_S \mathbf{G} \cdot d\mathbf{S} = -\oint_C (\mathbf{v} \times \mathbf{G}) \cdot d\mathbf{r}. \tag{10.16}$$

However, if $\mathbf{G}$ is time-dependent, then we must add to (10.16) the surface integral $\int (\partial \mathbf{G}/\partial t) \cdot d\mathbf{S}$, which brings us back to (10.15). (For a more detailed derivation, see Zangwill (2013).)

### *10.3.3 Deriving the Flux Rule from Faraday's Law*

Let us now combine this kinematic result with the differential form of Faraday's law,

$$\nabla \times \mathbf{E} = -\frac{\partial \mathbf{B}}{\partial t}, \tag{10.17}$$

where $\mathbf{E}$ is measured in the laboratory frame of reference. Integrating (10.17) over the open surface $S$, invoking Stokes' theorem and deploying (10.15) with $\mathbf{G} = \mathbf{B}$, we find

$$\oint_C \mathbf{E} \cdot d\mathbf{r} = -\int_S \frac{\partial \mathbf{B}}{\partial t} \cdot d\mathbf{S} = -\frac{d}{dt} \int_S \mathbf{B} \cdot d\mathbf{S} - \oint_C (\mathbf{v} \times \mathbf{B}) \cdot d\mathbf{r}, \tag{10.18}$$

where $\mathbf{v}(\mathbf{x}, t)$ is the prescribed velocity of the bounding curve $C$. Combining the two line integrals in (10.18) then yields

$$\oint_C (\mathbf{E} + \mathbf{v} \times \mathbf{B}) \cdot d\mathbf{r} = -\frac{d}{dt} \int_S \mathbf{B} \cdot d\mathbf{S}. \tag{10.19}$$

Evidently, this is very close to Faraday's flux rule (10.13), and so far we have used only Faraday's law in differential form, $\nabla \times \mathbf{E} = -\partial \mathbf{B}/\partial t$, plus some kinematics.

However, we are not quite there, since Faraday's flux rule is expressed in terms of an EMF, i.e. the forces on charges. So the final step is to invoke the idea of the Lorentz force $\mathbf{f} = q\mathbf{v} \times \mathbf{B}$ acting on a unit test charge that moves with the curve $C$. This tells us that the net force on the unit test charge is $\mathbf{E} + \mathbf{v} \times \mathbf{B}$. It follows that the EMF around the curve $C$ is, by definition,

$$\text{EMF} = \oint_C (\mathbf{E} + \mathbf{v} \times \mathbf{B}) \cdot d\mathbf{r} = \oint_C \mathbf{E}_{\text{rel}} \cdot d\mathbf{r}, \tag{10.20}$$

where $\mathbf{E}_{\text{rel}}$ is the electric field in a frame of reference moving with velocity $\mathbf{v}$, $\mathbf{E}_{\text{rel}} = \mathbf{E} + \mathbf{v} \times \mathbf{B}$. Combining (10.19) with (10.20) yields Faraday's flux rule:

$$\text{EMF} = \oint_C (\mathbf{E} + \mathbf{v} \times \mathbf{B}) \cdot d\mathbf{r} = \oint_C \mathbf{E}_{\text{rel}} \cdot d\mathbf{r} = -\frac{d}{dt} \int_S \mathbf{B} \cdot d\mathbf{S}. \tag{10.21}$$

We conclude that the flux rule can be derived from Faraday's law in differential form, provided that we use the Lorentz force to tell us how to calculate the force on a test charge moving with the curve $C$ and hence how to calculate the EMF. (Equivalently, we need the Lorentz force to tell us how $\mathbf{E}$ transforms from one frame of reference to another.) So Faraday's flux rule embodies both the force law $\mathbf{f} = q(\mathbf{E} + \mathbf{v} \times \mathbf{B})$ and Faraday's differential law, $\nabla \times \mathbf{E} = -\partial \mathbf{B}/\partial t$. From a classical point of view, it is a happy and convenient coincidence that the same flux rule works irrespective of whether the EMF comes from the Lorentz force or an induced electric field.

## 10.4 Applications of Faraday's Flux Rule

Let us now consider some simple examples of induced EMFs and of Faraday's flux rule.

We start with the rotating wire loop shown in Figure 10.5, which sits in an imposed horizontal magnetic field $\mathbf{B} = B\hat{\mathbf{e}}_x$. Suppose the rectangular loop has height $h$ and width $w$ and rotates about a vertical axis with an angular velocity $\boldsymbol{\Omega} = \Omega\hat{\mathbf{e}}_z$. Suppose also that the induced current is sufficiently weak that the imposed magnetic field is negligibly perturbed. Then, since the magnetic field is steady, we have $\nabla \times \mathbf{E} = 0$ and $\oint \mathbf{E} \cdot d\mathbf{r} = 0$. It follows that the EMF is purely motional, with no contribution from $\mathbf{E}$. We now take our closed curve $C$ to coincide with the conducting loop, so that we may combine (10.9) with the flux rule (10.21) to give

$$\text{EMF} = \oint_C (\mathbf{u} \times \mathbf{B}) \cdot d\mathbf{r} = \frac{1}{\sigma} \oint_C \mathbf{J} \cdot d\mathbf{r} = -\frac{d}{dt} \int_S \mathbf{B} \cdot d\mathbf{S}. \tag{10.22}$$

The instantaneous magnetic flux through the loop is evidently

$$\Phi = Bhw \cos\theta = Bhw \cos \Omega t, \tag{10.23}$$

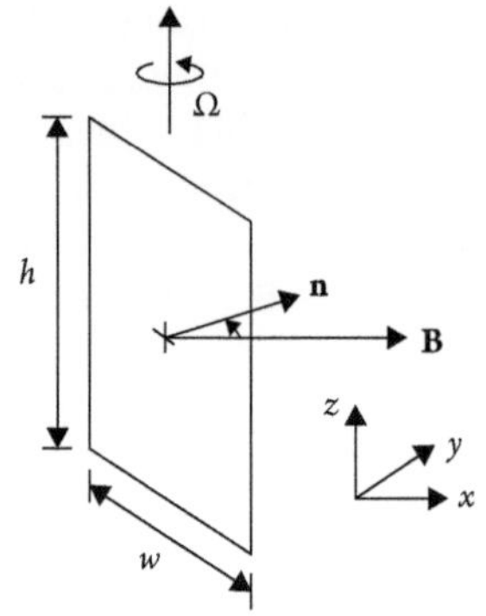


**Figure 10.5** A rectangular wire loop rotates about a vertical axis and is immersed in an imposed, horizontal magnetic field.

where $\theta$ is the angle between the normal to the loop and **B**. The flux rule now tells us that the induced EMF is

$$\text{EMF} = Bhw\Omega \sin \Omega t. \tag{10.24}$$

This is readily confirmed by direct calculation if we note that on either of the vertical sides of the loop we have

$$(\mathbf{u} \times \mathbf{B}) \cdot d\mathbf{r} = \mathbf{u} \cdot (\mathbf{B} \times d\mathbf{r}) = [(\Omega w/2) \sin \theta] B \, |d\mathbf{r}| ,$$

while $(\mathbf{u} \times \mathbf{B}) \cdot d\mathbf{r} = 0$ on the top and bottom of the loop. Integration then yields

$$\text{EMF} = \oint_C (\mathbf{u} \times \mathbf{B}) \cdot d\mathbf{r} = Bhw\Omega \sin \Omega t, \tag{10.25}$$

as required. We note in passing that the EMF is more readily obtained via the flux rule than by direct integration, at least in this particular example. The current density in the wire can now be found using (10.22), which gives

$$J = \frac{\sigma}{\ell} \text{EMF} = \frac{\sigma}{\ell} Bhw\Omega \sin \Omega t, \tag{10.26}$$

where $\ell = 2(h + w)$ is the length of the wire. If the wire has cross-sectional area $A$ and electrical resistance $R = \ell/\sigma A$, then the current $I$ flowing through the loop is evidently

$$I = JA = \frac{\sigma A}{\ell} \text{EMF} = \frac{\text{EMF}}{R}, \tag{10.27}$$

in accord with Ohm's law. We have an elementary form of alternating current generator.

As a second example, consider a rotating metal disc of radius $a$ that is threaded by an imposed magnetic field, as shown in Figure 10.6. This is known as Faraday's disc generator. The magnetic field is taken to be axisymmetric and a stationary wire connects the rim of the disc with its axis through sliding contacts at Y and Z. A current is driven through the wire

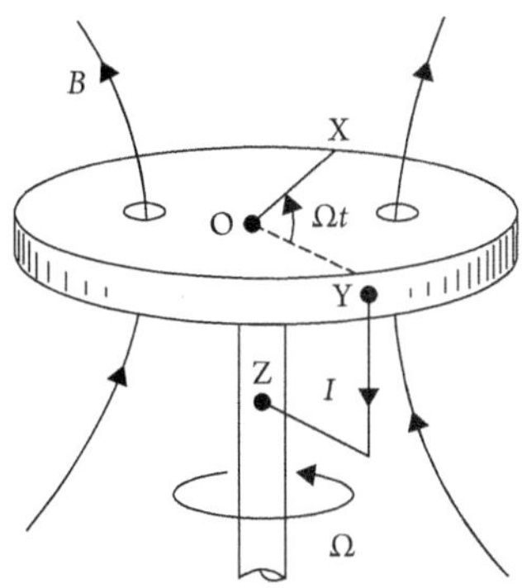


**Figure 10.6** Faraday's disc generator.

by a motional EMF within the disc and we wish to calculate that current. Once again, we shall assume that the induced current is sufficiently small for the imposed magnetic field to be negligibly perturbed. It follows from this assumption that $\nabla \times \mathbf{E} = 0$ and $\oint \mathbf{E} \cdot d\mathbf{r} = 0$, so that the EMF is purely motional. In this particular example the flux rule is less helpful and it is easier to calculate the motional EMF directly. Using cylindrical polar coordinates $(s, \theta, z)$ centred on the disc, we find that the azimuthal velocity of the disc, $\mathbf{u} = \Omega s \hat{\mathbf{e}}_\theta$, interacts with the vertical component of the magnetic field to give a radial Lorentz force (per unit charge) of $\mathbf{u} \times \mathbf{B} = \Omega s B_z \hat{\mathbf{e}}_s$. This drives a current radially outward in the disc, which returns in the current loop below. Integrating radially across the disc then gives a contribution to the EMF in the *positive radial direction* of

$$\int_0^X (\mathbf{u} \times \mathbf{B}) \cdot d\mathbf{r} = \int_0^a \Omega s B_z ds = \frac{\Omega}{2\pi} \int_0^a B_z 2\pi s \, ds = \frac{\Omega \Phi_0}{2\pi}, \tag{10.28}$$

where $\Phi_0$ is the total magnetic flux passing through the disc and the radial line OX rotates with the disc, as shown in Figure 10.6. Let us now assume that the resistance to current flow, $R$, is dominated by the wire YZ. In such cases we might anticipate that this radial EMF drives a current through the wire of $I = (\Omega \Phi_0 / 2\pi)/R$. This may be confirmed using (10.9), as we now show.

Consider the closed *material curve* $C$, labelled OXYZO in Figure 10.6. This consists of a rotating line OX painted on the disc, the azimuthal curve XY on the rim of the disc that links X to the sliding contact Y, and the return loop, YZO. Equation (10.9) applied to this material curve yields

$$\oint_C (\mathbf{u} \times \mathbf{B}) \cdot d\mathbf{r} = \frac{1}{\sigma} \oint_C \mathbf{J} \cdot d\mathbf{r}, \tag{10.29}$$

since $\oint \mathbf{E} \cdot d\mathbf{r} = 0$. Now there is no contribution to the motional EMF from the rim, XY, and so (10.28) and (10.29) combine to give the current density within the wire as

$$|\mathbf{J}| = \frac{\sigma}{\ell} |\text{EMF}| = \frac{\sigma}{\ell} \frac{\Omega \Phi_0}{2\pi},$$

where $\ell$ is the length of the wire. (We shall explain why we have placed a modulus sign on the EMF shortly.) If the wire has cross-sectional area $A$ and resistance $R = \ell/\sigma A$, then the current $I$ flowing through the loop is

$$I = |\mathbf{J}|A = \frac{\sigma A}{\ell}\frac{\Omega\Phi_0}{2\pi} = \frac{1}{R}\frac{\Omega\Phi_0}{2\pi}, \tag{10.30}$$

as anticipated above.

Let us now return to the flux rule (10.22) and to the calculation of the EMF. The expression $|\text{EMF}| = \Omega\Phi_0/2\pi$ can also be obtained directly from the flux rule applied to our material curve, although the argument is a little delicate. Suppose we take $t = 0$ to coincide with the instant where the rotating radial line embedded in the disc is coincident with the sliding contact on the rim of the disc. At this instant $C$ is a simple rectangle sitting in the $s$–$z$ plane and there is no flux through it. A short time later the radial line painted on the disc has swept out an angle $\Omega t$ and $C$ is now a three-dimensional curve, consisting of the rectangular return loop below and a portion of the disc of area $(\Omega t/2\pi)\pi a^2$. There is now a magnetic flux through the material curve $C$ of $\Phi = (\Omega t/2\pi)\Phi_0$, from which $d\Phi/dt = \Omega\Phi_0/2\pi$, at least for $\Omega t < 2\pi$. We now apply the flux rule to the material curve $C$. The right-hand rule then demands that we trace out $C$ in the sense OZYXO, i.e. we traverse the disc by starting at the rim and integrating radially *inward*. This is in the opposite direction to the Lorentz force acting on the free charges in the disc, and so the EMF determined in this way is numerically negative. The flux rule (10.22) then gives us

$$\text{EMF} = -\frac{d\Phi}{dt} = -\frac{\Omega\Phi_0}{2\pi}, \tag{10.31}$$

which is entirely consistent with our direct calculation of the EMF. So the flux rule does indeed work for a Faraday disc, at least up to the time $\Omega t = 2\pi$.

The delicacy arises when OX completes one revolution. How can we identify a material curve that evolves continuously in time as X passes over the sliding contact? One approach is to note that there is nothing special about this instant, since we chose the location of the radial line arbitrarily, and so we could all agree to blink as X passes the sliding contact and then restart the clock at $t = 0$ immediately afterwards. It is also instructive to consider what happens if we replace the conducting rim XY by a wire. Specifically, the azimuthal contribution to $C$ is taken to be an extension of the wire below, which is tied to the point X, pulled up from below, and continually wrapped around the rim of the disc. In this case the flux through $C$ rises continuously, and so (10.31) holds at all times.

## 10.5 The Flux Rule Applied to Material Circuits: Lenz's Law and Alfvén's Theorem

The examples above suggest that it is often convenient to take $C$ to be a material curve, i.e. to take $C$ to lie within a conductor and move with that conductor. So let us now return to (10.21) and focus on the particular case where the curve $C$ used to evaluate the EMF is a material curve always composed of the same atoms. This yields two important results: Lenz's law and Alfvén's theorem.

### 10.5.1 Lenz's Law

When $C$ is a material curve we have $\mathbf{v}(\mathbf{x}, t) = \mathbf{u}(\mathbf{x}, t)$, where $\mathbf{u}$ is the velocity field of the conducting medium. Moreover, Ohm's law tells us that

$$\mathbf{J} = \sigma \mathbf{E}_{\text{rel}} = \sigma(\mathbf{E} + \mathbf{u} \times \mathbf{B}), \tag{10.32}$$

and so $\mathbf{E}_{\text{rel}}$ in (10.21) may be written in terms of current density, $\mathbf{E}_{\text{rel}} = \mathbf{E} + \mathbf{u} \times \mathbf{B} = \mathbf{J}/\sigma$. As we have seen, Faraday's flux rule now adopts the particularly convenient form

$$\boxed{\frac{1}{\sigma} \oint_{C_m} \mathbf{J} \cdot d\mathbf{r} = -\frac{d}{dt} \int_S \mathbf{B} \cdot d\mathbf{S}.} \tag{10.33}$$

(From now on we shall place the subscript $m$ on $C$ whenever $C$ is a material curve, just to remind us that this is a special case of the flux rule.) Equation (10.33) often provides an efficient way to evaluate the current induced by a change of flux, and it works irrespective of whether the current is driven by a motional or a transformer EMF.

Note that the direction of the line integral in (10.33) is given through the right-hand rule. Consider, by way of illustration, the situation shown in Figure 10.3(a). Here the externally imposed magnetic field is from left to right and the relative orientation of $\mathbf{B}$ and $d\mathbf{r}$ in (10.33) is set by the right-hand rule. So, if the flux is falling, then (10.33) tells us that $\oint \mathbf{J} \cdot d\mathbf{r} > 0$ and hence the current flow is in the direction shown in Figure 10.3(a). Conversely, if the flux is rising, then (10.33) demands $\oint \mathbf{J} \cdot d\mathbf{r} < 0$ and the current flow is in the opposite direction to that shown in the figure. Note that, in either case, the additional magnetic field associated with the induced current has a direction such as to offset, at least to some degree, the change in the externally imposed field.

This is an example of what is known as *Lenz's law*, which states that *the direction of the induced current is always such as to tend to oppose any change in flux through a material circuit.* The rule applies equally to a motional EMF. Consider, for example, Figure 10.3(b), which shows a metal slide moving along conducting rails. It is readily confirmed that the Lorentz force acting on the free charges in the slide drives a current in the direction shown. The magnetic field associated with the induced current is then in the opposite direction to the externally applied magnetic field, and so tends to offset the rise in flux caused by the increasing area of the circuit. The nice thing about Lenz's law is that it is a *physical* statement, independent of sign conventions, and so it often provides a simple way of determining the direction of current flow.

### 10.5.2 Alfvén's Theorem

It is often useful to consider the idealised case in which the conducting medium is taken to be a perfect conductor, especially when dealing with certain astrophysical plasmas. So let us return to (10.33), only now we have in mind fluid conductors, like plasmas, as well as conventional solid conductors. In a perfect conductor we have $\sigma \to \infty$ and so (10.33) reduces to

$$\frac{d}{dt}\int_S \mathbf{B}\cdot d\mathbf{S} = 0. \tag{10.34}$$

This tells us that the magnetic flux though any material curve $C_m$ cannot change if the material that makes up the circuit is a perfect conductor. We can think of this as an extreme version of Lenz's law in which the magnetic field associated with the induced current perfectly offsets any change in the external flux through the material curve.

Consider, for example, Figure 10.7, in which a highly conducting ring is pulled through an imposed magnetic field. The lower image shows the net magnetic field, consisting of the sum of the externally imposed field and the magnetic field associated with the current induced in the ring. If the ring is a perfect conductor, then the total magnetic flux through it cannot change, and so as the ring moves to the right it appears to drag the magnetic field lines with it. It is as if the field lines cannot pass through the ring as it moves through the magnet. In fact, we show below that this is indeed the case: field lines cannot move relative to a perfectly conducting medium, but rather are frozen into the conducting medium in the sense that they move with the conductor.

There is something else of interest in Figure 10.7. Notice the curvature of the field lines caused by movement of the ring. From §8.6 we know that magnetic field lines behave like elastic bands in tension (Faraday's tension), and so the curvature shown in Figure 10.7

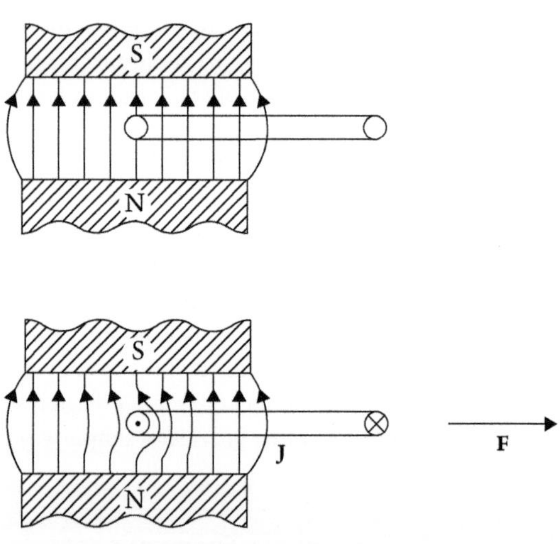


**Figure 10.7** Interaction of a magnetic field and a moving wire loop.

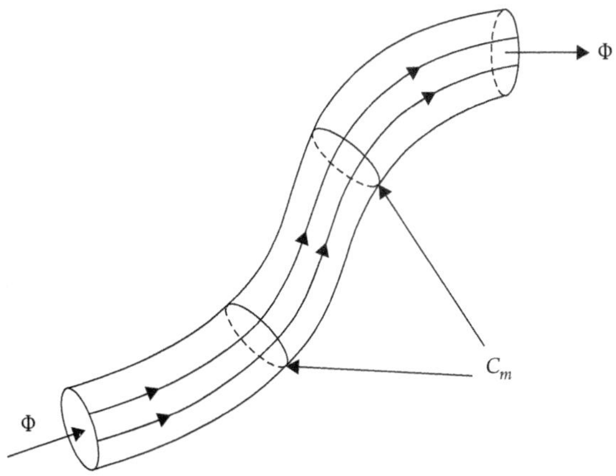


**Figure 10.8** A magnetic flux tube.

tells us that there is a Lorentz force $\mathbf{J} \times \mathbf{B}$ that acts to pull the ring back into the magnet. Thus a mechanical force $\mathbf{F}$ must be applied to the ring in order to drag it to the right. If this mechanical force were removed the ring, would spring back, overshoot its equilibrium position, and then oscillate. It is as if the ring were held in place by a magnetic spring. These kinds of oscillations are commonplace in plasmas threaded by a magnetic field and we will discuss this oscillation in more detail in the next section.

We shall now show that magnetic field lines are indeed frozen into a perfect conductor, as claimed above. In order to prove this result it is convenient to consider a conducting medium which extends across space, rather than the thin wire shown in Figure 10.7. Consider the isolated magnetic flux tube shown in Figure 10.8, which lies within a highly conducting medium, say a hot plasma. Let $\Phi_0$ be the magnetic flux in the tube and $C_m$ be a material curve that, at some initial instant, encircles the flux tube, sitting on its surface. If we treat the medium as a perfect conductor, then (10.34) tells us that the flux enclosed by $C_m$ will remain constant as the medium moves or deforms, and so the flux through $C_m$ stays as $\Phi_0$. Indeed, this is true of all material curves that encircle the flux tube at the initial instant. This suggests that *every* material curve that encircles the tube at $t = 0$ continues to encircle the flux tube for all time, which is possible only if the tube itself moves with the medium, as if frozen into it. This, in turn, implies that every magnetic field line in a perfectly conducting medium moves with that medium, since we could let the flux tube have a vanishingly small cross-section.

We have arrived at *Alfvén's theorem*, which states that, in a perfectly conducting medium, (i) the magnetic flux linking a material curve moving with the medium is conserved and (ii) the magnetic field lines are frozen into the medium, in the sense that they move with the medium. Alfvén's theorem plays a central role in, for example, magnetohydrodynamics and we shall return to this topic in Chapter 20.

## 10.6 The Oscillating Slide—A Poor Man's Alfvén Wave

Consider the apparatus shown in Figure 10.9, which is taken from Davidson (2017). This consists of a horizontal conducting circuit threaded by a uniform vertical magnetic field $\mathbf{B}_0$. The circuit is composed of a frictionless slide that is free to move between two parallel rails. We shall now show that, just like the conducting ring in Figure 10.7, if the slide is perturbed, and the apparatus has a high conductivity, the slide vibrates as if held in place by a magnetic spring. Our aim is to use the flux rule to quantify this oscillation.

To simplify the analysis, we shall suppose that the depth of the apparatus is much greater than its lateral dimensions $L$ and $W$ and take the rails and slide to be made from the same material and to have a common thickness $\Delta$, with $\Delta << L, W$. Now suppose that at some initial instant the slide is given some forward motion $\mathbf{u}_0$. The resulting motional EMF will drive a current density $\mathbf{J}$ in the direction shown in Figure 10.10, which, in turn, induces a magnetic field $\mathbf{B}_i$. The field $\mathbf{B}_i$ is directed downwards in accordance with Lenz's law and, as in a solenoid, it is small outside the apparatus and relatively uniform within it. Moreover, the analysis in §7.4.1 tells us it has a magnitude of $B_i = |\mathbf{B}_i| \approx \mu_0 \Delta J$.

We now apply the flux rule in the form of (10.33), where the material circuit $C$ comprises the slide and the return path for $\mathbf{J}$. When combined with $B_i = \mu_0 \Delta J$ this yields the equation

$$\frac{d}{dt}[LW(B_0 - B_i)] = \frac{\ell J}{\sigma} = \frac{\ell B_i}{\sigma \Delta \mu_0} = \frac{RB_i}{\mu_0}, \tag{10.35}$$

where $\ell = 2(L+W)$, $R = \ell/\sigma\Delta$ and $\Phi = (B_0 - B_i)LW$ are the length, resistance and flux linkage of the conducting circuit. Next we note that the Lorentz force $\mathbf{J} \times \mathbf{B}$ integrated through the slide gives a force per unit area of $\mathbf{F} = -J\Delta(B_0 - B_i/2)\,\hat{\mathbf{e}}_x$, where $B_0 - B_i/2$ is the average field within the slide. This gives the equation of motion for the slide as

$$\rho\Delta\frac{d^2L}{dt^2} = -J\Delta(B_0 - B_i/2) = -\frac{B_i}{\mu_0}(B_0 - B_i/2), \tag{10.36}$$

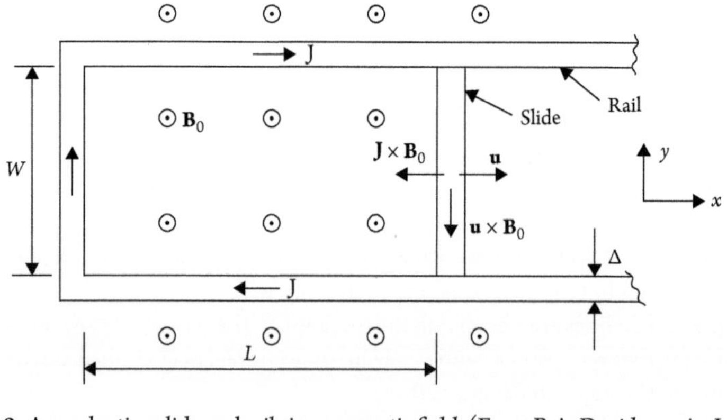


**Figure 10.9** A conducting slide and rails in a magnetic field. (From P. A. Davidson. *An Introduction to Magnetohydrodynamics*, 2nd ed. Cambridge University Press, 2017.)

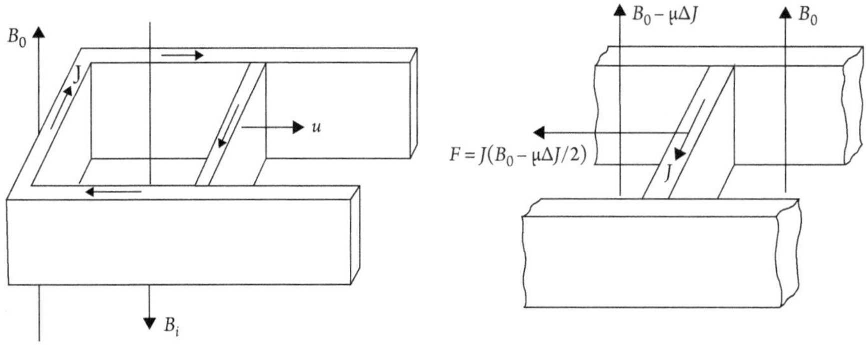


**Figure 10.10** Induced magnetic field and Lorentz force on slide. (From P. A. Davidson. *An Introduction to Magnetohydrodynamics*, 2nd ed. Cambridge University Press, 2017.)

where $\rho$ is the density of the metal. These expression, which are sufficient to determine the two unknowns $L$ and $B_i = \mu_0 \Delta J$, are most conveniently rewritten as

$$\frac{d}{dt}[L(B_0 - B_i)] = \left(\frac{R}{\mu_0 W u_0}\right) u_0 B_i \tag{10.37}$$

and

$$\rho \Delta \frac{d^2 L}{dt^2} = \frac{(B_0 - B_i)^2}{2\mu_0} - \frac{B_0^2}{2\mu_0}. \tag{10.38}$$

From §8.6 we recognise the right-hand side of (10.38) as the difference in magnetic pressure either side of the slide—a negative quantity. More importantly, $R/\mu_0 W u_0$ in (10.37) provides a convenient dimensionless measure of the resistivity of the circuit.

The low-conductivity limit of these equations is discussed in Exercises 7.1 and 10.1; here we focus on the high-conductivity limit. For $R << \mu_0 W u_0$ the right-hand side of (10.37) may be neglected and so the flux $\Phi$ linking the current path is conserved. This is an example of Alfvén's theorem. In such cases we may look for solutions of (10.38) of the form $L = L_0 + \eta$, with $\eta << L_0$, and (10.38) then reduces to (see Exercise 10.1)

$$\frac{d^2 \eta}{dt^2} + \frac{B_0^2}{\rho \mu_0 \Delta L_0} \eta = 0. \tag{10.39}$$

Thus, when the conductivity of the circuit is high, the slide oscillates in an elastic manner, and it does so for precisely the same reason as the ring in Figure 10.7.

Consider the case where the depth and width of the slide are similar, as in Figure 10.10. The field lines are frozen into the slide (Alfvén's theorem again), so when it moves to the right it drags the field lines with it. Since field lines act like elastic bands in tension, they resist this lateral displacement and decelerate the slide. Eventually the slide comes to a halt and reverses direction. When it reaches its equilibrium position, the magnetic restoring force falls to zero, but the inertia of the slide pushes it past the equilibrium point and the cycle now starts in reverse. Elastic oscillations follow. It is the combination of Faraday's tension in the magnetic field lines, the inertia of the slide and the freezing together of the magnetic field with the highly conducting body that gives rise to the oscillation.

Equation (10.39) tells us that the frequency of these oscillations is proportional to $B_0/\sqrt{\rho\mu_0}$, which is known as the Alfvén speed. Actually, we have been here before. We noted in §8.6 that a common feature of astrophysical plasmas is the appearance of Alfvén waves, which are transverse oscillations that propagate along magnetic field lines at the Alfvén speed. We shall discuss Alfvén waves in detail in Chapter 21.

## 10.7 Faraday's Law Expressed in Terms of the Vector Potential

We close this chapter by recalling that Faraday's law may be rewritten in terms of the vector potential **A**, defined by $\nabla \times \mathbf{A} = \mathbf{B}$. In particular, on substituting **A** for **B** in Faraday's equation, we obtain

$$\nabla \times \mathbf{E} = -\nabla \times \left( \frac{\partial \mathbf{A}}{\partial t} \right), \tag{10.40}$$

which uncurls to give

$$\boxed{\mathbf{E} = -\frac{\partial \mathbf{A}}{\partial t} - \nabla \varphi,} \tag{10.41}$$

for some scalar potential $\varphi$. This turns out to be a particularly useful formulation for many purposes. However, as discussed in Chapter 3, the precise interpretation of the scalar potential $\varphi$ depends on our choice of the gauge for **A**, i.e. how we choose to define $\nabla \cdot \mathbf{A}$. In practice, only two gauges are in common use.

For cases where we may ignore displacement currents, such as quasi-steady magnetic fields, the Ampère–Maxwell equation reverts to Ampère's law. In such situations the most convenient choice of the gauge is the Coulomb gauge, $\nabla \cdot \mathbf{A} = 0$. The divergence of (10.41) then demands $\nabla^2 \varphi = -\nabla \cdot \mathbf{E}$, and Gauss' law tells us that $\varphi$ is governed by the electrostatic Poisson equation $\nabla^2 \varphi = -\rho_e/\varepsilon_0$, where $\rho_e$ is the charge density. However, when we cannot ignore the displacement current in the Ampère–Maxwell equation, such as in electromagnetic radiation, it turns out that a better choice is the Lorenz gauge,

$$\nabla \cdot \mathbf{A} = -\frac{1}{c^2} \frac{\partial \varphi}{\partial t}, \tag{10.42}$$

where $c$ is the speed of light. As discussed in §7.9, $\varphi$ is then governed by a wave equation, rather than Poisson's equation. We shall return to (10.41) in Chapter 13, where we discuss the importance of the choice of gauge and its implications for $\varphi$ and **A**.

* * *

This concludes our review of Faraday's law and its relationship to Faraday's flux rule. Readers who seek an alternative perspective will find a spirited and informative discussion in Griffiths (2013).

## EXERCISES

10.1 Consider the high-resistance case of the slide and rails shown in Figure 10.9, in which **J** is low and $B_i << B_0$. Show that, in such cases, (10.37) and (10.38) yield

$$\frac{du}{dt} = -\frac{\sigma B_0^2}{\rho}\frac{W}{2(L+W)}u = -\frac{u}{\tau},$$

where $u$ is the speed of the slide and $\tau$ is defined by the equation above. If changes in $L$ may be neglected in the term $(L+W)$, we have $u = u_0 e^{-t/\tau}$, as discussed in Exercise 7.1. In the high-conductivity limit, (10.37) reduces to $(L_0+\eta)(B_0 - B_i) \approx L_0 B_0$, where $L = L_0 + \eta$. Use this to eliminate $B_0 - B_i$ from (10.38) and show that, for $\eta << L_0$,

$$\frac{d^2\eta}{dt^2} + \frac{B_0^2}{\rho\mu_0 \Delta L_0}\eta = 0.$$

10.2 A common illustration of electromagnetic induction is to let a pendulum composed of a thin light rod attached to a metal disc swing between the poles of a magnet, with the metal disc aligned with the faces of the magnet poles. If the disc is not too conducting and its momentum not too high, it is observed that the pendulum slows down on entering the magnet, coming to a halt on a timescale $\tau = (\sigma B_0^2/\rho)^{-1}$, as in Exercise 10.1. Explain what is happening. What do you think will happen if the disc is slotted like a fork? What do you think will happen if the disc is superconducting?

10.3 Consider the cylindrical bar magnet shown in Figure 9.13(a). Suppose it moves from left to right with constant speed $u$, passing through a copper ring of slightly larger radius. In doing so it induces an EMF in the ring. Sketch the variation of EMF with time.

10.4 A thin aluminium ring sits on top of an iron-cored solenoid whose axis is vertical. The ring and the solenoid have the same radius and are coaxial. The ring is free to move and rests on the top of the solenoid. Initially the solenoid is off. When it is turned on, the ring is seen to jump upward. Use Lenz's law to explain what is happening. What would happen if a thin radial cut were made through the ring and the experiment repeated?

10.5 A rectangular wire loop of dimensions $h$ and $w$ is coplanar with a long straight wire that carries a constant current $I$. The sides, of length $h$, are parallel to the wire and the instantaneous distance from the wire to the closest side of the loop is $r$. The loop recedes from the wire with a constant speed $u$. Show that the EMF induced in the loop is

$$\text{EMF} = \frac{\mu_0 I(hw)u}{2\pi r(r+w)}.$$

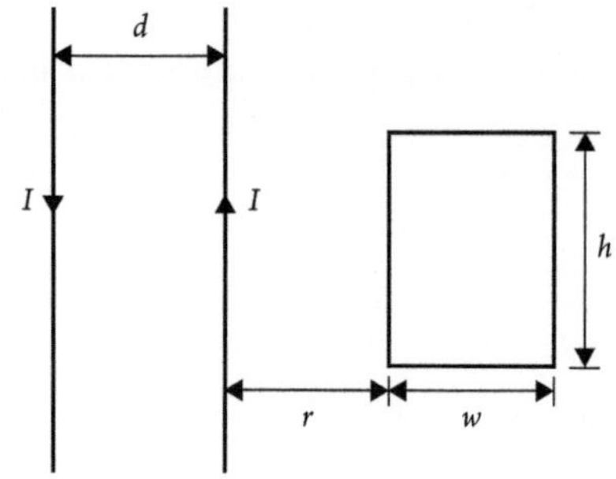


**Figure 10.11** An EMF is generated in a rectangular loop by two adjacent wires.

10.6 Two parallel wires are separated by a distance $d$ and carry equal and opposite currents $I(t)$. A rectangular wire loop of dimensions $h$ and $w$ is coplanar with the wires and a distance $r$ from the closest wire, as shown in Figure 10.11. Show that the EMF induced in the loop by the current is

$$\oint \mathbf{E} \cdot d\mathbf{r} = -\frac{\mu_0 h}{2\pi}\frac{dI}{dt} \ln\left[\frac{(r+w)(r+d)}{r(r+d+w)}\right].$$

10.7 A small circular coil of $N$ turns, radius $a$, and electrical resistance $R$ is orientated with its axis parallel to a magnetic field $\mathbf{B}$. The coil is suddenly 'flipped' so that its axis is reversed and a current $I$ is induced in the coil in accordance with EMF $= IR$. Show that the net charge that passes through the coil, $Q$, is

$$Q = \frac{2\pi a^2 NB}{R}.$$

## REFERENCES

Davidson P.A., 2017, *Introduction to Magnetohydrodynamics*, 2nd ed. Cambridge University Press.
Einstein, A., 1954, *Relativity, the Special and the General Theory*, 15th ed. Methuen.
Griffiths, D. J., 2013, *Introduction to Electrodynamics*, 4th ed. Pearson.
Zangwill, A., 2013, *Modern Electrodynamics*. Cambridge University Press.

# 11

# Quasi-Static Magnetic Fields

## *Magnetic Energy and Inductance*

Heaviside's [polemics] can be placed somewhere between brilliant and obnoxious, depending on one's point of view. At least he was not dull; indeed G. M. Minchin referred to Heaviside as 'the Walt Whitman of English Physics'.

M. J. Crowe, 1994

## 11.1 The Idea of a Quasi-Static Magnetic Field

In this chapter we shall focus on magnetic fields that vary relatively slowly in time. To be more precise, we shall assume that the timescale for changes in $\mathbf{B}$ is much longer than the time taken for electromagnetic waves to propagate across the domain of interest. (In practice, of course, this allows $\mathbf{B}$ to change relatively rapidly from a mechanical perspective.) We may then take the speed of light to be effectively infinite, so that the retarded potentials of §7.9 revert to the instantaneous inversion integrals of the quasi-steady Poisson equations for $\mathbf{A}$ and $\varphi$. Additionally, we shall assume that the charge density (both surface charges and internal charge) is very small, as is invariably the case for good conductors, unless charges are deliberately introduced onto the surface of those conductors. As discussed in §§3.1 and 3.2, we are then at liberty to take $\nabla \cdot \mathbf{J} = 0$ and to ignore the displacement current that gives rise to the phenomenon of electromagnetic waves. In short, we adopt the pre-Maxwell form of Ampère's law. Such a system of equations, known as the *quasi-magnetostatic approximation*, is appropriate for much of electrical engineering and for MHD. It is similar to, but distinct from, *quasi-electrostatics*, in which the surface charges *are* important, as discussed in §11.5.

For simplicity, we shall set aside dielectrics and magnetic materials, so that we may ignore polarization current $\mathbf{J}_{\text{pol}}$ and bound current, $\mathbf{J}_{\text{mag}}$. The only current density of interest is then the conduction current associated with the free charges in conductors, $\mathbf{J} = \mathbf{J}_f$. In the absence of magnetic materials and dielectrics, our governing equations are

$$\nabla \times \mathbf{B} = \mu_0 \mathbf{J}, \qquad \nabla \cdot \mathbf{J} = 0,$$

$$\nabla \times \mathbf{E} = -\frac{\partial \mathbf{B}}{\partial t}, \qquad \nabla \cdot \mathbf{B} = 0,$$

along with Ohm's law $\mathbf{J} = \sigma(\mathbf{E} + \mathbf{u} \times \mathbf{B})$ for the conduction current. Within this framework, Faraday's equation leads directly to two important and related concepts, namely those of magnetic energy and inductance. We shall discuss both of these in turn, bringing them together towards the end of the chapter.

We introduced the idea of magnetic energy in §3.5, where we asserted that $\mathbf{B}^2/2\mu_0$ is the magnetic energy density. Subsequently, in §8.4.3, we used energy conservation to derive an expression for the total energy required to assemble a set of steady currents loops and showed it to be

$$\boxed{U_B = \frac{1}{2}\int_{V_\infty} \mathbf{J}\cdot\mathbf{A}\,dV.} \tag{11.1}$$

For *steady* currents this is equivalent to a magnetic energy density of $\mathbf{B}^2/2\mu_0$, because Ampère's law allows us to rewrite (11.1) as

$$U_B = \frac{1}{2\mu_0}\int_{V_\infty} \mathbf{A}\cdot\nabla\times\mathbf{B}\,dV = \frac{1}{2\mu_0}\int_{V_\infty} \mathbf{B}^2 dV - \frac{1}{2\mu_0}\oint_{S_\infty} (\mathbf{A}\times\mathbf{B})\cdot d\mathbf{S} \tag{11.2}$$

and the surface integral vanishes as the surface recedes to infinity. However, for unsteady currents, (11.1) and (11.2) are not the same thing and so we might ask: which is the more fundamental form of magnetic energy? To that end we shall now redevelop the entire subject from first principles. This will lead us directly to the idea of the Poynting energy flux, which quantifies how electromagnetic energy is transferred from place to place.

The existence of such a flux was discovered independently by John Poynting and by the brilliant, if stubbornly independent, Oliver Heaviside. However, as was often the case with the maverick Heaviside, history chose not to credit Heaviside with the idea. In fact, relatively few of Heaviside's extraordinary discoveries in mathematical physics now bear his name, and after a life of exceptional achievement, his name has been largely relegated to the humble step-change in a function! (See Nahin (2002) for the history of Heaviside's life and achievements, while we provide a brief sketch in Appendix III.)

## 11.2 Magnetic Energy

### *11.2.1 The Rate of Working of the Electric Field on a Conductor*

Before discussing magnetic energy, it will be useful to review the work done by an electric field on a conductor. Recall that a charged particle moving with velocity $\mathbf{u}_q$ and carrying a charge $q$ experiences an electromagnetic force

$$\mathbf{f} = q(\mathbf{E} + \mathbf{u}_q \times \mathbf{B}). \tag{11.3}$$

The rate of working of this force on the charges within a unit volume of conducting material is

$$\sum \mathbf{f} \cdot \mathbf{u}_q = \mathbf{E} \cdot \sum q\mathbf{u}_q = \mathbf{E} \cdot \mathbf{J}, \tag{11.4}$$

where the current density $\mathbf{J} = \sum q\mathbf{u}_q$ in a conductor is dominated by the conduction current of the free charges. So, as noted in §7.2.2, the rate of working of the electric field on a conductor is $\mathbf{J} \cdot \mathbf{E}$ per unit volume. In an electrical circuit this energy is often replenished by an external source, such as a battery, which also does work on the charge carriers. However, in the absence of an external source, this energy must be supplied by the electromagnetic field itself.

In a stationary conductor the energy transferred to the free charges is lost by the charge carriers through inelastic collisions and ends up as heat in the conductor. Ohm's law in a stationary conductor, $\mathbf{J} = \sigma\mathbf{E}$, then gives $\mathbf{J}^2/\sigma$ as the rate of generation of heat per unit volume. This is Joule heating, or Ohmic dissipation, and it represents the rate at which electromagnetic energy is irreversibly lost to heat as a result of the flow of current. In a moving conductor, however, there is the possibility that some of the electrical power ends up as bulk kinetic energy in the conductor. This is apparent from Ohm's law, which yields

$$\mathbf{J} \cdot \mathbf{E} = \frac{\mathbf{J}^2}{\sigma} - \mathbf{J} \cdot (\mathbf{u} \times \mathbf{B}).$$

Rearranging the triple product gives us

$$\mathbf{J} \cdot \mathbf{E} = \frac{\mathbf{J}^2}{\sigma} + \mathbf{u} \cdot (\mathbf{J} \times \mathbf{B}), \tag{11.5}$$

where the second term on the right is the rate of working of the Lorentz force $\mathbf{J} \times \mathbf{B}$ acting on the conductor. This term may be positive or negative. When positive, it represents the conversion of electromagnetic energy into mechanical energy, as in the case of an electric motor. Conversely, when $\mathbf{u} \cdot (\mathbf{J} \times \mathbf{B}) < 0$, it represents the conversion of mechanical energy into electromagnetic energy, as in a dynamo.

So far we have ignored external sources, such as batteries. If external sources of current are present, then the net force acting on a test charge $q$ moving with velocity $\mathbf{u}_q$ must be generalised to read

$$\mathbf{f} = \mathbf{f}_s + q(\mathbf{E} + \mathbf{u}_q \times \mathbf{B}),$$

where the subscript $s$ stands for *source*. If we are dealing with batteries, say, then $\mathbf{f}_s$ is a chemical force. In such cases Ohm's law generalises to

$$\mathbf{J} = \sigma(\mathbf{F}_s + \mathbf{E} + \mathbf{u} \times \mathbf{B}),$$

where $\mathbf{F}_s = \mathbf{f}_s/q$. Equation (11.5) then becomes

$$\mathbf{J} \cdot \mathbf{E} + \mathbf{J} \cdot \mathbf{F}_s = \frac{\mathbf{J}^2}{\sigma} + \mathbf{u} \cdot (\mathbf{J} \times \mathbf{B}), \tag{11.6}$$

where the left-hand side now represents the work done on the charge carriers by both the electric field and the external source.

### 11.2.2 A Quasi-Steady Magnetic Energy Equation

We shall now develop an energy equation for quasi-steady magnetic fields. This is approximate, rather than exact, in that we shall neglect the displacement current in the Ampère–Maxwell equation, in line with the spirit of quasi-magnetostatics. We shall also, for the time being, ignore external sources of energy, such as batteries. Consider the dot product of Faraday's law with **B**, combined with the dot product of Ampère's law with **E**,

$$\frac{\partial}{\partial t}\left(\frac{\mathbf{B}^2}{2\mu_0}\right) + \mathbf{J}\cdot\mathbf{E} = -\frac{1}{\mu_0}(\mathbf{B}\cdot\nabla\times\mathbf{E} - \mathbf{E}\cdot\nabla\times\mathbf{B}),$$

from which

$$\boxed{\frac{\partial}{\partial t}\left(\frac{\mathbf{B}^2}{2\mu_0}\right) = -\mathbf{J}\cdot\mathbf{E} - \frac{1}{\mu_0}\nabla\cdot(\mathbf{E}\times\mathbf{B}).} \tag{11.7}$$

(quasi-steady magnetic fields)

This integrates over a volume $V$ to give

$$\boxed{\frac{d}{dt}\int_V \frac{\mathbf{B}^2}{2\mu_0}dV = -\int_V \mathbf{J}\cdot\mathbf{E}\,dV - \oint_S \frac{\mathbf{E}\times\mathbf{B}}{\mu_0}\cdot d\mathbf{S},} \tag{11.8}$$

(energy equation for quasi-steady magnetic fields)

where $S$ is the surface that encloses the volume $V$. This is our energy equation, in which we *define* $\mathbf{B}^2/2\mu_0$ to be the magnetic energy density.

Consider first the case of a vacuum, where there are no currents. Then (11.8) simplifies to

$$\frac{d}{dt}\int_V \frac{\mathbf{B}^2}{2\mu_0}dV = -\oint_S \frac{\mathbf{E}\times\mathbf{B}}{\mu_0}\cdot d\mathbf{S}, \tag{11.9}$$

where the vector $\mathbf{E}\times\mathbf{B}/\mu_0$ is called the *Poynting vector*. In view of our definition of magnetic energy, conservation of energy tells us that the integral on the right must represent the *net* rate at which electromagnetic energy flows out of $V$ through the closed surface $S$. This suggests, *but does not prove*, that the Poynting vector $\mathbf{E}\times\mathbf{B}/\mu_0$ represents the outward flux of electromagnetic energy per unit area at each and every point on the surface $S$.

In a conductor, on the other hand, (11.5) and (11.8) combine to give

$$\boxed{\frac{d}{dt}\int_V \frac{\mathbf{B}^2}{2\mu_0}dV = -\frac{1}{\sigma}\int_V \mathbf{J}^2 dV - \int_V \mathbf{u}\cdot(\mathbf{J}\times\mathbf{B})\,dV - \oint_S \frac{\mathbf{E}\times\mathbf{B}}{\mu_0}\cdot d\mathbf{S}.} \tag{11.10}$$

(energy equation for quasi-steady magnetic fields in conducting material)

(It is worth remembering that we are ignoring external sources of current here.) There are now two additional terms on the right. Evidently, the loss of magnetic energy represented by the first integral is the Ohmic dissipation within $V$. The second term on the right, on the other hand, is the rate of working of the Lorentz force $\mathbf{J} \times \mathbf{B}$ acting on the conductor. It may be positive or negative and represents an exchange of electromagnetic and mechanical energy, as discussed above. So, in words, (11.10) says that the magnetic energy within $V$ may fall because some magnetic energy is converted into heat, some is converted into mechanical energy, and some flows out of $V$ through the bounding surface $S$. On the other hand, if $\mathbf{u} \cdot (\mathbf{J} \times \mathbf{B}) < 0$, there is a conversion of mechanical energy into electromagnetic energy, and this may cause the magnetic energy to rise.

### 11.2.3 *What Happened to the Electric Energy?*

At this point one might ask: what happened to the electric energy in (11.8) and (11.10)? That is, in Chapter 5 we showed that the energy density of an electrostatic field may be written in the form $\varepsilon_0 \mathbf{E}^2/2$, and so we might have expected such a term to be present in a slowly varying electromagnetic field. In this respect it is instructive to compare (11.8) and (11.10) with the more general energy equations derived in §3.5. These are (3.42) and (3.44), which can be written as

$$\boxed{\frac{d}{dt}\int_V \left(\frac{\mathbf{B}^2}{2\mu_0} + \frac{\varepsilon_0 \mathbf{E}^2}{2}\right) dV = -\int_V \mathbf{J} \cdot \mathbf{E}\, dV - \oint_S \left(\frac{\mathbf{E} \times \mathbf{B}}{\mu_0}\right) \cdot d\mathbf{S},} \tag{11.11}$$

(exact energy equation for an electromagnetic field)

$$\boxed{\frac{d}{dt}\int_V \left(\frac{\mathbf{B}^2}{2\mu_0} + \frac{\varepsilon_0 \mathbf{E}^2}{2}\right) dV = -\frac{1}{\sigma}\int_V \mathbf{J}^2 dV - \int_V \mathbf{u} \cdot (\mathbf{J} \times \mathbf{B})\, dV - \oint_S \frac{\mathbf{E} \times \mathbf{B}}{\mu_0} \cdot d\mathbf{S}.} \tag{11.12}$$

(exact energy equation for an electromagnetic field in conducting material)

No restriction has been placed on the rate of change of the magnetic field in (11.11) and (11.12) and so, in their derivation, the displacement current has been retained in the Ampère–Maxwell equation:

$$\nabla \times \mathbf{B} = \mu_0 \left(\mathbf{J} + \varepsilon_0 \frac{\partial \mathbf{E}}{\partial t}\right). \tag{11.13}$$

Actually, a brief perusal of §3.5 is sufficient to confirm that the derivation of the exact expressions (11.11) and (11.12) is almost identical to that of the quasi-steady equations (11.8) and (11.10). The only difference is that (11.13) replaces Ampère's law in the analysis, and this has resulted in the electric energy appearing on the left.

So we conclude that the electric energy of an electromagnetic field is associated with the displacement current in the Ampère–Maxwell equation. It is negligible by comparison with the magnetic energy in the quasi-magnetostatic approximation, because the charge density *inside* a conductor is so small, which is why the electric energy is missing from (11.8) and (11.10). However, neither the displacement current nor the electric energy may be neglected if we deliberately introduce charges onto the *surface* of a conductor, as we shall discus in §11.5. So we shall have to reinstate the electric energy when we come to consider circuits that include capacitors.

### 11.2.4 Allowing for External Energy Sources

Let now consider how our energy equation changes if we allow for external sources of power, such as batteries. The expression (11.8),

$$\frac{d}{dt}\int_V \frac{\mathbf{B}^2}{2\mu_0}\,dV = -\int_V \mathbf{J}\cdot\mathbf{E}\,dV - \oint_S \frac{\mathbf{E}\times\mathbf{B}}{\mu_0}\cdot d\mathbf{S}, \tag{11.14}$$

is unchanged by the presence of an external source, since it follows directly from the laws of Faraday and Ampère. However, when we substitute for $\mathbf{J}\cdot\mathbf{E}$ in (11.14), we must now use (11.6), rather than (11.5). So let us write

$$P_s = \int_V \mathbf{J}\cdot\mathbf{F}_s\,dV \tag{11.15}$$

for the power supplied by the external source. It then follows that (11.10) is modified to read

$$\boxed{\frac{d}{dt}\int_V \frac{\mathbf{B}^2}{2\mu_0}\,dV = P_s - \frac{1}{\sigma}\int_V \mathbf{J}^2 dV - \int_V \mathbf{u}\cdot(\mathbf{J}\times\mathbf{B})\,dV - \oint_S \frac{\mathbf{E}\times\mathbf{B}}{\mu_0}\cdot d\mathbf{S}.} \tag{11.16}$$

(energy equation for quasi-steady magnetic fields in conducting material)

When the **E** and **B** fields decay sufficiently rapidly with distance from their sources, and the volume in (11.16) is taken to be all of space, the surface flux of the Poynting vector is zero. In such cases (11.16) simplifies to

$$P_s = \frac{d}{dt}\int_{V_\infty} \frac{\mathbf{B}^2}{2\mu_0}\,dV + \frac{1}{\sigma}\int_{V_\infty} \mathbf{J}^2 dV + \int_{V_\infty} \mathbf{u}\cdot(\mathbf{J}\times\mathbf{B})\,dV, \tag{11.17}$$

which says that the power from the external source can go into increasing the magnetic energy or compensating for the Joule heating, or be converted into mechanical energy (provided the rate of working of the Lorentz force on the conductor is positive).

For filamentary circuits we have $\mathbf{J}\,dV = I\,d\mathbf{r}$, where $I$ is the current in the circuit and $d\mathbf{r}$ part of that circuit. In such cases the power from the external source is often written as

$$P_s = I\int \mathbf{F}_s\cdot d\mathbf{r} = I\int \frac{\mathbf{f}_s}{q}\cdot d\mathbf{r}. \tag{11.18}$$

We now recall from (10.11) that, in the presence of external sources, the EMF around a circuit is defined as

$$\text{EMF} = \oint_C \frac{\mathbf{f}_s}{q} \cdot d\mathbf{r} + \oint_C (\mathbf{E} + \mathbf{v} \times \mathbf{B}) \cdot d\mathbf{r}, \tag{11.19}$$

and so the power delivered from the source is just the EMF of the source times the current in the circuit, as expected.

### 11.2.5 *An Ambiguity in the Poynting Energy Flux*

The divergence of the Poynting energy flux $\mathbf{E} \times \mathbf{B}/\mu_0$ is evidently crucial to ensuring that energy is conserved in (11.7). However, one should be careful not to insist that $\mathbf{E} \times \mathbf{B}/\mu_0$ uniquely determines the energy flux at *every* point in space. The point is that we can add to $\mathbf{E} \times \mathbf{B}/\mu_0$ any solenoidal vector field we wish, and this leaves the energy balance in both (11.7) and (11.8) unchanged. Thus any vector field of the form $\mathbf{E} \times \mathbf{B}/\mu_0 + \nabla \times \mathbf{G}$ can also lay claim to being an electromagnetic energy flux, in the sense that it balances the books, so to speak, in both (11.7) and (11.8).

Consider, by way of an example, the case of a long straight wire of radius $a$ and carrying a steady current of $I = J\pi a^2$, as shown in Figure 11.1. The electric and magnetic fields at the edge of the wire are $\mathbf{E} = \mathbf{J}/\sigma$ and $B = \mu_0 I/2\pi a$, and so the Poynting flux $\mathbf{E} \times \mathbf{B}/\mu_0$ is directed radially inward and has a magnitude of $JI/2\pi a\sigma$. When integrated over the surface of a cylindrical control volume of unit length, we find a net energy flux radially into the wire of $I^2/\pi a^2\sigma = I^2 R$, where $R$ is the resistance per unit length of the wire. Of course, this is exactly what the energy equation (11.10) demands,

$$\text{flux of energy} = \oint_S \frac{\mathbf{E} \times \mathbf{B}}{\mu_0} \cdot d\mathbf{S} = -\frac{1}{\sigma} \int_V \mathbf{J}^2 dV = -I^2 R, \tag{11.20}$$

where the surface integral is negative because $\mathbf{E} \times \mathbf{B}$ points radially inward and $d\mathbf{S}$ points outward.

Now suppose that we adopt as our energy flux the alternative expression

$$\frac{1}{\mu_0}[\mathbf{E} \times \mathbf{B} + \nabla \times (\varphi \mathbf{B})],$$

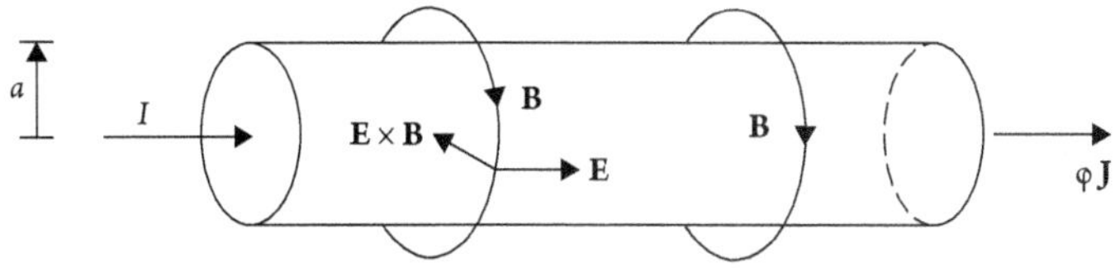


**Figure 11.1** A wire carries a current $I$. There is a radially inward Poynting energy flux.

where $\varphi$ is the electrostatic potential. Since this is of the form $\mathbf{E}\times\mathbf{B}/\mu_0+\nabla\times\mathbf{G}$, it is equally valid from the point of view of global energy conservation, in the sense that the extra term leaves (11.7) and (11.8) unaltered. Expanding $\nabla\times(\varphi\mathbf{B}/\mu_0)$, invoking Ampère's laws, and then Faraday's law in the form of (10.41), it is readily confirmed that

$$\frac{1}{\mu_0}[\mathbf{E}\times\mathbf{B}+\nabla\times(\varphi\mathbf{B})]=\varphi\mathbf{J}+\frac{\mathbf{B}}{\mu_0}\times\frac{\partial\mathbf{A}}{\partial t}. \tag{11.21}$$

For the case of a wire carrying a steady current, the electromagnetic energy flux per unit area is now $\varphi\mathbf{J}$. Of course, this is directed *along* the wire, rather than radially inward! Clearly, since $\nabla\times(\varphi\mathbf{B}/\mu_0)$ has zero divergence, the closed surface integrals of $\mathbf{E}\times\mathbf{B}/\mu_0$ and $\varphi\mathbf{J}$ are identical for a steady current and so this new definition must also satisfy

$$\text{flux of energy}=\oint_S(\varphi\mathbf{J})\cdot d\mathbf{S}=-\frac{1}{\sigma}\int_V\mathbf{J}^2dV=-I^2R. \tag{11.22}$$

This tells us that, just like the Poynting flux, $\varphi\mathbf{J}$ must integrate over a cylindrical control volume of radius $a$ to give a net energy flux into the wire of $I^2R$. Indeed, this may be confirmed by direct calculation of the surface integral in (11.22).

So, should we think of the electromagnetic energy flux as being radially inward towards the axis of the wire or directed axially along the wire itself? The answer is that, as far as energy conservation in a *quasi-steady* magnetic field is concerned, it does not matter. More generally, however, relativistic considerations go some way to establishing the uniqueness of $\mathbf{E}\times\mathbf{B}/\mu_0$ as an energy flux density (Jackson, 1999).

## 11.3 Inductance

### *11.3.1 Mutual Inductance and Neumann's Equation*

We now introduce the important concept of inductance, starting with the mutual inductance of two filamentary current loops. The point is this: if there two adjacent wire loops, $C_1$ and $C_2$, with a current $I_1$ flowing in $C_1$, then the magnetic field established by $C_1$ sets up a flux $\Phi_2$ through $C_2$, as shown in Figure 11.2. Evidently $\Phi_2$ is proportional to $I_1$, and so we can write $\Phi_2=MI_1$, where the constant of proportionality $M$ is called the *mutual inductance* of the two loops. We seek an explicit expression for $M$.

Our starting point is the Biot–Savart law written in terms of the vector potential $\mathbf{A}$. As is usual for quasi-steady fields, we shall use the Coulomb gauge for $\mathbf{A}$, $\nabla\cdot\mathbf{A}=0$, so that Ampère's law yields the Poisson equation $\nabla^2\mathbf{A}=-\mu_0\mathbf{J}$. As discussed in §7.3.2, this Poisson equation inverts in an infinite domain to give the Biot–Savart-like integral

$$\mathbf{A}(\mathbf{x})=\frac{\mu_0}{4\pi}\int\frac{\mathbf{J}(\mathbf{x}')}{|\mathbf{r}|}dV',\qquad \mathbf{r}=\mathbf{x}-\mathbf{x}'. \tag{11.23}$$

Note that, to get to (11.23), we have used Ampère's law rather than the Ampère–Maxwell equation, so we are restricted to quasi-steady fields in which the displacement current can be neglected.

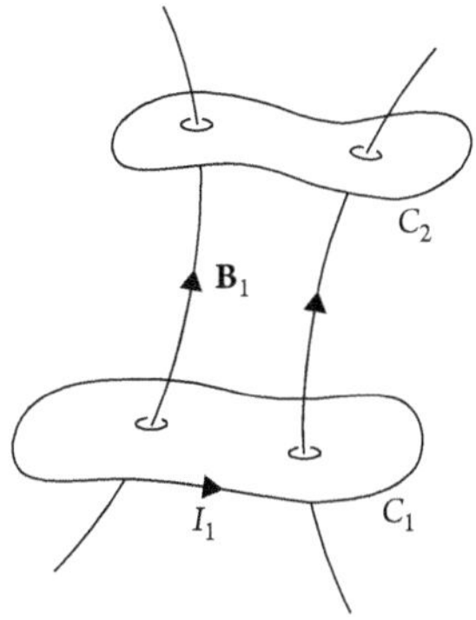


**Figure 11.2** A current $I_1$ flowing in loop $C_1$ establishes a magnetic field $\mathbf{B}_1$, which has a flux of $\Phi_2$ through $C_2$.

The idea of inductance is normally applied to filamentary current loops. However, (11.23) is readily rewritten in terms of such loops, since $\mathbf{J}'\,dV' = I'\,d\mathbf{l}'$, where $I'$ is the current in the loop and $d\mathbf{l}'$ is part of the circuit C. In such cases (11.23) gives the vector potential as

$$\mathbf{A}(\mathbf{x}) = \frac{\mu_0 I'}{4\pi} \oint_C \frac{d\mathbf{l}'}{|\mathbf{x}-\mathbf{x}'|}. \tag{11.24}$$

Consider now two such circuits, $C_1$ and $C_2$, spanned by surfaces $S_1$ and $S_2$ and with a current of $I_1$ flowing in $C_1$. The magnetic field established by $C_1$, call it $\mathbf{B}_1$, sets up a flux through $C_2$ as shown in Figure 11.2. This flux can be written in terms of $\mathbf{A}_1$ as follows:

$$\Phi_2 = \int_{S_2} \mathbf{B}_1 \cdot d\mathbf{S} = \int_{S_2} (\nabla \times \mathbf{A}_1) \cdot d\mathbf{S} = \oint_{C_2} \mathbf{A}_1 \cdot d\mathbf{l}_2. \tag{11.25}$$

However, $\mathbf{A}_1$ is given by (11.24) as

$$\mathbf{A}_1(\mathbf{x}) = \frac{\mu_0 I_1}{4\pi} \oint_{C_1} \frac{d\mathbf{l}_1}{|\mathbf{x}-\mathbf{x}_1|}, \tag{11.26}$$

where $\mathbf{x}_1$ locates part of loop $C_1$. So the flux $\Phi_2$ through $C_2$ can be written as the double line integral

$$\Phi_2 = \frac{\mu_0 I_1}{4\pi} \oint_{C_2} \oint_{C_1} \frac{d\mathbf{l}_1 \cdot d\mathbf{l}_2}{|\mathbf{x}_2-\mathbf{x}_1|}, \tag{11.27}$$

where $\mathbf{x}_2$ locates part of $C_2$. Noting that the flux through $C_2$ is proportional to $I_1$, we now define the mutual inductance of the two loops, denoted by $M_{21}$, to be that constant of proportionality. Thus, by definition,

$$\Phi_2 = M_{21} I_1, \tag{11.28}$$

and from (11.27) we obtain an explicit expression for the mutual inductance of the two loops:

$$M_{21} = \frac{\mu_0}{4\pi} \oint_{C_2} \oint_{C_1} \frac{d\mathbf{l}_1 \cdot d\mathbf{l}_2}{|\mathbf{x}_1 - \mathbf{x}_2|}. \tag{11.29}$$

This is known as *Neumann's equation*. Note that $M_{21}$ is a purely geometrical object.

The symmetry between 1 and 2 in (11.29) is striking and this tells us that if the current were to flow in loop 2, rather than loop 1, and we defined $M_{12}$ through $\Phi_1 = M_{12} I_2$, then we would have $M_{12} = M_{21}$. We conclude that, rather remarkably, the flux through $C_2$ due to a current flowing in $C_1$ is equal to the flux through $C_1$ due to the same current flowing in $C_2$. For this reason it is conventional to drop the subscript on $M$ and write $M_{12} = M_{21} = M$, where $M = \Phi_2/I_1 = \Phi_1/I_2$.

### *11.3.2 Faraday's Flux Rule Revisited and Self-Inductance*

The idea of inductance almost always goes hand-in-hand with Faraday's flux rule, and indeed it is only when combined with the flux law that we see its real value. Suppose that the current in loop $C_1$ is changing, so that the flux through $C_2$ also varies. Then Faraday's flux rule, combined with (11.28), tells us that the induced EMF in $C_2$ is

$$\text{EMF}_2 = -\frac{d\Phi_2}{dt} = -M\frac{dI_1}{dt}, \tag{11.30}$$

where we evaluate the EMF along the material curve $C_2$. Ohm's law in the form of (10.9) then gives us

$$EMF_2 = \frac{1}{\sigma} \oint_{C_2} \mathbf{J}_2 \cdot d\mathbf{l}_2 = I_2 \frac{\ell_2}{\sigma S_2} = -M\frac{dI_1}{dt}, \tag{11.31}$$

where $\ell_2$ and $S_2$ are the length and cross-sectional area of the wire in loop 2. We conclude that

$$I_2 R_2 = -M\frac{dI_1}{dt}, \tag{11.32}$$

where $R_2 = \ell_2/\sigma S_2$ is the resistance of loop 2. Thus we can find the current induced in loop 2 if we know the rate of change of current in loop 1.

Now, when the current flowing in $C_1$ changes, it not only alters the flux through $C_2$, but also changes the flux through $C_1$. This leads to the concept of *self-inductance*, denoted by $L$. As a natural extension to (11.28), we *define* the self-inductance of a current loop $C_1$ through the expression

$$\Phi_1 = L I_1. \tag{11.33}$$

Explicit expressions for $L$ are readily obtained from the Biot–Savart law, though sometimes the calculation is a little tedious. Simple examples of calculating $L$ are given in Exercises

11.2 and 11.3. A trickier calculation is that for a thin circular current loop or radius $a$ and cross-sectional area $S = \pi b^2, b << a$. It turns out this has a self-inductance of (see Jackson, 1999)

$$L = \mu_0 a \left[ \frac{1}{4} + \ln \left( \frac{8a}{e^2 b} \right) \right].$$

Returning to (11.33), Faraday's flux rule applied to $C_1$ gives us

$$\text{EMF}_1 = -\frac{d\Phi_1}{dt} = -L\frac{dI_1}{dt}. \tag{11.34}$$

Note that this induced EMF, which will tend to alter the current in loop $C_1$, has the opposite sign to $dI_1/dt$. Thus, in accordance with Lenz's law, the induced EMF is such as to resist any change in the current $I_1$, and hence any change in the magnetic flux through $C_1$. So, loosely speaking, inductance is the electrical analogue of mechanical inertia, resisting changes to current in the same way as mass resists changes to velocity.

Consider, for example, a resistive wire loop C that carries an initial current $I_0$, maintained by a battery. Suppose that, at $t = 0$, the battery is removed, so that the current in the loop starts to fall by Ohmic dissipation (Figure 11.3). A natural question to ask is: how fast will the current decline? As we shall now show, we can answer that question using either the language of inductance or the language of diffusion. Indeed, for this simple geometry, it turns out that these are two sides of the same coin.

Let us start by using the language of inductance. From (10.9) combined with (11.34), we have

$$\text{EMF} = \frac{1}{\sigma} \oint_C \mathbf{J} \cdot d\mathbf{l} = I\frac{\ell}{\sigma S} = -L\frac{dI}{dt}, \tag{11.35}$$

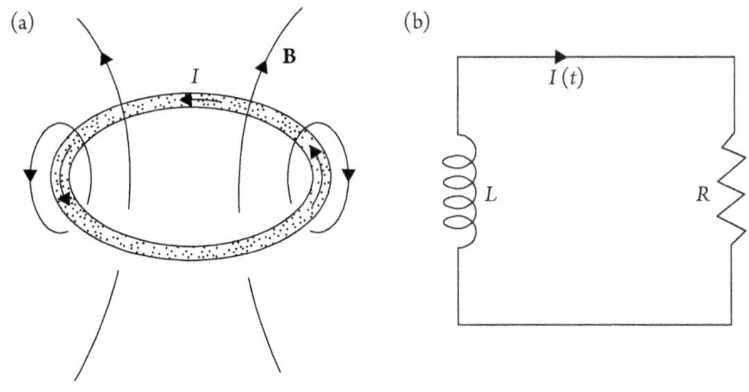


**Figure 11.3** (a) A wire loop has an initial current $I_0$, which then falls by Joule dissipation. (b) The equivalent circuit is a simple $LR$ circuit.

where $\ell$ and $S$ are the length and cross-sectional area of the wire in the loop. We conclude that

$$IR = -L\frac{dI}{dt}, \tag{11.36}$$

where $R = \ell/\sigma S$ is the resistance of the wire loop. In the terminology of electrical circuits (see Chapter 12), we have an $LR$ circuit whose current declines as

$$I = I_0 \exp(-t/\tau), \quad \tau = L/R. \tag{11.37}$$

The characteristic decay time for $I$ is controlled by the balance between $L$, which provides the inertia that resists the fall in current, and $R$, which promotes the decline through Ohmic dissipation.

It is instructive to estimate $\tau = L/R$ for the simple case of a thin circular current loop of radius $a$ and cross-sectional area $S = \pi b^2$. Here $R = 2a/\sigma b^2$ and we have already seen how to calculate $L$. Thus we have,

$$\tau = \frac{\sigma \mu_0 b^2}{2}\left[\frac{1}{4} + \ln\left(\frac{8a}{e^2 b}\right)\right] = \frac{b^2}{2\lambda}\left[\frac{1}{4} + \ln\left(\frac{8a}{e^2 b}\right)\right].$$

This makes explicit the physical process behind the timescale $\tau$. The point is that, as discussed in §2.2.1, magnetic fields (or equivalently currents) diffuse through stationary conductors just like heat, with a diffusivity $\lambda = 1/\sigma\mu_0$. This can be seen by combining the laws of Faraday and Ampère with $\mathbf{J} = \sigma\mathbf{E}$, which yields the diffusion equation,

$$\frac{\partial \mathbf{B}}{\partial t} = \lambda \nabla^2 \mathbf{B}, \quad \text{or} \quad \frac{\partial \mathbf{J}}{\partial t} = \lambda \nabla^2 \mathbf{J}. \tag{11.38}$$

We also know from §2.2.1 that the characteristic distance a substance diffuses in a time $t$, the diffusion length, is $\ell \sim 2\sqrt{\alpha t}$, where $\alpha$ is the appropriate diffusivity. (See, for example, the expression (2.33).) So, for $\mathbf{B}$ or $\mathbf{J}$, the time to diffuse a distance $b$ (i.e. the radius of the wire) is of the order of $t \sim b^2/4\lambda$, which is more or less the expression above for $\tau$. In short, the timescale for the decay of current in this resistive loop is the time taken for the current to diffuse out of the wire. We shall expand on this point in §11.4. (See also Exercise 11.9 for an alternative derivation of $\tau \sim \sigma\mu_0 b^2$.)

This simple example illustrates something else of interest. Note that (11.36) may be rewritten as

$$\frac{d}{dt}\left(\frac{1}{2}LI^2\right) = -I^2 R.$$

We recognise the term on the right as the Ohmic dissipation in the wire and ask what supplies the energy for that dissipation. Since the battery has been removed, the only possibility is magnetic energy, which suggests that the magnetic energy associated with a filamentary current loop is $U_B = LI^2/2$. We shall confirm that this is indeed the case in §11.3.3.

Before leaving this simple example, it is worth considering what happens if, once the current has fallen to zero, the battery is turned back on with its initial voltage of $V_0 = I_0 R$. We proceed as follows. In the presence of a battery (11.19) tells us that the EMF in the circuit is

$$\text{EMF} = \oint_C \mathbf{F}_s \cdot d\mathbf{r} + \oint_C \mathbf{E} \cdot d\mathbf{r},$$

where $\mathbf{F}_s = \mathbf{f}_s/q$ is the chemical force per unit charge in the battery. We also know from §11.2.1 that Ohm's law must be adjusted to read $\mathbf{J} = \sigma(\mathbf{F}_s + \mathbf{E})$. Since the voltage rise across the battery is $V_0 = \int \mathbf{F}_s \cdot d\mathbf{r}$, this tells us that

$$\text{EMF} = \frac{1}{\sigma}\oint_C \mathbf{J} \cdot d\mathbf{l} = \oint_C \mathbf{F}_s \cdot d\mathbf{r} + \oint_C \mathbf{E} \cdot d\mathbf{r} = V_0 - L\frac{dI}{dt}, \tag{11.39}$$

or equivalently,

$$\frac{1}{\sigma}\oint_C \mathbf{J} \cdot d\mathbf{l} = I\frac{\ell}{\sigma S} = V_0 - L\frac{dI}{dt}.$$

So our governing equation after the battery is turned back on is the same as (11.36), except for the inclusion of the term $V_0$,

$$IR = V_0 - L\frac{dI}{dt}. \tag{11.40}$$

This has the solution

$$I = I_0[1 - \exp(-t/\tau)], \quad \tau = L/R,$$

and so once again we see a response on the timescale $\tau$.

### *11.3.3 Magnetic Energy in Terms of Current and Inductance*

We shall now show that there is a close link between the idea of inductance and the energy of a quasi-static magnetic field. We start by returning to §8.4.3, where we saw that, for steady currents, the magnetic energy density $\mathbf{B}^2/2\mu_0$ can be rewritten as $\mathbf{J} \cdot \mathbf{A}/2$. That is, the magnetic energy can be written as

$$U_B = \frac{1}{2\mu_0}\int_{V_\infty} \mathbf{B}^2 dV = \frac{1}{2\mu_0}\int_{V_\infty} \mathbf{A} \cdot \nabla \times \mathbf{B}\, dV + \frac{1}{2\mu_0}\oint_{S_\infty} (\mathbf{A} \times \mathbf{B}) \cdot d\mathbf{S},$$

where the surface integral vanishes as the surface recedes to infinity. Thus we can use Ampère's law to rewrite the magnetic energy of a quasi-static field as

$$U_B = \frac{1}{2}\int_{V_\infty} \mathbf{A} \cdot \mathbf{J}\, dV. \tag{11.41}$$

For a single filamentary current loop $C$ carrying a current $I$, we have $\mathbf{J}\,dV = I\,d\mathbf{r}$, and so this becomes

$$U_B = \frac{1}{2} I \oint_C \mathbf{A} \cdot d\mathbf{r} = \frac{1}{2} I \int_S (\nabla \times \mathbf{A}) \cdot d\mathbf{S} = \frac{1}{2} I \int_S \mathbf{B} \cdot d\mathbf{S} = \frac{1}{2} I \Phi, \tag{11.42}$$

where $\Phi$ is the flux through $C$. So we conclude that the magnetic energy associated with a single filamentary current loop can be written in the particularly simple form

$$\boxed{U_B = \frac{1}{2} I \Phi = \frac{1}{2} L I^2,} \tag{11.43}$$

as suggested in §11.3.2.

When there are many such current loops $C_1, C_2, \ldots, C_N$, carrying currents $I_1, I_2, \ldots$, we proceed as follows. Generalising (11.42), we have

$$U_B = \frac{1}{2} \int_{V_\infty} \mathbf{A} \cdot \mathbf{J}\, dV = \frac{1}{2} \sum_N I_i \oint_{C_i} \mathbf{A} \cdot d\mathbf{r} = \frac{1}{2} \sum_N I_i \int_{S_i} \mathbf{B} \cdot d\mathbf{S} = \frac{1}{2} \sum_N I_i \Phi_i. \tag{11.44}$$

However, the flux though $C_1$ is $\Phi_1 = L_1 I_1 + M_{12} I_2 + \cdots$, and so

$$\begin{aligned} U_B = & \frac{I_1}{2} (L_1 I_1 + M_{12} I_2 + M_{13} I_3 + M_{14} I_4 + \cdots + M_{1N} I_N) \\ & + \frac{I_2}{2} (M_{21} I_1 + L_2 I_2 + M_{23} I_3 + M_{24} I_4 + \cdots + M_{2N} I_N) \\ & + \frac{I_3}{2} (M_{31} I_1 + M_{32} I_2 + L_3 I_3 + M_{34} I_4 + \cdots + M_{3N} I_N) \\ & + \cdots \end{aligned}$$

which is equivalent to

$$U_B = \frac{1}{2} \sum_{i=1}^{N} L_i I_i^2 + \frac{1}{2} \sum_{i=1}^{N} \sum_{j=1, j \neq i}^{N} M_{ij} I_i I_j. \tag{11.45}$$

For example, if there are just two current loops, then, using $M_{12} = M_{21} = M$, we have

$$\boxed{U_B = \frac{1}{2} L_1 I_1^2 + M I_1 I_2 + \frac{1}{2} L_2 I_2^2.} \tag{11.46}$$

We have been here before. The expression for the mutual magnetic energy of the two current loops is, in effect, (8.39), while that for the self-energy of the loops is (8.42).

Note that (11.46) may be rewritten as

$$U_B = \frac{1}{2}L_1\left(I_1 + \frac{MI_2}{L_1}\right)^2 + \frac{1}{2}\left(L_2 - \frac{M^2}{L_1}\right)I_2^2, \tag{11.47}$$

and, since $U_B$ must be positive for all possible values of $I_1$ and $I_2$, including $L_1I_1 = -MI_2$, we have the inequality $M < \sqrt{L_1L_2}$.

### 11.3.4 *Magnetic Forces in Terms of Inductance*

It turns out that the concept of inductance may be usefully combined with the idea of virtual work to calculate the forces and torques exerted by one current loop on another. We illustrate the general idea with a simple example.

Consider two circular current loops $C_1$ and $C_2$ carrying currents $I_1$ and $I_2$ that share a common axis, as shown in Figure 11.4. To focus our thoughts, suppose that the direction of the current is the same in the two loops. Since parallel currents attract, the net Lorentz force $\mathbf{F}_{\text{mag}} = \int \mathbf{J} \times \mathbf{B}\,dV$ acting on either loop tends to pull it towards the other loop. Suppose mechanical forces $\mathbf{F}_{\text{mech}} = -\int \mathbf{J} \times \mathbf{B}\,dV$, equal and opposite to $\mathbf{F}_{\text{mag}}$, are applied to both loops to keep them stationary. We now slowly apply a virtual displacement $d\mathbf{r}$ to the loop on the left, $C_1$, *while holding the currents in both loops constant.* Since we insist that the currents are held fixed during this virtual displacement, we must allow for batteries in both loops, whose function is to keep the currents constant.

Let the direction of $d\mathbf{r}$ be such as to reduce the gap between the loops. Then the virtual displacement $d\mathbf{r}$ acts to increase (i) the mutual inductance of the two loops, (ii) the flux through each loop and (iii) the mutual magnetic energy of the two loops (i.e. the magnetic energy that arises from the overlap of the two magnetic fields). Also, let $\delta U_{\text{mech}} = \mathbf{F}_{\text{mech}} \cdot d\mathbf{r}$ be the work done by the mechanical force acting on loop $C_1$ during its virtual displacement and let $\delta U_B$ be the change in magnetic energy arising from the same displacement. Clearly $\delta U_{\text{mech}} < 0$, since $d\mathbf{r}$ and $\mathbf{F}_{\text{mech}}$ are in opposite directions. Moreover, from (11.46) we have

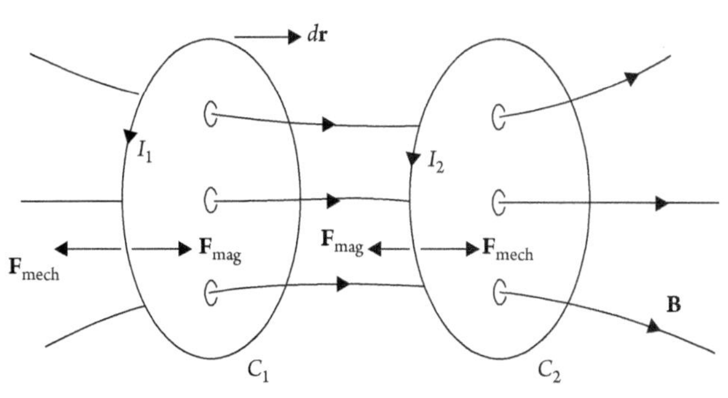


**Figure 11.4** Two circular current loops share a common axis. A virtual displacement $d\mathbf{r}$ is applied to the loop on the left while holding the currents in both loops constant.

$$\delta U_B = \delta\left(\frac{1}{2}L_1 I_1^2 + M I_1 I_2 + \frac{1}{2}L_2 I_2^2\right) = I_1 I_2 \delta M, \tag{11.48}$$

so that only the mutual magnetic energy changes. Now the changes in flux through the two loops are $\delta\Phi_1 = I_2\,\delta M$ and $\delta\Phi_2 = I_1\,\delta M$, and it follows that

$$\delta U_B = I_1 I_2\,\delta M = I_1\,\delta\Phi_1 = I_2\,\delta\Phi_2, \tag{11.49}$$

where $\delta\Phi_1$, $\delta\Phi_2$, $\delta M$ and $\delta U_B$ are all positive.

Finally, we need to consider the work done by the batteries during the virtual displacement. Consider, for example, loop 1. Since $\delta\Phi_1 > 0$, we have $d\Phi_1/dt > 0$ during the application of $d\mathbf{r}$, and so the EMF induced by Faraday's flux rule is negative and acts to oppose the flow of current in loop 1 (Lenz's law). In order to keep the current in loop 1 constant, the voltage of the battery must increase by an amount $d\Phi_1/dt$. This, in turn, entails a rise in the battery output power, over and above that required to compensate for the Joule dissipation associated with $I_1$. This rise in battery power is $I_1\,d\Phi_1/dt$, and if the virtual displacement is applied over a time interval $\delta t$, then the additional work done by the battery is $\int (I_1\,d\Phi_1/dt)\,dt = I_1\,\delta\Phi_1$. However, (11.49) tells us that this is simply $\delta U_B$. A similar argument may be applied to the battery in loop 2 to give the electrical work as $I_2\,\delta\Phi_2 = \delta U_B$. We conclude that the total work done by both batteries is

$$\delta U_{\text{battery}} = I_1\,\delta\Phi_1 + I_2\,\delta\Phi_2 = 2\,\delta U_B. \tag{11.50}$$

We may now apply conservation of energy. The total external work done during the virtual displacement, both electrical and mechanical, is evidently $2\delta U_B + \delta U_{\text{mech}}$, and this must equal the rise in magnetic energy:

$$\delta U_B = \delta U_{\text{battery}} + \delta U_{\text{mech}} = 2\delta U_B + \delta U_{\text{mech}}. \tag{11.51}$$

So we conclude that $\delta U_{\text{mech}} = -\delta U_B$. That is to say, the work done by the external mechanical force that holds loop 1 in place is

$$\delta U_{\text{mech}} = \mathbf{F}_{\text{mech}} \cdot d\mathbf{r} = -\delta U_B.$$

Since $\delta U_B$ is positive, this is consistent with $\delta U_{\text{mech}} < 0$. However, $\mathbf{F}_{\text{mech}}$ is equal and opposite to the magnetic force exerted on loop 1 by loop 2, $\mathbf{F}_{\text{mag}}$, and so we conclude that

$$\boxed{\mathbf{F}_{\text{mag}} \cdot d\mathbf{r} = \delta U_B = I_1 I_2 \delta M.} \tag{11.52}$$

Note that, since $\mathbf{F}_{\text{mag}}$ and $d\mathbf{r}$ are in the same direction, this is consistent with $\delta M$ being positive. Note also that $\mathbf{F}_{\text{mag}}$ and $d\mathbf{r}$ need not be parallel and so this is often rewritten in component form as

$$\boxed{(\mathbf{F}_{\text{mag}})_i = I_1 I_2 \frac{\partial M}{\partial x_i}.}$$

Perhaps some comments are in order at this point. First, a similar analysis involving incremental rotations can be applied to evaluate magnetic torques transmitted between loops. Second, instead of holding the currents constant during the virtual displacement, we could have elected to hold the fluxes through the loops constant. The analysis then unfolds in a somewhat different way, as discussed in Exercise 11.6. Third, the analysis above is essentially the same as that presented in §8.4.1, where we showed that a small magnetic dipole $\mathbf{m}$ moving through a background magnetic field $\mathbf{B}_0$ satisfies $U_B = -U_{\text{mech}} = \mathbf{m} \cdot \mathbf{B}_0$. Fourth, although we have considered a specific example, (11.52) is quite general, provided that *all currents are held constant* during the virtual displacement. So, if current loops are allowed to translate, rotate or deform in response to the Lorentz forces acting on them, the magnetic energy rises in accordance with $\delta U_B = \mathbf{F}_{\text{mag}} \cdot d\mathbf{r}$. This is exactly what we found for a magnetic dipole in Chapter 8, where the dipole $\mathbf{m}$ tends to align with $\mathbf{B}_0$.

## 11.4 Magnetic Diffusion Revisited

Magnetic diffusion, which we introduced in §2.2.1, is the natural process of adjustment during transients in stationary conductors. The governing equation is obtained by combining the laws of Faraday and Ampère with $\mathbf{J} = \sigma \mathbf{E}$, where $\sigma$ is treated as spatially uniform. This yields

$$\boxed{\frac{\partial \mathbf{B}}{\partial t} = \lambda \nabla^2 \mathbf{B}, \qquad \lambda = \frac{1}{\sigma \mu_0},} \tag{11.53}$$

where $\lambda$ is the *magnetic diffusivity*. Thus we see that a quasi-static magnetic field diffuses through a conductor, just like heat. Such diffusion processes are important because, amongst other things, they lead to the idea of a *skin depth* for oscillating fields, and because they sometimes control the timescale for DC transients in simple *LR* circuits.

Let us start with the notion of a skin depth. As we saw in §2.2.1, the characteristic distance a substance diffuses in a time $t$ is $\ell \sim 2\sqrt{\alpha t}$, where $\alpha$ is the appropriate diffusivity. This is known as the *diffusion length*. Now consider an oscillatory diffusion process, such as the diurnal fluctuation in air temperature above the ground. Such fluctuations diffuse only a distance $\delta \sim \sqrt{\alpha/\varpi}$ into the ground before the fluctuation reverses sign, $\alpha$ being the thermal diffusivity of the ground and $\varpi$ the frequency of the daily temperature cycle. So the fluctuations in air temperature penetrate only a distance $\delta \sim \sqrt{\alpha/\varpi}$ into the ground. Likewise, oscillating magnetic fields diffuse only a distance $\delta \sim \sqrt{\lambda/\varpi}$ into a stationary conductor, which is known as the *magnetic skin depth*.

Turning now to DC transients in *LR* circuits, magnetic diffusion often controls the response time, $\tau = L/R$, when the magnetic field in the inductor is enclosed by a conducting surface. Let us explain why, taking as an example a long, tightly wound solenoid of circular cross section. Let the solenoid have $N$ turns per unit length, $N_t$ turns in total and a radius of $a$, while the winding consist of a wire of rectangular cross section of radial thickness $c$ and axial width $b = 1/N$. Then the total resistance of the winding is evidently $R = N_t(2\pi a)/\sigma(bc)$, while in Exercise 11.2 it is shown that the inductance of the solenoid is $L = \mu_0 \pi a^2 N N_t$.

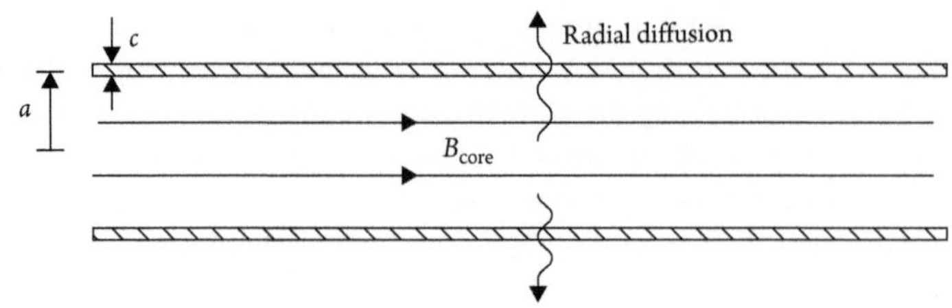


**Figure 11.5** The magnetic flux decays by diffusing radially outward through the winding.

According to (11.37), the timescale for transients in the solenoid, regarded as an *LR* circuit, is then $\tau = L/R = ac/2\lambda$, which is independent of $N$.

So how should we interpret the timescale $\tau = ac/2\lambda$ and what has this to do with magnetic diffusion? Suppose that the winding initially carries a steady current of $I_0$, but at $t = 0$ the power is switched off. The circuit remains intact, however, and so the current decays gradually through Ohmic dissipation in accordance with (11.37). For $t > 0$ we have a uniform axial magnetic field confined to a cylinder of radius $a$ by a conducting annular shell of thickness $c$, as shown in Figure 11.5. In such a situation the timescale $\tau = ac/2\lambda$ turns out to be the time taken for the magnetic flux within the solenoid to diffuse radially outward through the winding, as we now discuss.

Within the solenoid we have $\mathbf{B} = B_z(s,t)\hat{\mathbf{e}}_z$, so let us rewrite the axial component of (11.53) in the winding as

$$\frac{\partial B_z}{\partial t} = -\nabla \cdot (\text{diffusive flux of } B_z) = -\nabla \cdot \left( -\lambda \frac{\partial B_z}{\partial s} \hat{\mathbf{e}}_s \right). \tag{11.54}$$

Then the rate of radial diffusion of the magnetic field through the conducting annulus is, per unit area of the annulus, $-\lambda \partial B_z/\partial s = \lambda B_{\text{core}}/c$, $B_{\text{core}}(t)$ being the axial field inside the solenoid. The rate of loss of magnetic flux from the core by diffusion through the annulus is then obtained from the volume integral of (11.54), as described in Exercise 11.11:

$$\frac{d}{dt}(\pi a^2 B_{\text{core}}) = -\lambda \frac{B_{\text{core}}(t)}{c}(2\pi a).$$

Evidently this does indeed yield a timescale of $\tau = ac/2\lambda$!

## 11.5 Quasi-magnetostatics versus Quasi-electrostatics

Throughout this chapter we have been concerned with the evolution of magnetic fields induced in conductors and in free space by *slowly varying currents*. Because **B** varies slowly, we have neglected the displacement current, $\varepsilon_0 \partial \mathbf{E}/\partial t$ in the Ampère–Maxwell equation and adopted instead Ampère's law, which is strictly valid only for steady situations. So our governing equations are

$$\nabla \times \mathbf{B} = \mu_0 \mathbf{J}, \qquad \nabla \cdot \mathbf{J} = 0,$$
$$\nabla \times \mathbf{E} = -\frac{\partial \mathbf{B}}{\partial t}, \qquad \nabla \cdot \mathbf{B} = 0,$$

along with Ohm's law $\mathbf{J} = \sigma(\mathbf{E} + \mathbf{u} \times \mathbf{B})$. Note that Ampère's law demands $\nabla \cdot \mathbf{J} = 0$, in apparent contradiction to the charge conservation equation

$$\nabla \cdot \mathbf{J} = -\frac{\partial \rho_e}{\partial t}. \tag{11.55}$$

It is worth taking a moment to examine the limitations of this approximation, check that the resulting equations are self-consistent and compare them with those of quasi-electrostatics.

The key point is this: in good conductors, such as metals, the charge density within the interior of the conductor is tiny and so the right-hand side of (11.55) is negligible, giving $\nabla \cdot \mathbf{J} = 0$. We demonstrated this for the particular case of a stationary conductor in Chapter 3, where charge conservation, Ohm's law and Gauss' law, $\nabla \cdot \mathbf{E} = \rho_e/\varepsilon_0$, combine to yield

$$\frac{\partial \rho_e}{\partial t} = -\nabla \cdot \mathbf{J} = -\nabla \cdot (\sigma \mathbf{E}) = -\frac{\sigma}{\varepsilon_0}\rho_e. \tag{11.56}$$

This is the charge relaxation equation, whose solution is

$$\rho_e = \rho_e(0)\exp(-t/\tau_e), \quad \tau_e = \varepsilon_0/\sigma, \tag{11.57}$$

where $\tau_e$ is the charge relaxation time. For metallic conductors $\tau_e$ is tiny, around $10^{-14}$ s, and so any excess charge density that initially lies in the interior of a conductor moves extremely rapidly to its surface. It follows that $\rho_e$ is almost always zero inside stationary conductors and so $\nabla \cdot \mathbf{J} = 0$ within their interiors.

If the various parts of the conductor move through a magnetic field with a velocity $\mathbf{u}(\mathbf{x}, t)$, then it is easy to show that the right-hand side of (11.55) is still negligible. Ohm's law now becomes $\mathbf{J} = \sigma(\mathbf{E} + \mathbf{u} \times \mathbf{B})$ and (11.56) generalises to

$$\frac{\partial \rho_e}{\partial t} + \frac{\rho_e}{\tau_e} = -\sigma \nabla \cdot (\mathbf{u} \times \mathbf{B}). \tag{11.58}$$

In this case a small but finite charge can build up within the interior of the conductor. However, we are interested in events that take place on a timescale much longer than $\tau_e$ and so we may drop $\partial \rho_e/\partial t$ from (11.58), to give $\rho_e = -\sigma \tau_e \nabla \cdot (\mathbf{u} \times \mathbf{B})$. But dropping $\partial \rho_e/\partial t$ in (11.58) is equivalent to ignoring $\partial \rho_e/\partial t$ in (11.55) and so once again we conclude that the charge density $\rho_e$ is tiny and charge conservation reduces to the statement that $\nabla \cdot \mathbf{J} = 0$.

In summary then, unless we deliberately introduce charges onto the surface of the conductor, the charge density will be negligible everywhere. Note that dropping $\partial \rho_e/\partial t$ in (11.55) is entirely equivalent to dropping the displacement current $\varepsilon_0 \partial \mathbf{E}/\partial t$ in the Ampère–Maxwell equation

$$\nabla \times \mathbf{B} = \mu_0 \left( \mathbf{J} + \varepsilon_0 \frac{\partial \mathbf{E}}{\partial t} \right). \tag{11.59}$$

The point is that the displacement current is introduced into (11.59) specifically to make it consistent with the charge conservation equation (11.55). That is, since the left-hand side of (11.59) has zero divergence, so must the right. Using Gauss' law, the divergence of the right-hand side is

$$\nabla\cdot\mathbf{J}+\varepsilon_0\frac{\partial}{\partial t}\nabla\cdot\mathbf{E}=\nabla\cdot\mathbf{J}+\frac{\partial\rho_e}{\partial t}, \tag{11.60}$$

which is indeed zero in the light of the charge conservation equation. However, if the charge density is so tiny that charge conservation reduces to $\nabla\cdot\mathbf{J}=0$, then there is no need to introduce the displacement current into the Ampère–Maxwell equation, since its right-hand side is already solenoidal. Indeed, we can check directly that $\varepsilon_0\partial\mathbf{E}/\partial t << \mathbf{J}$ in the quasi-magnetostatic limit. Let $\varpi$ be a characteristic frequency for **B** and **E**. Then

$$\frac{\varepsilon_0\partial\mathbf{E}/\partial t}{|\mathbf{J}|}\sim\frac{\varepsilon_0\varpi E}{\sigma E}\sim\varpi\tau_e << 1, \tag{11.61}$$

and so dropping $\partial\rho_e/\partial t$ in (11.55) is entirely equivalent to dropping $\varepsilon_0\partial\mathbf{E}/\partial t$ in the Ampère–Maxwell equation.

Another consequence of the charge density being very small is that the electric energy is negligible, which is why this energy is absent in the quasi-static energy equations (11.8) and (11.10). In fact, it is left as an exercise for the reader to confirm that the ratio of the electric to magnetic energy is $O(\varpi^2\ell^2/c^2) << 1$, where $\ell$ is a typical length and $c$ the speed of light. So the quasi-magnetostatic equations are self-consistent.

However, there is a second kind of quasi-static system, *also satisfying* $\varpi\ell << c$, which is governed by a different set of equations. This is known as the *quasi-electrostatic regime*. Suppose we deliberately introduce charges onto the surface of the conductors and then slowly move those charges in space. In general currents will flow to satisfy charge conservation, and this is taken to be the dominant source of current. To leading order we are back with electrostatics—hence the name quasi-electrostatics. Because we cannot now neglect the surface charges, we must ensure that charge conservation is strictly satisfied, so **J** is no longer solenoidal. This, in turn, requires that we retain the displacement current in the Ampère–Maxwell equation. The simplification introduced by demanding that the charges move only *slowly* in space is that $\partial\rho_e/\partial t$, and hence **J** as estimated from (11.55), is relatively small. The Ampère–Maxwell equation now tells us that **B** is also relatively small, and an order-of-magnitude analysis then confirms that $\partial\mathbf{B}/\partial t$ may be neglected in Faraday's law. (The details are spelt out in Exercise 11.7.) So the governing equations for quasi-electrostatic fields turn out to be charge conservation plus

$$\nabla\times\mathbf{E}=0,\qquad\nabla\cdot\mathbf{E}=\frac{\rho_e}{\varepsilon_0}, \tag{11.62}$$

$$\nabla\times\mathbf{B}=\mu_0\left(\mathbf{J}+\varepsilon_0\frac{\partial\mathbf{E}}{\partial t}\right),\qquad\nabla\cdot\mathbf{B}=0. \tag{11.63}$$

These equations describe, for example, the steady charging of a parallel-plate capacitor. They are quite different to the quasi-magnetostatic equations considered in this chapter, and so it is important to distinguish between these two regimes. To that end, let us explore in a little more detail the similarities and differences between these two systems.

The absence of $\partial\mathbf{B}/\partial t$ in (11.62) means that, just like the quasi-magnetostatic equation (11.53), the equations (11.62) and (11.63) do *not* constitute a wave-bearing system, despite the retention of the displacement current in (11.63). Indeed, it is clear from (11.62) that $\mathbf{E}$ is simply determined by the usual static Poisson equation for the electrostatic potential. Moreover, the curl of (11.63),

$$\nabla^2\mathbf{B} = -\mu_0\nabla\times\mathbf{J}, \tag{11.64}$$

inverts to give

$$\mathbf{B}(\mathbf{x},t) = \frac{\mu_0}{4\pi}\int\frac{\nabla'\times\mathbf{J}(\mathbf{x}',t)}{|\mathbf{r}|}d\mathbf{x}', \quad \mathbf{r} = \mathbf{x}-\mathbf{x}', \tag{11.65}$$

which converts to the Biot–Savart law, as shown in §7.8.1. So, as in quasi-magnetostatics, the distribution of $\mathbf{B}(\mathbf{x}, t)$ is given by the Biot–Savart law. (See also Exercise 11.10).

Note that we can develop an energy equation for quasi-electrostatics by taking the dot product of $\mathbf{E}$ with the Ampère–Maxwell equation and then invoking $\nabla\times\mathbf{E} = 0$. The resulting expression is identical to the exact equation (11.11), but with the magnetic energy removed. In differential form this quasi-static energy equation is

$$\frac{\partial}{\partial t}(\varepsilon_0\mathbf{E}^2/2) = -\mathbf{J}\cdot\mathbf{E} - \frac{1}{\mu_0}\nabla\cdot(\mathbf{E}\times\mathbf{B}). \tag{11.66}$$

Comparing this with (11.7), we see that electrical energy has replaced magnetic energy.

In summary, then, in quasi-magnetostatics we neglect $\varepsilon_0\partial\mathbf{E}/\partial t$ in the Ampère–Maxwell equation but retain $\partial\mathbf{B}/\partial t$ in Faraday's law. By contrast, in quasi-electrostatics we neglect $\partial\mathbf{B}/\partial t$ in Faraday's law but retain $\varepsilon_0\partial\mathbf{E}/\partial t$ in the Ampère–Maxwell equation. Moreover, the energy equation of quasi-magnetostatics contains magnetic energy but neglects electric energy, whereas the energy equation for quasi-electrostatics retains electric energy but neglects magnetic energy. Both systems satisfy $\varpi\ell << c$.

Evidently it is crucial to distinguish between these two quasi-steady systems, although this is sometimes a rather subtle issue. Perhaps the simplest way to distinguish between the two regimes is to note that in quasi-electrostatics the primary source of the fields is an imposed slowly evolving charge distribution, which establishes an electric field via Gauss' law. Any current that happens to be present is there primarily to accommodate the redistribution of charge in accordance with charge conservation. This constitutes a rather weak current and so the magnetic energy in quasi-electrostatics is relatively small. By contrast, in quasi-magnetostatics the primary source of the fields is an imposed slowly varying current distribution that sets up a magnetic field in accordance with Ampère's law. The charge distribution is assumed to be weak on the understanding that no additional charges have been placed on the surfaces of the conductors. The magnetic energy then dominates. This distinction between quasi-electrostatics and quasi-magnetostatics is discussed in, for example, Zangwill (2013).

## EXERCISES

11.1 Consider a long solenoid of circular cross section that has $N_1$ turns per unit length and carries a current $I$. A second, smaller solenoid sits inside the larger one, coaxial with it and with $N_t$ turns in total. Show that their mutual inductance is $M = \mu_0 S N_1 N_t$, where $S$ is the cross-sectional area of the smaller solenoid.

11.2 Consider a long solenoid of circular cross section with $N$ turns per unit length and $N_t$ turns in total. It has a cross-sectional area $S$. Show that the self-inductance of the solenoid is $L = \mu_0 S N N_t$.

11.3 Consider a toroidal coil constructed from a long thin solenoid of square cross section of width $W$ and current $I$ whose centreline is bent into a circle of radius $R$, with $R >> W$. Cylindrical polar coordinates $(s, \theta, z)$ are centred on the toroidal coil such that the centreline of the coil is $s = R$. Show that the magnetic field within the interior of the toroidal coil is $\mu_0 N_t I / 2\pi s$, where $N_t$ is the total number of turns. Hence show that the self-inductance of the toroidal coil is

$$L = \frac{\mu_0 N_t^2 W}{2\pi} \ln \frac{R + W/2}{R - W/2}.$$

11.4 Consider a long solenoid of circular cross section with $N$ turns per unit length and $N_t$ turns in total. It has a cross-sectional area of $S$ and carries a current $I$. Starting from the energy density $\mathbf{B}^2/2\mu_0$, show that the energy in the solenoid is $U_B = \mu_0 I^2 N N_t S/2$. Now show that the same expression comes from (i) $U_B = LI^2/2$ and (ii) $U_B = \int (\mathbf{A} \cdot \mathbf{J}/2)\, dV$.

11.5 Consider two long coaxial solenoids of circular cross section. The outer one carries a current $I_1$ and has $N_1$ turns per unit length, while the inner one has cross-sectional area $S$, a current of $I_2$ and $N_2$ turns per unit length. The current in the two coils flow in the same direction and the inner coil lies partially inside the larger one but extends beyond it. Use the energy method of §11.3.4 to show that the axial force exerted by the larger coil on the smaller one has magnitude $F = \mu_0 I_1 I_2 N_1 N_2 S$.

11.6 Show that the magnetic energy of two current loops can be written as

$$U_B = \frac{\Phi_1^2}{2L_1} + \frac{\Phi_1 \Phi_2}{M} + \frac{\Phi_2^2}{2L_2}.$$

Now repeat the energy analysis of §11.3.4, only holding the fluxes through the loops constant during the virtual displacement, rather than the current. Show that (11.49) now reads $\delta U_B = -I_1 I_2\, \delta M$, but that energy conservations leaves $\mathbf{F}_{\text{mag}} \cdot d\mathbf{r} = I_1 I_2\, \delta M$ unchanged.

11.7 Consider the quasi-electrostatic regime in which the current distribution is dictated by charge conservation in the form of (11.55). Show that, if $\varpi$ is a typical frequency for $\mathbf{E}$ and $\rho_e$, then charge conservation gives $J \sim \varpi \ell \rho_e$, while Gauss' law demands $E \sim \ell \rho_e / \varepsilon_0$. Hence confirm that the displacement current is of the same order as the conduction current and that the Ampère–Maxwell equation requires $B \sim \mu_0 \varpi \ell^2 \rho_e$. Now combine the laws of Faraday and Gauss to give

$$|\nabla \times \mathbf{E}| \sim \mu_0 \varpi^2 \ell^2 \rho_e \sim (\varpi \ell / c)^2 |\nabla \cdot \mathbf{E}| << |\nabla \cdot \mathbf{E}|$$

and explain why $\partial \mathbf{B}/\partial t$ must be neglected in (11.62).

11.8 Apply the modified Poynting flux defined by (11.21), $\varphi \mathbf{J} + \mathbf{B} \times (\partial \mathbf{A}/\partial t)/\mu_0$, to the wire shown in Figure 11.1 for the case of an *unsteady* current in the wire. Show that the term $\mathbf{B} \times (\partial \mathbf{A}/\partial t)/\mu_0$ feeds the rate of change of magnetic energy surrounding the wire.

11.9 There is a third way of deriving the estimate $\tau = L/R \sim \sigma \mu_0 b^2$ for the decay time of the magnetic field shown in Figure 11.3. The energy equation (11.10) gives us

$$\frac{d}{dt} \int_{V_\infty} \frac{\mathbf{B}^2}{2\mu_0} dV = -\frac{1}{\sigma} \int_{V_{\text{wire}}} \mathbf{J}^2 dV = -\frac{I^2 (2\pi a)}{\sigma (\pi b^2)},$$

where $V_\infty$ is all of space. Taking $\mathbf{B}$ near the wire to be of order $B \sim \mu_0 I / 2\pi b$ and assuming that the field near the wire dominates the magnetic energy, show that the net magnetic energy of the wire loop is of the order of $\mu I^2 a$. Hence show that $\tau \sim \sigma \mu_0 b^2$. Actually, a better estimate of $\tau \sim \sigma \mu_0 b^2 \ln(a/b)$ is obtained if we note that the magnetic energy of the circular wire loop is $U_B = LI^2/2$, or $U_B \approx (I^2/2)\mu a \ln(8a/e^2 b)$.

11.10 Combine (7.58), which follows directly from the static Biot–Savart law, with charge conservation to show that

$$\begin{aligned} \nabla \times \mathbf{B} &= \mu_0 \mathbf{J}(\mathbf{x}) + \frac{\mu_0}{4\pi} \nabla \left( \int \frac{\nabla' \cdot \mathbf{J}'}{|\mathbf{r}|} d\mathbf{x}' \right) \\ &= \mu_0 \left\{ \mathbf{J}(\mathbf{x}) - \varepsilon_0 \frac{\partial}{\partial t} \nabla \left[ \frac{1}{4\pi \varepsilon_0} \int \frac{\rho_e(\mathbf{x}')}{|\mathbf{r}|} d\mathbf{x}' \right] \right\}. \end{aligned}$$

Now use the electrostatic equations (11.62) to show that, for the quasi-electrostatic regime, this yields

$$\nabla \times \mathbf{B} = \mu_0 \left( \mathbf{J} - \varepsilon_0 \frac{\partial}{\partial t} \nabla \varphi_s \right) = \mu_0 \left( \mathbf{J} + \varepsilon_0 \frac{\partial \mathbf{E}}{\partial t} \right),$$

where $\varphi_s$ is the electrostatic potential. Of course, this is (11.63), and so this merely serves to show the self-consistency of the quasi-electrostatic system in which $\mathbf{B}$ is determined by (11.65) and hence by the Biot–Savart law. Some authors use this result to suggest that the Ampère–Maxwell equation is implied by a combination of electrostatics, the static Biot–Savart law and charge conservation. Do you agree?

11.11 (a) Consider a magnetic field in a stationary conductor whose magnetic diffusivity $\lambda = 1/\sigma\mu_0$ is spatially varying but uniform and equal to $\lambda_S$ on a closed surface $S$ within the conductor. Combine the laws of Faraday and Ampère with $\mathbf{J} = \sigma\mathbf{E}$ to show that

$$\frac{\partial \mathbf{B}}{\partial t} = -\nabla \times (\lambda \nabla \times \mathbf{B}).$$

Integrate this over the volume enclosed by $S$ to give

$$\frac{d}{dt}\int \mathbf{B}\, dV = \lambda_S \oint (\nabla \times \mathbf{B}) \times d\mathbf{S}$$

and show that this may be rewritten as

$$\frac{d}{dt}\int B_i\, dV = -\oint (-\lambda_S \nabla B_i)\cdot d\mathbf{S} = -\oint \dot{\mathbf{q}}_i \cdot d\mathbf{S}.$$

Interpret the surface integral in terms of a diffusive flux of magnetic field, $\dot{\mathbf{q}}_i = -\lambda_S \nabla B_i$.

(b) Consider the solenoid shown in Figure 11.5, of radius $a$ and winding thickness $c$. Use Faraday's law, combined with Ohm's law, to show that the variation of magnetic flux within the core, $d\Phi/dt$, induces a current density at the inner surface of the winding of

$$J_\theta = -\frac{\sigma}{2\pi a}\frac{d\Phi}{dt} = -\frac{\sigma a}{2}\frac{dB_{\text{core}}}{dt}.$$

Now use Ampère's law to show that the diffusive flux of $B_z$ at the surface of the winding is

$$\dot{q}_z = -\lambda\frac{\partial B_z}{\partial s} = -\frac{1}{2\pi a}\frac{d\Phi}{dt} = -\frac{a}{2}\frac{dB_{\text{core}}}{dt}.$$

The gradient in $B_z$ across the annulus is linear, $\partial B_z/\partial s = -B_{\text{core}}/c$. Confirm that

$$\frac{d}{dt}(\pi a^2 B_{\text{core}}) = -\lambda\frac{B_{\text{core}}(t)}{c}(2\pi a),$$

which has a decay time of $\tau = ac/2\lambda$. Finally, deduce the same result directly from part (a) by assigning a finite but very low conductivity (a very high diffusivity) to the air core.

## REFERENCES

Jackson, J. D., 1999, *Classical Electrodynamics*, 3rd ed. Wiley.

Nahin P. J., 2002, *Oliver Heaviside: The Life, Work and Times of an Electrical Genius of the Victorian Age*. Johns Hopkins University Press.

Zangwill, A., 2013, *Modern Electrodynamics*. Cambridge University Press.

# 12

# Transient and AC Circuits

> We do not dwell in the Palace of Truth... the well in which Truth is said to reside is really a bottomless pit.
>
> Oliver Heaviside, 1893

Circuit theory, which is of course crucial to much of electrical engineering, is often ignored or given only a cursory glance in books on electrodynamics. To a large extent this is natural, since the centre of mass in electrodynamics is the theory of fields, whereas circuit theory deals with idealised circuit elements that have been reduced to zero-dimensional lumped parameters that are mere shadows of their three-dimensional reality. However, having established the concepts of charge conservation, EMFs, capacitors and inductors, i.e. most of the core elements of simple circuit theory, it seems somehow churlish not to bring these together and discuss what is a topic of immense practical importance. So we do this now, but in a way that does not stray too far from Maxwell's equations.

## 12.1 Idealised Components in Circuit Theory

The starting point for circuit theory is to introduce those idealised lumped parameters that serve as zero-order approximations to real devices, such as inductors and capacitors. Let us start with the notion of an *ideal inductor*. This is a conductor whose magnetic field, like that of a solenoid, is concentrated close to the inductor, so that its stray magnetic field is weak and there is little interaction with other circuit elements. We shall also assume that it has negligible charge density (both volume charge and surface charge), which is true of devices such as solenoids, provided that they operate at a moderately low frequency. To be more precise, if $\varpi$ is a characteristic frequency for the magnetic field, we shall assume that $\varpi\ell << c$, where $\ell$ is a typical lengthscale of the inductor and $c$ the speed of light. This then allows us to neglect any surface charges on the inductor (see Exercise 12.1), and since $\rho_e$ is tiny, the charge conservation equation, $\nabla\cdot\mathbf{J} = -\partial\rho_e/\partial t$, reduces to $\nabla\cdot\mathbf{J} = 0$. In short, an ideal inductor belongs to the regime of *quasi-magnetostatics* in which the displacement current is neglected and $\nabla\cdot\mathbf{J} = 0$. As noted in §11.5, the governing equations of quasi-magnetostatics are

$$\nabla \times \mathbf{B} = \mu_0 \mathbf{J}, \quad \nabla \times \mathbf{E} = -\frac{\partial \mathbf{B}}{\partial t}, \tag{12.1}$$

along with $\nabla \cdot \mathbf{B} = 0$ and $\nabla \cdot \mathbf{J} = 0$.

Suppose such an inductor is connected to a battery whose terminal voltage, i.e. potential difference, is $V_0$. Then Faraday's law, combined with the generalised definition of EMF, (11.19), tells us that

$$\text{EMF} = \oint_C \mathbf{F}_s \cdot d\mathbf{r} + \oint_C \mathbf{E} \cdot d\mathbf{r} = V_0 - L\frac{dI}{dt}, \tag{12.2}$$

where $\mathbf{F}_s$ is the chemical force acting on the charges in the battery. Also, the generalised form of Ohm's law, $\mathbf{J} = \sigma(\mathbf{F}_s + \mathbf{E})$, yields

$$\text{EMF} = \frac{1}{\sigma}\oint_C \mathbf{J} \cdot d\mathbf{r} = IR, \tag{12.3}$$

where $R$ is the net resistance of the inductive loop. Combining the two, we obtain

$$\text{EMF} = V_0 - L\frac{dI}{dt} = IR, \tag{12.4}$$

which is (11.39). We read this as saying that the net EMF in the circuit, which comprises the battery EMF plus the so-called *back EMF* of the inductor, drives a current $I$ through the circuit. However, there is a twist to the story: by convention, an *ideal inductor* is taken to have zero resistance, so the inevitable resistance associated with any real inductor is represented as a separate circuit element in series with the ideal inductor, as shown as Figure 12.1. This is the simple $LR$ circuit in which $R$ represents all possible sources of resistance in the circuit, including that of the inductor and the connecting wires and any internal resistance within the battery. Such a circuit is governed by (12.4).

There is now a matter of convention. In circuit theory it is conventional to transfer $L\,dI/dt$ in (12.4) from the left side to the right and rewrite the equation as

$$V_0 = L\frac{dI}{dt} + IR. \tag{12.5}$$

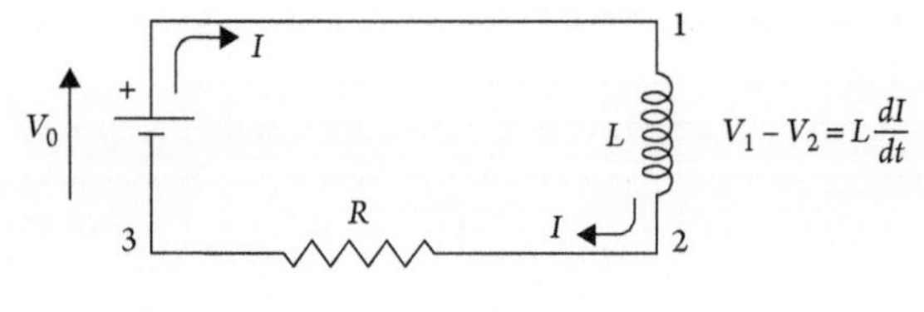


**Figure 12.1** A simple $LR$ circuit.

This is then interpreted as saying that the voltage rise (or increase in potential) across the terminals of the battery, $V_0 = V_1 - V_3$, is equal to the sum of the voltage drops (or drops in potential) across the inductor, $V_1 - V_2$, and the resistor, $V_2 - V_3 = IR$. By implication, $V_1 - V_2 = LdI/dt$. The step from (12.4) to (12.5) is, perhaps, trivial, but it is symbolic of the shift in language (we now talk of changes in voltage, rather than changes in potential) and modes of thought that arise as we move from electromagnetism to circuit theory.

Let us now consider the ideal capacitor. To focus thoughts, consider a circular, parallel-plate capacitor, as shown in Figure 12.2. First, we assume that the connecting wires and plates are perfect conductors and that there is no leakage of charge across the gap between the plates. Second, it is assumed that the gap is very narrow by comparison with the radius of the plates, so that the electric field is more or less confined to the region between the plates, with all the field lines that leave one plate reaching the other. There is then little interaction with other parts of the circuit and also equal and opposite charges on the two plates. Third, it is assumed that $\varpi\ell << c$, where $\varpi$ is a characteristic frequency for the electric field and $\ell$ is a typical lengthscale, say the radius of the capacitor plates. In such cases $\partial\mathbf{B}/\partial t$ can be neglected in Faraday's law, $\nabla \times \mathbf{E} \approx 0$, as discussed in §11.5. We are now in the regime of *quasi-electrostatics*, rather than quasi-magnetostatics. Since $\partial\rho_e/\partial t$ is now finite, charge conservation is retained in its full form, $\nabla \cdot \mathbf{J} = -\partial\rho_e/\partial t$, and for consistency the displacement current must be retained in the Ampère–Maxwell equation:

$$\nabla \times \mathbf{B} = \mu_0 \left(\mathbf{J} + \varepsilon_0 \frac{\partial \mathbf{E}}{\partial t}\right). \tag{12.6}$$

Now suppose that $Q$ is the net charge on the positive plate, and $I$ the current flowing onto that plate, as shown in Figure 12.2. Then $\nabla \cdot \mathbf{J} = -\partial\rho_e/\partial t$ integrated over the positive plate becomes $I = dQ/dt$. Moreover, by definition, the voltage drop across the capacitor, $V_{\text{cap}}$, is related to the charge $Q$ by $Q = CV_{\text{cap}}$, where $C$ is the capacitance. So an ideal capacitor is characterised by the simple relationship

$$I = C\frac{dV_{\text{cap}}}{dt}. \tag{12.7}$$

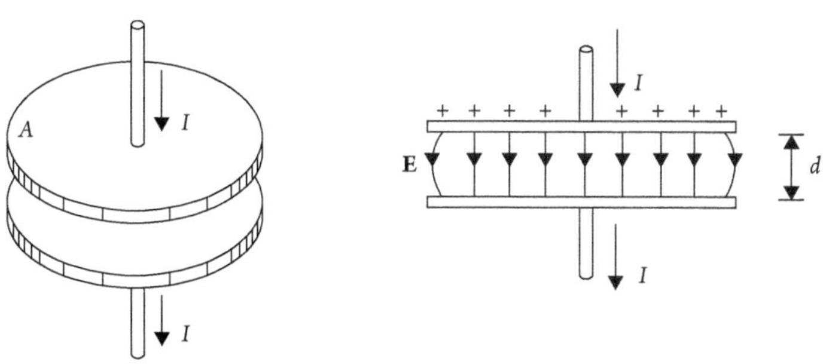


**Figure 12.2** A parallel-plate capacitor.

Thus, for example, if the inductor in Figure 12.1 were replaced by a capacitor, then (12.5) would be become $V_0 = V_{\rm cap} + IR$, and on differentiation this yields

$$R\frac{dI}{dt} + \frac{1}{C}I = 0.$$

We shall now explore the consequences of coupling ideal capacitors with ideal inductors.

## 12.2 An Example of Transient Response: The *RLC* Damped Oscillator

Perhaps the simplest circuit to exhibit non-trivial behaviour is the ubiquitous *RLC* circuit shown in Figure 12.3. A resistor $R$, an ideal capacitor $C$ and an ideal inductor $L$ are connected in series by perfectly conducting wires. The voltage drops across the various circuit elements must sum to zero, and so the current $I$ in the circuit is governed by

$$L\frac{dI}{dt} + IR + V_{\rm cap} = 0, \tag{12.8}$$

where $V_{\rm cap} = V_1 - V_2$. Differentiating with respect to time, and using (12.7) to evaluate the voltage drop across the capacitor, yields

$$L\frac{d^2I}{dt^2} + R\frac{dI}{dt} + \frac{1}{C}I = 0. \tag{12.9}$$

To within an arbitrary choice of phase (or origin in time), this has the solution

$$I = I_0 \exp(-t/2\tau)\cos(\varpi t), \tag{12.10}$$

where $\tau = L/R$ is the decay time introduced in (11.37) and the frequency of the oscillation is given by

$$\varpi^2 = \frac{1}{LC} - \left(\frac{1}{2\tau}\right)^2. \tag{12.11}$$

Evidently we have a damped oscillator. In practice the damping is usually relatively weak and so the resonant frequency is simply $\varpi \approx 1/\sqrt{LC}$.

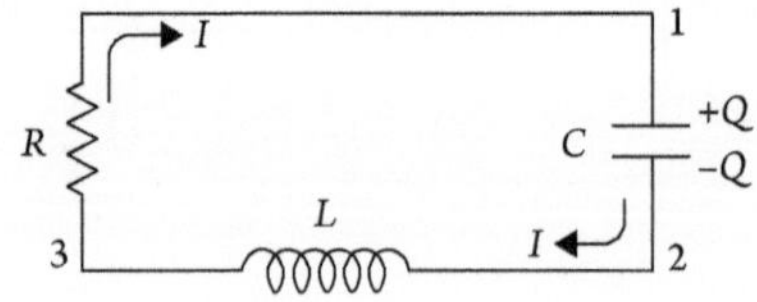


**Figure 12.3** An *RLC* circuit.

Note that $CV_{\text{cap}}^2/2$ is the electrical energy in the capacitor, as can be seen from

$$U_E = \frac{1}{2}\varepsilon_0 \int \mathbf{E}^2 d\mathbf{x} = \frac{1}{2}\varepsilon_0 \left(\frac{V_{\text{cap}}}{d}\right)^2 Ad = \frac{1}{2}\left(\frac{\varepsilon_0 A}{d}\right) V_{\text{cap}}^2, \tag{12.12}$$

where $A$ is the area of either of the capacitor plates and $d$ the gap between the plates. Moreover, the product of $I$ with (12.8) yields the energy equation

$$\frac{d}{dt}\left(\frac{LI^2}{2} + \frac{CV_{\text{cap}}^2}{2}\right) = -I^2 R, \tag{12.13}$$

which tells us that the sum of the magnetic energy in the inductor and the electrical energy in the capacitor decays at a rate set by the Ohmic dissipation.

If the damping is small, then $I$ and $V_{\text{cap}}$, and hence the magnetic energy and electric energy, are $\pi/2$ out of phase. Physically what happens is that energy sloshes back and forth between the capacitor and the inductor, and hence between the electric field and the magnetic field. At one moment the current is zero and the voltage across the capacitor at a maximum, so the capacitor is fully charged and all of the energy is electrical. The capacitor then discharges and $\pi/2$ later in the cycle the current is a maximum and the voltage across the capacitor is zero. All of the energy is now magnetic. The back EMF across the inductor now starts to charge the capacitor in the opposite direction, and so it goes on.

It is interesting that oscillations occur in a circuit composed of an ideal inductor, which is governed by the quasi-magnetostatic equations, and an ideal capacitor, which is governed by the quasi-electrostatic approximation. The quasi-magnetostatic approximation allows for oscillations, such as the Alfvén waves of §8.6 or the oscillating slide of §10.6, only if parts of the conducting circuit are allowed to move. If the conductor is stationary, then the absence of the displacement current in the Ampère–Maxwell equation means that the magnetic field is governed by the diffusion equation (11.38) and so no oscillations can arise. Likewise, as emphasised in §11.5, the quasi-electrostatic equations do *not* constitute a wave-bearing system, because $\partial\mathbf{B}/\partial t$ is missing from Faraday's law. So, when taken *individually* (and assuming that there is no movement of the conducting material), the equations of quasi-magnetostatics and quasi-electrostatics cannot support oscillatory behaviour. However, when *combined* in this *RLC* circuit, these two systems of equations do indeed yield oscillations.

## 12.3 Multi-loop Networks: Kirchhoff's Rules for Ideal Circuits

The analysis of more complicated circuits is facilitated by the application of *Kirchhoff's circuit rules*, as we now discuss. Consider, by way of an example, the circuit shown in Figure 12.4. While charge may accumulate on the plates of the capacitor, in an idealised circuit no charge is allowed to accumulate at *nodes* like X and Y. It follows that the instantaneous current

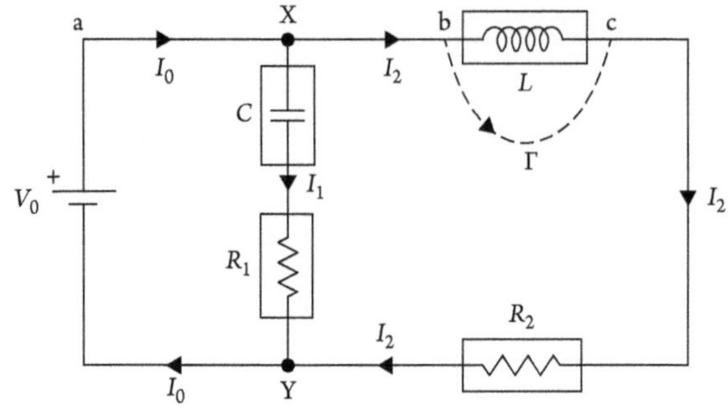


**Figure 12.4** A two-loop circuit.

flowing into either node in Figure 12.4 must equal the instantaneous current flowing out. So, at all times we have $I_0 = I_1 + I_2$ in this circuit. This is a manifestation of the fact that $\nabla \cdot \mathbf{J} = 0$ at all points outside the box enclosing the capacitor in Figure 12.4.

Next we note that the magnetic field associated with the ideal inductor $L$ is assumed to be restricted to the box enclosing that inductor. More generally, in an ideal circuit the magnetic field outside the inductors is assumed to be negligible. So, outside the box enclosing $L$ in Figure 12.4, Faraday's law reduces to $\nabla \times \mathbf{E} = 0$ and we may write $\mathbf{E} = -\nabla V$, $V$ being the voltage (or electrostatic potential). In integral form we have $\oint \mathbf{E} \cdot d\mathbf{r} = 0$, provided that the line integral avoids the inductor in Figure 12.4. It follows that the sum of the voltage differences around *any* closed loop that avoids the inductor must be zero. For example, consider the external loop comprising the current path c–Y–a–X–b plus the dashed curve Γ that goes from b to c. Then $\oint \mathbf{E} \cdot d\mathbf{r} = 0$ applied to that path yields, by analogy to (12.5),

$$V_0 = L\frac{dI_2}{dt} + I_2 R_2. \tag{12.14}$$

Equally we could apply this rule to the internal paths consisting of a–X–Y–a, or c–Y–X–b plus the dashed curve Γ.

We have arrived at Kirchhoff's rules for ideal circuits. These assert that

(i) the net current flowing into a node equals the net current flowing out of that node;
(ii) the sum of the voltage differences around any closed loop is zero.

In fact, we were already using the second of these rules when writing down (12.5) and (12.8). The use of Kirchhoff's rules allows ideal circuits of some complexity to be analysed with relative ease.

## 12.4 AC Networks: Impedances and Equivalent Circuits

We now turn to circuits with *alternating current* (AC) voltage sources that oscillate at some prescribed frequency $\varpi$. The first thing to notice is that our ideal circuit elements, as well as

Kirchhoff's rules, are all linear, reflecting the underlying linearity of Maxwell's equations. This tells us that the governing equations for any ideal circuit will be a linear system of ordinary differential equations. This, in turn, tells us that, in general, the net response of a forced ideal circuit is the superposition of (i) a transient response, which depends on the initial conditions and typically dies down with time if the circuit is dissipative, and (ii) a steady-state sinusoidal response, which is independent of the initial conditions and oscillates at the same frequency as the voltage source. We shall consider only the long-term response of AC circuits in this section.

If all quantities oscillate at the same frequency $\varpi$, then it is natural to write $I(t) = \hat{I}\exp(j\varpi t)$, $V(t) = \hat{V}\exp(j\varpi t)$ and so on, where the real part is understood, $\hat{I}$ and $\hat{V}$ are complex amplitudes, and $\varpi$ is taken to be real. For an inductor the voltage drop $V_{\text{ind}} = LdI/dt$ then becomes $\hat{V}_{\text{ind}} = (j\varpi L)\hat{I}$, while for a capacitor $\hat{I} = (j\varpi C)\hat{V}_{\text{cap}}$. In general, for any circuit element we may write $\hat{V} = Z\hat{I}$, where the complex number $Z$ is known as the *impedance* of that element. (As with much of the language now used to describe AC circuits, this term was coined by Heaviside.) Thus we have $Z = j\varpi L$ for an inductor, $Z = (j\varpi C)^{-1}$ for a capacitor and $Z = R$ for a resistor, as shown in Figure 12.5.

Using Kirchhoff's rules, it is easy to show that impedances may be combined just like resisters in a direct current (DC) circuit. Thus two impedances $Z_1$ and $Z_2$ in series add to give an equivalent impedance $Z_{\text{eq}} = Z_1 + Z_2$, while two impedances in parallel combine as

$$\frac{1}{Z_{\text{eq}}} = \frac{1}{Z_1} + \frac{1}{Z_2}. \tag{12.15}$$

The repeated application of these addition rules to a circuit with a single voltage source often, but not always, allows a complicated network of impedances to be reduced to a single equivalent impedance in series with that voltage source. There are, however, occasions where the repeated application of the impedance addition rules does not allow one to combine all of the individual impedances into one single equivalent impedance (see the example in Exercise 12.7). Nevertheless, when a circuit has a single voltage source $\hat{V}_0$, the linearity of the governing equations tells us that we can always write $\hat{V}_0 = Z_{\text{eq}}\hat{I}_0$ for some complex number $Z_{\text{eq}}$, where $\hat{I}_0$ is the current through the source. So we conclude that any two-terminal network of inductors, capacitors and resistors (which is connected to a single voltage source) can be replaced by a single equivalent impedance $Z_{\text{eq}}$, as shown

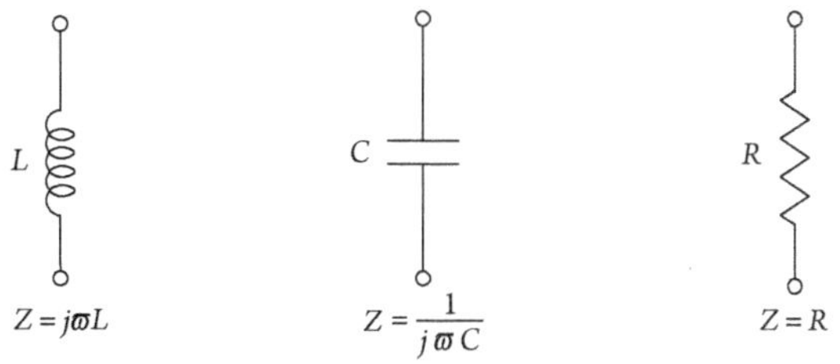


**Figure 12.5** The impedances of an inductor, a capacitor and a resistor.

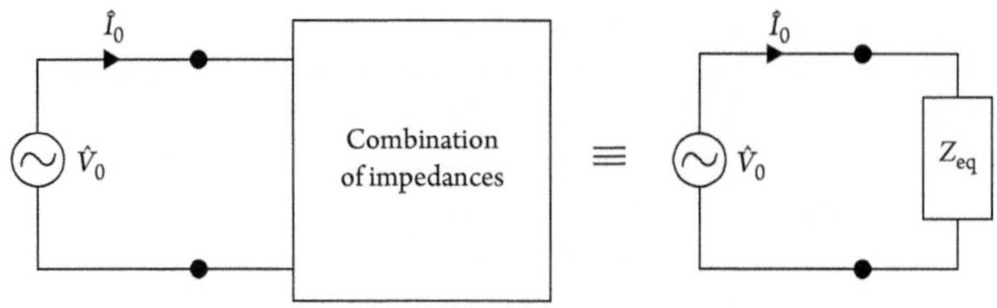


**Figure 12.6** A two-terminal network of inductors, capacitors and resistors can always be replaced by a single equivalent impedance $Z_{eq}$.

schematically in Figure 12.6. This equivalent impedance can always be found through the application of Kirchhoff's rules, even in those cases where the rules for combining impedances yield no useful result.

The linearity of the governing equations for ideal networks lends itself to a number of other useful theorems, such as *Thévenin's theorem*, which states that any two-terminal network of linear elements and energy sources can be replaced, as far as the outside word is concerned, by an ideal voltage source in series with a single impedance. Moreover, the *reciprocal theorem* states that, in any linear network, if a voltage $V$ applied in branch 1 causes a current $I$ to flow in branch 2, then voltage $V$ applied in branch 2 will cause the same current $I$ to flow in branch 1. This is illustrated in Exercise 12.8.

## 12.5 Ladder Networks: A Glimpse at Filters and Transmission Lines

Sometimes networks are built up through the repeated addition of a simple building block to yield a long repetitive chain of elements. These are called ladder networks and often it is convenient to take the chain of elements to be infinitely long. We shall discuss such networks here on the assumption that a steady-state AC response has been achieved and so we may use the idea of impedances.

A typical example of a semi-infinite network is shown in Figure 12.7, where $Z_1$ is the series impedance and $Z_2$ the shunt impedance. To focus our thoughts, we shall consider the case where $Z_1$ and $Z_2$ are purely imaginary, and so they may be capacitors or inductors, or perhaps some combination of the two. There are then no dissipative elements anywhere in the network. Let us now assume that the network has a well-defined equivalent impedance $Z_0$, which is known as the *characteristic impedance* of the infinite network. This impedance may be determined using a simple trick. Suppose that we add one more section to the front of the network, as shown in Figure 12.7(b). Then the net impedance will be unchanged and so (12.15) applied to Figure 12.7(b) yields

$$Z_0 = Z_1 + \frac{1}{1/Z_2 + 1/Z_0}, \tag{12.16}$$

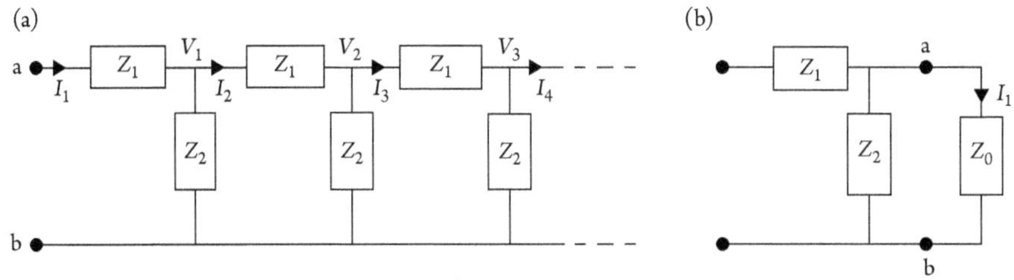


**Figure 12.7** (a) A simple ladder network of characteristic impedance $Z_0$. (b) An equivalent circuit for the ladder network with one more section added to the front.

from which we obtain

$$Z_0 = Z_1/2 \pm \sqrt{Z_1 Z_2 + Z_1^2/4}. \tag{12.17}$$

Of course, on physical grounds there cannot be two distinct characteristic impedances, and we shall see how to choose the appropriate sign in (12.17) shortly.

Given $Z_0$, we can determine the ratio of voltage drops across the $(n+1)$th and $n$th shunt elements, which will provide us with some insight as to how the network behaves. Since the net impedance to the right of the $n$th shunt element is $Z_0$, we have $\hat{V}_n = Z_0 \hat{I}_{n+1}$. Moreover, the voltage drop across the series element that lies between the $n$th and $(n+1)$th shunts is $\hat{V}_n - \hat{V}_{n+1} = Z_1 \hat{I}_{n+1}$. Combining these expressions gives us $\hat{V}_n - \hat{V}_{n+1} = Z_1 \hat{V}_n / Z_0$, from which

$$\frac{\hat{V}_{n+1}}{\hat{V}_n} = \frac{Z_0 - Z_1}{Z_0} = \frac{\sqrt{Z_1 Z_2 + Z_1^2/4} \mp Z_1/2}{\sqrt{Z_1 Z_2 + Z_1^2/4} \pm Z_1/2}. \tag{12.18}$$

Since $Z_1$ and $Z_2$ are imaginary, $\sqrt{Z_1 Z_2 + Z_1^2/4}$ is either strictly real or purely imaginary. Consider first the case where it is real, which we label as $R_0 = \sqrt{Z_1 Z_2 + Z_1^2/4}$. Then we have

$$Z_0 = Z_1/2 + R_0, \quad \left|\hat{V}_{n+1}\right| = \left|\hat{V}_n\right|, \tag{12.19}$$

so that the characteristic impedance looks like an inductor or a capacitor in series with a resistor. Note that we have had to drop the lower sign in (12.17) in order to avoid the unphysical situation in which there is a negative resistance creating energy within the circuit. Note also that there is no attenuation in voltage as we move along the network.

The other case of $\sqrt{Z_1 Z_2 + Z_1^2/4} = jX_0$, where $X_0$ is real, leads to a purely imaginary characteristic impedance and to

$$\frac{\hat{V}_{n+1}}{\hat{V}_n} = \frac{X_0 - |Z_1|/2}{X_0 + |Z_1|/2}. \tag{12.20}$$

Note that, once again, we have had to drop one of the signs in (12.17) and (12.18) in order to avoid the unphysical situation in which the voltage grows indefinitely as we move along the network. Equation (12.20) tells us that the voltage now decays exponentially with distance from the terminals.

A common network is one in which the series impedance is taken to be an inductor, $Z_1 = j\varpi L$, and the shunt impedance a capacitor, $Z_2 = (j\varpi C)^{-1}$, as shown in Figure 12.8. As we shall see in Chapter 14, this constitutes a simple model of an ideal transmission line. The expressions above then become

$$Z_0 = j\varpi L/2 + R_0, \quad R_0 = \sqrt{(L/C) - (\varpi L/2)^2}, \quad |\hat{V}_{n+1}| = |\hat{V}_n|, \tag{12.21}$$

for $\varpi^2 < 4/LC$, and

$$Z_0 = j\varpi L/2 + jX_0, \quad X_0 = \sqrt{(\varpi L/2)^2 - (L/C)}, \quad \frac{\hat{V}_{n+1}}{\hat{V}_n} = \frac{X_0 - \varpi L/2}{X_0 + \varpi L/2}, \tag{12.22}$$

for $\varpi^2 > 4/LC$. The frequency $\varpi_0 = 2/\sqrt{LC}$ is called the *cut-off frequency*.

Equations (12.21) and (12.22) constitute an example of what is called a *low-pass filter*, since frequencies lower than the cut-off pass down the network without any attenuation of voltage, whereas frequencies higher than $\varpi_0$ are blocked. By way of contrast, it is readily confirmed that swapping the $L$'s and $C$'s in the network, so that $Z_1 = (j\varpi C)^{-1}$ and $Z_2 = j\varpi L$, yields a *high-pass filter*, as discussed in Exercise 12.2.

One of the extraordinary features of (12.21) is that the characteristic impedance includes an effective resistance. This means that the generator in Figure 12.8 thinks it is feeding a combination of a resistor and an inductor, and so it continually loses energy to the network. However, there are no dissipative elements anywhere within the network, and hence no Ohmic dissipation. We seem to have reached a paradox: where does the energy passed from the generator to the network go? The answer is rather subtle. We have already mentioned that the ladder network in Figure 12.8 is often used as a model of an ideal transmission line, and we shall see in Chapter 14 that waves propagate along transmission lines. These waves can be initiated at the terminals shown on the left of the figure and then travel out along the network, carrying energy with them. If the transmission line is of finite length, there can be reflections at the far end, and ultimately wave energy propagates in both directions. (The possibility of energy propagation in two directions explains the two signs in (12.17).) However, for a semi-infinite ladder network any energy put into the transmission line on the left just propagates out to infinity. In this way the generator can continually feed energy into the network without there being any Ohmic dissipation within the ladder.

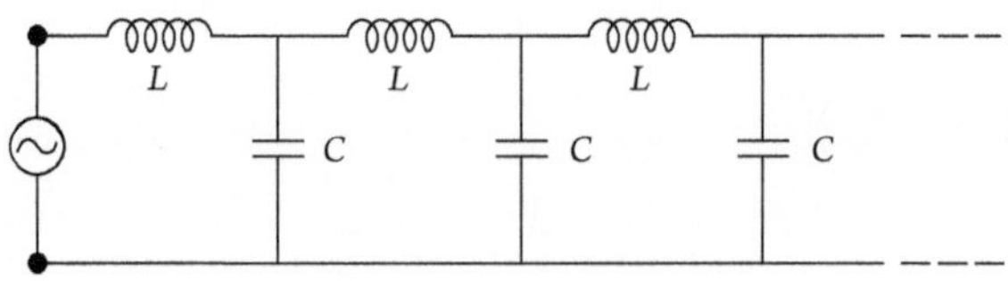


**Figure 12.8** A ladder network composed of an inductor and a capacitor is often used as a model of an ideal transmission line.

Actually this argument, while essentially correct, is a little glib in two respects. First, if energy continually propagates out to infinity, then at any instant there are parts of the network that have been energised and parts that have not. We have then violated one of our starting assumptions, which was that a steady-state AC response has been achieved within the network. Second, a transmission line has a continuous, rather than a discrete, distribution of impedances. So the $L$'s and $C$'s in the model of an ideal transmission line are usually written as $L = L'\,\delta x$ and $C = C'\,\delta x$, where $L'$ and $C'$ are the inductance and capacitance per unit length and $\delta x$ a short part of the line. We must then take the limit of $\delta x \to 0$, in which case we find $\varpi_0 \to \infty$ while (12.21) simplifies to $Z_0 = R_0 = \sqrt{L'/C'}$. So, for an ideal transmission line the characteristic impedance is the pure resistance, $Z_0 = \sqrt{L'/C'}$. If the ladder network of a transmission line is terminated after a finite distance with such a resistance, then the generator cannot detect that the network is in fact finite, and so there are no reflections from the far end.

## 12.6 Leaving the Quasi-Static Regime

So far we have limited ourselves to the quasi-static regime characterised by $\varpi\ell << c$, $\ell$ being a typical geometric lengthscale. Before leaving the topic of circuit theory, we consider what happens to our idealised circuit elements as the frequency is ramped up beyond the quasi-static limit. We shall see that, as the frequency rises, inductors start to exhibit some of the characteristics of capacitors, while capacitors exhibit some of properties of inductors. In short, the notion of pure inductors and capacitors starts to break down. Let us start by returning to the circular parallel-plate capacitor shown in Figure 12.2.

### *12.6.1 A Parallel-Plate Capacitor at High Frequencies*

As long as $\varpi\ell << c$, the electric field between the plates of the capacitor shown in Figure 12.2 satisfies the electrostatic equation $\nabla^2\mathbf{E} = 0$. Moreover, if we take the plates to be very broad relative to the gap $d$, the charge spreads out evenly over the plates and the electric field is more or less uniform at each instant, pointing in the axial direction. However, if the timescale of the electric field becomes comparable to the time of flight of electromagnetic waves across the capacitor plates, then we no longer satisfy the quasi-electrostatic approximation outlined in §11.5. In particular, we can no longer justify neglecting the term $\partial\mathbf{B}/\partial t$ in Faraday's law. In such a situation the electric field ceases to be uniform between the plates, as we shall now show.

When $\varpi\ell \sim c$, we must reinstate $\partial\mathbf{B}/\partial t$ in Faraday's law and adopt the full set of Maxwell equations,

$$\nabla\times\mathbf{E} = -\frac{\partial\mathbf{B}}{\partial t}, \qquad \nabla\cdot\mathbf{E} = \frac{\rho_e}{\varepsilon_0}, \tag{12.23}$$

$$\nabla\times\mathbf{B} = \mu_0\left(\mathbf{J}+\varepsilon_0\frac{\partial\mathbf{E}}{\partial t}\right), \quad \nabla\cdot\mathbf{B} = 0. \tag{12.24}$$

If we focus on the gap between the plates, then we have $\mathbf{J} = 0$ and $\rho_e = 0$, so that $\mathbf{E}$ now satisfies the equation

$$\nabla^2 \mathbf{E} = -\nabla \times (\nabla \times \mathbf{E}) = \nabla \times \frac{\partial \mathbf{B}}{\partial t} = \mu_0 \varepsilon_0 \frac{\partial^2 \mathbf{E}}{\partial t^2} = \frac{1}{c^2}\frac{\partial^2 \mathbf{E}}{\partial t^2}.$$

Let us focus on the axial component of $\mathbf{E}$ and adopt cylindrical polar coordinates $(s, \theta, z)$. For a fixed frequency $\varpi$, it is natural to write $E_z(\mathbf{x}, t) = \hat{E}(\mathbf{x}) \exp(j\varpi t)$ and our wave equation simplifies to

$$\nabla^2 \hat{E} + \frac{\varpi^2}{c^2}\hat{E} = 0. \tag{12.25}$$

We now take advantage of axial symmetry and assume that, in the narrow gap approximation, $\hat{E}(\mathbf{x})$ is a function of radius only. Our governing equation then becomes

$$\frac{1}{s}\frac{d}{ds}\left(s\frac{d\hat{E}}{ds}\right) + \frac{\varpi^2}{c^2}\hat{E} = 0,$$

whose solution is

$$\hat{E}(s) = \hat{E}_0 J_0(\varpi s/c), \tag{12.26}$$

where $J_0$ is the zero-order Bessel function of the first kind. Evidently $E_z$ is no longer uniform, but rather varies with radius. Note, however, that if the plate radius $a$ satisfies $\varpi a << c$, then we have $J_0 \approx 1$ and the field is again uniform. In short, we recover the quasi-static field in the limit of $\varpi a << c$, as we must.

From a knowledge of the shape of the function $J_0(x)$, we see that the electric field is a maximum at the centre of the capacitor plates and starts to fall off as we move radially outward. The reason for this particular profile can be understood if we keep the frequency modest and consider a small deviation from a uniform electrostatic field. In such cases we may, to leading order in $\varpi a/c$, take $\mathbf{E}$ to be uniform between the plates, and the Ampère–Maxwell equation becomes

$$\nabla \times \mathbf{B} = \mu_0 \varepsilon_0 \frac{\partial \mathbf{E}_0}{\partial t} = \frac{1}{c^2}\frac{\partial \mathbf{E}_0}{\partial t}, \tag{12.27}$$

where $\mathbf{E}_0$ is the uniform electrostatic field. This then gives rise to an azimuthal magnetic field with circular field lines centred on the capacitor. Indeed, from (12.24) we see that this magnetic field has exactly the same spatial structure as that generated by a uniform current density flowing along a straight wire. Invoking Stokes' theorem and applying (12.27) in integral form to a circular field line of radius $s$, we find

$$2\pi s B_\theta = \frac{1}{c^2}\frac{\partial E_0}{\partial t}\pi s^2,$$

or equivalently

$$B_\theta = \frac{s}{2c^2}\frac{\partial E_0}{\partial t}. \tag{12.28}$$

So far we have not left the confines of quasi-electrostatics. In order to do this, we turn to Faraday's equation and note that this oscillating magnetic field induces a second electric field that constitutes a correction to $\mathbf{E}_0$. Let us denote this correction by $\mathbf{E}^{\mathrm{corr}}$. In the narrow-gap approximation Faraday's law then becomes

$$\frac{\partial E_z^{\mathrm{corr}}}{\partial s} = \frac{\partial B_\theta}{\partial t},$$

and on substituting for $B_\theta$ we obtain

$$E_z^{\mathrm{corr}} = \frac{s^2}{4c^2}\frac{\partial^2 E_0}{\partial t^2} = -\frac{\varpi^2 s^2}{4c^2}E_0. \tag{12.29}$$

Note that this correction to $\mathbf{E}_0$ gets larger as we increase the frequency, or to be more precise, as we increase $\varpi a/c$. The total axial electric field is now

$$E_z = E_0\left(1 - \frac{\varpi^2 s^2}{4c^2}\right), \tag{12.30}$$

which constitutes the first two terms in a power-law expansion of (12.26). So we conclude that the deviation from a uniform axial field arises from the induced electric field associated with Faraday's law.

In summary, then, in addition to the conventional electrostatic field that we expect to see in a capacitor, the displacement current $\varepsilon_0 \partial \mathbf{E}/\partial t$ produces an azimuthal magnetic field, just as if a real current were flowing between the two plates (Figure 12.9). This oscillating magnetic field then induces a second electric field, which, in turn, perturbs the electrostatic field. In short, our capacitor is starting to exhibit some of the properties of an inductor, with the induced electric field playing an important role.

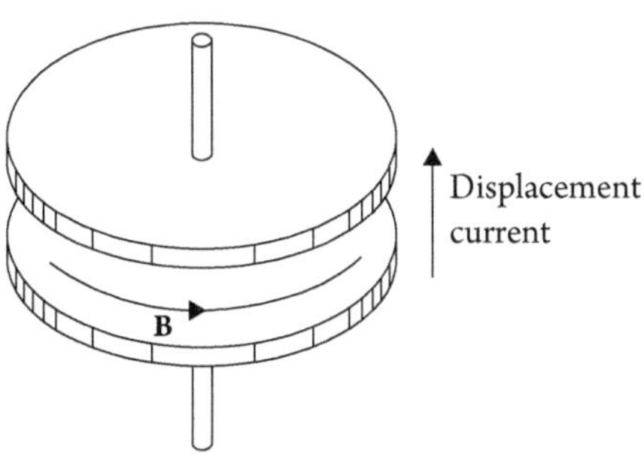


**Figure 12.9** The displacement current and associated magnetic field in a capacitor.

As the frequency increases, so the inductive characteristics of the capacitor become more evident. This is clear from the Ampère–Maxwell equation, which tells us that $|\mathbf{B}| \sim \varpi a\,|\mathbf{E}|/c^2$, and so the magnitude of the magnetic field, and of the associated induced electric field, grows as we increase $\varpi a/c$. For example, $J_0(x)$ changes sign at $x = 2.405$, and so (12.26) tells us that the instantaneous electric field changes direction within the capacitor whenever $\varpi a/c > 2.405$. At some point it ceases to be very meaningful to refer to the device as a capacitor, since this name carries with it much of the baggage of quasi-static circuit theory. Rather, we have two parallel plates exhibiting some interesting electromagnetic phenomena. We shall take up this story again in Chapter 14 when we discuss resonant cavities.

### 12.6.2 *A Solenoid at High Frequencies*

We conclude with a few remarks about what happens to an inductor as we increase $\varpi \ell/c$ beyond the quasi-static limit. The discussion is brief and qualitative. In part, this is because the geometry of most inductors is more complicated than that of the capacitor shown in Figure 12.9, and so a quantitative analysis of high-frequency inductors tends to lack the relative simplicity of that given in §12.6.1 for high-frequency capacitors.

Consider an inductor in the form of a solenoid. At low frequencies the surface charges and displacement current $\varepsilon_0 \partial \mathbf{E}/\partial t$ are negligible in the solenoid, for the reasons set out in Exercise 12.1. However, as $\varpi \ell/c$ is increased, Faraday's law tells us that the intensity of the electric field grows relative to the magnetic field according to $|\mathbf{E}| \sim \varpi \ell\,|\mathbf{B}|$. The surface charges and displacement current also increase, with the displacement current being proportional to $\mathbf{E}$. Moreover, the potential difference between adjacent turns of the solenoid creates an electric field not unlike that which would exist if a capacitor were placed between each turn in the coil (sometimes called the proximity effect). As seen from its terminals, the solenoid no longer behaves like an ideal inductor, but rather as if each turn of the solenoid were composed of an inductor in parallel with a so-called *parasitic capacitor*. Of course, the impedance of a capacitor falls off with frequency, as $1/\varpi C$, while that of an inductor rises as $\varpi L$. So, as the frequency is increased, less and less current is drawn through the wire loop in each turn and more 'jumps' across the parasitic capacitor in parallel with that loop. At high frequencies, then, the terminal impedance of the inductor looks quite unlike that of its ideal counterpart.

* * *

This concludes our brief excursion into circuit theory. As seen from the vantage point of field theory, it can seem a little simplistic, yet it works astonishingly well most of the time (provided $\varpi$ is not too high). Interested readers will find a gentle introduction in Feynman, Leighton & Sands (1964), while Grant & Phillips (1990) offer a more detailed account.

## EXERCISES

12.1 Let $\varpi$ be the characteristic frequency for an electromagnetic field and $\ell$ be a typical geometric lengthscale. In the *quasi-magnetostatic approximation* the magnitude of $\mathbf{E}$ in *free space* is set by Faraday's law. Show that $|\mathbf{E}| \sim \varpi \ell |\mathbf{B}|$ in such cases and hence confirm that the relative magnitude of the displacement current $\varepsilon_0 \partial \mathbf{E}/\partial t$ to $\mu_0^{-1} \nabla \times \mathbf{B}$ in the Ampère–Maxwell equation is $\varpi^2 \ell^2 / c^2$, where $c$ is the speed of light. Clearly we may ignore displacement currents when $\varpi \ell << c$, which is the quasi-magnetostatic approximation. Neglecting the displacement current in the Ampère–Maxwell equation implies that we must also ignore the charge density in the charge conservation equation $\nabla \cdot \mathbf{J} = -\partial \rho_e / \partial t$, as discussed in §11.5. So we then have $\nabla \cdot \mathbf{J} = 0$, which is consistent with Ampère's law.

12.2 Consider swapping the $L$'s and $C$'s in the network shown in Figure 12.8, so that $Z_1 = (j\varpi C)^{-1}$ and $Z_2 = j\varpi L$. Show that the arguments of §12.5 follow through more or less unchanged provided that we exchange $1/\varpi C$ for $\varpi L$. Hence show that this network constitutes a high-pass filter.

12.3 A switch is opened in a circuit containing a large inductance and carrying a steady current. The current in the open circuit tries to drop to zero but the energy in the inductor, $LI^2/2$, cannot suddenly vanish. What do you think happens? [*Hint:* sparks are often seen to fly in such circumstances.]

12.4 A charged capacitor $C$ of plate area $A_C$ and gap $d$ is connected to a wire resistor $R$ of length $\ell$, conductivity $\sigma$ and cross-sectional area $A_w$. The capacitor discharges through the wire resistor, governed by

$$R\frac{dI}{dt} + \frac{1}{C} I = 0,$$

where $I$ is the instantaneous current in the resistor. Show that the characteristic discharge time $\tau_C = RC$ is equal to $(\varepsilon_0/\sigma)(A_C \ell / A_w d)$. We recognise $\varepsilon_0/\sigma$ as the charge relaxation time of (3.12). The question is where the geometric factor $\ell A_C / d A_w$ comes from. Let $E_C$ be the electric field in the capacitor and $E_w$ the electric field in the wire, which is related to $I$ through Ohm's law, $E_w = \sigma^{-1}(I/A_w)$. Show that the quasi-electrostatic approximation in the form of $\nabla \times \mathbf{E} = 0$ demands that $E_C d = E_w \ell$, while Gauss' law relates $E_C$ to the instantaneous charge on the capacitor, $Q$, by $E_C = Q/\varepsilon_0 A_C$. (You may assume that $d << \ell$, and hence $E_C >> E_w$, when applying Gauss' law.) Hence show that

$$I = A_w \sigma E_w = \sigma A_w \frac{d}{\ell} E_C = \sigma A_w \frac{d}{\ell} \frac{Q}{\varepsilon_0 A_C} = \frac{Q}{\tau_C}.$$

Combine this with charge conservation, $I = -dQ/dt$, to deduce the characteristic discharge time.

12.5 A transmission line is modelled using the ladder network of Figure 12.8, but with a resistor $R_1 = R_1' \delta x$ in series with the inductor $L = L' \delta x$ and a second resistor $R_2 = R_2' \delta x$ in parallel with the capacitor $C = C' \delta x$. Find the characteristic impedance of the network.

12.6 Consider the ladder network of Figure 12.7 for the particular case where $Z_1$ is a series combination of an inductor and a capacitor, $Z_1 = j\varpi L_1 + (j\varpi C_1)^{-1}$, and $Z_2$ is a parallel combination of another inductor and capacitor, $Z_2^{-1} = j\varpi C_2 + (j\varpi L_2)^{-1}$. Show that this acts as a *band-pass filter*, blocking both low and high frequencies. Explain this behaviour in terms of the results of §12.5 and Exercise 12.2 by considering the limits of low and high frequency and approximating $Z_1$ and $Z_2$ by their dominant impedances.

12.7 Figure 12.10 shows a *bridge circuit*. The equivalent impedance for this network cannot be found using the impedance addition rules, but nevertheless the circuit may be analysed using Kirchhoff's rules. The bridge is normally operated so that the current through $Z_5$ is zero. Show that this requires $Z_1/Z_2 = Z_3/Z_4$ and that under these conditions the equivalent impedance of the circuit is

$$Z_{\text{eq}} = \frac{(Z_1 + Z_2)Z_3}{Z_1 + Z_3} = \frac{(Z_1 + Z_2)Z_4}{Z_2 + Z_4}.$$

12.8 We may illustrate the reciprocal theorem for ideal linear networks using the circuit shown in Figure 12.11. For the circuit on the left, show that the equivalent impedance, as seen from terminals a–b, is

$$Z_{\text{eq}} = \frac{Z_1Z_2 + Z_1Z_3 + Z_2Z_3}{Z_2 + Z_3},$$

and hence the current flowing in $Z_2$ is

$$\hat{I}_2 = \frac{Z_3}{Z_2 + Z_3}\hat{I}_0 = \frac{Z_3}{Z_1Z_2 + Z_1Z_3 + Z_2Z_3}\hat{V}_0.$$

Now show that the same current flows in $Z_1$ for the circuit on the right.

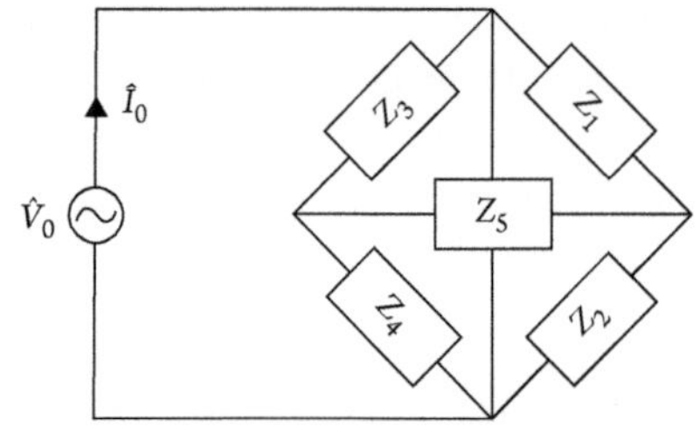


**Figure 12.10** An example of a bridge circuit (or Wheatstone bridge).

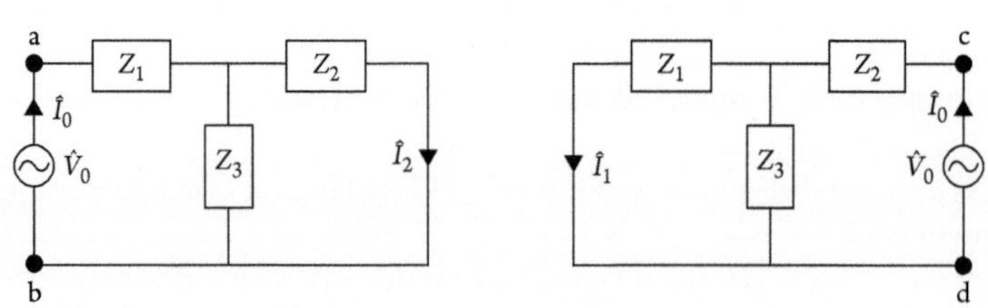


**Figure 12.11** A simple network that illustrates the reciprocal theorem.

## REFERENCES

Feynman, R. P., Leighton, R. B. & Sands, M., 1964, *The Feynman Lectures on Physics,* Volume II. Addison-Wesley.

Grant, I. S. & Phillips, W.R., 1990, *Electromagnetism,* 2nd ed. Wiley.

# 13

# Static Versus Dynamic Fields

## *Maxwell's Equations*

One scientific epoch ended and another began with James Clerk Maxwell.

Albert Einstein

We have finally reached the centre of mass of this textbook. Our task in this chapter is to bring together all of the fundamental laws of electrodynamics, starting with the static laws of Gauss and Ampère, and ending up with Maxwell's equations and their solution through the use of retarded potentials. In effect, we shall return to the overview offered in Chapter 3, where we started our journey. However, we do so now with the benefit of having discussed in some depth electrostatics, magnetostatics, Faraday's law, potentials and electromagnet energy. This will take us to the pinnacle of electrodynamics, from which we will be able to survey vast tracts of land, leading us into many interesting applications in later chapters, such as waves and radiation. For simplicity, we shall leave aside dielectric and magnetic materials and focus instead on electromagnetic fields in vacuum or in good electrical conductors. We shall return to dielectric and magnetic materials in Chapter 18.

## 13.1 A Recapitulation of the Laws of Electrostatics and Quasi-Magnetostatics

Let us start by assembling the key results from previous chapters. In Chapter 4 we noted that an *electrostatic* field is governed by

$$\nabla \times \mathbf{E}_s = 0, \qquad \nabla \cdot \mathbf{E}_s = \frac{\rho_e}{\varepsilon_0}, \tag{13.1}$$

where the second of these equations is Gauss' law, $\rho_e(\mathbf{x})$ is the charge density and the subscript $s$ is to remind us that the electric field is strictly static. Given the distribution of $\rho_e$ and suitable boundary conditions, this uniquely determines $\mathbf{E}_s$, since the requirements for a vector field to be uniquely determined are that its divergence and curl be known.

Since $\mathbf{E}_s$ is irrotational, it is conventional to introduce the electrostatic potential $\varphi_s$, defined by $\mathbf{E}_s = -\nabla\varphi_s$. Evidently this scalar potential is governed by the Poisson equation

$$\nabla^2\varphi_s = -\frac{\rho_e}{\varepsilon_0}, \tag{13.2}$$

and in §2.4.2 we showed that, in an infinite domain, this may be inverted using the Green's function solution,

$$\varphi_s(\mathbf{x}) = \frac{1}{4\pi\varepsilon_0}\int\frac{\rho_e(\mathbf{x}')}{|\mathbf{r}|}d\mathbf{x}', \qquad \mathbf{r} = \mathbf{x} - \mathbf{x}'. \tag{13.3}$$

Here the dummy variable $\mathbf{x}'$ allows us to sample the charge at different points.

Next, in Chapter 5, we noted that the mechanical work required to assemble a localised distribution of static charges in an infinite domain is

$$U_E = \frac{1}{2}\int_{V_\infty}\rho_e\varphi_s\,d\mathbf{x}. \tag{13.4}$$

Substituting for $\rho_e$ using (13.2), invoking the identity $\varphi_s\nabla^2\varphi_s = \nabla\cdot(\varphi_s\nabla\varphi_s) - (\nabla\varphi_s)^2$ and using Gauss' theorem to convert the divergence into a surface integral, this may be rewritten as

$$U_E = \frac{\varepsilon_0}{2}\int_{V_\infty}\mathbf{E}^2\,d\mathbf{x}. \tag{13.5}$$

Either of these expressions for $U_E$ represents the energy of an electrostatic field. Moreover, in Chapter 3, we claimed that (13.5), though not (13.4), represents the energy of any electric field, be it static or non-static. We shall return to this last point in §13.5.

Turning to magnetostatics, in Chapter 7 we saw that a *steady* magnetic field is governed by

$$\nabla\cdot\mathbf{B} = 0, \qquad \nabla\times\mathbf{B} = \mu_0\mathbf{J}, \tag{13.6}$$

the second of which is Ampère's law. Given the distribution of $\mathbf{J}$, and suitable boundary conditions, this uniquely determines $\mathbf{B}$. The solenoidal nature of $\mathbf{B}$ allows us to introduce the vector potential $\mathbf{A}$, defined, in part, by $\nabla\times\mathbf{A} = \mathbf{B}$. However, to unambiguously define $\mathbf{A}$, we must specify both its divergence and its curl, and in magnetostatics the so-called *Coulomb gauge* is invariably used, which takes the simplest option of $\nabla\cdot\mathbf{A} = 0$. Substituting for $\mathbf{B}$ in Ampère's law then gives us the Poisson equation

$$\nabla^2\mathbf{A} = -\mu_0\mathbf{J}, \tag{13.7}$$

which inverts like (13.2) to yield

$$\mathbf{A}(\mathbf{x}) = \frac{\mu_0}{4\pi}\int \frac{\mathbf{J}(\mathbf{x}')}{|\mathbf{r}|}d\mathbf{x}', \qquad \mathbf{r} = \mathbf{x} - \mathbf{x}'. \tag{13.8}$$

Noting that $\nabla \times (\mathbf{J}'/|\mathbf{r}|) = \nabla(1/|\mathbf{r}|) \times \mathbf{J}', \mathbf{J}'$ being independent of $\mathbf{x}$, and that

$$\nabla\left(\frac{1}{|\mathbf{r}|}\right) = -\frac{\mathbf{r}}{|\mathbf{r}|^3}, \tag{13.9}$$

this yields the Biot–Savart law,

$$\mathbf{B}(\mathbf{x}) = \frac{\mu_0}{4\pi}\int \frac{\mathbf{J}(\mathbf{x}') \times \mathbf{r}}{|\mathbf{r}|^3}d\mathbf{x}', \quad \mathbf{r} = \mathbf{x} - \mathbf{x}'. \tag{13.10}$$

Note that, as always, $\mathbf{x}'$ locates the source, in this case $\mathbf{J}$, while $\mathbf{x}$ is the point of measurement. The displacement vector $\mathbf{r}$ then takes us from the source to the observer of $\mathbf{B}$, as shown in Figure 13.1.

Next, in Chapter 8, we noted that the mechanical work required to assemble a localised distribution of steady currents in an infinite domain is

$$U_B = \frac{1}{2}\int_{V_\infty} \mathbf{J} \cdot \mathbf{A}\, dV. \tag{13.11}$$

Ampère's law allows us to rewrite (13.11) as

$$U_B = \frac{1}{2\mu_0}\int_{V_\infty} \mathbf{A} \cdot \nabla \times \mathbf{B}\, dV, \tag{13.12}$$

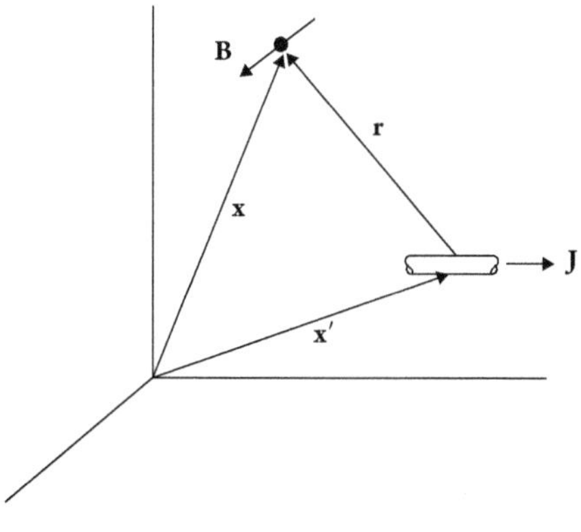


**Figure 13.1** Coordinate system used in the Biot–Savart law.

and the identity $\nabla \cdot (\mathbf{A} \times \mathbf{B}) = \mathbf{B}^2 - \mathbf{A} \cdot \nabla \times \mathbf{B}$ then gives us

$$U_B = \frac{1}{2\mu_0} \int_{V_\infty} \mathbf{B}^2 \, dV, \tag{13.13}$$

since the integral of $\nabla \cdot (\mathbf{A} \times \mathbf{B})$ converts to a surface integral whose integrand is zero at infinity. Either of (13.11) or (13.13) may be used to evaluate the magnetic energy of a steady magnetic field. More generally, in Chapter 3 we claimed that (13.13), though not (13.11), represents the energy of *any* magnetic field, static or non-static.

Next, in Chapter 10 we departed from statics and introduced Faraday's law,

$$\nabla \times \mathbf{E} = -\frac{\partial \mathbf{B}}{\partial t}, \tag{13.14}$$

which replaces the first of the two equations in (13.1). This uncurls to give

$$\mathbf{E} = -\frac{\partial \mathbf{A}}{\partial t} - \nabla \varphi, \tag{13.15}$$

where $\mathbf{A}$ is a vector potential, $\nabla \times \mathbf{A} = \mathbf{B}$ and $\varphi$ is some scalar potential. Note that $\mathbf{A}$ in (13.15) need *not* have zero divergence, i.e. $\mathbf{A}$ is not restricted to the Coulomb gauge. Note also that we have dropped the subscript $s$ on $\varphi$ since this is not, in general, an electrostatic potential governed by (13.2).

In Chapters 11 and 12 we reconciled the static equation of Ampère with the dynamic equation of Faraday by entering the world of quasi-magnetostatics. That is, we argued that if currents change slowly enough, we may retain Ampère's law and combine it with Faraday's law and Ohm's law to give the quasi-static system

$$\nabla \times \mathbf{B} = \mu_0 \mathbf{J}, \qquad \nabla \cdot \mathbf{J} = 0, \tag{13.16}$$

$$\nabla \times \mathbf{E} = -\frac{\partial \mathbf{B}}{\partial t}, \qquad \nabla \cdot \mathbf{B} = 0, \tag{13.17}$$

along with Ohm's law, $\mathbf{J} = \sigma(\mathbf{E} + \mathbf{u} \times \mathbf{B})$, $\mathbf{u}$ being the local velocity of the conducting medium. Note that the solenoidal nature of $\mathbf{J}$ and $\partial \mathbf{B}/\partial t$ follows from Ampère's and Faraday's laws, respectively, since $\nabla \cdot [\nabla \times (\cdot)] = 0$.

We also argued in Chapters 11 and 12 that the quasi-magnetostatic approximation is valid provided that no additional charges are added to the surfaces of conductors (so we exclude capacitors) and provided that $\varpi \ell << c$, where $\varpi$ is a characteristic frequency for the magnetic field, $\ell$ a characteristic lengthscale and $c$ the speed of light. However, at the end of Chapter 12 we saw how this system can break down if we introduce surface charges, such as in a capacitor, or if $\varpi \ell \sim c$. In both cases the fly in the ointment is charge conservation, and the reconciliation of this inconsistency leads us to Maxwell's equations.

## 13.2 Charge Conservation, the Displacement Current and Maxwell's Equations

We noted in Chapter 3 that the two fields $\rho_e$ and $\mathbf{J}$ are not, in general, independent. Rather, they are related by the principle of conservation of charge. This requires that for any volume $V$ enclosed by a surface $S$,

$$\oint_S \mathbf{J} \cdot d\mathbf{S} = -\frac{\partial}{\partial t} \int_V \rho_e \, dV \tag{13.18}$$

or, in differential form,

$$\nabla \cdot \mathbf{J} = -\frac{\partial \rho_e}{\partial t}, \tag{13.19}$$

which is the charge conservation equation. We see immediately that there is an inconsistency between (13.19) and $\nabla \cdot \mathbf{J} = 0$, and hence between (13.19) and Ampère's law. This does not rear its head in quasi-magnetostatics, because, inside good conductors, the charge density is very small and so a good approximation to charge conservation is simply $\nabla \cdot \mathbf{J} = 0$. However, at higher frequencies, or if surface charges are introduced, we cannot ignore the charge density and so we have a problem.

Maxwell resolved this dilemma by adding a second term to the right of Ampère's law,

$$\nabla \times \mathbf{B} = \mu_0 \left( \mathbf{J} + \varepsilon_0 \frac{\partial \mathbf{E}}{\partial t} \right). \tag{13.20}$$

This is the Ampère–Maxwell equation, and the additional term $\varepsilon_0 \partial \mathbf{E}/\partial t$ is the so-called displacement current. Charge conservation is now satisfied because, recalling that $\nabla \cdot \nabla \times (\cdot) = 0$, and using Gauss' law (13.1), the divergence of (13.20) yields

$$\nabla \cdot \mathbf{J} = -\varepsilon_0 \frac{\partial}{\partial t} \nabla \cdot \mathbf{E} = -\frac{\partial \rho_e}{\partial t}, \tag{13.21}$$

which is just (13.19).

At this point two natural question arise. (i) Are there any other features of the static and quasi-static equations that need to be changed when looking at high-frequency applications. (ii) Can we construct a self-consistent system of equations valid under all circumstances? Let us therefore return to §13.1 and run through the various laws. (We shall leave to one side the expressions for energy, which we shall discuss in §13.5.) Starting with electrostatics, clearly we must drop the first of (13.1), since it contravenes Faraday's law, but all of the experimental evidence suggests that Gauss' law has universal validity, so we may keep that. Since $\mathbf{E}$ is no longer guaranteed to be irrotational, we must also drop the Poisson equation for the electrostatic potential, as well as its static inversion formula, (13.3). Moving to magnetostatics and (13.6), the experimental evidence suggests that the expression $\nabla \cdot \mathbf{B} = 0$

is universally valid, and we have already seen how to generalise Ampère's law by adding the displacement current to the right-hand side. Since Ampère's law needs modification, we must drop the Poisson equation for **A**, (13.7), as well as its static inversion formula, (13.8). We must also abandon the Biot–Savart law, (13.10), since it follows from (13.8). On the other hand, it turns out that Faraday's law remains valid under all conditions, which is consistent with the universal validity of $\nabla\cdot\mathbf{B}=0$.

So let us see what we are left with. For materials that are neither dielectric nor magnetic, we have

$$\nabla\cdot\mathbf{E}=\frac{\rho_e}{\varepsilon_0},\ \text{(Gauss' law)}\qquad \nabla\times\mathbf{E}=-\frac{\partial\mathbf{B}}{\partial t},\ \text{(Faraday's law)} \tag{13.22, 13.23}$$

$$\nabla\cdot\mathbf{B}=0,\ \text{(}\mathbf{B}\text{ is solenoidal)}\qquad \nabla\times\mathbf{B}=\mu_0\left(\mathbf{J}+\varepsilon_0\frac{\partial\mathbf{E}}{\partial t}\right).\ \text{(Ampère–Maxwell law)} \tag{13.24, 13.25}$$

These four equation are known collectively as Maxwell's equations and, although they all appear in Maxwell's 1873 treatise, they were first collected together and written down in (more or less) this form by Heaviside in 1885. To obtain all of electrodynamics, we must add to these four equations two more laws:

$$\mathbf{f}=q(\mathbf{E}+\mathbf{u}_q\times\mathbf{B}),\ \text{(force on charge } q\text{)}\qquad \nabla\cdot\mathbf{J}=-\frac{\partial\rho_e}{\partial t}.\ \text{(charge conservation)} \tag{13.26, 13.27}$$

where $\mathbf{u}_q$ is the velocity of the charge $q$.

Perhaps some comments are in order at this point. First, Maxwell's equations yield *both* the divergence and the curl of **E** and **B**, albeit in an interlinked manner. It follows that they uniquely determine these two fields, provided of course that suitable boundary conditions are specified and the distributions of the sources $\rho_e$ and **J** are known. Second, we typically add to this system the empirical Ohm's law, $\mathbf{J}=\sigma(\mathbf{E}+\mathbf{u}\times\mathbf{B})$, which tells us that the current density in a conductor is (usually) proportional to the force acting on the free charges. Third, these laws are not all independent. For example, (13.24) follows, more or less, from the divergence of Faraday's law, while the divergence of the Ampère–Maxwell equation, combined with Gauss' law, yields the charge conservation equation. Fourth, since the **E** and **B** fields are coupled, we should not think of them as separate entities, but rather as two manifestations of a single electromagnetic field. Finally, since $\nabla\cdot\mathbf{B}=0$ remains valid, we may still introduce a vector potential for **B**, satisfying $\nabla\times\mathbf{A}=\mathbf{B}$. We may then uncurl Faraday's law to give

$$\mathbf{E}=-\frac{\partial\mathbf{A}}{\partial t}-\nabla\varphi, \tag{13.28}$$

for some scalar potential $\varphi$. The divergence of **A** appearing in (13.28) is then chosen as a matter of convenience, and need not be set to zero. Indeed, as we shall see shortly,

$\nabla \cdot \mathbf{A}$ is rarely set to zero when studying waves and radiation, although the Coulomb gauge $\nabla \cdot \mathbf{A} = 0$ is usually used in quasi-static problems. Since the divergence of (13.28), combined with Gauss' law, demands

$$\nabla^2 \varphi + \frac{\partial}{\partial t} \nabla \cdot \mathbf{A} = -\frac{\rho_e}{\varepsilon_0}, \tag{13.29}$$

the precise definition and meaning of $\varphi$ in (13.28) can be known only after the gauge (i.e. the divergence) of $\mathbf{A}$ has been specified.

## 13.3 How the Displacement Current Works

Let us see how the new term in the Ampère–Maxwell equation works. Consider the simple problem shown in Figure 13.2 of a long straight vertical wire that is charging a parallel-plate capacitor. Suppose that the current in the wire is $I$, the charge on the upper plate $Q$ and the area of each of the capacitor plates $A$. We now apply Gauss' law to a control volume that encloses the upper capacitor plate. Ignoring edge effects, we may take $\mathbf{E}$ to be uniform within the capacitor and negligible outside. Gauss' law then tells us that the electric field within the capacitor points downward and has a magnitude of $E = Q/\varepsilon_0 A$. Moreover, charge conservation demands that $I = dQ/dt$, and, combining these two statements, we conclude that

$$I = \varepsilon_0 A \frac{dE}{dt}. \tag{13.30}$$

Next, consider the Ampère–Maxwell equation integrated over a horizontal curve $C$ that is circular, of radius $r$, concentric with the wire and lies well above the capacitor. If $S$ is *any* open surface that spans the curve $C$, then we have

$$\oint_C \mathbf{B} \cdot d\mathbf{l} = \mu_0 \int_S \mathbf{J} \cdot d\mathbf{S} + \mu_0 \int_S \varepsilon_0 \frac{\partial \mathbf{E}}{\partial t} \cdot d\mathbf{S}. \tag{13.31}$$

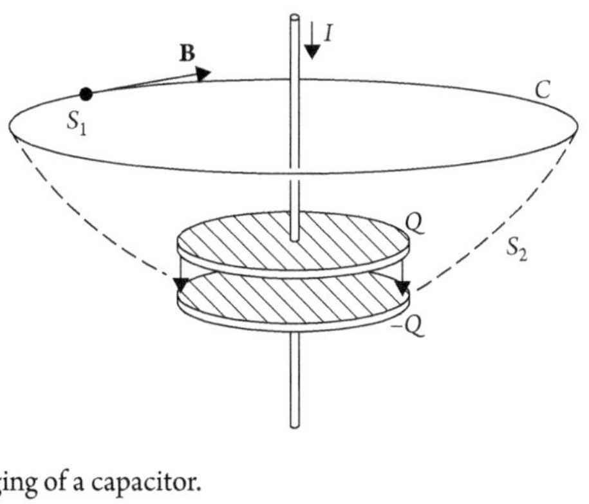


**Figure 13.2** The charging of a capacitor.

We consider two particular cases: $S_1$, which is coplanar with the curve $C$, and $S_2$, which stretches all the way down to the capacitor and passes between the capacitor plates. Since the electric field is more or less confined to the gap between the capacitor plates, for the case of $S_1$ (13.31) reduces to

$$\oint_C \mathbf{B} \cdot d\mathbf{l} = \mu_0 \int_S \mathbf{J} \cdot d\mathbf{S} = \mu_0 I, \tag{13.32}$$

from which we get the usual expression $B = \mu_0 I / 2\pi r$. For $S_2$, however, the first integral on the right of (13.31) is zero, and so Ampère's law gives the incorrect result $\int \mathbf{B} \cdot d\mathbf{l} = 0$. By comparison, the Ampère–Maxwell equation, combined with (13.30), yields

$$\oint_C \mathbf{B} \cdot d\mathbf{l} = \mu_0 \int_S \varepsilon_0 \frac{\partial \mathbf{E}}{\partial t} \cdot d\mathbf{S} = \mu_0 \varepsilon_0 \frac{\partial E}{\partial t} A = \mu_0 I, \tag{13.33}$$

which is consistent with (13.32). So, as the current density $\mathbf{J}$ passing through the first surface integral in (13.31) disappears, the contribution from the displacement current in the second surface integral grows to compensate. It is as if the displacement current were a real current that completes the circuit.

The essential point is this. Gauss' law and charge conservation combine to yield

$$\nabla \cdot \mathbf{J} = -\frac{\partial \rho_e}{\partial t} = -\frac{\partial}{\partial t}(\varepsilon_0 \nabla \cdot \mathbf{E}),$$

and so we have

$$\nabla \cdot \left( \mathbf{J} + \varepsilon_0 \frac{\partial \mathbf{E}}{\partial t} \right) = 0. \tag{13.34}$$

By replacing $\mathbf{J}$ in Ampère's law by $\mathbf{J} + \varepsilon_0 \partial \mathbf{E} / \partial t$ we ensure that the right-hand side of the Ampère–Maxwell equation is solenoidal and so integrates to give the same value for all surfaces $S$ that span the curve $C$ in (13.31).

## 13.4 Maxwell's Equations in Vacuum and the Speed of Light

The introduction of Maxwell's displacement current leads to the entirely new phenomenon of electromagnetic waves, thus forming a direct link between electromagnetism and light. This becomes clear if we write down Maxwell's equations for a vacuum:

$$\nabla \cdot \mathbf{E} = 0, \qquad \nabla \times \mathbf{E} = -\frac{\partial \mathbf{B}}{\partial t}, \tag{13.35}$$

$$\nabla \cdot \mathbf{B} = 0, \qquad \nabla \times \mathbf{B} = \mu_0 \varepsilon_0 \frac{\partial \mathbf{E}}{\partial t}. \tag{13.36}$$

Taking the curl of the Ampère–Maxwell equation and eliminating **E** using Faraday's law yields

$$\frac{\partial^2 \mathbf{B}}{\partial t^2} = c^2 \nabla^2 \mathbf{B}, \tag{13.37}$$

which describes the propagation of electromagnetic waves travelling at the speed of light, $c = 1/\sqrt{\mu_0 \varepsilon_0}$ . Similarly, taking the curl of Faraday's law and eliminating **B** using the Ampère–Maxwell equation gives us

$$\frac{\partial^2 \mathbf{E}}{\partial t^2} = c^2 \nabla^2 \mathbf{E}. \tag{13.38}$$

Again we have waves traveling at the speed of light.

The simplest class of waves in three dimensions are one-dimensional plane waves. Moreover, the easiest way to satisfy $\nabla \cdot \mathbf{E} = 0$ and $\nabla \cdot \mathbf{B} = 0$ is to have both **E** and **B** transverse to the direction of propagation. So let us look for solutions of the form

$$\mathbf{E}(z,t) = (E_x, E_y, 0), \qquad \mathbf{B}(z,t) = (B_x, B_y, 0). \tag{13.39}$$

Then Faraday's law plus the Ampère–Maxwell equation give us

$$\hat{\mathbf{e}}_z \times \frac{\partial \mathbf{E}}{\partial z} = -\frac{\partial \mathbf{B}}{\partial t}, \quad \hat{\mathbf{e}}_z \times \frac{\partial \mathbf{B}}{\partial z} = \frac{1}{c^2}\frac{\partial \mathbf{E}}{\partial t}. \tag{13.40}$$

Application of the operator $(\partial/\partial z)\hat{\mathbf{e}}_z \times$ to either of the two expressions in (13.40) yields, when combined with its companion equation, a one-dimensional wave equation for **E** or **B**:

$$\frac{\partial^2 \mathbf{E}}{\partial t^2} = c^2 \frac{\partial^2 \mathbf{E}}{\partial z^2}, \qquad \frac{\partial^2 \mathbf{B}}{\partial t^2} = c^2 \frac{\partial^2 \mathbf{B}}{\partial z^2}. \tag{13.41}$$

Of course, we could have got to (13.41) by simply substituting the trial solution (13.39) directly into the wave equations (13.37) and (13.38), but we shall find the intermediate result, (13.40), useful. As discussed in §2.3.4, such wave equations are *non-dispersive* and so d'Alembert's solution applies. We conclude, therefore, that the general solution of (13.41) is

$$\mathbf{E}(z,t) = \mathbf{E}_f(z - ct) + \mathbf{E}_b(z + ct), \tag{13.42}$$

$$\mathbf{B}(z,t) = \mathbf{B}_f(z - ct) + \mathbf{B}_b(z + ct), \tag{13.43}$$

where the subscripts $f$ and $b$ stand for forward- and backward-travelling waves and $\mathbf{E}_f$, $\mathbf{E}_b$, $\mathbf{B}_f$ and $\mathbf{B}_b$ are arbitrary functions of their arguments. When combined with (13.40), we find that the **E** and **B** fields in this transverse wave are related by

$$\hat{\mathbf{e}}_z \times \mathbf{E}_f = c\mathbf{B}_f, \quad \hat{\mathbf{e}}_z \times \mathbf{E}_b = -c\mathbf{B}_b, \tag{13.44}$$

so that the electric and magnetic field lines are perpendicular to each other, as well as to the direction of propagation of the wave.

Finally, if we retain the current density **J** in the Ampère–Maxwell equation before taking its curl, then (13.37) takes the form of an inhomogeneous wave equation

$$\frac{\partial^2 \mathbf{B}}{\partial t^2} = c^2\nabla^2\mathbf{B} + \frac{1}{\varepsilon_0}\nabla\times\mathbf{J}. \tag{13.45}$$

Evidently the term $\nabla\times\mathbf{J}$ acts a source of electromagnetic waves. It is informative to rewrite (13.45) as

$$\nabla^2\mathbf{B} - \frac{1}{c^2}\frac{\partial^2 \mathbf{B}}{\partial t^2} = -\mu_0\nabla\times\mathbf{J}, \tag{13.46}$$

and compare it with the curl of Ampère's law, which excludes displacement currents:

$$\nabla^2\mathbf{B} = -\mu_0\nabla\times\mathbf{J}. \tag{13.47}$$

Clearly Maxwell's displacement current has transformed a static Poisson equation into an inhomogeneous wave equation for electromagnetic waves. This is what is so exciting about Maxwell's displacement current, and one of the reasons why many see Maxwell's equations as one of the most important scientific advances of the nineteenth century. However, it was to take over 20 years after Maxwell first suggested the existence of such waves, in 1865, before the world would be convinced by the experiments of Hertz in 1888.

## 13.5 Energy Relationships and the Poynting Flux

Let us now return to the topic of electromagnetic energy, first introduced in Chapter 3. There are two points we wish to make. First, when working with static fields, we found convenient expressions for the energy of both an electric and a magnetic field, namely (13.5) and (13.13). However, we arrived at these expressions using ideas that are valid only for static fields, and we need to know if they generalise to time-dependent electromagnetic fields. Second, (13.5) suggests that the energy density of an electric field is $\varepsilon_0\mathbf{E}^2/2$, while (13.13) suggests that the energy density of a magnetic field is $\mathbf{B}^2/2\mu_0$. The implication is that, *at each point in space*, we can attribute to an electromagnetic field an *energy density* of

$$\text{energy density} = \frac{\mathbf{B}^2}{2\mu_0} + \frac{\varepsilon_0\mathbf{E}^2}{2}. \tag{13.48}$$

If we are to insist that we can attribute a definite energy density to an electromagnetic field at each point in space, then conservation of energy takes on two distinct aspects. On the one hand, there are global statements, such as any net loss of electromagnetic energy must reappear as, say, mechanical energy or heat. On the other hand, even when the global electromagnetic energy is conserved, its spatial distribution may change with time. In such a case we must invoke the idea of a *flux* (or flow) of electromagnetic energy from one point to another and find an expression for this energy flux. This is a concept that exercised both John Henry Poynting and Oliver Heaviside, who independently came to the same conclusion, as described below.

Let us start with the conversion of electromagnetic energy into other forms, which takes us back to §11.2.1. A charged particle moving with velocity $\mathbf{u}_q$ and carrying a charge $q$ experiences an electromagnetic force $\mathbf{f} = q(\mathbf{E} + \mathbf{u}_q \times \mathbf{B})$, and the rate of working of this force on the charges within a unit volume of conducting material is

$$\sum_q \mathbf{f} \cdot \mathbf{u}_q = \mathbf{E} \cdot \sum_q q\mathbf{u}_q = \mathbf{E} \cdot \mathbf{J}. \tag{13.49}$$

So, as noted in §11.2.1, the rate of working of the electromagnetic field on a conductor is $\mathbf{J} \cdot \mathbf{E}$ per unit volume. In a stationary conductor the energy transferred to the free charges is lost by the charge carriers through inelastic collisions and ends up as heat in the conductor. Ohm's law in a stationary conductor, $\mathbf{J} = \sigma\mathbf{E}$, then gives $\mathbf{J}^2/\sigma$ as the rate of generation of heat per unit volume, that is, the Ohmic dissipation. In a moving conductor, however, there is the possibility that some of the electrical power ends up as bulk kinetic energy in the conductor. This is apparent from the generalised version of Ohm's law, $\mathbf{J} = \sigma(\mathbf{E} + \mathbf{u} \times \mathbf{B})$, which yields

$$\mathbf{J} \cdot \mathbf{E} = \frac{\mathbf{J}^2}{\sigma} - \mathbf{J} \cdot (\mathbf{u} \times \mathbf{B}) = \frac{\mathbf{J}^2}{\sigma} + \mathbf{u} \cdot (\mathbf{J} \times \mathbf{B}). \tag{13.50}$$

The last term on the right is the rate of working of the Lorenz force per unit volume, $\mathbf{J} \times \mathbf{B}$, acting on the conductor. This term may be positive or negative, the former representing the conversion of electromagnetic energy into mechanical energy, and the latter the conversion of mechanical energy into electrical energy.

Let us now try to develop an energy equation for an electromagnetic field. First we take the dot product of $\mathbf{B}$ with Faraday's law, and the dot product of $\mathbf{E}$ with the Ampère–Maxwell equation. Adding the two expressions yields

$$\frac{\partial}{\partial t}\left(\frac{\mathbf{B}^2}{2\mu_0} + \frac{\varepsilon_0 \mathbf{E}^2}{2}\right) = -\mathbf{J} \cdot \mathbf{E} - \frac{1}{\mu_0}(\mathbf{B} \cdot \nabla \times \mathbf{E} - \mathbf{E} \cdot \nabla \times \mathbf{B}), \tag{13.51}$$

from which

$$\boxed{\frac{\partial}{\partial t}\left(\frac{\mathbf{B}^2}{2\mu_0} + \frac{\varepsilon_0 \mathbf{E}^2}{2}\right) = -\mathbf{J} \cdot \mathbf{E} - \frac{1}{\mu_0}\nabla \cdot (\mathbf{E} \times \mathbf{B}).} \tag{13.52}$$

This is our energy equation. We now use (13.48) to define the electromagnetic energy density, consistent with what we know about static fields. Then the left-hand side of (13.52) is the *local* rate of change of electromagnetic energy and the first term on the right is the *local* rate of loss of electromagnetic energy to heat and to mechanical energy. This suggests that the second term on the right might represent the flux of electromagnetic energy from place to place.

That this is indeed the case is most easily seen by integrating (13.52) over some control volume $V$. This gives

$$\boxed{\frac{d}{dt}\int_V \left(\frac{\mathbf{B}^2}{2\mu_0} + \frac{\varepsilon_0 \mathbf{E}^2}{2}\right) dV = -\int_V \mathbf{J} \cdot \mathbf{E}\, dV - \mu_0^{-1} \oint_S (\mathbf{E} \times \mathbf{B}) \cdot d\mathbf{S},} \tag{13.53}$$

where $S$ is the surface that encloses the volume $V$. In a vacuum this simplifies to

$$\frac{d}{dt}\int_V \left(\frac{\mathbf{B}^2}{2\mu_0}+\frac{\varepsilon_0\mathbf{E}^2}{2}\right)dV = -\oint_S \frac{\mathbf{E}\times\mathbf{B}}{\mu_0}\cdot d\mathbf{S}. \tag{13.54}$$

Conservation of energy tells us that the integral on the right must represent the rate at which electromagnetic energy flows out of $V$ through the surface $S$. The vector $\mathbf{E}\times\mathbf{B}/\mu_0$ is called the *Poynting vector*, and (13.54) suggests that this vector represents the flux of electromagnetic energy per unit area leaving through the surface $S$. Actually, as noted in §11.2.5, *any* vector field of the form $\mathbf{E}\times\mathbf{B}/\mu_0+\nabla\times\mathbf{G}$ can also lay claim to being an electromagnetic energy flux, in the sense that it balances the books in both (13.52) and (13.54). However, this freedom in the choice of energy flux is somewhat constrained by relativistic considerations and usually ignored (although see the example in §11.2.5).

In a conductor, by contrast, the generalised form of Ohm's law, $\mathbf{J}=\sigma(\mathbf{E}+\mathbf{u}\times\mathbf{B})$, gives us

$$\boxed{\frac{d}{dt}\int_V \left(\frac{\mathbf{B}^2}{2\mu_0}+\frac{\varepsilon_0\mathbf{E}^2}{2}\right)dV = -\frac{1}{\sigma}\int_V \mathbf{J}^2 dV - \int_V \mathbf{u}\cdot(\mathbf{J}\times\mathbf{B})\,dV - \oint_S \frac{\mathbf{E}\times\mathbf{B}}{\mu_0}\cdot d\mathbf{S}.} \tag{13.55}$$

In (13.55) there are two additional integrals on the right. The first is the irreversible Ohmic dissipation, while the second is the rate of working of the Lorentz force on the conductor, which may be positive or negative. As noted above, this represents an exchange of electromagnetic and mechanical energy.

## 13.6 The Lorenz Gauge and Wave Equations for the Potentials

When we introduced the 'uncurled' version of Faraday's law,

$$\mathbf{E} = -\frac{\partial\mathbf{A}}{\partial t} - \nabla\varphi, \tag{13.56}$$

we noted that we are at liberty to choose the divergence (gauge) of $\mathbf{A}$ in more or less any way we see fit; it is largely a matter of convenience. In magnetostatics, where Ampère's law demands that $\mathbf{J}$ be solenoidal, the simplest choice, $\nabla\cdot\mathbf{A}=0$, is universally adopted. This is the Coulomb gauge. However, when displacement currents are retained and $\mathbf{J}$ is no longer solenoidal, it is often prudent to choose an alternative definition of $\nabla\cdot\mathbf{A}$. We shall now discuss how to choose the optimal gauge for the vector potential when considering time-dependent phenomena, such as waves.

Let us start by using (13.56) to substitute for **E** in the Ampère–Maxwell equation,

$$\nabla \times \nabla \times \mathbf{A} = \mu_0 \mathbf{J} - \frac{1}{c^2}\frac{\partial}{\partial t}\left(\frac{\partial \mathbf{A}}{\partial t} + \nabla\varphi\right). \tag{13.57}$$

The identity $\nabla^2\mathbf{A} = \nabla(\nabla\cdot\mathbf{A}) - \nabla \times \nabla \times \mathbf{A}$ then yields the inelegant equation

$$\nabla^2\mathbf{A} - \nabla(\nabla\cdot\mathbf{A}) - \frac{1}{c^2}\frac{\partial^2\mathbf{A}}{\partial t^2} = -\mu_0\mathbf{J} + \frac{1}{c^2}\frac{\partial}{\partial t}\nabla\varphi. \tag{13.58}$$

Note that this involves both the vector potential **A** and the scalar potential $\varphi$. Moreover, the divergence of (13.56) demands that

$$\nabla^2\varphi + \frac{\partial}{\partial t}\nabla\cdot\mathbf{A} = -\frac{\rho_e}{\varepsilon_0}, \tag{13.59}$$

which gives us a second equation relating the scalar and vector potentials. Evidently the two potentials are coupled. We may, however, tidy things up considerably if we adopt the so-called *Lorenz gauge*,

$$\boxed{\nabla\cdot\mathbf{A} = -\frac{1}{c^2}\frac{\partial\varphi}{\partial t}.} \tag{13.60}$$

Our two equations for the vector and scalar potentials then decouple to give

$$\boxed{\nabla^2\mathbf{A} - \frac{1}{c^2}\frac{\partial^2\mathbf{A}}{\partial t^2} = -\mu_0\mathbf{J},} \tag{13.61}$$

$$\boxed{\nabla^2\varphi - \frac{1}{c^2}\frac{\partial^2\varphi}{\partial t^2} = -\frac{\rho_e}{\varepsilon_0}.} \tag{13.62}$$

The vector and scalar potentials are now governed by simple non-dispersive wave equations, with $\mu_0\mathbf{J}$ and $\rho_e/\varepsilon_0$ acting as source terms. Note that these source terms are coupled through charge conservation, rewritten in the form

$$\nabla\cdot(\mu_0\mathbf{J}) = -\frac{1}{c^2}\frac{\partial(\rho_e/\varepsilon_0)}{\partial t}.$$

This mirrors precisely the relationship between the two potentials demanded by the Lorenz gauge.

Within this framework it becomes clear that information propagates away from the sources $\mu_0\mathbf{J}$ and $\rho_e/\varepsilon_0$ at the speed of light, carried off by electromagnetic waves. This is so physically compelling that it seems somehow inevitable that we must adopt the Lorenz gauge for non-static applications. We might compare (13.61) and (13.62) with the equivalent quasi-static results, (13.2) and (13.7), in which we used the Coulomb gauge $\nabla \cdot \mathbf{A} = 0$. These are the Poisson equations,

$$\nabla^2 \mathbf{A} = -\mu_0 \mathbf{J}, \tag{13.63}$$

$$\nabla^2 \varphi_s = -\frac{\rho_e}{\varepsilon_0}. \tag{13.64}$$

Here there is no time delay between fluctuations in the sources and the corresponding changes in the potentials, no matter how remote the sources may be. This is consistent with the notion that quasi-static fields correspond to a system in which information travels infinitely fast. Indeed, for $c \to \infty$ the Lorenz gauge reverts to the Coulomb gauge, while the wave equations (13.61) and (13.62) simplify to the static equations (13.63) and (13.64).

Perhaps two footnotes are in order here. First, it is clear from the definitions $\mathbf{B} = \nabla \times \mathbf{A}$ for $\mathbf{A}$ and

$$\mathbf{E} = -\frac{\partial \mathbf{A}}{\partial t} - \nabla \varphi$$

for $\varphi$ that the fields $\mathbf{B}$ and $\mathbf{E}$ remain unchanged under the transformation

$$\mathbf{A}' = \mathbf{A} + \nabla \Lambda, \qquad \varphi' = \varphi - \frac{\partial \Lambda}{\partial t}. \tag{13.65}$$

This is known as a *gauge transformation*. This ambiguity in the value of $\mathbf{A}$ is equivalent to the freedom in the choice of $\nabla \cdot \mathbf{A}$ that we exploited in writing down (13.60). Second, some authors prefer to stay with the Coulomb gauge, $\nabla \cdot \mathbf{A} = 0$, even in time-dependent situations. Expression (13.59) then tells us that the scalar potential is governed by the static Poisson equation

$$\nabla^2 \varphi = -\frac{\rho_e}{\varepsilon_0}, \tag{13.66}$$

while $\mathbf{A}$ obeys

$$\nabla^2 \mathbf{A} - \frac{1}{c^2}\frac{\partial^2 \mathbf{A}}{\partial t^2} = -\mu_0 \mathbf{J} + \frac{1}{c^2}\frac{\partial}{\partial t}\nabla \varphi. \tag{13.67}$$

The instantaneous adjustment of $\varphi$ throughout all space triggered by changes in $\rho_e$ seems at odds with causality, but supporters of this scheme note that it is necessary only that the observable fields $\mathbf{E}$ and $\mathbf{B}$ obey causality and so this need not be a problem. However, we shall stay with the Lorenz gauge in this text since it places the two potentials on an equal footing and because both potentials then behave in a simple, intuitive way, as we now discuss.

## 13.7 Inverting the Wave Equations: Retarded Potentials

In Chapter 2 we saw how to invert a steady Poisson equation using the Green's function solution. In the case of the two static potentials the inversion integrals are (13.3) and (13.8):

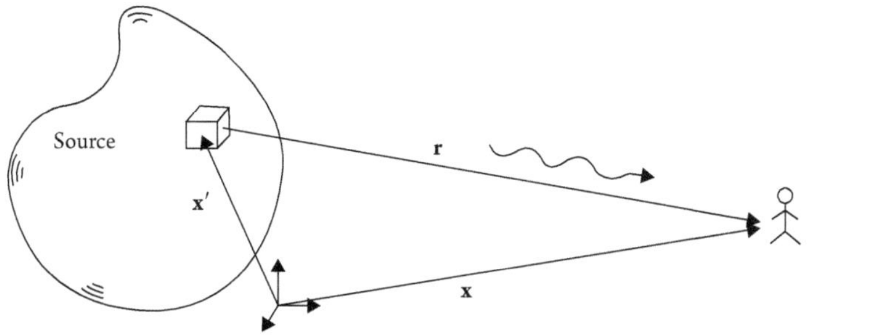


**Figure 13.3** The displacement vector $\mathbf{r} = \mathbf{x} - \mathbf{x}'$ takes us from the source at $\mathbf{x}'$ to the observer at $\mathbf{x}$.

$$\varphi_s(\mathbf{x},t) = \frac{1}{4\pi\varepsilon_0}\int \frac{\rho_e(\mathbf{x}',t)}{|\mathbf{r}|}d\mathbf{x}', \tag{13.68}$$

$$\mathbf{A}(\mathbf{x},t) = \frac{\mu_0}{4\pi}\int \frac{\mathbf{J}(\mathbf{x}',t)}{|\mathbf{r}|}d\mathbf{x}', \tag{13.69}$$

the second of which is equivalent to the Biot–Savart law. Note that, as always, we use $\mathbf{x}'$ to locate the source $\rho_e$ or $\mathbf{J}$, and $\mathbf{x}$ is the point of measurement of the field $\phi_s$ or $\mathbf{A}$. The displacement vector $\mathbf{r} = \mathbf{x} - \mathbf{x}'$ then takes us from the source to the observer, as shown in Figure 13.3. Note also that, by including time in these expressions, we explicitly affirm that the fields $\phi_s$ and $\mathbf{A}$ are evaluated at the same instant as their sources $\rho_e$ and $\mathbf{J}$.

We also learned how to invert the inhomogeneous wave equation in Chapter 2, albeit in the simpler context of acoustics. So we may hijack the inversion integral (2.83) for acoustic waves given in §2.5.1 and apply it directly to (13.61) and (13.62). The results are

$$\boxed{\varphi(\mathbf{x},t) = \frac{1}{4\pi\varepsilon_0}\int \frac{\rho_e(\mathbf{x}',t')}{|\mathbf{r}|}d\mathbf{x}',} \tag{13.70}$$

$$\boxed{\mathbf{A}(\mathbf{x},t) = \frac{\mu_0}{4\pi}\int \frac{\mathbf{J}(\mathbf{x}',t')}{|\mathbf{r}|}d\mathbf{x}',} \tag{13.71}$$

where $t' = t - r/c$ is the retarded time and the corresponding potentials are known as the *retarded potentials*. This simple, elegant generalisation of (13.68) and (13.69) makes it clear that information can be transmitted no faster than it can be carried away in the form of electromagnetic waves. That is to say, a fluctuation in $\rho_e$ or $\mathbf{J}$ at position $\mathbf{x}'$ is not felt immediately at location $\mathbf{x}$, but rather a time $|\mathbf{r}|/c$ later, which is the time of flight of waves travelling at the speed of light. These two integrals, which we first introduced in Chapter 3, will provide the starting point for much of our subsequent discussion.

## 13.8 From Retarded Potentials to Retarded Fields

The simplicity and beauty of (13.70) and (13.71) is striking. However, when we come to calculate the corresponding observable fields $\mathbf{E}$ and $\mathbf{B}$, life gets a little more complicated, as we shall now show. The reason for this complication is that to get **E** and **B** from $\phi$ and **A**, we need to take derivatives with respect to **x**. However, **x** appears not just in $\mathbf{r} = \mathbf{x} - \mathbf{x}'$, but also in $t' = t - r/c$. So the process of differentiation rapidly becomes somewhat messy and this is reflected in the slightly inelegant expressions for **E** and **B** that emerge. However, the end result is rather interesting, since it shows that time-dependent current distributions throw out an electromagnetic field that falls off as $|\mathbf{r}|^{-1}$ at large distances, which is a much slower decline than a static electric field, where $|\mathbf{E}|_\infty \sim |\mathbf{r}|^{-2}$, or a static magnetic field, for which $|\mathbf{B}|_\infty \sim |\mathbf{r}|^{-3}$. So, at large distances from a source, it is the time-dependent nature of the sources that really matters, since this dominates the far field.

We shall now find the general form of **E** and **B** corresponding to a given localised distribution of $\rho_e$ and **J**. Let us start with a couple of algebraic details. If $\mathbf{x}'$ is held constant while **x** is varied, then

$$\frac{\partial t'}{\partial x_i} = -\frac{1}{c}\frac{\partial |\mathbf{r}|}{\partial x_i} = -\frac{1}{c}\frac{\mathbf{r}}{|\mathbf{r}|}, \tag{13.72}$$

from which we obtain

$$\nabla \rho_e(\mathbf{x}', t') = -\frac{1}{c}\frac{\partial \rho_e(\mathbf{x}', t')}{\partial t'}\frac{\mathbf{r}}{|\mathbf{r}|}, \tag{13.73}$$

and

$$\nabla \times \mathbf{J}(\mathbf{x}', t') = -\frac{1}{c}\frac{\mathbf{r}}{|\mathbf{r}|} \times \frac{\partial \mathbf{J}(\mathbf{x}', t')}{\partial t'}, \tag{13.74}$$

where the operator $\nabla$ acts on **x** but not on $\mathbf{x}'$. We can now obtain **B** from (13.71) as follows. First we differentiate (13.71) with respect to **x** while keeping $\mathbf{x}'$ constant:

$$\mathbf{B}(\mathbf{x}, t) = \nabla \times \mathbf{A}(\mathbf{x}, t) = \frac{\mu_0}{4\pi}\int \left[\nabla\left(\frac{1}{|\mathbf{r}|}\right) \times \mathbf{J}(\mathbf{x}', t') + \left(\frac{1}{|\mathbf{r}|}\right)\nabla \times \mathbf{J}(\mathbf{x}', t')\right] d\mathbf{x}'.$$

Next we evaluate $\nabla(1/|\mathbf{r}|)$ from (13.9) and $\nabla \times \mathbf{J}(\mathbf{x}', t')$ using (13.74). This yields

$$\boxed{\mathbf{B}(\mathbf{x}, t) = \frac{\mu_0}{4\pi}\int \left\{\left[\mathbf{J}(\mathbf{x}', t') + \frac{|\mathbf{r}|}{c}\frac{\partial \mathbf{J}(\mathbf{x}', t')}{\partial t'}\right] \times \frac{\mathbf{r}}{|\mathbf{r}|^3}\right\} d\mathbf{x}'.} \tag{13.75}$$

Note that we have acquired an extra term, over and above that expected from the Biot–Savart law, because $\nabla \times \mathbf{J}' = 0$ in magnetostatics, yet $\nabla \times \mathbf{J}(\mathbf{x}', t') \neq 0$ in (13.75). Note also that the time-dependent term falls off with distance from the source as $B \sim r^{-1}$, and that this dominates the far field.

Similarly, combining (13.56) with (13.70) and (13.71), we have

$$\mathbf{E}(\mathbf{x}, t) = -\frac{\partial \mathbf{A}}{\partial t} - \nabla\varphi = -\frac{\mu_0}{4\pi}\int \frac{1}{|\mathbf{r}|}\frac{\partial \mathbf{J}(\mathbf{x}', t')}{\partial t} d\mathbf{x}' - \frac{1}{4\pi\varepsilon_0}\int \nabla\left(\frac{\rho_e(\mathbf{x}', t')}{|\mathbf{r}|}\right) d\mathbf{x}'.$$

Both $\mathbf{x}$ and $\mathbf{x}'$ are held constant when differentiating with respect to $t$ in the first integral, and so we may replace $\partial t$ by $\partial t'$ on the right. Moreover, we can use (13.73) to rewrite the integrand in the second integral as

$$\nabla\left(\frac{\rho_e(\mathbf{x}',t')}{|\mathbf{r}|}\right) = \rho_e\left(\mathbf{x}',t'\right)\nabla\left(\frac{1}{|\mathbf{r}|}\right) - \frac{1}{c}\frac{\partial\rho_e(\mathbf{x}',t')}{\partial t'}\frac{\mathbf{r}}{|\mathbf{r}|^2}.$$

Our expression for $\mathbf{E}$ now becomes

$$\mathbf{E}(\mathbf{x},t) = -\frac{\mu_0}{4\pi}\int\frac{1}{|\mathbf{r}|}\frac{\partial\mathbf{J}(\mathbf{x}',t')}{\partial t'}d\mathbf{x}' - \frac{1}{4\pi\varepsilon_0}\int\left[\nabla\left(\frac{1}{|\mathbf{r}|}\right)\rho_e\left(\mathbf{x}',t'\right) - \frac{1}{c}\frac{\partial\rho_e(\mathbf{x}',t')}{\partial t'}\frac{\mathbf{r}}{|\mathbf{r}|^2}\right]d\mathbf{x}'.$$

The final step is to combine the two integrals, while noting that $c = 1/\sqrt{\mu_0\varepsilon_0}$, and use (13.9) to evaluate $\nabla(1/|\mathbf{r}|)$. This yields

$$\boxed{\mathbf{E}(\mathbf{x},t) = \frac{1}{4\pi\varepsilon_0}\int\left\{\left[\rho_e\left(\mathbf{x}',t'\right) + \frac{|\mathbf{r}|}{c}\frac{\partial\rho_e(\mathbf{x}',t')}{\partial t'}\right]\frac{\mathbf{r}}{|\mathbf{r}|^3} - \frac{1}{c^2}\frac{1}{|\mathbf{r}|}\frac{\partial\mathbf{J}(\mathbf{x}',t')}{\partial t'}\right\}d\mathbf{x}'.} \qquad (13.76)$$

As with $\mathbf{B}$, the time-dependent terms dominate the far field and fall off with distance from the source as $|\mathbf{E}|_\infty \sim |\mathbf{r}|^{-1}$.

Comparing (13.75) and (13.76) with (13.70) and (13.71), we have lost all the elegant simplicity of the retarded potentials. Still, the end result is useful. These expressions for $\mathbf{E}$ and $\mathbf{B}$ have been given various names by different authors, depending on whom they think should get the credit for them. For example, several authors refer to them as the Jefimenko equations, although rather similar equations predate Jefimenko's publication of them in 1966. Note that the appearance of retarded time in these expressions is a direct consequence of the wave equations (13.37) and (13.38) for $\mathbf{B}$ and $\mathbf{E}$. Note also that, although we have derived these expressions from the retarded potentials, which in turn rely on the Lorenz gauge, they are in fact independent of the gauge used to define the potentials, since they refer only to $\mathbf{E}$ and $\mathbf{B}$.

Perhaps the most striking feature of these equations is that the far-field expressions $|\mathbf{B}|_\infty \sim |\mathbf{r}|^{-1}$ and $|\mathbf{E}|_\infty \sim |\mathbf{r}|^{-1}$ mean that the Poynting vector falls off as $\mathbf{E}_\infty \times \mathbf{B}_\infty/\mu_0 \sim |\mathbf{r}|^{-2}$, provided that the current density is time-dependent. This, in turn, offers the possibility that the flux of electromagnetic energy through a spherical surface that encloses the current might remain finite and of order one as the radius of the sphere tends to infinity. So, for example, an oscillating current source can radiate a finite amount of energy out to infinity.

## 13.9 A Summary of the Equations of Electrodynamics

We have covered a lot of ground in this chapter, so let us now summarise the key results. Following Feynman, Leighton & Sands (1964), we shall find it convenient to tabulate the fundamental equations in a way that emphasises the difference between the static results and the more general dynamic equations.

**Table 13.1** Comparison of the static and dynamic equations for **E** and **B**. The displacement vector $\mathbf{r}=\mathbf{x}-\mathbf{x}'$ takes us from a source at $\mathbf{x}'$ to the point of measurement of the fields at $\mathbf{x}$, and $t'=t-r/c$ is the retarded time.

| Static equations | True in general | |
|---|---|---|
| $\nabla\cdot\mathbf{E}=\frac{\rho_e}{\varepsilon_0}$ | $\boxed{\nabla\cdot\mathbf{E}=\frac{\rho_e}{\varepsilon_0}}$ | (Gauss' law) |
| $\nabla\times\mathbf{E}=0$ | $\boxed{\nabla\times\mathbf{E}=-\frac{\partial\mathbf{B}}{\partial t}}$ | (Faraday's law) |
| $\mathbf{E}=-\nabla\varphi$ | $\mathbf{E}=-\frac{\partial\mathbf{A}}{\partial t}-\nabla\varphi$ | |
| $\mathbf{E}(\mathbf{x},t)=\frac{1}{4\pi\varepsilon_0}\int\left[\frac{\rho_e(\mathbf{x}',t)\mathbf{r}}{\lvert\mathbf{r}\rvert^3}\right]d\mathbf{x}'$ | $\mathbf{E}(\mathbf{x},t)=\frac{1}{4\pi\varepsilon_0}\int\left\{\left[\rho_e(\mathbf{x}',t')+\frac{\lvert\mathbf{r}\rvert}{c}\frac{\partial\rho_e(\mathbf{x}',t')}{\partial t'}\right]\frac{\mathbf{r}}{\lvert\mathbf{r}\rvert^3}-\frac{1}{c^2\lvert\mathbf{r}\rvert}\frac{\partial\mathbf{J}(\mathbf{x}',t')}{\partial t'}\right\}d\mathbf{x}'$ | |
| $\nabla\cdot\mathbf{B}=0$ | $\boxed{\nabla\cdot\mathbf{B}=0}$ | (absence of magnetic monopoles) |
| $\mathbf{B}=\nabla\times\mathbf{A}$ | $\mathbf{B}=\nabla\times\mathbf{A}$ | |
| $\nabla\times\mathbf{B}=\mu_0\mathbf{J}$ | $\boxed{\nabla\times\mathbf{B}=\mu_0\left(\mathbf{J}+\varepsilon_0\frac{\partial\mathbf{E}}{\partial t}\right)}$ | (Ampère–Maxwell equation) |
| $\nabla\cdot\mathbf{J}=0$ | $\nabla\cdot\mathbf{J}=-\frac{\partial\rho_e}{\partial t}$ | |
| $\mathbf{B}(\mathbf{x},t)=\frac{\mu_0}{4\pi}\int\frac{\mathbf{J}(\mathbf{x}',t)\times\mathbf{r}}{\lvert\mathbf{r}\rvert^3}d\mathbf{x}'$ | $\mathbf{B}(\mathbf{x},t)=\frac{\mu_0}{4\pi}\int\left\{\left[\mathbf{J}(\mathbf{x}',t')+\frac{\lvert\mathbf{r}\rvert}{c}\frac{\partial\mathbf{J}(\mathbf{x}',t')}{\partial t'}\right]\times\frac{\mathbf{r}}{\lvert\mathbf{r}\rvert^3}\right\}d\mathbf{x}'$ | |
| $U=\int_V\left(\frac{\mathbf{B}^2}{2\mu_0}+\frac{\varepsilon_0\mathbf{E}^2}{2}\right)dV$ | $U=\int_V\left(\frac{\mathbf{B}^2}{2\mu_0}+\frac{\varepsilon_0\mathbf{E}^2}{2}\right)dV$ | |

**Table 13.2** Comparison of the static and dynamic equations for the potentials $\phi$ and **A**. The vector $\mathbf{r}=\mathbf{x}-\mathbf{x}'$ takes us from a source to the fields, and $t'$ is the retarded time.

| Static equations | True in general |
|---|---|
| $\nabla^2\varphi=-\frac{\rho_e}{\varepsilon_0}$ | $\nabla^2\varphi-\frac{1}{c^2}\frac{\partial^2\phi}{\partial t^2}=-\frac{\rho_e}{\varepsilon_0}$ |
| $\varphi(\mathbf{x},t)=\frac{1}{4\pi\varepsilon_0}\int\frac{\rho_e(\mathbf{x}',t)}{\lvert\mathbf{r}\rvert}d\mathbf{x}'$ | $\varphi(\mathbf{x},t)=\frac{1}{4\pi\varepsilon_0}\int\frac{\rho_e(\mathbf{x}',t')}{\lvert\mathbf{r}\rvert}d\mathbf{x}'$ |
| $\nabla\cdot\mathbf{A}=0$ | $\nabla\cdot\mathbf{A}=-\frac{1}{c^2}\frac{\partial\varphi}{\partial t}$ |
| $\nabla^2\mathbf{A}=-\mu_0\mathbf{J}$ | $\nabla^2\mathbf{A}-\frac{1}{c^2}\frac{\partial^2\mathbf{A}}{\partial t^2}=-\mu_0\mathbf{J}$ |
| $\mathbf{A}(\mathbf{x},t)=\frac{\mu_0}{4\pi}\int\frac{\mathbf{J}(\mathbf{x}',t)}{\lvert\mathbf{r}\rvert}d\mathbf{x}'$ | $\mathbf{A}(\mathbf{x},t)=\frac{\mu_0}{4\pi}\int\frac{\mathbf{J}(\mathbf{x}',t')}{\lvert\mathbf{r}\rvert}d\mathbf{x}'$ |

Let us start with the equations governing the fields **E** and **B**. Table 13.1 shows the static equations on the left and the more general dynamic equations on the right. As in the main text, $\mathbf{r}=\mathbf{x}-\mathbf{x}'$ takes us from a source, $\mu_0\mathbf{J}$ or $\rho_e/\varepsilon_0$, to the point of measurement of the fields (see Figure 13.3), and $t'=t-r/c$ is the retarded time. The expression for electromagnetic energy in the table is valid for both static and dynamic fields, though the equivalent expressions written in terms of potentials, (13.4) and (13.11), are valid *only* for static fields. Maxwell's equations are boxed on the right for emphasis.

Table 13.2 shows the same thing for the scalar and vector potentials $\varphi$ and **A**. We have used the Coulomb gauge on the left and the Lorenz gauge on the right, consistent with the development in this chapter.

## 13.10 The Electromagnetic Stress Tensor and Field Momentum

We close this chapter with a discussion of the Maxwell stress tensor and momentum conservation. In §5.5 we noted that we could write the Coulomb force in an electrostatic field as

$$\rho_e E_i = \frac{\partial \tau_{ij}^E}{\partial x_j}, \qquad \tau_{ij}^E = \varepsilon_0 \left( E_i E_j - \frac{1}{2}\mathbf{E}^2 \delta_{ij} \right), \tag{13.77}$$

where $\tau_{ij}^E$ is the *electric stress tensor*. However, the derivation of (13.77) relies on the fact that $\nabla \times \mathbf{E} = 0$ in a static field, which is not the case for an unsteady field. On the other hand, in §8.6 we noted that the $i$th component of the Lorentz force in magnetostatics can be written as

$$(\mathbf{J} \times \mathbf{B})_i = \frac{\partial \tau_{ij}^M}{\partial x_j}, \qquad \tau_{ij}^M = \frac{B_i B_j}{\mu_0} - \frac{\mathbf{B}^2}{2\mu_0}\delta_{ij}, \tag{13.78}$$

where $\tau_{ij}^M$ is the *magnetic stress tensor*. However, the derivation of (13.78) hinges on the absence of displacement currents in quasi-static magnetic fields.

It is natural to ask what these stress tensors look like for the most general case. Our starting point is to use Gauss' law and the Ampère–Maxwell equation to rewrite the net electromagnetic force as

$$\rho_e \mathbf{E} + \mathbf{J} \times \mathbf{B} = \varepsilon_0 (\nabla \cdot \mathbf{E})\mathbf{E} + \frac{1}{\mu_0}(\nabla \times \mathbf{B}) \times \mathbf{B} - \varepsilon_0 \frac{\partial \mathbf{E}}{\partial t} \times \mathbf{B}. \tag{13.79}$$

This can be rearranged using Faraday's law to give

$$\rho_e \mathbf{E} + \mathbf{J} \times \mathbf{B} = \varepsilon_0 [(\nabla \cdot \mathbf{E})\mathbf{E} + (\nabla \times \mathbf{E}) \times \mathbf{E}] + \frac{1}{\mu_0}(\nabla \times \mathbf{B}) \times \mathbf{B} - \varepsilon_0 \frac{\partial}{\partial t}(\mathbf{E} \times \mathbf{B}).$$

However, for any vector field **C**, we have the identity

$$[(\nabla \cdot \mathbf{C})\mathbf{C} + (\nabla \times \mathbf{C}) \times \mathbf{C}]_i = \frac{\partial}{\partial x_j}(C_i C_j) - \frac{\partial}{\partial x_i}(\mathbf{C}^2/2),$$

and so we conclude that

$$\boxed{\rho_e(\mathbf{E})_i + (\mathbf{J} \times \mathbf{B})_i = \frac{\partial}{\partial x_j}(\tau_{ij}^E + \tau_{ij}^M) - \varepsilon_0 \frac{\partial}{\partial t}(\mathbf{E} \times \mathbf{B})_i,} \tag{13.80}$$

where $\tau_{ij}^E$ and $\tau_{ij}^M$ are defined by (13.77) and (13.78). On the right of (13.80) we have the divergence of the *Maxwell stress tensor* $\tau_{ij}^{\text{Max}} = \tau_{ij}^E + \tau_{ij}^M$, which perhaps comes as no surprise. However, time dependence has introduced a second and unexpected term on the right, which is proportional to the time derivative of the Poynting vector.

Next, consider an electromagnetic field acting on a region of space $V$ that contains some charges in motion. Newton's second law tells us that

$$\frac{d\mathbf{p}_{\text{mech}}}{dt} = \int_V (\rho_e \mathbf{E} + \mathbf{J} \times \mathbf{B})\, dV, \tag{13.81}$$

where $\mathbf{p}_{\text{mech}}$ is the net linear momentum of the charges. We now encounter a curiosity: $\rho_e \mathbf{E} + \mathbf{J} \times \mathbf{B}$ does not, in general, obey Newton's third law in an *unsteady* field. For example, two moving electrons need not exert equal and opposite forces on each other. This gives rise to profound difficulties when applying momentum conservation to multi-body problems in electrodynamics, since the forces of interaction need not cancel. Luckily, however, this difficulty can be circumvented. Using (13.80), we rewrite (13.81) as

$$\frac{d}{dt}\left[\mathbf{p}_{\text{mech}} + \int_V (\varepsilon_0 \mathbf{E} \times \mathbf{B})\, dV\right]_i = \oint_S (\tau_{ij}^E + \tau_{ij}^M)\, dS_j. \tag{13.82}$$

Now it turns out that, just as we can associate energy with an electromagnetic field, so we can also associate momentum with that field. In particular, the linear momentum density is defined as $\varepsilon_0 \mathbf{E} \times \mathbf{B}$ (see e.g. Griffiths, 2013), and so (13.82) reads

$$\boxed{\frac{d}{dt}(\mathbf{p}_{\text{mech}} + \mathbf{p}_{\text{EM}})_i = \oint_S \tau_{ij}^{\text{Max}} dS_j.} \tag{13.83}$$

Here the left of (13.83) is the rate of change of total linear momentum (mechanical plus electromagnetic) inside the surface $S$. Note that $\mathbf{p}_{\text{mech}} + \mathbf{p}_{\text{EM}}$ is conserved as $S$ recedes to infinity, provided that the Maxwell stresses acting on $S$ fall to zero. Equation (13.83) is now interpreted as saying that the rate of change of total linear momentum inside $S$ is equal to the *momentum flux* through $S$, as represented by the Maxwell stress tensor acting on the boundary. In effect, we have chosen to associate part of the body force $\rho_e \mathbf{E} + \mathbf{J} \times \mathbf{B}$ with the field momentum inside $V$ and part with a momentum flux through $S$. The advantage of this interpretation is that adjacent regions of space that share a common boundary *exchange* linear momentum through their common surface integrals of the Maxwell stress, and so linear momentum is neither created nor destroyed through their interaction. We are now in a position to apply the law of momentum conservation, despite the failure of Newton's third law when applied to the electromagnetic body force.

Note that field momentum is not just some mathematical trick, invoked to keep the book-keeping in order; it really exists, as evidenced by radiation pressure and as shown by the following example, adapted from Feynman, Leighton & Sands (1964). Consider a

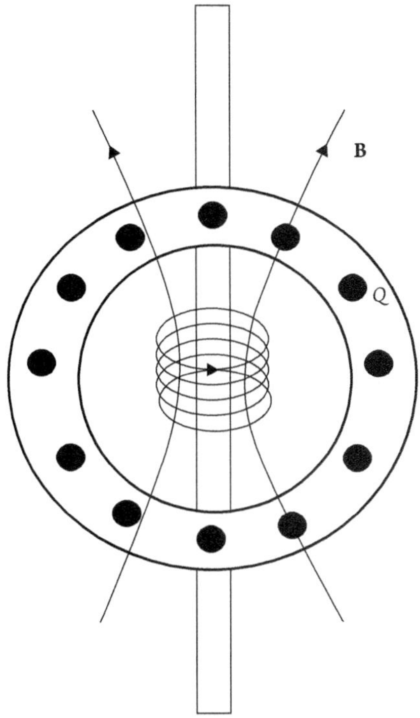


**Figure 13.4** Charges are uniformly embedded over the surface of a plastic sphere and a solenoid carrying a current is placed near its centre. (Reproduced from P. A. Davidson. *An Introduction to Magnetohydrodynamics*, 2nd ed. Cambridge University Press, 2017.)

stationary spherical plastic shell that is mounted on a frictionless spindle and is free to rotate. A roughly uniform charge distribution is embedded in the surface of the shell and gives rise to a radial electric field. Also, a solenoid is placed near the centre of the sphere, the axis of the solenoid being parallel to the rotation axis, and a current flows through the solenoid, creating an axisymmetric magnetic field (Figure 13.4). We now (somehow) arrange for the current to fall to zero. As the magnetic field declines, Faraday's law tells us that there is an induced electric field that is azimuthal, taking the form of rings that are concentric with the rotation axis. This induced electric field now acts on the charges to produce a torque on the sphere, causing it to spin up. At the end of this process we have acquired some mechanical angular momentum, and conservation of angular momentum tells us that this must have come from somewhere—but from where? In fact, it is easy to confirm that the final mechanical angular momentum was originally stored as field momentum in the static **E** and **B** fields at $t = 0$. That is, as **B** falls off, the torque acting on the plastic shell is

$$\mathbf{T} = \int \rho_e \mathbf{x} \times \mathbf{E}\, dV = -\int \rho_e \mathbf{x} \times \left(\frac{\partial \mathbf{A}}{\partial t}\right) dV = -\frac{d}{dt}\int \rho_e \mathbf{x} \times \mathbf{A}\, dV,$$

and so the final mechanical angular momentum is

$$\mathbf{L}_{\text{mech}} = \int\limits_{t=0} \rho_e \mathbf{x} \times \mathbf{A}\, dV = \varepsilon_0 \int\limits_{t=0} (\nabla \cdot \mathbf{E}) s A_\theta \, dV \, \hat{\mathbf{e}}_z = -\varepsilon_0 \int\limits_{t=0} \mathbf{E} \cdot \nabla(s A_\theta) dV \, \hat{\mathbf{e}}_z, \quad (13.84)$$

in $(s, \theta, z)$ coordinates. This is clearly equal to the initial field momentum, as given by

$$\mathbf{L}_{\text{EM}} = \varepsilon_0 \int \mathbf{x} \times (\mathbf{E} \times \mathbf{B})\, dV = \varepsilon_0 \int s(E_z B_s - E_s B_z)\, dV \, \hat{\mathbf{e}}_z, \quad (13.85)$$

and so there is a direct conversion of field momentum into mechanical momentum.

* * *

This concludes our discussion of Maxwell's equations. We shall devote the remainder of this textbook to unpicking the physical content of these equations, examining a range of applications, and exploring the relationship between Maxwell's equations and the special theory of relativity. Readers seeking more details of Maxwell's equations, and their solution through the use of retarded potentials, will find excellent accounts in Feynman, Leighton & Sands (1964) and Griffiths (2013).

## EXERCISES

13.1 A wire of infinite length and zero radius lies on the $z$ axis and carries no current for $t < 0$ and a steady current of $I$ for $t > 0$. Show that, for $s < ct$, $\mathbf{A}$ in cylindrical polar coordinates is given by

$$\mathbf{A}(s,t) = \frac{\mu_0 I}{4\pi} \hat{\mathbf{e}}_z \int_{-\hat{z}}^{\hat{z}} \frac{dz}{\sqrt{z^2 + s^2}}, \quad \hat{z}^2 + s^2 = (ct)^2.$$

Now confirm that, for $s < ct$,

$$\mathbf{E}(s,t) = -\frac{\mu_0 I}{2\pi} \frac{c}{\sqrt{(ct)^2 - s^2}} \hat{\mathbf{e}}_z, \quad \mathbf{B}(s,t) = \frac{\mu_0 I}{2\pi s} \frac{ct}{\sqrt{(ct)^2 - s^2}} \hat{\mathbf{e}}_\theta,$$

while $\mathbf{E} = \mathbf{B} = 0$ for $s > ct$. Notice that, for $s < ct$, $\hat{\mathbf{e}}_r \times \mathbf{E}(s,t) = (s/t)\mathbf{B}(s,t)$, and for $s << ct$ we recover the static field $|\mathbf{B}| = \mu_0 I / 2\pi s$, while the electric field falls off with time as $|\mathbf{E}| = \mu_0 I / 2\pi t$. Plot $\mathbf{A}(s)$ and show that the $s$-derivatives of $\mathbf{A}$ are discontinuous at $s = ct$. Confirm that this leads to $\delta(s - ct)$ corrections to the expressions above for $\mathbf{E}$ and $\mathbf{B}$.

13.2 Steady currents and static charges located near the origin establish a field. Show that the flux of energy through successive spherical surfaces falls to zero at large distances from the origin. Consider the energy balance for the annular volume defined by two concentric spherical surfaces of large but finite radius and show that the flux of energy through both spherical surfaces must be zero. (No mathematics is required.)

13.3 Calculate the Poynting flux and electromagnetic energy density for the **E** and **B** fields of Exercise 13.1. Show that the energy flux through a cylinder of radius $s$ and unit height is, for $s < ct$,

$$\mu_0^{-1}\oint(\mathbf{E}\times\mathbf{B})\cdot d\mathbf{S} = \frac{\mu_0 I^2}{2\pi}\frac{c^2 t}{(ct)^2 - s^2},$$

while the electromagnetic energy density is

$$\text{energy density} = \frac{\mu_0 I^2}{8\pi^2 s^2}\frac{(ct)^2 + s^2}{(ct)^2 - s^2},$$

also for $s < ct$. Confirm that these are consistent with the energy equation (13.54).

13.4 Consider a localised distribution of current. Show that (13.75) may be rewritten as

$$\mathbf{B}(\mathbf{x},t) = \frac{\mu_0}{4\pi}\int\left[\mathbf{J}(\mathbf{x}',t') + (t-t')\frac{\partial \mathbf{J}(\mathbf{x}',t')}{\partial t'}\right]\times\frac{\mathbf{r}}{|\mathbf{r}|^3}\,d\mathbf{x}'.$$

Now suppose that the current density varies slowly with time and consider the Taylor expansion

$$\mathbf{J}(\mathbf{x}',t) = \mathbf{J}(\mathbf{x}',t') + (t-t')\frac{\partial \mathbf{J}(\mathbf{x}',t')}{\partial t'} + O((t-t')^2).$$

Hence show that

$$\mathbf{B}(\mathbf{x},t) = \frac{\mu_0}{4\pi}\int\left[\mathbf{J}(\mathbf{x}',t)\times\frac{\mathbf{r}}{|\mathbf{r}|^3}\right]d\mathbf{x}' + O((t-t')^2).$$

Evidently the Biot–Savart law is accurate to second order in $t - t' = r/c$.

13.5 Consider a localised distribution of charge and current. Show that (13.76) may be rewritten as

$$\mathbf{E}(\mathbf{x},t) = \frac{1}{4\pi\varepsilon_0}\int\left\{\left[\rho_e(\mathbf{x}',t') + (t-t')\frac{\partial\rho_e(\mathbf{x}',t')}{\partial t'}\right]\frac{\mathbf{r}}{|\mathbf{r}|^3} - (t-t')^2\frac{1}{|\mathbf{r}|^3}\frac{\partial \mathbf{J}(\mathbf{x}',t')}{\partial t'}\right\}d\mathbf{x}'.$$

Now suppose that the charge distribution and current density vary slowly with time and consider the Taylor expansion

$$\rho_e(\mathbf{x}',t) = \rho_e(\mathbf{x}',t') + (t-t')\frac{\partial\rho_e(\mathbf{x}',t')}{\partial t'} + O((t-t')^2).$$

Hence show that

$$\mathbf{E}(\mathbf{x},t) = \frac{1}{4\pi\varepsilon_0}\int\left[\frac{\rho_e(\mathbf{x}',t)\mathbf{r}}{|\mathbf{r}|^3}\right]d\mathbf{x}' + O((t-t')^2).$$

Once again, the static result is accurate to second order in $t - t' = r/c$.

## REFERENCES

Feynman, R. P., Leighton, R. B. & Sands, M., 1964, *The Feynman Lectures on Physics*, Volume II. Addison-Wesley.

Griffiths, D. J., 2013, *Introduction to Electrodynamics*, 4th ed. Pearson.

Jefimenko, O. D., 1966, *Electricity and Magnetism*. Appleton-Century-Crofts.

# PART III
## Special Topics

# 14

# Confined Waves

## *Transmission Lines, Waveguides and Resonant Cavities*

Heaviside may have been a very snuffy lower-middle-class Prometheus, but at least he had snatched a piece of fire for mankind. If the vultures of poverty and the sense of persecution were gnawing at his liver, he shared with Prometheus the sense of having performed a godlike feat.

Nobert Wiener (*Invention*, 1993)

## 14.1 The Transmission of Energy and Information at High Frequencies

The wires connecting circuit components or devices operating at low frequencies, say 50 or 60 Hz, can be reasonably treated as ideal connections in the sense that they have negligible inductance, capacitance or resistance. Such connections transmit information virtually instantaneously (at least on the timescale of the circuit) and cause little loss of energy. As the frequency of the circuit increases, however, the connecting leads start to exhibit some distributed inductance and capacitance, as discussed in §12.6. In particular, parasitic capacitance develops between adjacent leads, while wire pairs can acquire a degree of self-inductance. This, in turn, can result in noticeable delays in the transmission of information along the connections, as well as possible reflections from discontinuities within, or at the ends of, the cables. Distributed impedance becomes particularly noticeable as the frequency approaches $\varpi \sim c/\ell$, where $\ell$ is a typical dimension of the circuit, say the length of the connecting wires, and $c$ the speed of light. Put another way, the distributed impedance of the connecting wires is important whenever the period, $2\pi/\varpi$, is comparable to the time of flight of light waves crossing the circuit. Moreover, as we increase $\varpi$, the connecting cables increasingly radiate waves, causing a loss of energy from the circuit as well as unwanted crosstalk between adjacent circuits.

In summary, then, at high frequencies the connecting cables must be carefully designed to minimise radiation, and once we reach a frequency of $\varpi \sim c/\ell$, it is essential to explicitly acknowledge that a cable possesses distributed capacitance and inductance. The modelling of distributed impedance within cables, and its consequences for the one-dimensional transmission of signals or energy, is the subject of *transmission line theory*.

From a practical point of view, the need to minimise radiation from connecting cables, as well as energy dissipation within connections, has given rise to a range of strategies. The simplest approach to radiation loss is to use unshielded twisted wire pairs, which is effective for controlling radiation losses for frequencies up to a few kilohertz and was traditionally used for local telephone connections. More generally, coaxial cables are commonly used, particularly for frequencies ranging from a few kilohertz up to several hundred megahertz. At yet higher frequencies (microwave frequencies) the coaxial cable as a means of carrying power has tended to give way to the so-called *waveguide*, at least when short distances are involved. For example, in a microwave oven a waveguide carries the microwaves from the wave source to the oven cavity. Finally, at very high frequencies, such as the infrared or visible spectrum, information is transferred along optical fibres.

Let us start at modest frequencies. A transmission line consists of two parallel conductors whose cross-sectional geometry is constant along the line and whose function is to transmit energy or information from a source to a remote load or device. The generic geometry of such a line is shown schematically in Figure 14.1(a). One general feature of a transmission line is that the **E** and **B** fields turn out to be normal to the axis, and so waves in a transmission line are called *transverse electromagnetic waves*, or TEM waves.

We shall take as our prototype transmission line the ubiquitous coaxial cable, as first patented by Oliver Heaviside in April 1880. In its simplest form this consists of a central conducting wire (the so-called active wire) enclosed by a thin conducting cylindrical sheath that is concentric with the inner wire (Figure 14.1(b)). The annular gap between the wire and the outer sheath is filled with a dielectric and the entire geometry is two-dimensional and axisymmetric. The outer conducting sheath acts as a return path for current and also minimises radiation by confining the electromagnetic field to the annular gap surrounding the inner wire. Because of the high frequency involved, the skin effect requires that the currents be confined to the outer surface of the active wire and the inner surface of the outer sheath.

At yet higher frequencies (say a few gigahertz), the loss of energy in coaxial cables can be relatively large. In such cases waveguides, as conceived and analysed by Rayleigh in 1897,

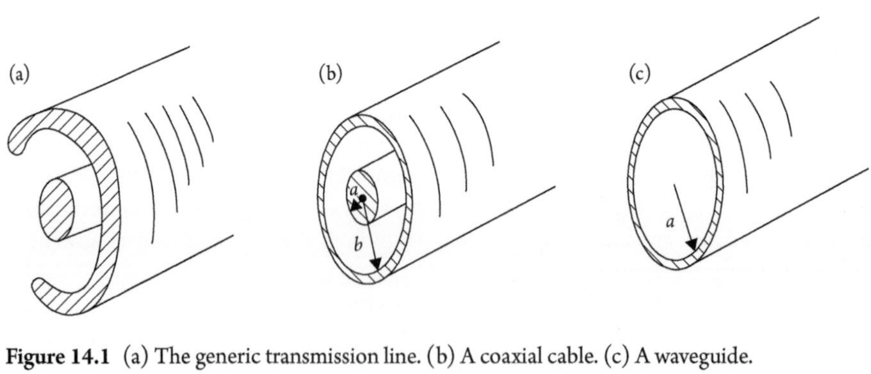


**Figure 14.1** (a) The generic transmission line. (b) A coaxial cable. (c) A waveguide.

are often used to channel electromagnetic waves from a source to a device. In effect, the inner wire is removed from a coaxial cable, leaving only the outer conducting tube, whose cross section may be circular or rectangular (Figure 14.1(c)). This then acts as a conduit for electromagnetic waves, in much the same way as an organ pipe guides acoustic waves along its length. Waveguides differ from transmission lines in that they do *not* support TEM waves, and there is a lower bound on the admissible frequency, called the *cut-off frequency*. This limits waveguides to microwave frequencies (1–100 GHz).

In this chapter we focus on transmission lines (especially coaxial cables) and waveguides, but touch briefly on resonant cavities and optical fibres. The aim is to establish the main physical ideas without drifting too far into technological issues, such as reflections and losses. We start by returning to Chapter 12 and to ladder networks.

## 14.2 Transmission Lines and TEM Waves

### *14.2.1 The Equivalent Circuit for Ideal Transmission Lines: Heaviside's Equations*

Let us see if we can use the quasi-static ideas of Chapter 12 to come up with the equivalent circuit for a transmission line consisting of a coaxial cable. (We shall allow for arbitrary cross-sectional shapes in §14.2.3.) To keep the algebra simple, we shall restrict ourselves to an ideal transmission line in which the resistance is neglected and take the dielectric to be air, or some other material with a relative permittivity of unity. We shall also assume that the conductivity of the wires is large enough for currents to be confined to thin surface layers on the outside of the inner wire and the inside of the outer wire.

Consider the semi-infinite ladder network introduced in §12.5 and shown in Figure 14.2(a). The distributed self-inductance of the coaxial cable, measured per unit length of the cable, is denoted by $L$, while the distributed capacitance between the active and outer wires is $C$, again measured per unit length. (We shall see how to calculate $L$ and $C$ for a coaxial cable shortly.) In Chapter 12 we restricted the analysis of this network to AC signals and used the rules for adding impedances to show that the terminal impedance of such an semi-infinite network, called the *characteristic impedance*, is a pure resistance of magnitude $Z_0 = \sqrt{L/C}$. Here we shall take a different approach, making no particular assumption about the nature of the input waveform. Rather, we shall follow Heaviside and develop partial differential equations for the current and voltage at any instant and at any location in the network.

Let $I(z,t)$ be the current in the central active wire, $V(z,t)$ the voltage difference between the active wire and the outer return wire, and $Q(z,t)$ the charge per unit length on the surface of the central wire. These are all functions of both time and distance along the cable. Note that, if we take the outer wire to be grounded, then $V$ is simply the potential of the active wire. Note also that, at any instant and at any given location, an equal and opposite current flows in the outer wire. This follows from the fact that the magnetic field is trapped in the annulus between the two wires and so the net current flowing in both wires must sum to zero, otherwise Ampère's circuital law would predict a magnetic field external to the cable.

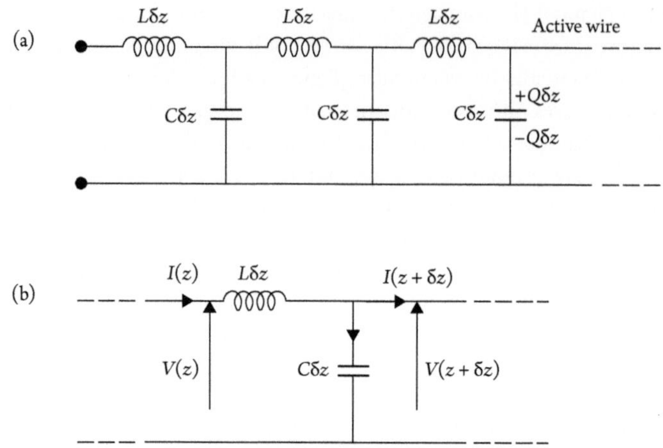


**Figure 14.2** (a) A ladder network composed of inductors and capacitors provides a model for an ideal transmission line. (b) Current and voltage changes for one step in the ladder.

Now consider one step in the ladder network of length $\delta z$, as shown in Figure 14.2(b). The input current is $I(z,t)$, the output current is $I(z+\delta z,t) = I(z,t)+\delta I$, and charge conservation demands that the rate of change of charge on the positive capacitor plate (or central wire) be $(\partial Q/\partial t)\,\delta z = -\delta I$. Of course, by definition of $C$, we have $Q = CV$, and so we conclude that

$$C\frac{\partial V}{\partial t} = -\frac{\partial I}{\partial z}. \tag{14.1}$$

Moreover, the voltage drop across the inductor, $-\delta V$, is given by $(L\delta z)\partial I/\partial t$, as shown in (12.5). So we have the second expression

$$L\frac{\partial I}{\partial t} = -\frac{\partial V}{\partial z}. \tag{14.2}$$

These are known as the *telegrapher's equations* and were first derived by Heaviside. When combined, they yield one-dimensional wave equations for both $I$ and $V$:

$$\boxed{\frac{\partial^2 I}{\partial t^2} = \frac{1}{LC}\frac{\partial^2 I}{\partial z^2}, \qquad \frac{\partial^2 V}{\partial t^2} = \frac{1}{LC}\frac{\partial^2 V}{\partial z^2}.} \tag{14.3}$$

As discussed in §2.1.3, the phase speed of waves governed by such an equation is $1/\sqrt{LC}$, which is independent of wavenumber, i.e. the waves are non-dispersive.

Let us now determine $L$ and $C$ for a coaxial cable, again using quasi-static theory. By symmetry, the electric field is radial and the magnetic field azimuthal, as shown in Figure 14.3. Ampère's circuital law and the integral version of Gauss' law applied to a circle of radius $s$ then tell us that

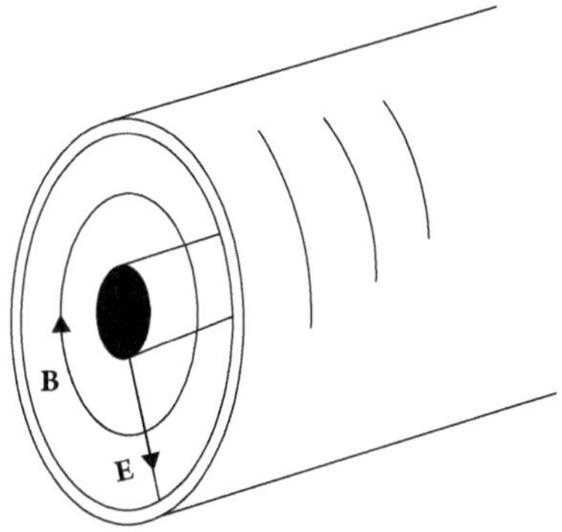


**Figure 14.3** The directions of **B** and **E** in a coaxial cable.

$$B_\theta = \frac{\mu_0 I}{2\pi s}, \qquad E_s = \frac{Q/\varepsilon_0}{2\pi s}, \tag{14.4}$$

in cylindrical polar coordinates $(s, \theta, z)$. The magnetic flux in the annulus, per unit length of the cable, and the radial gradient in electrostatic potential $\varphi$ are then given by

$$\Phi = \frac{\mu_0 I}{2\pi} \ln(b/a), \qquad \frac{\partial \varphi}{\partial s} = -\frac{Q/\varepsilon_0}{2\pi s}, \tag{14.5}$$

where $a$ and $b$ are the inner and outer radii of the annulus (see Figure 14.1(b)) and $E_s = -\partial\varphi/\partial s$. Since $L$, $C$ and $V$ are defined as $L = \Phi/I$, $C = Q/V$ and $V = \varphi(a) - \varphi(b)$, we find

$$L = \frac{\mu_0}{2\pi} \ln(b/a), \qquad C^{-1} = \frac{1}{2\pi\varepsilon_0} \ln(b/a). \tag{14.6}$$

So we conclude that, for a coaxial cable at modest frequencies, say $\varpi << c/b$,

$$\frac{1}{LC} = \frac{1}{\mu_0 \varepsilon_0} = c^2 \tag{14.7}$$

and

$$\boxed{\frac{\partial^2 I}{\partial t^2} = c^2 \frac{\partial^2 I}{\partial z^2}, \qquad \frac{\partial^2 V}{\partial t^2} = c^2 \frac{\partial^2 V}{\partial z^2},} \tag{14.8}$$

where $c$ is the speed of light. We have reached the extraordinary conclusion that electromagnetic waves propagate along the cable at the speed of light, confined to the annulus. The governing wave equation is exactly as for unbounded electromagnetic waves, except that the propagation is strictly one-dimensional. Note that the electric and magnetic fields in these waves are normal to the direction of propagation, and so they are transverse electromagnetic waves, or TEM waves for short.

As discussed in §2.3.4, such waves are *non-dispersive* and so d'Alembert's solution applies. We conclude that an ideal transmission line supports travelling waves of the form (see (2.18))

$$I(z,t) = I_f(z - ct) + I_b(z + ct), \tag{14.9}$$

$$V(z,t) = V_f(z - ct) + V_b(z + ct), \tag{14.10}$$

where $I_f$, $I_b$, $V_f$ and $V_b$ are arbitrary functions of their arguments. These are pulses of arbitrary profile that travel at the speed $c$ without change of shape. Here $I_f$ and $V_f$ are forward-travelling waves while $I_b$ and $V_b$ are backward-travelling waves. The conventional travelling wave solution, $\cos(k_z z - \varpi t)$, is just one particular case in which $\varpi = \pm ck_z$ is the *dispersion relationship* for our system. The relationship between $I_f$ and $V_f$, or $I_b$ and $V_b$, comes from combining (14.9) and (14.10) with either (14.1) or (14.2). This yields

$$V_f = Z_0 I_f, \qquad V_b = -Z_0 I_b, \tag{14.11}$$

where $Z_0 = \sqrt{L/C}$ is the characteristic impedance introduced in §12.5.

We shall say more about these equations and their solutions in §14.2.4. First, however, we want to show that Heaviside's equations for a coaxial cable do not rely on the idea of an equivalent circuit, but rather follow directly and simply from Maxwell's equations. Moreover, we shall show that the wave equations (14.3) and (14.8) apply equally to transmission lines of *arbitrary* cross section.

### 14.2.2 *Deriving the Ideal Transmission Line Equations from Maxwell's Equations*

The concepts of inductance and capacitance are really constructs rooted in quasi-static circuit theory. In particular, as discussed in §§11.5 and 12.1, the concepts of inductance and capacitance are normally invoked by applying the equations of quasi-magnetostatics and quasi-electrostatics to *distinct* regions of space, the former to inductors and the latter to capacitors. Yet in a transmission line the inductance and capacitance are spatially overlapping. So let us see if we can get the key expressions of §14.2.1 directly from Maxwell's equations, without the need to invoke the concepts of inductance and capacitance. As before, we shall restrict ourselves to an ideal coaxial transmission line, take the dielectric to have a relative permittivity of unity and assume that the conductivity of the wires is large enough for the currents to be confined to thin surface layers on the outside of the inner wire and the inside of the outer wire.

Let us start with the full set of Maxwell equations,

$$\nabla \times \mathbf{E} = -\frac{\partial \mathbf{B}}{\partial t}, \qquad \nabla \cdot \mathbf{E} = \frac{\rho_e}{\varepsilon_0}, \tag{14.12}$$

$$\nabla \times \mathbf{B} = \mu_0 \left( \mathbf{J} + \varepsilon_0 \frac{\partial \mathbf{E}}{\partial t} \right), \quad \nabla \cdot \mathbf{B} = 0. \tag{14.13}$$

As before, we shall assume that the electric field is radial and the magnetic field azimuthal. The integral of the axial component of the Ampère–Maxwell equation and the integral version of Gauss' law, both applied to a circle of radius $s$, then yield exactly the same equations as before,

$$B_\theta(s,z) = \frac{\mu_0 I}{2\pi s}, \quad E_s(s,z) = \frac{Q/\varepsilon_0}{2\pi s}. \tag{14.14}$$

(See, by way of comparison, (14.4).) These satisfy $\nabla \cdot \mathbf{B} = 0$ and $\nabla \cdot \mathbf{E} = 0$ in the annulus, as they must. Moreover, the charge conservation equation, $\nabla \cdot \mathbf{J} = -\partial \rho_e / \partial t$, integrated over a short length of the inner wire, yields

$$\frac{\partial I}{\partial z} = -\frac{\partial Q}{\partial t}, \tag{14.15}$$

which is nothing more than (14.1), with $Q = CV$. Next, applying Faraday's law, we find

$$\frac{\partial E_s}{\partial z} = -\frac{\partial B_\theta}{\partial t}, \tag{14.16}$$

which now gives us four equations for four unknowns: $I$, $Q$, $E_s$ and $B_\theta$. (It is left as an exercise for the reader to confirm that the Ampère–Maxwell equation in the annulus is automatically satisfied by (14.14) and (14.15), reflecting a certain redundancy in Maxwell's equations.) Finally, substituting for $E_s$ and $B_\theta$ in (14.16) using (14.14) yields

$$\frac{\partial I}{\partial t} = -c^2 \frac{\partial Q}{\partial z}, \tag{14.17}$$

which is really (14.2), with $c = 1/\sqrt{LC}$. We have now recovered the original system of equations for a coaxial cable. Specifically, we have recovered the *telegrapher's equations* (14.1) and (14.2), along with $c = 1/\sqrt{LC}$. Finally, combining (14.15) with (14.17) yields the now familiar one-dimensional wave equation,

$$\frac{\partial^2 I}{\partial t^2} = c^2 \frac{\partial^2 I}{\partial z^2}. \tag{14.18}$$

So what have we gained from this second look at the coaxial cable? The primary advantage of this second derivation is that we do not need to invoke the quasi-static ideas of inductance and capacitance. Of course, this may be seen as either an advantage or a disadvantage, depending on how heavily one's intuition rests on such concepts.

### *14.2.3 Generalising the Ideal Transmission Line Equations to Arbitrary Cross-sections*

We shall now show that the analysis given in the previous two sections generalises to transmission lines of arbitrary geometry, such as that shown in Figure 14.1(a). As before, we take the dielectric to be air and the transmission lines to be perfectly conducting. Our starting point is, once again, Maxwell's equations.

The key to this generalisation is to look for transverse electromagnetic waves, or TEM waves, of the form

$$\mathbf{B}(x,y,z) = (B_x, B_y, 0), \quad \mathbf{E}(x,y,z) = (E_x, E_y, 0), \tag{14.19}$$

where time dependence is understood. Maxwell's equations applied to free space then reduce to

$$\nabla_\perp \times \mathbf{E} + \hat{\mathbf{e}}_z \times \frac{\partial \mathbf{E}}{\partial z} = -\frac{\partial \mathbf{B}}{\partial t}, \qquad \nabla_\perp \cdot \mathbf{E} = 0, \tag{14.20}$$

$$\nabla_\perp \times \mathbf{B} + \hat{\mathbf{e}}_z \times \frac{\partial \mathbf{B}}{\partial z} = \frac{1}{c^2}\frac{\partial \mathbf{E}}{\partial t}, \qquad \nabla_\perp \cdot \mathbf{B} = 0, \tag{14.21}$$

where the operator $\nabla_\perp = \nabla - \hat{\mathbf{e}}_z(\partial/\partial z)$ involves only derivatives in $x$ and $y$. The first of the equations in either (14.20) or (14.21) naturally decomposes into axial and transverse components. The axial terms yield the two-dimensional electrostatic and magnetostatic systems,

$$\boxed{\nabla_\perp \times \mathbf{E} = 0, \qquad \nabla_\perp \cdot \mathbf{E} = 0,} \tag{14.22}$$

(2D electrostatics)

$$\boxed{\nabla_\perp \times \mathbf{B} = 0, \qquad \nabla_\perp \cdot \mathbf{B} = 0,} \tag{14.23}$$

(2D magnetostatics)

which are decoupled from each other. On the other hand, the transverse components give us

$$\hat{\mathbf{e}}_z \times \frac{\partial \mathbf{E}}{\partial z} = -\frac{\partial \mathbf{B}}{\partial t}, \qquad \hat{\mathbf{e}}_z \times \frac{\partial \mathbf{B}}{\partial z} = \frac{1}{c^2}\frac{\partial \mathbf{E}}{\partial t}. \tag{14.24}$$

Application of the operator $(\partial/\partial z)\hat{\mathbf{e}}_z\times$ to the either of the two expressions in (14.24) yields, when combined with its companion equation, a one-dimensional wave equation for **E** or **B**:

$$\boxed{\frac{\partial^2 \mathbf{E}}{\partial t^2} = c^2 \frac{\partial^2 \mathbf{E}}{\partial z^2}, \qquad \frac{\partial^2 \mathbf{B}}{\partial t^2} = c^2 \frac{\partial^2 \mathbf{B}}{\partial z^2}.} \tag{14.25}$$

Equations (14.22), (14.23) and (14.25) have the reassuring ring of generality (we have not yet said anything about the geometry of the conductors), and they tell us a great deal about TEM waves in transmission lines.

First, (14.22) and (14.23) tell us that the transverse structure of these waves is determined by the two-dimensional *electrostatic* and *magnetostatic* problems defined by the transverse geometry of the two conducting wires. The coaxial cable is just a special case that happens to be readily analysed. This explains why the quasi-static circuit analysis of §14.2.1 is consistent with the discussion in §14.2.2, based on the full set of Maxwell's equations. Second, (14.25) demands that all such fields propagate axially as one-dimensional waves at the speed of light. Note that the general form of these non-dispersive waves is, by analogy with (14.9) and (14.10),

$$\mathbf{E}(x,y,z,t) = \mathbf{E}_f(x,y,z-ct) + \mathbf{E}_b(x,y,z+ct), \tag{14.26}$$

$$\mathbf{B}(x,y,z,t) = \mathbf{B}_f(x,y,z-ct) + \mathbf{B}_b(x,y,z+ct). \tag{14.27}$$

When combined with (14.24), we find that the **E** and **B** fields in a TEM wave are related by

$$\hat{\mathbf{e}}_z \times \mathbf{E}_f = c\mathbf{B}_f, \qquad \hat{\mathbf{e}}_z \times \mathbf{E}_b = -c\mathbf{B}_b, \tag{14.28}$$

so that the **E** and **B** fields are everywhere perpendicular to each other, as well as to the direction of propagation of the wave.

It remains only to find the transverse structure of these waves. Since (14.22) and (14.23) constitute a two-dimensional system, at least for a given $z$ and $t$, we shall now suppress the variables $z$ and $t$ and focus on a given transverse plane at a given instant. The natural first step is to use (14.22) and (14.23) to introduce a two-dimensional vector potential for **B** and a two-dimensional scalar potential for **E**:

$$\mathbf{B} = \nabla_\perp \times (A\hat{\mathbf{e}}_z), \qquad \mathbf{E} = -\nabla_\perp \varphi. \tag{14.29}$$

Equations (14.22) and (14.23) tell us that both $A$ and $\varphi$ are harmonic, in the sense that

$$\nabla_\perp^2 A = 0, \quad \nabla_\perp^2 \varphi = 0. \tag{14.30}$$

The function $A$ is known as the *flux function* for **B**, with magnetic field lines being curves of constant $A$ (see §7.4.2). Moreover, the magnetic flux between two such field lines is equal to the difference in their respective values of $A$, as can be seen by integrating (7.30). Since the two wires are ideal conductors, **B** must be parallel to the surfaces of the wires, say $S_1$ and $S_2$, while **E** is normal to those surfaces. It follows that both $A$ and $\varphi$ are constant on $S_1$ and $S_2$. This, in turn, tells us that the solutions of (14.30) for $A$ and $\varphi$ are identical when normalised by their respective differences on the two wire surfaces.

Now consider the generic transmission line shown in Figure 14.4, and let $\Delta A = A_2 - A_1$ and $\Delta\varphi = \varphi_2 - \varphi_1$ where 2 labels the outer wire and 1 the inner wire. Suppose the outer wire is earthed and the inner one held at the voltage $V$. Then we have $\Delta A = \Phi$ and $\Delta\varphi = -V$, where $\Phi$ is the magnetic flux between the two wires (in the clockwise direction). We therefore have

$$\frac{\varphi(x,y)}{V} = -\frac{A(x,y)}{\Phi}. \tag{14.31}$$

Thus the equipotential surfaces are coincident with the magnetic field lines, which guarantees that **E** and **B** are mutually perpendicular at all points. To find the transverse structure of

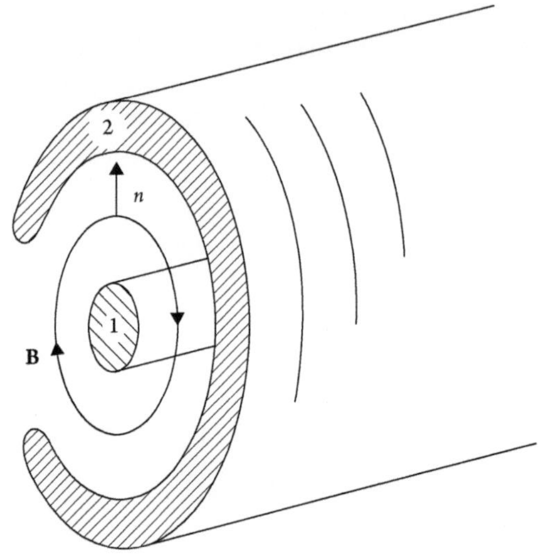


**Figure 14.4** The generic geometry of a transmission line.

either **E** or **B**, one simply needs to solve (14.30) subject to a given value of $\Phi$ or $V$. Moreover, (14.31) combined with the laws of Gauss and Ampère–Maxwell, and applied to the volume enclosed by a magnetic field line surrounding the inner wire, gives us

$$\frac{Q}{\varepsilon_0} = \oint \mathbf{E} \cdot d\mathbf{S} = -\oint \frac{\partial \varphi}{\partial n}\, dS = \frac{V}{\Phi} \oint \frac{\partial A}{\partial n}\, dS = \frac{V}{\Phi} \oint \mathbf{B} \cdot d\mathbf{r} = \frac{V}{\Phi} \mu_0 I.$$

Here $I$ is the current in the inner wire, $Q$ is the charge per unit length on the surface of the same wire, $n$ is an outward-pointing coordinate normal to the magnetic field line and $d\mathbf{r}$ is part of that field line. If we now define a capacitance and inductance (per unit length in $z$) though the usual expressions $Q = CV$ and $\Phi = LI$, we have

$$\frac{1}{LC} = \frac{1}{\mu_0 \varepsilon_0} = c^2, \tag{14.32}$$

which generalises (14.7) for the coaxial cable. In short, although we found it convenient to work with a coaxial cable in §§14.2.1 and 14.2.2, all of the essential results are quite general.

### *14.2.4 Footnotes on Transmission Lines: Ohmic Losses, Energy and Reflections*

Let us now return to the telegrapher's equations of §14.2.1, safe in the knowledge that they are not restricted to a coaxial cable. There are a number of final points that are worth making. First, if the dielectric is not air, and has a relative permittivity other than unity, all that is required is that we replace $\varepsilon_0$ by $\varepsilon = \varepsilon_r \varepsilon_0$. Second, it is relatively easy to allow for Ohmic dissipation in the conducting wires by adopting the equivalent circuit shown in Figure 14.5, which has a resistance per unit length, $R$, in series with $L$. This is discussed in

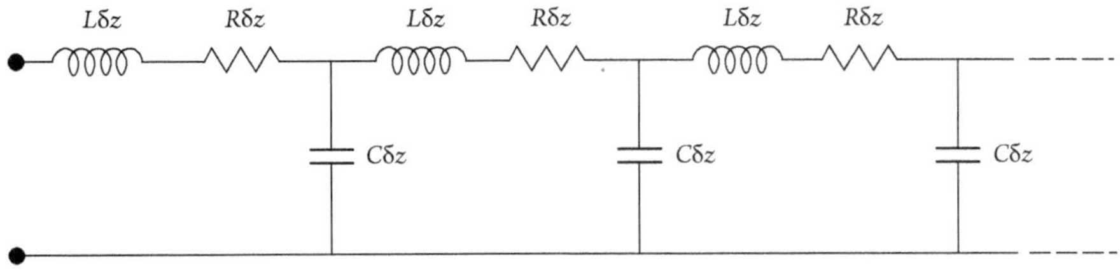


**Figure 14.5** A network including resistors provides a model for a real transmission line.

Exercise 14.1, where the characteristic decay length for waves is shown to be $c(L/R)$. Third, the ideal telegrapher's equations lend themselves to a simple energy equation. Taking the product of (14.1) with $V$, and of (14.2) with $I$, yields

$$\frac{\partial(IV)}{\partial z} = -\frac{\partial}{\partial t}\left(\frac{1}{2}CV^2 + \frac{1}{2}LI^2\right), \tag{14.33}$$

which is clearly an energy balance, with $CV^2/2$ and $LI^2/2$ being the electric and magnetic energy per unit length of the line. So, for example, if we have a semi-infinite transmission line running from 0 to $\infty$, with no signal at $z=\infty$, then (14.33) yields

$$[IV]_0 = \frac{d}{dt}\int_0^\infty \left(\frac{1}{2}CV^2 + \frac{1}{2}LI^2\right) dz = \frac{d}{dt}(U_E + U_B). \tag{14.34}$$

This tells us that the terminal input power $[IV]_0$ is equal to the rate of change of electric and magnetic energy in the line. The equivalent equation for the resistive network of Figure 14.5 is (see Exercise 14.2),

$$\frac{\partial(IV)}{\partial z} = -\frac{\partial}{\partial t}\left(\frac{1}{2}CV^2 + \frac{1}{2}LI^2\right) - I^2R. \tag{14.35}$$

The final point we wish to make concerns the reflection of waves at the end of a finite-length transmission line. As with any wave-bearing system, reflections can be important, and in the case of transmission lines, reflections from the end of the line are often undesirable. To avoid such reflections, it is important to match the impedance of the load to the characteristics of the line. We touched on this in §12.5, where we noted that there will be no reflections in an ideal transmission line provided that the impedance of the load equals the characteristic impedance of the line, $Z_0 = \sqrt{L/C}$. The point is this: the net impedance of a semi-infinite line is $Z_0$, so if a line is suddenly terminated by a load of this impedance, it is impossible to tell from signals received at the input terminal that the line has in fact been terminated, i.e. the load is interpreted as the remainder of the line.

Confirmation of this statement usually proceeds as follows. For an AC circuit (14.9) and (14.10) become, using the complex notation of §12.4,

$$I(z,t) = \hat{I}_f \exp j(\varpi t - k_z z) + \hat{I}_b \exp j(\varpi t + k_z z), \tag{14.36}$$

$$V(z,t) = \hat{V}_f \exp j(\varpi t - k_z z) + \hat{V}_b \exp j(\varpi t + k_z z), \tag{14.37}$$

where the subscripts $f$ and $b$ now indicate forward and reflected waves, $\varpi = ck_z$ and the real part is understood. Also, from (14.11), we have $\hat{V}_f = Z_0 \hat{I}_f$ and $\hat{V}_b = -Z_0 \hat{I}_b$, and so

$$Z_0 I(z,t) = \hat{V}_f \exp j(\varpi t - k_z z) - \hat{V}_b \exp j(\varpi t + k_z z). \tag{14.38}$$

Now suppose the line starts at $z = -L$ and is terminated at $z = 0$ with a load of impedance $Z_L$. Then $V(0,t) = (\hat{V}_f + \hat{V}_b)e^{j\varpi t}$ and $Z_0 I(0,t) = (\hat{V}_f - \hat{V}_b)e^{j\varpi t}$, and so $Z_L$ is given by

$$Z_L = Z_0 \frac{\hat{V}_f + \hat{V}_b}{\hat{V}_f - \hat{V}_b}. \tag{14.39}$$

Evidently, taking $Z_L = Z_0$ ensures that there is no reflected wave, as anticipated above.

## 14.3 Waveguides

### *14.3.1 Two More Types of Waves: TM and TE Waves*

A generic waveguide is shown in Figure 14.6. In effect, it is a hollow metal tube that channels electromagnetic waves along its axis, rather like an organ pipe guides acoustic waves along its length. Of course, we know that electromagnetic waves can propagate through such a tube, because we can see down the tube. Barring some changes in notation, the analysis of waveguides given below follows closely that of Rayleigh (1897), who laid the foundations of the subject.

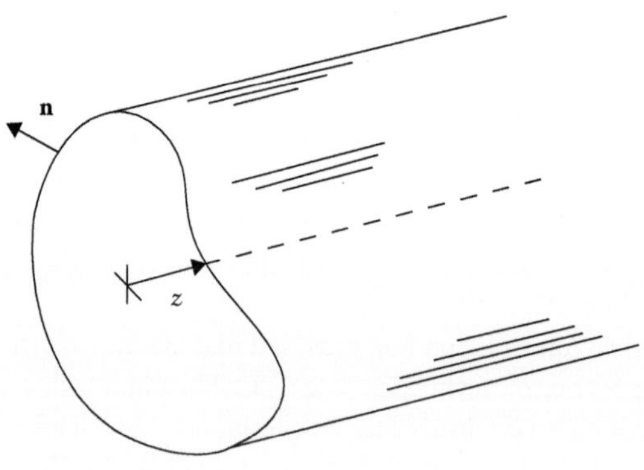


**Figure 14.6** The generic geometry of a waveguide.

The first point to note about waveguides is that they *cannot* support the TEM waves that we have been discussing in the context of transmission lines. This is easily seen from (14.30). Solving Laplace's equation in a closed two-dimensional domain subject to specified values of $A$ or $\varphi$ on an outer boundary has a unique solution, and if $A$ and $\varphi$ are constant on the boundary, then that unique solution is the degenerate one of constant $A$ and $\varphi$ in the interior. This, in turn, demands that both **E** and **B** be zero. So we must look for more complex wave-like solutions of the free-space Maxwell equations,

$$\nabla \times \mathbf{E} = -\frac{\partial \mathbf{B}}{\partial t}, \qquad \nabla \cdot \mathbf{E} = 0, \tag{14.40}$$

$$\nabla \times \mathbf{B} = \frac{1}{c^2}\frac{\partial \mathbf{E}}{\partial t}, \qquad \nabla \cdot \mathbf{B} = 0. \tag{14.41}$$

We shall restrict ourselves to sinusoidal waves, and take the lateral boundaries to be perfectly conducting and the dielectric to be air. Our starting point is the three-dimensional wave equations

$$\frac{\partial^2 \mathbf{E}}{\partial t^2} = c^2\nabla^2\mathbf{E}, \qquad \frac{\partial^2 \mathbf{B}}{\partial t^2} = c^2\nabla^2\mathbf{B}, \tag{14.42}$$

which we must solve subject to the lateral boundary conditions $\mathbf{B}\cdot\mathbf{n} = 0, \mathbf{E}\times\mathbf{n} = 0$, where $\mathbf{n}$ is an outward-pointing unit normal at the inside surface of the bounding wall. Since both **E** and **B** are zero inside the conducting wall, the jump in the normal component of **E** at the surface of the wall is accommodated by surface charges, in accordance with Gauss' law, while the jump in the tangential component in **B** gives rise to surface currents. We shall look for travelling-wave solutions of the form

$$\mathbf{E} = \hat{\mathbf{E}}(x,y)\exp j(k_z z \mp \varpi t), \quad \mathbf{B} = \hat{\mathbf{B}}(x,y)\exp j(k_z z \mp \varpi t), \tag{14.43}$$

so that our wave equations become

$$\nabla_\perp^2\hat{\mathbf{E}} + \left(\frac{\varpi^2}{c^2} - k_z^2\right)\hat{\mathbf{E}} = 0, \quad \nabla_\perp^2\hat{\mathbf{B}} + \left(\frac{\varpi^2}{c^2} - k_z^2\right)\hat{\mathbf{B}} = 0, \tag{14.44}$$

where both $\varpi$ and $k_z$ are taken to be positive and $\nabla_\perp = \nabla - \hat{\mathbf{e}}_z(\partial/\partial z)$.

Now, we know that $E_z$ and $B_z$ cannot both be zero, since there are no TEM waves in a waveguide, but we are at liberty to look for solutions in which $B_z = 0$ and $E_z \neq 0$, which are called *transverse magnetic* (or TM) waves. Then $\nabla_\perp \cdot \hat{\mathbf{B}} = 0$ allows us to introduce a two-dimensional solenoidal vector potential such that

$$\hat{\mathbf{B}} = \nabla_\perp \times [\hat{A}(x,y)\hat{\mathbf{e}}_z], \tag{14.45}$$

where the *flux function* $\hat{A}$ is governed by

$$\nabla_\perp^2\hat{A} + \left(\frac{\varpi^2}{c^2} - k_z^2\right)\hat{A} = 0. \tag{14.46}$$

The boundary conditions on $\mathbf{B}$ are then satisfied by demanding that $\hat{A} = 0$ on the lateral boundary $S$. Solving (14.46) subject to this boundary condition constitutes an eigenvalue problem for $\hat{A}$ in which the eigenvalues are all real and positive (see Exercise 14.6). If we denote the resulting eigenvalues by $\delta_n^2$, then we have

$$\nabla_\perp^2 \hat{A} + \delta_n^2 \hat{A} = 0, \quad \hat{A}|_S = 0, \quad \varpi^2 = c^2(\delta_n^2 + k_z^2). \tag{14.47}$$

Note that all of the geometric information associated with the lateral boundary is now tied up in the eigenvalues.

The corresponding $\mathbf{E}$ field for TM waves is given by the Ampère–Maxwell equation in the form

$$\hat{\mathbf{E}} = \mp \frac{k_z c^2}{\varpi} (\hat{\mathbf{e}}_z \times \hat{\mathbf{B}} - j k_z^{-1} \delta_n^2 \hat{A} \hat{\mathbf{e}}_z). \tag{14.48}$$

The key point about (14.48) is that it automatically satisfies the boundary conditions for $\mathbf{E}$, since $\hat{A} = 0$ and $\mathbf{B} \cdot \mathbf{n} = 0$ on the boundary. So we conclude that waveguides do indeed support TM waves, with the dispersion relationship $\varpi^2 = c^2(\delta_n^2 + k_z^2)$. Note that there is a lower bound on frequency, $\varpi_{\min} = c\delta_{\min}$, which is known as the *cut-off frequency*. Note also that (14.48) yields

$$\hat{\mathbf{E}}_\perp = j \frac{k_z}{\delta_n^2} \nabla_\perp \hat{E}_z, \quad \hat{E}_z = \pm j \frac{c^2 \delta_n^2}{\varpi} \hat{A}, \tag{14.49}$$

which gives $\nabla_\perp \times \hat{\mathbf{E}}_\perp = 0$ and also explains why, in many texts, $\hat{E}_z$ is used as a proxy for the flux function $\hat{A}$.

Let us now turn to the opposite case, in which $E_z = 0$ and $B_z \neq 0$, which are called *transverse electric* (or TE) waves. Then $\nabla_\perp \cdot \hat{\mathbf{E}} = 0$, which allow us to introduce a second vector potential such that

$$\hat{\mathbf{E}} = \nabla_\perp \times [\hat{C}(x,y) \hat{\mathbf{e}}_z], \tag{14.50}$$

which is also governed by

$$\nabla_\perp^2 \hat{C} + \left( \frac{\varpi^2}{c^2} - k_z^2 \right) \hat{C} = 0. \tag{14.51}$$

The boundary conditions on $\mathbf{E}$ are then satisfied by demanding that the normal derivative of $\hat{C}$ be zero on the boundary, $\partial \hat{C}/\partial n|_S = 0$. Solving (14.51) subject to this boundary condition constitutes a second eigenvalue problem. If we denote these eigenvalues by $\gamma_n^2$, then we have

$$\nabla_\perp^2 \hat{C} + \gamma_n^2 \hat{C} = 0, \quad \left. \frac{\partial \hat{C}}{\partial n} \right|_S = 0, \quad \varpi^2 = c^2(\gamma_n^2 + k_z^2). \tag{14.52}$$

The magnetic field is now given by Faraday's law in the form

$$\hat{\mathbf{B}} = \pm \frac{k_z}{\varpi} (\hat{\mathbf{e}}_z \times \hat{\mathbf{E}} - j k_z^{-1} \gamma_n^2 \hat{C} \hat{\mathbf{e}}_z). \tag{14.53}$$

Crucially, (14.53) automatically satisfies the boundary conditions on $\mathbf{B}$, and so we conclude that waveguides also support TE waves, with the dispersion relationship $\varpi^2 = c^2(\gamma_n^2 + k_z^2)$. Once again, there is a lower bound on frequency, $\varpi_{\min} = c\gamma_{\min}$. Indeed, it is this lower bound that tends to restrict waveguides to microwave frequencies. Finally, from (14.53) we have

$$\hat{\mathbf{B}}_\perp = j\frac{k_z}{\gamma_n^2}\nabla_\perp \hat{B}_z, \qquad \hat{B}_z = \mp j\frac{\gamma_n^2}{\varpi}\hat{C}, \tag{14.54}$$

which tells us that $\nabla_\perp \times \hat{\mathbf{B}}_\perp = 0$, while also explaining why $\hat{B}_z$ is often used as a proxy for $\hat{C}$ in TE waves. All of the results above are exclusively those of Rayleigh (1897).

### *14.3.2 Mode Shapes in Rectangular and Circular Waveguides*

It remains to solve the eigenvalue problems (14.47) and (14.52) in order to pin down the frequencies in these dispersion relationships. Here we shall restrict ourselves to the simplest transverse geometries, namely rectangles and circles, starting with the former. Suppose that $0 \le x \le a$ and $0 \le y \le b$, with $b < a$. Then the eigenfunctions for TM and TE waves that satisfy the boundary conditions $\hat{A}|_S = 0$ and $\partial\hat{C}/\partial n|_S = 0$ are clearly

$$\hat{A} \sim \sin\left(\frac{m\pi x}{a}\right)\sin\left(\frac{n\pi y}{b}\right), \quad \hat{C} \sim \cos\left(\frac{m\pi x}{a}\right)\cos\left(\frac{n\pi y}{b}\right), \tag{14.55}$$

for any positive integers $m$ and $n$. In either case the eigenvalues are

$$\delta_{mn}^2 = \gamma_{mn}^2 = \left(\frac{m\pi}{a}\right)^2 + \left(\frac{n\pi}{b}\right)^2. \tag{14.56}$$

Note that the TM mode vanishes if either $m$ or $n$ is zero, whereas the TE mode remains finite if one, but not both, of $m$ or $n$ is zero. So, as Rayleigh (1897) noted, the lowest frequency corresponds to the $\text{TE}_{10}$ mode, which leads to a cut-off frequency of $\varpi_{\min} = c\pi/a$. It is common practice to operate a rectangular waveguide just above this cut-off frequency so that *only* the $\text{TE}_{10}$ mode can be excited (the so-called *single-mode* regime). The shape of this mode is shown schematically in Figure 14.7.

For circular waveguides the eigenmodes with azimuthal wavenumber $m$ are governed by the axisymmetric equation

$$\frac{1}{s}\frac{d}{ds}\left(s\frac{d\hat{A}}{ds}\right) + \left(\delta_n^2 - \frac{m^2}{s^2}\right)\hat{A} = 0, \tag{14.57}$$

or the equivalent expression involving $\hat{C}$ and $\gamma_n^2$. This has the solution $\hat{A} \sim J_m(\delta_n s)$ for TM modes and $\hat{C} \sim J_m(\gamma_n s)$ for TE waves, where $J_m$ is the usual Bessel function. If the waveguide has radius $a$, then the eigenvalues of the TM modes must satisfy $J_m(\delta_n a) = 0$, while those for TE modes must satisfy $J'_m(\gamma_n a) = 0$, where the prime indicates a derivative. Let us denote the $n$th zero of $J_m$ by $j_{mn}$ and that of $J'_m$ by $j'_{mn}$. These zeros are widely tabulated and it turns out that the smallest one is $j'_{11} = 1.841$. So the cut-off frequency for a circular waveguide

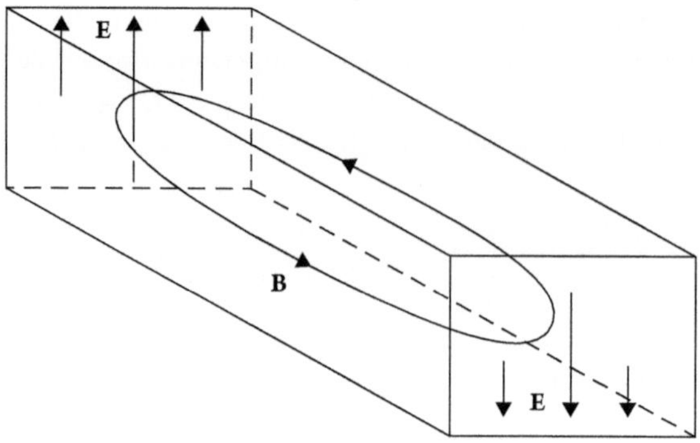


**Figure 14.7** The $TE_{10}$ mode in a rectangular waveguide.

corresponds to a $TE_{11}$ mode with an azimuthal wavenumber of unity and $\varpi_{min} = j'_{11}c/a$. Again, this result is due to Rayleigh (1897).

### *14.3.3 The Group Velocity in Waveguides*

We close our discussion of waveguides with a few comments about the propagation of energy in them. The dispersion relationship $\varpi^2 = c^2(\delta_n^2 + k_z^2)$ tells us that the phase velocity $c_p = \varpi / k_z$ is a function of $k_z$ and so these waves are dispersive, i.e. Fourier modes with different wavenumbers propagate at different speeds. It is a general feature of wave-bearing systems that the speed at which a slowly modulated wave packet travels, which is the speed at which information and energy disperse from a localised source, is given by the group velocity $c_g = d\varpi / dk_z$. (If you are not familiar with the concept of group velocity, see the discussion in §2.3.5.) For waveguides the group velocity is

$$c_g = c\frac{k_z}{\sqrt{k_z^2 + \delta_n^2}}, \quad \text{or} \quad c_g = c\frac{k_z}{\sqrt{k_z^2 + \gamma_n^2}}. \tag{14.58}$$

Note that $c_g < c$, as it must be. Equation (14.58) tells us that the energy belonging to the large wavenumbers propagates fastest, while the energy of the long-wavelength waves propagates more slowly. So a wave pulse composed of many wavenumbers will spread out (disperse) as it moves down the waveguide.

Suppose that our pulse is composed only of waves corresponding to the fundamental mode, say the $TE_{10}$ mode in a rectangular waveguide or the $TE_{11}$ mode for a circular guide. Then we can write our dispersion relationship as $\varpi^2 = \varpi_{min}^2 + c^2k_z^2$, and the corresponding forms for the phase and group velocities are

$$c_p = \frac{\varpi}{k_z} = \frac{c}{\sqrt{1 - (\varpi_{min}/\varpi)^2}}, \qquad c_g = \frac{d\varpi}{dk_z} = c\sqrt{1 - (\varpi_{min}/\varpi)^2}. \tag{14.59}$$

This shows explicitly how the speed of propagation of energy falls to zero as the frequency drops down to the cut-off frequency. Note that (14.59) yields the surprisingly simple relationship $c_g c_p = c^2$, with $c_g < c < c_p$. The group velocity is an important physical quantity, not just because it controls the rate of propagation of information along a waveguide, but also because the rate at which energy propagates down a guide is the product of the energy density and the group velocity.

## 14.4 Resonant Cavities

The fact that waves may propagate in both directions in a waveguide tells us that standing waves are a possibility, and indeed are inevitable if the guide is closed at both ends. This takes us to the topic of resonant cavities: enclosures bounded by highly conducting walls within which electromagnetic fields may be trapped and their energy stored. Let us take as a simple example a cylindrical cavity of length $L$ and arbitrary cross section, in other words a waveguide terminated by plane perpendicular walls at $z = 0$ and $z = L$. We may now construct standing wave solutions for the resonant cavity by superimposing the travelling waves of §14.3.1. As these solutions already satisfy the lateral boundary conditions, our task is to find suitable combinations of travelling waves that satisfy the conditions at $z = 0$ and $z = L$. As before, we take the boundaries to be perfectly conducting.

We start with TM waves, in which $B_z = 0$. In this case the additional boundary conditions to be satisfied are $\mathbf{E}_\perp = 0$ at $z = 0$ and $z = L$. Let us try adding a forward- and a backward-travelling wave of the same amplitude, say

$$\mathbf{B} = \frac{1}{2}\hat{\mathbf{B}}(x,y)\left[\exp j(k_z z - \varpi t) + \exp j(k_z z + \varpi t)\right].$$

Recall that actual solutions are obtained from the complex representation of §14.3.1 by taking the real part. So the expression above becomes

$$\mathbf{B} = \hat{\mathbf{B}}(x,y)\cos(k_z z)\cos(\varpi t) = \nabla_\perp \times \left[\hat{A}(x,y)\hat{\mathbf{e}}_z\right]\cos(k_z z)\cos(\varpi t), \tag{14.60}$$

while (14.48) yields the corresponding form for the electric field as

$$\mathbf{E}_\perp = -\frac{k_z c^2}{\varpi}(\hat{\mathbf{e}}_z \times \hat{\mathbf{B}})\sin(k_z z)\sin(\varpi t), \tag{14.61}$$

$$E_z = \frac{c^2 \delta_n^2}{\varpi}\hat{A}(x,y)\cos(k_z z)\sin(\varpi t). \tag{14.62}$$

Provided that $k_z = \pi p/L$, where $p$ is an integer, (14.61) satisfies the boundary conditions on $\mathbf{E}_\perp$ at $z = 0$ and $z = L$. The dispersion relationship (14.47) now tells us that the resonant frequencies are

$$\varpi_{mnp}^2 = c^2\left[\delta_{mn}^2 + \left(\frac{\pi p}{L}\right)^2\right], \tag{14.63}$$

where $\delta_{mn}^2$ is given by (14.56) for a rectangular cross section. For a circular cross section of radius $a$, on the other hand, we have $\delta_{mn} = j_{mn}/a$ (with $m = 0, 1, 2, \ldots$ and $n = 1, 2, 3, \ldots$), where $j_{mn}$ is the $n$th root of $J_m(x) = 0$. Thus the resonant frequencies for TM modes in a circular cylinder are

$$\varpi^{\mathrm{TM}}_{mnp} = c\sqrt{\left(\frac{j_{mn}}{a}\right)^2 + \left(\frac{\pi p}{L}\right)^2}. \tag{14.64}$$

The lowest TM mode corresponds to $m = 0$, $n = 1$, $p = 0$, in which **E** is purely axial, **B** and **E** are both uniform in $z$, and $\varpi^{TM}_{010} = cj_{01}/a = 2.405c/a$. This $\mathrm{TM}_{010}$ mode is shown in Figure 14.8. It is reminiscent of the high-frequency parallel-plate capacitor discussed in §12.6.1.

For TE waves, in which $E_z = 0$, the additional boundary conditions to be satisfied are $B_z = 0$ and $\mathbf{E}_\perp = 0$ at $z = 0$ and $z = L$. This time we must subtract a forward- and a backward-travelling wave of the same amplitude:

$$\mathbf{E} = \frac{1}{2}\hat{\mathbf{E}}(x,y)\left[\exp j(k_z z - \varpi t) - \exp j(k_z z + \varpi t)\right].$$

Taking the real part, we obtain

$$\mathbf{E} = \hat{\mathbf{E}}(x,y)\sin(k_z z)\sin(\varpi t) = \nabla_\perp \times \left[\hat{C}(x,y)\hat{\mathbf{e}}_z\right]\sin(k_z z)\sin(\varpi t), \tag{14.65}$$

while (14.53) shows the corresponding solution for the magnetic field to be

$$\mathbf{B}_\perp = \frac{k_z}{\varpi}(\hat{\mathbf{e}}_z \times \hat{\mathbf{E}})\cos(k_z z)\cos(\varpi t), \tag{14.66}$$

$$B_z = \frac{\gamma_n^2}{\varpi}\hat{C}(x,y)\sin(k_z z)\cos(\varpi t). \tag{14.67}$$

Once again, this satisfies the boundary conditions provided that $k_z = \pi p/L$. The dispersion relationship (14.52) now tells us that the resonant frequencies are

$$\varpi^2_{mnp} = c^2\left[\gamma^2_{mn} + \left(\frac{\pi p}{L}\right)^2\right], \tag{14.68}$$

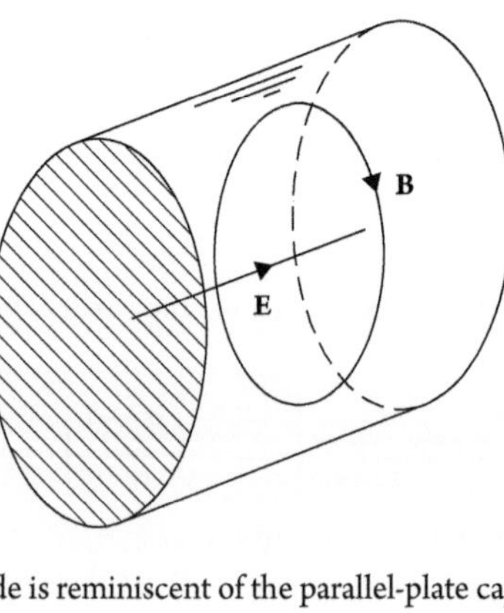


**Figure 14.8** The $\mathrm{TM}_{010}$ mode is reminiscent of the parallel-plate capacitor of §12.6.1.

where $\gamma_{mn}^2$ is given by (14.56) for a rectangular cross section. For a circular cross section, on the other hand, we have $\gamma_{mn} = j'_{mn}/a$ (with $m = 0, 1, 2, \dots$ and $n = 1, 2, 3, \dots$), where $j'_{mn}$ is the $n$th zero of $J'_m$. So the resonant frequencies for TE modes in a circular cylinder are

$$\varpi_{mnp}^{\mathrm{TE}} = c\sqrt{\left(\frac{j'_{mn}}{a}\right)^2 + \left(\frac{\pi p}{L}\right)^2}. \tag{14.69}$$

The lowest TE mode corresponds to $m = 1$, $n = 1$, $p = 1$, and this $\mathrm{TE}_{111}$ mode has a resonant frequency of

$$\varpi_{111}^{\mathrm{TE}} = c\sqrt{\left(\frac{1.84}{a}\right)^2 + \left(\frac{\pi}{L}\right)^2}. \tag{14.70}$$

This is as far as we shall take the subject of guides and cavities. However, it is worth noting that, although we often think of waveguides and resonant cavities as part of the electrical engineer's toolkit, nature also provides us with some intriguing examples. For example, the spherical gap between the surface of the earth and the ionosphere acts as an (imperfect) resonant cavity, trapping low-frequency waves (see e.g. Jackson, 1999).

We have made only modest inroads into the important topics of metallic waveguides and cavities, but the interested reader will find many more details in Rayleigh (1897), Smythe (1969) and Jackson (1999). If allowances are made for notation, and the fact that vector calculus was in its infancy in 1897, Rayleigh is particularly lucid.

## 14.5 An Aside: Optical Fibres

It would be odd to end a chapter on transmission lines and metallic waveguides without commenting on optical fibres, since these have come to dominate high-speed telecommunications over the last few decades. From a pedagogical point of view, such comments should really come after our discussion in Chapter 19 of the reflection of waves at interfaces, since optical fibres rely on the principle of total internal reflection. However, it seems more appropriate that we introduce these fibres here.

An optical fibre is a thin thread of transparent dielectric, known as the core fibre, enclosed by an outer layer of cladding of lower refractive index (Figure 14.9). Such fibres are usually constructed from silica ($SiO_2$), although occasionally plastic is used for short runs where attenuation is less important. Losses in optical fibres are relatively low, at least when silica is used, and so thin fibres can be used to send signals over many tens of kilometres. Fibre optics also have the advantage of being immune from electromagnetic interference and crosstalk, are able to carry a particularly high information density, and provide a secure means of transmitting information. Pure silica has a refractive index of around $n = 1.45$ in the visible spectrum, but this can be raised in the core fibre by doping the glass with oxides such as germanium oxide ($GeO_2$). The core fibre is typically between 4 and 100 $\mu$m in diameter and visible or near-infrared light with a wavelength around 1 $\mu$m is used to transmit the information down the line.

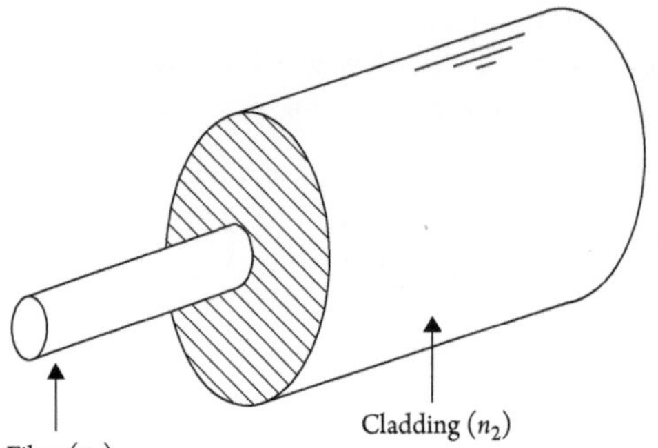


**Figure 14.9** Schematic of an optical fibre.

The number of independent modes that can be sent down a given fibre is usually expressed in terms of the so-called *fibre parameter*

$$V = ka\sqrt{1 - \left(\frac{n_2}{n_1}\right)^2}, \tag{14.71}$$

where $a$ is the radius of the core fibre, $n_1$ and $n_2$ are the refractive indices of the core and cladding, respectively, and $k = 2\pi/\lambda$ is the wavenumber. It turns out that, for so-called step-index fibres, only one mode may be transmitted when $V$ is less than 2.405 (the first zero of $J_0$), whereas for $V \gg 1$ the number of modes is approximately $N \approx V^2/2$ (Jackson, 1999). The choice of wavelength in (14.71) is dictated in large part by the absorption spectrum of silica. For example, there is relatively low attenuation in the near-infrared windows $\lambda \approx 1.25$–$1.35\,\mu$m and $\lambda \approx 1.5$–$1.6\,\mu$m, which makes these popular for long-distance communication (Keiser, 2011), while $\lambda \approx 0.8$–$0.9\,\mu$m is more common for short distances because of the availability of cheap optical sources at these wavelengths, although the losses in this range are higher.

The smaller-diameter optical fibres, say 4–8 $\mu$m in diameter, can be thought of as dielectric waveguides of circular cross section, described by the kind of analysis presented in §14.3. Usually these operate in the *single-mode regime*, as do metallic waveguides. However, fibres whose core diameters are much greater than the wavelength of the light, say 50–100 $\mu$m, operate in the *multi-mode regime*, capable of transmitting many hundreds of modes. These are more simply described using the ray tracing methods of geometrical optics, as discussed below. Such fibres rely on the fact that rays travelling down the line are perfectly reflected each time they strike the cladding, provided that they have a small angle of propagation relative to the axis of the fibre. So the rays that are almost parallel to the axis undergo multiple reflections and cannot escape from the fibre.

Perhaps it is worth recalling Snell's law at this point. Consider a light ray travelling through a medium of refractive index $n_1$, which strikes the interface to a second medium of lower refractive index $n_2$. According to Snell's law, the wave vector **k** bends away from the normal as light passes through the interface, as shown in Figure 14.10(a). Suppose that $\phi_I$ and $\phi_T$

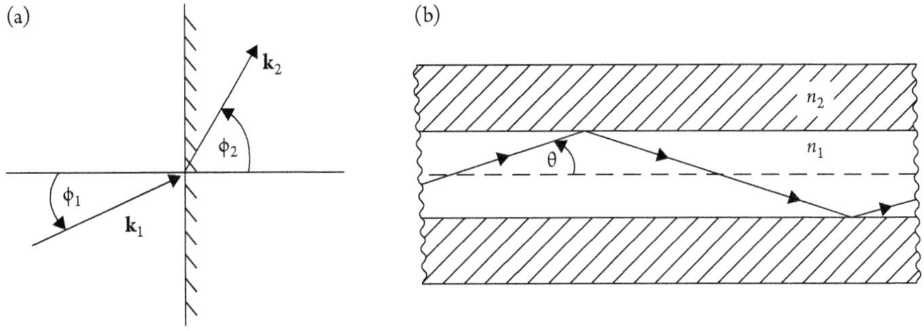


**Figure 14.10** (a) Refraction of a light ray at a plane interface. (b) Reflections at the interface of the fibre and the cladding in a multi-mode fibre.

are the angles to the normal made by the incident and transmitted waves. Then Snell's law tells us that $n_1 \sin\phi_I = n_2 \sin\phi_T$, so at the critical angle of $\phi_I = \phi_{\rm crit} = \sin^{-1}(n_2/n_1)$ the transmitted wave just grazes the interface, i.e. $\phi_T = 90°$. For $\phi_I > \sin^{-1}(n_2/n_1)$ there is no refracted ray, merely a reflected one, a phenomenon known as *total internal reflection*. (Actually, as discussed in Chapter 19, the field in medium 2 is not completely zero for $\phi_I > \phi_{\rm crit}$, but rather takes the form of an *evanescent wave* that decays exponentially over a short distance from the interface.) In terms of the complimentary angle parallel to the interface, $\theta = 90° - \phi$, we get total internal reflection whenever the incident wave adopts an angle less than $\theta_{\rm crit} = \cos^{-1}(n_2/n_1)$. Thus rays in an optical fibre will experience multiple internal reflections provided that their angle of propagation relative to the axis of the fibre is less than $\theta_{\rm crit} = \cos^{-1}(n_2/n_1)$. This is shown in Figure 14.10(b). In practice, propagation is typically restricted to a cone angle somewhat less than $\theta_{\rm crit} \sim 10°$.

The construction shown in Figure 14.10(b) is called a *step-index* fibre because of the discontinuous jump in refractive index at the interface of the core fibre and the cladding. However, superior properties, such as reduced dispersion, can be realised if there is a gradual decrease in refractive index with radius, $n(s)$, which is referred to as a *graded-index* fibre. Such multi-mode fibres are more difficult to analyse, requiring advanced methods from geometrical optics, such as the eikonal approximation.

When dealing with very thin fibres, say cores of 4–20 $\mu$m diameter, the methods of ray tracing cease to be valid and we must revert to Maxwell's equations, just as we did for metallic waveguides. There are then similarities to the analysis of §14.3 in that propagation occurs only for certain discrete transverse modes, and for a given mode there is a unique axial wavenumber corresponding to a given frequency. However, the situation is complicated by the fact that we must match solutions across the interface between the fibre and the cladding, whereas with metal waveguides we simply impose boundary conditions at the lateral edges of the guide. When the radius of the fibre is comparable to the wavelength of the light to be transmitted, then only the lowest transverse mode will propagate. These are called *single-mode fibres* and they typically have a core diameter of around 4–8 $\mu$m. Single-mode fibres tend to be used for communication links in excess of 1 km, whereas multi-mode fibres are preferred for shorter-distance links, or when a particularly high power is to be transmitted.

Although graded-index fibres tend to be restricted to multi-mode communication, they can in principle be used for single-mode fibres. A modal analysis of graded-index fibres is, of course, more complicated than that of step-index fibres. Since $n^2(s) = \varepsilon/\varepsilon_0$ (see Chapter 19), the governing equation for **E** becomes

$$\nabla^2\mathbf{E} = -\nabla\times\nabla\times\mathbf{E} + \nabla(\nabla\cdot\mathbf{E}) = \frac{n^2(s)}{c^2}\frac{\partial^2\mathbf{E}}{\partial t^2} + \nabla\left[\nabla\cdot\left(\frac{\mathbf{D}}{\varepsilon}\right)\right]. \tag{14.72}$$

Moreover, as there are no free charges, we have $\nabla\cdot\mathbf{D} = \rho_f = 0$, and so this yields the *inhomogeneous* wave equation

$$\nabla^2\mathbf{E} - \frac{n^2(s)}{c^2}\frac{\partial^2\mathbf{E}}{\partial t^2} = -\nabla\left(\mathbf{E}\cdot\frac{\nabla\varepsilon}{\varepsilon}\right). \tag{14.73}$$

Solving (14.73) is far from straightforward and the corresponding mode shapes are typically rather complicated. For example, unless we impose azimuthal symmetry, the resulting waves are not pure TE or TM modes, but rather a more complicated form of hybrid wave. Such modes are discussed in Jackson (1999).

That is all we shall say about optical fibres; in effect, we have done little more than acknowledged that they exist. Because of their technological importance, there is a large literature devoted to their theory, as well as to practical aspects of their use. Interested readers are directed to the somewhat mathematical introduction in Jackson (1999) and to the more extensive and qualitative treatment by Keiser (2011).

## EXERCISES

14.1 Consider the dissipative transmission line shown in Figure 14.5. Show that the telegrapher's equations become

$$C\frac{\partial V}{\partial t} = -\frac{\partial I}{\partial z}, \qquad L\frac{\partial I}{\partial t} + RI = -\frac{\partial V}{\partial z},$$

and so voltage fluctuations are governed by

$$\frac{\partial^2 V}{\partial t^2} + \frac{R}{L}\frac{\partial V}{\partial t} = \frac{1}{LC}\frac{\partial^2 V}{\partial z^2}.$$

Confirm that waves are dissipated over a characteristic distance of $2c\tau$, where $\tau = L/R$.

14.2 Use the telegrapher's equations of Exercise 14.1 to derive the energy equation

$$\frac{\partial(IV)}{\partial z} = -\frac{\partial}{\partial t}\left(\frac{1}{2}CV^2 + \frac{1}{2}LI^2\right) - I^2R.$$

Hence show that the input power to a semi-infinite line is

$$[IV]_0 = \frac{d}{dt}\int_0^\infty \left(\frac{1}{2}CV^2 + \frac{1}{2}LI^2\right) dz + \int_0^\infty I^2 R\, dz.$$

14.3 Use the telegrapher's equations for an ideal transmission line to show that

$$\frac{\partial(\Phi Q)}{\partial t} = -\frac{\partial}{\partial z}\left(\frac{1}{2}CV^2 + \frac{1}{2}LI^2\right),$$

where $\Phi$ and $Q$ are the flux and charge per unit length. Hence show that, for a semi-infinite transmission line,

$$\left[\frac{1}{2}CV^2 + \frac{1}{2}LI^2\right]_z = \frac{d}{dt}\int_z^\infty \Phi Q\, dz.$$

14.4 Use the telegrapher's equations for an ideal transmission line, along with (14.32), to show that

$$\frac{\partial P}{\partial t} = -c^2\frac{\partial e}{\partial z}, \qquad \frac{\partial e}{\partial t} = -\frac{\partial P}{\partial z},$$

where $P = IV$ and $e$ is the electromagnetic energy per unit length of the line. Hence show that $P$ and $e$ both obey the one-dimensional wave equation and propagate at the speed $c$.

14.5 Consider an ideal transmission line of characteristic impedance $Z_0$ terminated with an impedance $Z_L$. Show that the ratio of the reflected to the incident voltage, $\rho_L$, is

$$\rho_L = \frac{\hat{V}_b}{\hat{V}_f} = \frac{Z_L - Z_0}{Z_L + Z_0}.$$

Hence show that, if the line is terminated with an open circuit, we have $\rho_L = 1$ and the voltage is a standing wave with a maximum at the reflection point. Also show that, if the line is terminated with a closed circuit (zero impedance), we have $\rho_L = -1$ and the voltage is a standing wave with a node at the reflection point.

14.6 Use (14.47), plus Gauss' theorem, to express the two-dimensional integral of $\hat{\mathbf{B}}^2$ over the cross section of a waveguide in terms of the two-dimensional integral of $\hat{A}^2$. Hence show that the eigenvalues for TM waves are all real and positive. Now use (14.52) to show that the eigenvalues for TE waves are also real and positive.

14.7 Let subscripts $n$ and $m$ indicate two independent TM modes in a waveguide. Use (14.47), plus Gauss' theorem, to express the two-dimensional integral of $\hat{\mathbf{B}}_n \cdot \hat{\mathbf{B}}_m$ over the cross-section of the waveguide in terms of the two-dimensional integral of $\hat{A}_n\hat{A}_m$. Hence show that two independent TM modes are orthogonal in the sense that the two-dimensional integral of $\hat{\mathbf{B}}_n \cdot \hat{\mathbf{B}}_m$ is zero for $n \neq m$. Now use (14.52) to show that TE modes are orthogonal in the sense that the two-dimensional integral of $\hat{\mathbf{E}}_n \cdot \hat{\mathbf{E}}_m$ is zero for $n \neq m$.

## REFERENCES

Jackson, J. D., 1999, *Classical Electrodynamics*, 3rd ed. Wiley.

Rayleigh, Lord, 1897, On the passage of electric waves through tubes, or the vibrations of dielectric cylinders. *Phil. Mag.*, **43**, 125–132.

Smythe, W. R., 1969, *Static and Dynamic Electricity*, 3rd ed. McGraw-Hill.

Keiser, G., 2011, *Optical Fiber Communications*, 4th ed. McGraw-Hill.

# 15

# Maxwell's Equations in Free Space I

## *The Propagation of Waves*

> The tendency of modern physics is to resolve the whole material world into waves, and nothing but waves. These waves are of two kinds, bottled-up waves, which we call matter, and un-bottled waves, which we call radiation or light. If annihilation of matter occurs, the process is merely that of un-bottling imprisoned waves... These concepts reduce the whole universe to a world of light, potential or existent, so that the whole story of its creation can be told with perfect accuracy and completeness in the six words, 'God said, "Let there be light." '.
>
> Sir James Jeans, 1930

In this chapter we consider how electromagnetic waves travel through free space. We are less interested in the sources of these waves, which are discussed in Chapter 16, and focus instead on their intrinsic properties, such as their transverse structure and their ability to transport energy. We shall postpone our discussion of waves in matter until Chapter 19.

## 15.1 The Electromagnetic Spectrum

Before discussing the electrodynamics of waves, perhaps it is useful to note the range of frequencies and wavelengths commonly encountered in physics and engineering. Maxwell's equations place no limitation on the wavelength $\lambda$, but in practice this tends to range from many tens of kilometres for long radio waves, down to around $10^{-15}$ m for high-energy gamma rays in cosmic radiation. The corresponding range of frequencies is $f = 10^4$ Hz to $f = 10^{24}$ Hz, where $f = c/\lambda$ and the speed of light in vacuum is $c = 1/\sqrt{\mu_0 \varepsilon_0} = 2.998 \times 10^8$ m/s. So both frequency and wavelength can vary by 20 orders of magnitude. The visible spectrum, on the other hand, is rather narrow, ranging from $\lambda = 7.6 \times 10^{-7}$ m and $f = 3.9 \times 10^{14}$ Hz for the longest visible red light to $\lambda = 4.0 \times 10^{-7}$ m and $f = 7.5 \times 10^{14}$ Hz for the shortest visible blue light. Below the visible spectrum, as measured by frequency, we have radio waves, microwaves and the infrared spectrum, while above visible light comes

**Table 15.1** Some common types of electromagnetic waves

| Wavelength (m) | Type of waves | Frequency (Hz) |
|---|---|---|
| $10^{-12}$ | Gamma rays | $3\times10^{20}$ |
| $10^{-11}$ | Gamma rays, X-rays | $3\times10^{19}$ |
| $10^{-10}$ | X-rays | $3\times10^{18}$ |
| $10^{-9}$ | X-rays | $3\times10^{17}$ |
| $10^{-8}$ | X-rays, ultraviolet | $3\times10^{16}$ |
| $10^{-7}$ | Ultraviolet | $3\times10^{15}$ |
| $10^{-6}$ | Visible | $3\times10^{14}$ |
| $10^{-5}$ | Infrared | $3\times10^{13}$ |
| $10^{-4}$ | Infrared | $3\times10^{12}$ |
| $10^{-3}$ | Infrared | $3\times10^{11}$ |
| $10^{-2}$ | Microwaves | $3\times10^{10}$ |
| $10^{-1}$ | Microwaves, radar | $3\times10^{9}$ |
| 1 | Microwaves, radar | $3\times10^{8}$ |
| 10 | Radio, TV | $3\times10^{7}$ |
| $10^{2}$ | Radio | $3\times10^{6}$ |

ultraviolet radiation, X-rays and gamma rays. The frequency and wavelength of various types of common waves are noted in Table 15.1.

## 15.2 Maxwell's Equations in Vacuum

### 15.2.1 *Retarded Potentials and Wave Propagation*

In §13.7 we noted that, provided we use the Lorenz gauge to determine the divergence of the vector potential,

$$\nabla\cdot\mathbf{A} = -\frac{1}{c^2}\frac{\partial\varphi}{\partial t}, \tag{15.1}$$

the scalar and vector potentials are governed by the inhomogeneous wave equations

$$\nabla^2\varphi - \frac{1}{c^2}\frac{\partial^2\varphi}{\partial t^2} = -\frac{\rho_e}{\varepsilon_0}, \tag{15.2}$$

$$\nabla^2\mathbf{A} - \frac{1}{c^2}\frac{\partial^2\mathbf{A}}{\partial t^2} = -\mu_0\mathbf{J}, \tag{15.3}$$

where $\mu_0\mathbf{J}$ and $\rho_e/\varepsilon_0$ act as sources of waves. In the same section we saw how to invert these wave equations to give the retarded potentials

$$\varphi(\mathbf{x},t) = \frac{1}{4\pi\varepsilon_0}\int\frac{\rho_e(\mathbf{x}',t')}{|\mathbf{r}|}d\mathbf{x}', \tag{15.4}$$

$$\mathbf{A}(\mathbf{x},t) = \frac{\mu_0}{4\pi}\int\frac{\mathbf{J}(\mathbf{x}',t')}{|\mathbf{r}|}d\mathbf{x}', \tag{15.5}$$

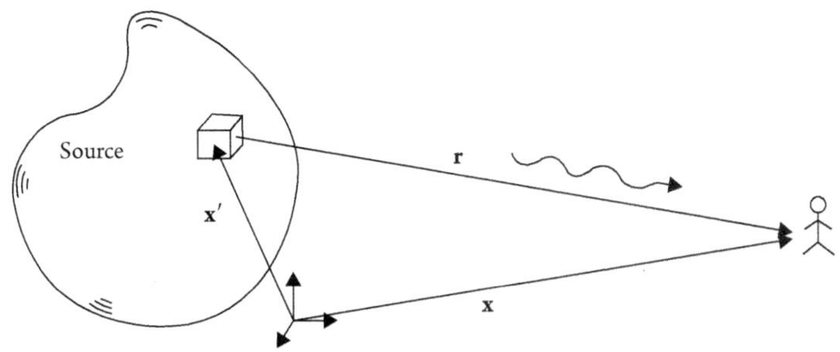


**Figure 15.1** The vector $\mathbf{r} = \mathbf{x} - \mathbf{x}'$ takes us from the source at $\mathbf{x}'$ to the observer at $\mathbf{x}$.

where $t' = t - r/c$ is the retarded time and $\mathbf{r} = \mathbf{x} - \mathbf{x}'$ takes us from the source to the point of observation (Figure 15.1). We emphasised that these inversion integrals are identical to those for the corresponding static Poisson equations for $\varphi$ and $\mathbf{A}$, except that $t$ is replaced by $t'$ in the integrand. This reflects the finite rate at which information propagates. Indeed, if we let $c \to \infty$ in (15.4) and (15.5), we recover the strictly static potentials.

In Chapter 16, where we discuss radiation, we shall use (15.4) and (15.5) to relate wave fields to their sources. In particular, we shall look at the asymptotic structure of wave fields at large distances from their sources, in the so-called *radiation zone*, and discuss how that structure changes with the nature of the source. In this chapter, by contrast, we set aside the sources and focus on the intrinsic properties of electromagnetic waves as they propagate through free space. So we shall consider the free-space form of Maxwell's equations in which the sources are absent:

$$\nabla \times \mathbf{E} = -\frac{\partial \mathbf{B}}{\partial t}, \quad \nabla \cdot \mathbf{E} = 0, \tag{15.6}$$

$$\nabla \times \mathbf{B} = \frac{1}{c^2}\frac{\partial \mathbf{E}}{\partial t}, \quad \nabla \cdot \mathbf{B} = 0. \tag{15.7}$$

Taking the curl of the Ampère–Maxwell equation and eliminating $\mathbf{E}$ using Faraday's law, or taking the curl of Faraday's law and eliminating $\mathbf{B}$ using the Ampère–Maxwell equation, yields the non-dispersive wave equations

$$\frac{\partial^2 \mathbf{B}}{\partial t^2} = c^2\nabla^2\mathbf{B}, \quad \frac{\partial^2 \mathbf{E}}{\partial t^2} = c^2\nabla^2\mathbf{E}. \tag{15.8}$$

We shall work with the potentials $\varphi$ and $\mathbf{A}$, using (15.2) and (15.3), as well as with the fields $\mathbf{E}$ and $\mathbf{B}$, adopting the Lorenz gauge throughout. We start in §15.3 with plane waves, in which the propagation is rectilinear. We then consider spherical wave propagation in §15.4 and conclude with fully three-dimensional waves and §15.5.

When dealing with three-dimensional waves it is often convenient to find ways of replacing the vector wave equation by an equivalent scalar wave equation. It is useful in this regard to note that (15.8) may be combined with $\nabla \cdot \mathbf{E} = 0$ and $\nabla \cdot \mathbf{B} = 0$ to yield the scalar wave equations

$$\frac{\partial^2}{\partial t^2}(\mathbf{x}\cdot\mathbf{B}) = c^2\nabla^2(\mathbf{x}\cdot\mathbf{B}), \tag{15.9}$$

$$\frac{\partial^2}{\partial t^2}(\mathbf{x}\cdot\mathbf{E}) = c^2\nabla^2(\mathbf{x}\cdot\mathbf{E}), \tag{15.10}$$

as discussed in Exercise 15.1. We shall return to (15.9) and (15.10) when we discuss the Debye potentials in §15.5.

### 15.2.2 *Symmetries in the Free-Space Maxwell Equations*

Maxwell's equations are, of course, invariant under translation and rotation in space, as well as translation in time. Before turning to waves, it will prove useful to note two additional symmetries associated with the free-space Maxwell equations. First, Maxwell's equations in vacuum are reversible in time, in the sense that they are unaltered under the transformation $t \to -t$, $\mathbf{B} \to -\mathbf{B}$. So, for every solution we find of Maxwell's equations, there is another solution in which $t$ is replaced by $-t$. When working with waves triggered by localised sources, *causality* can usually be used to eliminate one of the two possible solutions.

The second symmetry is less obvious but equally important. If we introduce the auxiliary fields $\mathbf{E}^* = c\mathbf{B}$ and $\mathbf{B}^* = -\mathbf{E}/c$, then it is readily confirmed that $\mathbf{E}^*$ and $\mathbf{B}^*$ also satisfy Maxwell's equations in vacuum. This means that every time we find a wave-like solution of Maxwell's equations in free space, we know immediately that a second class of waves exists in which the roles of $\mathbf{E}$ and $\mathbf{B}$ are reversed. This will prove particularly convenient when working with three-dimensional waves, where each solution of the wave equations has a partner system that differs in a non-trivial way from the original. We shall take advantage of this symmetry, which is sometimes called *dual symmetry*, when discussing the potentials of Hertz and Debye in §15.5. Note that dual symmetry does not apply when the sources $\rho_e$ and $\mathbf{J}$ are reinstated in Maxwell's equations, owing to the absence of magnetic monopoles.

## 15.3 Plane Waves

We start, then, with the simplest of all possible electromagnetic waves: one-dimensional plane waves, first introduced in §13.4. So let us return to the results of that section.

### 15.3.1 *TEM Waves and D'Alembert Revisited*

The simplest way to ensure that $\nabla\cdot\mathbf{E} = 0$ and $\nabla\cdot\mathbf{B} = 0$ are satisfied in electromagnetic waves is to have both $\mathbf{E}$ and $\mathbf{B}$ transverse to the direction of propagation. This suggests that we look for plane wave solutions of the form

$$\mathbf{E}(z,t) = (E_x, E_y, 0), \qquad \mathbf{B}(z,t) = (B_x, B_y, 0), \tag{15.11}$$

from which Faraday's law and the Ampère–Maxwell equation give us

$$\hat{\mathbf{e}}_z \times \frac{\partial \mathbf{E}}{\partial z} = -\frac{\partial \mathbf{B}}{\partial t}, \quad \hat{\mathbf{e}}_z \times \frac{\partial \mathbf{B}}{\partial z} = \frac{1}{c^2}\frac{\partial \mathbf{E}}{\partial t}. \tag{15.12}$$

Application of $(\partial/\partial z)\hat{\mathbf{e}}_z\times$ to these equations yields one-dimensional wave equations for both **E** and **B**:

$$\frac{\partial^2 \mathbf{E}}{\partial t^2} = c^2 \frac{\partial^2 \mathbf{E}}{\partial z^2}, \qquad \frac{\partial^2 \mathbf{B}}{\partial t^2} = c^2 \frac{\partial^2 \mathbf{B}}{\partial z^2}. \tag{15.13}$$

Such wave equations are non-dispersive and so d'Alembert's solution applies. It follows that the general solution of (15.13) is

$$\boxed{\mathbf{E}(z,t) = \mathbf{E}_f(z-ct) + \mathbf{E}_b(z+ct),} \tag{15.14}$$

$$\boxed{\mathbf{B}(z,t) = \mathbf{B}_f(z-ct) + \mathbf{B}_b(z+ct),} \tag{15.15}$$

where $\mathbf{E}_f$, $\mathbf{E}_b$ $\mathbf{B}_f$ and $\mathbf{B}_b$ are arbitrary functions of their arguments and the subscripts $f$ and $b$ stand for forward- and backward-travelling waves, respectively.

Adopting the terminology of Chapter 14, these are *transverse electromagnetic waves*, or *TEM waves* for short. When combined with (15.12), our d'Alembert solutions yield

$$\boxed{c\mathbf{B}_f = \hat{\mathbf{e}}_z \times \mathbf{E}_f, \quad c\mathbf{B}_b = -\hat{\mathbf{e}}_z \times \mathbf{E}_b.} \tag{15.16}$$

Evidently the electric and magnetic field lines are perpendicular to each other, as well as to the direction of propagation of the wave.

Perhaps the simplest illustration of these waves is the somewhat artificial problem of a current sheet of infinite extent that is suddenly 'turned on' at $t = 0$. Suppose that the sheet is located at the $x$–$y$ plane, $z = 0$, and that the imposed current density is

$$\mathbf{J}(z,t) = K(t)\delta(z)\hat{\mathbf{e}}_x, \tag{15.17}$$

where $K(t)$ is a Heaviside step function. As in Chapter 7, we integrate through the current sheet and rewrite (15.17) as $\mathbf{K} = K(t)\hat{\mathbf{e}}_x$, where **K** is the surface current density in A/m. The geometry is as shown in Figure 15.2.

In §7.4 we showed that the static magnetic field associated with a steady current sheet of this type is $\mathbf{B} = \mp(\mu_0/2)K\hat{\mathbf{e}}_y$, where the negative sign applies above the current sheet and the positive sign below. When considering the dynamic response of the magnetic field to such a current sheet being turned on, it is necessary only to replace $t$ by the retarded time $t' = t - z/c$. It follows that, above the current sheet,

$$\mathbf{B}(z,t) = -(\mu_0/2)K(t-z/c)\hat{\mathbf{e}}_y, \tag{15.18}$$

so that **B** extends only a distance $z = ct$ from the sheet. The relationship $c\mathbf{B}_f = \hat{\mathbf{e}}_z \times \mathbf{E}_f$ then gives the electric field as

$$\mathbf{E}(z,t) = -\frac{K(t-z/c)}{2\varepsilon_0 c}\hat{\mathbf{e}}_x. \tag{15.19}$$

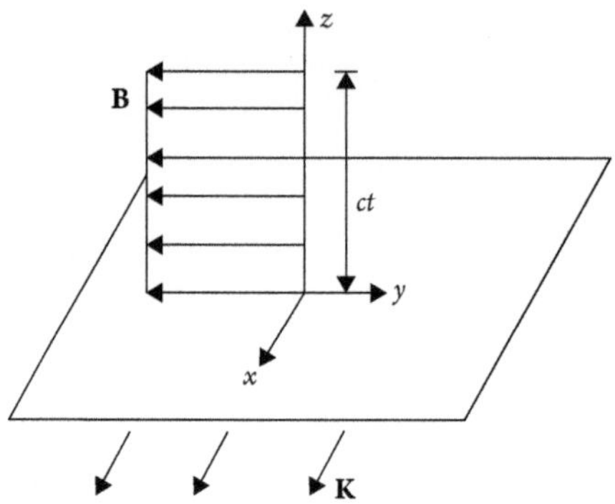


**Figure 15.2** A uniform current sheet is confined to the $x$–$y$ plane and turned on at $t = 0$. It induces a magnetic field in the $y$ direction that extends a distance $z = ct$ from the sheet.

### *15.3.2 Monochromatic Plane Waves, Polarisation and the Poynting Flux*

The most important special case of d'Alembert's solution is a monochromatic sinusoidal wave,

$$\mathbf{E}(\mathbf{x},t) = \hat{\mathbf{E}} \exp j(\mathbf{k}\cdot\mathbf{x} - \varpi t), \quad \mathbf{k}\cdot\hat{\mathbf{E}} = 0, \tag{15.20}$$

$$\mathbf{B}(\mathbf{x},t) = \hat{\mathbf{B}} \exp j(\mathbf{k}\cdot\mathbf{x} - \varpi t), \quad \mathbf{k}\cdot\hat{\mathbf{B}} = 0, \tag{15.21}$$

where the real part is understood and the amplitudes $\hat{\mathbf{E}}$ and $\hat{\mathbf{B}}$ may be complex. Evidently $c = \varpi/|\mathbf{k}|$ and it is readily confirmed that, by analogy with (15.16),

$$\boxed{c\hat{\mathbf{B}} = \hat{\mathbf{e}}_k \times \hat{\mathbf{E}},} \tag{15.22}$$

where $\hat{\mathbf{e}}_k$ is a unit vector in the direction of travel of the wave, $\hat{\mathbf{e}}_k = \mathbf{k}/|\mathbf{k}|$. The electric and magnetic fields are clearly perpendicular to each other, as well as to the direction of propagation of the wave, i.e. this is a TEM wave, as illustrated in Figure 15.3.

If $\hat{\mathbf{E}}$ and $\hat{\mathbf{B}}$ are both real, then $\mathbf{E}$ lies in a single plane, as does $\mathbf{B}$, and the $\mathbf{E}$ and $\mathbf{B}$ fields are in phase. Such a plane wave is said to be *linearly polarised* and, by convention, the *plane of polarisation* is taken to be the plane in which $\mathbf{E}$ and $\mathbf{k}$ both lie. Such a wave is illustrated in Figure 15.3. However, more complicated arrangements are possible. For example, suppose we superimpose two such plane waves which propagate in the same direction but are 90° out of phase. For example, if the waves propagate in the positive $z$ direction, we might take $\hat{\mathbf{E}} = E_0(\hat{\mathbf{e}}_x - j\hat{\mathbf{e}}_y)$, from which

$$\mathbf{E}(\mathbf{x},t) = E_0[\cos(kz - \varpi t)\hat{\mathbf{e}}_x + \sin(kz - \varpi t)\hat{\mathbf{e}}_y]. \tag{15.23}$$

Equation (15.22) then gives $c\hat{\mathbf{B}} = E_0(j\hat{\mathbf{e}}_x + \hat{\mathbf{e}}_y)$ and so the corresponding magnetic field is

$$c\mathbf{B}(\mathbf{x},t) = E_0[-\sin(kz - \varpi t)\hat{\mathbf{e}}_x + \cos(kz - \varpi t)\hat{\mathbf{e}}_y]. \tag{15.24}$$

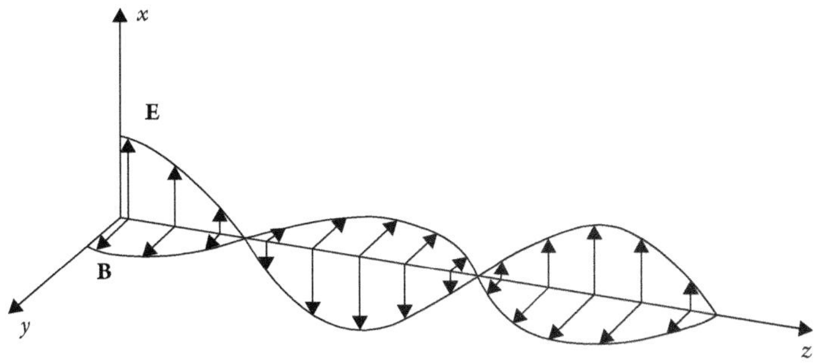


**Figure 15.3** A linearly polarized plane wave.

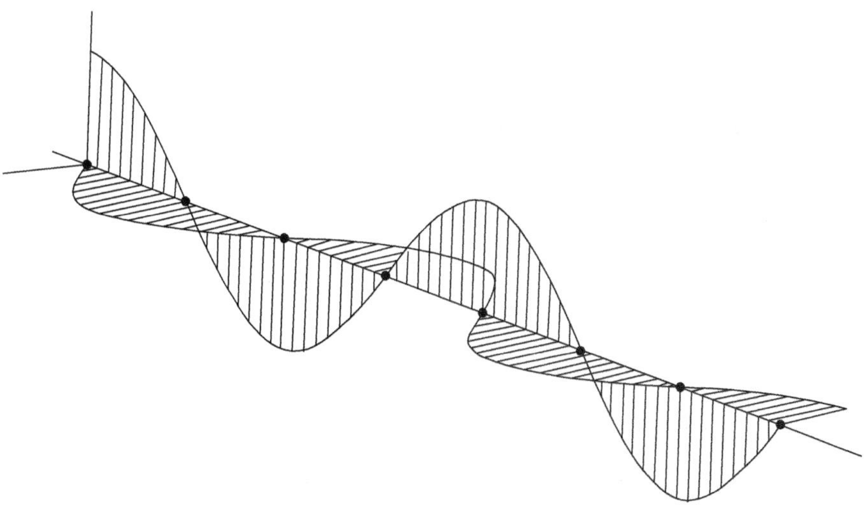

**Figure 15.4** The electric field of the circularly polarised wave (15.23).

Now suppose we focus on the location $z = 0$. Then, $\mathbf{E} = E_0(\hat{\mathbf{e}}_x \cos \varpi t - \hat{\mathbf{e}}_y \sin \varpi t)$ and $\mathbf{E}$ is seen to be a vector of fixed magnitude that rotates about the $z$ axis with an angular velocity that is antiparallel to $\mathbf{k}$. Conversely, if we take a snapshot of the wave at $t = 0$, we find $\mathbf{E} = E_0(\hat{\mathbf{e}}_x \cos kz + \hat{\mathbf{e}}_y \sin kz)$, and once again $\mathbf{E}$ spirals around the $z$ axis, this time as $z$ increases. Such a wave is said to be *circularly polarised*, and the electric field of a circularly polarised wave is shown in Figure 15.4. It is readily confirmed that (15.23) has negative helicity, $\mathbf{E} \cdot \nabla \times \mathbf{E} < 0$, whereas the choice $\hat{\mathbf{E}} = E_0(\hat{\mathbf{e}}_x + j\hat{\mathbf{e}}_y)$ gives a wave of positive helicity $\mathbf{E} \cdot \nabla \times \mathbf{E} > 0$.

Note that more complex forms of polarisation are possible. For example, if we replace (15.23) by

$$\mathbf{E}(\mathbf{x}, t) = E_1 \cos(kz - \varpi t)\hat{\mathbf{e}}_x + E_2 \sin(kz - \varpi t)\hat{\mathbf{e}}_y, \quad E_1 \neq E_2,$$

then the wave is said to be *elliptically polarised.* One might even consider cases where $E_1 \neq E_2$ and the phase difference is other than 90°. However, we shall not pause to discuss these more intricate cases, but rather refer the reader to Jackson (1999).

We close this section with a comment about energy transport by one-dimensional waves. Let us rewrite (15.16) as $c\mathbf{B} = \hat{\mathbf{e}}_c \times \mathbf{E}$, where $\hat{\mathbf{e}}_c = \pm\hat{\mathbf{e}}_z$ is a unit vector in the direction of propagation of the wave, equal to $\hat{\mathbf{e}}_k = \mathbf{k}/|\mathbf{k}|$ for sinusoidal waves. Evidently the Poynting vector for one-dimensional waves of the d'Alembert type is

$$\boxed{\frac{\mathbf{E}\times\mathbf{B}}{\mu_0} = \frac{|\mathbf{E}|^2}{\mu_0 c}\hat{\mathbf{e}}_c = \frac{c\,|\mathbf{B}|^2}{\mu_0}\hat{\mathbf{e}}_c.} \tag{15.25}$$

As might be expected, the energy flux is in the direction of $\hat{\mathbf{e}}_c$, i.e. the direction of propagation of the wave. Note that (15.25) includes sinusoidal waves as a special case. On averaging over a cycle for sinusoidal waves, and using $\langle\sim\rangle$ for that average, we find

$$\frac{\langle\mathbf{E}\times\mathbf{B}\rangle}{\mu_0} = \frac{\langle\mathbf{E}^2\rangle}{2\mu_0 c}\hat{\mathbf{e}}_c = \frac{c\langle\mathbf{B}^2\rangle}{2\mu_0}\hat{\mathbf{e}}_c. \tag{15.26}$$

## 15.4 Spherical Waves Far from Their Sources: A Glimpse at Radiation

Perhaps the next simplest type of wave is one that propagates in the radial direction. In some sense this is also a form of one-dimensional wave, although its behaviour is somewhat different to that of a plane wave. As we shall see when we discuss radiation in Chapter 16, this type of spherical wave is typical of the leading-order contribution to the far-field radiation generated by a localised source. So let us now consider spherical waves at large distances from the origin, assumed to be generated by a compact localised source located near the origin. By large distances from the origin we mean that $|\mathbf{x}| >> \lambda$, so that spherical surfaces look 'almost flat' on the scale of the wavelength of the waves. Since $\lambda \sim c/\varpi$, this approximation amounts to the assumption that $r = |\mathbf{x}| >> c/\varpi$.

### 15.4.1 *Radially Propagating Waves Far from an Oscillating Electric Dipole*

Consider the vector potential

$$\mathbf{A}(r,t) = \frac{G(r,t)}{r}\hat{\mathbf{e}}_z, \quad r = |\mathbf{x}|. \tag{15.27}$$

Evidently $A_z$ is governed by the spherically symmetric wave equation,

$$\frac{\partial^2}{\partial t^2}\left(\frac{G}{r}\right) = c^2\frac{1}{r^2}\frac{\partial}{\partial r}\left[r^2\frac{\partial}{\partial r}\left(\frac{G}{r}\right)\right].$$

This, in turn, yields

$$\frac{\partial^2 G}{\partial t^2} = c^2 \frac{\partial^2 G}{\partial r^2}, \tag{15.28}$$

which admits the d'Alembert solutions $G = G(r \pm ct)$. Moreover, since we have in mind a localised source near the origin, causality demands that we discard the incoming waves, leaving us with the outward-propagating wave

$$\mathbf{A} = \frac{G(kr - \varpi t)}{r} \hat{\mathbf{e}}_z, \quad \varpi = ck. \tag{15.29}$$

We may now determine the scalar potential from the Lorenz gauge (15.1). In spherical polar coordinates $(r,\theta,\phi)$, (15.1) yields

$$\frac{1}{c^2}\frac{\partial \varphi}{\partial t} = -\nabla \cdot \mathbf{A} = -\cos\theta \frac{\partial}{\partial r}\left(\frac{G}{r}\right) \approx -\frac{\cos\theta}{r}\frac{\partial G}{\partial r} = -k\frac{\cos\theta}{r}\frac{\partial G}{\partial kr} = k\frac{\cos\theta}{r}\frac{\partial G}{\partial \varpi t},$$

where we have used $kr >> 1$ in order to drop the term $\cos\theta G/r^2$. We conclude that, to leading order in $(kr)^{-1}$,

$$\varphi = c\frac{\cos\theta}{r} G(kr - \varpi t). \tag{15.30}$$

The **E** and **B** fields now follow. In particular, after some algebra, (15.29) and (15.30) yield

$$\mathbf{B} = \nabla \times \mathbf{A} = -\sin\theta \frac{\partial}{\partial r}\left(\frac{G}{r}\right) \hat{\mathbf{e}}_\phi \approx -\frac{\sin\theta}{r}\frac{\partial G}{\partial r}\hat{\mathbf{e}}_\phi \tag{15.31}$$

and

$$\mathbf{E} = -\frac{\partial \mathbf{A}}{\partial t} - \nabla\varphi = c\cos\theta \frac{G}{r^2}\hat{\mathbf{e}}_r - c\sin\theta \frac{\partial}{\partial r}\left(\frac{G}{r}\right)\hat{\mathbf{e}}_\theta \approx -c\frac{\sin\theta}{r}\frac{\partial G}{\partial r}\hat{\mathbf{e}}_\theta, \tag{15.32}$$

where we have used $\partial G/\partial t = -c\partial G/\partial r$ in evaluating $\partial \mathbf{A}/\partial t$ and $kr >> 1$ on the right of (15.31) and (15.32). Note that, to leading order in $(kr)^{-1}$, **B** and **E** are normal to $\hat{\mathbf{e}}_r$, the direction of propagation of the wave. These are therefore (approximate) TEM waves, in the sense that the leading-order fields are perpendicular to each other and also to $\hat{\mathbf{e}}_r$. In summary then, for $kr >> 1$ we have

$$\boxed{\mathbf{B} = -\frac{\varpi}{c}\frac{\sin\theta}{r} G'(kr - \varpi t)\hat{\mathbf{e}}_\phi, \qquad \mathbf{E} = -\varpi\frac{\sin\theta}{r} G'(kr - \varpi t)\hat{\mathbf{e}}_\theta,} \tag{15.33}$$

where the prime on $G$ indicates differentiation with respect to $kr - \varpi t$. From these expressions, we obtain

$$\boxed{c\mathbf{B} = \hat{\mathbf{e}}_r \times \mathbf{E}} \tag{15.34}$$

and a Poynting vector

$$\boxed{\frac{\mathbf{E}\times\mathbf{B}}{\mu_0} = \frac{|\mathbf{E}|^2}{\mu_0 c}\hat{\mathbf{e}}_r = \frac{c\,|\mathbf{B}|^2}{\mu_0}\hat{\mathbf{e}}_r.} \tag{15.35}$$

So, as we might have anticipated, the energy flux is radially outward.

In Chapter 16 we shall use (15.4) and (15.5) to show that, to leading order in $(kr)^{-1}$, these are precisely the wave fields generated at large distances from a compact oscillating *electric dipole* $\mathbf{p} = p_0\hat{\mathbf{e}}_z \cos\varpi t$. It necessary only to take

$$G(kr - \varpi t) = \frac{\mu_0 p_0 \varpi}{4\pi}\sin(kr - \varpi t). \tag{15.36}$$

Thus, for an oscillating electric dipole we have

$$\mathbf{A} = \frac{\mu_0 p_0 \varpi}{4\pi r}\sin(kr - \varpi t)\,\hat{\mathbf{e}}_z, \tag{15.37}$$

$$\mathbf{B} = -\frac{\mu_0 p_0 \varpi^2}{4\pi c}\frac{\sin\theta}{r}\cos(kr - \varpi t)\,\hat{\mathbf{e}}_\phi, \tag{15.38}$$

$$\mathbf{E} = -\frac{\mu_0 p_0 \varpi^2}{4\pi}\frac{\sin\theta}{r}\cos(kr - \varpi t)\,\hat{\mathbf{e}}_\theta \tag{15.39}$$

and

$$\frac{\mathbf{E}\times\mathbf{B}}{\mu_0} = \frac{\mu_0 p_0^2 \varpi^4}{(4\pi)^2 c}\left(\frac{\sin\theta}{r}\right)^2 \cos^2(kr - \varpi t)\,\hat{\mathbf{e}}_r. \tag{15.40}$$

Note that the field lines for **B** are azimuthal, encircling the axis of the electric dipole. The electric field, on the other hand, has a more complicated structure, as shown schematically in Figure 15.5. Note also that the energy flux is zero in the polar directions and has a maximum on the equatorial plane. Finally, we note that the energy flux rises sharply with frequency, as $\mathbf{E}\times\mathbf{B} \sim \varpi^4$.

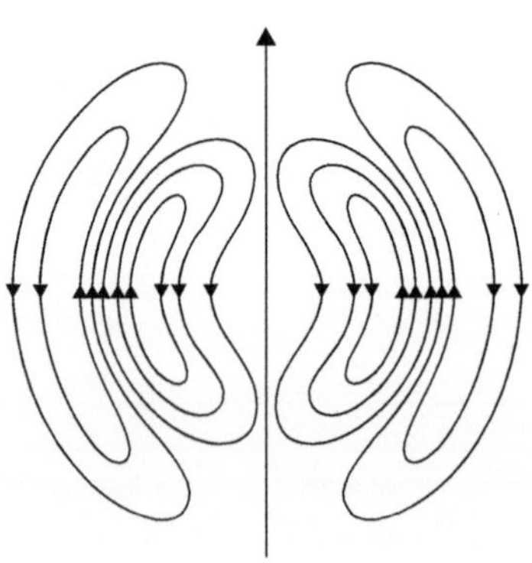

**Figure 15.5** Schematic of the electric field associated with an oscillating electric dipole.

### 15.4.2 *Radially Propagating Waves Far from an Oscillating Magnetic Dipole*

We now consider the vector potential

$$\mathbf{A}(\mathbf{x},t) = \sin\theta \frac{G(r,t)}{r}\hat{\mathbf{e}}_\phi, \tag{15.41}$$

also expressed in spherical polar coordinates $(r,\theta,\phi)$. Since $\mathbf{A}$ has zero divergence, the Lorenz gauge tells us that we may drop the scalar potential and take $\varphi = 0$. Thus we are considering a wave source near the origin consisting of currents but no charge density.

When due attention is paid to the derivatives of the unit vector $\hat{\mathbf{e}}_\phi$, it is straightforward, though tedious, to show that

$$\nabla^2\mathbf{A} = \frac{\sin\theta}{r}\left(\frac{\partial^2 G}{\partial r^2} - 2\frac{G}{r^2}\right)\hat{\mathbf{e}}_\phi,$$

and so the wave equation for $\mathbf{A}$ reduces to

$$\frac{\partial^2 G}{\partial t^2} = c^2\left(\frac{\partial^2 G}{\partial r^2} - 2\frac{G}{r^2}\right). \tag{15.42}$$

The solution of (15.42) is discussed in Exercise (15.5), although for the present purposes we need only note that

$$\frac{\partial^2 G}{\partial t^2} \approx c^2\frac{\partial^2 G}{\partial r^2}$$

for $kr >> 1$. Once again $G$ admits the d'Alembert solutions $G = G(r \pm ct)$, with causality demanding that we discard the incoming waves. We conclude that, for $kr >> 1$,

$$\mathbf{A} = \frac{\sin\theta}{r}G(kr - \varpi t)\,\hat{\mathbf{e}}_\phi, \quad \varpi = ck. \tag{15.43}$$

In Chapter 16 we shall use (15.5) to show that this is the vector potential at large distances from an oscillating *magnetic dipole* located at the origin, $\mathbf{m} = m_0\hat{\mathbf{e}}_z\cos\varpi t$. It is necessary only to take

$$G(kr - \varpi t) = \frac{\mu_0 m_0 \varpi}{4\pi c}\sin(kr - \varpi t). \tag{15.44}$$

As with the wave field in §15.4.1, the resulting waves are perpendicular to each other and also to $\hat{\mathbf{e}}_r$, and so they are TEM waves, at least to leading order in $(kr)^{-1}$. They also satisfy $c\mathbf{B} = \hat{\mathbf{e}}_r \times \mathbf{E}$, just like the far field of an electric dipole. So, once again, the Poynting vector is

$$\frac{\mathbf{E}\times\mathbf{B}}{\mu_0} = \frac{|\mathbf{E}|^2}{\mu_0 c}\hat{\mathbf{e}}_r = \frac{c\,|\mathbf{B}|^2}{\mu_0}\hat{\mathbf{e}}_r. \tag{15.45}$$

## 15.5 Three-Dimensional Waves: The Potentials of Hertz and Debye

We now turn to the difficult task of describing fully three-dimensional waves. If we are far enough away from the sources that create the wave field, then we might expect such disturbances to look increasingly like the radially propagating waves of §15.4. However, closer to those sources, the waves will have a more complex structure and so a more general framework is required. We now introduce two strategies that, historically, have been adopted for precisely this purpose. These are the potentials of *Hertz* and *Debye*.

The first, as the name suggests, came very early, being introduced by Hertz in 1888. In effect, this replaces $\varphi$ and $\mathbf{A}$ by two related potentials $\mathbf{\Pi}_e$ and $\mathbf{\Pi}_m$, which are known as the *electric and magnetic Hertz vectors*. In the absence of sources these new potentials are governed by

$$\nabla_\mu^2 \mathbf{\Pi}_e = 0, \quad \nabla_\mu^2 \mathbf{\Pi}_m = 0, \qquad \nabla_\mu^2 = \nabla^2 - \frac{1}{c^2}\frac{\partial^2}{\partial t^2}, \tag{15.46}$$

where the wave operator $\nabla_\mu^2$ is called the *d'Alembertian*. At first sight these offer no advantage over $\varphi$ and $\mathbf{A}$. However, when sources in the form of time-dependent electric and magnetic dipoles are introduced, the advantage of Hertz's potentials becomes clear, since they capture the wave field with elegant simplicity. The second approach has its roots in a 1909 paper by Debye and starts with the general observation that *any* solenoidal vector field can be formally decomposed into *poloidal and toroidal components*, each of which is uniquely specified by a scalar field, say $P$ and $T$. When this decomposition is applied to $\mathbf{B}$ in the free-space Maxwell equations, $P$ and $T$ are found to be governed by $\nabla_\mu^2 P = \nabla_\mu^2 T = 0$. These are *Debye's magnetic and electric scalar potentials*, and they have the advantage that we need only solve the scalar wave equation, rather than (15.46). The algebra that follows is not for the faint-hearted, but the end result is worthwhile.

### *15.5.1 The Hertz Vectors*

In the absence of the magnetic Hertz vector $\mathbf{\Pi}_m$, which we shall define shortly, the *electric Hertz vector* $\mathbf{\Pi}_e$ is defined by

$$\mathbf{A} = \frac{1}{c^2}\frac{\partial \mathbf{\Pi}_e}{\partial t}, \quad \nabla \cdot \mathbf{\Pi}_e = -\varphi. \tag{15.47}$$

This clearly satisfies the Lorenz gauge and uniquely determines $\mathbf{\Pi}_e$ in terms of $\varphi$ and $\mathbf{A}$. Moreover, the wave equation for $\mathbf{A}$ tell us that, in the absences of sources, $\nabla_\mu^2 \mathbf{\Pi}_e = 0$. The fields $\mathbf{E}$ and $\mathbf{B}$ are then given by

$$\mathbf{B} = \nabla \times \mathbf{A} = \frac{1}{c^2}\frac{\partial}{\partial t}\nabla \times \mathbf{\Pi}_e, \tag{15.48}$$

$$\mathbf{E} = -\nabla\varphi - \frac{\partial \mathbf{A}}{\partial t} = \nabla(\nabla \cdot \mathbf{\Pi}_e) - \frac{1}{c^2}\frac{\partial^2 \mathbf{\Pi}_e}{\partial t^2} = \nabla \times \nabla \times \mathbf{\Pi}_e. \tag{15.49}$$

In summary, then, any solution $\mathbf{\Pi}_e$ of the wave equation $\nabla^2_\mu \mathbf{\Pi}_e = 0$ yields a legitimate electromagnetic field through (15.48) and (15.49).

Now $\mathbf{E}^* = -c\mathbf{B}$ and $\mathbf{B}^* = \mathbf{E}/c$ also satisfy Maxwell's equations in vacuum, and so there is a mirror solution to (15.48) and (15.49). If we replace $\mathbf{\Pi}_e/c$ by $\mathbf{\Pi}_m$, this is

$$\mathbf{E} = -\frac{\partial \mathbf{\Lambda}_m}{\partial t}, \qquad \mathbf{B} = \nabla \times \mathbf{\Lambda}_m, \quad \mathbf{\Lambda}_m = \nabla \times \mathbf{\Pi}_m, \tag{15.50}$$

where $\mathbf{\Pi}_m$ also satisfies the wave equation: $\nabla^2_\mu \mathbf{\Pi}_m = 0$. So any solution $\mathbf{\Pi}_m$ of the wave equation $\nabla^2_\mu \mathbf{\Pi}_m = 0$ yields a second electromagnetic field through (15.50). Uncurling the expression for $\mathbf{B}$ in (15.50) gives $\mathbf{A} = \mathbf{\Lambda}_m - \nabla\psi$ for some $\psi$. However, for reasons that will become apparent, this second electromagnetic field is usually restricted to problems in which the wave source consists of currents only, there being no charge density. The retarded potential (15.4) then tells us that $\varphi = 0$ and the Lorenz gauge demands $\nabla \cdot \mathbf{A} = 0$. So it is conventional to suppress $\psi$ and take $\mathbf{A} = \mathbf{\Lambda}_m = \nabla \times \mathbf{\Pi}_m$. This then serves to define the curl of the *magnetic Hertz vector* $\mathbf{\Pi}_m$, in the absence of $\mathbf{\Pi}_e$.

We might now ask: may we combine these expressions in an unambiguous way to yield a general representation of any electromagnetic field? This is discussed by Stratton (1941), who proposes

$$\boxed{\mathbf{A} = \frac{1}{c^2}\frac{\partial \mathbf{\Pi}_e}{\partial t} + \mathbf{\Lambda}_m, \quad \varphi = -\nabla \cdot \mathbf{\Pi}_e, \quad \mathbf{\Lambda}_m = \nabla \times \mathbf{\Pi}_m.} \tag{15.51}$$

This satisfies the Lorenz gauge, and the corresponding $\mathbf{B}$ and $\mathbf{E}$ fields are evidently

$$\mathbf{B} = \nabla \times \mathbf{A} = \frac{1}{c^2}\frac{\partial}{\partial t}\nabla \times \mathbf{\Pi}_e + \nabla \times \mathbf{\Lambda}_m, \tag{15.52}$$

$$\mathbf{E} = -\nabla\varphi - \frac{\partial \mathbf{A}}{\partial t} = \nabla(\nabla \cdot \mathbf{\Pi}_e) - \frac{1}{c^2}\frac{\partial^2 \mathbf{\Pi}_e}{\partial t^2} - \frac{\partial \mathbf{\Lambda}_m}{\partial t}. \tag{15.53}$$

Note that $\mathbf{\Pi}_e$ and $\mathbf{\Lambda}_m$ in (15.51) are ambiguous, as we can add $\nabla \times \mathbf{G}$ to $\mathbf{\Pi}_e$, for some $\mathbf{G}$, while subtracting $c^{-2}\partial(\nabla \times \mathbf{G})/\partial t$ from $\mathbf{\Lambda}_m$, leaving (15.51) unaltered. We shall exploit this gauge-like freedom shortly. Since $\nabla^2_\mu$ commutes with both $\nabla \cdot (\sim)$ and $\nabla \times (\sim)$ (see Exercise 15.1), (15.2), (15.3) and (15.51) yield, in the absence of sources,

$$\nabla \cdot (\nabla^2_\mu \mathbf{\Pi}_e) = 0, \qquad \frac{1}{c^2}\frac{\partial}{\partial t}(\nabla^2_\mu \mathbf{\Pi}_e) + \nabla \times (\nabla^2_\mu \mathbf{\Pi}_m) = 0. \tag{15.54}$$

The first of these may be rewritten as $\nabla^2_\mu \mathbf{\Pi}_e = \nabla \times$ (some vector), or

$$\nabla^2_\mu \mathbf{\Pi}_e = \nabla \times (\nabla^2_\mu \mathbf{G}) = \nabla^2_\mu (\nabla \times \mathbf{G}),$$

for some $\mathbf{G}$. We now exploit the gauge-like freedom in $\mathbf{\Pi}_e$ and *elect* to take $\mathbf{G} = 0$. This gives us $\nabla^2_\mu \mathbf{\Pi}_e = 0$ and hence, from (15.54),

$$\boxed{\nabla^2_\mu \mathbf{\Pi}_e = 0, \qquad \nabla^2_\mu \mathbf{\Pi}_m = 0,} \tag{15.55}$$

whose **B** and **E** fields are

$$\boxed{\mathbf{B} = \frac{1}{c^2}\frac{\partial}{\partial t}\nabla \times \mathbf{\Pi}_e + \nabla \times \nabla \times \mathbf{\Pi}_m,} \tag{15.56}$$

$$\boxed{\mathbf{E} = \nabla \times \nabla \times \mathbf{\Pi}_e - \frac{\partial}{\partial t}\nabla \times \mathbf{\Pi}_m.} \tag{15.57}$$

Note that, when $\mathbf{\Pi}_e = 0$, all of the information about the fields is contained in $\mathbf{\Pi}_m$. In such cases the divergence of **A** is zero and so the Lorenz gauge requires that $\varphi = 0$. It follows from (15.4) and (15.5) that, when $\mathbf{\Pi}_e = 0$, the wave source consists of currents only, there being no charge density. So we associated $\mathbf{\Pi}_m$ with non-electric sources, such as oscillating magnetic dipoles.

In order to illustrate the utility of the Hertz potentials, consider a compact magnetic dipole **m**, sitting at the origin. We seek a representation of the fields in terms of $\mathbf{\Pi}_m$ alone, and so take $\mathbf{\Pi}_e = 0$. In magnetostatics the vector potential of a point dipole is

$$\mathbf{A} = \nabla \times \left(\frac{\mu_0 \mathbf{m}}{4\pi r}\right),$$

and since $\mathbf{\Pi}_m$ is defined by $\mathbf{A} = \nabla \times \mathbf{\Pi}_m$, we have, for *quasi-static fields*,

$$\mathbf{\Pi}_m^{\text{static}}(\mathbf{x}, t) = \frac{\mu_0 \mathbf{m}(t)}{4\pi r}. \tag{15.58}$$

The Laplacian of this is

$$\nabla^2 \mathbf{\Pi}_m^{\text{static}} = -\mu_0 \mathbf{m}\delta(\mathbf{x}),$$

since $\nabla^2 r^{-1} = -4\pi\delta(\mathbf{x})$ (see §2.1.4). If the dipole now oscillates, then $\mathbf{\Pi}_m$ is controlled by the analogous wave equation,

$$\nabla_\mu^2 \mathbf{\Pi}_m = -\mu_0 \mathbf{m}(t)\delta(\mathbf{x}) \tag{15.59}$$

(see Jackson, 1999), and the theory of retarded potentials tells us that the solution to this wave equation is simply the quasi-static solution in which $t$ is replaced by retarded time. So the Hertz vector associated with $\mathbf{m}(t)$ is

$$\boxed{\mathbf{\Pi}_m(\mathbf{x}, t) = \frac{\mu_0 \mathbf{m}(t - r/c)}{4\pi r}.} \tag{15.60}$$

For $\varpi r/c >> 1$, (13.74) then yields

$$\mathbf{A} = \nabla \times \mathbf{\Pi}_m = \frac{\mu_0 \dot{\mathbf{m}}(t - r/c)}{4\pi r c} \times \hat{\mathbf{e}}_r, \tag{15.61}$$

where a dot over a quantity indicates a time derivative. Similarly, if we have an electric dipole at the origin, $\mathbf{p}(t)$, then it may be shown that $\mathbf{\Pi}_m = 0$ while $\mathbf{\Pi}_e$ is governed by (Stratton, 1941)

$$\nabla_{\mu}^2 \mathbf{\Pi}_e = -\varepsilon_0^{-1} \mathbf{p}(t)\delta(\mathbf{x}),$$

The solution of this wave equation is

$$\boxed{\mathbf{\Pi}_e(\mathbf{x},t) = \frac{\mathbf{p}(t - r/c)}{4\pi\varepsilon_0 r},} \tag{15.62}$$

as discussed in Exercise 15.4. Moreover, using (15.51) and (13.74), the corresponding $\mathbf{A}$ and $\mathbf{B}$ fields for $\varpi r/c >> 1$ are

$$\mathbf{A}(\mathbf{x},t) = \frac{\mu_0 \dot{\mathbf{p}}(t - r/c)}{4\pi r}, \quad \mathbf{B} = \nabla \times \mathbf{A} = \frac{\mu_0 \ddot{\mathbf{p}}(t - r/c)}{4\pi r c} \times \hat{\mathbf{e}}_r. \tag{15.63}$$

If we now use superposition, it is clear from (15.60) and (15.62) that the Hertz vectors provide an efficient representation of the wave field from a random collection of dipole sources. This lends support to the suggestion of Stratton (1941) that (15.51) provides a general representation of free-space fields, at least for dipolar sources.

### *15.5.2 The Poloidal–Toroidal Decomposition of Solenoidal Vector Fields*

We now discuss the *poloidal–toroidal (P–T)* decomposition of a solenoidal vector field, partly as it leads to the Debye potentials and partly because it is important in its own right. This decomposition is described in Chandrasekhar (1961) and Moffatt (1978).

Any solenoidal field of the form

$$\mathbf{B}_T = \nabla \times (\mathbf{x}T) = \nabla T \times \mathbf{x} \tag{15.64}$$

is referred to as *toroidal*. Clearly, since $\mathbf{x} \cdot \mathbf{B}_T = 0$, toroidal field lines sit on spherical surfaces. Moreover, any vector field of the form

$$\mathbf{B}_P = \nabla \times \nabla \times (\mathbf{x}P) = \nabla \times (\nabla P \times \mathbf{x}) \tag{15.65}$$

is referred to as *poloidal*. For axisymmetric fields described in cylindrical polar coordinates, $\mathbf{B}_P$ sits in the $s$–$z$ plane while $\mathbf{B}_T$ is azimuthal. Now expanding $\nabla \times (\nabla P \times \mathbf{x})$ in (15.65) gives

$$\mathbf{B}_P = -(\nabla^2 P)\mathbf{x} + \nabla(P + \mathbf{x} \cdot \nabla P), \tag{15.66}$$

as discussed in Exercise 15.2. It follows that

$$\mathbf{x} \cdot \mathbf{B}_P = -(\nabla^2 P)\mathbf{x}^2 + \mathbf{x} \cdot \nabla(P + \mathbf{x} \cdot \nabla P) = -L^2 P, \tag{15.67}$$

where $L^2$ is the operator defined in spherical polar coordinates by

$$\frac{1}{r^2}L^2\psi = \nabla^2\psi - \frac{1}{r^2}\frac{\partial}{\partial r}\left(r^2\frac{\partial\psi}{\partial r}\right). \tag{15.68}$$

In effect, $r^{-2}L^2$ is the Laplacian expanded in spherical polar coordinates and with the radial derivatives subtracted out. (Beware that some authors define $r^{-2}L^2$ as minus (15.68).)

Clearly the curl of a toroidal field is poloidal. Less obviously, the curl of a poloidal field is toroidal, since

$$\nabla \times \mathbf{B}_P = \nabla \times [-(\nabla^2 P)\mathbf{x}]. \tag{15.69}$$

Now suppose that we combine these expressions and write our solenoidal field as

$$\mathbf{B} = \mathbf{B}_P + \mathbf{B}_T = \nabla \times \nabla \times (\mathbf{x}P) + \nabla \times (\mathbf{x}T). \tag{15.70}$$

Then

$$\nabla \times \mathbf{B} = \nabla \times [-(\nabla^2 P)\mathbf{x}] + \nabla \times \nabla \times (\mathbf{x}T), \tag{15.71}$$

and so

$$\mathbf{x}\cdot\mathbf{B} = -L^2 P, \quad \mathbf{x}\cdot\nabla\times\mathbf{B} = -L^2 T. \tag{15.72}$$

We now make two important observations. First, a solenoidal vector field $\mathbf{B}$ is uniquely determined if $\mathbf{x}\cdot\mathbf{B}$ and $\mathbf{x}\cdot\nabla\times\mathbf{B}$ are specified, i.e. the radial components of the field and its curl are known (Bouwkamp & Casimir, 1954). This means that, provided (15.72) can be inverted to determine $P$ and $T$, these two functions uniquely determine the solenoidal field, and a decomposition of the field into poloidal and toroidal parts is always possible. Second, the inversion of (15.72) is, in fact, readily achieved, as we now discuss.

The operator $L^2$ involves only the $\theta$ and $\phi$ coordinates and has the property that

$$L^2 S_n + n(n+1)S_n = 0, \tag{15.73}$$

where $S_n$ are the *surface harmonics* of degree $n$,

$$S_n(\theta,\phi) = \sum_{m=0}^{n} A_n^m P_n^m(\cos\theta)\exp(jm\phi), \tag{15.74}$$

$A_n^m$ are arbitrary amplitudes and $P_n^m$ are *associated Legendre polynomials* (see e.g. Moffatt, 1978). Note that the functions $P_n^m$ may be derived from the Legendre polynomials of Appendix I using

$$P_n^m(x) = (-1)^m(1-x^2)^{m/2}\frac{d^m}{dx^m}P_n(x). \tag{15.75}$$

Now, surface harmonics form a complete orthogonal set on a sphere, and so it is possible to expand $\mathbf{x}\cdot\mathbf{B}$ and $\mathbf{x}\cdot\nabla\times\mathbf{B}$ using surface harmonics as a basis and with coefficients that depend on $r$. The condition (15.73) then allows both equations in (15.72) to be inverted term by term and the inversion is complete.

In summary, then, the decomposition of a smooth solenoidal vector field $\mathbf{B}$ into poloidal and toroidal parts is always possible, and the defining scalars, $P$ and $T$, are closely related to $\mathbf{x} \cdot \mathbf{B}$ and $\mathbf{x} \cdot \nabla \times \mathbf{B}$. We now apply this to Maxwell's equations.

### 15.5.3 *The P–T Decomposition Applied to the Free-Space Equations*

Let us now move from kinematics to electrodynamics. Given

$$\nabla \times \mathbf{B} = \nabla \times [-(\nabla^2 P)\mathbf{x}] + \nabla \times \nabla \times (\mathbf{x}T), \tag{15.76}$$

the Ampère–Maxwell equation in free space becomes

$$\frac{\partial \mathbf{E}}{\partial t} = c^2\{\nabla \times [-(\nabla^2 P)\mathbf{x}] + \nabla \times \nabla \times (\mathbf{x}T)\}, \tag{15.77}$$

from which

$$\frac{\partial}{\partial t}\nabla \times \mathbf{E} = c^2\{\nabla \times \nabla \times [-(\nabla^2 P)\mathbf{x}] + \nabla \times [-(\nabla^2 T)\mathbf{x}]\}.$$

On the other hand, Faraday's law tells us that

$$\frac{\partial}{\partial t}\nabla \times \mathbf{E} = -\frac{\partial^2 \mathbf{B}}{\partial t^2} = -[\nabla \times \nabla \times (\ddot{P}\mathbf{x}) + \nabla \times (\ddot{T}\mathbf{x})],$$

and equating the two yields

$$\nabla \times \nabla \times [\mathbf{x}(\ddot{P} - c^2\nabla^2 P)] + \nabla \times [\mathbf{x}(\ddot{T} - c^2\nabla^2 T)] = 0. \tag{15.78}$$

This might be compared with the original kinematic decomposition:

$$\mathbf{B} = \nabla \times \nabla \times (\mathbf{x}P) + \nabla \times (\mathbf{x}T). \tag{15.79}$$

Now, just as $\mathbf{B} = 0$ ensures $P = T = 0$, through the inversion of (15.72), so (15.78) ensures

$$\ddot{P} = c^2\nabla^2 P, \quad \ddot{T} = c^2\nabla^2 T. \tag{15.80}$$

Our scalar fields $P$ and $T$ have acquired some dynamical significance. Note that $\mathbf{E}$ now takes the form

$$\dot{\mathbf{E}} = -\nabla \times (\mathbf{x}\ddot{P}) + c^2\nabla \times \nabla \times (\mathbf{x}T). \tag{15.81}$$

### 15.5.4 From the P–T Decomposition to the Debye Potentials

We are finally in a position to introduce the Debye potentials. If we introduce the scalar function $Q$ defined by $\partial Q/\partial t = cT$, then the Debye potentials are $P$ and $Q$, with $P$ being the *magnetic Debye potential* and $Q$ the *electric Debye potential*. From (15.79)–(15.81), we see that the most general representation of **B** and **E** in terms of Debye potentials is

$$\boxed{\mathbf{B} = \nabla \times \nabla \times (\mathbf{x}P) + \frac{1}{c}\frac{\partial}{\partial t}\nabla \times (\mathbf{x}Q),} \tag{15.82}$$

$$\boxed{\mathbf{E} = -\frac{\partial}{\partial t}\nabla \times (\mathbf{x}P) + c\nabla \times \nabla \times (\mathbf{x}Q),} \tag{15.83}$$

where

$$\boxed{\ddot{P} = c^2\nabla^2 P, \quad \ddot{Q} = c^2\nabla^2 Q.} \tag{15.84}$$

Note that the potentials are related to the radial components of **B** and **E** through (15.72), which gives us

$$\mathbf{x}\cdot\mathbf{B} = -L^2 P, \quad \mathbf{x}\cdot\mathbf{E} = -L^2(cQ). \tag{15.85}$$

Of course, $\mathbf{x}\cdot\mathbf{B}$ and $\mathbf{x}\cdot\mathbf{E}$ are themselves governed by the scalar wave equation, as noted in (15.9) and (15.10). If we compare (15.82)–(15.84) with (15.55)–(15.57), we see that there is a close relationship between the Debye potentials and the Hertz vectors, with

$$\nabla \times \mathbf{\Pi}_m = \nabla \times (\mathbf{x}P), \qquad \nabla \times \mathbf{\Pi}_e = c\nabla \times (\mathbf{x}Q) \tag{15.86}$$

being a *particular case* of the Hertz vectors. So there are many points of contact between these two, nominally distinct, approaches.

To illustrate the utility of the Debye potentials, let us consider two special cases. First, suppose that the sources involve only current (there are no free charges) and that the toroidal field is zero. Then we have $\varphi = 0$ and $\nabla\cdot\mathbf{A} = 0$, and the **B** and **E** fields are given by the magnetic Debye potential as

$$\mathbf{B} = \nabla \times \nabla \times (\mathbf{x}P), \qquad \mathbf{E} = -\nabla\dot{P} \times \mathbf{x}. \tag{15.87}$$

The electric field satisfies $\mathbf{x}\cdot\mathbf{E} = 0$ and so the **E**-lines lie on spherical surfaces. In cases where the waves propagate radially outward from a localised source, these constitute transverse electric (TE) waves. Indeed, for a compact *magnetic* dipole sitting at the origin, combining (15.61) with (15.86) yields a far-field electric field

$$\mathbf{E} = -\nabla\dot{P} \times \mathbf{x} = -\frac{\mu_0\ddot{\mathbf{m}}(t - r/c)}{4\pi rc} \times \hat{\mathbf{e}}_r, \quad \varpi r/c >> 1. \tag{15.88}$$

The opposite case is where the poloidal field is zero:

$$\mathbf{B} = c^{-1}\nabla \times (\mathbf{x}\dot{Q}), \quad \mathbf{E} = c\nabla \times \nabla \times (\mathbf{x}Q). \tag{15.89}$$

Note that we may move from one solution to the other, say from $P$ to $Q$, using the *dual symmetry transformation* introduced in §15.2.2: $\mathbf{E}^* = c\mathbf{B}$ and $\mathbf{B}^* = -\mathbf{E}/c$, with $Q^* = P$. Note also that $\mathbf{x} \cdot \mathbf{B} = 0$ in (15.89), and so the **B**-lines now lie on spherical surfaces. In cases where the waves propagate radially outward from the origin, these constitute transverse magnetic (TM) waves. Indeed, for a compact *electric* dipole sitting at the origin, (15.63) yields

$$c\mathbf{B} = \nabla \dot{Q} \times \mathbf{x} = \frac{\mu_0 \ddot{\mathbf{p}}(t - r/c)}{4\pi r} \times \hat{\mathbf{e}}_r, \qquad \varpi r/c >> 1. \tag{15.90}$$

The nice thing about the Debye potentials is that any solution of the scalar wave equation yields a possible wave field. Perhaps the simplest type of solution to look for consists of spherical waves of the form

$$P, Q = f(r) S(\theta, \phi) \exp(-j\varpi t). \tag{15.91}$$

As discussed in §15.4, these are the sorts of waves one expects to see if the source sits near the origin and we are a long way from the source. After a little algebra, separation of variables applied to (15.91) yields

$$L^2 S + n(n+1) S = 0, \tag{15.92}$$

$$r^2 f''(r) + 2r f'(r) + [(rk)^2 - n(n+1)] f = 0, \tag{15.93}$$

where $k = \varpi/c$. We recognise from (15.73) that $S = S_n$, the surface harmonics of degree $n$. Of more interest is the radial equation, whose solutions are the spherical Bessel functions of the first, second and third kinds. As shown in Bouwkamp & Casimir (1954), the most appropriate Bessel function in this case is the spherical Bessel function of the third kind, $f(r) = h_n^{(1)}(kr)$, also called the spherical Hankel function. These functions take the form

$$h_0^{(1)}(x) = -j \frac{\exp jx}{x}, \qquad h_1^{(1)}(x) = -j \frac{\exp jx}{x^2} - \frac{\exp jx}{x}, \tag{15.94}$$

and so on. For large distances from the origin, $kr >> 1$, the Hankel function $h_n^{(1)}(kr)$ is

$$h_n^{(1)}(kr) \sim (-j)^{n+1} (kr)^{-1} \exp(jkr),$$

at least to leading order in $(kr)^{-1}$. This may be combined with the $\exp(-j\varpi t)$ contribution to (15.91) to yield radially propagating waves of the general form discussed in §15.4 (see Exercise 15.5).

The solution to the more general case of a cluster of multipole sources is set out in Bouwkamp & Casimir (1954). In brief, given a set of sources, Bouwkamp & Casimir provide a systematic way of determining $\mathbf{x} \cdot \mathbf{B}$ and $\mathbf{x} \cdot \mathbf{E}$ at all points outside a sphere that encloses those sources. They then use (15.85) to relate $P$ and $Q$, and hence **E** and **B**, to the sources. This strategy is discussed at length in Jackson (1999).

## EXERCISES

15.1 Show that the identity

$$\mathbf{x} \cdot \nabla^2 \mathbf{D} = \nabla^2 (\mathbf{x} \cdot \mathbf{D}) - 2\nabla \cdot \mathbf{D}$$

holds for any vector field $\mathbf{D}$. Hence deduce the scalar wave equations (15.9) and (15.10) from the vector wave equations (15.8). Also confirm that $\nabla^2$ commutes with both $\nabla \cdot (\sim)$ and $\nabla \times (\sim)$, and hence show that (15.54) follows from the wave equations for the potentials $\varphi$ and $\mathbf{A}$.

15.2 Expand $\mathbf{B}_P = \nabla \times (\nabla P \times \mathbf{x})$ to show that a poloidal magnetic field satisfies

$$\mathbf{B}_P = 2\nabla P - (\nabla^2 P)\mathbf{x} + (\mathbf{x} \cdot \nabla)\nabla P.$$

Now confirm the identity $\nabla(\mathbf{x} \cdot \nabla P) = (\mathbf{x} \cdot \nabla)\nabla P + \nabla P$ and use it to verify (15.66).

15.3 Find the $\mathbf{E}$ and $\mathbf{B}$ fields corresponding to the vector potential (15.43) and show that, to leading order in $(kr)^{-1}$, they satisfy $c\mathbf{B} = \hat{\mathbf{e}}_r \times \mathbf{E}$.

15.4 Use (4.24) to show that, for a quasi-steady electric dipole at the origin, the electric Hertz vector is

$$\mathbf{\Pi}_e^{\text{static}}(\mathbf{x}, t) = \frac{\mathbf{p}(t)}{4\pi\varepsilon_0 r}.$$

Hence confirm that

$$\nabla^2 \mathbf{\Pi}_e^{\text{static}} = -\frac{\mathbf{p}\delta(\mathbf{x})}{\varepsilon_0}$$

and discuss the relationship between these equations and (15.62), i.e.

$$\nabla_\mu^2 \mathbf{\Pi}_e = -\frac{\mathbf{p}(t)\delta(\mathbf{x})}{\varepsilon_0}, \quad \mathbf{\Pi}_e(\mathbf{x}, t) = \frac{\mathbf{p}(t - r/c)}{4\pi\varepsilon_0 r}.$$

15.5 Consider solutions of (15.42) of the form $G(r,t) = \hat{G}(r)\exp(-j\varpi t)$ and show that one solution is $\hat{G} \sim krh_1^{(1)}(kr)$, where $h_1^{(1)}(kr)$ is the spherical Hankel function of order one. Now consider the surface harmonic $S_1(\theta) = A_1^0 P_1^0(\cos\theta) = A_1^0 \cos\theta$ and show that the corresponding solution of (15.91) for the magnetic Debye potential is

$$P = h_1^{(1)}(kr) S_1(\theta) \exp(-j\varpi t).$$

Confirm that this is the oscillating magnetic dipole of (15.42) and that, for $kr >> 1$,

$$P \sim \cos\theta \frac{\cos(kr - \varpi t)}{kr}.$$

## REFERENCES

Bouwkamp, C. J. & Casimir, H. B. G., 1954, On multipole expansions in the theory of electromagnetic radiation. *Physica*, **20**, 539–554.

Chandrasekhar, S., 1961, *Hydrodynamic and Hydromagnetic Stability*. Oxford University Press.

Jackson, J. D., 1999, *Classical Electrodynamics*, 3rd ed. Wiley.

Moffatt, H. K., 1978, *Magnetic Field Generation in Electrically Conducting Fluids*. Cambridge University Press.

Stratton, J. A., 1941, *Electromagnetic Theory*. McGraw-Hill.

# 16

# Maxwell's Equations in Free Space II

## *Radiation*

'Mine is a long and a sad tale!' said the Mouse, turning to Alice, and sighing. 'It *is* a long tail, certainly', said Alice, looking down with wonder at the Mouse's tail; 'but why do you call it sad?'

Lewis Carroll (*Alice's Adventures in Wonderland*)

We now turn to the topic of radiation, which in electrodynamics is usually taken to mean the behaviour of waves at large distances from a localised source and in particular the transmission of electromagnetic energy out to infinity.

## 16.1 What is a Radiation Field in Electrodynamics?

In §13.8 we showed that the **E** and **B** fields corresponding to localised distributions of $\rho_e$ and **J** are

$$\mathbf{E}(\mathbf{x},t) = \frac{1}{4\pi\varepsilon_0}\int\left\{\left[\rho_e(\mathbf{x}',t') + \frac{|\mathbf{r}|}{c}\frac{\partial\rho_e(\mathbf{x}',t')}{\partial t'}\right]\frac{\mathbf{r}}{|\mathbf{r}|^3} - \frac{1}{c^2}\frac{1}{|\mathbf{r}|}\frac{\partial\mathbf{J}(\mathbf{x}',t')}{\partial t'}\right\}d\mathbf{x}', \tag{16.1}$$

$$\mathbf{B}(\mathbf{x},t) = \frac{\mu_0}{4\pi}\int\left\{\left[\mathbf{J}(\mathbf{x}',t') + \frac{|\mathbf{r}|}{c}\frac{\partial\mathbf{J}(\mathbf{x}',t')}{\partial t'}\right]\times\frac{\mathbf{r}}{|\mathbf{r}|^3}\right\}d\mathbf{x}'. \tag{16.2}$$

Here, as in earlier chapters, the displacement vector $\mathbf{r} = \mathbf{x} - \mathbf{x}'$ takes us from the source at $\mathbf{x}'$ to the observer at $\mathbf{x}$, as shown in Figure 16.1, and $t'$ is the retarded time at the source, $t' = t - |\mathbf{r}|/c$.

Evidently, a time-dependent current produces **E** and **B** fields that tail off as $|\mathbf{r}|^{-1}$. This is a slow decline in comparison with an electrostatic field, where $|\mathbf{E}|_\infty \sim |\mathbf{r}|^{-2}$, or a magnetostatic field, for which $|\mathbf{B}|_\infty \sim |\mathbf{r}|^{-3}$. At large distances from a source, then, it is the time-dependent contribution to the source that matters, as this casts the longest shadow.

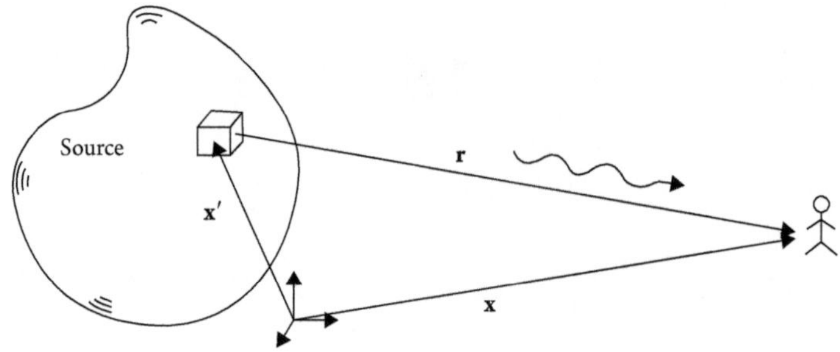


**Figure 16.1** The vector $\mathbf{r} = \mathbf{x} - \mathbf{x}'$ takes us from the source at $\mathbf{x}'$ to the observer at $\mathbf{x}$.

The far-field scalings $|\mathbf{B}|_\infty \sim |\mathbf{r}|^{-1}$ and $|\mathbf{E}|_\infty \sim |\mathbf{r}|^{-1}$ for a time-dependent source imply that, in general, the Poynting vector falls as $\mathbf{E}_\infty \times \mathbf{B}_\infty/\mu_0 \sim |\mathbf{r}|^{-2}$. This, in turn, tells us that the flux of electromagnetic energy through a spherical surface that encloses the source can remain finite and of order one as the radius of that sphere tends to infinity. In short, an oscillating current can radiate a finite amount of energy out to infinity. The far field of a source in which $|\mathbf{B}|_\infty \sim |\mathbf{r}|^{-1}$ and $|\mathbf{E}|_\infty \sim |\mathbf{r}|^{-1}$ is referred to as a *radiation field*.

In this chapter we shall examine the process by which waves carry energy out to infinity, focusing exclusively on the far field, called the *radiation zone*. One of our main conclusions will be that, in order for energy to be despatched to infinity, it is necessary for the charges that generate the field to *accelerate*. We shall start by looking at a compact oscillating dipole, either a magnetic or an electric dipole, and then move to the more general case of a compact source of arbitrary distribution. Throughout, we restrict ourselves to non-relativistic charges, i.e. to charges whose velocity is much less than *c*.

Perhaps we should say now exactly what we mean by the terms *compact source* and *far field*. Suppose the source of the waves has a characteristic lengthscale $\ell$, the waves emitted from the source have a wavelength $\lambda$ and the source is centred on the origin. Then a source is said to be compact if $\ell \ll \lambda$, and the far field is that region of space for which $|\mathbf{x}| \gg \lambda$. Thus the far field of a compact source satisfies the double inequality $\ell \ll \lambda \ll |\mathbf{x}|$, or, equivalently,

$$\ell \ll c/\varpi \ll |\mathbf{x}|, \tag{16.3}$$

where $\varpi$ is frequency. We shall apply these limits repeatedly in the next three sections.

## 16.2 The Magnetic Dipole

We start by considering a compact oscillating magnetic dipole, centred on the origin. Such a dipole turns out to be less efficient at radiating energy than its electric counterpart, but it is simpler to analyse. Of course, we already know from our discussion of the magnetic Hertz vector what the far field looks like, but it is instructive to derive those results directly from (15.5), the retarded potential for $\mathbf{A}$. So we shall now find the far-field vector potential for a simple magnetic dipole and then derive the $\mathbf{E}$ and $\mathbf{B}$ fields from that potential.

### 16.2.1 The Far-Field Potential

Consider a circular wire loop of radius $a$ carrying a current $I(t) = I_0 \cos \varpi t$, as shown in Figure 16.2. Since $\nabla \cdot \mathbf{J} = 0$, there are no free charges and so the scalar potential is zero. The vector potential, on the other hand, is

$$\mathbf{A}(\mathbf{x},t) = \frac{\mu_0}{4\pi} \int \frac{\mathbf{J}(\mathbf{x}',t')}{|\mathbf{r}|} \, d\mathbf{x}' = \frac{\mu_0}{4\pi} \int \frac{I_0 \cos(\varpi t')\, \hat{\mathbf{e}}'_\phi}{|\mathbf{r}|} a \, d\phi', \tag{16.4}$$

where $\mathbf{r} = \mathbf{x} - \mathbf{x}'$ and $t' = t - |\mathbf{r}|/c$ is the retarded time. Our task now is to find the asymptotic form of the integral on the right of (16.4) for large $|\mathbf{x}|$ and small $a$.

We shall adopt both Cartesian and spherical polar coordinates $(r, \theta, \phi)$, centred on the loop, with the loop in the $x-y$ plane. To focus our thoughts, we consider an observer sitting above the $y$ axis with a position vector $\mathbf{x} = |\mathbf{x}|\, \hat{\mathbf{e}}_r = |\mathbf{x}| \cos\theta\, \hat{\mathbf{e}}_z + |\mathbf{x}| \sin\theta\, \hat{\mathbf{e}}_y$, while a segment of the current loop is located by $\mathbf{x}' = a \cos\phi'\, \hat{\mathbf{e}}_x + a \sin\phi'\, \hat{\mathbf{e}}_y$, as shown in Figure 16.3. Evidently

$$|\mathbf{r}|^2 = |\mathbf{x}|^2 - 2\mathbf{x} \cdot \mathbf{x}' + a^2 = |\mathbf{x}|^2 - 2a\,|\mathbf{x}| \sin\theta \sin\phi' + a^2,$$

from which, for $|\mathbf{x}| \gg a$,

$$|\mathbf{r}| = |\mathbf{x}|\,[1 - (a/|\mathbf{x}|) \sin\theta \sin\phi']. \tag{16.5}$$

We now evaluate the various terms in the integrand of (16.4), starting with $I(t') = I_0 \cos(\varpi t')$. From (16.5) we may rewrite this as

$$I(t') = I_0 \cos \varpi t' = I_0 \cos[\varpi(t - |\mathbf{x}|/c) + (\varpi a/c) \sin\theta \sin\phi'].$$

Expanding the cosine as a function of its two arguments, and taking advantage of the fact that $\varpi a/c \ll 1$ to linearize the resulting trigonometric functions of $\varpi a/c$, yields

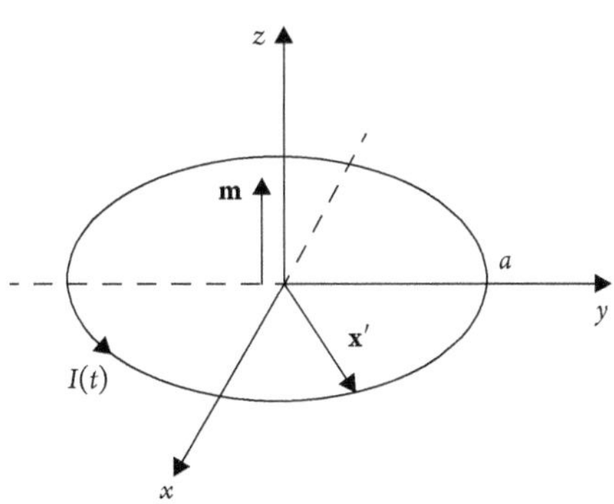


**Figure 16.2** An oscillating magnetic dipole consisting of a wire loop of radius $a$.

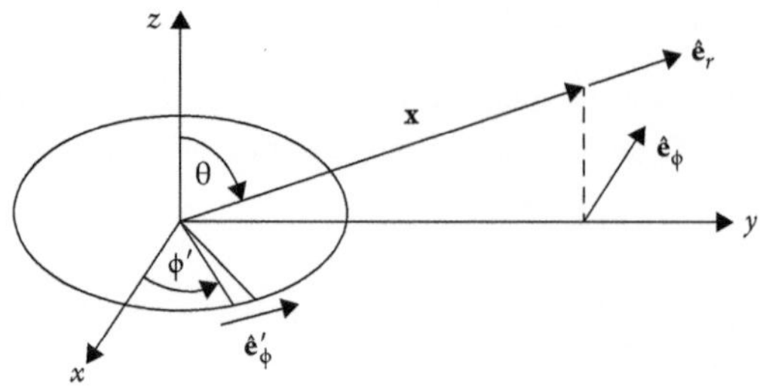


**Figure 16.3** The coordinate systems used to describe the fields of a magnetic dipole.

$$I_0 \cos \varpi t' = I_0 [\cos(k|\mathbf{x}| - \varpi t) + (\varpi a/c) \sin\theta \sin\phi' \sin(k|\mathbf{x}| - \varpi t)], \tag{16.6}$$

where $k = \varpi/c$ is the wavenumber.

Next we note that $\hat{\mathbf{e}}'_\phi = -\sin\phi' \hat{\mathbf{e}}_x + \cos\phi' \hat{\mathbf{e}}_y$. However, symmetry tells us that the contribution to (16.4) from $\cos\phi' \hat{\mathbf{e}}_y$ for $x' > 0$ is exactly cancelled by the corresponding contribution from $x' < 0$. On the other hand, the contributions from $\sin\phi' \hat{\mathbf{e}}_x$ for positive and negative $x'$ are equal and additive. It follows from (16.5) and (16.6) that

$$\mathbf{A}(\mathbf{x}, t) = -\frac{2a\mu_0 I_0}{4\pi |\mathbf{x}|} \int_{-\pi/2}^{\pi/2} \frac{\cos(k|\mathbf{x}| - \varpi t) + (\varpi a/c) \sin\theta \sin\phi' \sin(k|\mathbf{x}| - \varpi t)}{1 - (a/|\mathbf{x}|) \sin\theta \sin\phi'} \sin\phi' \, d\phi' \, \hat{\mathbf{e}}_x.$$

We now use $|\mathbf{x}| \gg a$ to rewrite $[1 - (a/|\mathbf{x}|) \sin\theta \sin\phi']^{-1}$ as $1 + (a/|\mathbf{x}|) \sin\theta \sin\phi'$, gather terms in $\sin^n \phi'$ and note that all terms that are odd in $\sin\phi'$ integrate to zero. This yields

$$\mathbf{A}(\mathbf{x}, t) = \frac{\mu_0 m_0}{4\pi |\mathbf{x}|^2} \sin\theta \left[ \cos(k|\mathbf{x}| - \varpi t) + \frac{\varpi |\mathbf{x}|}{c} \sin(k|\mathbf{x}| - \varpi t) \right] \hat{\mathbf{e}}_\phi, \tag{16.7}$$

where $m_0 = \pi a^2 I_0$ is the amplitude of the dipole moment. The first term on the right falls off as $|\mathbf{x}|^{-2}$. In the limit of $\varpi \to 0$, this becomes the quasi-static vector potential of a point dipole. The second term, however, falls off as $|\mathbf{x}|^{-1}$ and so dominates the far field. For $\varpi |\mathbf{x}|/c \gg 1$, this gives the vector potential in the radiation zone as

$$\mathbf{A}(\mathbf{x}, t) = \frac{\mu_0 \varpi m_0}{4\pi c} \frac{\sin\theta}{|\mathbf{x}|} \sin(k|\mathbf{x}| - \varpi t) \, \hat{\mathbf{e}}_\phi, \quad \nabla \cdot \mathbf{A} = 0, \tag{16.8}$$

which might be compared with (15.61). This is a monochromatic wave travelling radially outward from its source at the speed $c$. Our task now is to obtain the far-field forms of **E** and **B** from (16.8).

### 16.2.2 The Electromagnetic Field and Poynting Energy Flux

The electric and magnetic fields may be obtained from the vector potential according to

$$\mathbf{E}(\mathbf{x},t) = -\frac{\partial \mathbf{A}}{\partial t}, \quad \mathbf{B}(\mathbf{x},t) = \nabla \times \mathbf{A}. \tag{16.9}$$

Evidently $\nabla \cdot \mathbf{E} = 0$ and

$$\boxed{\mathbf{E}(\mathbf{x},t) = \frac{\mu_0 \varpi^2 m_0}{4\pi c} \frac{\sin\theta}{|\mathbf{x}|} \cos(k|\mathbf{x}| - \varpi t)\, \hat{\mathbf{e}}_\phi,} \tag{16.10}$$

which is azimuthal and so normal to the position vector **x**. Moreover, it falls off as $|\mathbf{x}|^{-1}$, as required by a radiation field. The corresponding *solenoidal* magnetic field is, to leading order in $c/\varpi|\mathbf{x}|$ and $a/|\mathbf{x}|$,

$$\boxed{\mathbf{B}(\mathbf{x},t) = -\frac{\mu_0 \varpi^2 m_0}{4\pi c^2} \left[ \frac{\sin\theta}{|\mathbf{x}|} \cos(k|\mathbf{x}| - \varpi t)\, \hat{\mathbf{e}}_\theta - \frac{2c}{\varpi} \frac{\cos\theta}{|\mathbf{x}|^2} \sin(k|\mathbf{x}| - \varpi t)\, \hat{\mathbf{e}}_r \right].} \tag{16.11}$$

The dominant component is now $\hat{\mathbf{e}}_\theta$. Like **E**, this component is normal to **x** and falls off as $|\mathbf{x}|^{-1}$. If we ignore the weak radial contribution to **B**, then (16.10) and (16.11) between them represent a transverse electromagnetic (TEM) wave, as shown in Figure 16.4.

Ignoring the weak radial contribution to **B**, the dominant **E** and **B** fields are related by $c\mathbf{B} = \hat{\mathbf{e}}_r \times \mathbf{E}$, where $\hat{\mathbf{e}}_r = \mathbf{x}/|\mathbf{x}|$, and so the Poynting vector is

$$\boxed{\frac{\mathbf{E} \times \mathbf{B}}{\mu_0} = \frac{|\mathbf{E}|^2}{\mu_0 c} \hat{\mathbf{e}}_r = \frac{\mu_0 \varpi^4 m_0^2}{(4\pi)^2 c^3} \frac{\sin^2\theta}{|\mathbf{x}|^2} \cos^2(k|\mathbf{x}| - \varpi t)\, \hat{\mathbf{e}}_r.} \tag{16.12}$$

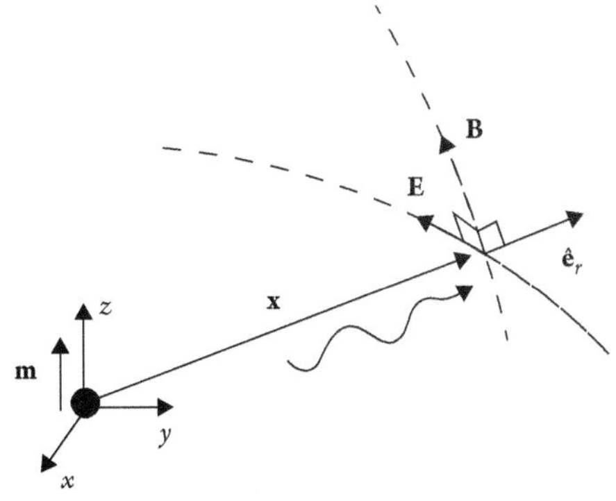


**Figure 16.4** The radiation field of an oscillating magnetic dipole is a TEM wave.

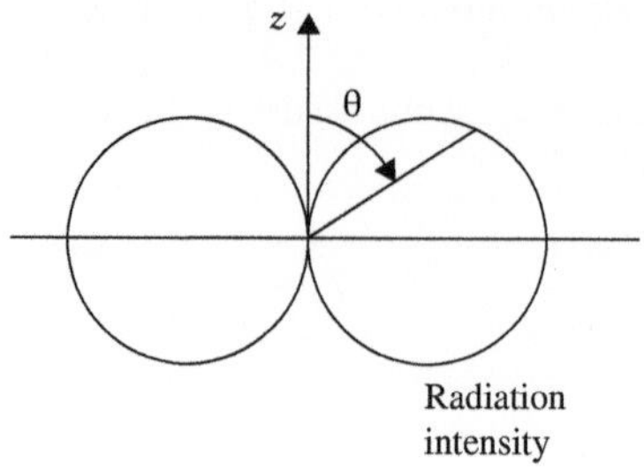


**Figure 16.5** The energy flux of a magnetic dipole is greatest in the equatorial plane and zero in the polar directions.

As we might have anticipated, this energy flux is directed radially outward. Note that the flux depends on the polar angle $\theta$, being greatest in the equatorial plane and zero in the polar directions, as shown in Figure 16.5. Moreover, on integrating (16.12) over the surface $|\mathbf{x}| = R$, we find

$$\frac{1}{\mu_0}\oint \mathbf{E}\times\mathbf{B}\cdot d\mathbf{S} = \frac{\mu_0\varpi^4 m_0^2}{6\pi c^3}\cos^2(k|\mathbf{x}| - \varpi t), \tag{16.13}$$

where we have used $\int(\sin\theta/|\mathbf{x}|)^2\, dS = 8\pi/3$. This is independent of $R$, as required by a radiation field.

## 16.3 The Hertzian Dipole

We now consider a compact oscillating *electric dipole*, centred on the origin. This is often called the Hertzian dipole. Physically, we take it to be a charge $-Q(t)$ located at $\mathbf{x} = -\mathbf{d}/2$ with a second charge $Q(t)$ at $\mathbf{x} = \mathbf{d}/2$, giving a net dipole moment of $\mathbf{p}(t) = Q(t)\mathbf{d}$. The two charges are connected by a wire whose function is to pass a current $I(t) = dQ/dt$ between them, thus ensuring that charge conservation is satisfied. To focus our thoughts, we take $\mathbf{d} = d\hat{\mathbf{e}}_z$, so that the two charges sit on the $z$ axis either side of the origin, and we employ spherical polar coordinates $(r,\theta,\phi)$, as shown in Figure 16.6. Finally, we restrict ourselves to harmonic oscillations in which $\mathbf{p}(t) = p_0\cos\varpi t\,\hat{\mathbf{e}}_z$, $Q(t) = Q_0\cos\varpi t$ and $d \ll c/\varpi$.

### *16.3.1 The Far-Field Potentials*

We have already deduced the far-field of a compact electric dipole using the electric Hertz vector. (See the discussion in §15.5.1.) However, it is instructive to derive the same results by more conventional means. So, following Griffiths (2013), we now find the far-field vector and scalar potentials and then derive the **E** and **B** fields from those potentials. Let $\mathbf{r}_\pm = \mathbf{x} \mp \mathbf{d}/2$ be the displacement vector that takes us from the top/bottom charge to the observer and let $t'_\pm = t - |\mathbf{r}_\pm|/c$ be the corresponding retarded time at the top/bottom charge. Squaring $\mathbf{r}_\pm = \mathbf{x} \mp \mathbf{d}/2$ gives

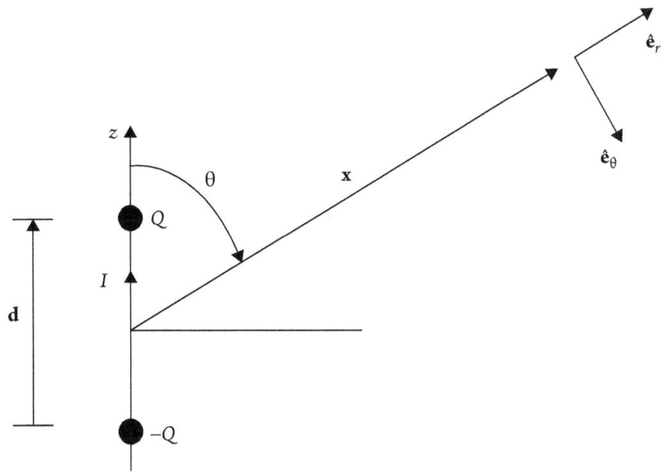


**Figure 16.6** An oscillating electric dipole consists of two charges and a connecting wire.

$$|\mathbf{r}_\pm|^2 = |\mathbf{x}|^2 \mp \mathbf{d}\cdot\mathbf{x} + (d/2)^2 = |\mathbf{x}|^2 \mp d\,|\mathbf{x}|\cos\theta + (d/2)^2, \tag{16.14}$$

and in the far field, where $|\mathbf{x}| \gg d$, this yields,

$$|\mathbf{r}_\pm| = |\mathbf{x}|\,[1 \mp (d/2\,|\mathbf{x}|)\cos\theta], \quad |\mathbf{r}_\pm|^{-1} = |\mathbf{x}|^{-1}\,[1 \pm (d/2\,|\mathbf{x}|)\cos\theta]. \tag{16.15}$$

Turning now to the scalar potential, we have

$$\varphi(\mathbf{x},t) = \frac{1}{4\pi\varepsilon_0}\int \frac{\rho_e(\mathbf{x}',t')}{|\mathbf{r}|}\,d\mathbf{x}' = \frac{Q_0}{4\pi\varepsilon_0}\left(\frac{\cos\varpi t'_+}{|\mathbf{r}_+|} - \frac{\cos\varpi t'_-}{|\mathbf{r}_-|}\right), \tag{16.16}$$

where (16.15) gives us

$$\frac{\cos\varpi t'_\pm}{|\mathbf{r}_\pm|} = |\mathbf{x}|^{-1}\left(1 \pm \frac{d}{2\,|\mathbf{x}|}\cos\theta\right)\cos\left[\varpi\,(t - |\mathbf{x}|/c) \pm \frac{\varpi d}{2c}\cos\theta\right]. \tag{16.17}$$

The expression (16.17) is messy, but it may be simplified for a compact source by using $\varpi\,d/c \ll 1$. The strategy is as follows. We expand the cosine on the right as a function of its two arguments, take advantage of the fact that $\varpi\,d/c \ll 1$ to linearize the resulting trigonometric functions of $\varpi\,d/c$, and finally multiply out the brackets while discarding a term involving the product of the two small parameters $\varpi\,d/c$ and $d/|\mathbf{x}|$. This yields

$$\frac{\cos\varpi\,t'_\pm}{|\mathbf{r}_\pm|} = \frac{\cos(k\,|\mathbf{x}| - \varpi\,t)}{|\mathbf{x}|} \pm \frac{d}{2\,|\mathbf{x}|^2}\cos\theta\left[\cos(k\,|\mathbf{x}| - \varpi\,t) + \frac{\varpi\,|\mathbf{x}|}{c}\sin(k\,|\mathbf{x}| - \varpi\,t)\right],$$

where $k = \varpi/c$ is the wavenumber. It follows that, for $d \ll |\mathbf{x}|$ and $\varpi d/c \ll 1$,

$$\varphi(\mathbf{x},t) = \frac{p_0 \cos\theta}{4\pi\varepsilon_0 |\mathbf{x}|^2}\left[\cos(k|\mathbf{x}| - \varpi t) + \frac{\varpi\,|\mathbf{x}|}{c}\sin(k|\mathbf{x}| - \varpi t)\right]. \tag{16.18}$$

The first term on the right of (16.18) falls off as $|\mathbf{x}|^{-2}$. For $\varpi \to 0$ this becomes the quasi-static Coulomb potential for a point dipole. The second term, however, falls off as $|\mathbf{x}|^{-1}$ and so dominates the far field. For $\varpi\,|\mathbf{x}|/c \gg 1$, this gives the far-field scalar potential as

$$\varphi(\mathbf{x},t) = \frac{p_0}{4\pi\varepsilon_0}\frac{\varpi}{c\,|\mathbf{x}|}\cos\theta\sin(k|\mathbf{x}| - \varpi t). \tag{16.19}$$

The determination of the far-field vector potential is more straightforward. It turns out that, to obtain an estimate for $\mathbf{A}$ that is asymptotically consistent with (16.18), it is necessary only to make the most obvious first-order approximations in (13.71), with retarded time evaluated at the centre of the dipole:

$$\begin{aligned}\mathbf{A}(\mathbf{x},t) &= \frac{\mu_0}{4\pi}\int \frac{\mathbf{J}(\mathbf{x}',t')}{|\mathbf{r}|}\,d\mathbf{x}' = \frac{\mu_0 I(t')\mathbf{d}}{4\pi\,|\mathbf{x}|} = -\frac{\mu_0\varpi\,\mathbf{p}_0}{4\pi\,|\mathbf{x}|}\sin\varpi t' \\ &= -\frac{\mu_0\varpi\,\mathbf{p}_0}{4\pi\,|\mathbf{x}|}\sin[\varpi(t - |\mathbf{x}|/c)].\end{aligned}$$

It is convenient to rewrite this using $\hat{\mathbf{e}}_z = \hat{\mathbf{e}}_r\cos\theta - \hat{\mathbf{e}}_\theta\sin\theta$, from which

$$\mathbf{A}(\mathbf{x},t) = \frac{\mu_0\varpi\,p_0}{4\pi\,|\mathbf{x}|}(\cos\theta\,\hat{\mathbf{e}}_r - \sin\theta\,\hat{\mathbf{e}}_\theta)\sin(k|\mathbf{x}| - \varpi t). \tag{16.20}$$

It is readily confirmed that (16.18) and (16.20) jointly satisfy the Lorenz gauge.

Our task now is to obtain the far-field forms of $\mathbf{E}$ and $\mathbf{B}$ from (16.19) and (16.20). We shall retain terms of order $|\mathbf{x}|^{-1}$ only, dispensing with higher-order terms.

### 16.3.2 *The Electromagnetic Field and Poynting Energy Flux*

Let us start with the electric field,

$$\mathbf{E}(\mathbf{x},t) = -\frac{\partial\mathbf{A}}{\partial t} - \nabla\varphi. \tag{16.21}$$

For small $c/\varpi\,|\mathbf{x}|$, (16.19) gives

$$\nabla\varphi = \frac{\varpi^2 p_0}{4\pi\varepsilon_0 c^2}\frac{\cos\theta}{|\mathbf{x}|}\cos(k|\mathbf{x}| - \varpi t)\,\hat{\mathbf{e}}_r, \tag{16.22}$$

while (16.20) tells us that

$$\frac{\partial\mathbf{A}}{\partial t} = -\frac{\mu_0\varpi^2 p_0}{4\pi\,|\mathbf{x}|}(\cos\theta\,\hat{\mathbf{e}}_r - \sin\theta\,\hat{\mathbf{e}}_\theta)\cos(k|\mathbf{x}| - \varpi t). \tag{16.23}$$

Combining the two yields the far-field electric field:

$$\boxed{\mathbf{E}(\mathbf{x},t) = -\frac{\mu_0 \varpi^2 p_0}{4\pi} \frac{\sin\theta}{|\mathbf{x}|} \cos(k|\mathbf{x}| - \varpi t)\,\hat{\mathbf{e}}_\theta.} \tag{16.24}$$

Note that $\mathbf{E}$ is normal to $\mathbf{x}$, the position vector of the observer, and falls off as $|\mathbf{x}|^{-1}$, as a radiation field should. Note also that, somewhat disturbingly, $\nabla \cdot \mathbf{E} \neq 0$ in (16.24). This is because we have ignored a weak radial contribution to $\mathbf{E}$ of order $|\mathbf{x}|^{-2}$, which is equivalent to the second term on the right of (16.11) and whose primary function is to ensure that the $\mathbf{E}$-lines are closed in the $r$–$\theta$ plane.

The magnetic field can now be found from $\mathbf{B}(\mathbf{x},t) = \nabla \times \mathbf{A}$, which yields

$$\boxed{\mathbf{B}(\mathbf{x},t) = -\frac{\mu_0 \varpi^2 p_0}{4\pi c} \frac{\sin\theta}{|\mathbf{x}|} \cos(k|\mathbf{x}| - \varpi t)\hat{\mathbf{e}}_\phi,} \tag{16.25}$$

to leading order in $c/\omega|\mathbf{x}|$. This is equivalent to (15.63) of §15.5.1, obtained using the electric Hertz vector. Note that the magnetic field falls off as $|\mathbf{x}|^{-1}$, as it must. Note also that a faster way to obtain $\mathbf{E}$, which bypasses $\varphi$ completely, is to take the curl of (16.25) and then use the Ampère–Maxwell law to get $\mathbf{E}$. Finally, note that we may shift from the magnetic to the electric dipole fields using the dual symmetry transformation of §15.2.2:

$$\mathbf{E}^* = c\mathbf{B}, \quad \mathbf{B}^* = -\mathbf{E}/c, \quad p_0 = m_0/c. \tag{16.26}$$

Between them, (16.24) and (16.25) represent a monochromatic TEM wave travelling radially outward from its source at the speed $c$. The $\mathbf{B}$-lines are azimuthal, encircling the dipole $\mathbf{p}$, while the $\mathbf{E}$-lines are poloidal, as shown in Figure 16.7.

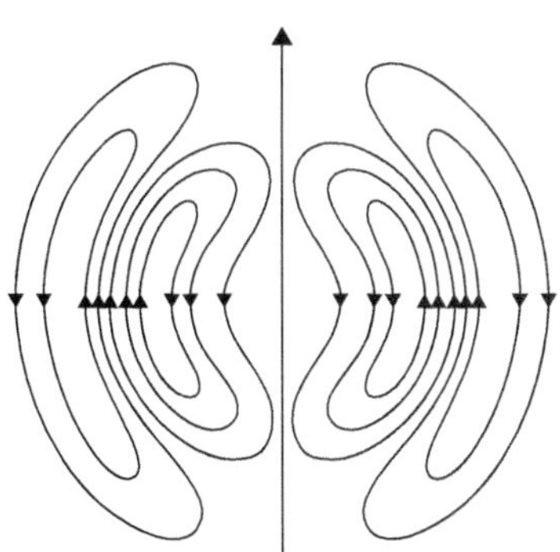

**Figure 16.7** The electric field, including the weak radial component, of an electric dipole.

As with the magnetic dipole, the dominant **E** and **B** fields are related by $c\mathbf{B} = \hat{\mathbf{e}}_r \times \mathbf{E}$, and so the radiated energy per unit area is

$$\boxed{\frac{\mathbf{E}\times\mathbf{B}}{\mu_0} = \frac{|\mathbf{E}|^2}{\mu_0 c}\hat{\mathbf{e}}_r = \frac{\mu_0 \varpi^4 p_0^2}{(4\pi)^2 c}\frac{\sin^2\theta}{|\mathbf{x}|^2}\cos^2(k|\mathbf{x}| - \varpi t)\,\hat{\mathbf{e}}_r.} \tag{16.27}$$

The energy flux is radial and is proportional to the *fourth power* of $\varpi$. It depends on the polar angle $\theta$ and has the same spatial distribution as a magnetic dipole, being greatest on the equatorial plane and zero in the polar directions. On integrating (16.27) over the spherical surface $|\mathbf{x}| = R$, we find

$$\frac{1}{\mu_0}\oint \mathbf{E}\times\mathbf{B}\cdot d\mathbf{S} = \frac{\mu_0 \varpi^4 p_0^2}{6\pi c}\cos^2(k|\mathbf{x}| - \varpi t). \tag{16.28}$$

Note that the ratio of this radiated energy to that of the equivalent magnetic dipole, assuming the dipole currents and sizes to be comparable, is

$$\frac{\text{radiation from electric dipole}}{\text{radiation from magnetic dipole}} = \frac{p_0^2 c^2}{m_0^2} = \frac{Q_0^2 d^2 c^2}{\pi^2 a^4 I_0^2} \sim \frac{c^2}{\varpi^2 d^2} \gg 1. \tag{16.29}$$

We conclude that a magnetic dipole is much less efficient at radiating electromagnetic energy out to infinity.

## 16.4 The Far Field of an Arbitrary Compact Source

So far we have restricted ourselves to a small dipole source, either a magnetic dipole or an electric dipole, which varies harmonically in time. We now consider a localised source centred on the origin and consisting of currents and charges that vary in space and time in an arbitrary manner, subject only to charge conservation. However, we shall continue to take the source to be compact, so that (16.3) gives us $\varpi\ell/c \ll 1$, where $\varpi$ and $\ell$ are a characteristic frequency and lengthscale of the source. This, in turn, tells us that the velocity of the charges is much less than $c$, and so we may ignore corrections that arise when charges travel at relativistic speeds. (We shall postpone the discussion of such corrections until Chapter 17.)

Let us see if we can guess what the far field of a compact source looks like from what we know already. We might anticipate that the far field of a compact source can be approximated by a combination of a magnetic dipole and an electric dipole. Moreover, in §15.5.1 we saw that the magnetic Hertz vector associated with a magnetic dipole at the origin, $\mathbf{m}(t)$, is

$$\mathbf{\Pi}_m(\mathbf{x},t) = \frac{\mu_0 \mathbf{m}(t - |\mathbf{x}|/c)}{4\pi|\mathbf{x}|}.$$

In the far field, $\varpi\,|\mathbf{x}|/c \gg 1$, this leads to a vector potential

$$\mathbf{A}(\mathbf{x},t) = \nabla \times \Pi_m = \frac{\mu_0 \dot{\mathbf{m}}(t - |\mathbf{x}|/c)}{4\pi\,|\mathbf{x}|\,c} \times \hat{\mathbf{e}}_r, \tag{16.30}$$

where $\hat{\mathbf{e}}_r = \mathbf{x}/|\mathbf{x}|$ and the dot over a quantity indicates its time derivative. Similarly, for an electric dipole at the origin, $\mathbf{p}(t)$, the electric Hertz vector is

$$\Pi_e(\mathbf{x},t) = \frac{\mathbf{p}(t - |\mathbf{x}|/c)}{4\pi\varepsilon_0\,|\mathbf{x}|},$$

giving **A** and **B** fields in the far field, $\varpi\,|\mathbf{x}|/c \gg 1$, of

$$\boxed{\mathbf{A}(\mathbf{x},t) = \frac{\mu_0 \dot{\mathbf{p}}(t - |\mathbf{x}|/c)}{4\pi\,|\mathbf{x}|}, \qquad \mathbf{B}(\mathbf{x},t) = \frac{\mu_0}{4\pi\,|\mathbf{x}|\,c}\ddot{\mathbf{p}}(t - |\mathbf{x}|/c) \times \hat{\mathbf{e}}_r.} \tag{16.31}$$

Now we have already seen that the far field of a magnetic dipole is, in general, very weak in comparison with that of an electric dipole, which suggests that we may neglect (16.30) in comparison with (16.31). Moreover, in the far field of the dipoles examined so far, we have always found $\mathbf{E} = c\mathbf{B} \times \hat{\mathbf{e}}_r$. *If* this turns out to be a general result, then (16.31) suggests

$$\boxed{\mathbf{E}(\mathbf{x},t) = \frac{\mu_0}{4\pi\,|\mathbf{x}|}\hat{\mathbf{e}}_r \times [\hat{\mathbf{e}}_r \times \ddot{\mathbf{p}}(t - |\mathbf{x}|/c)].} \tag{16.32}$$

We shall now show that, except for cases in which $\mathbf{p} \approx 0$, (16.31) and (16.32) do indeed dominate the far field of a compact source. We follow, more or less, Griffiths (2013).

### 16.4.1 The Far-Field Potentials

Consider a source centred on the origin of characteristic size $\ell$, as shown in Figure 16.8. Let us start with the retarded scalar potential,

$$\varphi(\mathbf{x},t) = \frac{1}{4\pi\varepsilon_0}\int \frac{\rho_e(\mathbf{x}',t')}{|\mathbf{r}|}\,d\mathbf{x}', \quad \mathbf{r} = \mathbf{x} - \mathbf{x}', \tag{16.33}$$

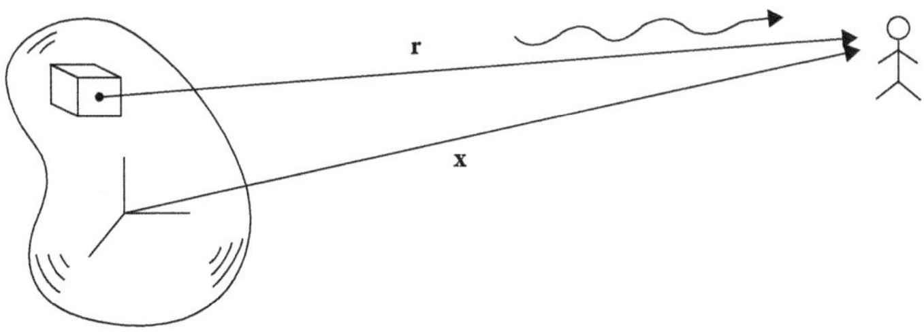


**Figure 16.8** A compact source centred at the origin of characteristic size $\ell$.

where $t' = t - |\mathbf{r}|/c$ is the retarded time. Since $|\mathbf{r}|^2 = |\mathbf{x}|^2 - 2\mathbf{x}' \cdot \mathbf{x} + |\mathbf{x}'|^2$, we have, in the far field,

$$|\mathbf{r}| \approx |\mathbf{x}| \left(1 - \frac{\mathbf{x}' \cdot \mathbf{x}}{|\mathbf{x}|^2}\right), \quad |\mathbf{x}| \gg \ell. \tag{16.34}$$

It follows that

$$\rho_e(\mathbf{x}', t - |\mathbf{r}|/c) \approx \rho_e(\mathbf{x}', t_0' + \mathbf{x}' \cdot \mathbf{x}/c|\mathbf{x}|), \tag{16.35}$$

where $t_0' = t - |\mathbf{x}|/c$ is the *retarded time at the origin.* We now Taylor-expand $\rho_e$ in time,

$$\rho_e(\mathbf{x}', t_0' + \mathbf{x}' \cdot \mathbf{x}/|\mathbf{x}|c) \approx \rho_e(\mathbf{x}', t_0') + \frac{\partial \rho_e(\mathbf{x}', t_0')}{\partial t} \frac{\mathbf{x}' \cdot \mathbf{x}}{|\mathbf{x}|c},$$

where we are free to ignore higher-order terms on the grounds that $\varpi \ell/c \ll 1$. Thus, in the far field, we have the approximation

$$\varphi(\mathbf{x}, t) \approx \frac{1}{4\pi\varepsilon_0 |\mathbf{x}|} \int \frac{\rho_e(\mathbf{x}', t_0') + (\mathbf{x}' \cdot \mathbf{x}/|\mathbf{x}|\, c)\dot{\rho}_e(\mathbf{x}', t_0')}{1 - \mathbf{x}' \cdot \mathbf{x}/|\mathbf{x}|^2}\, d\mathbf{x}',$$

where the dot over a quantity again indicates its time derivative. We now expand terms and discard a contribution that is the product of the two small parameters $\varpi \ell/c$ and $\ell/|\mathbf{x}|$. This yields

$$\varphi(\mathbf{x}, t) \approx \frac{1}{4\pi\varepsilon_0 |\mathbf{x}|} \left[ Q_0 + \frac{\mathbf{x}}{|\mathbf{x}|^2} \cdot \int \mathbf{x}' \rho_e(\mathbf{x}', t_0')\, d\mathbf{x}' + \frac{\mathbf{x}}{|\mathbf{x}|\, c} \cdot \int \mathbf{x}' \dot{\rho}_e(\mathbf{x}', t_0')\, d\mathbf{x}' \right],$$

where $Q_0$ is the total charge, a constant. Using (5.37) as the definition of the electric dipole moment $\mathbf{p}(t)$, this yields

$$\varphi(\mathbf{x}, t) = \frac{1}{4\pi\varepsilon_0 |\mathbf{x}|} \left[ Q_0 + \frac{\mathbf{x} \cdot \mathbf{p}(t_0')}{|\mathbf{x}|^2} + \frac{\mathbf{x} \cdot \dot{\mathbf{p}}(t_0')}{c\,|\mathbf{x}|} \right], \tag{16.36}$$

where we have replaced $\approx$ by $=$ on the understanding that these are the dominant terms for small $\varpi \ell/c$ and $\ell/|\mathbf{x}|$. The first two terms on the right of (16.36) represent the static monopole and dipole potentials of the source in limit of $\varpi \to 0$. The final term in (16.36), by contrast, contributes to the radiation field.

Tuning now to the vector potential, we have the approximation

$$\mathbf{A}(\mathbf{x}, t) = \frac{\mu_0}{4\pi} \int \frac{\mathbf{J}(\mathbf{x}', t')}{|\mathbf{r}|}\, d\mathbf{x}' \approx \frac{\mu_0}{4\pi\,|\mathbf{x}|} \int \mathbf{J}(\mathbf{x}', t_0')\, d\mathbf{x}'. \tag{16.37}$$

However, charge conservation tells us that

$$J_i = \mathbf{J} \cdot \nabla x_i = \nabla \cdot (x_i \mathbf{J}) - x_i \nabla \cdot \mathbf{J} = \nabla \cdot (x_i \mathbf{J}) + x_i \left( \frac{\partial \rho_e}{\partial t} \right),$$

and on integration the divergence converts to a surface integral, which vanishes for a localised source. We conclude that, to leading order in $\varpi\ell//c$ and $\ell/|\mathbf{x}|$,

$$\boxed{\mathbf{A}(\mathbf{x},t) = \frac{\mu_0\dot{\mathbf{p}}(t_0')}{4\pi|\mathbf{x}|}, \quad t_0' = t - |\mathbf{x}|/c,} \tag{16.38}$$

which is indeed the vector potential of (16.31). It is readily confirmed that (16.36) and (16.38) are asymptotically consistent, in the sense that they satisfy the Lorenz gauge.

Actually, a more careful analysis of (16.37), in which $t'$ is not replaced by $t_0'$, but rather by $t_0' + \mathbf{x}' \cdot \mathbf{x}/c|\mathbf{x}|$, and $\mathbf{J}$ is then Taylor-expanded in time, yields (see Exercise 16.1),

$$\boxed{\mathbf{A}(\mathbf{x},t) = \frac{\mu_0\dot{\mathbf{p}}(t_0')}{4\pi|\mathbf{x}|} + \frac{\mu_0\dot{\mathbf{m}}(t_0')}{4\pi|\mathbf{x}|\,c} \times \hat{\mathbf{e}}_r + \text{electric quadrupole term.}} \tag{16.39}$$

This now captures the magnetic dipole contribution of (16.30). However, provided $\mathbf{p}$ is not unusually small, (16.29) allows us to revert to (16.38), to leading order in $\varpi\ell/c$.

Our task now is to determine the far-field forms of $\mathbf{E}$ and $\mathbf{B}$ from (16.36) and (16.38), retaining only terms of order $1/|\mathbf{x}|$.

### *16.4.2 The Electromagnetic Field and Poynting Energy Flux*

Let us start with the magnetic field. From (16.38) we have

$$\mathbf{B}(\mathbf{x},t) = \nabla \times \mathbf{A} = \frac{\mu_0}{4\pi}\nabla \times \frac{\dot{\mathbf{p}}(t_0')}{|\mathbf{x}|} = \frac{\mu_0}{4\pi}\left[\left(\nabla\frac{1}{|\mathbf{x}|}\right) \times \dot{\mathbf{p}}(t_0') + \frac{1}{|\mathbf{x}|}\nabla \times \dot{\mathbf{p}}(t_0')\right].$$

For $\varpi|\mathbf{x}|/c \gg 1$, the dominant term comes from differentiating $\dot{\mathbf{p}}(t_0')$, rather than $1/|\mathbf{x}|$, and so we shall retain only the last term on the right,

$$\mathbf{B}(\mathbf{x},t) = \frac{\mu_0}{4\pi|\mathbf{x}|}\nabla \times \dot{\mathbf{p}}(t_0') = \frac{\mu_0}{4\pi|\mathbf{x}|}(\nabla t_0') \times \ddot{\mathbf{p}}(t_0').$$

However, $\nabla t_0' = -\mathbf{x}/c|\mathbf{x}| = -\hat{\mathbf{e}}_r/c$, and so we deduce that the magnetic field in the radiation zone is

$$\mathbf{B}(\mathbf{x},t) = \frac{\mu_0}{4\pi c|\mathbf{x}|}\ddot{\mathbf{p}}(t - |\mathbf{x}|/c) \times \hat{\mathbf{e}}_r. \tag{16.40}$$

We have arrived back at (16.31), with a magnetic field that is normal to the position vector $\mathbf{x}$ and falls off as $1/|\mathbf{x}|$. If we adopt spherical polar coordinates with the $z$ axis aligned with $\ddot{\mathbf{p}}$, then (16.40) becomes

$$\boxed{\mathbf{B}(\mathbf{x},t) = \frac{\mu_0 \sin\theta}{4\pi c|\mathbf{x}|}\ddot{p}(t - |\mathbf{x}|/c)\hat{\mathbf{e}}_\phi,} \tag{16.41}$$

which reduces to (16.25) for a sinusoidal Hertzian dipole.

Turning to the electric field, we have

$$\mathbf{E}(\mathbf{x},t) = -\frac{\partial \mathbf{A}}{\partial t} - \nabla\varphi = -\frac{\mu_0 \ddot{\mathbf{p}}(t_0')}{4\pi|\mathbf{x}|} - \nabla\varphi, \tag{16.42}$$

and so it is necessary only to find $\nabla\varphi$. Since only the final term in (16.36) contributes to the radiation field, we write

$$\nabla\varphi = \frac{1}{4\pi\varepsilon_0 c}\nabla\left[\frac{\mathbf{x}\cdot\dot{\mathbf{p}}(t_0')}{|\mathbf{x}|^2}\right].$$

Once again, the dominant term comes from differentiating $\dot{\mathbf{p}}(t_0')$, rather than $1/|\mathbf{x}|$. So, in the far field, the chain rule yields

$$\nabla\varphi = \frac{\mathbf{x}\cdot\ddot{\mathbf{p}}(t_0')\nabla t_0'}{4\pi\varepsilon_0 c|\mathbf{x}|^2} = -\frac{\mu_0\mathbf{x}\cdot\ddot{\mathbf{p}}(t_0')}{4\pi|\mathbf{x}|^2}\hat{\mathbf{e}}_r,$$

which combines with (16.42) to give

$$\mathbf{E}(\mathbf{x},t) = \frac{\mu_0}{4\pi|\mathbf{x}|}\hat{\mathbf{e}}_r \times [\hat{\mathbf{e}}_r \times \ddot{\mathbf{p}}(t-|\mathbf{x}|/c)]. \tag{16.43}$$

Alternatively, we might note that the curl of (16.40) is

$$\nabla\times\mathbf{B}(\mathbf{x},t) = \frac{\mu_0}{4\pi c^2|\mathbf{x}|}[\dddot{\mathbf{p}}(t-|\mathbf{x}|/c)\times\hat{\mathbf{e}}_r]\times\hat{\mathbf{e}}_r,$$

which may be combined with the Ampère–Maxwell law to also give (16.43). Either way, we have arrived back at (16.32), which we first obtained using the Hertz vectors.

Like $\mathbf{B}$, the electric field in the radiation zone is normal to the position vector $\mathbf{x}$ and falls off as $1/|\mathbf{x}|$. If we adopt spherical polar coordinates with the $z$ axis aligned with $\ddot{\mathbf{p}}(t_0')$, then (16.43) becomes

$$\boxed{\mathbf{E}(\mathbf{x},t) = \frac{\mu_0\sin\theta}{4\pi|\mathbf{x}|}\ddot{p}(t-|\mathbf{x}|/c)\hat{\mathbf{e}}_\theta,} \tag{16.44}$$

which reduces to (16.24) for a sinusoidal dipole. We conclude that a compact source generates a TEM wave (at least to leading order in $1/|\mathbf{x}|$) and that wave is of the d'Alembert type. As always, the dominant $\mathbf{E}$ and $\mathbf{B}$ fields in the radiation zone satisfy $c\mathbf{B} = \hat{\mathbf{e}}_r \times \mathbf{E}$, and so the radiated energy per unit area is

$$\boxed{\frac{\mathbf{E}\times\mathbf{B}}{\mu_0} = \frac{|\mathbf{E}|^2}{\mu_0 c}\hat{\mathbf{e}}_r = \frac{\mu_0}{(4\pi)^2 c}\frac{\sin^2\theta}{|\mathbf{x}|^2}[\ddot{p}(t_0')]^2\hat{\mathbf{e}}_r.} \tag{16.45}$$

Clearly, (16.27) is just a special case of this.

**Table 16.1** The dominant **E** and **B** fields in the radiation zone of a compact source

| | |
|---|---|
| Compact magnetic dipole at the origin, | $\mathbf{B}(\mathbf{x},t) = -\frac{\mu_0 \varpi^2 m_0}{4\pi c^2}\frac{\sin\theta}{\lvert\mathbf{x}\rvert}\cos(k\lvert\mathbf{x}\rvert - \varpi t)\,\hat{\mathbf{e}}_\theta$ |
| $\mathbf{m}(t) = m_0 \cos\varpi t\,\hat{\mathbf{e}}_z$ | $\mathbf{E}(\mathbf{x},t) = \frac{\mu_0 \varpi^2 m_0}{4\pi c}\frac{\sin\theta}{\lvert\mathbf{x}\rvert}\cos(k\lvert\mathbf{x}\rvert - \varpi t)\,\hat{\mathbf{e}}_\phi$ |
| Compact electric dipole at the origin, | $\mathbf{B}(\mathbf{x},t) = -\frac{\mu_0 \varpi^2 p_0}{4\pi c}\frac{\sin\theta}{\lvert\mathbf{x}\rvert}\cos(k\lvert\mathbf{x}\rvert - \varpi t)\,\hat{\mathbf{e}}_\phi$ |
| $\mathbf{p}(t) = p_0 \cos\varpi t\,\hat{\mathbf{e}}_z$ | $\mathbf{E}(\mathbf{x},t) = -\frac{\mu_0 \varpi^2 p_0}{4\pi}\frac{\sin\theta}{\lvert\mathbf{x}\rvert}\cos(k\lvert\mathbf{x}\rvert - \varpi t)\,\hat{\mathbf{e}}_\theta$ |
| Arbitrary compact source at the origin with electric dipole moment $\mathbf{p}(t)$ | $\mathbf{B}(\mathbf{x},t) = \frac{\mu_0}{4\pi c\lvert\mathbf{x}\rvert}\ddot{\mathbf{p}}(t - \lvert\mathbf{x}\rvert/c) \times \hat{\mathbf{e}}_r$ |
| | $\mathbf{E}(\mathbf{x},t) = \frac{\mu_0}{4\pi\lvert\mathbf{x}\rvert}\hat{\mathbf{e}}_r \times [\hat{\mathbf{e}}_r \times \ddot{\mathbf{p}}(t - \lvert\mathbf{x}\rvert/c)]$ |
| | $\frac{\mathbf{E}\times\mathbf{B}}{\mu_0} = \frac{\mu_0}{(4\pi)^2 c}\frac{\sin^2\theta}{\lvert\mathbf{x}\rvert^2}[\ddot{p}(t_0')]^2\hat{\mathbf{e}}_r$ |

### 16.4.3 *A Comparison of Sources*

We have covered a lot of ground in the last few sections and so perhaps it is now appropriate to compare the dominant **E** and **B** fields in the radiation zones of these various compact sources. These are summarised in Table 16.1, where we use spherical polar coordinates $(r,\theta,\phi)$ centred on the source, with unit vectors $(\hat{\mathbf{e}}_r, \hat{\mathbf{e}}_\theta, \hat{\mathbf{e}}_\phi)$. In each case the **E** and **B** fields are linked by $c\mathbf{B} = \hat{\mathbf{e}}_r \times \mathbf{E}$, which yields a TEM wave that propagates radially outward from the source at the speed $c$. Note that we may shift from the magnetic dipole to the electric dipole fields using the dual symmetry transformation of §15.2.2: $\mathbf{E}^* = c\mathbf{B}$ and $\mathbf{B}^* = -\mathbf{E}/c$, with $p_0 = m_0/c$. Note also that the radiated energy rises sharply with frequency, as $\varpi^4$, which will turn out to be important for light scattering processes, as we shall discuss in Chapter 19.

## 16.5 Larmor's Radiation Equation

Let us suppose that our compact source consists of a negative charge of fixed magnitude $-q$ located at the origin and an equal but opposite charge $q$ located at the point $\mathbf{d}(t)$, as shown in Figure 16.9. Then we have an electric dipole of magnitude $\mathbf{p}(t) = q\mathbf{d}(t)$. Evidently $\ddot{\mathbf{p}}(t_0') = q\mathbf{a}_q(t_0')$, where $\mathbf{a}_q$ is the acceleration of the moving charge. The radiated energy per unit area, which we may associate with the moving charge, is then given by (16.45),

$$\frac{\mathbf{E}\times\mathbf{B}}{\mu_0} = \frac{\mu_0 q^2 a_q^2(t_0')}{(4\pi)^2 c}\frac{\sin^2\theta}{\lvert\mathbf{x}\rvert^2}\hat{\mathbf{e}}_r, \tag{16.46}$$

where we have taken the $z$ axis to coincide with $\ddot{\mathbf{d}}(t_0')$. Integrating over a sphere of large radius $\lvert\mathbf{x}\rvert = R$, and noting that $\int(\sin\theta/\lvert\mathbf{x}\rvert)^2\,dS = 8\pi/3$, we find the radiated power to be

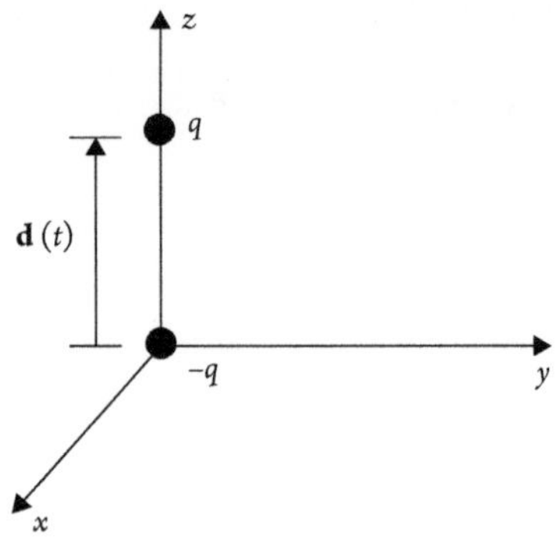


**Figure 16.9** An electric dipole consists of a stationary charge and a moving charge.

$$\boxed{P_q = \frac{1}{\mu_0}\oint \mathbf{E}\times\mathbf{B}\cdot d\mathbf{S} = \frac{\mu_0 q^2 a_q^2(t-|\mathbf{x}|/c)}{6\pi c}.} \tag{16.47}$$

This is *Larmor's radiation equation* of 1897, which makes it explicit that the radiation of energy out to infinity requires charges to *accelerate*. Note that, because the dipole is compact, we have $\varpi\, d/c \ll 1$, where $\varpi$ is a characteristic frequency. It follows that (16.47) is restricted to charges moving at velocities much less than the speed of light. In Chapter 17 we shall see how to generalise (16.47) for relativistic charges.

## 16.6 An Aside: Why Don't Atoms Collapse?

As a simple illustration of Larmor's formula, consider an electron spinning around the nucleus of a hydrogen atom in a circular orbit of radius $R$, as shown in Figure 16.10. The electron has mass $m_e$, charge $-q_e$ and speed $u_q$. The radial force balance for the charge is evidently

$$\frac{1}{4\pi\varepsilon_0}\frac{q_e^2}{R^2} = m_e\frac{u_q^2}{R} = m_e a_q,$$

from which we can find the radiated energy:

$$P_q = \frac{\mu_0}{6\pi c}\left(\frac{q_e^3}{4\pi\varepsilon_0 R^2 m_e}\right)^2.$$

Now the potential energy of the charge is $\mathrm{PE} = -q_e^2/4\pi\varepsilon_0 R$, whose magnitude is twice the kinetic energy

$$\mathrm{KE} = \frac{1}{2}m_e u_q^2 = \frac{1}{2}\frac{q_e^2}{4\pi\varepsilon_0 R}.$$

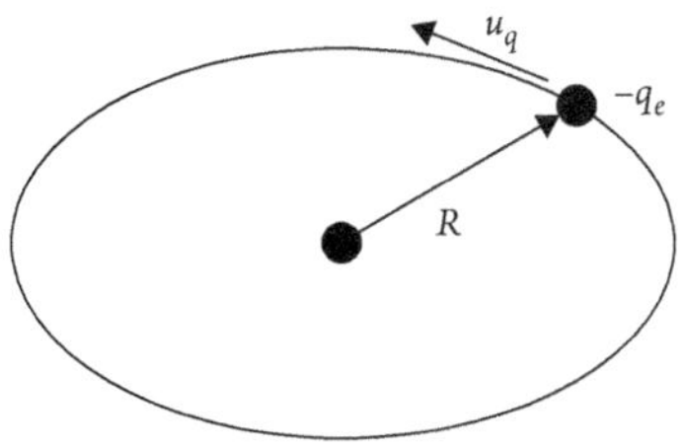


**Figure 16.10** An electron has an orbit of radius $R$ around the nucleus of a hydrogen atom.

The total energy of the charge, kinetic plus potential, is therefore equal to minus its kinetic energy, a result that is characteristic of bodies held in a circular orbit by a central force. If this energy diminishes at a rate set by the radiation, then we have

$$\frac{d}{dt}(\mathrm{PE}+\mathrm{KE}) = \frac{q_e^2\dot{R}}{8\pi\varepsilon_0 R^2} = -\frac{\mu_0}{6\pi c}\left(\frac{q_e^3}{4\pi\varepsilon_0 R^2 m_e}\right)^2. \tag{16.48}$$

The electron is therefore predicted to spiral inward at a rate

$$\dot{R} = -\frac{\mu_0 q_e^4}{12\pi^2\varepsilon_0 c R^2 m_e^2}. \tag{16.49}$$

If correct, this would give the hydrogen atom a lifespan of $\sim 10^{-11}$ s, a failure of classical electrodynamics that greatly bothered Heaviside: 'The radiation of energy is very rapid—so great as seeming to shut out the possibility of anything more than momentary persistence. . . .' It would take the advent of quantum mechanics to resolve this paradox.

## 16.7 The Centre-Fed Linear Antenna

We conclude our brief survey of radiation by turning to a topic of immense practical importance: that of antennas. Entire books are devoted to this complicated subject, so we can do little justice to it here. We shall consider only the simple geometry of a *centre-fed linear antenna*, as shown in Figure 16.11. This is a thin wire or tube, energised by a coaxial cable at its centre. It behaves a bit like an open-ended transmission line, so standing waves are set up in which the current goes to zero at both ends of the antenna.

Perhaps the first point to emphasise is that this is *not* a compact source, since the wavelength of the radiation is comparable to the length of the antenna, $d$. Second, for most antennas one cannot easily predict in advance the shape of the current distribution, so for the purposes of computing radiation fields one typically starts with an empirical estimate of the current distribution. For the centre-fed linear antenna oscillating at a single frequency $\varpi = kc$, it is observed that a reasonable approximation to the current is

$$I(z,t) = \hat{I}(z)\cos\varpi t = I_0 \sin(kd/2 - k|z|)\cos\varpi t. \tag{16.50}$$

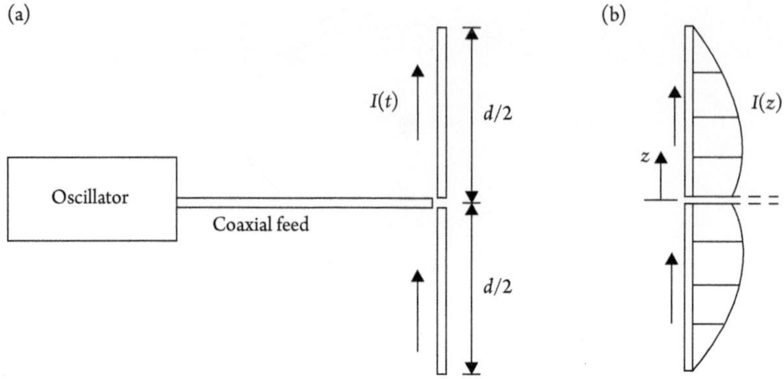


**Figure 16.11** (a) A linear antenna energised at its centre. (b) The current distribution.

Here the antenna sits on the $z$ axis, centred at the origin. Note that $I$ falls to zero at the ends of the antenna, $z = \pm d/2$, as shown schematically in Figure 16.11(b) for the case of $d = 2\lambda/3$, $\lambda$ being the wavelength. A common special case is $kd = \pi$, or equivalently $d = \lambda/2$, which is known as the *half-wave antenna*.

Let us see if we can find the **E** and **B** fields in the radiation zone associated with such a spatially extended distribution of current. Our starting point is

$$\mathbf{A}(\mathbf{x},t) = \frac{\mu_0}{4\pi}\int \frac{\mathbf{J}(\mathbf{x}',t')}{|\mathbf{r}|}\,d\mathbf{x}' = \frac{\mu_0 I_0}{4\pi}\int_{-d/2}^{d/2} \frac{\sin(kd/2 - k|z'|)\cos\varpi t'}{|\mathbf{r}|}\,dz\,\hat{\mathbf{e}}_z. \tag{16.51}$$

We now use the fact that $|\mathbf{x}| \gg d$ to simplify this integral in two ways. First, we replace $|\mathbf{r}|$ by $|\mathbf{x}|$ in the denominator. Second, we invoke (16.34), from which $t' = t_0' + z'z/c|\mathbf{x}|$, where $t_0' = t - |\mathbf{x}|/c$ is the retarded time at the centre of the antenna. This gives us

$$\mathbf{A}(\mathbf{x},t) = \frac{\mu_0 I_0}{4\pi|\mathbf{x}|}\int_{-d/2}^{d/2} \sin(kd/2 - k|z'|)\cos(\varpi t_0' + kz'z/|\mathbf{x}|)\,dz'\,\hat{\mathbf{e}}_z. \tag{16.52}$$

Evaluating this integral is somewhat tedious, though straightforward. The end result is

$$\mathbf{A}(\mathbf{x},t) = \frac{\mu_0 I_0 \hat{\mathbf{e}}_z}{2\pi k|\mathbf{x}|}\left\{\frac{\cos[(kd/2)\cos\theta] - \cos(kd/2)}{\sin^2\theta}\right\}\cos(k|\mathbf{x}| - \varpi t), \tag{16.53}$$

where $\theta$ is the polar angle shown in Figure 16.12.

The corresponding magnetic field can be found from $\mathbf{B} = \nabla\times\mathbf{A}$. Neglecting terms smaller than $1/|\mathbf{x}|$, we find

$$\mathbf{B}(\mathbf{x},t) = \frac{\mu_0 I_0}{2\pi|\mathbf{x}|}\left\{\frac{\cos[(kd/2)\cos\theta] - \cos(kd/2)}{\sin\theta}\right\}\sin(k|\mathbf{x}| - \varpi t)\,\hat{\mathbf{e}}_\phi. \tag{16.54}$$

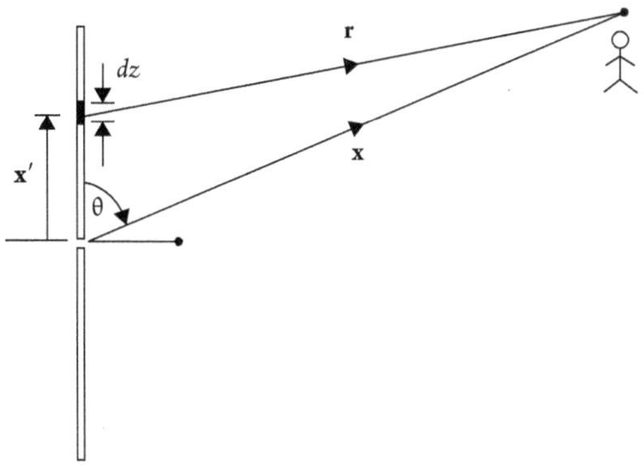


**Figure 16.12** The coordinate system used to evaluate the radiation zone of an antenna.

Taking the curl once more, and again neglecting terms smaller than $1/|\mathbf{x}|$, yields

$$\frac{1}{c^2}\frac{\partial \mathbf{E}}{\partial t} = \nabla \times \mathbf{B} = -\frac{\mu_0 I_0 k}{2\pi|\mathbf{x}|}\left\{\frac{\cos[(kd/2)\cos\theta] - \cos(kd/2)}{\sin\theta}\right\}\cos(k|\mathbf{x}| - \varpi t)\,\hat{\mathbf{e}}_\theta, \tag{16.55}$$

from which

$$\mathbf{E}(\mathbf{x}, t) = \frac{\mu_0 I_0 c}{2\pi|\mathbf{x}|}\left\{\frac{\cos[(kd/2)\cos\theta] - \cos(kd/2)}{\sin\theta}\right\}\sin(k|\mathbf{x}| - \varpi t)\,\hat{\mathbf{e}}_\theta. \tag{16.56}$$

Evidently, just as with a compact source, we have $c\mathbf{B} = \hat{\mathbf{e}}_r \times \mathbf{E}$. So, as usual, the Poynting vector is radial and of magnitude $|\mathbf{E}|^2/\mu_0 c$, but its spatial structure is now complicated.

For the particular case of the half-wave antenna, however, the currents and charge density in the antenna resemble those of a simple electric dipole (see Exercise 16.2). In particular, when $kd = \pi$, the $\mathbf{B}$, $\mathbf{E}$ and $|\mathbf{E} \times \mathbf{B}|$ fields are given by

$$\mathbf{B}(\mathbf{x}, t) = \frac{\mu_0 I_0}{2\pi}\frac{\sin\theta}{|\mathbf{x}|}\left\{\frac{\cos[(\pi/2)\cos\theta]}{\sin^2\theta}\right\}\sin(k|\mathbf{x}| - \varpi t)\,\hat{\mathbf{e}}_\phi, \tag{16.57}$$

$$\mathbf{E}(\mathbf{x}, t) = \frac{\mu_0 I_0 c}{2\pi}\frac{\sin\theta}{|\mathbf{x}|}\left\{\frac{\cos[(\pi/2)\cos\theta]}{\sin^2\theta}\right\}\sin(k|\mathbf{x}| - \varpi t)\,\hat{\mathbf{e}}_\theta, \tag{16.58}$$

$$\frac{\mathbf{E} \times \mathbf{B}}{\mu_0} \sim \frac{\sin^2\theta}{|\mathbf{x}|^2}\left\{\frac{\cos[(\pi/2)\cos\theta]}{\sin^2\theta}\right\}^2 \sin^2(k|\mathbf{x}| - \varpi t)\,\hat{\mathbf{e}}_r. \tag{16.59}$$

Crucially, the term in the curly brackets in (16.57)–(16.59) is only a weak function of the polar angle and indeed close to unity. So these expressions resemble the equivalent distributions for a compact electric dipole, as given by (16.24)–(16.27).

The angular dependence of the Poynting vector in (16.59) is shown in Figure 16.13(a). As with the Hertzian dipole, it is dumbbell-like in shape, being maximal on the equatorial plane and zero on the polar axis. However, as $kd$ increases beyond $\pi$, the radiation pattern

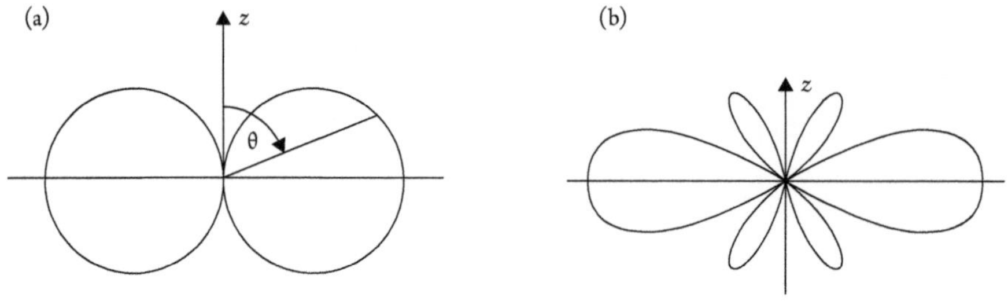


**Figure 16.13** Schematic of the radiation patterns for (a) a half-wave antenna and (b) an antenna where $d$ is somewhat larger than $\lambda$.

of the antenna becomes increasingly complicated, exhibiting multiple lobes, as indicated schematically in Figure 16.13(b) and as discussed in Exercise 16.4.

Axial symmetry demands that the radiation field of a single linear antenna be omnidirectional in the equatorial plane. However, the radiation pattern can be focused in a particular direction by using more than one antenna, i.e. by using an *antenna array*. The general idea is to take advantage of superposition and arrange for the radiation fields from the various antennas to cancel in those directions where no radiation is wanted, and interfere constructively in those directions where radiation is required. This can be done partly through the positioning of the various antenna elements, and partly through the choice of the relative phases of the currents in the antennas.

As a simple example, consider two identical half-wave antennas that lie side by side, are separated by a distance $\lambda/2$ and point in the vertical direction. For example, suppose that their centres lie at $(x, y, z) = (\pm\lambda/4, 0, 0)$, the antennas run from $z = -\lambda/4$ to $z = \lambda/4$ and the currents in the two antennas are in phase. Then the two radiation fields differ only through a phase lag in the $x$ direction, and so on the $x$ axis the two radiation fields are everywhere out of phase and cancel, while on the $y$ axis they are additive. In short, the radiation pattern in a horizontal plane is focused in the $y$ direction.

More generally, the use of multiple antennas, whose relative phases and positions are cunningly chosen, allow narrowly focused radiation patterns to be engineered. We shall not pause to discuss the radiation fields of such antenna arrays, but rather direct the reader to Lorrain & Corson (1970), who discuss antennas and antenna arrays at length.

## EXERCISES

16.1 Use charge conservation to show that

$$2x_j\dot{J}_i = (x_j\dot{J}_i - x_i\dot{J}_j) + x_i x_j \ddot{\rho}_e + \nabla \cdot (x_i x_j \dot{\mathbf{J}}).$$

Hence derive the vector potential for a compact source, as given by (16.39). (*Hint:* replace $t'$ in (16.37) by $t'_0 + \mathbf{x}' \cdot \mathbf{x}/c\,|\mathbf{x}|$, then Taylor-expand $\mathbf{J}$ in time, as was done for

$\rho_e$ in the build-up to (16.36), and finally use the expression above to split the integrand involving $\dot{\mathbf{J}}$ into symmetric and antisymmetric parts.) Show that the antisymmetric part of the equation above leads to the magnetic dipole term in (16.39), while the symmetric part yields the electric quadrupole contribution. Find the electric quadrupole contribution.

16.2 Find the free charge density in a centre-fed half-wave linear antenna, starting from (16.50). Show that the charge density in the antenna resembles that of a simple electric dipole and that the electric dipole moment of the free charge is

$$p = \frac{2I_0 d}{\varpi \pi} \sin \varpi t = p_0 \sin \varpi t.$$

Hence show that the magnetic field for a half-wave antenna, (16.57), can be rewritten as

$$\mathbf{B}(\mathbf{x}, t) = \frac{\mu_0 \varpi^2 p_0}{4\pi c} \frac{\sin\theta}{|\mathbf{x}|} \left\{ \frac{\cos[(\pi/2)\cos\theta]}{\sin^2\theta} \right\} \sin(k|\mathbf{x}| - \varpi t)\, \hat{\mathbf{e}}_\phi.$$

Plot the term in the curly brackets as a function of polar angle and show that it is close to unity. Compare this expression for $\mathbf{B}$ with (16.25), the field of a point electric dipole.

16.3 Use charge conservation to show that

$$\dot{J}_i = \dot{\mathbf{J}} \cdot \nabla x_i = \nabla \cdot (x_i \dot{\mathbf{J}}) + x_i \ddot{\rho}_e.$$

Now derive (16.40), the magnetic field in the radiation zone of a compact source, directly from Jefimenko's equation (16.2).

16.4 Sketch the dependence of the Poynting flux on polar angle for a centre-fed linear antenna operating at $d = \lambda$, $d = 3\lambda/2$ and $d = 5\lambda/2$. Confirm that the number of lobes is equal to the number of half-wavelengths that can fit into the antenna.

16.5 Consider a centre-fed linear antenna in the compact limit of $kd \to 0$. Show that the free charge density in the antenna is that of a point electric dipole of magnitude $p = (I_0 d^2/4c) \sin \varpi t$. Hence confirm that, in the limit of $kd \to 0$, the vector potential for a linear antenna, (16.53), reduces to that for a compact source:

$$\mathbf{A}(\mathbf{x}, t) = \frac{\mu_0 \dot{\mathbf{p}}(t_0')}{4\pi |\mathbf{x}|}, \quad t_0' = t - |\mathbf{x}|/c.$$

## REFERENCES

Griffiths, D. J., 2013, *Introduction to Electrodynamics*, 4th ed. Pearson.

Lorrain, P. & Corson, D., 1970, *Electromagnetic Fields and Waves*, 2nd ed. W. H. Freeman.

# 17

# Maxwell's Equations in Free Space III

## *The Fields of Moving Charges*

Whether good mathematicians, when they die, go to Cambridge, I do not know.

Oliver Heaviside (*Nature*, 1903)

We now consider the electromagnetic field established by point charges moving in an arbitrary manner. In many respects we could have included this material in the previous chapter, since we shall be concerned largely with the radiation fields far from the charges. However, the theory for point charges tends to follow its own particular internal logic and so, in practice, it helps to treat it separately from the dipole radiation of Chapter 16. In particular, we shall find it convenient to adopt a Lagrangian description of the moving sources, rather than the Eulerian formulation used to date. Moreover, we shall consider for the first time charges moving at relativistic velocities, and this introduces a phenomenon that, although implicit in the equations of Chapter 13, is effectively new.

We shall find it convenient to make a small change in notation in this chapter. In earlier chapters the presence of a prime on a variable played one of two distinct roles. On the one hand the prime in $\mathbf{x}'$ indicated that this locates part of a source, with the displacement vector $\mathbf{r} = \mathbf{x} - \mathbf{x}'$ taking us from a source at $\mathbf{x}'$ to an observer at $\mathbf{x}$. On the other hand, the prime in $t' = t - |\mathbf{r}|/c$ indicates retarded time. In order to avoid possible confusion, in this chapter we shall use a prime strictly to indicate a variable evaluated at retarded time, and use $\mathbf{x}''$ as the position vector that locates part of a source. So the displacement vector $\mathbf{r} = \mathbf{x} - \mathbf{x}''$ takes us from a source at $\mathbf{x}''$ to an observer at $\mathbf{x}$.

There are two distinct, yet equivalent, ways of deriving the fields of point charges in motion. The most common approach is to start by determining the potentials for a moving point charge from (13.70) and (13.71), which leads to the so-called *Liénard–Wiechert potentials.* The $\mathbf{E}$ and $\mathbf{B}$ fields then follow after some algebra. One technical difficulty here is that the moving charges are described in a Lagrangian framework, while the fields they create are necessarily expressed in an Eulerian form, and this makes for rather a lot of algebra. An alternative strategy is to start with the Jefimenko expressions for $\mathbf{E}$ and $\mathbf{B}$, (13.75) and (13.76), and then adapt those to the special case of moving point charges. This takes us back to Heaviside's discoveries in 1904, and to what are now called the Heaviside–Feynman equations. We shall present both formulations in what follows. Let us start, however, by returning to §2.5.2, and to moving sources in acoustics.

## 17.1 The Potentials for a Point Charge Executing Arbitrary Motion

### *17.1.1 Moving Sources in Acoustics Revisited*

Let us return to (2.87), which governs pressure waves travelling at the speed of sound, $a$, and generated by an acoustic source per unit volume, $s(\mathbf{x}, t)$:

$$\nabla^2 p - \frac{1}{a^2}\frac{\partial^2 p}{\partial t^2} = -s(\mathbf{x}, t). \tag{17.1}$$

Note that the source in (17.1) is written as an *Eulerian field* $s(\mathbf{x}, t)$, in which $\mathbf{x}$ and $t$ are taken to be *independent variables*. Now suppose that $s(\mathbf{x}, t)$ is uniform within a moving volume of fixed shape, has overall strength $S = \int s\, d\mathbf{x}$ and moves along a known trajectory $\mathbf{x}_0''(t)$, where $\mathbf{x}_0''(t)$ locates the centre of the moving source. We shall also assume that it is a *compact source*, that is, its size is much smaller than the wavelength of the waves it generates. The vectors $\mathbf{x}_0''(t)$ and $\mathbf{u}(t) = d\mathbf{x}_0''/dt$ now provide a *Lagrangian* description of the source (Figure 7.1). (Those readers unfamiliar with the difference between Eulerian and Lagrangian descriptions of motion may wish to consult §20.2.1.)

Now we can solve (17.1) in the usual way using retarded potentials, and so if the source is small enough one might expect the pressure field $p(\mathbf{x}, t)$ to be given by

$$p(\mathbf{x}, t) = \frac{1}{4\pi}\int \frac{s(\mathbf{x}'', t')}{|\mathbf{r}|}\, d\mathbf{x}'' = \frac{1}{4\pi\left|\mathbf{r}_0'\right|}\int s(\mathbf{x}'', t')\, d\mathbf{x}'' = \frac{S}{4\pi\left|\mathbf{r}_0'\right|}, \tag{17.2}$$

where $\mathbf{r}_0' = \mathbf{x} - \mathbf{x}_0''(t')$ and the prime on $\mathbf{r}_0'$ indicates that $\mathbf{r}_0$ is evaluated at the retarded time $t' = t - \left|\mathbf{r}_0'\right|/a$ (see Figure 17.1). However, as noted in §2.5.2, this is *not* the case, the correct solution being

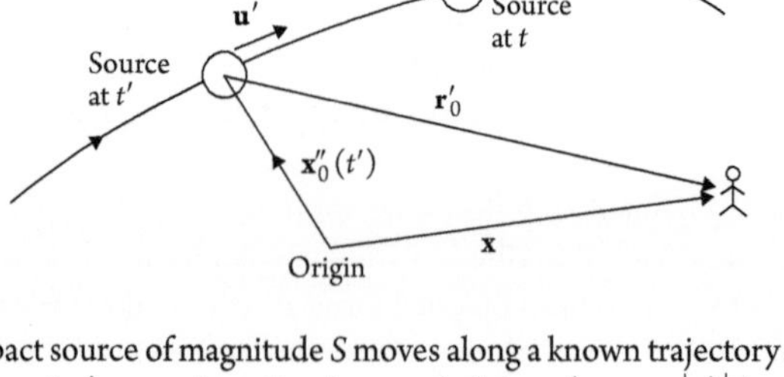


**Figure 17.1** A compact source of magnitude $S$ moves along a known trajectory $\mathbf{x}_0''(t)$ with velocity $\mathbf{u}(t)$. $\mathbf{r}_0' = \mathbf{x} - \mathbf{x}_0''(t')$ and $\mathbf{u}'$ are evaluated at the retarded time $t' = t - \left|\mathbf{r}_0'\right|/a$.

$$p(\mathbf{x},t) = \frac{1}{4\pi\left|\mathbf{r}_0'\right|}\int s(\mathbf{x}'',t')\,d\mathbf{x}'' = \frac{S}{4\pi\left|\mathbf{r}_0'\right|\kappa_a'}, \quad \kappa_a' = 1 - \frac{\mathbf{u}'\cdot\mathbf{r}_0'}{a\left|\mathbf{r}_0'\right|}, \tag{17.3}$$

where $\mathbf{u}' = \mathbf{u}(t')$ is the velocity of the source at the retarded time $t' = t - \left|\mathbf{r}_0'\right|/a$.

The reason for this is rather subtle and is related to the fact that, although the source is compact, it still has a finite volume, say $V_S$. It follows that, for a given time $t$ at the observer, we must evaluate contributions to the integral in (17.3) from different parts of the source at *different retarded times*. By way of contrast, $S$ is defined by $S = \int s(\mathbf{x},t_0)\,d\mathbf{x}$, which is evaluated at a *fixed* time, $t_0$. To focus our thoughts, suppose the source moves towards the observer. Then there is a difference between the time at which we first start evaluating contributions to the integral in (17.3) from the furthest parts of the source and the time at which we finish adding contributions to the integral from the closest parts of the source. Moreover, the source moves towards the observer during that time period. Thus the source sweeps through, or occupies, a volume somewhat larger than $V_S$ during the period over which the integral in (17.3) is evaluated, and it turns out that the effective volume of integration is $V_S/\kappa_a'$. If we approximate $|\mathbf{r}|$ by $\left|\mathbf{r}_0'\right|$ in the denominator of the left-hand integral in (17.2), on the assumption that $\left|\mathbf{x}'' - \mathbf{x}_0''\right| << \left|\mathbf{r}_0'\right|$, then the integral takes the form

$$p(\mathbf{x},t) = \frac{1}{4\pi}\int \frac{s(\mathbf{x}'',t')}{|\mathbf{r}|}\,d\mathbf{x}'' = \frac{S/\kappa_a'}{4\pi\left|\mathbf{r}_0'\right|} = \frac{S}{4\pi\kappa_a'\left|\mathbf{r}_0'\right|}, \tag{17.4}$$

which is (17.3). Notice that the size of the source does not enter into the final result when expressed in terms of $S$, and so this volumetric correction persists no matter how small we make $V_S$.

### *17.1.2 The Liénard–Wiechert Potentials for a Moving Point Charge*

Of course, what is true in acoustics is also true in electromagnetic radiation. Suppose we have a point charge $q$ that follows the trajectory $\mathbf{x}_0''(t)$ with velocity $\mathbf{u}(t) = d\mathbf{x}_0''/dt$. Then there is a charge density $\rho_e(\mathbf{x},t) = q\delta(\mathbf{x} - \mathbf{x}_0''(t))$, as well as a current density $\mathbf{J}(\mathbf{x},t) = q\mathbf{u}(t)\delta(\mathbf{x} - \mathbf{x}_0''(t))$. By analogy with (17.3), the potentials (13.70) and (13.71),

$$\varphi(\mathbf{x},t) = \frac{1}{4\pi\varepsilon_0}\int \frac{\rho_e(\mathbf{x}'',t')}{|\mathbf{r}|}\,d\mathbf{x}'', \quad \mathbf{A}(\mathbf{x},t) = \frac{\mu_0}{4\pi}\int \frac{\mathbf{J}(\mathbf{x}'',t')}{|\mathbf{r}|}\,d\mathbf{x}'', \tag{17.5}$$

now become

$$\boxed{\varphi(\mathbf{x},t) = \frac{1}{4\pi\varepsilon_0}\frac{q}{|\mathbf{r}'|\,\kappa'}, \qquad \mathbf{A}(\mathbf{x},t) = \frac{\mu_0}{4\pi}\frac{q\mathbf{u}'}{|\mathbf{r}'|\,\kappa'} = \frac{\mathbf{u}'}{c^2}\varphi(\mathbf{x},t).} \tag{17.6}$$

Here $\mathbf{r} = \mathbf{x} - \mathbf{x}_0''(t)$ takes us from the charge to the observer, the vectors $\mathbf{r}' = \mathbf{x} - \mathbf{x}_0''(t')$ and $\mathbf{u}' = \mathbf{u}(t')$ are evaluated at the retarded time $t' = t - \left|\mathbf{r}_0'\right|/c$ (see Figure 17.2), and $\kappa' = 1 - \mathbf{u}'\cdot\mathbf{r}'/c\left|\mathbf{r}'\right|$ is the volumetric correction introduced in §17.1.1. The expressions (17.6) are the potentials of Liénard and Wiechert.

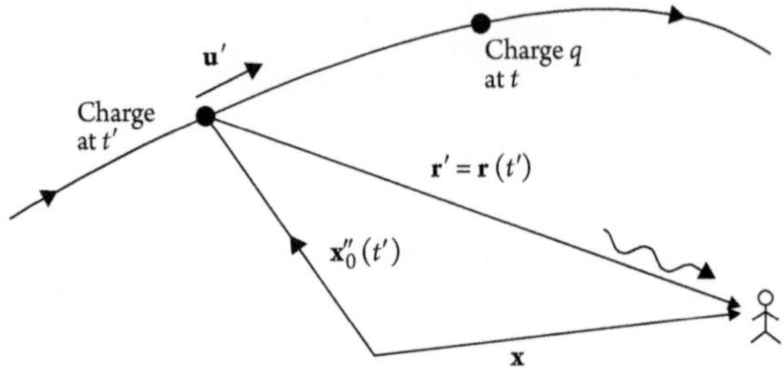


**Figure 17.2** A point charge of magnitude $q$ moves along a known trajectory $\mathbf{x}_0''(t)$ with velocity $\mathbf{u}(t)$. $\mathbf{r}' = \mathbf{x} - \mathbf{x}_0''(t')$ and $\mathbf{u}'$ are evaluated at the retarded time $t' = t - |\mathbf{r}'|/c$.

Note that $\kappa'$ is *not* a relativistic correction, as is evident from the fact that it arises in acoustics. Note also that, if we write $r'(t') = |\mathbf{r}'|$, then the retarded time is now determined implicitly though the expression

$$t = t' + r'(t')/c. \tag{17.7}$$

## 17.2 From the Liénard–Wiechert Potentials to the Electromagnetic Field

### 17.2.1 *Kinematic Preliminaries*

We now wish to determine the $\mathbf{E}$ and $\mathbf{B}$ fields from the Liénard–Wiechert potentials. In principle this is straightforward, but in practice it is messy. The main complication is that our fields are necessarily expressed in Eulerian form

$$\mathbf{E}(\mathbf{x}, t) = -\frac{\partial \mathbf{A}}{\partial t} - \nabla\varphi, \quad \mathbf{B}(\mathbf{x}, t) = \nabla \times \mathbf{A}, \tag{17.8}$$

where $\mathbf{x}$ and $t$ are taken to be *independent* variables, while the motion of our charge is described in a Lagrangian framework, in which the position of the charge is a prescribed function of $t$. This difference in viewpoints leads to a number of complications, such as retarded time being determined indirectly through the expression $t = t' + r'(t')/c$. In order to ease the derivation of the $\mathbf{E}$ and $\mathbf{B}$ fields in the following sections, we shall first establish a number of simple kinematic results.

We start by deriving expressions for the derivative with respect to $t'$ of various kinematic quantities associated with the charge at the retarded time $t'$. For reasons that will become apparent, when taking derivatives with respect to $t'$ we shall take the location of the observer, $\mathbf{x}$, to be *fixed*. For example, we might take the observer to be located at the origin, in which case $\mathbf{r}' = \mathbf{x} - \mathbf{x}_0''(t')$ becomes (minus) the position vector of the charge at the retarded time. We start by noting that

$$\frac{dr'}{dt'} = \frac{1}{2r'}\frac{d}{dt'}(\mathbf{r}'\cdot\mathbf{r}') = \frac{\mathbf{r}'}{r'}\cdot\frac{d}{dt'}(\mathbf{x}-\mathbf{x}_0''(t')) = -\frac{\mathbf{u}'\cdot\mathbf{r}'}{r'}. \tag{17.9}$$

It follows that $t = t' + r'(t')/c$ differentiates to give

$$\frac{dt}{dt'} = 1 + \frac{1}{c}\frac{dr'}{dt'} = 1 - \frac{\mathbf{u}'\cdot\mathbf{r}'}{cr'} = \kappa'. \tag{17.10}$$

Similarly, we have

$$\frac{d}{dt'}(r'\kappa') = \frac{d}{dt'}(r' - \mathbf{u}'\cdot\mathbf{r}'/c) = -\frac{\mathbf{u}'\cdot\mathbf{r}'}{r'} - \frac{1}{c}[\mathbf{a}'\cdot\mathbf{r}' - (\mathbf{u}')^2], \tag{17.11}$$

where $\mathbf{a}'$ is the acceleration of the charge at the retarded time $t'$. We shall make use of these expressions shortly, but perhaps it is worth reiterating now that $\mathbf{x}$ is held constant when taking derivatives in (17.9)–(17.11).

Next we seek expressions for $\nabla r'$ and $\nabla(\mathbf{r}'\cdot\mathbf{u}')$. Here the operator $\nabla$ represents derivatives with respect to $\mathbf{x}$, the position of the observer, while holding $t$, but *not* $t'$, fixed. In short, this is the same operator that appears in (17.8). We proceed as follows. For $\nabla r'$ we have the identity

$$r'\nabla r' = \nabla(\mathbf{r}'\cdot\mathbf{r}'/2) = (\mathbf{r}'\cdot\nabla)\mathbf{r}' + \mathbf{r}'\times(\nabla\times\mathbf{r}'),$$

where $\mathbf{r}' = \mathbf{x} - \mathbf{x}_0''(t')$, and so

$$(\mathbf{r}'\cdot\nabla)\mathbf{r}' = (\mathbf{r}'\cdot\nabla)[\mathbf{x}-\mathbf{x}_0''(t')] = \mathbf{r}' - (\mathbf{r}'\cdot\nabla t')\mathbf{u}'$$

and

$$\begin{aligned}\mathbf{r}'\times\{\nabla\times[\mathbf{x}-\mathbf{x}_0''(t')]\} &= -\mathbf{r}'\times[\nabla\times\mathbf{x}_0''(t')] = -\mathbf{r}'\times(\nabla t'\times\mathbf{u}')\\ &= (\mathbf{r}'\cdot\nabla t')\mathbf{u}' - (\mathbf{r}'\cdot\mathbf{u}')\nabla t'.\end{aligned}$$

Gathering terms yields

$$r'\nabla r' = \mathbf{r}' - (\mathbf{r}'\cdot\mathbf{u}')\nabla t'. \tag{17.12}$$

Moreover, since we are holding $t$ constant, $t = t' + r'(t')/c$ differentiates to give $c\nabla t' = -\nabla r'$, which combines with (17.12) to yield the particularly simple expression

$$\kappa' r'\nabla r' = \mathbf{r}'. \tag{17.13}$$

Next we consider $\nabla(\mathbf{r}'\cdot\mathbf{u}')$. Once again we start with an identity, in this case

$$\nabla(\mathbf{r}'\cdot\mathbf{u}') = \mathbf{r}'\cdot\nabla\mathbf{u}' + \mathbf{u}'\cdot\nabla\mathbf{r}' + \mathbf{r}'\times\nabla\times\mathbf{u}' + \mathbf{u}'\times\nabla\times\mathbf{r}'.$$

This expands to give

$$\nabla(\mathbf{r}'\cdot\mathbf{u}') = (\mathbf{r}'\cdot\nabla t')\mathbf{a}' + \mathbf{u}'\cdot\nabla[\mathbf{x}-\mathbf{x}_0''(t')] + \mathbf{r}'\times(\nabla t'\times\mathbf{a}') + \mathbf{u}'\times\nabla\times[\mathbf{x}-\mathbf{x}_0''(t')],$$

from which

$$\nabla(\mathbf{r}'\cdot\mathbf{u}') = (\mathbf{r}'\cdot\nabla t')\mathbf{a}' + \mathbf{u}' - (\mathbf{u}'\cdot\nabla t')\mathbf{u}' + \mathbf{r}'\times(\nabla t'\times\mathbf{a}') - \mathbf{u}'\times(\nabla t'\times\mathbf{u}').$$

Expanding the vector triple products and gathering terms, we find

$$\nabla(\mathbf{r}'\cdot\mathbf{u}') = \mathbf{u}' + [\mathbf{r}'\cdot\mathbf{a}' - (\mathbf{u}')^2]\nabla t'. \tag{17.14}$$

Finally we recall that, with $t$ held constant, we have $c\nabla t' = -\nabla r'$ and $\kappa' r'\nabla r' = \mathbf{r}'$, and so (17.14) becomes

$$\nabla(\mathbf{r}'\cdot\mathbf{u}') = \mathbf{u}' - [\mathbf{r}'\cdot\mathbf{a}' - (\mathbf{u}')^2]\frac{\mathbf{r}'}{c\kappa' r'}. \tag{17.15}$$

Armed with these (admittedly uninspiring) kinematic results, we may now find the **E** and **B** fields from the Liénard–Wiechert potentials.

### 17.2.2 *The Electric Field for a Moving Point Charge*

Let us start with the electric field **E**. Substituting the Liénard–Wiechert potentials into (17.8), we obtain

$$\mathbf{E}(\mathbf{x},t) = -\frac{\partial\mathbf{A}}{\partial t} - \nabla\varphi = -\frac{q}{4\pi\varepsilon_0}\left[\frac{1}{c^2}\frac{\partial}{\partial t}\left(\frac{\mathbf{u}'}{r'\kappa'}\right) + \nabla\left(\frac{1}{\kappa' r'}\right)\right]. \tag{17.16}$$

Consider first the partial time derivative of **A**. Here **x**, the position of the observer, is held fixed and so we may avail ourselves of (17.10):

$$\frac{\partial}{\partial t}\left(\frac{\mathbf{u}'}{r'\kappa'}\right) = \frac{dt'}{dt}\frac{d}{dt'}\left(\frac{\mathbf{u}'}{r'\kappa'}\right) = \frac{1}{\kappa'}\left[\frac{\mathbf{a}'}{r'\kappa'} - \frac{\mathbf{u}'}{(r'\kappa')^2}\frac{d}{dt'}(r'\kappa')\right].$$

We now substitute for the last term on the right using (17.11), which yields

$$\frac{\partial}{\partial t}\left(\frac{\mathbf{u}'}{r'\kappa'}\right) = \frac{1}{c(r'\kappa')^3}\{cr'(r'\kappa')\mathbf{a}' + [c\mathbf{u}'\cdot\mathbf{r}' + r'(\mathbf{a}'\cdot\mathbf{r}' - \mathbf{u}'^2)]\mathbf{u}'\}. \tag{17.17}$$

Next, consider the gradient on the right of (17.16), which we may rewrite as

$$\nabla\left(\frac{1}{\kappa' r'}\right) = -\frac{1}{(\kappa' r')^2}\nabla(\kappa' r') = -\frac{1}{(\kappa' r')^2}\nabla[r' - (\mathbf{r}'\cdot\mathbf{u}')/c].$$

Using (17.13) and (17.15) to substitute for the two gradients, we find

$$\nabla\left(\frac{1}{\kappa' r'}\right) = \frac{1}{c^2(\kappa' r')^3}[c(\kappa' r')\mathbf{u}' - (\mathbf{r}'\cdot\mathbf{a}' + c^2 - \mathbf{u}'^2)\mathbf{r}']. \tag{17.18}$$

It only remains to combine (17.17) and (17.18) and gather terms. After a little algebra, we obtain

$$\boxed{\mathbf{E} = \frac{q}{4\pi\varepsilon_0 c^2}\frac{(c^2 - \mathbf{u}'^2)(\mathbf{r}' - r'\mathbf{u}'/c) + \mathbf{r}' \times [(\mathbf{r}' - r'\mathbf{u}'/c) \times \mathbf{a}']}{(\kappa' r')^3}.} \tag{17.19}$$

Some authors prefer to write this as

$$\boxed{\mathbf{E} = \frac{q}{4\pi\varepsilon_0 c^2}\left\{\frac{1}{(\kappa r)^3}\{(c^2 - \mathbf{u}^2)(\mathbf{r} - r\mathbf{u}/c) + \mathbf{r} \times [(\mathbf{r} - r\mathbf{u}/c) \times \mathbf{a}]\}\right\}_{\text{ret}},} \tag{17.20}$$

where the subscript 'ret' implies that everything inside the large curly brackets is evaluated at the retarded time $t'$.

The first contribution to (17.20),

$$\mathbf{E}_u = \frac{q}{4\pi\varepsilon_0}\left\{\frac{1 - \mathbf{u}^2/c^2}{(\kappa r)^3}(\mathbf{r} - r\mathbf{u}/c)\right\}_{\text{ret}}, \tag{17.21}$$

does not depend on $\mathbf{a}'$ and is called the *generalised Coulomb field*. It reduces to the static Coulomb field for $u' << c$ and falls off as $r'^{-2}$. If the acceleration is zero, then we have $\mathbf{r} = \mathbf{r}' - r'\mathbf{u}'/c$, as discussed in Exercise 17.5. In such cases we are led to the remarkable conclusion that $\mathbf{E}$ points radially outward from the *instantaneous* position of the charge, not the retarded position, and in this sense we may regard such a field as being permanently attached to the charge, dragged along by it. The second contribution,

$$\mathbf{E}_a = \frac{q}{4\pi\varepsilon_0 c^2}\left\{\frac{1}{(\kappa r)^3}\mathbf{r} \times [(\mathbf{r} - r\mathbf{u}/c) \times \mathbf{a}]\right\}_{\text{ret}}, \tag{17.22}$$

is perpendicular to $\mathbf{r}'$. It is called the *radiation field* since it falls off as $r'^{-1}$, yielding a finite Poynting flux for $r' \to \infty$. It is *not* attached to the charge, in the sense that a burst of acceleration gives rise to a spherical pulse of radiation that propagates out to infinity, centred on the retarded location of the acceleration. We discuss this in §17.2.5 and in Exercises 17.1 and 17.2.

### 17.2.3 The Magnetic Field and Poynting Vector

Let us now find the equivalent magnetic field from (17.6),

$$\mathbf{B}(\mathbf{x}, t) = \nabla \times \mathbf{A} = \frac{1}{c^2}\nabla \times [\mathbf{u}'\varphi(\mathbf{x}, t)] = \frac{1}{c^2}(\nabla\varphi \times \mathbf{u}' + \varphi\nabla \times \mathbf{u}'). \tag{17.23}$$

Now we have already found $\nabla\varphi$ in the process of evaluating $\mathbf{E}$, and so (17.18) gives us

$$\nabla\varphi = \frac{q}{4\pi\varepsilon_0}\nabla\left(\frac{1}{\kappa' r'}\right) = \frac{q}{4\pi\varepsilon_0}\frac{1}{c^2(\kappa' r')^3}[c(\kappa' r')\mathbf{u}' - (\mathbf{r}' \cdot \mathbf{a}' + c^2 - \mathbf{u}'^2)\mathbf{r}'].$$

Moreover, with $t$ constant, we have, from (17.13), $c\nabla t' = -\nabla r'$ and $\kappa' r' \nabla r' = \mathbf{r}'$, and so

$$\nabla \times \mathbf{u}' = \nabla t' \times \frac{d\mathbf{u}'}{dt'} = \nabla t' \times \mathbf{a}' = -\frac{1}{c\kappa' r'}\mathbf{r}' \times \mathbf{a}'.$$

Substituting into (17.23) and gathering terms yields

$$\boxed{\mathbf{B} = \frac{\mu_0 q}{4\pi c^2}\frac{1}{(\kappa' r')^3}\{(c^2 - \mathbf{u}'^2)\mathbf{u}' + [(\mathbf{r}' \cdot \mathbf{a}')\mathbf{u}' + c\kappa' r' \mathbf{a}']\} \times \mathbf{r}',} \tag{17.24}$$

or equivalently

$$\boxed{\mathbf{B} = \frac{\mu_0 q}{4\pi c^2}\left\{\frac{1}{(\kappa r)^3}\{(c^2 - \mathbf{u}^2)\mathbf{u} + [(\mathbf{r} \cdot \mathbf{a})\mathbf{u} + c\kappa r \mathbf{a}]\} \times \mathbf{r}\right\}_{\text{ret}}.} \tag{17.25}$$

As with $\mathbf{E}$, it is convenient to split this into two terms. The first contribution to (17.25),

$$\mathbf{B}_u = \frac{\mu_0 q}{4\pi}\left\{\frac{1 - \mathbf{u}^2/c^2}{(\kappa r)^3}\mathbf{u} \times \mathbf{r}\right\}_{\text{ret}}, \tag{17.26}$$

does not depend on $\mathbf{a}'$ and falls off as $r'^{-2}$. It does not contribute to the radiation field. The second contribution,

$$\mathbf{B}_a = \frac{\mu_0 q}{4\pi c^2}\left\{\frac{1}{(\kappa r)^3}[(\mathbf{r} \cdot \mathbf{a})\mathbf{u} + c\kappa r \mathbf{a}] \times \mathbf{r}\right\}_{\text{ret}}, \tag{17.27}$$

falls off as $r'^{-1}$ and so *does* contribute to the radiation field at infinity. It is readily confirmed that $\mathbf{B}_u$ and $\mathbf{B}_a$ are related to $\mathbf{E}_u$ and $\mathbf{E}_a$ by

$$\boxed{c\mathbf{B}_u = \hat{\mathbf{e}}'_r \times \mathbf{E}_u, \qquad c\mathbf{B}_a = \hat{\mathbf{e}}'_r \times \mathbf{E}_a,} \tag{17.28}$$

where $\hat{\mathbf{e}}_r$ is the unit vector $\hat{\mathbf{e}}_r = \mathbf{r}/r$, and $\hat{\mathbf{e}}'_r$ is the same unit vector at the retarded time $t'$. Evidently $c\mathbf{B} = \hat{\mathbf{e}}'_r \times \mathbf{E}$, and so $\mathbf{B}$ is perpendicular to both $\mathbf{E}$ and $\mathbf{r}'$, the displacement vector that takes us from the charge at time $t'$ to the observer. The Poynting vector for the radiation field at large distances is

$$\boxed{\frac{\mathbf{E}_a \times \mathbf{B}_a}{\mu_0} = \frac{\mathbf{E}_a^2}{c\mu_0}\hat{\mathbf{e}}'_r.} \tag{17.29}$$

### 17.2.4 The Case of Uniform Motion and a Glimpse at the Lorentz Transformation

We now illustrate the Liénard–Wiechert potentials and fields by considering two simple examples. We start with the simplest configuration of all—that of a charge moving along the $x$ axis with constant velocity $\mathbf{u}$, as shown in Figure 17.3. The potential $\varphi$ of the charge is

$$\varphi(\mathbf{x},t) = \frac{1}{4\pi\varepsilon_0}\frac{q}{r'\kappa'} = \frac{q}{4\pi\varepsilon_0(r' - ur'_x/c)}. \tag{17.30}$$

If we want to express $\varphi$ in terms of $\mathbf{x}$ and $t$, we need to find $r'(\mathbf{x},t)$, which in turn requires $t'(\mathbf{x},t)$. This may be found by combining the definitions $t' = t - |\mathbf{r}'|/c$ and $\mathbf{r}' = \mathbf{x} - ut'\hat{\mathbf{e}}_x$ to give

$$r'^2 = c^2(t-t')^2 = (x-ut')^2 + y^2 + z^2.$$

Solving for $t'$ then yields

$$(1-u^2/c^2)t' = t - ux/c^2 \pm c^{-1}\sqrt{(x-ut)^2 + (1-u^2/c^2)(y^2+z^2)}. \tag{17.31}$$

However, the factor $r' - ur'_x/c$ in the denominator of (17.30) can be rewritten as

$$r' - ur'_x/c = c(t-t') - u(x-ut')/c = c[t - (1-u^2/c^2)t' - ux/c^2],$$

and so (17.31) tells us that

$$r' - ur'_x/c = \sqrt{(x-ut)^2 + (1-u^2/c^2)(y^2+z^2)}. \tag{17.32}$$

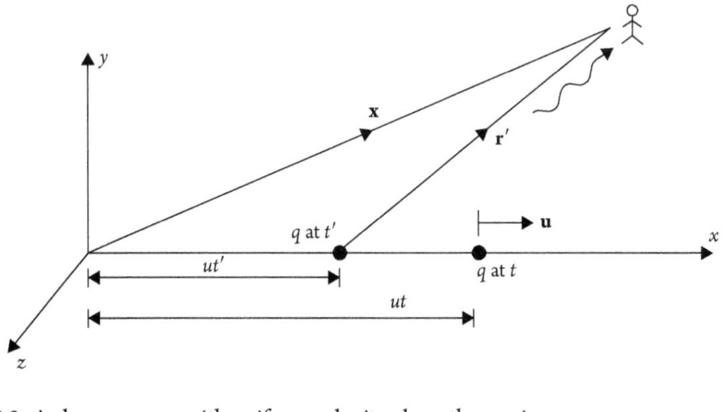


**Figure 17.3** A charge moves with uniform velocity along the $x$ axis.

We conclude that the potential of the moving point charge is

$$\boxed{\varphi(\mathbf{x},t)=\frac{q/\sqrt{1-u^2/c^2}}{4\pi\varepsilon_0\sqrt{\left[(x-ut)/\sqrt{1-u^2/c^2}\right]^2+y^2+z^2}}.} \tag{17.33}$$

Note, however, that in a coordinate system $\hat{S}$ that is centred on, and moves with, the charge, the potential is simply

$$\varphi(\mathbf{x},t)=\frac{q}{4\pi\varepsilon_0\sqrt{\hat{x}^2+\hat{y}^2+\hat{z}^2}}.$$

Focusing on the denominator of (17.33), it is as if the same event observed in the original frame $S$ and the moving frame $\hat{S}$ requires that the spatial coordinates transform as

$$\hat{x}=(x-ut)/\sqrt{1-u^2/c^2},\quad \hat{y}=y,\quad \hat{z}=z, \tag{17.34}$$

which is different from the Galilean transformation, $\hat{x}=x-ut$, $\hat{y}=y$, $\hat{z}=z$. This is our first glimpse at the Lorentz transformation of special relativity, which we shall discuss in Chapter 22.

Let us now turn to the Liénard–Wiechert fields **E** and **B**, starting with (17.19) for the electric field:

$$\mathbf{E}=\frac{q}{4\pi\varepsilon_0}\frac{(1-u^2/c^2)(\mathbf{r}'-r'\mathbf{u}/c)}{(r'-ur'_x/c)^3}. \tag{17.35}$$

Once again, we want to express this in terms of $\mathbf{x}$ and $t$. Let us start with the numerator. Combining $t'=t-|\mathbf{r}'|/c$ with $\mathbf{r}'=\mathbf{x}-ut'\hat{\mathbf{e}}_x$ gives us $\mathbf{r}'-r'\mathbf{u}/c=\mathbf{x}-\mathbf{u}t$, where $\mathbf{x}-\mathbf{u}t$ is the displacement vector that takes from the charge at time $t$ to the observer at $\mathbf{x}$. Moreover, (17.32) allows us to rewrite the denominator in terms of $\mathbf{x}$ and $t$, and so our expression for **E** becomes

$$\mathbf{E}(\mathbf{x},t)=\frac{q}{4\pi\varepsilon_0}\frac{(1-u^2/c^2)(\mathbf{x}-\mathbf{u}t)}{[(x-ut)^2+(1-u^2/c^2)(y^2+z^2)]^{3/2}}. \tag{17.36}$$

This is most conveniently expressed as

$$\boxed{\mathbf{E}(\mathbf{x},t)=\frac{q}{4\pi\varepsilon_0}\frac{(1-u^2/c^2)(\mathbf{x}-\mathbf{u}t)}{|\mathbf{x}-\mathbf{u}t|^3\,(1-u^2\sin^2\theta/c^2)^{3/2}},} \tag{17.37}$$

where $\theta$ is the angle between $\mathbf{x}-\mathbf{u}t$ and $\mathbf{u}$.

There are two striking features of (17.37). First, the field is not spherically symmetric, but rather most intense in directions normal to $\mathbf{u}$, as illustrated in Figure 17.4(a). Specifically, the field aligned with $\mathbf{u}$ is reduced by a factor of $1-u^2/c^2$, while that normal to $\mathbf{u}$ is enhanced by a factor of $1/\sqrt{1-u^2/c^2}$. Second, the **E**-lines point radially outward from the *instantaneous position* of the charge, *not* the retarded position. This is a quite remarkable result, since the

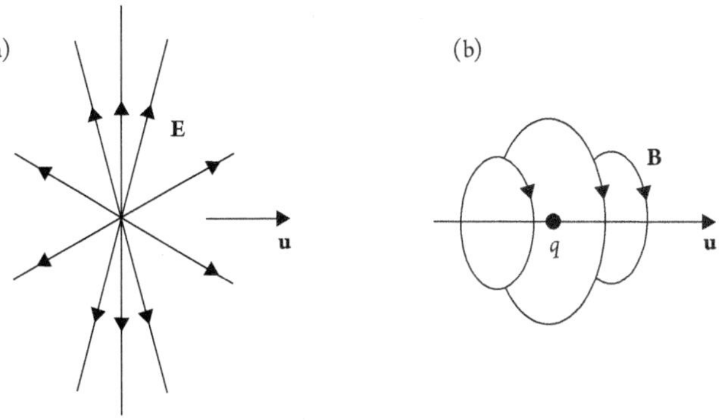


**Figure 17.4** The electromagnetic field of a charge moving with uniform velocity $\mathbf{u}$: (a) the electric field; (b) the magnetic field.

information received at point $\mathbf{x}$ at time $t$ came from the charge at its retarded position! The reason for this strange behaviour is closely related to Exercises 13.4 and 13.5.

The magnetic field is most readily obtained from $\mathbf{E}$ using $c\mathbf{B} = \hat{\mathbf{e}}'_r \times \mathbf{E}$. Noting that $\mathbf{r}' - r'\mathbf{u}/c = \mathbf{x} - \mathbf{u}t$, and hence

$$\hat{\mathbf{e}}'_r = (\mathbf{x} - \mathbf{u}t)/r' + \mathbf{u}/c, \tag{17.38}$$

we find,

$$\boxed{\mathbf{B} = \frac{1}{c^2}\mathbf{u} \times \mathbf{E}.} \tag{17.39}$$

The magnetic field lines encircle the charge in the azimuthal direction, as shown in Figure 17.4(b).

### 17.2.5 *The Sudden Deceleration of a Moving Charge: Bremsstrahlung*

We now consider what happens if our moving charge is suddenly subject to an abrupt deceleration, reducing its velocity to zero over a short period of time. This leads to so-called *bremsstrahlung*, the German for 'braking radiation'.

Suppose the charge stops abruptly at time $t = 0$ and that the origin of coordinates is chosen to be the location of the stationary charge for $t > 0$. Far from the stationary charge an observer at $\mathbf{x}$ does not know that the charge has stopped, because the information has not yet reached that point. So for distances greater than $|\mathbf{x}| = ct$ the field is that of a moving charge, with the electric field lines pointing radially outward from the position the charge would have had if it had not stopped. In short, for $|\mathbf{x}| > ct$ the $\mathbf{E}$-lines point radially outward from the location $\mathbf{x}_0 = \mathbf{u}t$. Close to the stationary charge, however, an observer *does* know that the charge has stopped and so we find a static Coulomb field in which the $\mathbf{E}$-lines point radially outward from the origin. So we have an inner field centred on the origin and an outer field centred on the 'downstream' point $\mathbf{x}_0 = \mathbf{u}t$. It turns out that the two fields connect through

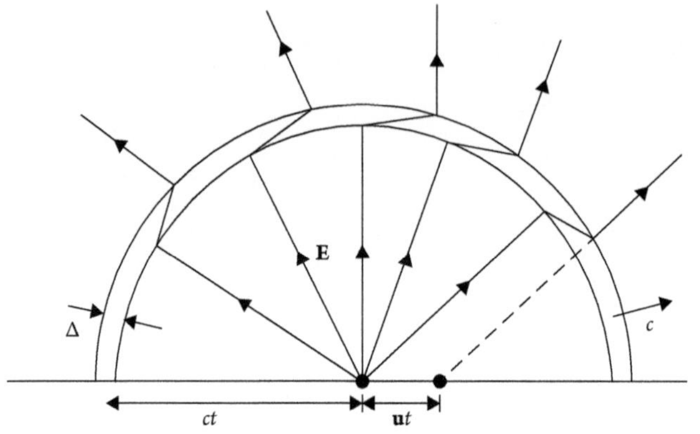


**Figure 17.5** The electric field of a charge that had been moving with uniform velocity **u** and is suddenly stopped at time $t = 0$.

a thin shell of radius $|\mathbf{x}| = ct$, with the thickness of the spherical shell dictated by the time taken to decelerate the charge. The situation is shown in Figure 17.5, with the dominant field in the shell being *normal* to **x**. As time passes, the spherical shell $r = ct$ sweeps radially outward at speed $c$, carrying the information that the charge is stationary and leaving a static Coulomb field in its wake. This shell is, in fact, a wave front, triggered by the deceleration of the charge and carrying an intense transverse electromagnetic field out to infinity (Exercise 17.1).

## 17.3 The Heaviside–Feynman Equations for the Field of a Point Charge

We now consider an alternative way of deriving the field of a moving point charge. This will lead us to the Heaviside–Feynman equations, which look very different to the Liénard–Wiechert fields, though of course they are equivalent. Our starting point is the Jefimenko equations (13.75) and (13.76):

$$\mathbf{E}(\mathbf{x},t) = \frac{1}{4\pi\varepsilon_0}\int\left\{\left[\rho_e(\mathbf{x}'',t') + \frac{|\mathbf{r}|}{c}\frac{\partial\rho_e(\mathbf{x}'',t')}{\partial t'}\right]\frac{\mathbf{r}}{|\mathbf{r}|^3} - \frac{1}{c^2}\frac{1}{|\mathbf{r}|}\frac{\partial\mathbf{J}(\mathbf{x}'',t')}{\partial t'}\right\}d\mathbf{x}'', \tag{17.40}$$

$$\mathbf{B}(\mathbf{x},t) = \frac{\mu_0}{4\pi}\int\left\{\left[\mathbf{J}(\mathbf{x}'',t') + \frac{|\mathbf{r}|}{c}\frac{\partial\mathbf{J}(\mathbf{x}'',t')}{\partial t'}\right]\times\frac{\mathbf{r}}{|\mathbf{r}|^3}\right\}d\mathbf{x}'', \tag{17.41}$$

where $\mathbf{r} = \mathbf{x} - \mathbf{x}''$. Since $t' = t - |\mathbf{x} - \mathbf{x}''|/c$, and both $\mathbf{x}$ and $\mathbf{x}''$ are held constant when differentiating the *Eulerian* fields $\rho_e(\mathbf{x}'',t')$ and $\mathbf{J}(\mathbf{x}'',t')$ with respect to time, we may replace $\partial/\partial t'$ by $\partial/\partial t$ in the integrals. This gives us

$$\mathbf{E}(\mathbf{x},t) = \frac{1}{4\pi\varepsilon_0}\int \frac{\rho_e(\mathbf{x}'',t')\mathbf{r}}{|\mathbf{r}|^3}\,d\mathbf{x}'' + \frac{1}{4\pi\varepsilon_0}\frac{\partial}{\partial t}\int\left[\frac{\rho_e(\mathbf{x}'',t')\mathbf{r}}{c\,|\mathbf{r}|^2} - \frac{\mathbf{J}(\mathbf{x}'',t')}{c^2\,|\mathbf{r}|}\right] d\mathbf{x}'', \tag{17.42}$$

$$\mathbf{B}(\mathbf{x},t) = \frac{\mu_0}{4\pi}\int \frac{\mathbf{J}(\mathbf{x}'',t')\times\mathbf{r}}{|\mathbf{r}|^3}\,d\mathbf{x}'' + \frac{\mu_0}{4\pi}\frac{\partial}{\partial t}\int \frac{\mathbf{J}(\mathbf{x}'',t')\times\mathbf{r}}{c\,|\mathbf{r}|^2}\,d\mathbf{x}'', \tag{17.43}$$

which are easier to evaluate for point charges.

Now consider a point charge $q$ that follows the trajectory $\mathbf{x}_0''(t)$ with velocity $\mathbf{u}(t) = d\mathbf{x}_0''/dt$. Then the charge and current density are given by $\rho_e(\mathbf{x},t) = q\delta(\mathbf{x}-\mathbf{x}_0''(t))$ and $\mathbf{J}(\mathbf{x},t) = q\mathbf{u}(t)\delta(\mathbf{x}-\mathbf{x}_0''(t))$. Taking into account the volumetric correction introduced in §17.1.2, $\kappa' = 1-\mathbf{u}'\cdot\mathbf{r}'/c\,|\mathbf{r}'|$, the integrals may be evaluated to give the fields in terms of Lagrangian time derivatives (in which $\mathbf{x}$ is assumed to be held fixed):

$$\mathbf{E} = \frac{q}{4\pi\varepsilon_0}\left[\frac{\mathbf{r}'}{\kappa' r'^3} + \frac{1}{c}\frac{d}{dt}\left(\frac{\mathbf{r}'}{\kappa' r'^2}\right) - \frac{1}{c^2}\frac{d}{dt}\left(\frac{\mathbf{u}'}{r'\kappa'}\right)\right], \tag{17.44}$$

$$\mathbf{B} = \frac{\mu_0 q}{4\pi}\left[\frac{\mathbf{u}'\times\mathbf{r}'}{\kappa' r'^3} + \frac{1}{c}\frac{d}{dt}\left(\frac{\mathbf{u}'\times\mathbf{r}'}{\kappa' r'^2}\right)\right], \tag{17.45}$$

where $\mathbf{r}' = \mathbf{x}-\mathbf{x}_0''(t')$ and $\mathbf{u}' = \mathbf{u}(t')$ are evaluated at the retarded time $t' = t - |\mathbf{r}'|/c$.

We may now use the results of §17.2.1 to simplify these expressions. The first step is to use (17.10) to rewrite the time derivatives in terms of $t'$:

$$\mathbf{E} = \frac{q}{4\pi\varepsilon_0}\left[\frac{\mathbf{r}'}{\kappa' r'^3} + \frac{1}{c\kappa'}\frac{d}{dt'}\left(\frac{\mathbf{r}'}{\kappa' r'^2}\right) + \frac{1}{c^2\kappa'}\frac{d}{dt'}\left(\frac{1}{r'\kappa'}\frac{d\mathbf{r}'}{dt'}\right)\right], \tag{17.46}$$

$$\mathbf{B} = \frac{\mu_0 q}{4\pi}\left[\frac{\mathbf{u}'\times\mathbf{r}'}{\kappa' r'^3} + \frac{1}{c\kappa'}\frac{d}{dt'}\left(\frac{1}{r'}\frac{\mathbf{u}'\times\mathbf{r}'}{\kappa' r'}\right)\right], \tag{17.47}$$

from which we find

$$\mathbf{E} = \frac{q}{4\pi\varepsilon_0}\left\{\frac{\mathbf{r}'}{\kappa' r'^3} + \frac{1}{c\kappa'}\frac{d}{dt'}\left[\left(1+\frac{1}{c}\frac{dr'}{dt'}\right)\frac{\mathbf{r}'}{\kappa' r'^2}\right] + \frac{1}{c^2\kappa'}\frac{d}{dt'}\left[\frac{d(\mathbf{r}'/r')}{\kappa'\,dt'}\right]\right\}, \tag{17.48}$$

$$\mathbf{B} = \frac{\mu_0 q}{4\pi}\left[\left(1-\frac{1}{c\kappa'}\frac{dr'}{dt'}\right)\frac{\mathbf{u}'\times\mathbf{r}'}{\kappa' r'^3} + \frac{1}{c\kappa' r'}\frac{d}{dt'}\left(\frac{\mathbf{u}'\times\mathbf{r}'}{\kappa' r'}\right)\right]. \tag{17.49}$$

However, (17.9) tells us that $dr'/dt' = -c(1-\kappa')$ and so our expressions for $\mathbf{E}$ and $\mathbf{B}$ simplify to

$$\mathbf{E} = \frac{q}{4\pi\varepsilon_0}\left\{\frac{\mathbf{r}'}{r'^3} + \frac{r'}{c\kappa'}\frac{d}{dt'}\left(\frac{\mathbf{r}'}{r'^3}\right) + \frac{1}{c^2\kappa'}\frac{d}{dt'}\left[\frac{d(\mathbf{r}'/r')}{\kappa'\,dt'}\right]\right\}, \tag{17.50}$$

$$\mathbf{B} = \frac{\mu_0 q}{4\pi}\left[\frac{\mathbf{u}'\times\mathbf{r}'}{\kappa'^2 r'^3} + \frac{1}{c\kappa' r'}\frac{d}{dt'}\left(\frac{\mathbf{u}'\times\mathbf{r}'}{\kappa' r'}\right)\right]. \tag{17.51}$$

Finally, reverting back to derivatives with respect to $t$ yields

$$\boxed{\mathbf{E} = \frac{q}{4\pi\varepsilon_0}\left[\frac{\mathbf{r}'}{r'^3} + \frac{r'}{c}\frac{d}{dt}\left(\frac{\mathbf{r}'}{r'^3}\right) + \frac{1}{c^2}\frac{d^2}{dt^2}\left(\frac{\mathbf{r}'}{r'}\right)\right],} \tag{17.52}$$

$$\boxed{\mathbf{B} = \frac{\mu_0 q}{4\pi}\left[\frac{\mathbf{u}'\times\mathbf{r}'}{\kappa'^2 r'^3} + \frac{1}{cr'}\frac{d}{dt}\left(\frac{\mathbf{u}'\times\mathbf{r}'}{\kappa' r'}\right)\right].} \tag{17.53}$$

These are the Heaviside–Feynman equations. Note that they are simpler and more compact than (17.19) and (17.24), and so invite physical interpretation.

Consider first the equation for $\mathbf{E}$, which was derived by Heaviside in 1904 and then rediscovered in Feynman, Leighton & Sands (1964). Since $r'/c = t - t'$, it may be rewritten as

$$\mathbf{E} = \frac{q}{4\pi\varepsilon_0}\left[\frac{\mathbf{r}'}{r'^3} + (t-t')\frac{d}{dt}\left(\frac{\mathbf{r}'}{r'^3}\right) + \frac{1}{c^2}\frac{d^2}{dt^2}\left(\frac{\mathbf{r}'}{r'}\right)\right].$$

The first term on the right is the *retarded* Coulomb field. Now suppose the motion of the charge is such that $\mathbf{E}(\mathbf{x}, \mathrm{t})$ varies slowly in time. Then the Taylor expansion

$$\frac{\mathbf{r}(t)}{r^3} = \frac{\mathbf{r}(t')}{r'^3} + (t-t')\frac{d}{dt'}\left[\frac{\mathbf{r}(t')}{r'^3}\right] + O((t-t')^2)$$

tells us that, for $u << c$ and $\kappa \approx 1$,

$$\mathbf{E} = \frac{q}{4\pi\varepsilon_0}\left[\frac{\mathbf{r}}{r^3} + \frac{1}{c^2}\frac{d^2}{dt^2}\left(\frac{\mathbf{r}'}{r'}\right)\right] + O((t-t')^2). \tag{17.54}$$

The first term on the right of (17.54) is now the *instantaneous* Coulomb field, centred on the instantaneous location of $q$. Evidently, this contribution to $\mathbf{E}$ remains attached to the charge, being dragged along by it. This extraordinary lack of retardation is closely related to (17.37) and is discussed in Exercise 17.5. So, if the field is changing slowly, then departures from the instantaneous Coulomb field are largely contained in the acceleration term, and it is this which gives rise to the radiation field at large distances from the charge. Indeed, a little algebra is sufficient to confirm that, to leading order in $1/r$,

$$\frac{\mu_0 q}{4\pi}\frac{d^2}{dt^2}\left(\frac{\mathbf{r}'}{r'}\right) = \mathbf{E}_a = \frac{\mu_0 q}{4\pi}\left\{\frac{\hat{\mathbf{e}}_r \times [(\hat{\mathbf{e}}_r - \mathbf{u}/c)\times\mathbf{a}]}{\kappa^3 r}\right\}_{\text{ret}} = \frac{\mu_0 q}{4\pi r'}\left(\frac{d^2\mathbf{r}'}{dt^2} - \frac{d^2 r'}{dt^2}\hat{\mathbf{e}}'_r\right).$$

The middle term is (17.22), while the term on the far right is due to Heaviside (1904) and is an astonishingly simple representation of the radiation field. This contribution to $\mathbf{E}$ is *not* attached to the charge, but rather despatches energy out to infinity as spherical waves.

Turning to the magnetic field, we may use $dr'/dt' = -c(1-\kappa')$ to write (17.53) as

$$c\mathbf{B} = \frac{q}{4\pi\varepsilon_0}\left(\frac{\mathbf{r}'}{r'}\right)\times\left[\frac{r'}{c}\frac{d}{dt}\left(\frac{\mathbf{r}'}{r'^3}\right) + \frac{1}{c^2}\frac{d^2}{dt^2}\left(\frac{\mathbf{r}'}{r'}\right)\right]. \tag{17.55}$$

This may be combined with (17.52) to give the inevitable relationship $c\mathbf{B} = \hat{\mathbf{e}}_r' \times \mathbf{E}$. Finally, note that Heaviside's analysis predates the Jefimenko equations by 62 years.

## 17.4 Radiation Patterns and a Second Look at Larmor's Radiation Equation

We close with a discussion of the radiation field of an accelerating charge. As we have seen, the electric field in the radiation zone is

$$\mathbf{E}_a(\mathbf{x},t) = \frac{\mu_0 q}{4\pi}\left\{\frac{\hat{\mathbf{e}}_r \times [(\hat{\mathbf{e}}_r - \mathbf{u}/c)\times \mathbf{a}]}{\kappa^3 r}\right\}_{\text{ret}}, \tag{17.56}$$

while the corresponding $\mathbf{B}$ field satisfies $c\mathbf{B}_a = \hat{\mathbf{e}}_r' \times \mathbf{E}_a$. Since $\mathbf{E}_a$ is normal to $\mathbf{r}'$, this yields a Poynting flux

$$\frac{\mathbf{E}_a \times \mathbf{B}_a}{\mu_0} = \frac{|\mathbf{E}_a|^2}{\mu_0 c}\hat{\mathbf{e}}_r' = \frac{\mu_0 q^2}{(4\pi)^2 c}\left\{\frac{\{\hat{\mathbf{e}}_r \times [(\hat{\mathbf{e}}_r - \mathbf{u}/c)\times \mathbf{a}]\}^2}{\kappa^6 r^2}\right\}_{\text{ret}}\hat{\mathbf{e}}_r'. \tag{17.57}$$

Evidently, energy radiates radially outward from the position the charge had at time $t'$.

Consider first the case of a charge for which $|\mathbf{u}| << c$. Then $\kappa' \approx 1$ and (17.56) simplifies to

$$\mathbf{E}_a(\mathbf{x},t) = \frac{\mu_0 q}{4\pi}\left\{\frac{\hat{\mathbf{e}}_r \times (\hat{\mathbf{e}}_r \times \mathbf{a})}{r}\right\}_{\text{ret}} = \frac{\mu_0 q}{4\pi}\left\{\frac{|\mathbf{a}|\sin\theta}{r}\hat{\mathbf{e}}_\theta\right\}_{\text{ret}},$$

where we have used spherical polar coordinates aligned with $\mathbf{a}'$. The energy flux per unit area is then

$$\frac{\mathbf{E}_a \times \mathbf{B}_a}{\mu_0} = \frac{\mu_0 q^2}{(4\pi)^2 c}\left\{\frac{a^2 \sin^2\theta}{r^2}\right\}_{\text{ret}}\hat{\mathbf{e}}_r'.$$

This is the familiar dipole radiation pattern of (16.46) and Figure 16.5, consisting of an axisymmetric lobe centred on the equatorial plane with zero radiation along $\pm a$. We now integrate over a sphere of large radius $r' = R$, centred on the charge at the retarded time $t'$. Noting that $\int (\sin\theta/r)^2\, dS = 8\pi/3$, we find the radiated power is

$$\boxed{P_q = \frac{1}{\mu_0}\oint \mathbf{E}\times\mathbf{B}\cdot d\mathbf{S} = \frac{\mu_0 q^2 a^2(t')}{6\pi c}.} \tag{17.58}$$

We have arrived back at Larmor's radiation equation (16.47), which is valid for charges whose speed is much less than $c$.

The case where the velocity of the charge is a significant fraction of $c$ is altogether more difficult. In part this is because the form of the Poynting flux is more complicated, and in part because the energy flux measured (or received) in a time interval $dt$ by a remote observer is not the same as the energy flux *emitted* in a time interval $dt'$ adjacent to the charge, the two

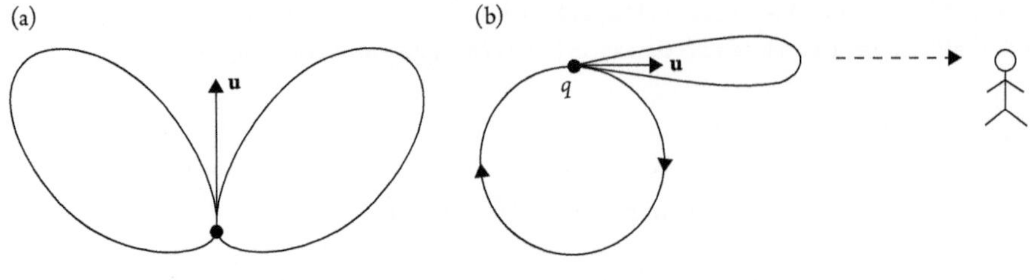


**Figure 17.6** Relativistic radiation field for (a) **u** parallel to **a** and (b) circular motion.

differing by a factor of $dt/dt' = \kappa'$. So, to get the rate of *emission* of energy from the charge, we must multiply (17.57), the energy flux seen by a remote observer, by $\kappa'$ before integrating over a sphere. The integration itself is complicated, but it turns out that the generalised version of Larmor's radiation equation is (see e.g. Zangwill, 2013)

$$\boxed{P_{\text{emitted}} = \frac{1}{\mu_0}\oint \mathbf{E}\times\mathbf{B}\cdot d\mathbf{S} = \frac{\mu_0 q^2\gamma^6}{6\pi c}\left\{a^2 - \frac{|\mathbf{u}\times\mathbf{a}|^2}{c^2}\right\}_{\text{ret}},} \tag{17.59}$$

where $\gamma = 1/\sqrt{1-\mathbf{u}'^2/c^2}$. Note that the factor of $\gamma^6$ leads to a large amplification of the radiation at relativistic speeds.

The particular case where **a** and **u** are collinear is of some interest. It has the simpler radiation pattern

$$\frac{\mathbf{E}_a \times \mathbf{B}_a}{\mu_0} = \frac{\mu_0 q^2}{(4\pi)^2 c}\left\{\frac{[\hat{\mathbf{e}}_r \times (\hat{\mathbf{e}}_r \times \mathbf{a})]^2}{\kappa^6 r^2}\right\}_{\text{ret}} \hat{\mathbf{e}}'_{r'}, \tag{17.60}$$

which differs from the non-relativistic case by the factor of $\kappa^{-6}$ only. The most striking feature here is the strong tendency for the term $\kappa' = 1 - \mathbf{u}'\cdot\mathbf{r}'/c\left|\mathbf{r}'\right|$ to focus the radiation ahead of, and diminish the radiation behind, the moving charge (see Exercise 17.3). So the radiation pattern remains axisymmetric, but it is now dominated by a radiation cone ahead of the moving charge, as shown in Figure 17.6(a). For high velocities, $u \to c$, this radiation cone becomes highly elongated in the direction of **u**.

The case of circular motion, where **a** and **u** are perpendicular, is discussed in Exercise 17.4. Here the radiation field for $u \to c$ is dominated by a single beam aligned with the instantaneous velocity, as shown in Figure 17.6(b). A distant observer therefore sees periodic blips of radiation as the beam sweeps over their location, rather like a rotating searchlight. This is *synchrotron radiation* and it is typical of the radiation from relativistic electrons spiralling around magnetic field lines. Synchrotron radiation is commonly observed in astrophysics, generating much of the radio emissions from distant galaxies.

* * *

This concludes our brief introduction to the fields of moving charges. It is tricky topic, in part because the algebraic details can get rather involved. Readers who hanker after a more leisurely introduction to the Liénard–Wiechert potentials could do worse than consult

Griffiths (2013) or Feynman, Leighton & Sands (1964). Griffiths also provides a detailed derivation of the Liénard–Wiechert fields **E** and **B**, as do Panofsky & Phillips (1962). Finally, Purcell (2011) offers a simple overview of bremsstrahlung, while the Heaviside–Feynman equations are discussed in Jackson (1999) and Zangwill (2013).

## EXERCISES

17.1 Consider a charge that is initially moving with constant velocity $\mathbf{u}$ and then abruptly decelerates to $\mathbf{u} = 0$ as it passes the origin, as shown in in Figure 17.5. The thickness of the spherical shell in Figure 17.5 is of order $\Delta \sim c\tau$, where $\tau$ is the short time it takes to decelerate the charge. Suppose that $u << c$. Then the electric field at large distances is, according to (17.37),

$$\mathbf{E} = \frac{q}{4\pi\varepsilon_0} \frac{\mathbf{x} - \mathbf{u}t}{|\mathbf{x} - \mathbf{u}t|^3}, \qquad |\mathbf{x}| > ct + \Delta,$$

while that close to the charge is

$$\mathbf{E} = E_r \hat{\mathbf{e}}_r = \frac{q}{4\pi\varepsilon_0} \frac{\hat{\mathbf{e}}_r}{|\mathbf{x}|^2}, \qquad |\mathbf{x}| < ct - \Delta.$$

Figure 17.7 shows an electric field line that makes an angle of $\theta$ relative to $\mathbf{u}$. Using spherical polar coordinates aligned with $\mathbf{u}$, the field line is deflected forward a distance $ut \sin\theta$ in the (minus) $\theta$ direction as it passes through the shell, and then continues radially outward, but now centred on the point $\mathbf{x} = \mathbf{u}t$.

Show that the ratio of the transverse to radial field components within the shell scales as $E_\theta / E_r \sim -\, ut \sin\theta / c\tau$. Hence show that $E_\theta$ within the shell falls off with distance as

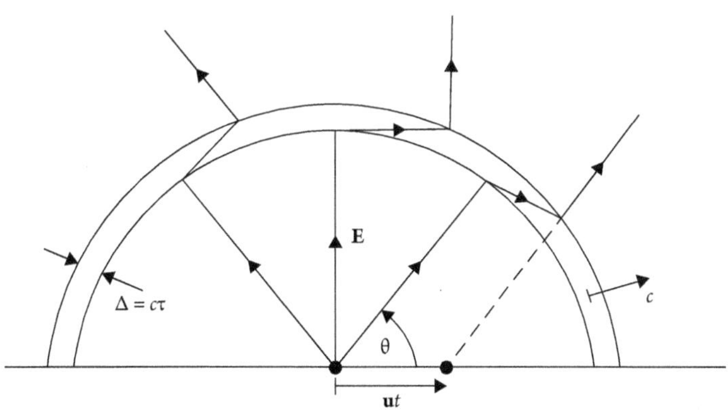


**Figure 17.7** The electric field of a charge that had been moving with uniform velocity $\mathbf{u} = u\hat{\mathbf{e}}_z$ and is suddenly stopped at time $t = 0$.

$$E_\theta \sim -\frac{q}{4\pi\varepsilon_0}\frac{u\sin\theta}{c^2\tau(ct)} \sim -\frac{q}{4\pi\varepsilon_0}\frac{u\sin\theta}{c^2\tau\,|\mathbf{x}|} \sim \frac{\mu_0 q}{4\pi}\frac{a\sin\theta}{|\mathbf{x}|},$$

where $a = -u/\tau$. This estimate might be compared with (17.56) applied to a slowly moving charge:

$$\mathbf{E}_a(\mathbf{x},t) = \frac{\mu_0 q}{4\pi}\left\{\frac{\hat{\mathbf{e}}_r \times (\hat{\mathbf{e}}_r \times \mathbf{a})}{r}\right\}_{\text{ret}} = \frac{\mu_0 q}{4\pi}\left\{\frac{a\sin\theta}{r}\hat{\mathbf{e}}_\theta\right\}_{\text{ret}}.$$

Thus the shell is in fact a wave front, initiated by the deceleration of the charge. In such an acceleration-induced wave $\mathbf{E}$ is predominantly normal to $\mathbf{x}$ and falls off as $1/|\mathbf{x}|$, as described by (17.56).

17.2 Consider a charge that is initially stationary and located at the origin. At $t = 0$ it is abruptly accelerated up to a velocity $\mathbf{u}$, which it maintains for $t > 0$. Determine $\mathbf{E}$ close to the instantaneous location of the charge and also for $|\mathbf{x}| > ct$, taking $u << c$. As in Exercise 17.1, the jump from the inner to the outer field occurs within a thin spherical shell centred on $|\mathbf{x}| = ct$ and of thickness $\Delta \sim c\tau$. Sketch the electric field lines and estimate the strength of the transverse electric field and Poynting vector within the thin shell. Hence estimate the energy radiated out to infinity.

17.3 A charge accelerates in a straight line. Adopting spherical polar coordinates aligned with the motion, use (17.60) to find the angular dependence of the Poynting flux. Sketch this angular variation for $u \to c$ and find the angles of the maximum Poynting vector as a function of $\gamma$.

17.4 Use (17.57) to find the radiation pattern for a charge moving in a circular orbit at a constant speed $u$. (*Hint:* use spherical polar coordinates aligned with the instantaneous direction of $\mathbf{u}$, and with $\mathbf{a}$ aligned with the $x$ axis.) For the case of $u \to c$, show that the radiation is focused into a single beam that points in the instantaneous direction of $\mathbf{u}$.

17.5 Confirm that, in the absence of charge acceleration, $\mathbf{r}$ and $\mathbf{r}'$ are related by $\mathbf{r} = \mathbf{r}' - r'\mathbf{u}'/c$. Hence show that, when $\mathbf{a} = 0$, the Liénard–Wiechert expression for $\mathbf{E}$, (17.19), simplifies to

$$\mathbf{E} = \frac{q(1-\mathbf{u}^2/c^2)}{4\pi\varepsilon_0}\frac{(\mathbf{r}' - r'\mathbf{u}'/c)r'^3}{[\mathbf{r}'\cdot(\mathbf{r}' - r'\mathbf{u}'/c)]^3} = \frac{q(1-\mathbf{u}^2/c^2)}{4\pi\varepsilon_0}\frac{\mathbf{r}}{r^3\cos^3\alpha},$$

where $\alpha$ is the angle between $\mathbf{r}$ and $\mathbf{r}'$. You may find it useful to note that $r'^2\kappa' = \mathbf{r}'\cdot(\mathbf{r}' - r'\mathbf{u}'/c)$. Note that the electric field points radially outward from the *instantaneous* position of the charge, not the retarded position. Show that this expression for $\mathbf{E}$ is identical to (17.37) and reduces to a static Coulomb field for $u << c$.

Now show that, when $\mathbf{a} = 0$ and $u << c$, the first two terms in the Heaviside–Feynman expression for $\mathbf{E}$, (17.44), can be written as

$$\frac{q}{4\pi\varepsilon_0}\left[\frac{\mathbf{r}'}{\kappa' r'^3} + \frac{1}{c}\frac{d}{dt}\left(\frac{\mathbf{r}'}{\kappa' r'^2}\right)\right] = \frac{q}{4\pi\varepsilon_0}\frac{(\mathbf{r}' - r'\mathbf{u}'/c)r'^3}{[\mathbf{r}'\cdot(\mathbf{r}' - r'\mathbf{u}'/c)]^3} = \frac{q}{4\pi\varepsilon_0}\frac{\mathbf{r}}{r^3\cos^3\alpha},$$

which again points radially outward from the instantaneous position of the charge. It is this which lies behind (17.54).

The absence of retardation in the Liénard–Wiechert equation (17.37), as well as the Heaviside–Feynman expression above, goes back to the cancelation of terms in the Jefimenko equations (17.40) and (17.41) for slowly varying fields, as discussed in Exercises 13.4 and 13.5.

17.6 Deduce the Liénard–Wiechert expression for **B**, (17.25), starting from the equivalent Heaviside–Feynman expression, (17.53). Show that, for the particular case of $\mathbf{a} = 0$, the Liénard–Wiechert expression simplifies to

$$\mathbf{B} = \frac{\mu_0 q(1 - \mathbf{u}^2/c^2)}{4\pi} \frac{\mathbf{u} \times \mathbf{r}}{r^3 \cos^3 \alpha}.$$

(As in Exercise 17.5, you may find it useful to note that $r'^2 \kappa' = \mathbf{r}' \cdot (\mathbf{r}' - r'\mathbf{u}'/c)$ and $\mathbf{r} = \mathbf{r}' - r'\mathbf{u}'/c$.) Note that the orientation of **B** is determined by the instantaneous position of the charge. Finally, confirm that this expression for **B** satisfies (17.39):

$$\mathbf{B} = \frac{1}{c^2}\mathbf{u} \times \mathbf{E}.$$

## REFERENCES


Feynman, R. P., Leighton, R. B. & Sands, M., 1964, *The Feynman Lectures on Physics*, Volume II. Addison-Wesley.

Heaviside, O., 1904, The radiation from an electron moving in an elliptic, or any other orbit. *Nature*, **69**, 342–343.

Griffiths, D. J., 2013, *Introduction to Electrodynamics*, 4th ed. Pearson.

Jackson, J. D., 1999, *Classical Electrodynamics*, 3rd ed. Wiley.

Panofsky, W. K. H. & Phillips, M., 1962, *Classical Electricity and Magnetism*, 2nd ed. Addison-Wesley.

Purcell, E., 2011, *Electricity and Magnetism*, 2nd ed. Cambridge University Press.

Zangwill, A., 2013, *Modern Electrodynamics*. Cambridge University Press.

# 18

# Maxwell's Equations in Dielectric and Magnetic Materials

The first time a French reader opens Maxwell's book, his admiration is tempered with a feeling of uneasiness, and often of distrust.

Henri Poincaré (*Science and Hypothesis*, 1905)

We now return to dielectric and magnetic materials, which were first discussed in Chapters 6 and 9 in the context of steady fields. Our aim is to establish the macroscopic form of Maxwell's equations appropriate for matter, along with the corresponding boundary conditions. This will then form the basis for our discussion of a variety of phenomena in Chapter 19, such as the reflection, refraction and dispersion of electromagnetic waves. We start, however, by recalling how static fields behave in dielectric and magnetic materials, assembling the key results of Chapters 6 and 9.

## 18.1 A Summary of the Behaviour of Static Fields in Matter

Let us start with dielectrics. The process of electronic polarisation allows a mean electric field to induce molecular dipoles within a dielectric, electric dipoles that are partially aligned with the imposed field. Moreover, if the molecules have an intrinsic dipole moment (i.e. they are polar molecules) and are free to rotate, then these polar molecules will have their dipole moment partially aligned with **E** through orientation polarisation. Either way, the molecular dipoles tend to align with **E** and we define the *electric polarisation vector* **P** to be the mean dipole moment per unit volume, $\mathbf{P}\,dV = \sum \mathbf{p}_{\text{mol}}$, where $\mathbf{p}_{\text{mol}}$ are the molecular dipoles in the volume element $dV$. This macroscopic, or locally averaged, field is a function of position, as well as of the applied electric field.

As we noted in Chapter 6, gradients in **P** are associated with the accumulation of bound charges, both volume charges and surface charges, according to

$$\rho_b = -\nabla \cdot \mathbf{P}, \quad \sigma_b = \mathbf{P} \cdot \mathbf{n}, \tag{18.1}$$

where **n** is the unit normal to the surface of the dielectric. These are not imaginary charges, but rather represent the genuine accumulation of bound charges either within the dielectric or on its surface. Since Gauss' law applies to all charge, irrespective of whether it is free charge or bound charge, we have

$$\nabla \cdot \mathbf{E} = \frac{\rho_e}{\varepsilon_0} = \frac{\rho_f + \rho_b}{\varepsilon_0}, \tag{18.2}$$

where $\rho_f$ is the free charge density. Moreover, since the bound charge density is given by $\rho_b = -\nabla \cdot \mathbf{P}$, Gauss' law may be rewritten as $\nabla \cdot (\varepsilon_0 \mathbf{E} + \mathbf{P}) = \rho_f$. This is usually expressed in terms of the auxiliary field

$$\mathbf{D} = \varepsilon_0 \mathbf{E} + \mathbf{P}, \tag{18.3}$$

called the *electric displacement vector*. Evidently, Gauss' law rewritten in terms of **D** takes a particularly simple form:

$$\nabla \cdot \mathbf{D} = \rho_f. \tag{18.4}$$

We also noted in Chapter 6 that charge conservation in the form

$$\nabla \cdot \mathbf{J} = -\frac{\partial \rho_e}{\partial t} \tag{18.5}$$

applies separately to both the free charges $\rho_f$, where **J** is then the conduction current associated with $\rho_f$, and to the bound charges $\rho_b$, which leads to the additional expression

$$\nabla \cdot \mathbf{J}_{\text{pol}} = -\frac{\partial \rho_b}{\partial t}. \tag{18.6}$$

Here $\mathbf{J}_{\text{pol}}$ is known as the *polarisation current density*, and it represents the current associated with movement of the bound charges—a kind of conduction current arising from the distortion or displacement of the polarised molecules. Since $\rho_b = -\nabla \cdot \mathbf{P}$, we may rewrite (18.6) as

$$\nabla \cdot \left( \mathbf{J}_{\text{pol}} - \frac{\partial \mathbf{P}}{\partial t} \right) = 0. \tag{18.7}$$

Of course, *one* particular solution of (18.7) is

$$\mathbf{J}_{\text{pol}} = \frac{\partial \mathbf{P}}{\partial t}, \tag{18.8}$$

although we shall see shortly that this is not the only possibility.

Finally we recall that, if the applied field is not too excessive, the degree of polarisation tends to be proportional to **E**, which leads to the idealisation of a *linear isotropic dielectric*. Any linear isotropic dielectric obeys the constitutive relationship

$$\mathbf{P} = \chi_e \varepsilon_0 \mathbf{E} = (\varepsilon_r - 1)\varepsilon_0 \mathbf{E}, \tag{18.9}$$

where $\chi_e = \varepsilon_r - 1$ is a dimensionless material constant called the *electric susceptibility* and $\varepsilon_r$ is known as the *dielectric constant* or the *relative permittivity*. In terms of **D**, this becomes

$$\mathbf{D} = \varepsilon_r \varepsilon_0 \mathbf{E} = \varepsilon \mathbf{E}, \tag{18.10}$$

where $\varepsilon = \varepsilon_r \varepsilon_0$ is the *permittivity* of the dielectric.

Let us now turn to magnetic materials and to the results of Chapter 9. We start with the *magnetisation vector* **M**, which is the magnetic dipole moment per unit volume, defined by $\mathbf{M}\,dV = \sum \mathbf{m}_{\text{mol}}$, where $\mathbf{m}_{\text{mol}}$ are the molecular dipoles in the volume $dV$. Note that, like **P**, the vector **M** is a macroscopic field. Note also that $\mathbf{m}_{\text{mol}}$, and hence **M**, may be a consequence of diamagnetism, paramagnetism or ferromagnetism—it does not matter which. In Chapter 9 we saw that the magnetisation vector is associated with a set of volumetric and surface current densities, according to

$$\mathbf{J}_{\text{mag}} = \nabla \times \mathbf{M}, \quad \mathbf{K}_{\text{mag}} = \mathbf{M} \times \mathbf{n}, \tag{18.11}$$

where **n** is a unit normal vector to the surface of the magnetised material. These are known as *bound currents* and, like **M**, they represent a macroscopic field.

From a macroscopic point of view, the bound currents act as real currents, at least as far as calculating the magnetic field is concerned, and so they must be included on the right-hand side of Ampère's law along with the conduction current. Ampère's law then becomes

$$\nabla \times \mathbf{B} = \mu_0(\mathbf{J}_f + \mathbf{J}_{\text{mag}}) = \mu_0(\mathbf{J}_f + \nabla \times \mathbf{M}),$$

where $\mathbf{J}_f$ is the conduction current associated with the free changes. It is now convenient to introduce the auxiliary field **H**, defined by

$$\mathbf{H} = \frac{1}{\mu_0}\mathbf{B} - \mathbf{M}, \tag{18.12}$$

so that Ampère's law take the particularly simple form

$$\nabla \times \mathbf{H} = \mathbf{J}_f. \tag{18.13}$$

This is a convenient expression in the sense that we usually know, or control, the conduction current, whereas we do not directly control the bound current.

In diamagnetic and paramagnetic materials the induced magnetisation **M** is observed to be proportional to the externally imposed magnetic field. Traditionally this has been expressed in the form

$$\mathbf{M} = \chi_m \mathbf{H}, \tag{18.14}$$

where $\chi_m$ is called the *magnetic susceptibility*, which is positive for paramagnetic materials and negative in diamagnetism. A material obeying the constitutive law (18.14) is said to be a linear isotropic medium. A soft ferromagnetic material in which the imposed field is not too large, and hysteresis may be tentatively ignored, is also sometimes approximated by

(18.14). However, a more realistic model for soft iron (which still ignores hysteresis) is to allow $\chi_m$ to be a function of $|\mathbf{H}|$. In any event, combining (18.12) and (18.14) gives us

$$\mathbf{B} = \mu_0(\mathbf{H}+\mathbf{M}) = \mu_0(1+\chi_m)\mathbf{H} = \mu_0\mu_r\mathbf{H} = \mu\mathbf{H}, \tag{18.15}$$

where $\mu = \mu_0(1+\chi_m)$ is the *permeability* of the medium and the related constant $\mu_r = 1+\chi_m$ is the *relative permeability*.

## 18.2 A Second Look at the Polarisation Current Density

We are now in a position move from static fields to dynamic fields, and hence write down the macroscopic version of Maxwell's equations appropriate to matter. First, however, there is a small discrepancy concerning the polarisation current that needs sorting out. We have seen that charge conservation applied to the bound current, along with the expression $\rho_b = -\nabla \cdot \mathbf{P}$, leads directly to

$$\nabla \cdot \left( \mathbf{J}_{\text{pol}} - \frac{\partial \mathbf{P}}{\partial t} \right) = 0. \tag{18.16}$$

Most authors then conclude that

$$\mathbf{J}_{\text{pol}} = \frac{\partial \mathbf{P}}{\partial t}, \tag{18.17}$$

the physical interpretation being that $\mathbf{P}$ can be increased by extending the gap between the charges in $\mathbf{p}_{\text{mol}}$, and this movement of charge constitutes a current. However, a more general solution of (18.16) is in fact

$$\mathbf{J}_{\text{pol}} = \frac{\partial \mathbf{P}}{\partial t} + \nabla \times \mathbf{M}_{\text{pol}}, \tag{18.18}$$

where the vector field $\mathbf{M}_{\text{pol}}$ is the vector potential for $\mathbf{J}_{\text{pol}} - \partial\mathbf{P}/\partial t$ and we have chosen the symbol $\mathbf{M}$ because of the similarity between (18.18) and $\mathbf{J}_{\text{mag}} = \nabla \times \mathbf{M}$. So what is $\mathbf{M}_{\text{pol}}$ and why do most authors set it to zero?

This question is tackled in, for example, Jackson (1999). It turns out that $\mathbf{M}_{\text{pol}}$ is associated with molecular magnetic dipoles of the form $\mathbf{m}_{\text{mol}} \sim \mathbf{p}_{\text{mol}} \times \mathbf{u}_{\text{mol}}$, where $\mathbf{u}_{\text{mol}}$ is the velocity of the molecules. It is not so hard to see how such magnetic dipoles might arise. Suppose that we have a small localised region of static charge $\rho_e$, which is set into motion with uniform velocity $\mathbf{u}$. Then (twice) the magnetic dipole moment of the moving charge distribution is

$$2\mathbf{m} = \int \mathbf{x} \times \mathbf{J}\, dV = \int \mathbf{x} \times (\rho_e \mathbf{u})\, dV = \int \rho_e \mathbf{x}\, dV \times \mathbf{u} = \mathbf{p} \times \mathbf{u}.$$

So molecular electric dipoles that drift can set up molecular magnetic dipoles, and on averaging over these magnetic dipoles we obtain an additional source of magnetisation.

It is this that lies behind $\mathbf{M}_{\text{pol}}$. However, if there is no bulk motion of the dielectric, then $\mathbf{u}_{\text{mol}}$ is both small and random. In such a situation the result of this averaging leaves $\mathbf{M}_{\text{pol}}$ negligible and (18.18) simply reverts to (18.17). On the other hand, if there *is* bulk motion of the dielectric, then $\mathbf{M}_{\text{pol}}$ can be significant, and it turns out that $\mathbf{M}_{\text{pol}} = \mathbf{P} \times \mathbf{u}_{\text{bulk}}$ where $\mathbf{u}_{\text{bulk}}(\mathbf{x}, t)$ is the macroscopic velocity field (see Fano, Chu & Adler, 1960). The expression (18.18) now becomes

$$\mathbf{J}_{\text{pol}} = \frac{\partial \mathbf{P}}{\partial t} + \nabla \times (\mathbf{P} \times \mathbf{u}_{\text{bulk}}). \tag{18.19}$$

If $\mathbf{u}_{\text{bulk}}$ is uniform, then the curl in (18.19) can be expanded to give

$$\mathbf{J}_{\text{pol}} = \left[\frac{\partial \mathbf{P}}{\partial t} + (\mathbf{u}_{\text{bulk}} \cdot \nabla)\mathbf{P}\right] - (\nabla \cdot \mathbf{P})\mathbf{u}_{\text{bulk}}.$$

Introducing the notation

$$\frac{D\mathbf{P}}{Dt} = \frac{\partial \mathbf{P}}{\partial t} + (\mathbf{u}_{\text{bulk}} \cdot \nabla)\mathbf{P}, \tag{18.20}$$

and recalling that $\rho_b = -\nabla \cdot \mathbf{P}$, we obtain

$$\mathbf{J}_{\text{pol}} = \frac{D\mathbf{P}}{Dt} + \rho_b \mathbf{u}_{\text{bulk}}. \tag{18.21}$$

The quantity $D\mathbf{P}/Dt$ is called the *convective derivative* of $\mathbf{P}$ and it gives the rate of change of $\mathbf{P}$ in a frame moving with velocity $\mathbf{u}_{\text{bulk}}$, as we shall discuss in Chapter 20. Notice that, if we change to a frame of reference moving with velocity $\mathbf{u}_{\text{bulk}}$, then (18.21) reverts to (18.17). Equation (18.21) says that the effect of introducing uniform bulk motion is to (i) supplement the polarisation current density by the convection current $\rho_b \mathbf{u}_{\text{bulk}}$ and (ii) demand that we use the relative time derivative $D\mathbf{P}/Dt$, rather than $\partial \mathbf{P}/\partial t$, to evaluate the contribution to $\mathbf{J}_{\text{pol}}$ from molecular fluctuations. The case of non-uniform bulk motion is discussed in Exercise 18.3.

## 18.3 The Macroscopic Maxwell Equations for Polarised and Magnetised Matter

We are now in a position to develop the macroscopic version of Maxwell's equations for polarised or magnetised matter. Our objective is to manipulate Maxwell's equations into a form in which only the free charges and associated conduction current appear explicitly. The bound charges and current, which we do not control, are then hidden in the auxiliary fields **D** and **H**. We may then close the system by using the constitutive equations that relate **D** and **H** to **E** and **B**, at least for linear materials.

We start by noting that Faraday's law requires no change, since it contains no sources. Moreover, the static approach to a macroscopic version of Gauss' law needs no modification when we move to dynamic fields. We merely introduce the electric displacement vector

$\mathbf{D} = \varepsilon_0\mathbf{E} + \mathbf{P}$ and rewrite Gauss' law in the form $\nabla \cdot \mathbf{D} = \rho_f$. So we shall focus our attention on the Ampère–Maxwell equation. When all possible sources of macroscopic current are accounted for, we have

$$\mathbf{J} = \mathbf{J}_f + \mathbf{J}_{\text{pol}} + \mathbf{J}_{\text{mag}} = \mathbf{J}_f + \frac{\partial \mathbf{P}}{\partial t} + \nabla \times \mathbf{M}_{\text{pol}} + \nabla \times \mathbf{M}, \tag{18.22}$$

where we have lumped any convection current in with the conduction current and labelled the sum as $\mathbf{J}_f$. So, taking the displacement current over to the left-hand side, the macroscopic version of the Ampère–Maxwell equation reads

$$\frac{1}{\mu_0}\nabla \times \mathbf{B} - \varepsilon_0\frac{\partial \mathbf{E}}{\partial t} = \mathbf{J}_f + \mathbf{J}_{\text{pol}} + \mathbf{J}_{\text{mag}} = \mathbf{J}_f + \frac{\partial \mathbf{P}}{\partial t} + \nabla \times (\mathbf{M}_{\text{pol}} + \mathbf{M}).$$

We now generalise our definition of $\mathbf{H}$ to include the magnetisation $\mathbf{M}_{\text{pol}}$,

$$\mathbf{H} = \frac{1}{\mu_0}\mathbf{B} - (\mathbf{M} + \mathbf{M}_{\text{pol}}), \tag{18.23}$$

noting that this reverts to the more usual definition when there is no bulk motion of the dielectric. Gathering terms now yields,

$$\nabla \times \mathbf{H} = \mathbf{J}_f + \frac{\partial \mathbf{D}}{\partial t}. \tag{18.24}$$

We conclude that the macroscopic Maxwell equations are

$$\boxed{\begin{array}{cc} \nabla \cdot \mathbf{D} = \rho_f, & \nabla \times \mathbf{E} = -\dfrac{\partial \mathbf{B}}{\partial t}, \\ \text{(Gauss' law)} & \text{(Faraday's law)} \end{array}} \tag{18.25}$$

$$\boxed{\begin{array}{cc} \nabla \cdot \mathbf{B} = 0, & \nabla \times \mathbf{H} = \mathbf{J}_f + \dfrac{\partial \mathbf{D}}{\partial t}. \\ \text{(}\mathbf{B}\text{ is solenoidal)} & \text{(Ampère-Maxwell law)} \end{array}} \tag{18.26}$$

In some engineering texts these are simply referred to as *the* Maxwell equations, the suggestion being that the free-space equations are just a special case. However, we shall take the position that

$$\nabla \cdot \mathbf{E} = \frac{\rho_e}{\varepsilon_0}, \quad \nabla \times \mathbf{E} = -\frac{\partial \mathbf{B}}{\partial t}, \tag{18.27}$$

$$\nabla \cdot \mathbf{B} = 0, \quad \nabla \times \mathbf{B} = \mu_0\left(\mathbf{J} + \varepsilon_0\frac{\partial \mathbf{E}}{\partial t}\right), \tag{18.28}$$

are the fundamental equations, while the macroscopic equations (18.25) and (18.26) are simply a rearrangement of (18.27) and (18.28) designed to hide the bound charges and current.

Note that the macroscopic equations are not closed as they stand, since they involve four unknown vector fields, **D**, **H**, **E** and **B**, and each of these fields requires that its divergence and curl be specified if it is to be uniquely determined. In order to close the system it is common to assume that the material is linear and isotropic, so that $\mathbf{P} = \chi_e \varepsilon_0 \mathbf{E}$ yields

$$\boxed{\mathbf{D} = \varepsilon_r \varepsilon_0 \mathbf{E} = \varepsilon \mathbf{E}, \quad \varepsilon_r = 1 + \chi_e,} \tag{18.29}$$

while $\mathbf{M} = \chi_m \mathbf{H}$ gives us

$$\mu_0(\mathbf{H} + \mathbf{M}) = \mu_0(1 + \chi_m)\mathbf{H} = \mu \mathbf{H}. \tag{18.30}$$

If there is no bulk motion of the dielectric, then $\mathbf{M}_{\text{pol}} \approx 0$ and our constitutive law for **H** is the familiar one

$$\boxed{\mathbf{B} = \mu_r \mu_0 \mathbf{H} = \mu \mathbf{H}, \quad \mu_r = 1 + \chi_m.} \tag{18.31}$$

In such a situation our macroscopic equations are simply

$$\nabla \cdot \mathbf{E} = \frac{\rho_f}{\varepsilon}, \quad \nabla \times \mathbf{E} = -\frac{\partial \mathbf{B}}{\partial t}, \tag{18.32}$$

$$\nabla \cdot \mathbf{B} = 0, \quad \nabla \times \mathbf{B} = \mu \left( \mathbf{J}_f + \varepsilon \frac{\partial \mathbf{E}}{\partial t} \right). \tag{18.33}$$

In effect, $(\varepsilon, \mu)$ and $(\rho_f, \mathbf{J}_f)$ replace $(\varepsilon_0, \mu_0)$ and $(\rho_e, \mathbf{J})$ in Maxwell's equations. If $\mathbf{M}_{\text{pol}} \neq 0$, or the medium is anisotropic, then these equations become more complicated, but we shall not pause to pursue those cases.

In the absence of sources, the governing equations for a linear isotropic material are $\nabla \cdot \mathbf{E} = 0$, $\nabla \cdot \mathbf{B} = 0$ and

$$\nabla \times \mathbf{E} = -\frac{\partial \mathbf{B}}{\partial t}, \quad \nabla \times \mathbf{B} = \mu \varepsilon \frac{\partial \mathbf{E}}{\partial t}. \tag{18.34}$$

These are identical to the vacuum equations except that $\varepsilon \mu$ replaces $\varepsilon_0 \mu_0$, and so the wave speed is now $\upsilon = 1/\sqrt{\varepsilon \mu}$. In optics, the ratio

$$n = c/\upsilon = \sqrt{\varepsilon \mu / \varepsilon_0 \mu_0} = \sqrt{\varepsilon_r \mu_r} \tag{18.35}$$

is known as the *refractive index*. It is larger than unity and so electromagnetic waves travel more slowly through matter. Both $n$ and $\varepsilon_r$ are functions of wave frequency, which will turn out to important for certain applications in Chapter 19.

## 18.4 Why Are There No Magnetic Charges?

Now might be a good time to raise the curious issue of the absence of magnetic monopoles. Heaviside was reluctant to declare absolutely that there was no such thing as magnetic charge, and so tended to reserve a place for magnetic monopoles when writing out the general form of Maxwell's equations, leaving it until the last moment before striking out such a term and drawing some conclusion. Many subsequent authors, such as Parker (2007), have shared Heaviside's sense of curiosity. Indeed, Parker muses whimsically at some length about the structure of an uninhabitable universe in which there *are* magnetic monopoles.

If magnetic charges did indeed exist, then Maxwell's equations might look something like

$$\nabla \cdot \mathbf{E} = \frac{\rho_e}{\varepsilon_0}, \qquad \nabla \times \mathbf{E} = -\mu_0 \mathbf{J}_m - \frac{\partial \mathbf{B}}{\partial t}, \tag{18.36}$$

$$\nabla \cdot \mathbf{B} = \mu_0 \rho_m, \quad \nabla \times \mathbf{B} = \mu_0 \left( \mathbf{J} + \varepsilon_0 \frac{\partial \mathbf{E}}{\partial t} \right), \tag{18.37}$$

where $\rho_m$ is the magnetic charge density and $\mathbf{J}_m$ the associated flux of magnetic charge:

$$\nabla \cdot \mathbf{J}_m = -\frac{\partial \rho_m}{\partial t}.$$

Such an idea appeals to the symmetrical roles played by **E** and **B** in (18.36) and (18.37), and indeed it was this symmetry that left Heaviside reluctant to let go of magnetic monopoles. While there is no obvious theoretical objection to a universe in which magnetic monopoles exist, and indeed some theories anticipate such monopoles, all attempts to date to detect magnetic charge have failed. Nevertheless, the quest for monopoles continues, in large part because Dirac managed to show that the existence of a magnetic monopole could explain the quantization of electrical charge (see e.g. the discussion in Jackson, 1999). So every now and then the search for magnetic charge is renewed, only to reconfirm its apparent absence. It seems that, for whatever reason, after the 'big bang' there were no magnetic charges, and ever since we have had a magnetic field that satisfies $\nabla \cdot \mathbf{B} = 0$.

## 18.5 Boundary Conditions for the Macroscopic Maxwell Equations

We now turn to the boundary conditions that must be satisfied at the interface between two different materials. These are obtained from the integral forms of the macroscopic Maxwell equations, which are

$$\oint \mathbf{D} \cdot d\mathbf{S} = Q_f, \quad \oint \mathbf{E} \cdot d\mathbf{r} = -\frac{\partial}{\partial t} \int \mathbf{B} \cdot d\mathbf{S}, \tag{18.38}$$

$$\oint \mathbf{B} \cdot d\mathbf{S} = 0, \quad \oint \mathbf{H} \cdot d\mathbf{r} = \int \mathbf{J}_f \cdot d\mathbf{S} + \frac{\partial}{\partial t} \int \mathbf{D} \cdot d\mathbf{S}, \tag{18.39}$$

where $Q_f$ is the free charge lying within the closed surface.

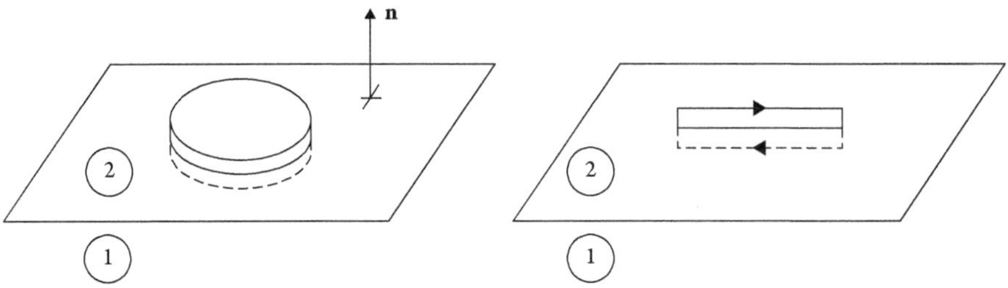


**Figure 18.1** The closed surface and closed contour to which the macroscopic Maxwell equations are applied to obtain boundary conditions at the interface.

The two closed surface integrals are now applied to a thin disc, or pillbox, of unit cross-sectional area that straddles the interface, as shown on the left of Figure 18.1. If the lower material is labelled 1 and the upper material 2, and **n** is a unit normal vector that takes us from material 1 to material 2, then we have

$$\boxed{(\mathbf{D}_2 - \mathbf{D}_1)\cdot\mathbf{n} = \sigma_f, \quad (\mathbf{B}_2 - \mathbf{B}_1)\cdot\mathbf{n} = 0,} \tag{18.40}$$

where $\sigma_f$ is the free surface charge. Evidently the normal component of **B** is continuous while that of **D** is discontinuous whenever there are surface charges.

Turning to the closed line integrals, we apply these to a thin rectangular loop of unit length that straddles the interface, as shown on the right of Figure 18.1. If the loop is thin enough, then we may ignore the contribution from the two time-dependent fluxes. The only surface integral we need worry about is that associated with the free surface current $\mathbf{K}_f$ and so our integral equations simplify to

$$\oint \mathbf{E}\cdot d\mathbf{r} = 0, \quad \oint \mathbf{H}\cdot d\mathbf{r} = \int \mathbf{K}_f \cdot d\mathbf{S}. \tag{18.41}$$

It is readily confirmed that these yield (see Exercise 18.1)

$$\boxed{\mathbf{n}\times(\mathbf{E}_2 - \mathbf{E}_1) = 0, \quad \mathbf{n}\times(\mathbf{H}_2 - \mathbf{H}_1) = \mathbf{K}_f.} \tag{18.42}$$

We conclude that the tangential component of **E** is continuous, while that of **H** is discontinuous whenever there is a surface current. The second equation in (18.42) is often rewritten as

$$(\mathbf{H}_2 - \mathbf{H}_1)_\parallel = \mathbf{K}_f \times \mathbf{n},$$

where the subscript $\parallel$ indicates that this applies to the components of **H** parallel to the interface.

For linear, isotropic media, these boundary conditions may be written in terms of **E** and **B** alone:

$$(\varepsilon_2\mathbf{E}_2 - \varepsilon_1\mathbf{E}_1)\cdot\mathbf{n} = \sigma_f, \quad (\mathbf{E}_2 - \mathbf{E}_1)_\parallel = 0, \tag{18.43}$$

$$(\mathbf{B}_2 - \mathbf{B}_1)\cdot\mathbf{n} = 0, \quad (\mu_2^{-1}\mathbf{B}_2 - \mu_1^{-1}\mathbf{B}_1)_\parallel = \mathbf{K}_f \times \mathbf{n}. \tag{18.44}$$

In the absence of surface charges and current, these simplify to

$$\boxed{\varepsilon_2(\mathbf{E}_2)_\perp = \varepsilon_1(\mathbf{E}_1)_\perp, \quad (\mathbf{E}_2)_\parallel = (\mathbf{E}_1)_\parallel,} \tag{18.45}$$

$$\boxed{(\mathbf{B}_2)_\perp = (\mathbf{B}_1)_\perp, \quad \frac{1}{\mu_2}(\mathbf{B}_2)_\parallel = \frac{1}{\mu_1}(\mathbf{B}_1)_\parallel,} \tag{18.46}$$

where the subscript $\perp$ indicates a component perpendicular to the interface. We shall make frequent use of these interfacial boundary conditions in Chapter 19.

## EXERCISES

18.1 Apply (18.41) to the loop shown in Figure 18.1, taking the unit vector $\mathbf{n}$ to point in the $z$ direction. By considering the loop to be first aligned with the $x$ axis, and then with the $y$ axis, show that $(H_2 - H_1)_x = K_y$ and $(H_2 - H_1)_y = -K_x$, and hence confirm that $\mathbf{n} \times (\mathbf{H}_2 - \mathbf{H}_1) = \mathbf{K}_f$.

18.2 Show that, in certain crystalline materials, there is *not* a unique decomposition of the charge into bound charge and free charge, so that the definition of $\mathbf{P}$ is often ambiguous. (*Hint:* group the charges into dipoles in different ways within the crystal.)

18.3 Combine (10.15), in which $\mathbf{P} = \mathbf{G}$ and $\mathbf{u}_{\text{bulk}}(\mathbf{x}, t) = \mathbf{v}(\mathbf{x}, t)$, with (18.19) to show that, if $S_m$ is any material surface in a translating or deforming dielectric (a surface always composed of the same atoms), then

$$\int_{S_m} \mathbf{J}_{\text{pol}} \cdot d\mathbf{S} = \frac{d}{dt} \int_{S_m} \mathbf{P} \cdot d\mathbf{S}.$$

This is the dynamic equivalent of the static result (18.17) and its physical interpretation is the same: if $\mathbf{P}$ in the vicinity of $S_m$ is to be increased, then the gap between charges in $\mathbf{p}_{\text{mol}}$ must be locally increased and this leads to a polarisation current across $S_m$.

18.4 According to (18.34), a plane wave will travel through a linear dielectric, say glass, at the speed $v = c/n$. Consider a monochromatic plane wave passing from a vacuum into a large glass block, the wave vector $\mathbf{k}$ being *normal* to the surface of the glass. Find the relationship between $\mathbf{E}$, $\mathbf{B}$ and $\mathbf{k}$ for the wave in the glass and show that, except for replacing $c$ by $c/n$, the wave behaves exactly like the incident vacuum wave. Of course, this is a consequence of the linearity of (18.34). From a physical point of view, however, it is extraordinary that, as the wave passes through the glass, the molecules are continually being polarised first one way and then the other, and when the fields of these polarised molecules are *superimposed* on the incident wave field travelling at the speed $c$, we obtain a new wave with the same frequency, *but a different speed*!

18.5 Use the results of Exercise 18.4 to show that boundary conditions (18.45) and (18.46) applied to the surface of the glass block take the form

$$\mathbf{E}_I + \mathbf{E}_R = \mathbf{E}_T, \quad c^{-1}\mathbf{E}_I - c^{-1}\mathbf{E}_R = \frac{n}{c}\mathbf{E}_T,$$

where the subscripts $I$, $R$ and $T$ stand for *incident, reflected* and *transmitted.* The glass has a refractive index $n = 1.6$. Find the ratio of the reflected to the transmitted electric field at the interface.

## REFERENCES

Fano, R. M., Chu, L. J. & Adler, R. B., 1960, *Electromagnetic Fields, Energy and Forces*. Wiley.
Jackson, J. D., 1999, *Classical Electrodynamics*, 3rd ed. Wiley.
Parker, E. N., 2007, *Conversations on Electric and Magnetic Fields in the Cosmos*. Princeton University Press.

# 19

# Plane Waves in Stationary Dielectrics and Conductors

Concerning vision alone, a separate science formed among philosophers, namely optics.... It is possible that some other science may be more useful, but no other science has so much sweetness and beauty of utility. Therefore it is the flower of the whole of philosophy and through it, and not without it, can the other sciences be known.

Roger Bacon (*Opus Majus*, 1268)

We now turn to electromagnetic waves in matter, particularly in dielectrics, where we touch on optics. We start with the governing equations and boundary conditions.

## 19.1 Electromagnetic Waves in Stationary Dielectrics

### 19.1.1 Governing Equations and Boundary Conditions

Our discussion of dielectrics will be restricted to homogeneous isotropic linear media in which $\mathbf{D} = \varepsilon_r\varepsilon_0\mathbf{E} = \varepsilon\mathbf{E}$ and $\mu_r = 1$. The governing equations for such a system were introduced in §18.3 and are

$$\nabla \cdot \mathbf{E} = 0, \quad \nabla \times \mathbf{E} = -\frac{\partial \mathbf{B}}{\partial t}, \tag{19.1}$$

$$\nabla \cdot \mathbf{B} = 0, \quad \nabla \times \mathbf{B} = \mu_0\varepsilon\frac{\partial \mathbf{E}}{\partial t}, \tag{19.2}$$

which yield the wave equations

$$\frac{\partial^2 \mathbf{E}}{\partial t^2} = \frac{1}{\mu_0\varepsilon}\nabla^2\mathbf{E}, \quad \frac{\partial^2 \mathbf{B}}{\partial t^2} = \frac{1}{\mu_0\varepsilon}\nabla^2\mathbf{B}. \tag{19.3}$$

These are identical to the vacuum equations except that $\varepsilon$ replaces $\varepsilon_0$, and so the wave speed is $v = 1/\sqrt{\varepsilon\mu_0}$, with a *refractive index*

$$n = c/v = \sqrt{\varepsilon_r}. \tag{19.4}$$

Note that the dielectric constant $\varepsilon_r$ is invariably a function of frequency, and this will become important in §19.1.5 when we come to discuss the refraction of dispersive waves. These equations support monochromatic plane waves of the form

$$\mathbf{E}(\mathbf{x},t) = \hat{\mathbf{E}} \exp j(\mathbf{k}\cdot\mathbf{x} - \varpi t), \quad \mathbf{k}\cdot\hat{\mathbf{E}} = 0, \tag{19.5}$$

$$\mathbf{B}(\mathbf{x},t) = \hat{\mathbf{B}} \exp j(\mathbf{k}\cdot\mathbf{x} - \varpi t), \quad \mathbf{k}\cdot\hat{\mathbf{B}} = 0, \tag{19.6}$$

where $v = \varpi/|\mathbf{k}|$. Either of (19.1) or (19.2) then gives

$$v\hat{\mathbf{B}} = \hat{\mathbf{e}}_k \times \hat{\mathbf{E}}, \tag{19.7}$$

where $\hat{\mathbf{e}}_k = \mathbf{k}/|\mathbf{k}|$ is a unit vector in the direction of travel of the wave.

In §18.5 we showed that, in the absence of surface charges and current, the interfacial boundary conditions between two dielectrics are

$$\varepsilon_2(\mathbf{E}_2)_\perp = \varepsilon_1(\mathbf{E}_1)_\perp, \quad (\mathbf{E}_2)_\parallel = (\mathbf{E}_1)_\parallel, \quad \mathbf{B}_2 = \mathbf{B}_1, \tag{19.8}$$

where the subscripts $\parallel$ and $\perp$ indicate components parallel and perpendicular to the interface. Armed with these governing equations and boundary conditions, we shall now explore the behaviour of waves as they pass from one dielectric into another.

### *19.1.2 The Transmission and Reflection of Normal Waves at Plane Interfaces*

Let us start with the simplest case of two semi-infinite dielectrics, labelled 1 and 2, which share the common interface $x = 0$. We shall consider a monochromatic wave that approaches the interface through medium 1 and is normal to the interface, $\mathbf{k} = k\hat{\mathbf{e}}_x$, as shown in Figure 19.1. We use the subscripts $I$, $R$ and $T$ to indicate *incident*, *reflected* and *transmitted* waves, and so the electric fields either side of the interface are

$$\mathbf{E}_1(\mathbf{x},t) = \hat{\mathbf{E}}_I \exp j(k_1 x - \varpi t) + \hat{\mathbf{E}}_R \exp j(k_1 x + \varpi t), \quad k_1 = \varpi/v_1, \tag{19.9}$$

$$\mathbf{E}_2(\mathbf{x},t) = \hat{\mathbf{E}}_T \exp j(k_2 x - \varpi t), \quad k_2 = \varpi/v_2, \tag{19.10}$$

with similar expressions for $\mathbf{B}$. Note that we take the amplitudes, such as $\hat{\mathbf{E}}_I$, to be real, so that the wave is linearly polarised. Note also that the frequency of the wave must be conserved as it crosses the interface in order that the exponentials in (19.9) and (19.10) can be matched at $x = 0$. The requirement that $v\hat{\mathbf{B}} = \hat{\mathbf{e}}_k \times \hat{\mathbf{E}}$ now demands

$$v_1(\hat{\mathbf{B}}_I + \hat{\mathbf{B}}_R) = \hat{\mathbf{e}}_x \times (\hat{\mathbf{E}}_I - \hat{\mathbf{E}}_R), \tag{19.11}$$

$$v_2\hat{\mathbf{B}}_T = \hat{\mathbf{e}}_x \times \hat{\mathbf{E}}_T, \tag{19.12}$$

since $\hat{\mathbf{e}}_k = \hat{\mathbf{e}}_x$ for the incident and transmitted waves and $\hat{\mathbf{e}}_k = -\hat{\mathbf{e}}_x$ for the reflected wave.

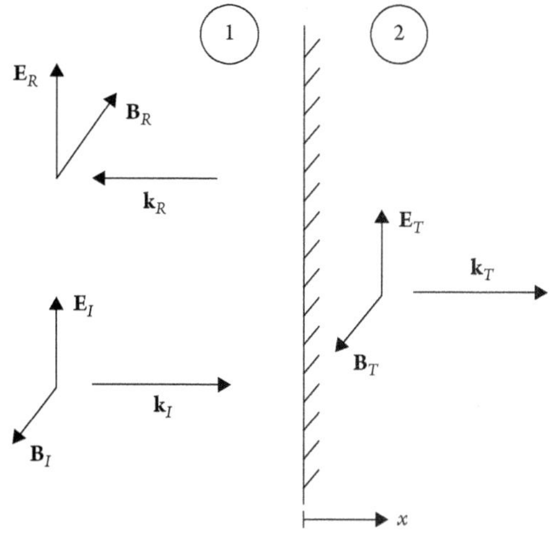


**Figure 19.1** A plane wave normal to the interface $x = 0$ is partially reflected and partially transmitted. The **E** and **B** fields lie in the $y$–$z$ plane.

Let us now consider the boundary conditions. First note that, since $\mathbf{k} \cdot \hat{\mathbf{B}} = 0$ and $\mathbf{k} \cdot \hat{\mathbf{E}} = 0$, both **B** and **E** lie in the $y$–$z$ plane and so the boundary conditions at the interface simplify to $\mathbf{E}_2 = \mathbf{E}_1$ and $\mathbf{B}_2 = \mathbf{B}_1$. The expressions (19.9)–(19.12) now yield

$$\hat{\mathbf{E}}_I + \hat{\mathbf{E}}_R = \hat{\mathbf{E}}_T, \quad \frac{1}{v_1}(\hat{\mathbf{E}}_I - \hat{\mathbf{E}}_R) = \frac{1}{v_2}\hat{\mathbf{E}}_T, \tag{19.13}$$

from which we obtain the amplitudes of the transmitted and reflected waves:

$$\hat{\mathbf{E}}_T = \frac{2n_1}{n_1 + n_2}\hat{\mathbf{E}}_I, \quad \hat{\mathbf{E}}_R = \frac{n_1 - n_2}{n_1 + n_2}\hat{\mathbf{E}}_I. \tag{19.14}$$

Note that the reflected wave may be in phase or 180° out of phase with the incident wave, depending on which medium has the higher refractive index.

We now turn from wave amplitudes to power. The time-averaged Poynting vector for the incident wave is

$$\frac{1}{\mu_0}\overline{\mathbf{E} \times \mathbf{B}} = \frac{\hat{E}_I^2}{2\mu_0 v_1}\hat{\mathbf{e}}_x = \frac{1}{2\mu_0 c} n_1 \hat{E}_I^2 \hat{\mathbf{e}}_x, \tag{19.15}$$

where the overbar indicates a time average. Of course, similar expressions apply to the reflected and transmitted waves. It follows that the ratio of the transmitted to the incident power, $T$, and that of the reflected to the incident power, $R$, are given by

$$T = \frac{4n_1 n_2}{(n_1 + n_2)^2}, \quad R = \frac{(n_1 - n_2)^2}{(n_1 + n_2)^2}, \tag{19.16}$$

where $T$ and $R$ are known as the *transmission* and *reflection coefficients*. Note that $T + R = 1$, as required by conservation of energy. Note also that there is negligible reflection of energy if the two materials have similar refractive indices.

We close this section with a comment, first raised in Exercise 18.4. Consider the case where medium 1 is a vacuum, so that we have a wave propagating though free space which then impacts on the surface of a dielectric, say glass. By virtue of superposition, the total electromagnetic field at each point, both inside and outside the dielectric, is the sum of the incident wave and the radiation fields of all the tiny molecular dipoles. This, in turn, tells us that the net field associated with all the molecular dipoles has three parts:

(i) a wave that propagates through the dielectric at speed $c$ and that is out of phase with the incident wave, thus annihilating the incident wave within the dielectric by destructive interference;

(ii) a wave that propagates backward thought the vacuum at speed $c$, which constitutes the reflected wave;

(ii) a wave that propagates through the dielectric at the lower speed $v_2$, which constitutes the transmitted wave.

It is extraordinary that the magnitudes and phases of the tiny molecular dipoles are so perfectly orchestrated as to produce these three systematic effects. This phenomenon is known in optics as the *Ewald–Oseen extinction theorem*.

### *19.1.3 The Refraction and Reflection of Oblique Waves at Plane Interfaces*

Let us now consider the trickier case of an oblique incident wave. As above, we consider two semi-infinite dielectrics that share the common interface $x = 0$, with a monochromatic wave that approaches the interface through medium 1, as shown in Figure 19.2. The incident, transmitted and reflected waves now take the form

$$\mathbf{E}_I(\mathbf{x},t) = \hat{\mathbf{E}}_I \exp j(\mathbf{k}_I \cdot \mathbf{x} - \varpi t), \quad v_1 \hat{\mathbf{B}}_I = \hat{\mathbf{e}}_I \times \hat{\mathbf{E}}_I, \tag{19.17}$$

$$\mathbf{E}_T(\mathbf{x},t) = \hat{\mathbf{E}}_T \exp j(\mathbf{k}_T \cdot \mathbf{x} - \varpi t), \quad v_2 \hat{\mathbf{B}}_T = \hat{\mathbf{e}}_T \times \hat{\mathbf{E}}_T, \tag{19.18}$$

$$\mathbf{E}_R(\mathbf{x},t) = \hat{\mathbf{E}}_R \exp j(\mathbf{k}_R \cdot \mathbf{x} - \varpi t), \quad v_1 \hat{\mathbf{B}}_R = \hat{\mathbf{e}}_R \times \hat{\mathbf{E}}_R, \tag{19.19}$$

where $\hat{\mathbf{e}}_I$, $\hat{\mathbf{e}}_T$ and $\hat{\mathbf{e}}_R$ are unit vectors in the directions of $\mathbf{k}_I$, $\mathbf{k}_T$ and $\mathbf{k}_R$. As before, we take the amplitudes to be real, so that the wave is linearly polarised.

In order to match the boundary conditions at the interface, the exponential terms in (19.17)–(19.19) must all be the same at the plane $x = 0$, and this in turn requires

$$(\mathbf{k}_I)_\parallel = (\mathbf{k}_R)_\parallel = (\mathbf{k}_T)_\parallel. \tag{19.20}$$

We conclude that $\mathbf{k}_I$, $\mathbf{k}_T$ and $\mathbf{k}_R$ are coplanar (the so-called *plane of incidence*) and we shall orientate our axes so that these wave vectors to all lie in the $x$–$y$ plane. Moreover, the dispersion relationship requires

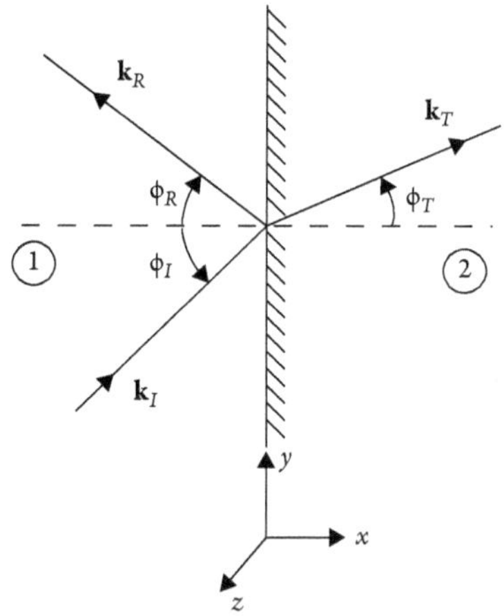


**Figure 19.2** An oblique wave approaches the interface $x = 0$ and is partially reflected and partially transmitted. The wave vectors all lie in the $x$–$y$ plane.

$$\varpi = k_I v_1 = k_R v_1 = k_T v_2, \tag{19.21}$$

or equivalently

$$k_I = k_R = \frac{n_1}{n_2} k_T, \tag{19.22}$$

which combines with (19.20) to yield

$$\frac{(k_I)_y}{k_I} = \frac{(k_R)_y}{k_R} = \frac{n_2}{n_1}\frac{(k_T)_y}{k_T}. \tag{19.23}$$

Evidently the angle of incidence equals the angle of reflection, $\phi_I = \phi_R$, while the angle of transmission is related to the angle of incidence by

$$\boxed{\frac{\sin\phi_T}{\sin\phi_I} = \frac{n_1}{n_2},} \tag{19.24}$$

all angles being measured relative to the normal to the interface (Figure 19.2). The bending of the light ray at the interface is referred to as *refraction* and (19.24) is called *Snell's law of refraction*, a cornerstone of geometrical optics.

Notice that, if $n_2 > n_1$, say light in air entering water or glass, the wave is refracted *towards* the normal, $\phi_I > \phi_T$. Notice also that we have yet to apply boundary conditions (19.8), which we now rewrite as

$$n_2^2(\mathbf{E}_2)_\perp = n_1^2(\mathbf{E}_1)_\perp, \quad (\mathbf{E}_2)_\parallel = (\mathbf{E}_1)_\parallel, \quad \mathbf{B}_2 = \mathbf{B}_1. \tag{19.25}$$

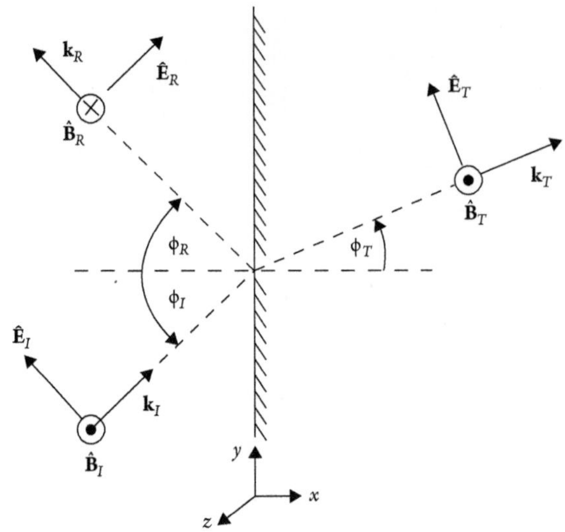


**Figure 19.3** A linearly polarised, oblique wave approaches the interface $x = 0$. The plane of polarisation is aligned with that of incidence.

For simplicity, we shall take the plane of polarisation (the plane containing **k** and **E**) to be aligned with the plane of incidence, so that **E** lies in the $x$–$y$ plane and **B** is in the $z$ direction, as shown in Figure 19.3. Equation (19.7) then gives us $v_1\hat{B}_I = \hat{E}_I$, $v_1\hat{B}_R = -\hat{E}_R$ and $v_2\hat{B}_T = \hat{E}_T$, where the positive directions for $\hat{B}_I$, $\hat{B}_R$ and $\hat{B}_T$ are taken to be in the positive $z$ direction, while those for $\hat{E}_I$, $\hat{E}_R$ and $\hat{E}_T$ are as shown in Figure 19.3. Our boundary conditions for **E** now become

$$(\hat{E}_I)_x + (\hat{E}_R)_x = \frac{n_2^2}{n_1^2}(\hat{E}_T)_x, \quad (\hat{E}_I)_y + (\hat{E}_R)_y = (\hat{E}_T)_y, \tag{19.26}$$

or equivalently

$$-\hat{E}_I \sin\phi_I + \hat{E}_R \sin\phi_R = -\frac{n_2^2}{n_1^2}\hat{E}_T \sin\phi_T, \tag{19.27}$$

$$\hat{E}_I \cos\phi_I + \hat{E}_R \cos\phi_R = \hat{E}_T \cos\phi_T, \tag{19.28}$$

while that for **B** can be rewritten as

$$\hat{E}_I - \hat{E}_R = \frac{n_2}{n_1}\hat{E}_T. \tag{19.29}$$

A little algebra now gives us

$$\hat{E}_I + \hat{E}_R = \frac{\cos\phi_T}{\cos\phi_I}\hat{E}_T = \alpha\hat{E}_T, \tag{19.30}$$

which defines $\alpha$, and

$$\boxed{\hat{E}_T = \frac{2}{\alpha + n_2/n_1}\hat{E}_I, \quad \hat{E}_R = \frac{\alpha - n_2/n_1}{\alpha + n_2/n_1}\hat{E}_I.} \tag{19.31}$$

These are *Fresnel's equations* for the particular case where the planes of incidence and polarisation are aligned. For many purposes it is preferable to express $\alpha$ in terms of the incident angle alone. To that end, we note that $\alpha^2$ can be written as

$$\alpha^2 = \frac{\cos^2\phi_T}{\cos^2\phi_I} = \frac{1-\sin^2\phi_T}{1-\sin^2\phi_I} = \frac{1-(n_1^2/n_2^2)\sin^2\phi_I}{1-\sin^2\phi_I}. \tag{19.32}$$

When combined with (19.31), this allows $\hat{E}_T/\hat{E}_I$ and $\hat{E}_R/\hat{E}_I$ to be expressed in terms of $\sin\phi_I$ and $n_2/n_1$ only. The net result is illustrated in Figure 19.4 for the particular case of waves passing from air ($n = 1$) into water, for which $n = 1.33$ in the visible spectrum. Note that $\hat{E}_T$ is always in phase with $\hat{E}_I$, whereas $\hat{E}_R$ is 180° out of phase with $\hat{E}_I$ at low to moderate angles of incidence and in phase for high angles of incidence.

It is instructive to consider certain special cases of (19.31) for $n_2 > n_1$. For normal incidence ($\phi_I = 0$) we have $\alpha = 1$ and in such cases we recover (19.14) from (19.31). At the other extreme, when the incident wave just grazes the surface ($\phi_I \to \pi/2$) we find $\alpha \to \infty$, and hence $\hat{E}_T = 0$ and $\hat{E}_R = \hat{E}_I$ In such a situation the wave is entirely reflected. A third important case is that in which $\alpha = n_2/n_1$, where (19.31) tells us that there is no reflected

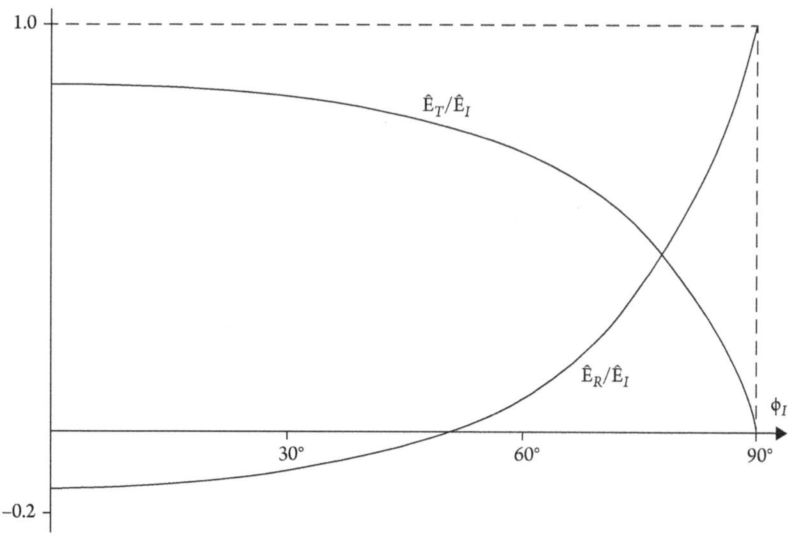


**Figure 19.4** Fresnel's equations for waves passing from air ($n = 1$) into water ($n = 1.33$). The plane of polarisation of the wave is aligned with the plane of incidence.

wave. The angle of incidence at which this happens is called the *Brewster angle* $\phi_B$ and it can be determined from (19.32), according to which

$$\alpha^2 = n_2^2/n_1^2 = \frac{1-(n_1^2/n_2^2)\sin^2\phi_B}{1-\sin^2\phi_B}.$$

This may be rearranged to give

$$\sin^2\phi_B = \frac{n_2^2/n_1^2}{1+n_2^2/n_1^2}, \tag{19.33}$$

from which we obtain $\phi_B = \tan^{-1}(n_2/n_1)$. For light passing from air into water ($n = 1.33$) this angle is around 53°.

### 19.1.4 Total Internal Reflection

There is one other important special case, which is connected to the phenomenon of *total internal reflection*. This occurs when the incident wave travels through a medium of relatively high refractive index and tries to exit into a dielectric of lower refractive index. For example, it can occur at the interface of glass and air when glass carries the incident wave. To understand total internal reflection, let us return to Snell's law and restrict ourselves to situations in which $n_2 < n_1$. Then

$$\frac{\sin\phi_T}{\sin\phi_I} = \frac{n_1}{n_2} > 1,$$

which tells us that the transmitted wave is refracted away from the normal and towards the interface. Evidently, for an angle of incidence equal to

$$\phi_I = \sin^{-1}(n_2/n_1), \tag{19.34}$$

the transmitted wave just grazes the interface. For larger angles of incidence, $\sin\phi_I > n_2/n_1$, there is no transmitted wave and indeed the analysis of §19.1.3 breaks down. This critical angle is, for glass in air, around 40°. For angles of incidence larger than the critical value an acceptable solution can be re-established by rewriting the wave vector in medium 2 as

$$\mathbf{k}_T = jk_x\hat{\mathbf{e}}_x + \mathbf{k}_\parallel, \tag{19.35}$$

which leads to a field in dielectric 2 that decays exponentially with distance from the interface, as $\exp(-k_x x)$. Such a field is called an *evanescent wave*.

Let us see if we can work out just how rapidly such an evanescent field decays with distance from the interface. Using (19.35), the dispersion relationship in dielectric 2 is readily shown to be

$$\mathbf{k}_\parallel^2 - k_x^2 = \mu_0\varepsilon_2\varpi^2 = (\varepsilon_r)_2\frac{\varpi^2}{c^2} = n_2^2\frac{\varpi^2}{c^2},$$

while that in dielectric 1 remains

$$\mathbf{k}_{\parallel}^2 = k_I^2 \sin^2 \varphi_I = n_1^2 \frac{\varpi^2}{c^2} \sin^2 \phi_I.$$

Since $\mathbf{k}_{\parallel}$ is continuous across the interface, this yields

$$k_x^2 = (n_1^2 \sin^2 \phi_I - n_2^2) \frac{\varpi^2}{c^2}, \tag{19.36}$$

which determines the rate of decay of the **E** and **B** fields in dielectric 2 according to $\exp(-k_x x)$. The principle of total internal reflection is put to good use in fibre optics, where a signal can be confined to a thin dielectric tube through multiple internal reflections, as discussed in Chapter 14.

### *19.1.5 Dispersive Waves and Rainbows*

Let us now turn to the phenomenon of *dispersion* in dielectrics. The key point is this. We know that white light passing from air into a glass prism will spread out into a spectrum of colours, with the blue light refracted more than the red. Moreover, each frequency refracts according to Snell's law, and so this tells us that the refractive index of glass must be an *increasing* function of frequency, at least in the visible spectrum. (Blue light, which refracts more, has a higher frequency than red light). It turns out that most transparent materials exhibit a similar frequency dependence of $n$, and this leads to many interesting phenomena, not the least of which are rainbows.

Historically, this phenomenon was characterised by the empirical equation

$$n(\lambda) = 1 + a\left(1 + \frac{b}{\lambda^2}\right),$$

where $\lambda$ is the wavelength in vacuum and the empirical coefficients $a$ and $b$ are positive. This is known as *Cauchy's equation* and $a$ and $b$ are called the coefficients of refraction and dispersion, respectively. In terms of frequency, this reads

$$n(\varpi) = 1 + a\left[1 + \frac{b}{(2\pi)^2} \frac{\varpi^2}{c^2}\right]. \tag{19.37}$$

In practice, (19.37) is often a reasonable approximation for the visible spectrum in transparent materials, and its general form, $n(\varpi) = 1 + a(1 + \beta\varpi^2)$, is in line with the predictions of simple models of atomic oscillations. For example, the *Lorentz oscillator model* of electronic polarisation, which mimics the way in which electrons in an atom oscillate in response to a fluctuating ambient electric field, predicts this kind of behaviour provided that there are no atomic resonances in the range of frequencies of interest (see e.g. Griffiths, 2013).

More generally, wave-bearing systems in which the phase velocity $c_p = \varpi/k$ is a function of frequency (or, equivalently, wavelength) are called *dispersive systems*. The name derives from the fact that the various spatial Fourier modes (various wavelengths) that make up a single disturbance all propagate away from the source of that disturbance at different speeds, and so a localised disturbance does not stay localised for long, but rather disperses (spreads out) as it propagates. For example, consider the ripples on the surface of a pond that follow the impact of a stone on the surface. As the ripples spread out from the point of impact, they occupy an ever greater area, because waves of slightly different wavelength propagate at slightly different speeds and so travel different distances in a given time (see Figure 2.14 in §2.3.5).

As discussed in §2.3.5, localised disturbances spread through a dispersive system in the form of wave packets, with the wave packets carrying information and energy away from the source of the disturbance at a velocity called the *group velocity*. In one-dimensional systems with a dispersion relationship of the form $\varpi = \varpi(k)$, the group velocity is $c_g = d\varpi/dk$ (see (2.60) in §2.3.5). For waves in a dielectric with a frequency-dependent refractive index $n(\varpi)$, we have $\varpi = \upsilon k = ck/n(\varpi)$, and so the phase velocity is $c_p = \varpi/k = c/n(\varpi)$, which is indeed a function of frequency. So we have a dispersive system whose group velocity can be obtained by differentiating $\varpi n(\varpi) = ck$ to give

$$\left(n + \varpi \frac{dn}{d\varpi}\right)\frac{d\varpi}{dk} = c,$$

from which

$$c_g = \frac{c}{n + \varpi\, dn/d\varpi}. \tag{19.38}$$

Since $n$ is normally an increasing function of frequency for transparent materials, at least for the visible spectrum, this tells us that the group velocity is usually smaller than the phase velocity. However, for certain rather limited frequency ranges (those frequencies at which atomic resonances occur and wave energy is absorbed by the medium), $n$ can be a sharply decreasing function of frequency, a behaviour that is known as *anomalous dispersion*. This can lead to $c_g$, as calculated from (19.38), exceeding $c$, in which case the concept of group velocity breaks down. It can also cause $n$ to dip down below unity, so that $c_p > c$. (Readers interested in more details may wish to consult Zangwill (2013).)

Perhaps the most obvious consequence of dispersive waves in a dielectric is the common rainbow. Often you will see a rainbow if you have your back to the sun and look upward at an angle of around 40° to the ground and into a region where water droplets are suspended in the atmosphere. Each water drop acts like a tiny prism, which first refracts the light through the drop and then reflects the incident sunlight back towards the observer. To understand why the colours in the rainbow run from red at the top to blue or violet at the bottom, and why $\sim 40°$ is often an optimum angle for seeing a rainbow, we must consider the action of a single droplet.

Suppose the sun is low in the sky and so the incident light approaching the droplet is roughly horizontal. A light ray entering the droplet is first refracted towards the normal, after which it is reflected, at least in part, from the back face of the drop. Finally the ray exits from the front surface of the drop, as illustrated in Figure 19.5. On leaving the drop, the light ray is again refracted, only this time away from the normal. Of course, the path taken through the

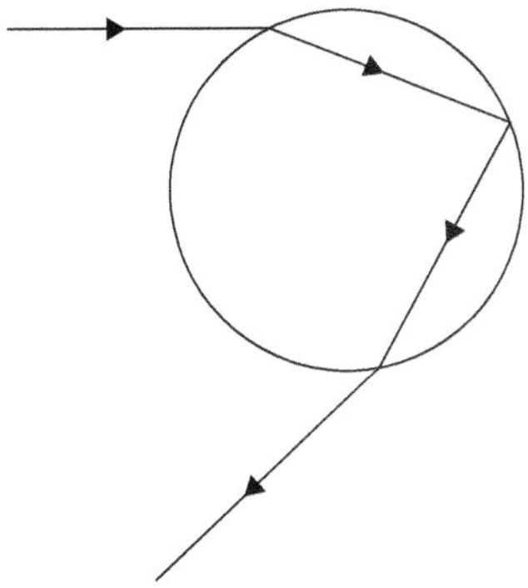

**Figure 19.5** A light ray passing from air into a water droplet is first refracted at the front of the drop and then reflected from its rear.

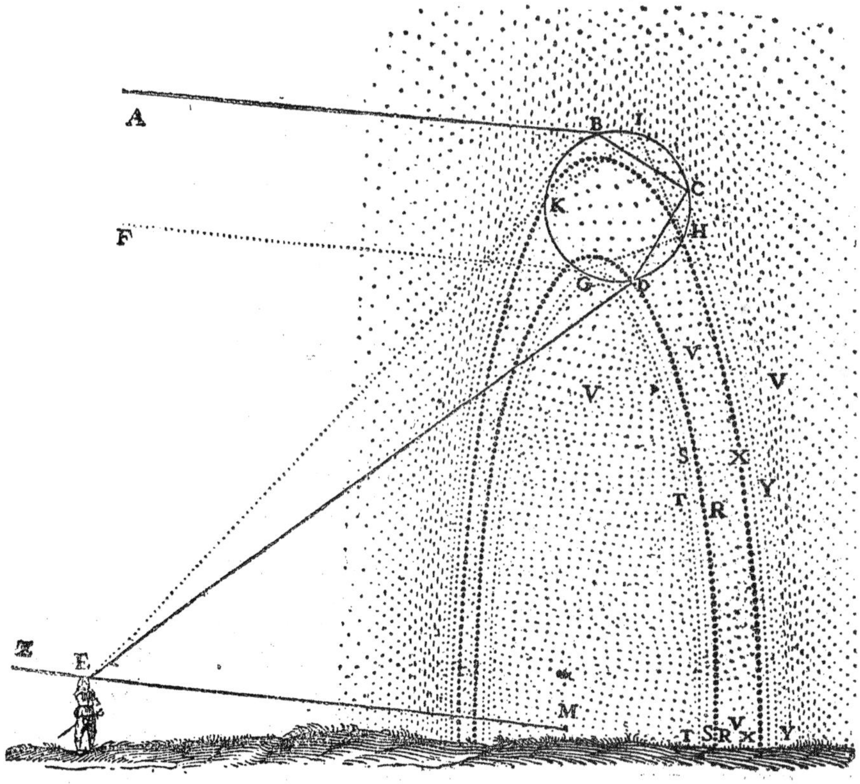


**Figure 19.6** An observer on the ground looking upward at an angle of 40° to the horizontal sees predominantly violet light reflected from water droplets, while at an angle of 42° the observer sees predominantly red light. From an original sketch by Descartes.

droplet is slightly different for different wavelengths, as the blue light ($n \approx 1.34$) is refracted more than the red ($n \approx 1.33$). Moreover, the path taken through the droplet depends on where the light ray hits the surface. However, in 1637 the French philosopher and scientist René Descartes computed thousands of different possible paths though a drop and showed that, on average, and for a given wavelength, there is a strong tendency for the light rays to follow a path close to that which gives the maximum overall deviation of the light ray to the horizontal. The angle of maximum possible deviation turns out to be 42° for red light, and 40° for violet.

It follows from the argument above that, if the sun is not too high in the sky, an observer on the ground looking upward at an angle of 40° will see predominantly blue or violet light reflected towards them. Moreover, the same observer looking slightly higher, at an angle of 42° to the horizontal, will see predominantly red light. This is illustrated in Figure 19.6, which is from an original sketch by Descartes.

Of course, there are many complications. For example, the sun's rays will not in general be horizontal. Moreover, a second bow may form at an angle of $\sim 51°$ as a result of two internal reflections in the water droplets. Finally, when the drops are relatively small, diffraction can play as large a role as dispersion and reflection. However, we shall not pause to discuss these complications, but rather refer the reader to Greenler (1980).

## 19.2 An Aside: The Scattering of Light

### *19.2.1 Rayleigh Scattering of Light and Why the Sky Is Blue*

In keeping with the meteorological theme above, we now ask why the sky is blue, a question to which Rayleigh gave an excellent answer in a series of papers during the period 1871–1899. Rayleigh was inspired by the experiments of the Irish physicist John Tyndall on the effect of small suspended particles on the passage of light, experiments that bore some resemblance to Leonardo da Vinci's observations, around 1500, on the scattering of light by wood smoke (Jackson, 1999). He was also unconvinced by the assertion of Clausius that the sky is blue as a result of multiple reflections from small 'water bubbles' suspended in the atmosphere. So Rayleigh set out to develop his own explanation, based on the scattering of light by tiny particles.

The key point is that light passing through air is scattered by small particles and molecules, especially nitrogen and oxygen molecules, a process now known as *Rayleigh scattering*. Moreover, as Rayleigh appreciated, this scattering is more pronounced for blue light than for red light. This, in turn, results in the diffuse blue light we see in the atmosphere at most times of day, except at sunset. Perhaps it is worth taking a moment to sketch out some of Rayleigh's ideas, although we shall omit the mathematical details in the interests of brevity.

Rayleigh scattering occurs when an electromagnetic wave excites molecules that are much smaller than the wavelength of the light. As the wave passes over the molecules, the fluctuating electric field in the wave induces electronic polarisation, causing the electrons to oscillate at the same frequency as the light. Each molecule then acts as an oscillating electric

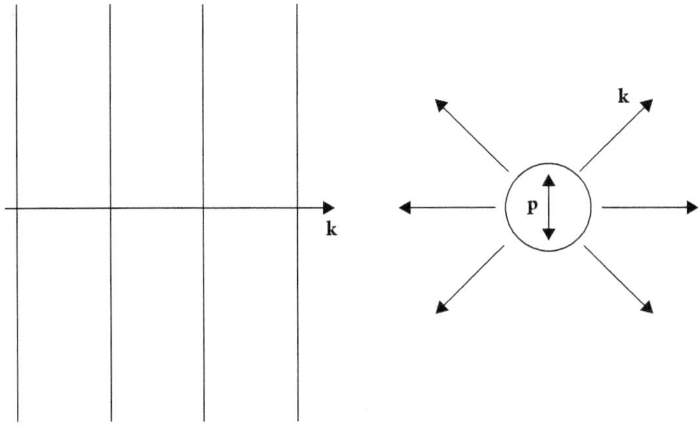


**Figure 19.7** The scattering of light by a molecule.

dipole of magnitude $\mathbf{p} = \alpha\varepsilon_0\mathbf{E}$, which radiates energy in the manner discussed in §16.3. It is this incoherent reradiated energy that appears as scattered light (Figure 19.7).

Crucially, the dipole radiation from the polarised molecules obeys Larmor's radiation equation for a compact source, (16.28), which in time-averaged form says that

$$\frac{1}{\mu_0}\oint \overline{\mathbf{E}\times\mathbf{B}}\cdot d\mathbf{S} = \frac{\mu_0\varpi^4 p_0^2}{12\pi c}, \tag{19.39}$$

the overbar being a time average. A decisive role is now played by the $\varpi^4$ dependence in (19.39), a dependence that is known as *Rayleigh's law*. Blue light, which has a higher frequency, is much more effectively scattered than red light. The net result is that the blue content of the sun's rays is repeatedly bounced around the atmosphere by the nitrogen and oxygen molecules, eventually becoming scattered across the sky. It then appears as a diffuse glow emanating from all directions. The red or yellow light, by contrast, is much less effectively scattered, and so direct radiation from the sun reaching the ground is significantly depleted in blue light but less so in red or yellow light. This is why the sun appears yellow, and why sunsets often have a red glow.

There are, of course, many complications, as is clear from the fact that the simple argument given above, based on the $\varpi^4$ dependence of light scattering, might be taken to support a violet sky, rather than a blue sky. Complications include the fact that solar radiation is not uniformly intense at all wavelengths in the visible spectrum, there is absorption of ultraviolet light by molecules in the atmosphere, and the human eye responds to light of different wavelengths in different ways. One curiosity is that, given the high density of scattering molecules, one might have expected a large degree of destructive interference of the scattered light, yet this is not the case. We shall not pause to explore these various subtleties, but readers interested in more details could do worse that consult Zangwill (2013), or even Rayleigh's original papers (Rayleigh, 1871, 1899).

### *19.2.2 Thomson Scattering by Free Charges*

In Rayleigh scattering the incident wave interacts with a neutral molecule whose centre of mass is assumed to remain stationary. The oscillating dipole then arises through electronic polarisation of the molecule by the incident electric field. There is a second common form of scattering, in which the incident wave interacts with a single free charge, say an electron, which then oscillates in space, radiating energy. This is called *Thomson scattering*. In order to keep the analysis simple, we shall assume that the amplitude of the induced oscillation is much smaller than the wavelength of the incident wave. This means that the oscillating charge may be treated as a compact source, of the type considered in §16.4, in which the motion of the charge is non-relativistic.

Suppose that the charge $q$ is centred at the origin, and has mass $m$ and position $\mathbf{x}(t)$. If the incident electric field is $\mathbf{E}_0 \exp j(\mathbf{k}\cdot\mathbf{x} - \varpi t)$, then $m\ddot{\mathbf{x}} = q\mathbf{E}$. This may be replaced by

$$m\ddot{\mathbf{x}} = q\mathbf{E}_0 \cos \varpi t$$

if the amplitude of the motion is small. It follows that

$$\mathbf{x}(t) = -\frac{q\mathbf{E}_0}{m\varpi^2} \cos \varpi t,$$

and so our oscillating charge represents a compact source, provided that

$$\frac{qE_0}{m\varpi} << c.$$

This also ensures that the velocity of the charge is non-relativistic. We may therefore determine the radiated power using the non-relativistic Larmor equation (17.58). This gives us

$$\frac{1}{\mu_0} \oint \overline{\mathbf{E}\times\mathbf{B}} \cdot d\mathbf{S} = \frac{\mu_0 q^4 E_0^2}{12\pi\, m^2 c}, \tag{19.40}$$

the overbar being a time average. Note that, unlike Rayleigh scattering, the re-radiated power is independent of frequency. The means that all wavelengths are equally scattered by free charges.

This is all we shall say about the scattering of light. There is a vast literature devoted to the subject, and to the associated phenomenon of diffraction, but readers will find convenient overviews in Jackson (1999) and Zangwill (2013), and an extensive discussion in Landau & Lifshitz (1960).

## 19.3 Waves in Stationary Conductors

### *19.3.1 Damped Waves and the Skin Depth*

Let us now turn to waves in good conductors, such as metals, in which there is no polarisation or magnetisation, $\varepsilon_r = \mu_r = 1$. In principle, we must now allow for both free charges and a conduction current, and so our governing equations are, notionally,

$$\nabla \cdot \mathbf{E} = \frac{\rho_e}{\varepsilon_0}, \quad \nabla \times \mathbf{E} = -\frac{\partial \mathbf{B}}{\partial t}, \tag{19.41}$$

$$\nabla \cdot \mathbf{B} = 0, \quad \nabla \times \mathbf{B} = \mu_0 \left( \mathbf{J} + \varepsilon_0 \frac{\partial \mathbf{E}}{\partial t} \right). \tag{19.42}$$

To these we must add Ohm's law, $\mathbf{J} = \sigma \mathbf{E}$, and charge conservation,

$$\nabla \cdot \mathbf{J} = -\frac{\partial \rho_e}{\partial t}. \tag{19.43}$$

However, as noted in Chapter 3, in good conductors the charge density is almost always zero and so (19.41) and (19.43) reduce to $\nabla \cdot \mathbf{E} = 0$ and $\nabla \cdot \mathbf{J} = 0$. Moreover, the displacement current in the Ampère–Maxwell equation was originally added to Ampère's law in order to ensure consistency with the charge conservation equation when $\rho_e$ is finite. That is to say, the divergence of the Ampère–Maxwell equation, combined with Gauss' law, yields (19.43). However, if $\rho_e$ is negligible, then we have the option of setting the displacement current to zero, since the divergence of Ampère's law (which has no displacement current) automatically yields $\nabla \cdot \mathbf{J} = 0$, consistent with charge conservation. So we might anticipate that the displacement current in (19.42) might also be neglected in conductors. We shall now show that this is indeed the case, at least for good conductors.

Let us start by returning to the observation in Chapter 3 that free charges may be neglected in the interior of a good conductor, such as a metal. We restrict the discussion to stationary conductors. The point is that charge conservation, Ohm's law and Gauss' law combine to yield

$$\frac{\partial \rho_e}{\partial t} = -\nabla \cdot \mathbf{J} = -\nabla \cdot (\sigma \mathbf{E}) = -\frac{\sigma}{\varepsilon_0} \rho_e.$$

This may be rewritten as

$$\frac{\partial \rho_e}{\partial t} + \frac{\rho_e}{\tau_e} = 0, \quad \tau_e = \frac{\varepsilon_0}{\sigma}, \tag{19.44}$$

whose solution is $\rho_e = \rho_e(0) \exp(-t/\tau_e)$. The quantity $\tau_e$ is called the charge relaxation time and for a typical metallic conductor it is very small, around $10^{-15}$ s. So we conclude that any excess charge density that initially lies within the interior of a conductor will move extremely rapidly to the surface under the action of the electrostatic repulsion forces. It follows that $\rho_e$ is almost always zero in stationary conductors, leaving $\nabla \cdot \mathbf{E} = 0$ and $\nabla \cdot \mathbf{J} = 0$.

Turning now to the Ampère–Maxwell equation, we first note that, when $\rho_e = 0$, both sides of Ampère's law (in which the displacement current is suppressed) are automatically solenoidal. There is therefore no compelling need to add a displacement current to Ampère's law, since the divergence of Ampère's law automatically yields $\nabla \cdot \mathbf{J} = 0$, consistent with charge conservation. Of course, there might be some other reason for introducing a displacement current, other than the requirement to be consistent with charge conservation. So let us allow for a finite displacement current and consider the ratio of that displacement current to the conduction current in the Ampère–Maxwell equation. This is of the order of

$$\frac{\varepsilon_0 \partial \mathbf{E}/\partial t}{|\mathbf{J}|} \sim \frac{\varepsilon_0 \partial \mathbf{E}/\partial t}{\sigma\,|\mathbf{E}|} \sim \varpi\,\tau_e, \tag{19.45}$$

where $\varpi$ is a typical frequency. We might take as our definition of a good conductor the condition that $\varpi\tau_e << 1$, so that the charge relaxation time is much shorter than any other timescale in the problem. (In this approximation we effectively exclude extremely fast processes.) We therefore conclude that in good conductors we may drop both the free charge density and the displacement current, leaving our governing equations as

$$\nabla\cdot\mathbf{E} = 0, \quad \nabla\times\mathbf{E} = -\frac{\partial\mathbf{B}}{\partial t}, \tag{19.46}$$

$$\nabla\cdot\mathbf{B} = 0, \quad \nabla\times\mathbf{B} = \mu_0\sigma\mathbf{E}. \tag{19.47}$$

Since we have lost one time derivative in our governing equations, we might anticipate that their solution will be very different to the waves discussed so far in this chapter. This is readily confirmed by taking the curl of Ampère's equation and eliminating **E** using Faraday's law. This yields, not a wave equation, but rather a diffusion equation:

$$\frac{\partial\mathbf{B}}{\partial t} = \lambda\nabla^2\mathbf{B}. \tag{19.48}$$

Here $\lambda = 1/\mu_0\sigma$ is called the *magnetic diffusivity*, which should not to be confused with wavelength! So magnetic fields in stationary conductors diffuse, rather than propagate as waves. Actually, we have met (19.48) before in §2.2.1, where we discussed its solution.

While it is true that we may obtain wave-like solutions to (19.48) if we impose oscillatory boundary conditions at the edge of the domain of interest, these are not really physical waves in any conventional sense, but rather heavily damped boundary oscillations. Suppose, for example, that we look for a one-dimensional wave-like solution of (19.48) of the form

$$\mathbf{B}(x,t) = \Re\{\hat{\mathbf{B}}(x)\exp(j\varpi t)\},$$

where $\Re$ indicates the real part of what follows and $\hat{\mathbf{B}}$ is complex and lies in the $y$–$z$ plane. Clearly

$$\hat{\mathbf{B}}''(x) = \frac{j\varpi}{\lambda}\hat{\mathbf{B}}$$

and hence

$$\hat{\mathbf{B}}''(x) = \left(\frac{1+j}{\delta}\right)^2\hat{\mathbf{B}}, \qquad \delta = \sqrt{\frac{2\lambda}{\varpi}}. \tag{19.49}$$

This has the solution

$$\mathbf{B}(x,t) = \mathbf{B}_0\exp\left(-\frac{x}{\delta}\right)\cos\left(\varpi t - \frac{x}{\delta}\right), \tag{19.50}$$

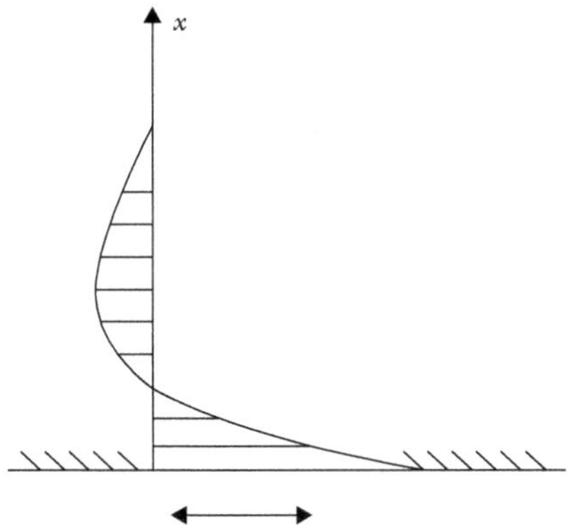


**Figure 19.8** The diffusive solution for a magnetic field oscillating with constant amplitude adjacent to a semi-infinite conducting solid.

which is illustrated in Figure 19.8. It represents a magnetic field oscillating with constant amplitude adjacent to a semi-infinite conducting solid. As noted in §2.2.1, the magnetic field penetrates a characteristic distance

$$\delta = \sqrt{\frac{2\lambda}{\varpi}} = \sqrt{\frac{2}{\sigma \mu_0 \varpi}} \tag{19.51}$$

into the conductor, which is referred to as the *skin depth*. In a good conductor operating at normal frequencies the skin depth is typically very thin, much smaller than the dimensions of the conductor. For example, in copper at $10^3$ Hz we have $\delta \approx 2.1$ mm, while at $10^6$ Hz the skin depth $\delta \approx 0.066$ mm.

Note that the electric field corresponding to (19.50) can be obtained from either Faraday's law or else (19.47). Either expression yields

$$|\mathbf{E}| \sim \varpi \delta \, |\mathbf{B}| \sim \sqrt{\lambda \varpi} \, |\mathbf{B}| \,. \tag{19.52}$$

So, in a good conductor, where $\delta$ is typically very small, the electric field is equally weak.

### 19.3.2 *Reflection at the Surface of a Good Conductor*

We conclude by considering the reflection of a plane wave from the surface of a good conductor. Consider the situation shown in Figure 19.1, where we take medium 1 to be a dielectric, say air, and medium 2 to be a good conductor, and a plane wave in the dielectric reflects from the surface of the conductor at normal incidence. Ohm's law demands that the surface current be zero, $\mathbf{K} = 0$, since a finite value of $\mathbf{K}$ would require an infinite electric field at the surface of the conductor. So the boundary conditions (18.43) and (18.44) reduce to

$$\mathbf{E}_2 = \mathbf{E}_1, \quad \mu_2^{-1}\mathbf{B}_2 = \mu_1^{-1}\mathbf{B}_1, \tag{19.53}$$

and the incident, reflected and transmitted electric fields satisfy

$$\mathbf{E}_I + \mathbf{E}_R = \mathbf{E}_T.$$

Now we have already seen that the electric field in a good conductor operating at normal frequencies is very small, $\mathbf{E}_T \approx 0$, and it follows that $\mathbf{E}_I \approx -\mathbf{E}_R$. In such cases the wave is perfectly reflected at the surface of the conductor and there is no transmitted wave. In short, good conductors behave almost like mirrors.

## EXERCISES

19.1 Derive the Fresnel equations, analogous to (19.31), for the particular case where the planes of incidence and polarisation are perpendicular.

19.2 Light passes through a plate glass window at normal incidence. The window is of thickness $d$ and the glass has a refractive index $n$. The refractive index of the air either side of the glass is equal to unity and the frequency of the light is $\varpi$. Show that the overall transmission coefficient is

$$T = \left[1 + \frac{(n^2-1)^2}{4n^2}\sin^2\left(\frac{n\varpi d}{c}\right)\right]^{-1}.$$

What are the maximum and minimum values of the transmission coefficient?

19.3 Light passes through a plate glass window at normal incidence and into a large tank of still water. The window is of thickness $d$ and the frequency of the light is $\varpi$. The refractive index of the air outside the tank is equal to unity, while the refractive indices of the glass and water are $n_1$ and $n_2$, respectively. Find an expression for the overall transmission coefficient in terms of $d$, $n_1$, $n_2$, $\varpi$ and $c$.

19.4 The refractive index of light in water at a frequency of $7 \times 10^{14}$ Hz is $n = 1.34$, while that at $4 \times 10^{14}$ Hz is $n = 1.33$. The variation of the refractive index with frequency obeys Cauchy's equation. Find the ratio of the phase velocity to the group velocity in water for a frequency of $7 \times 10^{14}$ Hz.

19.5 For very poor conductors we might retain the displacement current in (19.47), so the governing equations become $\nabla \cdot \mathbf{E} = 0$, $\nabla \cdot \mathbf{B} = 0$ and

$$\nabla \times \mathbf{E} = -\frac{\partial \mathbf{B}}{\partial t}, \quad \nabla \times \mathbf{B} = \mu_0 \sigma \mathbf{E} + \mu_0 \varepsilon \frac{\partial \mathbf{E}}{\partial t}.$$

Show that this results in damped travelling waves and find the phase velocity and rate of damping for a monochromatic one-dimensional wave. Show that, for extremely poor conductors, the skin depth is independent of frequency and inversely proportional to the electrical conductivity.

## REFERENCES

Greenler, R., 1980, *Rainbows, Halos, and Glories*. Cambridge University Press.
Griffiths, D. J., 2013, *Introduction to Electrodynamics*, 4th ed. Pearson.
Jackson, J. D., 1999, *Classical Electrodynamics*, 3rd ed. Wiley.
Landau, L. D. & Lifshitz, E. M., 1960, *Electrodynamics of Continuous Media* Pergamon Press.
Lord Rayleigh, 1871, On the light from the sky, its polarization and colour. *Phil. Mag.*, Ser. 4, **41**, 107–120, 274–279.
Lord Rayleigh, 1899, On the transmission of light through an atmosphere containing small particles in suspension, and on the origin of the blue of the sky. *Phil. Mag.*,. Ser. 5, **47**, 375–394.
Zangwill, A., 2013, *Modern Electrodynamics*. Cambridge University Press.

# 20

# Magnetohydrodynamics I

## *Governing Equations and Kinematic Theorems*

Anyone informed that the universe is expanding and contracting in pulsations of eighty billion years has the right to ask, 'What's in it for me?'.

Peter De Vries (*The Glory of the Hummingbird*, 1974)

We now turn to the topic of magnetohydrodynamics, or MHD for short, which is the study of the interaction between magnetic fields and conducting fluids, such as liquid metals and hot plasmas.

## 20.1 What is Magnetohydrodynamics (MHD)?

MHD is central to many terrestrial processes, such as plasma confinement in nuclear fusion reactors and the heating, stirring and levitation of liquid metals in materials processing. It also describes motion within the interiors of the planets and stars, and how that motion produces planetary or stellar magnetic fields through dynamo action. At larger scales it determines the behaviour of the solar wind as it spirals outward from the sun, and of the magnetosphere that protects the earth from that wind. It even controls the stability of the accretion discs that form around young and dying stars. In this chapter we establish the governing equations of MHD and explore some of their simpler properties, while in Chapter 21 we turn our attention to applications. To keep the exposition simple, we shall limit the discussion to *incompressible fluids*, which provides an adequate basis for introducing many of the observed phenomena.

In effect, MHD is the union of electrodynamics and fluid mechanics, describing the mutual interaction of magnetic fields and electrically conducting, but non-magnetic, fluids. We have already hinted at aspects of this interaction in earlier chapters, and perhaps it is worth recalling now the nature of this two-way coupling. In brief, the relative motion of a conductor and an imposed magnetic field gives rise to an EMF, $\mathbf{u} \times \mathbf{B}$, which in turn drives a current, as described by the generalised version of Ohm's law,

$$\mathbf{J} = \sigma(\mathbf{E} + \mathbf{u} \times \mathbf{B}), \tag{20.1}$$

where $\mathbf{u}(\mathbf{x}, t)$ is the velocity distribution in the conducting medium. This current then gives rise to a second magnetic field, which supplements the imposed field, and, as discussed in §10.5.2, the net effect is that the lines of force of the combined magnetic field appear to be swept along by the conductor. This combined magnetic field interacts with the current $\mathbf{J}$ to exert a Lorentz force on the conductor, $\mathbf{J} \times \mathbf{B}$, and this force also acts so as to inhibit the relative motion of the field and the conductor (again, see §10.5.2).

The nature of this mutual interaction can be most clearly understood using Faraday's flux rule, (10.13), coupled with the notion of Faraday's tension in the field lines, (8.56). Let us start by combining Faraday's flux rule with Ohm's law to give

$$\frac{d}{dt}\int_S \mathbf{B} \cdot d\mathbf{S} = -\oint_{C_m} (\mathbf{E} + \mathbf{u} \times \mathbf{B}) \cdot d\mathbf{r} = -\frac{1}{\sigma}\oint_{C_m} \mathbf{J} \cdot d\mathbf{r}, \tag{20.2}$$

where $C_m$ is a *material curve* that moves with the conductor and $S$ is any surface that spans $C_m$. Clearly, when the conductivity is sufficiently large, the magnetic flux linking any material curve is approximately conserved. As discussed in §10.5.2, this demands that a magnetic field be frozen into a highly conducting medium, in the sense that the atoms of the fluid are bound to the field lines. In short, the conductor drags the field along with it.

This is illustrated in Figure 20.1 for the case of a solid conductor. Here a highly conducting ring is pulled through an imposed magnetic field by a mechanical force $\mathbf{F}$. The net magnetic field consists of the sum of the externally imposed field and the magnetic field associated with the current induced in the ring. If the ring is highly conducting, then, according to (20.2), the total magnetic flux through it cannot change, and so as the ring moves it appears to drag the magnetic field lines along with it. In effect, the field lines passing through the ring are frozen into it.

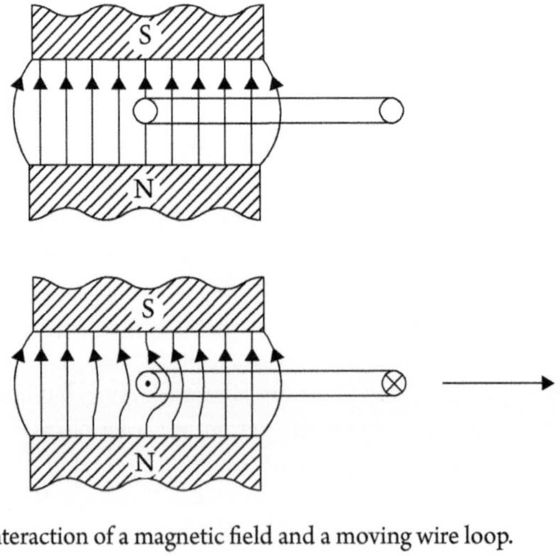


**Figure 20.1** The interaction of a magnetic field and a moving wire loop.

Let us now consider the role of the Lorentz force, or Faraday tension, in Figure 20.1. From §8.6 we know that magnetic field lines behave like elastic bands in tension (Faraday tension) and so the field-line curvature evident in Figure 20.1 tells us that the Lorentz force acts to pull the ring back into the magnet, countering the mechanical force **F**. So, in this simple example, both the induced magnetic field and the Faraday tension act to suppress the relative motion of the conductor and the net magnetic field.

Of course, from a purely electrodynamic perspective, what is true for a solid conductor must also be true for a conducting fluid. The only difference between the two is that the fluid has none of the rigidity of the solid and so responds to the Lorentz force in more subtle ways. So, one of the hallmarks of MHD is that there is a tendency to suppress the relative motion of a magnetic field and a conducting fluid, with the moving fluid tending to drag the field lines along with it, and the Lorentz force reacting back on the fluid, resisting relative motion. As we shall see, the degree to which the **B**-lines are effectively frozen into a fluid depends on a dimensionless measure of the conductivity.

MHD tends to be divided into its *kinematic* aspects, in which $\mathbf{u}(\mathbf{x}, t)$ is assumed to be prescribed and we seek the influence of a given velocity field on the magnetic field, and the full *dynamical* problem, where we also ask how the Lorentz force acts back on the fluid, thus altering $\mathbf{u}(\mathbf{x}, t)$. The first of these is strictly an exercise in electrodynamics, whereas the second, two-way coupling, takes us into the realm of fluid dynamics.

## 20.2 A Crash Course in Fluid Dynamics

As some physics syllabi touch on fluid mechanics only briefly, we now provide a short introduction to the subject. For simplicity, we temporarily set aside the Lorentz force.

### *20.2.1 An Eulerian Description of Motion and the Convective Derivative*

We start with the difference between Eulerian and Lagrangian descriptions of motion. In fluid mechanics it is conventional to represent the velocity or temperature fields in a fluid in precisely the same way as you would describe a magnetic field: we write $\mathbf{u}(\mathbf{x}, t)$ and $T(\mathbf{x}, t)$, with position and space considered as *independent* variables. This is known as an Eulerian formulation, to distinguish it from the Lagrangian description of motion used in conventional mechanics. Although an Eulerian formulation is quite natural, a difficulty arises because the laws of mechanics and thermodynamics are usually framed in terms of the rate of change of a quantity associated with a *given fluid particle.*

To illustrate the point, consider the case of a steady velocity field $\mathbf{u}(\mathbf{x})$. If we wish to apply $\mathbf{F} = m\mathbf{a}$ to a lump of fluid of mass $m$, then the acceleration that appears on the right is the rate of change of $\mathbf{u}$ following a fluid particle as it slides along a streamline. (Streamlines are to a velocity field $\mathbf{u}(\mathbf{x})$ what magnetic field lines are to the snapshot of a magnetic field $\mathbf{B}(\mathbf{x})$.) Evidently, $\mathbf{a}$ in Newton's second law is not $\partial\mathbf{u}/\partial t$, which is the rate of change of $\mathbf{u}$ at a fixed point in space, through which a succession of fluid particles will pass, but rather the rate of change of $\mathbf{u}$ associated with a given fluid particle as it moves through space. Such a derivative is written as $D\mathbf{u}/Dt$; a convention that was introduced by Stokes. Similarly, $DT/Dt$ is the rate

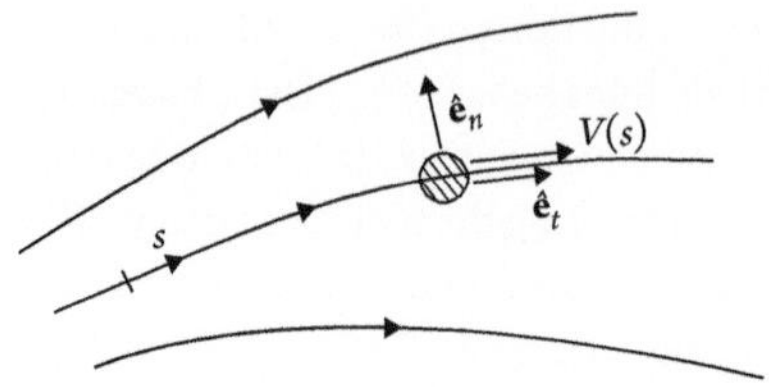


**Figure 20.2** The intrinsic coordinates used in (20.5).

of change of temperature following a given fluid particle. The operator $D(\cdot)/Dt$ is called the *convective derivative.*

We can get an explicit expression for the convective derivative as follows. Consider the change in temperature resulting from small changes in time and position, $\delta T = (\partial T/\partial t)\delta t + (\partial T/\partial x)\delta x + \cdots$. If we want to follow a fluid particle, then $\delta \mathbf{x}$ and $\delta t$ are not independent, but rather are related by $\delta \mathbf{x} = \mathbf{u}\delta t$. So, following a given element of fluid,

$$\delta T = \frac{\partial T}{\partial t}\delta t + \frac{\partial T}{\partial x}u_x \delta t + \cdots = \left(\frac{\partial T}{\partial t} + \mathbf{u}\cdot\nabla T\right)\delta t,$$

from which

$$\frac{DT}{Dt} = \frac{\partial T}{\partial t} + \mathbf{u}\cdot\nabla T. \tag{20.3}$$

Similarly, the acceleration of a lump of fluid is

$$\mathbf{a} = \frac{D\mathbf{u}}{Dt} = \frac{\partial \mathbf{u}}{\partial t} + (\mathbf{u}\cdot\nabla)\mathbf{u}. \tag{20.4}$$

When the flow is steady, this acceleration is most conveniently written in terms of intrinsic coordinates,

$$\mathbf{a} = (\mathbf{u}\cdot\nabla)\mathbf{u} = V\frac{\partial}{\partial s}(V\hat{\mathbf{e}}_t) = V\frac{\partial V}{\partial s}\hat{\mathbf{e}}_t - \frac{V^2}{R}\hat{\mathbf{e}}_n, \tag{20.5}$$

where $\mathbf{u} = V\hat{\mathbf{e}}_t$, $\hat{\mathbf{e}}_t$ is a unit vector tangential to a streamline, $\hat{\mathbf{e}}_n$ is the unit normal vector directed away from the local centre of curvature, and $s$ is a coordinate measured along the streamline, as shown in Figure 20.2. Also, $R$ is the radius of curvature of the streamline, defined in the usual way by $d\hat{\mathbf{e}}_t/ds = -\hat{\mathbf{e}}_n/R$.

### *20.2.2 More Kinematics: Mass Conservation, Spin and Deformation in a Fluid*

Mass conservation applied to a control volume fixed in space and with bounding surface $S$ gives us

$$\oint_S \rho\mathbf{u}\cdot d\mathbf{S} = -\frac{d}{dt}\int_V \rho\, dV,$$

where $\rho$ is the fluid density. Appling Gauss' theorem to the surface integral allows us to rewrite this in differential form,

$$\nabla \cdot (\rho \mathbf{u}) = -\frac{\partial \rho}{\partial t}, \tag{20.6}$$

or equivalently $D\rho/Dt + \rho \nabla \cdot \mathbf{u} = 0$. This reduces to

$$\boxed{\nabla \cdot \mathbf{u} = 0} \tag{20.7}$$

for an incompressible fluid, which is known as the *continuity equation*. In this chapter we shall restrict ourselves to incompressible fluids, in which case $\mathbf{u}$ is solenoidal.

We now turn to the topics of deformation and rotation in a fluid. Consider the relative motion of two adjacent fluid particles that are instantaneously located at points $\mathbf{x}$ and $\mathbf{x} + \delta\mathbf{x}$, with velocities $\mathbf{u}$ and $\mathbf{u} + \delta\mathbf{u}$. Then $\delta\mathbf{u}$ is the velocity at $\mathbf{x} + \delta\mathbf{x}$ as seen by an observer moving with the fluid at location $\mathbf{x}$. Clearly $\delta u_i = (\partial u_i/\partial x_j)\delta x_j$ and, for reasons that will become apparent, it is convenient to divide $\partial u_i/\partial x_j$ into symmetric and antisymmetric parts:

$$\delta u_i = \frac{\partial u_i}{\partial x_j}\delta x_j = \frac{1}{2}\left(\frac{\partial u_i}{\partial x_j} + \frac{\partial u_j}{\partial x_i}\right)\delta x_j + \frac{1}{2}\left(\frac{\partial u_i}{\partial x_j} - \frac{\partial u_j}{\partial x_i}\right)\delta x_j. \tag{20.8}$$

We now introduce the *rate-of-strain tensor* $S_{ij}$ and the *vorticity vector* $\boldsymbol{\omega}$, defined by

$$S_{ij} = \frac{1}{2}\left(\frac{\partial u_i}{\partial x_j} + \frac{\partial u_j}{\partial x_i}\right), \qquad \boldsymbol{\omega} = \nabla \times \mathbf{u}, \tag{20.9}$$

and so our expression for $\delta\mathbf{u}$ becomes

$$\delta\mathbf{u} = \delta\mathbf{u}^{(s)} + \delta\mathbf{u}^{(a)} = S_{ij}\,\delta x_j + \tfrac{1}{2}\boldsymbol{\omega} \times (\delta\mathbf{x}). \tag{20.10}$$

We shall see that the two terms on the right of (20.10) make fundamentally different contributions to the relative velocity $\delta\mathbf{u}$, with $\delta\mathbf{u}^{(s)}$ and $S_{ij}$ being associated with the deformation of fluid elements and $\delta\mathbf{u}^{(a)}$ (or equivalently $\boldsymbol{\omega}$) with the spin of fluid elements about their instantaneous centres.

Consider first the anti-symmetric contribution to (20.10), $\delta\mathbf{u}^{(a)} = \frac{1}{2}\boldsymbol{\omega} \times (\delta\mathbf{x})$. This represents rigid-body rotation about the point $\mathbf{x}$ with angular velocity $\boldsymbol{\omega}/2$. Evidently, as seen by an observer moving with the fluid at location $\mathbf{x}$, the relative velocity $\delta\mathbf{u}^{(a)}$ rotates fluid elements about $\mathbf{x}$ without causing any deformation of those elements. In words, the vorticity at location $\mathbf{x}$ and at time $t$, $\boldsymbol{\omega}(\mathbf{x}, t)$, is twice the angular velocity $\boldsymbol{\Omega}$ of a fluid element passing through the point $\mathbf{x}$ at time $t$, the angular velocity being measured about the centre of the element. In short, vorticity is all about the intrinsic spin of fluid lumps as they slide along streamlines, as indicated in Figure 20.3.

Consider now the symmetric contribution to $\partial u_i/\partial x_j$. Since $S_{ij}$ is symmetric, it may be put into diagonal form through an appropriate rotation of the coordinate system. The coordinate axes are then the principal axes of the tensor $S_{ij}$, which we label 1, 2 and 3. Let $\dot{\alpha}$,

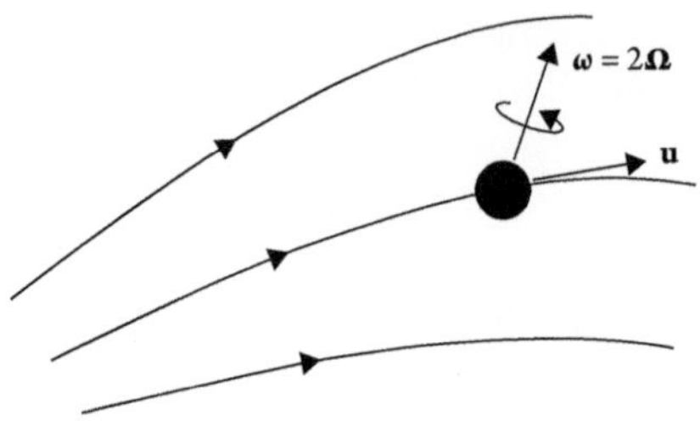


**Figure 20.3** The vorticity at location **x** and time $t$ is twice the angular velocity of a fluid element passing through the point **x** at time $t$.

$\dot{\beta}$ and $\dot{\gamma}$ be the three principal rates of strain. Then $\dot{\alpha} = \partial u_1/\partial x_1$, $\dot{\beta} = \partial u_2/\partial x_2$ and $\dot{\gamma} = \partial u_3/\partial x_3$, while (20.7) demands that $\dot{\alpha} + \dot{\beta} + \dot{\gamma} = 0$. If we temporarily set $\delta\mathbf{u}^{(a)}$ to zero, then, in coordinates aligned with the principal axes, we have $\delta\mathbf{u} = (\dot{\alpha}\,\delta x_1, \dot{\beta}\,\delta x_2, \dot{\gamma}\,\delta x_3)$. So, a short material line element oriented parallel to $x_1$ experiences a relative velocity field $\delta\mathbf{u} = (\dot{\alpha}\,\delta x_1, 0, 0)$, being stretched or compressed at a rate $\dot{\alpha}$ while remaining parallel to $x_1$. Likewise, line elements aligned with $x_2$ and $x_3$ stretch or contract, without rotation, at rates $\dot{\beta}$ and $\dot{\gamma}$. We conclude that, provided $\delta\mathbf{u}^{(a)} = 0$, a spherical blob of fluid centred on **x** deforms into an ellipsoid whose principal axes do not rotate. In short, $S_{ij}$ is associated with the deformation of fluid elements, as its name suggests.

Such deformations require stresses acting on fluid elements, and so, associated with $S_{ij}$, we have a stress tensor $\tau_{ij}$. A common constitutive law, called *Newton's law of viscosity*, proposes that $\tau_{ij}$ is proportional to $S_{ij}$:

$$\tau_{ij} = 2\rho\nu S_{ij} = \rho\nu\left(\frac{\partial u_i}{\partial x_j} + \frac{\partial u_j}{\partial x_i}\right), \tag{20.11}$$

where the constant of proportionality $\nu$ is the *kinematic viscosity* of the fluid. We shall return to (20.11) shortly, when we shall explain the logic behind (20.11).

### 20.2.3 *Dynamics at Last: The Navier–Stokes Equation and Energy Dissipation*

Let us turn from kinematics to dynamics. Consider a small blob of fluid of mass $\rho\,\delta V$. The net pressure force acting on this blob is, using a variant of Gauss' theorem,

$$\oint_S (-p)\,d\mathbf{S} = \int_{\delta V} (-\nabla p)\,dV = -(\nabla p)\delta V,$$

where the surface integral on the left is the sum of all the elementary pressure forces $p\,d\mathbf{S}$ acting on the surface $S$ of the blob. Newton's second law applied to this moving blob of fluid now gives us

$$(\rho\,\delta V)\frac{D\mathbf{u}}{Dt} = -(\nabla p)\delta V + \text{viscous forces}, \tag{20.12}$$

where we have still to determine the viscous term in (20.12).

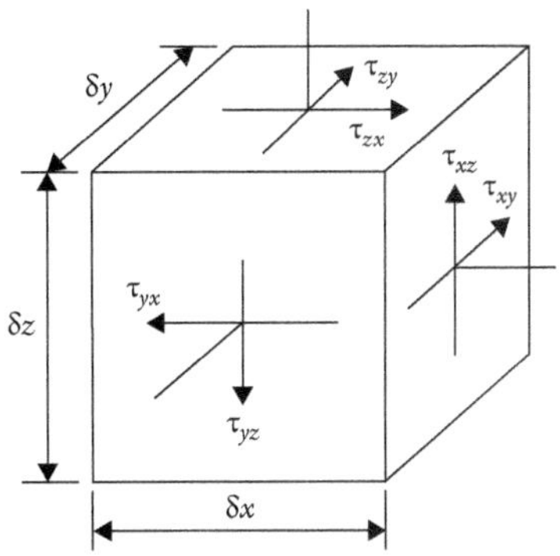


**Figure 20.4** The viscous stresses acting on a rectangular element.

Suppose the fluid lump is instantaneously in the form of a rectangular element $dy\,dx\,dz$, as shown in Figure 20.4. The stresses arising from the presence of viscosity comprise both shear stresses $\tau_{xy}$, $\tau_{yz}$ and so forth, as well as normal stresses, $\tau_{xx}$, $\tau_{yy}$ and $\tau_{zz}$, which supplement the normal stresses due to pressure. These viscous stresses can influence the motion of a fluid lump because any imbalance in stress will lead to a net viscous force acting on the element. For example, a difference between $\tau_{zx}$ at the top and bottom of the element will produce a net force in the $x$ direction of magnitude $f_x = (\delta\tau_{zx})\,dx\,dy$, which may also be written as $f_x = (\partial\tau_{zx}/\partial z)\,\delta V$. Gathering all such terms, we conclude that the net viscous force in the $x$ direction is

$$f_x = \left(\frac{\partial\tau_{xx}}{\partial x} + \frac{\partial\tau_{yx}}{\partial y} + \frac{\partial\tau_{zx}}{\partial z}\right)\delta V = \frac{\partial\tau_{jx}}{\partial x_j}\,\delta V.$$

Of course, similar expressions can be found for $f_y$ and $f_z$ and it is evident that the net viscous force in the $i$th direction acting on the rectangular element is $f_i = (\partial\tau_{ji}/\partial x_j)\,\delta V$. Our equation of motion now becomes

$$\rho\frac{D\mathbf{u}}{Dt} = -\nabla p + \frac{\partial\tau_{ji}}{\partial x_j}. \tag{20.13}$$

This is as far as Newton's second law will take us. If we are to make further progress, we need to somehow relate the viscous stresses to the velocity field, which is exactly what (20.11) achieves. To understand where this constitutive relationship comes from, consider the simple shear flow $\mathbf{u} = u_x(y)\hat{\mathbf{e}}_x$ shown in Figure 20.5. Here the fluid elements slide over each other and one measure of this rate of sliding is the angular distortion rate, $d\gamma/dt$, of an initially rectangular element. In this simple flow a shear stress $\tau_{yx}$ is required to maintain the relative sliding of the fluid layers. Moreover, it is reasonable to suppose that $\tau_{yx}$ is directly proportional to the rate of sliding $d\gamma/dt$, and we define the constant of proportionality to be $\rho\nu$. It follows that $\tau_{yx} = \rho\nu\,d\gamma/dt$. Moreover, it is clear from Figure 20.5 that $d\gamma/dt = \partial u_x/\partial y$, and so for this simple rectilinear shear flow we have

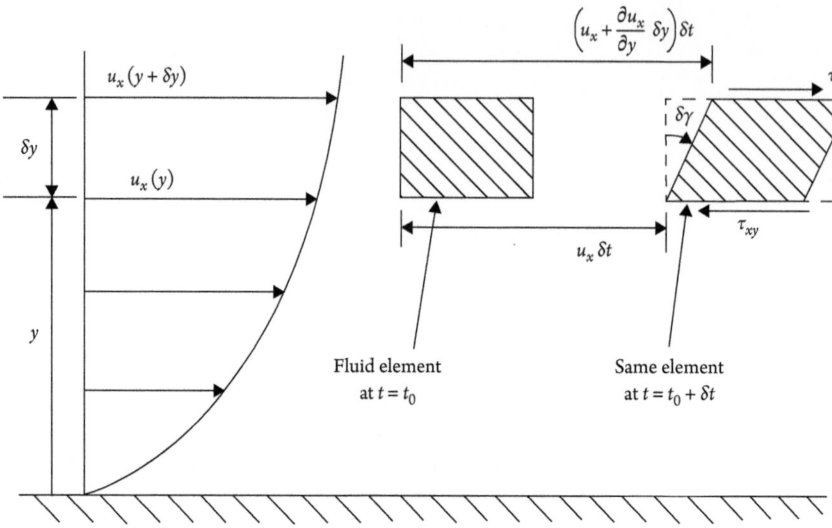


**Figure 20.5** The distortion of a fluid element in a parallel shear flow.

$$\tau_{yx} = \rho\nu\frac{\partial u_x}{\partial y}.$$

Of course, this is just a special case of our constitutive equation (20.11). This kind of argument may be generalised in an obvious way to justify (20.11) in three dimensions, and indeed Newton's law of viscosity is found to have very wide applicability.

Finally, we substitute (20.11) into our equation of motion (20.13) and invoke $\nabla\cdot\mathbf{u} = 0$. This yields

$$\boxed{\frac{\partial\mathbf{u}}{\partial t} + (\mathbf{u}\cdot\nabla)\mathbf{u} = -\nabla\left(\frac{p}{\rho}\right) + \nu\nabla^2\mathbf{u},} \tag{20.14}$$

which is the *Navier–Stokes equation*. The boundary condition on $\mathbf{u}$ corresponding to (20.14) is $\mathbf{u} = 0$ on any stationary solid surface, i.e. fluids 'stick' to surfaces. This is known as the *no-slip condition*. Sometimes it is useful to imagine a fluid with zero viscosity. Such fluids, which do not really exist, are governed by the *Euler equation*,

$$\boxed{\frac{D\mathbf{u}}{Dt} = -\nabla\left(\frac{p}{\rho}\right).} \tag{20.15}$$

The boundary condition appropriate to (20.15) is $\mathbf{u}\cdot d\mathbf{S} = 0$ at an impermeable surface.

The first, and most important, point to note about (20.14) is that it is nonlinear because of the term $(\mathbf{u}\cdot\nabla)\mathbf{u}$. This nonlinearity, which is not shared by Maxwell's equations, has profound implications for fluid mechanics. For example, relatively few non-trivial exact solutions of the steady Navier–Stokes equations have so far been found. More importantly, unless the viscosity is relatively large, so that there is a significant amount of viscous damping, the solutions of (20.14) are chaotic, a phenomenon we call turbulence. The usual dimensionless measure of the relative importance of viscosity is the *Reynolds number* $\mathrm{Re} = u\ell/\nu$, where $u$ and $\ell$ are characteristic velocity and length scales. Re is normally interpreted as a measure of the ratio of inertial to viscous forces.

We close this section with a comment about viscous dissipation. We can obtain an energy equation by taking the dot product of $\mathbf{u}$ with (20.13). Noting that

$$u_i \frac{\partial \tau_{ji}}{\partial x_j} = \frac{\partial}{\partial x_j}(\tau_{ji}u_i) - \tau_{ji}\frac{\partial u_i}{\partial x_j} = \frac{\partial}{\partial x_j}(\tau_{ji}u_i) - \tau_{ij}S_{ij}, \tag{20.16}$$

this gives

$$\frac{\partial}{\partial t}(\rho \mathbf{u}^2/2) = -\nabla \cdot [(\rho \mathbf{u}^2/2)\mathbf{u}] - \nabla \cdot (p\mathbf{u}) + \frac{\partial}{\partial x_j}(\tau_{ji}u_i) - \tau_{ij}S_{ij}. \tag{20.17}$$

When integrated over a control volume fixed in space, the divergences on the right become surface integrals representing the material transport of kinetic energy across the bounding surface and the rate of working of the pressure and viscous stresses on the same surface. Conservation of energy then tells us that the term $\tau_{ij}S_{ij}$ must represent an irreversible loss of mechanical energy to heat. It follows that the rate of increase of internal energy per unit volume due to viscous dissipation is $\tau_{ij}S_{ij} = 2\rho\nu S_{ij}S_{ij}$.

### 20.2.4 *A Transport Equation for Vorticity*

Note that the Navier–Stokes equation involves both the velocity and pressure fields. Moreover, these are not independent, since the divergence of (20.14) yields

$$\nabla^2\left(\frac{p}{\rho}\right) = -\nabla \cdot (\mathbf{u} \cdot \nabla \mathbf{u}), \tag{20.18}$$

which may be inverted in an infinite domain to give

$$p(\mathbf{x}) = \frac{\rho}{4\pi}\int \frac{[\nabla \cdot (\mathbf{u} \cdot \nabla \mathbf{u})]'}{|\mathbf{x} - \mathbf{x}'|}\, d\mathbf{x}'. \tag{20.19}$$

Evidently, the pressure field is a non-local function of the global velocity field. If we are to regard (20.14) as an evolution equation for $\mathbf{u}$, of the form $\partial \mathbf{u}/\partial t = F(\mathbf{u},p)$, then each time we calculate a new velocity field from the old, we must update the pressure field using (20.19). This can be cumbersome, and is made all the more awkward by the non-local nature of the pressure field. For this, and other, reasons, it is usually preferable to convert the Navier–Stokes equation into an evolution equation for vorticity, $\boldsymbol{\omega} = \nabla \times \mathbf{u}$. Of course, given $\boldsymbol{\omega}(\mathbf{x}, t)$, we may use the Biot–Savart law,

$$\mathbf{u}(\mathbf{x}, t) = \frac{1}{4\pi} \int \frac{\boldsymbol{\omega}(\mathbf{x}', t) \times \mathbf{r}}{r^3} d\mathbf{x}', \quad \mathbf{r} = \mathbf{x} - \mathbf{x}', \tag{20.20}$$

to invert $\boldsymbol{\omega} = \nabla \times \mathbf{u}$ and hence find $\mathbf{u}(\mathbf{x}, t)$.

In order to obtain the governing equation for $\boldsymbol{\omega}(\mathbf{x}, t)$, we use the vector identity $\nabla(\mathbf{u}^2/2) = (\mathbf{u} \cdot \nabla)\mathbf{u} + \mathbf{u} \times \boldsymbol{\omega}$ to rewrite (20.14) in the form

$$\frac{\partial \mathbf{u}}{\partial t} = \mathbf{u} \times \boldsymbol{\omega} - \nabla C + \nu \nabla^2 \mathbf{u}, \quad C = \frac{p}{\rho} + \frac{\mathbf{u}^2}{2}, \tag{20.21}$$

where $C$ is known as *Bernoulli's function*. We now take the curl of (20.21), which yields

$$\boxed{\frac{\partial \boldsymbol{\omega}}{\partial t} = \nabla \times (\mathbf{u} \times \boldsymbol{\omega}) + \nu \nabla^2 \boldsymbol{\omega}.} \tag{20.22}$$

Moreover, since $\mathbf{u}$ and $\boldsymbol{\omega}$ are both solenoidal, we have $\nabla \times (\mathbf{u} \times \boldsymbol{\omega}) = (\boldsymbol{\omega} \cdot \nabla)\mathbf{u} - (\mathbf{u} \cdot \nabla)\boldsymbol{\omega}$, and so (20.22) becomes

$$\boxed{\frac{D\boldsymbol{\omega}}{Dt} = (\boldsymbol{\omega} \cdot \nabla)\mathbf{u} + \nu \nabla^2 \boldsymbol{\omega}.} \tag{20.23}$$

This is our vorticity evolution equation. We can now advance $\boldsymbol{\omega}$ in time by writing (20.22) or (20.23) in the form $\partial \boldsymbol{\omega}/\partial t = F(\mathbf{u}, \boldsymbol{\omega})$, while using (20.20) to update the velocity along with the vorticity field.

Perhaps some comments are in order at this point. First, we have used $\nabla \cdot \boldsymbol{\omega} = 0$ to move from (20.22) to (20.23). This is not a consequence of incompressibility, but rather follows from the identity $\nabla \cdot [\nabla \times (\sim)] = 0$. Second, although we have said that $\boldsymbol{\omega}$ is related to the spin of fluid particles as they slide along streamlines, this has nothing to do with the *global* rotation in a flow. For example, the velocity $\mathbf{u} = (u_x(y), 0, 0)$ possesses vorticity, yet the streamlines are straight and parallel. On the other hand, the flow $\mathbf{u}(s) = (0, k/s, 0)$in $(s, \theta, z)$ coordinates has no vorticity, except for a singularity at $s = 0$, yet the streamlines are circular. This is illustrated in Figure 20.6. Third, since vorticity is a measure of spin, we might expect (20.23) to be related to angular momentum conservation. To see that this is indeed the case, consider a blob of fluid that is *instantaneously spherical*, as shown in Figure 20.3. Since $\boldsymbol{\omega}$ is twice the angular velocity of the sphere, the angular momentum of the blob is $\mathbf{H} = I\boldsymbol{\omega}/2$, where $I$ is its moment of inertia. Moreover, $\mathbf{H}$ can change as a result of the tangential surface stresses alone, the pressure having no influence as long as the blob is spherical. Therefore, at the instant the blob is spherical, we have

$$\frac{D\mathbf{H}}{Dt} = (\text{viscous torque on spherical element}),$$

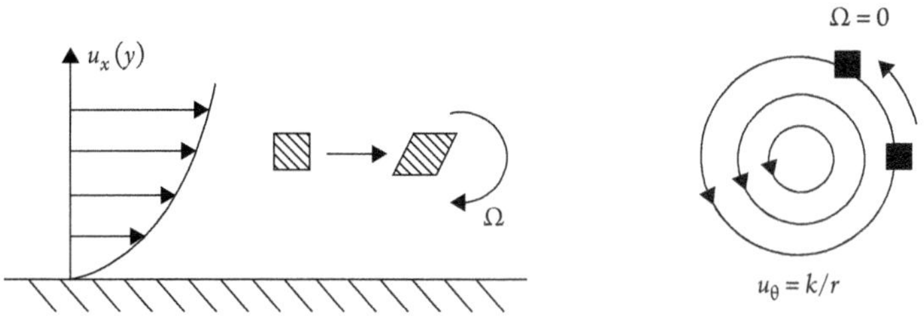


**Figure 20.6** Fluid elements on the left have vorticity, while those on the right do not.

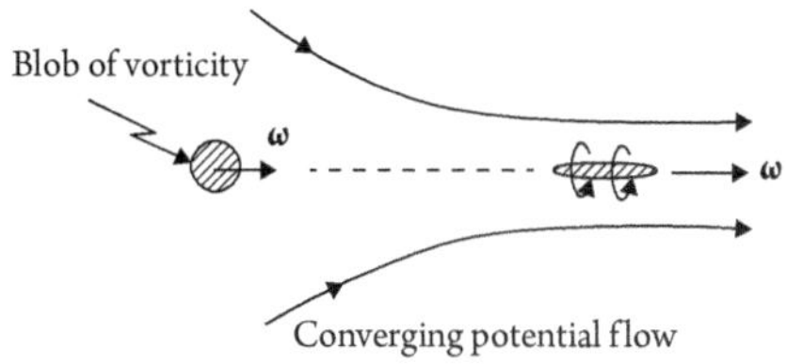


**Figure 20.7** Stretching a fluid element can intensify its vorticity.

or equivalently

$$\frac{D\boldsymbol{\omega}}{Dt} = -\frac{\boldsymbol{\omega}}{I}\frac{DI}{Dt} + 2\frac{\text{viscous torque}}{I}. \tag{20.24}$$

This is the physical interpretation of (20.23). In short, the vorticity can change in a blob of fluid because its moment of inertia changes or through the action of viscous stresses.

A comparison of (20.23) with (20.24) suggests that the first term on the right, $(\boldsymbol{\omega}\cdot\nabla)\mathbf{u}$, represents changes in the moment of inertia of a fluid blob, and indeed this is the case. Consider a thin *vortex tube*, that is, an aggregate of vortex lines, rather like a magnetic flux tube. If we let $u_{\parallel}$ be the component of velocity parallel to the tube and $s$ a coordinate measured along the tube, then $(\boldsymbol{\omega}\cdot\nabla)u_{\parallel} = |\boldsymbol{\omega}|\, du_{\parallel}/ds$. Moreover, the vortex tube is stretched (or compressed) whenever $du_{\parallel}/ds > 0$ (or $du_{\parallel}/ds < 0$). It follows that, when $(\boldsymbol{\omega}\cdot\nabla)u_{\parallel} > 0$, the vortex tube is stretched and thinned, causing a reduction in its moment of inertia and a consequent rise in angular velocity, in accordance with (20.24). So, the stretching of a fluid element in the direction of its vorticity causes it to spin up, as illustrated in Figure 20.7.

There is one special case where there is no vortex stretching (or compression), namely two-dimensional motion: $\mathbf{u}(x,y) = (u_x, u_y, 0)$, $\boldsymbol{\omega} = (0,0,\omega)$. Here (20.23) reduces to the simple scalar equation

$$\frac{D\omega}{Dt} = \nu\nabla^2\omega. \tag{20.25}$$

This might be compared with the governing equation for the temperature $T$ in a fluid,

$$\frac{DT}{Dt} = \alpha\nabla^2 T, \tag{20.26}$$

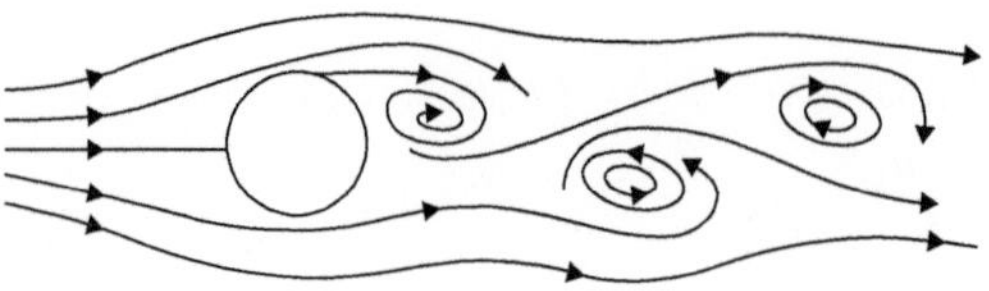

**Figure 20.8** Vortex shedding behind a cylinder.

where $\alpha$ is the thermal diffusivity. Equations of this type are referred to as *advection–diffusion equations*. Evidently, in planar flows, vorticity is swept around by the flow and diffuses, just like heat. Note that, within the interior of such a flow, neither diffusion nor advection (material transport) of a patch of vorticity can change the net amount of vorticity in that patch. Both advection and diffusion merely act to redistribute a given amount the vorticity in space. (Try proving this for yourself by integrating (20.25) over the vortex patch, applying Gauss' theorem to the viscous and nonlinear advection terms.)

In summary, then, vorticity may be advected (materially transported) and diffused. Moreover, in three dimensions it may be intensified or reduced through the stretching or compression of the vortex tubes. However, none of these processes can create vorticity in a fluid that had none to start with. One might ask, therefore, where the vorticity in the wake of the cylinder shown in Figure 20.8 has come from. Here the analogy between vorticity in a two-dimensional flow and heat transport is useful. Heat cannot be created or destroyed within the interior of a fluid by advection or diffusion—it can only be redistributed in space. However, if the cylinder in Figure 20.8 is heated, then heat will get into the fluid by diffusing in from the surface of the cylinder. The same is true of vorticity: in the absence of body forces, vorticity can only get into a fluid though the boundaries. Just as there are thin *thermal boundary layers* on the heated cylinder, through which heat diffuses into the fluid, so there are thin *mechanical boundary layers*, through which vorticity diffuses into the fluid. All of the vorticity downstream of the cylinder in Figure 20.8 started off in a thin boundary layer on the upstream surface of the cylinder.

### *20.2.5 Helmholtz's Laws of Inviscid Vortex Dynamics and Kinetic Helicity*

Sometimes it is allowable to neglect viscous stresses, and so we now turn to the somewhat idealised world of *inviscid* vortex dynamics. This may be couched in terms of Kelvin's 1869 theorem or, more comprehensively, in terms of Helmholtz's 1858 laws.

We start by revisiting the concept of a vortex tube. Since $\boldsymbol{\omega}$ is solenoidal, the flux of vorticity, $\Phi$, is constant along the length of such a tube, there being no flux through the sides. Often, $\Phi$ is known the strength of the vortex tube. A related quantity is the *circulation* $\Gamma$, defined as

$$\Gamma = \oint_C \mathbf{u} \cdot d\mathbf{l}. \tag{20.27}$$

If the curve $C$ encircles an isolated vortex tube, then Stokes' theorem tells us that $\Gamma = \Phi$.

Kelvin's theorem is expressed in terms of circulation. It says that, if $C_m(t)$ is a closed material curve, always composed of the same fluid particles, then

$$\Gamma = \oint_{C_m(t)} \mathbf{u} \cdot d\mathbf{l} = \text{constant}. \tag{20.28}$$

This theorem does not hold if the fluid is subject to a rotational body force, nor if viscous stresses are significant at any point on the curve $C_m$. The proof of Kelvin's theorem follows from (10.15), the kinematic equation

$$\frac{d}{dt}\int_S \mathbf{G} \cdot d\mathbf{S} = \int_S \left[\frac{\partial \mathbf{G}}{\partial t} - \nabla \times (\mathbf{u} \times \mathbf{G})\right] \cdot d\mathbf{S}. \tag{20.29}$$

If we take $S$ to be any surface that spans $C_m$, and write $\mathbf{G} = \boldsymbol{\omega}$, then (20.22) combined with (20.29) yields Kelvin's theorem:

$$\frac{d\Gamma}{dt} = \frac{d}{dt}\int_S \boldsymbol{\omega} \cdot d\mathbf{S} = 0. \tag{20.30}$$

Helmholtz's laws also apply to inviscid flow, and they state that, in the absence of rotational body forces:

(1) fluid particles that lie on a vortex line at some initial instant continue to lie on that vortex line for all time, i.e. vortex lines are frozen into the fluid like dye lines;

(2) the flux of vorticity along a vortex tube, $\Phi$, is the same at all cross sections of the tube and is independent of time.

We may prove Helmholtz's first law as follows. Let $d\mathbf{l}$ be a short material line that moves with the fluid, like a dyeline. Then the rate of change of $d\mathbf{l}$ is $\mathbf{u}(\mathbf{x} + d\mathbf{l}) - \mathbf{u}(\mathbf{x})$, where $\mathbf{x}$ and $\mathbf{x} + d\mathbf{l}$ mark the two ends of $d\mathbf{l}$. It follows that

$$\frac{D}{Dt}(d\mathbf{l}) = \mathbf{u}(\mathbf{x} + d\mathbf{l}) - \mathbf{u}(\mathbf{x}) = (d\mathbf{l} \cdot \nabla)\mathbf{u}, \tag{20.31}$$

which might be compared with the inviscid version of (20.23). It would seem that $\boldsymbol{\omega}$ and $d\mathbf{l}$ obey the same equation and so they must evolve in identical ways under the influence of $\mathbf{u}$. We conclude that vortex lines are frozen into the fluid, just like dye lines.

Turning to Helmholtz's second law, this follows from the solenoidal nature of $\boldsymbol{\omega}$ plus a combination of the first law and Kelvin's theorem. The fact that the vorticity flux is constant along a vortex tube follows directly from $\nabla \cdot \boldsymbol{\omega} = 0$, while the invariance of $\Phi$ comes from the fact that a vortex tube moves with the fluid (the first law) and so Kelvin's theorem may be applied to a material curve $C$ encircling the tube, to give $\Phi$ = constant.

Helmholtz's laws strongly constrain the evolution of an (idealised) inviscid fluid. As a simple example, consider two thin interlinked vortex tubes, as shown in Figure 20.9. Each

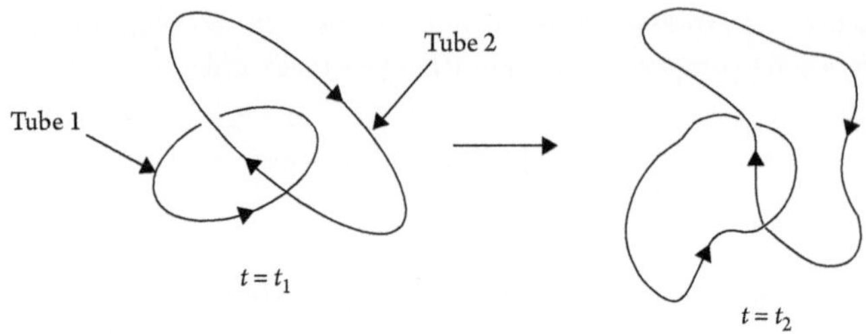


**Figure 20.9** Inviscid vortex tubes preserve their topology as they are swept around.

tube creates a velocity field via the Biot–Savart law, and so they advect each other in some complex way. However, no matter how complicated the motion, the tubes remain linked in the same manner for all time, and the strength of each tube is conserved.

This conservation of vortex-line topology in an inviscid fluid is captured by an invariant called the *kinetic helicity*. This is analogous to the magnetic helicity of (7.7) and is defined as

$$h_\omega = \int_{V_\omega} \mathbf{u} \cdot \boldsymbol{\omega} \, dV, \tag{20.32}$$

where $V_\omega$ is a *material volume* for which $\boldsymbol{\omega} \cdot d\mathbf{S} = 0$ on its surface $S_\omega$. We may confirm that $h_\omega$ is indeed an invariant by combining Euler's equation with the inviscid version of the vorticity equation (20.23):

$$\frac{D\mathbf{u}}{Dt} \cdot \boldsymbol{\omega} + \frac{D\boldsymbol{\omega}}{Dt} \cdot \mathbf{u} = -\nabla \left( \frac{p}{\rho} \right) \cdot \boldsymbol{\omega} + (\boldsymbol{\omega} \cdot \nabla \mathbf{u}) \cdot \mathbf{u}. \tag{20.33}$$

Given that $\boldsymbol{\omega}$ is solenoidal, this yields

$$\frac{D}{Dt}(\mathbf{u} \cdot \boldsymbol{\omega}) = \nabla \cdot \left[ \left( \frac{u^2}{2} - \frac{p}{\rho} \right) \boldsymbol{\omega} \right], \tag{20.34}$$

which may be integrated over a set of *material* volume elements $\delta V$, each of invariant volume, to give

$$\frac{d}{dt} \int_{V_\omega} (\mathbf{u} \cdot \boldsymbol{\omega}) \, dV = \int_{V_\omega} \frac{D(\mathbf{u} \cdot \boldsymbol{\omega})}{Dt} dV = \oint_{S_\omega} \left( \frac{u^2}{2} - \frac{p}{\rho} \right) \boldsymbol{\omega} \cdot d\mathbf{S} = 0, \tag{20.35}$$

as required.

The invariance of $h_\omega$ is a direct consequence of the inviscid conservation of vortex-line topology, as may be illustrated by returning to Figure 20.9. Suppose that $V_\omega$ is all of space. Then $h_\omega$ has a contribution from vortex tube 1, which has volume $V_1$, flux $\Phi_1$ and centreline

$C_1$, and another contribution from vortex tube 2, with volume $V_2$, flux $\Phi_2$ and centreline $C_2$. Denoting these contributions by $h_1$ and $h_2$, we have

$$h_1 = \int_{V_1} (\mathbf{u} \cdot \boldsymbol{\omega}) dV = \oint_{C_1} \mathbf{u} \cdot (\Phi_1 d\mathbf{l}) = \Phi_1 \oint_{C_1} \mathbf{u} \cdot d\mathbf{l}, \tag{20.36}$$

where we have used $\boldsymbol{\omega}\, dV = \Phi_1\, d\mathbf{l}$. However, provided that the loops are interlinked once, we have $\oint_{C_1} \mathbf{u} \cdot d\mathbf{l} = \pm\Phi_2$, where the positive sign corresponds to a right-handed linkage and the negative sign to a left-handed linkage. Evidently $h_1 = \pm\Phi_1\Phi_2$ and a similar calculation gives $h_2 = \pm\Phi_2\Phi_1$. It follows that

$$h_\omega = h_1 + h_2 = \pm 2\Phi_1\Phi_2. \tag{20.37}$$

Note that, if the tubes were unlinked, then $h_\omega$ would be zero. We conclude from (20.37) that the invariance of $h_\omega$ in this simple example is a direct consequence of the conservation of the vortex-line topology and of the fluxes $\Phi_1$ and $\Phi_2$.

### *20.2.6 Boundary Layers and Turbulence*

We conclude our brief introduction to fluid mechanics by noting that the kinematic viscosity of nearly all common fluids is very small, around $\nu \sim 10^{-6}\,\mathrm{m^2\,s^{-1}}$ for water and liquid metals and perhaps a little larger for air. An immediate consequence of the smallness of $\nu$ is that the Reynolds number $\mathrm{Re} = u\ell/\nu$ in most flows is very large, so that in the bulk of the fluid the viscous stresses are usually negligible by comparison with inertia. For example, if the cylinder in Figure 20.8 had a diameter $d = 1$ cm, and a modest cross flow $u = 10$ cm/s, then the Reynolds number would be $\mathrm{Re} = ud/\nu \sim 10^3$. If it were a chimney in a high wind, on the other hand, we might have $\mathrm{Re} \sim 10^7$.

It should be noted, however, that no matter how large Re might be, viscous stresses are rarely negligible very close to solid surfaces, as we now discuss. Away from solid surfaces we might treat a high-Re flow as nominally inviscid, and solve for the velocity using the Euler equation, say. The boundary condition appropriate for the Euler equation is $\mathbf{u} \cdot d\mathbf{S} = 0$ at an impermeable surface, which allows for a tangential slip velocity at the boundary. However, all real fluids are viscous and they must satisfy the stronger *no-slip* boundary condition on a solid surface. Close to a surface, therefore, the tangential velocity must drop from the inviscid slip value down to zero. The forces that achieve this sharp reduction in tangential velocity are the viscous stresses, and to accomplish this they must be able to compete with the inertial forces, despite the large nominal value of the Reynolds number $\mathrm{Re} = ud/\nu$. The way this is achieved is that the tangential velocity drops to zero across a very narrow surface layer, called a *boundary layer*, and the extreme thinness of this surface layer, $\delta$, allows the viscous stresses $\tau \sim \rho\nu u/\delta$ to achieve a value of order one despite the smallness of the viscosity. In short, adjacent to solid surfaces a thin boundary layer forms with large transverse gradients in velocity, and within this boundary layer the tangential velocity drops to zero in order to satisfy the no-slip condition. This large transverse gradient in velocity leads to large local values of vorticity, and in many ways these mechanical boundary layers may be thought of as diffusion layers for vorticity, with vorticity diffusing out into the flow.

The second main consequence of a low viscosity is that most flows are *turbulent*, with a chaotic velocity field. Turbulence arises when the viscous dissipation is too weak to damp out random perturbations, and this occurs for flow in a pipe or flow around a cylinder at values of $\mathrm{Re} = ud/\nu$ in excess of $\sim 10^3$. The critical value of Re at which turbulence first appears in a given flow depends on the precise geometry involved, although critical values tend to be in the range $10^2$–$10^4$, and this usually corresponds to a relatively low velocity. So, while the raindrops running down a window may be *laminar* (i.e. non-turbulent), the wind in the street outside will certainly not be.

In a turbulent flow one typically has a mean velocity field, which might be obtained from the actual velocity field by time-averaging, plus random fluctuations superimposed on the mean. Just as the random fluctuations of the molecules in a laminar flow give rise to the macroscopic phenomenon of viscous stresses, through the molecular transfer of momentum, so averaging over the random fluctuations in velocity in a turbulent flow give rise to additional stresses acting on the mean flow, called *Reynolds stresses*. There is then a two-way interaction between the mean flow and the turbulence, with the turbulence shaping the mean flow through the Reynolds stresses, and the mean flow generating and dispersing the turbulence. This is a difficult problem that remains largely unsolved, although a brave reader might consult, say, Frisch (1995) or Davidson (2015) for more details.

## 20.3 Stripping Down Maxwell's Equations for MHD

Let us now return to electrodynamics and establish the reduced form of Maxwell's equations required for MHD. The key point is that, in good conductors, such as liquid metals and hot plasmas, the charge density within the interior of the conductor is tiny. A number of simplifications follow directly from this, such as the following:

- charge conservation simplifies to $\nabla \cdot \mathbf{J} = 0$;
- the displacement current may be neglected in the Ampère–Maxwell equation;
- the Coulomb force is negligible by comparison with the Lorentz force $\mathbf{J} \times \mathbf{B}$;
- the electrostatic energy is negligible by comparison with the magnetic energy.

Let us start, therefore, with charge conservation and return to the discussion of §11.5.

### *20.3.1 The Charge Density and Electrostatic Energy are Negligible*

Before we tackle moving conductors, let us consider the case of a stationary conductor. Here charge conservation,

$$\nabla \cdot \mathbf{J} = -\frac{\partial \rho_e}{\partial t}, \tag{20.38}$$

combined with Gauss' law, $\nabla \cdot \mathbf{E} = \rho_e/\varepsilon_0$, and Ohm's law, $\mathbf{J} = \sigma \mathbf{E}$, yields

$$\frac{\partial \rho_e}{\partial t} = -\nabla \cdot \mathbf{J} = -\nabla \cdot (\sigma \mathbf{E}) = -\frac{\sigma}{\varepsilon_0}\rho_e. \tag{20.39}$$

Of course, this is the charge relaxation equation, whose solution is

$$\rho_e = \rho_e(0)\exp\left(-\frac{t}{\tau_e}\right), \quad \tau_e = \frac{\varepsilon_0}{\sigma}, \tag{20.40}$$

where $\tau_e$ is the charge relaxation time. For liquid metals $\tau_e$ is tiny, around $10^{-14}$ s, and so any excess charge is rapidly eradicated from the interior of the conductor.

Now suppose the conductor moves through a magnetic field with a velocity $\mathbf{u}(\mathbf{x}, t)$ that is much less than the speed of light, $c$. Then Ohm's law becomes $\mathbf{J} = \sigma(\mathbf{E} + \mathbf{u} \times \mathbf{B})$ and (20.39) generalises to

$$\frac{\partial \rho_e}{\partial t} + \frac{\rho_e}{\tau_e} = -\sigma \nabla \cdot (\mathbf{u} \times \mathbf{B}). \tag{20.41}$$

A small but finite charge can now build up within the interior of the conductor. However, MHD is non-relativistic, $u << c$, and so we are interested in events that take place on a timescale much longer than $\tau_e$ (see Exercise 20.1). That is to say, MHD is characterised by $\varpi\tau_e << 1$, where $\varpi$ is a typical frequency or inverse timescale. It follows that we may drop $\partial \rho_e/\partial t$ from (20.41), to give the pseudo-static result

$$\rho_e = -\varepsilon_0 \nabla \cdot (\mathbf{u} \times \mathbf{B}). \tag{20.42}$$

Now, dropping $\partial \rho_e/\partial t$ in (20.41) is equivalent to ignoring $\partial \rho_e/\partial t$ in the charge conservation equation, and so we conclude that the charge density $\rho_e$ is so small in MHD that charge conservation reduces to the statement that $\nabla \cdot \mathbf{J} = 0$. Moreover, ignoring $\partial \rho_e/\partial t$ in (20.38) is equivalent to dropping the displacement current $\varepsilon_0 \partial \mathbf{E}/\partial t$ from the Ampère–Maxwell equation,

$$\nabla \times \mathbf{B} = \mu_0\left(\mathbf{J} + \varepsilon_0 \frac{\partial \mathbf{E}}{\partial t}\right). \tag{20.43}$$

The point here is that the displacement current was introduced into (20.43) specifically to make it consistent with the charge conservation equation. However, if the charge density is so small that charge conservation reduces to $\nabla \cdot \mathbf{J} = 0$, then there is no need to introduce the displacement current into Ampère's equation, since its right-hand side is already solenoidal. Indeed, we can check directly that $\varepsilon_0 \partial \mathbf{E}/\partial t << \mathbf{J}$ in the MHD limit, since

$$\frac{\varepsilon_0\, \partial \mathbf{E}/\partial t}{|\mathbf{J}|} \sim \frac{\varepsilon_0 \varpi E}{\sigma E} \sim \varpi \tau_e << 1. \tag{20.44}$$

Another consequence of $\rho_e$ being very small is that the electric force $\rho_e \mathbf{E}$ is much weaker than the Lorentz force $\mathbf{J} \times \mathbf{B}$ and so can be neglected. That is, from (20.42) we have $\rho_e \sim \varepsilon_0 uB/\ell$, while Ohm's law requires $\mathbf{E} \sim \mathbf{J}/\sigma$, and so

$$\rho_e |\mathbf{E}| \sim (\varepsilon_0 uB/\ell)(J/\sigma) \sim \frac{u\tau_e}{\ell} JB \sim (\varpi \tau_e) JB << JB,$$

where $\ell$ is a typical length-scale for the flow. Thus, in the MHD approximation, the force law simplifies to $\mathbf{F} = \mathbf{J} \times \mathbf{B}$.

Since the Coulomb force is negligible, it should come as no surprise that the electrostatic energy may also be ignored by comparison with the magnetic energy. In fact, the ratio of the electric to magnetic energy density is readily shown to be $O(u^2/c^2) << 1$. This follows from application of Faraday's law:

$$\frac{\varepsilon_0 \mathbf{E}^2}{\mathbf{B}^2/\mu_0} \sim \frac{\varepsilon_0 \ell^2 (\nabla \times \mathbf{E})^2}{\mathbf{B}^2/\mu_0} \sim \frac{\varepsilon_0 \ell^2 (\varpi \mathbf{B})^2}{\mathbf{B}^2/\mu_0} \sim \frac{\ell^2 \varpi^2}{c^2} \sim \frac{u^2}{c^2} << 1.$$

In summary, then, although the charge density is finite in MHD, it is so small that it plays no significant role. It does not enter into the charge conservation equation, nor do we need it to calculate the electric force or electric energy, since these are both negligible. In fact, $\rho_e$ is significant only in Gauss' law, and so in MHD we simply ignore $\rho_e$ and drop Gauss' law. The divergence of $\mathbf{E}$ is then determined by a combination of $\nabla \cdot \mathbf{J} = 0$ and Ohm's law, $\mathbf{J} = \sigma(\mathbf{E} + \mathbf{u} \times \mathbf{B})$, which gives us

$$\nabla \cdot \mathbf{E} = -\nabla \cdot (\mathbf{u} \times \mathbf{B}). \tag{20.45}$$

Comparing this with (20.42) shows that (20.45) is equivalent to Gauss' law in the MHD approximation.

### 20.3.2 The Reduced Form of Maxwell's Equations for MHD

We are now in a position to summarise the reduced form of Maxwell's equations used in MHD. We have seen that charge conservation simplifies to $\nabla \cdot \mathbf{J} = 0$, while the displacement current is negligible in the Ampère–Maxwell equation, which then reduces to the differential form of Ampère's law. Finally, we have shown that, in the MHD regime, Gauss' law may be dropped because it is equivalent to a combination of Ohm's law and $\nabla \cdot \mathbf{J} = 0$, and in any event it involves a quantity, $\rho_e$, that is of no interest to us. Our reduced set of electrodynamic equations are then as follows:

- Faraday's law plus the solenoidal nature of **B**,

$$\boxed{\nabla \times \mathbf{E} = -\frac{\partial \mathbf{B}}{\partial t}, \qquad \nabla \cdot \mathbf{B} = 0;} \tag{20.46}$$

- Ampère's law plus charge conservation,

$$\boxed{\nabla \times \mathbf{B} = \mu_0 \mathbf{J}, \qquad \nabla \cdot \mathbf{J} = 0;} \tag{20.47}$$

- Ohm's law plus the Lorentz force per unit volume,

$$\boxed{\mathbf{J} = \sigma(\mathbf{E} + \mathbf{u} \times \mathbf{B}), \qquad \mathbf{F} = \mathbf{J} \times \mathbf{B}.} \tag{20.48}$$

Since we have dropped the displacement current, these are probably best described as the *pre-Maxwell equations* of electrodynamics.

There are, of course, auxiliary equations that come along with the vector potential **A**. These are

$$\boxed{\mathbf{B} = \nabla \times \mathbf{A}, \qquad \nabla \cdot \mathbf{A} = 0.} \tag{20.49}$$

Note that in MHD the Coulomb gauge is invariably used for **A**. Ampère's law now becomes the magnetostatic Poisson equation,

$$\nabla^2 \mathbf{A} = -\mu_0 \mathbf{J}, \tag{20.50}$$

which inverts in the usual way to give

$$\mathbf{A}(\mathbf{x}, t) = \frac{\mu_0}{4\pi} \int \frac{\mathbf{J}(\mathbf{x}', t)}{|\mathbf{r}|} d\mathbf{x}', \qquad \mathbf{r} = \mathbf{x} - \mathbf{x}'. \tag{20.51}$$

If we add to these the Navier–Stokes equation, with the Lorentz force incorporated on the right-hand side, then we have the full set of governing equations for incompressible MHD.

### 20.3.3 *An Evolution Equation for the Magnetic Field*

Combining the laws of Faraday, Ohm and Ampère, we obtain an expression relating **B** to **u**:

$$\frac{\partial \mathbf{B}}{\partial t} = -\nabla \times \mathbf{E} = -\nabla \times \left(\frac{\mathbf{J}}{\sigma} - \mathbf{u} \times \mathbf{B}\right) = \nabla \times \left(\mathbf{u} \times \mathbf{B} - \frac{\nabla \times \mathbf{B}}{\mu_0 \sigma}\right).$$

However **B** is solenoidal and so $\nabla \times \nabla \times \mathbf{B} = -\nabla^2 \mathbf{B}$. Our equation for **B** now simplifies to

$$\boxed{\frac{\partial \mathbf{B}}{\partial t} = \nabla \times (\mathbf{u} \times \mathbf{B}) + \lambda \nabla^2 \mathbf{B},} \tag{20.52}$$

where $\lambda = (\mu_0 \sigma)^{-1}$ is the *magnetic diffusivity* (not to be confused with wavelength). The expression (20.52) is sometimes called the *induction equation* and it is one of the key equations in MHD. In effect, it is a transport equation for **B**, in the sense that if **u** is known, then it dictates the evolution of **B** from prescribed initial conditions. Note that we may take advantage of the fact that both **u** and **B** are solenoidal to rewrite (20.52) in the more useful form

$$\boxed{\frac{D\mathbf{B}}{Dt} = (\mathbf{B} \cdot \nabla)\mathbf{u} + \lambda \nabla^2 \mathbf{B}.} \tag{20.53}$$

An important, if highly idealised, special case is that of a perfect conductor (i.e. a conductor of infinite conductivity), in which case $\lambda = 0$ and

$$\frac{D\mathbf{B}}{Dt} = (\mathbf{B} \cdot \nabla)\mathbf{u}. \tag{20.54}$$

If we compare (20.53) with (20.23), we discover something remarkable: $\mathbf{B}$ and $\boldsymbol{\omega}$ are governed by the same equation, at least to within a change in diffusivity! This allows us to transfer much of what we learn from the study of vortex dynamics to MHD, as we shall see. The MHD analogue of the hydrodynamic Reynolds number is evidently

$$R_m = \frac{u\ell}{\lambda} = u\ell\mu_0\sigma, \tag{20.55}$$

which provides a convenient measure of the ratio of advection to diffusion of magnetic flux. This dimensionless quantity is known as the *magnetic Reynolds number*.

## 20.4 Some Simple Kinematic Theorems

We now enter the realms of kinematic MHD and explore some of the consequences of the induction equation, starting with the analogy between vorticity and magnetic field transport. Throughout this section we shall assume that $\mathbf{u}$ is somehow prescribed.

### *20.4.1 The Analogy Between the Evolution of Magnetic Fields and Vorticity*

The similarity between (20.53) with (20.23) is striking. It tells us that, just like vorticity in pure hydrodynamics, magnetic fields in a conducting medium are materially advected by the term on the left of (20.53), while being diffused by the second term on the right. Indeed, if $\mathbf{u}$ and $\mathbf{B}$ are both two-dimensional fields, then the governing equation for the vector potential $\mathbf{A} = A(x,y)\hat{\mathbf{e}}_z$ turns out to be the simple advection–diffusion equation

$$\frac{DA}{Dt} = \lambda\nabla^2 A,$$

just like that for heat (see Exercise 20.3). Thus, for example, the magnetic field lines surrounding a long current-carrying wire in a cross flow have the same form as the isotherms surrounding a heated wire in a wind, as discussed in Exercise 20.4 and as illustrated in Figure 20.10.

Moreover, our study of vorticity tells us that, for three-dimensional problems, the first term on the right of (20.53) will act to intensify $\mathbf{B}$ whenever magnetic flux tubes are stretched, or diminish $\mathbf{B}$ if they are compressed. This amplification of magnetic fields by flux-tube stretching lies behind dynamo action in planetary cores and stars.

There are, however, differences between magnetic fields and vorticity. For example, in the absence of body forces, vorticity enters a fluid via the boundaries, just like heat, whereas

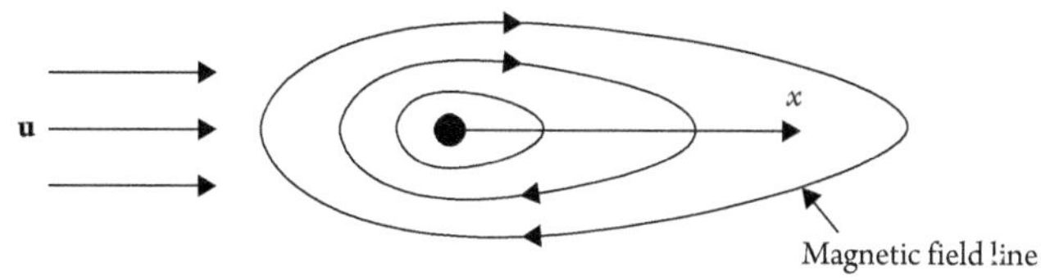


**Figure 20.10** The magnetic field of a wire carrying a current that sits in a cross flow. (From P. A. Davidson. *An Introduction to Magnetohydrodynamics*, 2nd ed. Cambridge University Press, 2017.)

in MHD one is typically interested in how a plasma or liquid metal interacts with a pre-existing magnetic field. Another difference is that **u** is functionally related to **ω** via the Biot–Savart law, whereas there is no equivalent kinematic relationship between **u** and **B**. There are therefore many more self-consistent solutions of the induction equation than the vorticity equation.

Despite these differences, the analogy is a powerful one. For example, theories of MHD turbulence in, say, the solar wind are in large part based on our understanding of conventional hydrodynamic turbulence. That is, we know a lot about how vorticity fields behave in chaotic velocity fields, and the analogy between the evolution of **ω** and **B** allows us to transfer some of that knowledge to MHD turbulence. However, the analogy is probably at its most powerful when considering ideal fluids, i.e. fluids in which $\nu = \lambda = 0$. Here the induction equation and the vorticity equation for strictly hydrodynamic flow are

$$\frac{D\mathbf{B}}{Dt} = (\mathbf{B} \cdot \nabla)\mathbf{u} \quad \text{and} \quad \frac{D\boldsymbol{\omega}}{Dt} = (\boldsymbol{\omega} \cdot \nabla)\mathbf{u}, \tag{20.56}$$

both of which take the same form as the governing equation for a short material line, such as a dye line:

$$\frac{D}{Dt}(d\mathbf{l}) = (d\mathbf{l} \cdot \nabla)\mathbf{u}. \tag{20.57}$$

It follows that, in a perfect conductor, the **B**-lines are frozen into the fluid, just like the vortex lines in an inviscid fluid (free of body forces). We might anticipate, therefore, that there are MHD analogues of Helmholtz's laws and of the conservation of kinetic helicity, and indeed there are. The analogue of Helmholtz's laws is *Alfvén's theorem* and the counterpart of the conservation of kinetic helicity is the *conservation of magnetic helicity*.

### *20.4.2 Perfect Conductors 1: Alfvén's Theorem Revisited*

We met Alfvén's theorem in §10.5.2. It states that, in a perfectly conducting medium,

(i) the magnetic field lines are frozen into the fluid, in the sense that they move with the fluid;

(ii) the magnetic flux linking a material curve moving with the medium is conserved.

These are the analogues of Helmholtz's two laws for an inviscid fluid. The second of these, which is equivalent to Kelvin's theorem in conventional hydrodynamics, follows from Faraday's flux rule plus Ohm's law in the form

$$\frac{d}{dt}\int_S \mathbf{B}\cdot d\mathbf{S} = -\oint_{C_m} (\mathbf{E}+\mathbf{u}\times\mathbf{B})\cdot d\mathbf{r} = -\frac{1}{\sigma}\oint_{C_m} \mathbf{J}\cdot d\mathbf{r}, \tag{20.58}$$

where $C_m$ is a closed *material curve*. The point is that, as $\sigma \to \infty$, the right-hand side of (20.58) must go to zero because Ampère's law demands that **J** remains finite. So we have

$$\frac{d}{dt}\int_S \mathbf{B}\cdot d\mathbf{S} = 0 \tag{20.59}$$

in a perfect conductor, as required by the second part of Alfvén's theorem.

The first part of the theorem then follows either from a comparison of (20.56) and (20.57), or from part (ii) using the following argument. Consider the isolated magnetic flux tube shown in Figure 20.11. Let $\Phi_0$ be the magnetic flux along the tube and let $C_m$ be a material curve that, at some initial instant, encircles the flux tube, sitting on its surface. If the medium is a perfect conductor, then (20.59) tells us that the flux enclosed by $C_m$ will remain constant as the medium moves and deforms, and so the flux through $C_m$ stays as $\Phi_0$. Moreover, this is true of all material curves that encircle the flux tube at $t = 0$. This suggests that every material curve that initially encircles the flux tube continues to do so for all time, which is possible only if the tube itself moves with the medium. Finally, if we let the flux tube have a vanishingly small cross-section, we conclude that every magnetic field line in a perfectly conducting fluid moves with that medium.

The fact that the **B**-lines are frozen into a fluid explains why stretching a flux tube intensifies the magnetic field. Consider a short portion of the flux tube shown in Figure

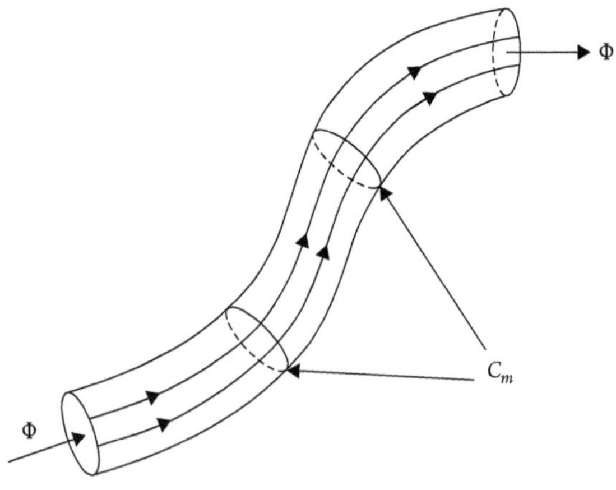


**Figure 20.11** A magnetic flux tube.

20.11. Suppose it has volume $\delta V = A\delta\ell$, where $A$ is the local cross-sectional area of the tube and $\delta\ell$ the length of the volume element. Since the fluid is incompressible, the material volume $\delta V$ is conserved, as is $\Phi_0 = BA$. It follows that $\Phi_0/\delta V = B/\delta\ell$ is an invariant of the volume element. In summary, then, if the element is stretched, its cross-sectional area falls to conserve volume and the magnetic field rises to conserve flux. Moreover, since $B/\delta\ell$ is conserved, a doubling of $\delta\ell$ will give rise to a doubling of $B$.

### 20.4.3 *Perfect Conductors 2: The Conservation of Magnetic Helicity*

We introduced *magnetic helicity* in §§7.1 and 8.8. It is defined as

$$h_m = \int_V \mathbf{A} \cdot \mathbf{B}\, dV, \tag{20.60}$$

where the volume $V$ is chosen such that $\mathbf{B} \cdot d\mathbf{S} = 0$ at all points on its surface. It is the MHD analogue of kinetic helicity and is often used as a topological measure of the degree of entanglement of magnetic flux tubes. To understand its significance, consider a magnetic field composed of two thin magnetic flux tubes of volumes $V_1$ and $V_2$, as shown in Figure 20.12. Let the tubes have centrelines $C_1$ and $C_2$, and magnetic fluxes $\Phi_1$ and $\Phi_2$. If $\delta V$ is a short portion of one of the flux tubes, of length $\delta\mathbf{l}$, then $\mathbf{B}\,\delta V = \Phi\,\delta\mathbf{l}$, and so the net magnetic helicity is

$$h_m = \int_{V_1} \mathbf{A} \cdot \mathbf{B}\, dV + \int_{V_2} \mathbf{A} \cdot \mathbf{B}\, dV = \Phi_1 \oint_{C_1} \mathbf{A} \cdot d\mathbf{l} + \Phi_2 \oint_{C_2} \mathbf{A} \cdot d\mathbf{l}.$$

However, provided the flux tubes are linked, Stokes' theorem yields

$$\oint_{C_1} \mathbf{A} \cdot d\mathbf{l} = \pm\Phi_2, \quad \oint_{C_2} \mathbf{A} \cdot d\mathbf{l} = \pm\Phi_1,$$

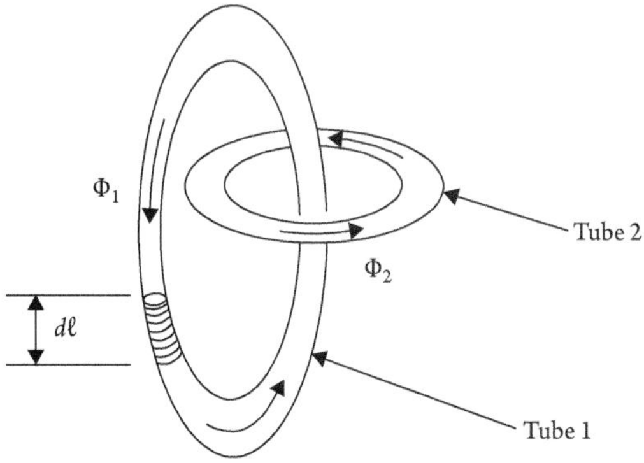


**Figure 20.12** Two interlinked magnetic flux tubes. (From P. A. Davidson. *An Introduction to Magnetohydrodynamics*, 2nd ed. Cambridge University Press, 2017.)

where the positive sign must be used for a right-handed linkage and the negative sign for a left-handed one. For this simple geometry, then, we have $h_m = \pm 2\Phi_1\Phi_2$, which might be compared with (20.37) for kinetic helicity. On the other hand, if the flux tubes are wound around each other $n$ times, then $h_m = \pm 2n\Phi_1\Phi_2$, with $h_m = 0$ for unlinked tubes.

It is clear from Alfvén's theorem that $h_m = \pm 2\Phi_1\Phi_2$ is an ideal invariant of this simple geometry, since $\Phi_1$ and $\Phi_2$ are conserved, as well as the linkage of the tubes. (Some diffusion is required to change the linkage.) The conservation of $h_m$ in more complicated geometries may be established as follows. We start by uncurling the ideal induction equation

$$\frac{\partial \mathbf{B}}{\partial t} = \nabla \times (\mathbf{u} \times \mathbf{B}), \tag{20.61}$$

which yields

$$\frac{\partial \mathbf{A}}{\partial t} = \mathbf{u} \times \mathbf{B} + \nabla f, \tag{20.62}$$

for some scalar $f$. These expressions may be combined to give

$$\frac{\partial}{\partial t}(\mathbf{A} \cdot \mathbf{B}) = \nabla \cdot (f\mathbf{B}) + \mathbf{A} \cdot [\nabla \times (\mathbf{u} \times \mathbf{B})],$$

which, with the help of the identity

$$\mathbf{A} \cdot \nabla \times (\mathbf{u} \times \mathbf{B}) = \nabla \cdot [\mathbf{A} \times (\mathbf{B} \times \mathbf{u})] = \nabla \cdot [(\mathbf{A} \cdot \mathbf{u})\mathbf{B} - (\mathbf{A} \cdot \mathbf{B})\mathbf{u}],$$

and $\nabla \cdot \mathbf{u} = 0$, becomes

$$\frac{D}{Dt}(\mathbf{A} \cdot \mathbf{B}) = \nabla \cdot [(f + \mathbf{A} \cdot \mathbf{u})\mathbf{B}]. \tag{20.63}$$

Finally, we integrate (20.63) over a material volume $V$ for which $\mathbf{B} \cdot d\mathbf{S} = 0$ on the bounding surface. If $V$ consists of the *material* volume elements $\delta V$, each of which satisfies $D(\delta V)Dt = 0$, we obtain

$$\frac{d}{dt}\int_V (\mathbf{A} \cdot \mathbf{B})\, dV = 0. \tag{20.64}$$

This establishes conservation of $h_m$ in a perfect conductor for fields of arbitrary complexity.

### *20.4.4 Azimuthal Field Generation by Differential Rotation*

We shall now allow for a small but finite magnetic diffusivity, so that we must consider the combined effects of advection and diffusion. We start with axisymmetric flows. An important phenomenon in planetary and stellar interiors is the effect of differential rotation on the dipolar magnetic field. For example, the sun exhibits a variation of surface rotation rate with latitude, and this spirals out an east–west field from the dipolar field. It turns out that this kind of differential rotation has a profound influence on the structure of planetary and stellar magnetic fields, as we shall see in Chapter 21.

Consider a rotating fluid that is confined to a sphere, say the core of a planet or a star. Suppose the fluid rotates non-uniformly, as $\mathbf{u} = (0, \Omega(s,z)s, 0)$ in polar coordinates, and possesses an axisymmetric poloidal magnetic field $\mathbf{B}_p(s,z) = (B_s, 0, B_z)$. To focus our thoughts, we take the fluid to rotate faster at the equator than at its poles. Then it is clear from Alfvén's theorem that the poloidal field lines will be advected as indicated in Figure 20.13, with the north–south field lines spiralled out as shown. This is called the $\Omega$-effect.

This process may be quantified using the azimuthal component of the induction equation,

$$\frac{\partial \mathbf{B}_\theta}{\partial t} = \nabla \times (\mathbf{u} \times \mathbf{B}_p) + \lambda \nabla^2 \mathbf{B}_\theta. \tag{20.65}$$

It is readily confirmed that

$$\nabla \times (\mathbf{u} \times \mathbf{B}_p) = [s(\mathbf{B}_p \cdot \nabla)\Omega]\hat{\mathbf{e}}_\theta,$$

which makes explicit the role played by non-uniform rotation in generating the azimuthal field. In the steady state we have a balance in (20.65) between differential rotation and diffusion, $\lambda \nabla^2 \mathbf{B}_\theta = -s(\mathbf{B}_p \cdot \nabla\Omega)\hat{\mathbf{e}}_\theta$, and if $\lambda$ is small, then a large azimuthal magnetic field may be generated by this mechanism, of the order of $B_\theta \sim R_m B_p$.

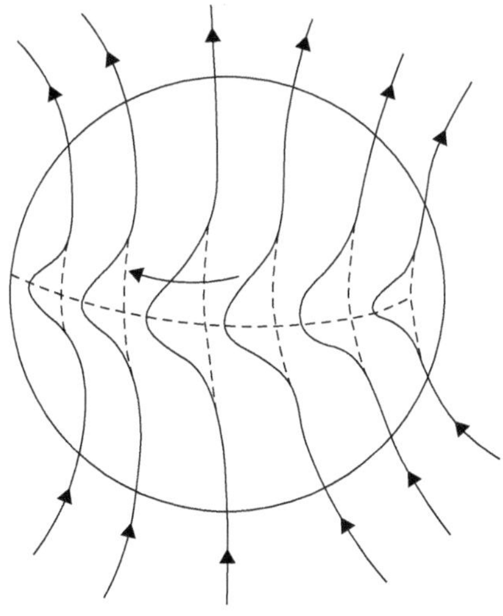

**Figure 20.13** The spiralling out of dipolar magnetic field lines by differential rotation. (From P. A. Davidson. *An Introduction to Magnetohydrodynamics*, 2nd ed. Cambridge University Press, 2017.)

### 20.4.5 *The Phenomenon of Flux Expulsion*

A second example of the combined effects of advection and diffusion is a phenomenon known as *flux expulsion*. Consider a steady two-dimensional flow consisting of a region of closed streamlines. Provided that $R_m = u\ell/\lambda$ is large, $\ell$ being a typical lengthscale, then any planar magnetic field that spans the flow is gradually eradicated, as shown in Figure 20.14. That is, magnetic flux is expelled from the region of closed streamlines.

A proof of the statement above proceeds as follows. We start by writing $\mathbf{B}$ in terms of the vector potential,

$$\mathbf{B} = \nabla \times (A\hat{\mathbf{e}}_z) = \left(\frac{\partial A}{\partial y}, -\frac{\partial A}{\partial x}, 0\right), \tag{20.66}$$

where $A$ is the *flux function* for the planar field $\mathbf{B}$. Similarly, we write

$$\mathbf{u} = \nabla \times (\psi\hat{\mathbf{e}}_z) = \left(\frac{\partial \psi}{\partial y}, -\frac{\partial \psi}{\partial x}, 0\right),$$

where $\psi$ is the analogous *streamfunction* for $\mathbf{u}$. These definitions ensure that both $\mathbf{u}$ and $\mathbf{B}$ are solenoidal, with each $\mathbf{B}$-line identified by a particular value of $A$ and each streamline by a given value of $\psi$ (see Exercise 20.2). It is readily confirmed that the induction equation now uncurls to give

$$\frac{\partial \mathbf{A}}{\partial t} = \mathbf{u} \times \mathbf{B} + \lambda \nabla^2 \mathbf{A},$$

from which we find (see Exercise 20.3)

$$\frac{DA}{Dt} = \lambda \nabla^2 A. \tag{20.67}$$

Thus $A$ satisfies an advection–diffusion equation, just like heat in a two-dimensional flow. When $R_m$ is large, and diffusion weak, we have $DA/Dt \approx 0$ and hence $A$ is *almost* materially conserved and so more or less constant along the streamlines, although its value may vary from streamline to streamline. Moreover, we shall now show that a small but finite amount of diffusion slowly eradicates cross-stream gradients in $A$ until such time as $A$ is perfectly uniform, resulting in $\mathbf{B} = 0$.

A proof of this last statement proceeds as follows. For $\lambda \to 0$, (20.67) becomes $DA/Dt = 0$, which simplifies to $\mathbf{u} \cdot \nabla A = 0$ for a steady magnetic field. This, in turn, requires that $A$ be constant along the streamlines, i.e. $A = A(\psi)$. Now suppose that $\lambda$ is not exactly zero, but rather small but finite. Then we still have $A \approx A(\psi)$, while the steady version of (20.67),

$$\nabla \cdot (A\mathbf{u}) = \lambda \nabla^2 A,$$

integrates over any two-dimensional volume $V_\psi$ which is bounded by a closed streamline $C_\psi$, to give

$$\lambda \int_{V_\psi} \nabla^2 A \, dV = 0. \tag{20.68}$$

**Figure 20.14** An example of flux expulsion at $R_m = 100$, based on computations by N. O. Weiss. The panels show the gradual expulsion of an initially uniform field by a clockwise vortex. (From D. J. Galloway & N. O. Weiss. Convection and magnetic fields in stars. *Astrophys. J.* **243**, 945–953 (1981), reproduced by permission of AAS.)

(The divergence on the left integrates to zero because of Gauss' theorem and because $\mathbf{u} \cdot d\mathbf{S} = 0$ on $C_\psi$.) Crucially, (20.68) holds for any finite value of $\lambda$, and in particular when $\lambda$ is infinitesimally small, so that $A \approx A(\psi)$. So let us now combine the integral constraint (20.68) with the high-$R_m$ approximation $A \approx A(\psi)$. Gauss' theorem gives us

$$\int_{V_\psi} \nabla^2 A \, dV = \int_{S_\psi} \nabla A \cdot d\mathbf{S} = A'(\psi) \int_{S_\psi} \nabla\psi \cdot d\mathbf{S} = A'(\psi) \int_{V_\psi} \nabla^2 \psi \, dV = -A'(\psi) \int_{V_\psi} \omega \, dV,$$

where $\omega = -\nabla^2 \psi$ is the vorticity and $S_\psi$ is the surface of $V_\psi$. So, in the high-$R_m$ regime, (20.68) becomes

$$\lambda A'(\psi) \oint_{C_\psi} \mathbf{u} \cdot d\mathbf{l} = 0. \tag{20.69}$$

We conclude that $A'(\psi) = 0$, since the line integral cannot be zero and $\lambda$ is finite, if small. Evidently the cross-stream gradient of $A$ is zero, so that $A$ must take the same value on all closed streamlines.

In summary, then, in a region of closed streamlines, and at high values of $R_m$, the flux function $A$ gradually diffuses across the streamlines until it is perfectly uniform, at which point we have $\mathbf{B} = 0$. It is this slow but inevitable cross-stream diffusion of $A$ that lies behind the expulsion of flux shown in Figure 20.14.

If we compare (20.67) with the vorticity equation for two-dimensional hydrodynamic flow, (20.25), or the advection–diffusion equation for heat, (20.26), i.e.

$$\frac{D\omega}{Dt} = \nu \nabla^2 \omega \quad \text{and} \quad \frac{DT}{Dt} = \alpha \nabla^2 T,$$

then we might anticipate that this phenomenon has its counterparts in both vortex dynamics and heat transfer, and indeed it does. In a two-dimensional hydrodynamic flow at high Re and low thermal diffusivity, the vorticity and temperature are perfectly uniform in regions of closed streamlines.

## 20.5 The Governing Equations of MHD

### 20.5.1 *Governing Equations and Dimensionless Groups*

The governing equations of incompressible MHD consist of the reduced form of Maxwell's equations,

$$\boxed{\nabla \times \mathbf{B} = \mu_0 \mathbf{J}, \qquad \nabla \times \mathbf{E} = -\frac{\partial \mathbf{B}}{\partial t},} \tag{20.70}$$

along with Ohm's law, $\mathbf{J} = \sigma(\mathbf{E} + \mathbf{u} \times \mathbf{B})$, and charge conservation, $\nabla \cdot \mathbf{J} = 0$. These combine to give the induction equation,

$$\boxed{\frac{D\mathbf{B}}{Dt} = (\mathbf{B} \cdot \nabla)\mathbf{u} + \lambda \nabla^2 \mathbf{B}.} \tag{20.71}$$

To this we must add the continuity equation, $\nabla \cdot \mathbf{u} = 0$, and the Navier–Stokes equation incorporating the Lorentz force,

$$\boxed{\frac{D\mathbf{u}}{Dt} = -\nabla\left(\frac{p}{\rho}\right) + \nu \nabla^2 \mathbf{u} + \rho^{-1}\mathbf{J} \times \mathbf{B}.} \tag{20.72}$$

The corresponding vorticity equation is

$$\boxed{\frac{D\boldsymbol{\omega}}{Dt} = (\boldsymbol{\omega} \cdot \nabla)\mathbf{u} + \nu \nabla^2 \boldsymbol{\omega} + \rho^{-1}\nabla \times (\mathbf{J} \times \mathbf{B}).} \tag{20.73}$$

Note that, since both $\rho$ and $\mu_0$ are constants in an incompressible flow, some authors find it convenient to work with the scaled magnetic field $\mathbf{h} = \mathbf{B}/\sqrt{\rho\mu_0}$, which has the dimensions of a *velocity*. The Lorentz force per unit mass is then $\mathbf{F} = (\nabla \times \mathbf{h}) \times \mathbf{h}$.

These equations are nonlinear, partly because the Navier–Stokes equation is nonlinear and partly because of the two-way coupling between $\mathbf{u}$ and $\mathbf{B}$. When dealing with nonlinear systems it is often helpful to identify the relevant dimensionless groups, since these can provide some guidance as to the dominant processes and balances in different regimes. We have already met two dimensionless groups, $\mathrm{Re} = u\ell/\nu$ and $R_m = u\ell/\lambda$, both of which are indicative of the relative magnitudes of advection and diffusion, in one case the diffusion of vorticity and in the other case the diffusion of a magnetic field. It is natural to ask if there are other dimensionless groups in MHD and, if so, what these represent.

Let us start by determining how many independent dimensionless groups there are. Consider the Navier–Stokes equation with the Lorentz force per unit mass written as $(\nabla \times \mathbf{h}) \times \mathbf{h}$. This introduces the physical variables $|\mathbf{h}|$, $u$, $\ell$ and $\nu$, where $|\mathbf{h}|$ is a typical scaled magnetic field and $u$ a characteristic velocity. If we now include the induction equation, we must add $\lambda$ to this list, giving us a total of five dimensional parameters. Since all diffusivities have the dimensions of $\mathrm{m}^2/\mathrm{s}$, these five parameters contain between them only two dimensions: those of length and time. The Buckingham Pi theorem now tells us that the number of independent dimensionless groups is five (parameters) minus two (dimensions), i.e. three groups. Clearly two of these are $R_m = u\ell/\lambda$ and $\mathrm{Re} = u\ell/\nu$, and, since $\mathbf{h}$ has the dimensions of a velocity, the third group can be taken as $\mathbf{h}\ell$ divided by a diffusivity. By convention, this third group is usually taken as $|\mathbf{h}|\,\ell/\sqrt{\nu\lambda}$, which is known as the *Hartmann number*. In terms of $\mathbf{B}$, this is $\mathrm{Ha} = B\ell\,(\sigma/\rho\nu)^{1/2}$. From these three independent dimensionless groups we can construct others. For example, the *magnetic Prandtl number* is $\nu/\lambda$, while the *Lundquist number* is $|\mathbf{h}|\,\ell/\lambda$, or $B\ell\sigma\sqrt{\mu_0/\rho}$.

When there is a strong background rotation $\boldsymbol{\Omega}$, it is often useful to move into a rotating frame of reference. The Coriolis force $2\mathbf{u} \times \boldsymbol{\Omega}$ then appears on the right of (20.72) and the introduction of $\boldsymbol{\Omega}$ means that there is one more dimensionless group. This is usually taken to be either the *Rossby number* $\mathrm{Ro} = u/\Omega\ell$, which is indicative of the ratio of inertial to Coriolis forces, or the *Ekman number* $\mathrm{Ek} = \nu/\Omega\ell^2$, which measures the ratio of viscous to Coriolis forces. Of course, $u$ is measured in the rotating frame.

### 20.5.2 *Energy Relationships*

We shall now show that the Lorentz force allows energy to be exchanged between the magnetic field and the fluid. We start by returning to §11.2.2, where we showed that the dot product of $\mathbf{B}$ with Faraday's law, combined with Ampère's law, yields

$$\frac{\partial}{\partial t}\left(\frac{\mathbf{B}^2}{2\mu_0}\right) = -\mathbf{J}\cdot\mathbf{E} - \nabla\cdot\left(\frac{\mathbf{E}\times\mathbf{B}}{\mu_0}\right). \tag{20.74}$$

On substituting for $\mathbf{E}$ using Ohm's law, and rearranging the triple product $\mathbf{J}\cdot(\mathbf{u}\times\mathbf{B})$, we obtain

$$\frac{\partial}{\partial t}\left(\frac{\mathbf{B}^2}{2\mu_0}\right) = -\frac{\mathbf{J}^2}{\sigma} - \nabla\cdot\left(\frac{\mathbf{E}\times\mathbf{B}}{\mu_0}\right) - (\mathbf{J}\times\mathbf{B})\cdot\mathbf{u}. \tag{20.75}$$

The left-hand side is the rate of change of electromagnetic energy, the electric energy being negligible, while the terms on the right represent Joule dissipation $\mathbf{J}^2/\sigma$, the redistribution of electromagnetic energy via the Poynting flux $\mathbf{E}\times\mathbf{B}/\mu_0$ and the rate of working of the Lorentz force $(\mathbf{J}\times\mathbf{B})\cdot\mathbf{u}$. When integrated over all space, the Poynting flux vanishes (there is no radiation field in MHD) and we obtain

$$\boxed{\frac{d}{dt}\int_{V_\infty}\frac{\mathbf{B}^2}{2\mu_0}\,dV = -\int_{V_\infty}(\mathbf{J}\times\mathbf{B})\cdot\mathbf{u}\,dV - \int_{V_\infty}\frac{\mathbf{J}^2}{\sigma}\,dV.} \tag{20.76}$$

Thus the magnetic energy can decline either through Joule dissipation or because the Lorentz force does work on the fluid.

Let us now turn to the Navier–Stokes equation. Taking the dot product of $\mathbf{u}$ with (20.72) yields

$$\frac{\partial}{\partial t}(\rho\mathbf{u}^2/2) = -\nabla\cdot[(\rho\mathbf{u}^2/2)\mathbf{u}] - \nabla\cdot(p\mathbf{u}) + (\mathbf{J}\times\mathbf{B})\cdot\mathbf{u} + \rho\nu\mathbf{u}\cdot\nabla^2\mathbf{u}. \tag{20.77}$$

We now integrate over all space, assume the motion is localised in space, and use (20.17) to rewrite the rate of working of the viscous forces in terms of viscous dissipation. This gives

$$\boxed{\frac{d}{dt}\int_{V_\infty}(\rho\mathbf{u}^2/2)\,dV = \int_{V_\infty}(\mathbf{J}\times\mathbf{B})\cdot\mathbf{u}\,dV - 2\rho\nu\int_{V_\infty}S_{ij}S_{ij}dV.} \tag{20.78}$$

Crucially, the rate of working of the Lorentz force appears with opposite signs in (20.76) and (20.78), confirming that it provides a mechanism for exchanging energy between the magnetic field and the fluid. When $(\mathbf{J} \times \mathbf{B}) \cdot \mathbf{u} > 0$ the magnetic field transfers energy to the fluid, as in an electric motor, while $(\mathbf{J} \times \mathbf{B}) \cdot \mathbf{u} < 0$ means that the fluid motion acts to increase the magnetic energy through flux-tube stretching, as in a dynamo.

### 20.5.3 *The Lorentz Force Revisited: Maxwell's Stresses Versus Faraday's Tension*

In MHD it frequently pays to rewrite the Lorentz force in terms of Maxwell stresses, or else Faraday's tension, so perhaps it is worth revisiting §8.6 at this point, reminding ourselves of how this works. The starting point is the Lorentz force in the form

$$\mathbf{J} \times \mathbf{B} = (\mathbf{B} \cdot \nabla)\left(\frac{\mathbf{B}}{\mu_0}\right) - \nabla\left(\frac{\mathbf{B}^2}{2\mu_0}\right), \tag{20.79}$$

where $\mathbf{B}^2/2\mu_0$ is the *magnetic pressure*, denoted by $p_m$. Provided there are no free surfaces, the role of $p_m$ is simply to change the pressure distribution in the fluid, with $p_m$ absent from the vorticity equation. The Lorentz force (20.79) is mechanically equivalent to the imaginary set of stresses

$$\tau_{ij}^M = \frac{B_i B_j}{\mu_0} - \frac{\mathbf{B}^2}{2\mu_0}\delta_{ij}, \tag{20.80}$$

in the sense that we can write

$$(\mathbf{J} \times \mathbf{B})_i = \frac{\partial \tau_{ij}^M}{\partial x_j}.$$

These are Maxwell's stresses. The utility of writing the Lorentz force in terms of $\tau_{ij}^M$ can be seen by integrating (20.79) over a control volume $V$. Invoking $\nabla \cdot \mathbf{B} = 0$ and Gauss' theorem, we find that the integrated effect of the Lorentz force can be represented by surface integrals alone,

$$\int_V \mathbf{J} \times \mathbf{B}\, dV = \oint \frac{\mathbf{B}}{\mu_0} \mathbf{B} \cdot d\mathbf{S} - \oint \frac{\mathbf{B}^2}{2\mu_0} d\mathbf{S}, \tag{20.81}$$

and these surface integrals represent the cumulative effect of the distributed stress system $\tau_{ij}^M$ acting over the surface of $V$.

We now turn to the representation of $\mathbf{J} \times \mathbf{B}$, or more precisely $(\mathbf{B} \cdot \nabla)\mathbf{B}/\mu_0$, in terms of Faraday's tension. The key is to write $(\mathbf{B} \cdot \nabla)\mathbf{B}/\mu_0$ in terms of intrinsic coordinates. By analogy with the derivation of (20.5), we have

$$(\mathbf{B} \cdot \nabla)\left(\frac{\mathbf{B}}{\mu_0}\right) = \frac{B}{\mu_0}\frac{dB}{ds}\hat{\mathbf{e}}_t - \frac{B^2}{\mu_0 R}\hat{\mathbf{e}}_n, \tag{20.82}$$

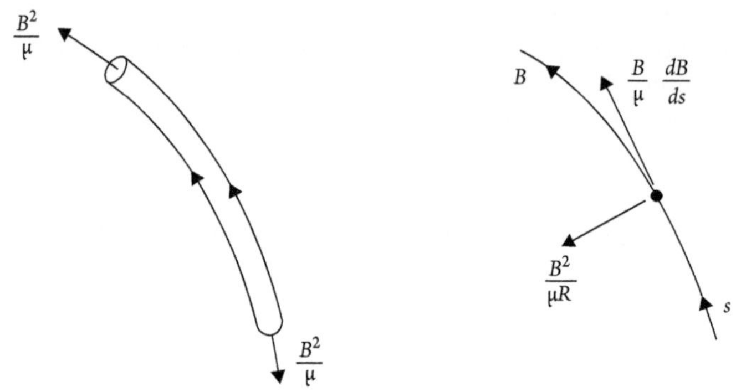


**Figure 20.15** The components of $(\mathbf{B}\cdot\nabla)\mathbf{B}/\mu_0$ acting on a flux tube can be interpreted as a consequence of Faraday's tension $B^2/\mu_0$ acting along the axis of that tube.

where $\hat{\mathbf{e}}_t$ is a unit vector tangential to the field line and $\hat{\mathbf{e}}_n$ is the unit normal vector. This immediately tells us that there is a component of the Lorentz force transverse to the magnetic field lines as a result of the curvature of those lines.

To understand the physical origin of (20.82), consider a thin isolated flux tube of cross-sectional area $A$ and carrying a magnetic flux $\Phi = B(s)A(s)$, as shown in Figure 20.15. The force $(\mathbf{B}\cdot\nabla)\mathbf{B}/\mu_0$ can be represented by the Maxwell stresses $B_iB_j/\mu_0$, and for our isolated flux tube this reduces to a tensile stress $B^2/\mu_0$ acting along the axis of the tube. This is Faraday's tension and it leads to a tensile force in the flux tube of $T(s) = (B^2/\mu_0)A$. Moreover, we know that a string with a tensile force $T$ experiences a force per unit length of $\mathbf{F} = (dT/ds)\hat{\mathbf{e}}_t - (T/R)\hat{\mathbf{e}}_n$, where $R$ is the local radius of curvature of the string. If we substitute $T = \Phi B/\mu_0$ into this expression, we conclude that our flux tube experiences a force

$$\mathbf{F} = \frac{1}{\mu_0}\frac{d(\Phi B)}{ds}\hat{\mathbf{e}}_t - \frac{\Phi B}{\mu_0 R}\hat{\mathbf{e}}_n = \frac{\Phi}{\mu_0}\frac{dB}{ds}\hat{\mathbf{e}}_t - \frac{B^2 A}{\mu_0 R}\hat{\mathbf{e}}_n, \tag{20.83}$$

and dividing through by $A$ yields (20.82). In short, the normal and tangential components of $\mu_0^{-1}(\mathbf{B}\cdot\nabla)\mathbf{B}$ in (20.82) can be thought of as arising from Faraday's tension $B^2/\mu_0$ acting along the axis of the flux tube, as shown in Figure 20.15.

Faraday's tension frequently provides a powerful means for picturing the effects of the Lorentz force. We may think of magnetic field lines as being in tension, carrying a tensile stress $B^2/\mu_0$, and exerting a force on the conductor, like stretched elastic bands.

## 20.6 Some Simple Dynamics: Alfvén Waves and Conservation of Cross Helicity

We shall now consider some simple applications of the governing equations, starting with Alfvén waves. We touched on Alfvén waves in §8.6. The underlying mechanism of these waves comes in two parts. On the one hand, Alfvén's theorem tells us that a highly conducting fluid tends to stick to the magnetic field lines, and so we may associate a certain amount of inertia with the **B**-lines. On the other hand, the field lines behave like an elastic band in tension, and if they are perturbed, then they push back on the fluid. Consider, for example, the situation shown in Figure 20.16, where the fluid is given an impulsive motion to the right. The initially straight field lines are pushed out of shape by the transverse movement of the medium, and the resulting tension in the field lines reacts back on the fluid, slowing it down. Eventually the fluid must come to rest, but the field lines are still bent, exerting a restoring force on the fluid. The fluid now reverses direction, carrying the field lines back towards the equilibrium position. However, as the field lines approach their initial position, the inertia of the fluid carries them past the equilibrium point and the lines start to bow out in the opposite direction. Clearly, oscillations will result.

Thus, just as the sideways force on a plucked string, $\mathbf{F}_n = -(T/R)\hat{\mathbf{e}}_n$, leads to waves with a wave speed $c = \sqrt{T/\rho_s}$ (where $\rho_s$ is the mass per unit length of the string), so the Faraday tension $B^2/\mu_0$ in a magnetic field line leads to oscillations of the conducting medium, with a wave speed $c = \sqrt{(B^2/\mu_0)/\rho}$. Alfvén waves play an important role in astrophysics, as we shall see in Chapter 21.

Let us now turn to so-called ideal MHD, which is an idealised system in which both $\nu$ and $\lambda$ are put equal to zero. There is an integral invariant of ideal MHD, over and above magnetic helicity, which is called *cross helicity*. This is defined as

$$h_c = \int \mathbf{u} \cdot \mathbf{B}\, dV, \tag{20.84}$$

where **B** is assumed to be localised in space and the integral encloses the magnetic field.

To prove that (20.84) is an ideal invariant, we must combine the ideal induction equation,

$$\frac{D\mathbf{B}}{Dt} = \mathbf{B} \cdot \nabla \mathbf{u}, \tag{20.85}$$

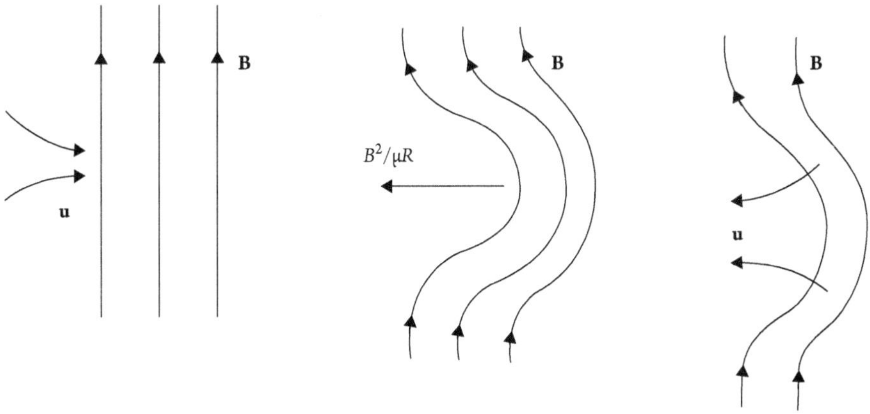


**Figure 20.16** In a highly conducting fluid the magnetic field lines behave like elastic bands frozen into the fluid. This is the origin of Alfvén waves.

with Euler's equation,

$$\rho\frac{D\mathbf{u}}{Dt} = -\nabla p + \mathbf{J}\times\mathbf{B}. \tag{20.86}$$

Between them, these yield

$$\frac{D}{Dt}(\mathbf{u}\cdot\mathbf{B}) = \nabla\cdot\left[\left(\frac{u^2}{2} - \frac{p}{\rho}\right)\mathbf{B}\right]. \tag{20.87}$$

We may now integrate (20.87) over any material volume that encloses $\mathbf{B}$ to show that $h_c$ is conserved, just as integrating (20.63) gives conservation of magnetic helicity.

The geometrical interpretation of the conservation of cross helicity can be clarified by considering a thin isolated flux tube that links with a thin isolated vortex tube, rather like the tubes shown in Figure 20.12. Using $\mathbf{B}\,\delta V = \Phi_B\,\delta\mathbf{l}$ to represent part of the flux tube, we have

$$h_c = \int_{V_B} \mathbf{u}\cdot\mathbf{B}\,dV = \Phi_B \oint_{C_B} \mathbf{u}\cdot d\mathbf{l}, \tag{20.88}$$

where $V_B$, $C_B$ and $\Phi_B$ are the volume, centreline and flux of the magnetic flux tube. However, provided the vortex tube and magnetic flux tube are linked, Stokes' theorem yields

$$\oint_{C_B} \mathbf{u}\cdot d\mathbf{l} = \pm\Phi_\omega,$$

where $\Phi_\omega$ is the flux of vorticity in the vortex tube and the positive (negative) sign must be used for a right-handed (left-handed) linkage. For this simple geometry, then, we have $h_c = \pm\Phi_B\Phi_\omega$, provided the two tubes are linked. Crucially, this linkage is conserved in an ideal fluid, as are $\Phi_B$ and $\Phi_\omega$, which explains the conservation of $h_c$. Cross helicity has recently received much attention in the context of astrophysical turbulence, as discussed in, for example, Davidson (2017).

* * *

This concludes our brief introduction to the governing equations of MHD. Readers who have not had occasion to study fluid mechanics in any depth will probably require a more gentle introduction to that offered in §20.2. The excellent textbook by Acheson (1990) is recommended. Regarding MHD, introductory texts accessible to undergraduates are rather thin on the ground, although Shercliff (1965) and Davidson (2017) may be worth a look.

## EXERCISES

20.1 Show that the charge relaxation time $\tau_e$ is related to the magnetic Reynolds number $R_m = u\ell/\lambda$ by $\varpi\tau_e = (u/c)^2/R_m$, where $\varpi \sim u/\ell$ is an inverse timescale of the motion. MHD is non-relativistic, $u << c$, and so, provided $R_m > u/c$, we have $\varpi\tau_e << 1$.

20.2 Show that the two-dimensional magnetic field

$$\mathbf{B} = \nabla \times (A\hat{\mathbf{e}}_z) = \left(\frac{\partial A}{\partial y}, -\frac{\partial A}{\partial x}, 0\right)$$

satisfies $\mathbf{B} \cdot \nabla A = 0$, where $A$ is the flux function. Hence confirm that each **B**-line is associated with a constant value of $A$. Similarly show that, if $\mathbf{u} = (\partial\psi/\partial y, -\partial\psi/\partial x, 0)$, then each streamline is associated with a constant value of the streamfunction, $\psi$.

20.3 Show that the induction equation uncurls to give

$$\frac{\partial \mathbf{A}}{\partial t} = \mathbf{u} \times \mathbf{B} + \lambda\nabla^2\mathbf{A} + \nabla f,$$

for some scalar $f$. (Remember that the Coulomb gauge is used in MHD.) Now show that, for a two-dimensional magnetic field extending to infinity and sitting in a coplanar flow, we are free to take $f = 0$. Finally confirm that, with $f = 0$, the vector potential for a planar magnetic field, $\mathbf{A} = A(x,y)\hat{\mathbf{e}}_z$, satisfies the simple advection–diffusion equation

$$\frac{DA}{Dt} = \lambda\nabla^2 A.$$

20.4 A long, straight wire of negligible cross-sectional area carries a current $I$ and sits in a uniform cross flow $u$, as shown in Figure 20.10. The magnetic field lines surrounding the wire are swept downstream, governed by

$$u\frac{\partial A}{\partial x} = \lambda\nabla^2 A,$$

where $\mathbf{B} = \nabla \times (A\hat{\mathbf{e}}_z)$, $x$ is in the streamwise direction and the origin is taken as the centre of the wire. This equation is the same as that governing the temperature field of a heated wire in a cross flow. As there is no geometric lengthscale in this problem, the only way of constructing a magnetic Reynolds number is to use the radial coordinate $s$ as the characteristic lengthscale, giving $R_m = \mu\sigma us = us/\lambda$. Near the wire we have a diffusive (or quasi-static) regime, with $R_m << 1$ and $\nabla^2 A \approx 0$, while at large distances from the wire we have $R_m >> 1$ and advection dominates. Show that the advection–diffusion equation for $A$ may be solved by looking for solutions of the form $A = g(x,y)\exp(ux/2\lambda)$, and that this yields $A = aK_0(us/2\lambda)\exp(ux/2\lambda)$, where $K_0$ is the usual modified Bessel function. The constant $a$ may be determined by matching this to the static solution at $s \to 0$, i.e. $A = (\mu_0 I/2\pi)\ln(1/s)$. Show that this gives $a = \mu_0 I/2\pi$.

20.5 Consider the isolated flux tube

$$\mathbf{B} = \frac{\Phi}{\pi\delta^2}\exp(-s^2/\delta^2)\,\hat{\mathbf{e}}_z,$$

where $\delta(t)$ is the instantaneous characteristic radius of the tube, $\Phi$ is the net magnetic flux along the tube and we use polar coordinates $(s,\theta,z)$. The flux tube sits in the irrotational straining flow $\mathbf{u} = (-\alpha s/2, 0, \alpha z)$, as shown in Figure 20.17(a), where $\alpha(t)$ is a prescribed strain rate. Show that this is a solution of the induction equation provided that $\delta$ satisfies

$$\frac{d\delta^2}{dt} + \alpha(t)\delta^2 = 4\lambda.$$

Also show that, for the case where $\alpha$ is constant, the solution for $\delta$ asymptotes to the steady value of $\delta_{\text{steady}} = \sqrt{4\lambda/\alpha}$, irrespective of its initial value. Use simple physical arguments to explain this behaviour.

20.6 The Cartesian analogue of Exercise 20.5 is the planar flux sheet

$$\mathbf{B} = \frac{\Phi}{\sqrt{\pi}\,\delta}\exp(-x^2/\delta^2)\,\hat{\mathbf{e}}_y,$$

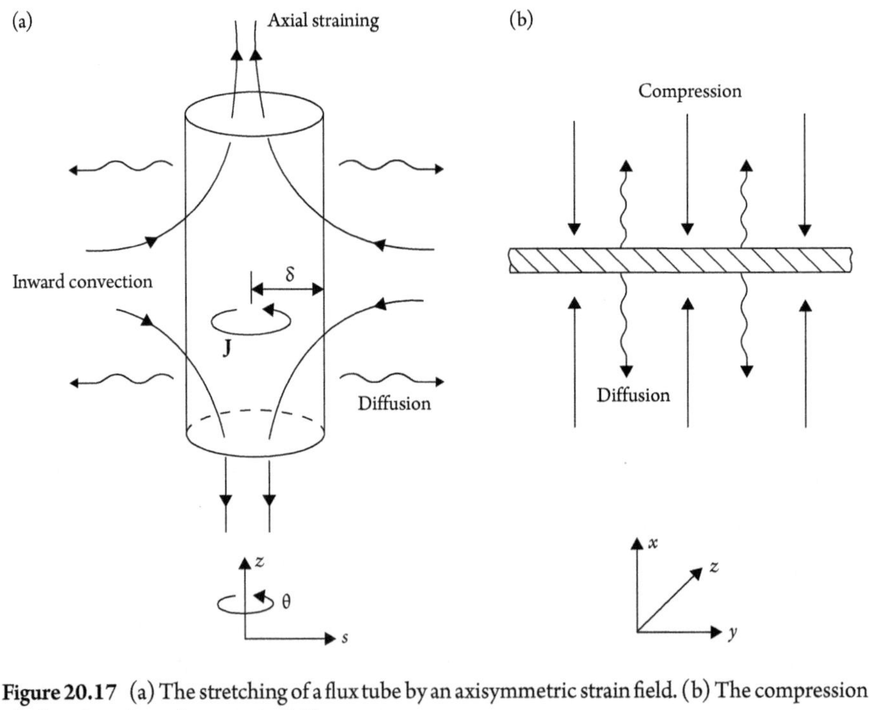


**Figure 20.17** (a) The stretching of a flux tube by an axisymmetric strain field. (b) The compression of a flux sheet by a planar strain field.

which is subject to the irrotational strain field $(u_x, u_y, 0) = (-\alpha x, \alpha y, 0)$, as shown in Figure 20.17(b). Show that this is a solution of the induction equation provided that $\delta$ satisfies

$$\frac{d\delta^2}{dt} + 2\alpha(t)\delta^2 = 4\lambda. \tag{20.89}$$

Now show that, when $\alpha$ is constant, the flux sheet always asymptotes to the equilibrium thickness $\delta = \sqrt{2\lambda/\alpha}$.

20.7 A perfectly conducting fluid is subject to an axisymmetric motion and contains an axisymmetric azimuthal magnetic field $\mathbf{B}_\theta$. Show that $B_\theta/r$ is materially conserved by each fluid element.

## REFERENCES


Acheson, D. J., 1990, *Elementary Fluid Dynamics*. Oxford University Press.

Davidson, P. A., 2015, *Turbulence: An Introduction for Scientists and Engineers*, 2nd ed. Oxford University Press.

Davidson P. A., 2017, *An Introduction to Magnetohydrodynamics*, 2nd ed. Cambridge University Press.

Frisch, U., 1995, *Turbulence*. Cambridge University Press.

Shercliff, J. A., 1965, *A Textbook of Magnetohydrodynamics*. Pergamon Press.

# 21

# Magnetohydrodynamics II

## *An Introduction to Plasma Containment, Planetary Dynamos and Stellar Magnetism*

> During Alfvén's visit he gave a lecture at the University of Chicago, which was attended by Fermi. As Alfvén described his work, Fermi nodded his head and said, 'Of course.' The next day the entire world of physics said, 'Oh, of course.'
>
> A. Dessler

Alfvén is the father of astrophysical MHD, although he had to fight many battles before his ideas gained wide acceptance. We now offer a short, largely qualitative, introduction to MHD in astrophysics and plasma containment, restricting ourselves to incompressible motion. We start with plasma confinement, which is an exercise in high-$R_m$ MHD.

## 21.1 The Magnetic Confinement of Fusion Plasmas

### *21.1.1 Toroidal Confinement in Tokomaks*

In order to achieve nuclear fusion, the colliding nuclei must have sufficient kinetic energy to overcome the Coulomb repulsion that tends to keep them apart. This, in turn, requires that the plasma be maintained at a high temperature, in excess of $10^8$ K if the reactants are deuterium and tritium, which is the reaction favoured by those seeking controlled fusion in the laboratory. Consequently, if we are to realise sustained nuclear fusion in a reactor, it is crucial that the hot plasma be kept away from the reactor walls, and magnetic confinement is the most obvious means of achieving this. Moreover, the plasma in a fusion reactor must be pressurised and held in a stable configuration for long periods of time (years, if the process is to be economic). Unfortunately, after half a century of concerted effort, confinement times have yet to exceed one second. (There are, however, ambitious programmes aimed at extending this.) The central problem is that a magnetically confined plasma is plagued by a myriad of instabilities, as we now discuss.

Let us start with some geometry. A plasma in equilibrium satisfies $\mathbf{J} \times \mathbf{B} = \nabla p$, from which we find

$$\mathbf{J} \cdot \nabla p = \mathbf{B} \cdot \nabla p = 0. \tag{21.1}$$

It follows that the field lines of **J** and **B** lie on constant-pressure surfaces and those surfaces must close within the plasma in an isolated system. So the **J** and **B** fields lie on a set of nested surfaces, as shown in Figure 21.1. Note that those surfaces cannot be simply connected, because any closed contour for which $\oint \mathbf{B} \cdot d\mathbf{r}$ is non-zero would require current flowing across that surface, which is forbidden by (21.1). Thus spherical surfaces are inadmissible and the simplest configuration is a set of nested tori. This is the geometry used in tokamaks, which form the basis of most attempts at plasma confinement. In such a confinement scheme one must arrange for the Lorentz force to point inwards, towards the so-called *magnetic axis*, which is the constant-pressure surface of zero surface area.

It is natural to adopt cylindrical polar coordinates $(s, \theta, z)$ centred on the torus. We shall use $R$ to denote the radius of the major axis and $a$ the radius of the minor axis, so that $R - a \leq s \leq R + a$, and $|z| \leq a$. Axisymmetric fields in cylindrical polar coordinates are usually decomposed into their $s$–$z$ and $\theta$ components, with the former denoted *poloidal* and the latter either *azimuthal* or *toroidal*. When discussing tokamaks, it is normal to adopt the term toroidal in preference to azimuthal. We shall stay with that convention, and so we shall talk of a poloidal–toroidal decomposition of **B** or **J**.

It is sometimes convenient to imagine a torus with a relatively large aspect ratio, $R >> a$, so that the geometry may be may be locally 'unwrapped' to form a cylinder of radius $a$, as in Figure 21.2. However, notational confusion can easily develop in such cases: what was a poloidal ($s$–$z$) field in the context of a torus is now apparently a toroidal (i.e. $\theta$) field in the context of a cylinder. In order to avoid such confusion, we shall always have in mind a torus, even when it is unwrapped to look like a cylinder. Thus, for example, toroidal **B** fields run parallel to the magnetic axis, supported by poloidal currents encircling the plasma. Conversely, a poloidal magnetic field wraps around the plasma in the $s$–$z$ plane, supported by toroidal currents flowing within the plasma, running parallel to the magnetic axis. (Recall that, for an axisymmetric field, the curl of a poloidal field is toroidal while the curl of the

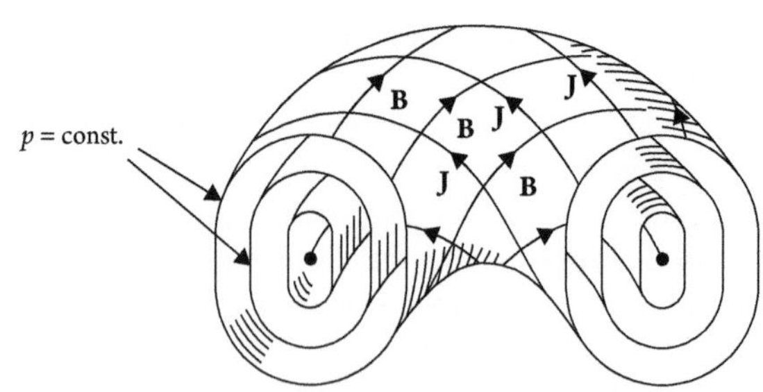


**Figure 21.1** In plasma confinement the **J**- and **B**-lines sit on a set of nested surfaces. The magnetic axis lies at the centre of those surfaces. (Reproduced from P. A. Davidson. *An Introduction to Magnetohydrodynamics*, 2nd ed. Cambridge University Press, 2017.)

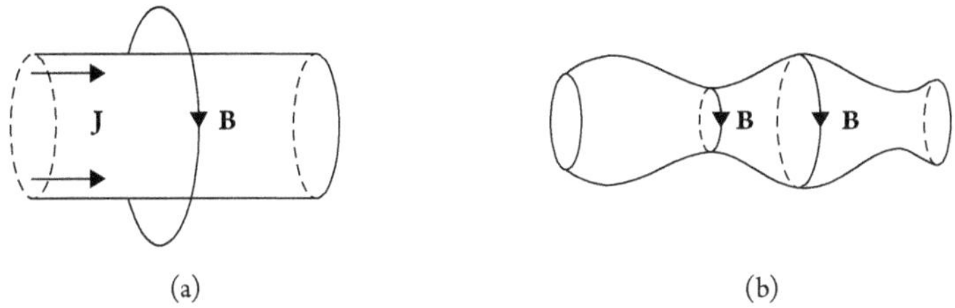


**Figure 21.2** (a) A poloidal **B** field, supported by toroidal surface currents, confines a plasma. (b) The sausage-mode instability. (Reproduced from P. A. Davidson. *An Introduction to Magnetohydrodynamics*, 2nd ed. Cambridge University Press, 2017.)

toroidal field is poloidal, and so Ampère's law tells us that toroidal currents support a poloidal magnetic field.)

As discussed in § 8.7, a plasma may be pressurised by inducing toroidal surface currents, which support a poloidal magnetic field encircling the torus, as shown in Figure 21.2(a). The magnetic pressure then pinches radially inward, confining the plasma. Unfortunately, this configuration is unstable to the so-called *sausage-mode instability*, as discussed in §8.7 and as shown in Figure 21.2(b). However, as we shall see, this instability may be neutralised by trapping a toroidal magnetic field within the plasma, running parallel to the magnetic axis. Indeed, the idea of using a combination of poloidal and toroidal magnetic fields is the basis of the plasma confinement scheme in tokamaks.

In a tokamak the toroidal **B** field is maintained by D-shaped superconducting coils arranged around the plasma, as shown in Figure 21.3, while the toroidal plasma current, which supports the poloidal **B** field, is induced (at least for short periods) by transformer action. In some sense, we may think of the poloidal magnetic field as pressurising the plasma, while the toroidal field provides stability. However, this is a little simplistic, since often the toroidal field provides much of the pressurisation by maintaining a higher field outside the plasma than inside. As we shall see, stability requires that the toroidal field always be dominant, typically ten times the poloidal field strength, and so the pitch of the helical field lines within the plasma is small. The success of a magnetic confinement scheme is often measured using the so-called $\beta$-ratio of the plasma, $\beta = p/(B^2/2\mu_0)$, where $p$ is the pressure on the magnetic axis and $B$ is the magnetic field external to the plasma. In tokamaks, $\beta$ rarely rises above 0.2 and is more typically around 0.03–0.1.

### *21.1.2 Sausage-Mode and Kink Instabilities*

In the absence of a toroidal magnetic field, the plasma in a torus is always unstable to the sausage-mode instability, as shown in Figure 21.2. That is, if $I$ is the net toroidal current flowing around the plasma loop, then the mean poloidal field at the surface of the plasma is $B_P = \mu I/2\pi a$. So, if the plasma pinches in at some point, conserving $I$ in the process, then the external poloidal field increases, along with the associated magnetic pressure. This accentuates the initial disturbance and drives an instability.

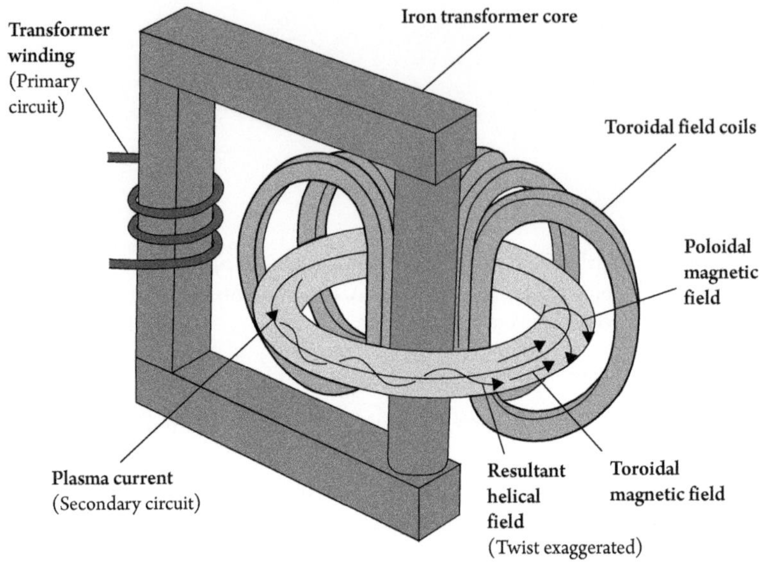


**Figure 21.3** Magnetic confinement in a tokamak uses both poloidal and toroidal fields.

As noted above, this instability may be neutralised by trapping a toroidal magnetic field $B_T$ within the plasma $B_T$. This field pushes radially outward from the magnetic axis to the plasma surface, with the outward Lorentz force, integrated from the axis to the surface, being the difference in magnetic pressures, $(B_T^2/2\mu_0)_{\rm axis} - (B_T^2/2\mu_0)_{\rm ext}$. (Here $B_{T,\rm ext}$ is the external toroidal field.) Consider the simplest case where the toroidal field is negligible outside the plasma and uniform within it. Then $B_T = \Phi/\pi a^2$ within the plasma, where $\Phi$ is the toroidal flux in the plasma loop. If the plasma pinches inward at some point, conserving $\Phi$ in the process, then the toroidal field will grow, along with the associated outward Lorentz force, and this counters the initial disturbance. Physically, the toroidal field stabilises the plasma to sausage-mode disturbances because such modes bend the toroidal field lines, increasing their magnetic energy, and work must be done against the Faraday tension within the field lines, a tension that tends to push the toroidal **B**-lines back towards their equilibrium position. A pseudo-static stability calculation is outlined in Exercise 21.1, which ignores any *external* toroidal field, as well as curvature associated with the major axis of the torus. This calculation shows that such a configuration is stable to sausage-mode disturbances provided that $B_T^2 > B_P^2/2$.

However, there is then a second problem. Even when stable to sausage-mode perturbations, such a configuration is unstable to disturbances in the form of the *kink instability*. Suppose that the plasma column starts to buckle, as shown in Figure 21.4. Then the external poloidal field lines are pressed together on the inside of the bend and spread out on the outer side, causing the external magnetic pressure to increase on the concave side of the column and reduce on the opposite side. This produces a sideways force that amplifies the initial perturbation. Luckily, the kink instability may be stabilised using an *external* toroidal field,

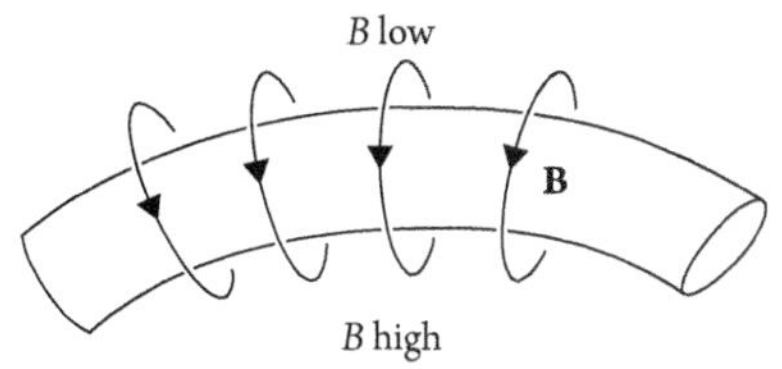


**Figure 21.4** The kink instability. (Reproduced from P. A. Davidson. *An Introduction to Magnetohydrodynamics*, 2nd ed. Cambridge University Press, 2017.)

essentially because the Faraday tension in the toroidal field lines resists bending. A simple stability analysis shows that avoidance of the kink instability requires $\left|B_P(a)/B_{T,\mathrm{ext}}\right| < 2\pi a/L$, where $L$ is the wavelength of the disturbance (see Boyd & Sanderson, 2003). In a tokamak $L$ cannot be greater than $2\pi R$, and so the criterion for avoiding the kink instability is $\left|B_P(a)/B_{T,\mathrm{ext}}\right| < a/R$. This is the reason why the toroidal field is dominant in a tokamak.

### *21.1.3 Interchange and Ballooning Instabilities*

The sausage-mode and kink instabilities are the simplest members of a complex family of magnetic instabilities that plague fusion plasmas. The problem is that magnetic fields provide stable support only when the Faraday tension resists bending of the field lines. However, modes for which $\mathbf{B}\cdot\mathbf{k} = 0$, $\mathbf{k}$ being a wave vector, do not bend the field lines and so the Faraday tension offers no resistance to such disturbances. These modes are called *interchange modes* and the sausage-mode instability is just one example.

Consider, by way of illustration, plasma in a cylindrical column carrying an axial current density that is a function of distance $r$ from the magnetic axis and which gives rise to circular **B**-lines, also a function of $r$. Now suppose that a sausage-mode disturbance perturbs the plasma, as shown in Figure 21.2. The stored magnetic energy of a thin circular flux loop of radius $r$ is $U_B = (B^2/2\mu_0)V = (\Phi^2/2\mu_0 V)(2\pi r)^2$, where $V$ is the volume of the loop and $\Phi$ its flux. Evidently, if the loop contacts while preserving its flux and volume, then its magnetic energy will fall, and this energy must reappear as kinetic energy. However, as some flux loops contract, reducing their energy, incompressibility demands that others expand, increasing their energy, and so the criterion for a net release of magnetic energy is more subtle. In fact, we must consider what happens when two adjacent loops at different radii exchange position while conserving their volume and flux. If the inner flux loop has radius $r_1$ and flux $\Phi_1$, and the outer one radius $r_2 = r_1 + \delta r$ and flux $\Phi_2$, then the net change in magnetic energy following an interchange of the two flux loops is

$$\delta U_B = \frac{(2\pi)^2}{2\mu_0 V}(\Phi_2^2 - \Phi_1^2)(r_1^2 - r_2^2) = -\frac{(2\pi)^2 r_1}{\mu_0 V}\frac{\Phi_2^2 - \Phi_1^2}{\delta r}(\delta r)^2. \tag{21.2}$$

However, since $B/r = 2\pi\Phi/V$, this becomes

$$\delta U_B = -\frac{rV}{\mu_0}\frac{d}{dr}\left(\frac{B}{r}\right)^2(\delta r)^2. \tag{21.3}$$

**Figure 21.5** Ballooning modes in a tokamak. (From R. Khan, N. Mizuguchi, N. Nakajima & T. Hayashi. Dynamics of the ballooning mode and the relationship to edge-localized modes in a spherical tokamak. *Phys. Plasmas* **14**, 062302 (2007), with permission.)

We conclude that magnetic energy is converted into kinetic energy, and an interchange instability can take hold, provided that $B^2/(\mu_0 r^2)$ increases with radius at the point where the perturbation occurs. This condition is satisfied when the magnetic energy density is localised at the outer surface of the plasma, as in Figure 21.2.

This type of analysis can be carried over to more complex geometries by using a variational method called the *energy principle*. It turns out that interchange modes grow unchecked whenever there is a source of magnetic energy to drive the instability, and this occurs when (i) the **B**-lines are free to contract and (ii) the magnetic energy density increases with distance from the local centre of curvature of the field lines (Davidson, 2017). In the simple case of a cylindrical column with circular **B**-lines, interchange modes are readily realised because of the symmetry of the base state. However, in more complex toroidal geometries, in which the field lines are helical, the requirement that $\mathbf{B} \cdot \mathbf{k} = 0$ at *all points* in the growing instability greatly restricts the possibility of interchange modes.

Indeed, to help suppress these instabilities in tokamaks, it is normal to vary the pitch of the helical **B**-lines from one constant-pressure surface to the next, so that if the plasma bulges outward at some point, it finds its way barred by surrounding field lines of a different inclination.

In a well-designed tokamak interchange modes are rare and attention transfers to the closely related *ballooning modes*. The difference between the two types of instability is that in the former we demand $\mathbf{B}\cdot\mathbf{k}=0$, so as to eliminate all field-line bending, while in the latter this is relaxed to $\mathbf{B}\cdot\mathbf{k}\approx 0$, so that some field-line bending is allowed. In both cases the instability is long and slender, localised along a flux tube (Figure 21.5). In ballooning modes, where $\mathbf{B}\cdot\mathbf{k}\neq 0$, some work must be done against the Faraday tension while bending the field lines, but this is more than compensated for by the freedom to focus the growing mode in those regions where magnetic energy is readily released by interchanging the field lines. Ballooning modes are prevalent in tokamaks.

Our brief (and qualitative) discussion of plasma containment has been confined to incompressible motion, but plasmas are, of course, compressible. However, the most unstable ballooning modes turn out to be the incompressible ones. The ballooning instability is characterised by the appearance of filamentary flux tubes near the outer edge of the plasma, as shown in Figure 21.5. They are an important member of the broader class of tokamak instabilities known as *edge-localised modes* and such instabilities have severely hampered attempts at sustained plasma confinement in fusion reactors. Readers interested in a more detailed discussion of plasma stability could do worse than consult Boyd & Sanderson (2003).

## 21.2 Alfvén Waves Revisited

### *21.2.1 Small-Amplitude Alfvén Waves*

We now turn to Alfvén waves. These rely on the fact that magnetic field lines behave like stretched elastic bands, carrying a Faraday tension $B^2/\mu_0$. Thus, if a magnetic field is pushed out of equilibrium, as in Figure 21.6, the resulting curvature of the field lines gives rise to a restoring force $B^2/\mu_0 R$, where $R$ is the radius of curvature of the field lines. Moreover, Alfvén's theorem tells us that, at high $R_m$, the conducting medium sticks to the **B**-lines, and so the field lines have some inertia. Oscillations are then inevitable.

Let us now quantify this. Suppose that we have a uniform magnetic field $\mathbf{B}_0$ in an inviscid fluid that is perturbed to $\mathbf{B}=\mathbf{B}_0+\mathbf{b}$ by a weak velocity field $\mathbf{u}$. Then the linearised form of the induction equation for a small-amplitude perturbation is

$$\frac{\partial \mathbf{b}}{\partial t}=(\mathbf{B}_0\cdot\nabla)\mathbf{u}+\lambda\nabla^2\mathbf{b}. \tag{21.4}$$

Moreover, the linearised momentum equation can be written in terms of Maxwell stresses as

$$\frac{\partial \mathbf{u}}{\partial t}=\frac{1}{\rho\mu_0}(\mathbf{B}_0\cdot\nabla)\mathbf{b}, \tag{21.5}$$

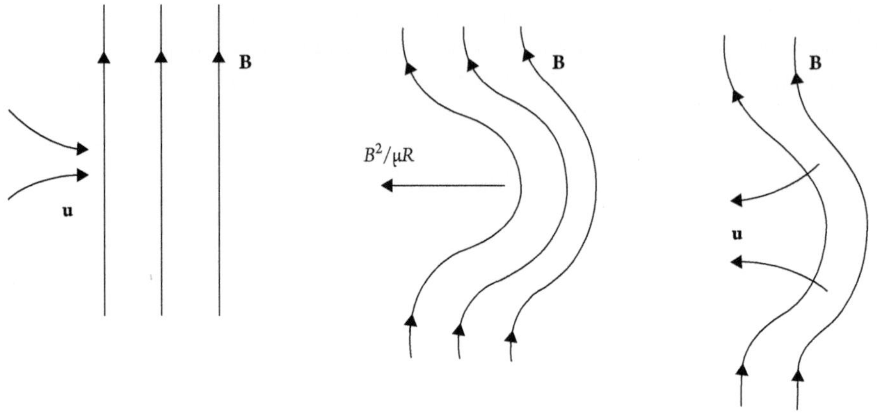


**Figure 21.6** The mechanism of Alfvén waves.

where the combined pressure term, including the magnetic pressure, $-\nabla(p+p_m)$, is zero because **u** and **b** are both solenoidal (see Exercise 21.2). Eliminating **b** now yields the wave equation,

$$\left(\frac{\partial}{\partial t} - \lambda\nabla^2\right)\frac{\partial \mathbf{u}}{\partial t} = \frac{1}{\rho\mu_0}(\mathbf{B}_0 \cdot \nabla)^2\mathbf{u}, \tag{21.6}$$

which supports travelling waves of the form $\mathbf{u} = \hat{\mathbf{u}}\exp[j(\mathbf{k}\cdot\mathbf{x} - \varpi t)]$. The corresponding dispersion relationship is

$$\varpi^2 + j\lambda k^2\varpi = v_a^2 k_{\|}^2, \tag{21.7}$$

where $\boldsymbol{v}_a = \mathbf{B}_0/\sqrt{\rho\mu_0}$ is the *Alfvén velocity* and $k_{\|}$ is the component of **k** parallel to $\mathbf{B}_0$. For a perfect conductor, $\lambda = 0$, we obtain $\varpi = \pm v_a k_{\|}$, which is a non-dispersive transverse wave with a phase velocity $v_a$. When $\lambda$ is small but finite, on the other hand, we obtain the high-$R_m$ result, $\varpi = \pm v_a k_{\|} - (\lambda k^2/2)j$, which is weakly damped.

For perfect conductors, our governing equations adopt a more symmetrical form when written in terms of the scaled magnetic field $\mathbf{h} = \mathbf{B}/\sqrt{\rho\mu_0}$:

$$\frac{\partial \mathbf{h}}{\partial t} = (\mathbf{h}_0 \cdot \nabla)\mathbf{u}, \quad \frac{\partial \mathbf{u}}{\partial t} = (\mathbf{h}_0 \cdot \nabla)\mathbf{h}, \tag{21.8}$$

where $\mathbf{h}_0 = \boldsymbol{v}_a$. Since Alfvén waves are non-dispersive, we expect (21.8) to support d'Alembert-like solutions, and indeed two general solutions of (21.8) are

$$\mathbf{u} = \mathbf{V}^F(\mathbf{x} - \mathbf{h}_0 t), \quad \mathbf{h} = \mathbf{h}_0 - \mathbf{V}^F(\mathbf{x} - \mathbf{h}_0 t), \tag{21.9}$$

$$\mathbf{u} = \mathbf{V}^B(\mathbf{x} + \mathbf{h}_0 t), \quad \mathbf{h} = \mathbf{h}_0 + \mathbf{V}^B(\mathbf{x} + \mathbf{h}_0 t). \tag{21.10}$$

Here $\mathbf{V}^F$ and $\mathbf{V}^B$ are arbitrary functions of their arguments and represent disturbances that travel along the $\mathbf{B}_0$-lines at the Alfvén velocity without change of shape.

It can be useful to rewrite these results in terms of the so-called *Elsasser fields* $\mathbf{v}^+ = \mathbf{u} + (\mathbf{h} - \mathbf{h}_0)$ and $\mathbf{v}^- = \mathbf{u} - (\mathbf{h} - \mathbf{h}_0)$, where $\mathbf{v}^- = 2\mathbf{V}^F$ is a forward-travelling wave and $\mathbf{v}^+ = 2\mathbf{V}^B$ is a backward-travelling wave. In terms of these new variables, (21.8) becomes

$$\frac{\partial \mathbf{v}^\pm}{\partial t} = \pm(\mathbf{h}_0 \cdot \nabla)\mathbf{v}^\pm. \tag{21.11}$$

### 21.2.2 *Finite-Amplitude Waves, Colliding Wave Packets, and Alfvénic Turbulence*

Somewhat surprisingly, the *fully nonlinear* governing equations for an ideal fluid,

$$\frac{\partial \mathbf{h}}{\partial t} + (\mathbf{u} \cdot \nabla)\mathbf{h} = (\mathbf{h} \cdot \nabla)\mathbf{u}, \tag{21.12}$$

$$\frac{\partial \mathbf{u}}{\partial t} + (\mathbf{u} \cdot \nabla)\mathbf{u} = -\nabla\left(\frac{p + p_m}{\rho}\right) + (\mathbf{h} \cdot \nabla)\mathbf{h}, \tag{21.13}$$

support *finite-amplitude* Alfvén waves. Specifically, (21.12) and (21.13) support finite-amplitude waves of the form (21.9) and (21.10), where $\mathbf{V}^F$ and $\mathbf{V}^B$ are now not only of arbitrary shape, but also of arbitrary amplitude. (The proof of this is left to the reader.)

A more symmetric form of the governing equations may be obtained by using the Elsasser variables $\mathbf{v}^\pm = \mathbf{u} \pm (\mathbf{h} - \mathbf{h}_0)$. The expressions (21.12) and (21.13) then become

$$\frac{\partial \mathbf{v}^\pm}{\partial t} + (\mathbf{v}^\mp \cdot \nabla)\mathbf{v}^\pm = \pm(\mathbf{h}_0 \cdot \nabla)\mathbf{v}^\pm - \nabla\left(\frac{p + p_m}{\rho}\right), \tag{21.14}$$

which reduces to (21.11) for small-amplitude waves of the form (21.9) or (21.10). Now consider some form of arbitrary motion, not necessarily wave-like, in which $\mathbf{v}^\pm$ are localised in space. We take the dot product of (21.14) with $\mathbf{v}^\pm$, integrate over all space, and use Gauss' theorem to convert all divergences into surface integrals, which vanish for a localised disturbance. Only the time derivative in (21.14) then survives and we conclude that the volume integrals of $(\mathbf{v}^+)^2$ and $(\mathbf{v}^-)^2$ constitute two invariants of ideal MHD. Given that

$$(\mathbf{v}^+)^2 + (\mathbf{v}^-)^2 = 2[\mathbf{u}^2 + (\mathbf{h} - \mathbf{h}_0)^2], \quad (\mathbf{v}^+)^2 - (\mathbf{v}^-)^2 = 4\mathbf{u} \cdot (\mathbf{h} - \mathbf{h}_0),$$

it is evident that the global conservation of $(\mathbf{v}^+)^2 + (\mathbf{v}^-)^2$ is nothing more than conservation of energy for the disturbance. The global conservation of $(\mathbf{v}^+)^2 - (\mathbf{v}^-)^2$, on the other hand, is a manifestation of the conservation of cross helicity of the disturbance, $h_c = \int \mathbf{u} \cdot \mathbf{b}\, dV$. (See § 20.6 for a discussion of the conservation of cross helicity.)

Note that the non-linear term in (21.14), $(\mathbf{v}^\mp \cdot \nabla)\mathbf{v}^\pm$, is non-zero only when oppositely travelling wave packets overlap, and so finite-amplitude Alfvén waves travelling in the same direction do not interact. It is interesting to speculate, therefore, as to what happens when two finite-amplitude wave packets travelling in opposite directions collide. Such events are thought to play an important role in solar-wind turbulence.

Let us start with an ideal fluid model. Suppose that $\mathbf{v}^+$ and $\mathbf{v}^-$ constitute two localised disturbances (wave packets) that are initially distinct, so that $\mathbf{v}^+ = 2\mathbf{V}^B$ and $\mathbf{v}^- = 2\mathbf{V}^F$. They then collide, and during the collision the non-linear term $(\mathbf{v}^\mp \cdot \nabla)\mathbf{v}^\pm$ comes into play and so the wave packets progressively distort each other, although the volume integrals of both $(\mathbf{v}^+)^2$ and $(\mathbf{v}^-)^2$ are conserved at all times. Eventually the wave packets pass through one another and emerge with new shapes. Nonlinear interactions then cease and the waves once again propagate without change of shape.

Consider now those times during which the two wave packets are non-overlapping, either before or after the collision. Then we have

$$\frac{1}{4}\int (\mathbf{v}^+)^2 dV = \frac{1}{2}\int \left(\mathbf{u}^2 + \frac{\mathbf{b}^2}{\rho\mu_0}\right) dV = \frac{1}{\sqrt{\rho\mu_0}}\int \mathbf{u}\cdot\mathbf{b}\, dV, \tag{21.15}$$

$$\frac{1}{4}\int (\mathbf{v}^-)^2 dV = \frac{1}{2}\int \left(\mathbf{u}^2 + \frac{\mathbf{b}^2}{\rho\mu_0}\right) dV = -\frac{1}{\sqrt{\rho\mu_0}}\int \mathbf{u}\cdot\mathbf{b}\, dV, \tag{21.16}$$

where we integrate over the forward- or backward-travelling waves only, as appropriate. However, throughout the collision the volume integrals of $(\mathbf{v}^+)^2$ and $(\mathbf{v}^-)^2$ are both conserved, and so the waves must emerge with their energy and cross helicity unchanged. In short, there is no exchange of energy or cross helicity between the wave packets.

Consider now the situation where the fluid has a small but finite magnetic diffusivity and there exists a random sea of wave-packets, some travelling parallel to the mean field $\mathbf{B}_0$ and some travelling antiparallel to $\mathbf{B}_0$. This is known as *Alfvénic turbulence.* Multiple collisions will occur and, as is typical with nonlinear interactions, it is observed that the energy in the wave packets becomes increasingly spread over a wider range of scales as a result of those collisions. Moreover, the magnetic and kinetic energy that is gradually passed down to small scales is eventually converted into heat within thin current sheets. Numerical simulations of such collisions tend to suggest that, in a typical collision, the weaker of $\mathbf{v}^+$ and $\mathbf{v}^-$ is more severely buffeted by the stronger of the two wave packets, and so passes it energy down to small scales more rapidly. After multiple collisions, then, the weaker wave packets tend to be eradicated. Thus, for example, if there is an excess of energy in the forward-travelling waves, then those wave packets will progressively eliminate the backward-travelling disturbances until all the wave packets propagate in the same direction. This process is known as *dynamic alignment,* and it suggests that any local dominance of forward- or backward-travelling waves will be progressively reinforced, until all the waves travel in the same direction. The nonlinear interactions then cease and we reach a state in which $\mathbf{u} = \mathbf{h} - \mathbf{h}_0$, or else $\mathbf{u} = -(\mathbf{h} - \mathbf{h}_0)$, which is known as an *Alfvénic state.*

This concludes our brief review of Alfvén waves. Readers interested in more details of the nonlinear interaction of such waves might wish to consult Moffatt (1978) or Davidson (2017).

## 21.3 Planetary Dynamos

The earth, Mercury, Jupiter and Saturn all maintain dipolar magnetic fields that are more or less aligned with the rotation axis of the planet. This, in turn, requires that there exist

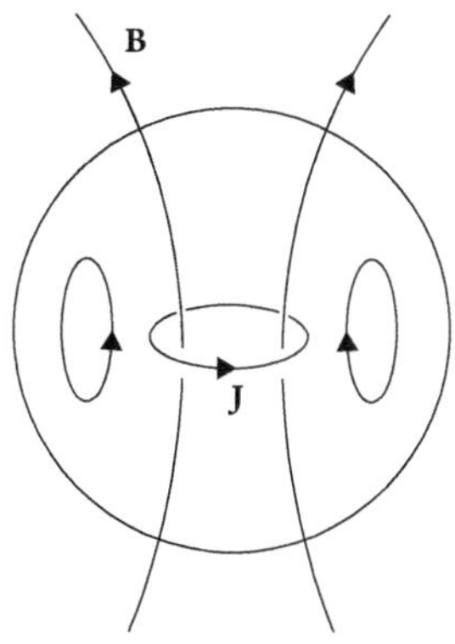


**Figure 21.7** A dipolar magnetic field requires azimuthal currents circulating in the core.

large-scale east–west currents circulating in the cores of these planets (Figure 21.7), and scientists have been speculating for over a century as to what maintains those currents in the face of the Ohmic losses. We now discuss this fascinating question, focusing primarily on the earth and using cylindrical polar coordinates $(s, \theta, z)$ that rotate with the mantle. However, perhaps it is only fair to warn readers that this is still an active area of research and that the precise mechanisms at work are still being debated.

### *21.3.1 The Need for a Dynamo Theory for the Earth*

As noted in § 7.7, the geomagnetic field is not the result of permanent magnetism, since the temperature at the centre of the earth is well above the Curie point, and indeed is similar to that at the surface of the sun. Nor is the earth's magnetic field the last vestige of some primordial field trapped within the conducting core during its formation, since such a fossil relic would have decayed long ago through Ohmic dissipation. To see that this is so, consider the energy equation (20.76) in the absence of fluid motion,

$$\frac{d}{dt}\int\limits_{V_\infty} \frac{\mathbf{B}^2}{2\mu_0}\, dV = \frac{dU_B}{dt} = -\int\limits_{V_C} \frac{\mathbf{J}^2}{\sigma}\, dV, \tag{21.17}$$

where $V_C$ is the electrically conducting core of the earth. Evidently, the magnetic energy decays at a rate set by the Ohmic dissipation in the earth's core. The characteristic decay time of such a field can be determined from an eigenvalue analysis of the diffusion equation, $\partial \mathbf{B}/\partial t = \lambda \nabla^2 \mathbf{B}$, coupled to an external potential field outside the iron core. This yields, for the slowest-decaying mode, a decay time $t_d = R_C^2/\lambda\pi^2$, where $\lambda \approx 2\text{m}^2/\text{s}$ and $R_C$ is the radius of the conducting core (see § 2.2.3). However, such a field decays on a timescale of around $10^4$ years, whereas the earth's magnetic field has existed for at least $10^9$ years. Sir Horace Lamb published this calculation in 1883, prompting a multitude of speculative theories of terrestrial magnetism. This included a thin magnetic crust, the thermoelectric effect, the Hall effect, rotating electrostatic charges and a proposed modification to Maxwell's equations. However, these were all quickly dismissed and it was Joseph Larmor who, in 1919, finally suggested the possibility of a fluid dynamo (in the context of the sun) maintained by convective motion.

So let us return to (20.76), but retain the possibility of fluid motion in the earth's iron core:

$$\frac{d}{dt}\int_{V_\infty}\frac{\mathbf{B}^2}{2\mu_0}dV = -\int_{V_C}(\mathbf{J}\times\mathbf{B})\cdot\mathbf{u}\,dV - \int_{V_C}\frac{\mathbf{J}^2}{\sigma}dV. \tag{21.18}$$

We conclude that the magnetic field may be maintained, despite the Joule dissipation, through the convective action of the conducting fluid in the core. That is to say, in the core of the earth there could exist a self-excited fluid dynamo that converts mechanical energy into magnetic energy, with motion of the conducing fluid across the magnetic field lines generating the currents that are required to maintain the magnetic field (Figure 21.8). In such a dynamo the conductor does work on the magnetic field by stretching the flux tubes, and at a rate that compensates for the Joule dissipation. All current theories of terrestrial magnetism take this as their starting point, with the convection driven partly by the temperature difference across the core and partly by the slow solidification of the solid inner core. (As the inner core slowly freezes, the pure iron solidifies first, leaving behind iron that is rich in an admixture of lighter impurities and hence buoyant.)

### 21.3.2 *The Structure of the Planets, Force Balances and Planetary Magnetic Fields*

Before discussing the geodynamo in detail, perhaps it is worth saying something about the interior of the planets and the dominant force balances in their conducting cores. The earth has a radius of 6370 km and consists of a thin outer crust, a rocky mantle and an iron-nickel core (which is ~92% iron). The core itself is divided into two parts. The solid inner core has a radius $R_i = 1220\,\text{km}$, while the liquid outer core has a radius $R_C = 3480\,\text{km}$ and contains

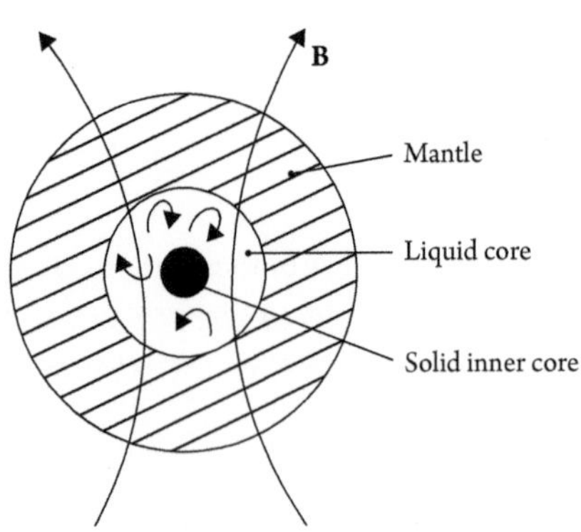


**Figure 21.8** Motion in the earth's core maintains the terrestrial magnetic field.

a weak admixture of lighter elements, such as sulphur, oxygen and silicon. Despite its high temperature, the inner core is growing at a rate of $\sim$1 mm/year, the solidification being driven by the high pressure at the centre of the earth.

The external magnetic field of the earth is dipolar, with a magnetic axis roughly aligned with the rotation axis. The dipole moment has a strength $m = 7.9 \times 10^{22}\,\mathrm{A\,m^2}$, from which we can estimate the average axial field in the core using

$$\mathbf{m} = \frac{1}{2}\int_{V_C} \mathbf{x} \times \mathbf{J}\, dV = \frac{3}{2\mu_0}\int_{V_C} \mathbf{B}\, dV. \tag{21.19}$$

This gives a mean axial field strength $\bar{B}_z = 3.75\,\mathrm{G}$. The dipole field reverses randomly, the reversals being relatively rapid, around $10^3 - 10^4$ years in duration, while the time between reversals is of the order of $10^6$ years. The reversal timescale is comparable to the magnetic diffusion time for the solid inner core, $t_d = R_i^2/\lambda\pi^2 \sim 2000$ years, which suggests that the reversal timescale might be set by the time taken for the magnetic field to diffuse out of the inner core.

In a frame of reference rotating with the earth a typical fluid velocity in the core can be estimated from the rate at which the magnetic field drifts across the core–mantle boundary, and this suggests $u \sim 0.2\mathrm{mm/s}$. Since $\lambda \approx 2\mathrm{m^2/s}$, we might estimate a large-scale magnetic Reynolds number to be

$$R_m = \frac{u(R_C - R_i)}{\lambda} \sim 200. \tag{21.20}$$

Let us now consider the relative magnitudes of the various forces acting on the fluid in the core. The angular velocity of the earth is $\Omega = 7.28 \times 10^{-5}\,\mathrm{s}^{-1}$ and the ratio of the inertial force $\mathbf{u} \cdot \nabla \mathbf{u}$ to the Coriolis force $2\mathbf{u} \times \boldsymbol{\Omega}$, which arises from the use of a rotating frame of reference, is usually estimated using the *Rossby number*

$$\mathrm{Ro} = \frac{u}{\Omega(R_C - R_i)}. \tag{21.21}$$

For the earth we have $\mathrm{Ro} \sim 10^{-6}$, and so $\mathbf{u} \cdot \nabla \mathbf{u}$ is negligible in the core, although this does not mean that we can neglect $\partial \mathbf{u}/\partial t$ in the governing equations. Turning to the viscous forces, the kinematic viscosity of liquid iron is $\nu \approx 10^{-6}\mathrm{m^2/s}$, which is less than the viscosity of air at room temperature. The relative magnitude of the viscous forces in a rapidly rotating sphere is usually measured using the Ekman number

$$\mathrm{Ek} = \frac{\nu}{\Omega R_C^2}, \tag{21.22}$$

which is $\mathrm{Ek} \sim 10^{-15}$ for the earth. This tells us that the viscous stresses are absolutely tiny by comparison with the Coriolis force, and so cannot play any significant role in the dynamics. Moreover, since the viscous stresses are negligible, the dissipation of energy must be almost entirely Ohmic.

Given that both $\mathbf{u} \cdot \nabla \mathbf{u}$ and the viscous stresses are much too small to play any important role, it is usually assumed that the primary force balance in the core is between buoyancy, which drives the flow, the Coriolis force, which shapes the motion, and the Lorentz force, which acts back on the fluid to dampen the motion. The dimensionless group usually used to estimate the ratio of the Lorentz to Coriolis forces is the *Elsasser number*

$$\Lambda = \frac{\sigma B^2}{\rho \Omega}. \tag{21.23}$$

The mean density in the liquid core is $\rho = 10.9 \times 10^3\,\mathrm{kg/m^3}$, and if we take $B \sim 4\,\mathrm{G}$, then we find $\Lambda = O(1)$. This suggests that there is an approximate balance between the Lorentz and Coriolis forces, and it seems plausible that the Lorentz, Coriolis and buoyancy forces are all of similar magnitudes.

Of course, it is not just the earth that maintains a magnetic field through dynamo action. Mercury, Jupiter, Saturn, Uranus, Neptune and Jupiter's largest moon, Ganymede, all appear to have dynamos in their cores. The external magnetic fields of the terrestrial planets, Mercury and the earth, and the gas giants, Jupiter and Saturn, are all rather similar, being dipolar and with a magnetic axis more or less aligned with the rotation axis. Even the normalised intensities of the magnetic fields within these planets are similar, as discussed below. This similarity is surprising, since the structures of these planets are quite different, as indicated in Figure 21.9. For example, the terrestrial planets have iron cores and rocky

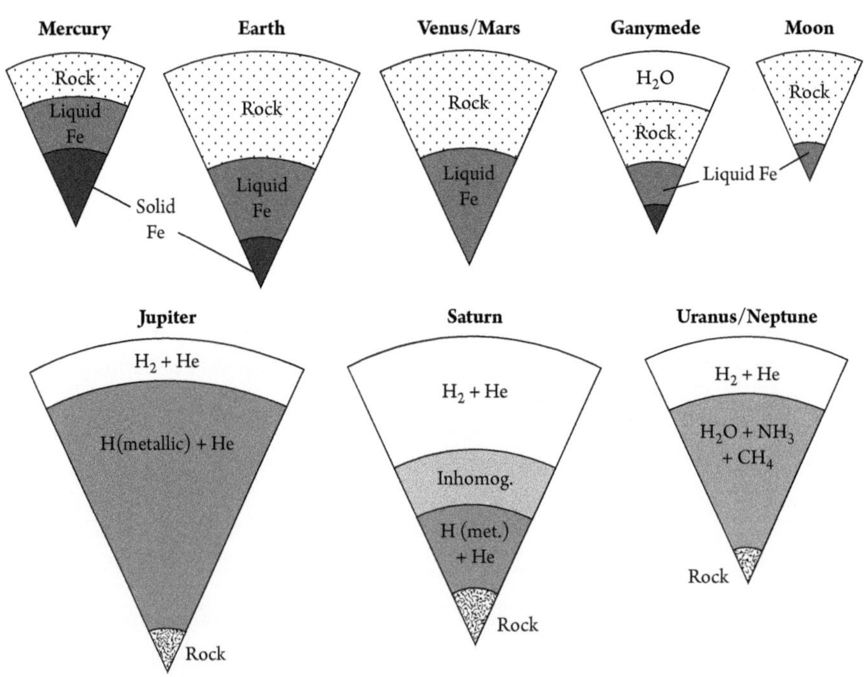


**Figure 21.9** Approximate structure of the planets. (Figure courtesy of Uli Christensen.)

mantles, while the gas giants are composed of hydrogen and helium, with the high pressure at their centres creating metallic hydrogen, which is electrically conducting. Note, however, that the observable magnetic fields of the ice giants, Uranus and Neptune, are very different to those of the other planets, being not particularly dipolar and with a strongly tilted magnetic axes.

From the point of view of dynamo theory, there are three quantities that we can measure with some degree of confidence in the planets. These are the size of the planet, from which we can estimate the size of the conducting core, $R_C$, how fast the planet rotates, $\Omega$, and the dipole moment of the magnetic field, $\mathbf{m}$. Moreover, from the dipole moment we can calculate the mean axial field in the conducting core, $\bar{B}_z$, using (21.19). From these three quantities we may construct a dimensionless measure of the core magnetic field,

$$\Pi_B = \frac{\bar{B}_z/\sqrt{\rho\mu_0}}{\Omega R_C}. \tag{21.24}$$

This dimensionless field strength may be used to compare the magnetic fields across the various planets. The results are shown in Table 21.1, where we have included for comparison a fully convective low-mass M-dwarf star (V374) which is thought to have a fluid dynamo similar to that in the gas giants. It is surprising how uniform $\Pi_B$ is across all of these objects, despite the wide range of sizes, rotation rates and interior structures (Davidson, 2016). This suggests some degree of uniformity in the nature of the dynamos operating in the planets, at least as far as the terrestrial planets and gas giants are concerned.

### 21.3.3 *Geophysical Constraints on Planetary Dynamo Theories*

Dynamo action is commonly observed in computer simulations of a rotating conducting fluid in a sphere. However, at the time of writing, the numerical simulations of dynamos cannot get close to a planet-like regime (Christensen, 2011), and so it is far from clear that

**Table 21.1** Properties of the planets. The mean axial magnetic fields in the conducting cores, $\bar{B}_z$, are obtained using (21.19).

| Planet/star | Rotation period (days) | Radius of planet/star ($10^3$ km) | Core radius $R_C$ ($10^3$ km) | Dipole moment ($10^{22}$ A m$^2$) | Mean $B_z$ in core (G) | $\frac{\bar{B}_z/\sqrt{\rho\mu_0}}{\Omega R_C} \times 10^6$ |
|---|---|---|---|---|---|---|
| Mercury | 58.6 | 2.44 | 1.8 | 0.004 | 0.014 | 5.5 |
| Earth | 1 | 6.37 | 3.48 | 7.9 | 3.7 | 13 |
| Jupiter | 0.413 | 69.9 | 55 | 150 000 | 18 | 5.2 |
| Saturn | 0.440 | 58.2 | 29 | 4600 | 3.8 | 2.3 |
| Uranus | 0.718 | 25.3 | ~18 | 390 | ~1.3 | ~1.0 |
| Neptune | 0.671 | 24.6 | ~20 | 200 | ~0.5 | ~0.3 |
| V374 Pegasi | 0.44 | 200 | — | — | ~800 | ~17 |

Source: P. A. Davidson. *An Introduction to Magnetohydrodynamics*, 2nd ed. Cambridge University Press, 2017.

the mechanisms underpinning the numerical dynamos are the same as those in a planet. It is important, therefore, to find ways of constraining theories of the geodynamo.

Perhaps the first point to note is that, with $\mathrm{Ek} \sim 10^{-15}$ for the earth, the viscous stresses are extremely small and so viscosity cannot play any significant role in the core, even when we allow for viscous boundary layers. Likewise, with $\mathrm{Ro} \sim 10^{-6}$, the inertial force $\mathbf{u} \cdot \nabla \mathbf{u}$ should play no role in a theory of planetary dynamos. Next we note that, despite having very different rotation rates, sizes and internal structures, the magnetic fields of Mercury, Jupiter, Saturn and the earth are all surprisingly similar, being dipolar and with a magnetic axis more or less aligned with the rotation axis. Even the field strengths, as measured by $\Pi_B$, are similar. Occam's razor then tends to suggest that the same basic dynamo mechanisms operate in all four planets, which in turn suggests that the mantle in the earth does not play a crucial role. Finally, the driving force for convection in the core is around $10^4$ times the critical value at which convection first appears (Christensen, 2011) and consequently the convection is strongly driven and likely to be highly chaotic. The current generation of numerical simulations, by contrast, are highly viscous and rather weakly driven. These various observations suggest that we should seek a theory of planetary dynamos that

(i) rests on simple and robust dynamical processes commonly found in a rapidly rotating fluid;
(ii) is not sensitive to thermal and mechanical boundary conditions and in particular does not rely on the presence of a mantle;
(iii) is consistent with the convection being strongly driven and highly chaotic;
(iv) does not rely on the existence of viscous stresses or the inertial force $\mathbf{u} \cdot \nabla \mathbf{u}$.

Theories of the geodynamo come in two parts. The first is *kinematic* and seeks *any* velocity field in a sphere that yields a dynamo, and many candidate velocity fields have been identified. The second task is to identify which of these kinematic dynamos is *dynamically* consistent and in line with the four criteria above. Unfortunately, there is still little agreement on this. While we have some understanding of the dynamos operating in the weakly driven, overly viscous numerical simulations, the gap between those and a planetary core is large and it is unclear how to transfer that understanding to the planets.

### 21.3.4 Four Kinematic Theorems

Let us start with some kinematics. There are four simple kinematic ideas that are central to dynamo theory:

(i) A *large magnetic Reynolds number* is needed for a sustained dynamo.
(ii) The need to conserve angular momentum in the presence of convection tends to drive differential rotation. As noted in § 20.4.4, such differential rotation can spiral out an east–west field from the dipole field, which is known as the $\Omega$-*effect*.
(iii) In 1934 Cowling discovered that an *axisymmetric dynamo is not possible*.

(iv) An evolution equation for the axial component of **B**, integrated over the core, shows that this field is maintained by a *non-axisymmetric azimuthal EMF.*

Let us start with the requirement for a large value of $R_m$. According to (21.18), in order to maintain a dynamo, we require

$$-\int_{V_C} \mathbf{u}\cdot(\mathbf{J}\times\mathbf{B})\,dV \geq \int_{V_C} \frac{\mathbf{J}^2}{\sigma}\,dV. \tag{21.25}$$

If we label the net Ohmic dissipation by $D$ and the rate of conversion of kinetic energy into magnetic energy by $\dot{W}$, then the ratio of $\dot{W}$ to $D$ is of the order of $R_m = u\ell/\lambda$ for some suitable $u$ and $\ell$. It follows that, for a sustained dynamo, $R_m$ must be reasonably large and in fact we can identify a minimum acceptable value of $R_m$ by placing bounds on the integrals in (21.25). For example, the Schwartz inequality yields

$$\mu_0^2\dot{W}^2 \leq u_{\max}^2\left[\int_{V_C}(\nabla\times\mathbf{B})\times\mathbf{B}\,dV\right]^2 \leq u_{\max}^2\int_{V_\infty}\mathbf{B}^2 dV\int_{V_C}(\nabla\times\mathbf{B})^2 dV, \tag{21.26}$$

where $u_{\max}$ is the maximum of **u**. Moreover, if currents are confined to a sphere, the calculus of variations gives us

$$\int_{V_\infty}\mathbf{B}^2 dV \leq \frac{R_C^2}{\pi^2}\int_{V_C}(\nabla\times\mathbf{B})^2 dV \tag{21.27}$$

(see Moffatt, 1978), from which

$$\left|\dot{W}\right| \leq \frac{u_{\max}R_C}{\pi\lambda}D. \tag{21.28}$$

If we base $R_m$ on $u_{\max}$ and $R_C$, then a necessary condition for sustained dynamo action is $R_m \geq \pi$. In fact, tighter bounds on $R_m$ exist and, in practice, it is found that dynamo action in a sphere is not possible unless $R_m$ exceeds a few multiples of 10.

Let us now turn to the $\Omega$-effect. Differential rotation is thought to be driven by buoyant upwellings which occur above and below the inner core, within the so-called *tangent cylinder*, an imaginary cylinder that is coaxial with the rotation axis and circumscribes the inner core. The idea is that rising buoyant fluid within the tangent cylinder will push material laterally outwards towards the mantle. In an inertial frame of reference fluid elements tend to conserve their angular momentum and, Lorentz forces apart, we might expect $(\mathbf{x}\times\mathbf{u})_z$ to be roughly conserved. Thus, as fluid moves out to a greater cylindrical radius, driven by the buoyant upwellings, the azimuthal velocity $u_\theta$ will tend to fall. So we might expect the fluid near the inner core to spin slightly faster than that near the mantle. In a frame of reference rotating with the mantle the fluid surrounding the inner core will then have $u_\theta > 0$. *If* this is all true, then an azimuthal field will be swept out from the dipole field, as shown in Figure 21.10.

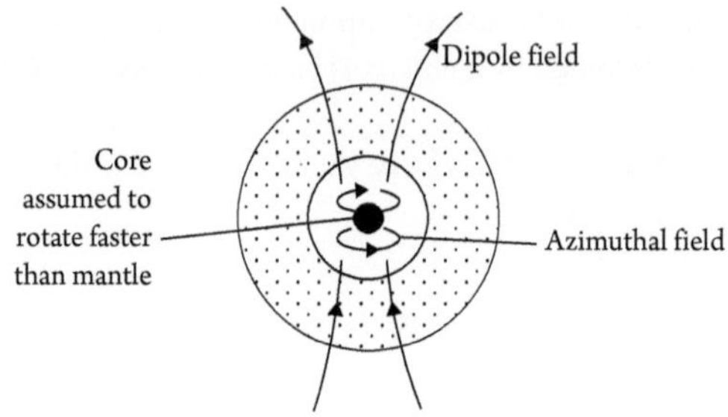


**Figure 21.10** In the Ω-effect, an east–west field is generated by differential rotation.

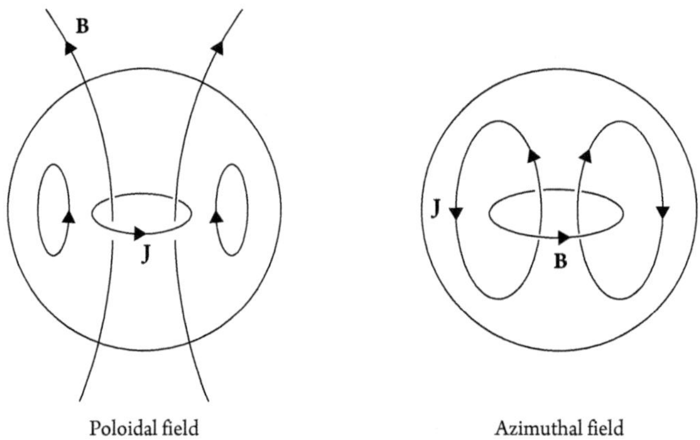


**Figure 21.11** Poloidal and azimuthal magnetic fields and their associated currents.

Let now quantify the Ω-effect. We shall adopt a frame of reference rotating with the earth and use cylindrical polar coordinates $(s,\theta,z)$ centred on the core. It is natural to divide $\mathbf{B}$ into its azimuthal, $\mathbf{B}_\theta = (0, B_\theta, 0)$, and poloidal, $\mathbf{B}_P = (B_s, 0, B_z)$, components. For axisymmetric fields, $\mathbf{B}_\theta$ and $\mathbf{B}_P$ are separately solenoidal and the current that supports $\mathbf{B}_P$ is azimuthal, while $\mathbf{B}_\theta$ is supported by poloidal currents (Figure 21.11).

The effect of differential rotation on the dipole is most easily seen if we assume axial symmetry. The azimuthal component of the induction equation is then

$$\frac{\partial \mathbf{B}_\theta}{\partial t} = \nabla \times (\mathbf{u}_\theta \times \mathbf{B}_P) + \nabla \times (\mathbf{u}_P \times \mathbf{B}_\theta) + \lambda \nabla^2 \mathbf{B}_\theta, \tag{21.29}$$

and, noting that

$$\nabla \times (\mathbf{u}_\theta \times \mathbf{B}_P) = s\mathbf{B}_P \cdot \nabla \left( \frac{u_\theta}{s} \right) \hat{\mathbf{e}}_\theta,$$

$$\nabla \times (\mathbf{u}_P \times \mathbf{B}_\theta) = -s\mathbf{u}_P \cdot \nabla \left( \frac{B_\theta}{s} \right) \hat{\mathbf{e}}_\theta,$$

this simplifies to

$$\frac{D}{Dt}\left(\frac{B_\theta}{s}\right) = \mathbf{B}_P \cdot \nabla\left(\frac{u_\theta}{s}\right) + \lambda s^{-2}\nabla_*^2(sB_\theta), \tag{21.30}$$

where $\nabla_*^2$ is the Stokes operator,

$$\nabla_*^2 = \frac{\partial^2}{\partial z^2} + s\frac{\partial}{\partial s}\left(\frac{1}{s}\frac{\partial}{\partial s}\right).$$

The interaction of the differential rotation, $\nabla(u_\theta/s)$, with the dipole field is now explicit. For example, suppose the dipole points to the north and $u_\theta/s$ is a maximum near the inner core. Then $\mathbf{B}_P \cdot \nabla(u_\theta/s)$, and hence $B_\theta$, will be negative in the north and positive in the south, as shown in Figure 21.10.

Balancing the two terms on the right of (21.30) suggests $B_\theta \sim [|\nabla(u_\theta/s)|\ell^3/\lambda]B_z$. If $R_m$, based on gradients in $u_\theta/s$, were large, then we would expect the east–west field to be much stronger than the dipole field. However, it is not so clear that the differential rotation is that strong in planetary cores. For example, in the (admittedly imperfect) numerical simulations of the geodynamo only a few of the simulations show a strong east–west field, while most show only weak differential rotation and an east–west field that is of the same order as $B_P$. The implication is that, in most of the numerical simulations, the differential rotation is largely suppressed by the poloidal magnetic field, which resists being sheared, clamping the flow into a state close to rigid-body rotation.

We now turn to Cowling's theorem, which states that an axisymmetric dynamo is not possible. The simplest way of seeing where this comes from is Cowling's *neutral-point argument*. Suppose that we look for a steady axisymmetric dynamo in which $\mathbf{B}$ and $\mathbf{u}$ are both poloidal, while $\mathbf{J}$ and $\nabla \times \mathbf{u}$ are necessarily azimuthal. Since the dynamo is steady, Faraday's law requires $\nabla \times \mathbf{E} = 0$ and Ohm's law reduces to $\mathbf{J} = \sigma(-\nabla V + \mathbf{u} \times \mathbf{B})$. The electrostatic potential is then governed by the divergence of Ohm's law,

$$\nabla^2 V = \nabla \cdot (\mathbf{u} \times \mathbf{B}) = \mathbf{B} \cdot \nabla \times \mathbf{u} - \mu_0 \mathbf{u} \cdot \mathbf{J} = 0.$$

This demands that $V = 0$, and hence Ohm's law simplifies to $\mathbf{J} = \sigma\, \mathbf{u} \times \mathbf{B}$. Crucially, in an axisymmetric poloidal field there is always at least one *neutral point*, N, at which $|\mathbf{B}| = 0$ and the $\mathbf{B}$-lines are locally closed in the $s$–$z$ plane, as shown in Figure 21.12. Ohm's law, $\mathbf{J} = \sigma\, \mathbf{u} \times \mathbf{B}$, requires that $\mathbf{J} = 0$ at the neutral point, yet Ampère's circuital law applied to a field line surrounding N demands that $\mathbf{J}$ is non-zero near N, which leads to a contradiction. The implication is that a steady axisymmetric dynamo is not possible. (Actually, this argument is a little simplistic, but see Exercise 21.4 for more details.)

A more general proof of Cowling's theorem proceeds as follows. The velocity and magnetic fields are taken to be axisymmetric, but we now allow for both poloidal and azimuthal contributions:

$$\mathbf{B}(s,z,t) = \mathbf{B}_P + \mathbf{B}_\theta, \quad \mathbf{u}(s,z,t) = \mathbf{u}_P + \mathbf{u}_\theta. \tag{21.31}$$

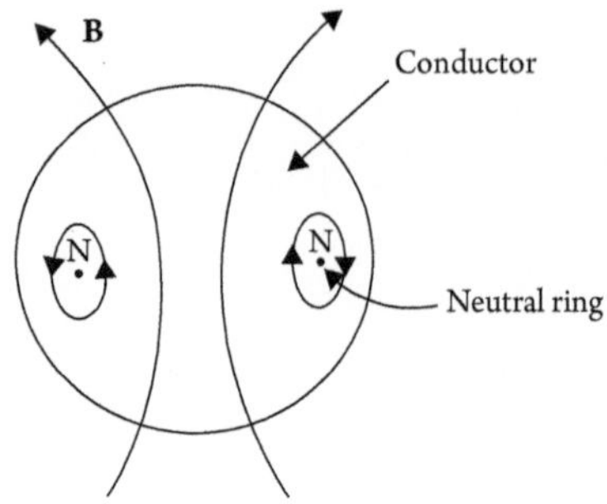


**Figure 21.12** The neutral ring in an axisymmetric poloidal field. (From P. A. Davidson. *Turbulence in Rotating, Stratified and Electrically Conducting Fluids.* Cambridge University Press, 2013.)

The induction equation can be similarly divided into poloidal and azimuthal parts, with the azimuthal component, (21.29), simplifying to (21.30). The poloidal component, on the other hand, is

$$\frac{\partial \mathbf{B}_P}{\partial t} = \nabla \times (\mathbf{u}_P \times \mathbf{B}_P) + \lambda \nabla^2 \mathbf{B}_P. \tag{21.32}$$

Since $\mathbf{B}_P$ is solenoidal, we can write it in terms of a vector potential: $\mathbf{B}_P = \nabla \times [(\chi/s)\,\hat{\mathbf{e}}_\theta]$, where $\chi$ is the *flux function* for $\mathbf{B}_P$. Uncurling (21.32) then yields, after a little algebra,

$$\frac{D\chi}{Dt} = \lambda \nabla_*^2 \chi, \tag{21.33}$$

where $\nabla_*^2$ is the Stokes operator. Crucially, (21.33) contains no source term and so $\chi$, and hence $\mathbf{B}_P$, must eventually diffuse away. As $\mathbf{B}_P$ vanishes, so also must $\mathbf{B}_\theta$, since there is then no east–west field generation by differential rotation, and hence no source term in (21.30). Once again, we see that an axisymmetric dynamo is not possible.

We now turn to our final kinematic result, in which we seek an evolution equation for the integrated axial field. In the process we shall discover the key to circumventing Cowling's theorem. We start with (21.19), which, using Ohm's law, can be rewritten as

$$\mathbf{m} = \frac{3}{2\mu_0} \int_{V_C} \mathbf{B}\, dV = \frac{\sigma}{2} \int_{V_C} \mathbf{x} \times (\mathbf{E} + \mathbf{u} \times \mathbf{B})\, dV. \tag{21.34}$$

Faradays law, plus the identity

$$\nabla \times [(\mathbf{x}^2/2)\mathbf{E}] = (\mathbf{x}^2/2)\nabla \times \mathbf{E} + \mathbf{x} \times \mathbf{E},$$

allows us to rewrite this in the form

$$\frac{3}{2\mu_0} \int_{V_C} \mathbf{B}\, dV = \frac{\sigma}{2} \int_{V_C} \mathbf{x} \times (\mathbf{u} \times \mathbf{B})\, dV - \frac{\sigma}{4} \int_{V_C} (R_C^2 - \mathbf{x}^2) \frac{\partial \mathbf{B}}{\partial t}\, dV. \tag{21.35}$$

This may be rearranged as

$$\frac{d}{dt}\int_{V_C}(R_C^2-\mathbf{x}^2)\mathbf{B}\,dV = 2\int_{V_C}\mathbf{x}\times(\mathbf{u}\times\mathbf{B})\,dV - 6\lambda\int_{V_C}\mathbf{B}\,dV, \tag{21.36}$$

with axial component

$$\frac{d}{dt}\int_{V_C}(R_C^2-\mathbf{x}^2)B_z\,dV = 2\int_{V_C}s(\mathbf{u}\times\mathbf{B})_\theta\,dV - 6\lambda\int_{V_C}B_z\,dV. \tag{21.37}$$

This tells us that the integrated axial field is maintained by the volume integral of $s(\mathbf{u}\times\mathbf{B})_\theta$. However, the first integral on the right of (21.37) is zero when the velocity and magnetic fields are axisymmetric. That is, if $\mathbf{B}$ is axisymmetric, then we have $\mathbf{B}=\mathbf{B}_P+\mathbf{B}_\theta$, where $\mathbf{B}_P=\nabla\times\mathbf{A}_\theta$. We then find that $s(\mathbf{u}\times\mathbf{B})_\theta=-\mathbf{u}_P\cdot\nabla(sA_\theta)$ and hence

$$\int_{V_C}s(\mathbf{u}\times\mathbf{B})_\theta\,dV = -\oint_{S_C}(sA_\theta)\mathbf{u}_P\cdot d\mathbf{S}=0, \tag{21.38}$$

where $S_C$ is the surface of $V_C$. We conclude that dynamo action in a sphere is not possible if $\mathbf{u}$ and $\mathbf{B}$ are both axisymmetric. Of course, this is Cowling's theorem.

Suppose now that $\mathbf{B}$ and $\mathbf{u}$ are non-axisymmetric, with $\mathbf{B}=\bar{\mathbf{B}}(s,z)+\mathbf{b}$ and $\mathbf{u}=\bar{\mathbf{u}}(s,z)+\mathbf{v}$, where $\bar{\mathbf{B}}$ and $\bar{\mathbf{u}}$ are azimuthal averages of $\mathbf{B}$ and $\mathbf{u}$. Since $\mathbf{b}$ and $\mathbf{v}$ have zero azimuthal means, the integrals of $s(\mathbf{v}\times\bar{\mathbf{B}})_\theta$ and $s(\bar{\mathbf{u}}\times\mathbf{b})_\theta$ are both zero and so (21.37) becomes

$$\frac{d}{dt}\int_{V_C}(R_C^2-\mathbf{x}^2)\bar{B}_z\,dV = 2\int_{V_C}s(\mathbf{v}\times\mathbf{b})_\theta\,dV - 6\lambda\int_{V_C}\bar{B}_z\,dV. \tag{21.39}$$

This tells us that the dipole field is supported by an azimuthal EMF associated with *non-axisymmetric* fluctuations in both $\mathbf{u}$ and $\mathbf{B}$. We shall now show that this EMF arises from the interaction of *helical* disturbances with the east–west field.

### 21.3.5 Parker's Lift-and-Twist Mechanism and Mean-Field Electrodynamics

Parker (1955) was the first to suggest that an important part of the dynamo cycle is the local EMF induced by small helical disturbances lifting and twisting the background field. We now describe Parker's all-important lift-and-twist mechanism, which is known as the $\alpha$-effect. To focus thoughts, we shall consider a single disturbance $\mathbf{v}$, with kinetic helicity density $\mathbf{v}\cdot\nabla\times\mathbf{v}$, spiralling through the east–west field, as shown in Figure 21.13.

Suppose we have a local helical disturbance with both axial velocity and axial vorticity, so that the kinetic helicity is $h_\omega \sim \upsilon_z\omega_z$. Recalling that $\boldsymbol{\omega}=\nabla\times\mathbf{v}$ is twice the angular velocity of a fluid particle, it is clear that $\mathbf{v}$ will tend to lift and twist the field lines. It seems plausible that the resulting perturbation in magnetic flux will be associated with a local current, or EMF, which has a component parallel to the background east–west field, as

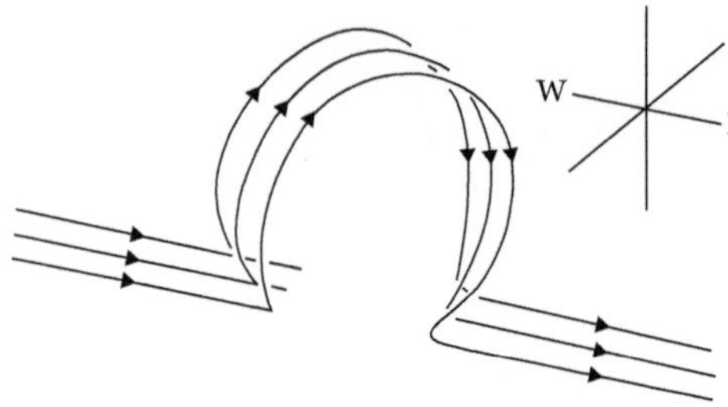


**Figure 21.13** A helical vortex lifts and twists the **B**-lines. (Adapted from T. G. Cowling. *Magnetohydrodynamics*. Interscience, 1957.)

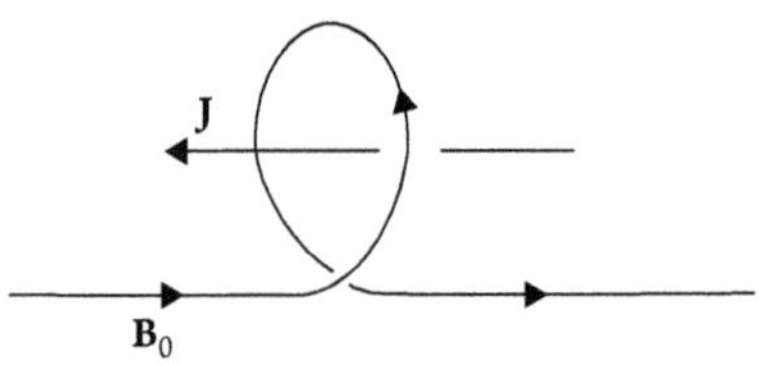


**Figure 21.14** A helical disturbance with positive (negative) kinetic helicity will tend to generate an EMF that is antiparallel (parallel) to the background field $\bar{\mathbf{B}}$.

shown in Figure 21.14. The linearity of Maxwell's equations then tells us that the resulting EMF will be proportional to the local background field $\bar{\mathbf{B}}$, and indeed we might expect it to be proportional to the product of the kinetic helicity and $\bar{\mathbf{B}}$. Moreover, if the twist of the field lines is right-handed, corresponding to positive kinetic helicity in the disturbance, it seems plausible from Figure 22.14 that the local current or EMF is antiparallel to $\bar{\mathbf{B}}$, whereas a left-handed twist (negative helicity) will tend to give rise to a local EMF that is parallel to $\bar{\mathbf{B}}$. Putting all of this together, we might expect the local EMF to be proportional to $-h_\omega \bar{\mathbf{B}}$. We shall see shortly that this is indeed the case.

Now, the small local EMF generated by a single disturbance would be of little interest if an adjacent disturbance generated an EMF of the opposite sign, since the two would tend to cancel. However, if the kinetic helicities of all of the disturbances were of the same sign in a given hemisphere, then all the EMFs would point in the same direction, adding together like batteries wired in series to drive a large-scale azimuthal current in the core, a current that can support a dipole field in accordance with (21.39). So Parker's lift-and-twist mechanism relies crucially on most of the disturbances in a given hemisphere having helicities of the same sign.

A formal calculation of the lift-and-twist EMF usually proceeds as follows. As before, we divide **B** and **u** into axisymmetric and non-axisymmetric parts: $\mathbf{B} = \bar{\mathbf{B}}(s,z) + \mathbf{b}$ and $\mathbf{u} = \bar{\mathbf{u}}(s,z) + \mathbf{v}$. Next, we assume that the horizontal scale of the disturbance is sufficiently small, say around $\ell = 10$ km in the case of the earth, that we may treat the background field as locally uniform. Some additional assumptions are then required in order to make progress, and these vary somewhat from author to author. In Davidson (2016) it is assumed that the disturbance is columnar, aligned with the rotation axis and of maximal helicity, assumptions that are inspired by the results of the numerical dynamos (see § 21.3.7). It is also assumed that, on the small scale of the disturbance, the local magnetic Reynolds number is of the order of unity or less, $R_m = \upsilon\ell/\lambda \leq 1$, which is consistent with $u(R_C - R_i)/\lambda \sim 200$ and $\ell \sim 10$km. Finally, it is assumed that there are multiple disturbances homogeneously distributed in space. Given all of these assumptions, it may be shown that

$$\langle \mathbf{v} \times \mathbf{b} \rangle = -\frac{\ell^2}{2\lambda} \langle \mathbf{v} \cdot \nabla \times \mathbf{v} \rangle \bar{\mathbf{B}}_\theta, \tag{21.40}$$

where the angled brackets represent a local spatial average. Note that the EMF is proportional to $-h_\omega \bar{\mathbf{B}}_\theta$, as anticipated above. Similar estimates are given by other authors. It has become conventional to write expressions like (21.40) in the form $\langle \mathbf{v} \times \mathbf{b} \rangle = \alpha \bar{\mathbf{B}}_\theta$, which is the origin of the term $\alpha$-effect. Note that $\alpha$ is a pseudoscalar with the dimensions of a velocity. Although we have focused on helical disturbances punching through the east–west field, a similar analysis may be done for any background magnetic field that is normal to the rotation axis, $\bar{\mathbf{B}}_\perp$, and we obtain (Davidson, 2016)

$$\langle \mathbf{v} \times \mathbf{b} \rangle = -\frac{\ell^2}{2\lambda} \langle \mathbf{v} \cdot \nabla \times \mathbf{v} \rangle \bar{\mathbf{B}}_\perp = \alpha \bar{\mathbf{B}}_\perp, \tag{21.41}$$

Since $\mathbf{b}$ and $\mathbf{v}$ have zero azimuthal means, the *azimuthally averaged* induction equation now becomes

$$\frac{\partial \bar{\mathbf{B}}}{\partial t} = \nabla \times (\bar{\mathbf{u}} \times \bar{\mathbf{B}}) + \nabla \times (\alpha \bar{\mathbf{B}}_\perp) + \lambda \nabla^2 \bar{\mathbf{B}}.$$

This is the equation of *mean-field electrodynamics,* whose poloidal and azimuthal components are

$$\frac{\partial \mathbf{B}_P}{\partial t} = \nabla \times (\mathbf{u}_P \times \mathbf{B}_P) + \nabla \times (\alpha \mathbf{B}_\theta) + \lambda \nabla^2 \mathbf{B}_P, \tag{21.42}$$

$$\frac{\partial \mathbf{B}_\theta}{\partial t} = \nabla \times (\mathbf{u}_\theta \times \mathbf{B}_P) + \nabla \times (\mathbf{u}_P \times \mathbf{B}_\theta) + \nabla \times (\alpha B_s \hat{\mathbf{e}}_s) + \lambda \nabla^2 \mathbf{B}_\theta, \tag{21.43}$$

where we have omitted the overbars for clarity. These equations are axisymmetric and are best rewritten as evolution equations for $\chi$, the flux function for $\mathbf{B}_P$, and the azimuthal field $B_\theta$:

$$\frac{D\chi}{Dt} = \alpha s B_\theta + \lambda \nabla_*^2 \chi, \tag{21.44}$$

$$\frac{D}{Dt}\left(\frac{B_\theta}{s}\right) = \mathbf{B}_P \cdot \nabla \left(\frac{u_\theta}{s}\right) + \frac{1}{s}\frac{\partial}{\partial z}(\alpha B_s) + \lambda s^{-2} \nabla_*^2 (s B_\theta). \tag{21.45}$$

These might be compared with (21.33) and (21.30). Note that $\alpha B_\theta$ now appears as a source of azimuthal current, and hence poloidal field. Similarly, $\alpha B_s$ is a source of poloidal current, and hence east–west field. It is readily confirmed that these axisymmetric equations support dynamo action, provided that $|\alpha| R_C/\lambda$ is large enough (Moffatt, 1978).

### 21.3.6 *Two Popular Kinematic Dynamo Cartoons for Planets and Stars*

Turbulence in a rapidly rotating fluid tends to be dominated by a random sea of columnar vortices aligned with the rotation axis (Davidson, 2013). Most numerical simulations of the geodynamo show precisely that, with the columnar vortices lying outside the tangent cylinder, surrounding the inner core. In the weakly driven, highly viscous simulations the columnar vortices are laminar, relatively broad and span the core from mantle to mantle, and are arranged in cyclone ($u_\theta > 0$)–anticyclone ($u_\theta < 0$) pairs. By way of contrast, in the more strongly driven simulations the columnar vortices are thin and chaotic, and often do not intersect the mantle, though they still tend to appear in cyclone–anticyclone pairs (Figure 21.15). In either case the columnar vortices are observed to be highly helical and to drive an $\alpha$-effect outside the tangent cylinder.

There are two popular cartoons of planetary and stellar dynamos based on mean-field electrodynamics. In $\alpha - \Omega$ dynamos, differential rotation converts the dipole into an east–west field. To complete the regenerative cycle $\mathbf{B}_P \to \mathbf{B}_\theta \to \mathbf{B}_P$, Parker's $\alpha$-effect is then used to convert $\mathbf{B}_\theta$ back into a dipole field. By way of contrast, in $\alpha^2$ dynamos, any differential rotation is incidental to dynamo action. Instead, helical columnar vortices generate $\mathbf{B}_\theta$ from the radial field via an $\alpha$-effect, and then convert $\mathbf{B}_\theta$ back into a poloidal field using a second $\alpha$-effect. Note that, in either case, we require that the helicity be of uniform sign (at least statistically) in each hemisphere, so that the local EMFs are additive. Let us consider these two cartoons in a little more detail. To focus our thoughts, we shall assume that the dipole points to the north.

In an $\alpha-\Omega$ dynamo, the $\Omega$-effect sweeps out an azimuthal field that is negative in the north and positive in the south, as shown in Figure 21.10. Moreover, in order to support the original dipole field, we require azimuthal currents that are positive in both hemispheres. Given that $\langle \mathbf{v} \times \mathbf{b} \rangle_\theta \sim -h_\omega \bar{\mathbf{B}}_\theta$, such a dynamo clearly requires a kinetic helicity density that is predominantly positive in the north and negative in the south. However, most numerical simulations of the geodynamo yield a helicity distribution of the opposite sign outside the tangent cylinder (where the $\alpha$-effect resides), and in any case they show only a weak

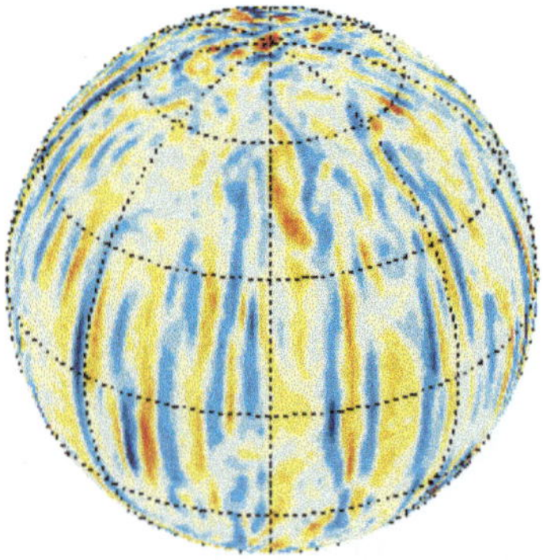

**Figure 21.15** The radial velocity near the mantle in a numerical simulation of the geodynamo. (Adapted from U. R. Christensen & J. Wicht. Numerical dynamo simulations. In: P. Olsen (ed.), *Core Dynamics* (*Treatise on Geophysics*, Volume 8) Elsevier, 2007, pp. 245–282.)

$\Omega$-effect. So these numerical dynamos are not of the $\alpha$-$\Omega$ type, but rather $\alpha^2$ dynamos. (An $\alpha$-$\Omega$ dynamo does, however, operate in the sun.)

So, in planetary dynamos, attention is focussed on the $\alpha$-effect associated with the columnar vortices. A typical example is shown in Figure 21.16, where, in line with most of the numerical simulations, the kinetic helicity outside the tangent cylinder is negative in the north and positive in the south. Here an $\alpha$-effect operating on the radial field $B_s$ drives a positive radial current in the north and south, and, since $\mathbf{J}_p$ is solenoidal, this establishes poloidal currents of quadrupole structure. Ampère's law then tells us that these give rise to an azimuthal **B** field that is positive in the north and negative in the south. A second $\alpha$-effect operating on the azimuthal **B** field then yields azimuthal currents that are positive both north and south, supporting the original dipole field.

Note that this kind of $\alpha^2$ dynamo relies crucially on a helicity distribution that is statistically antisymmetric about the equator. It is natural to ask what maintains this helicity and why it is spatially segregated, being predominantly positive in the south and negative in the north.

### *21.3.7 Where Does the Helicity Come From and Why is it Spatially Segregated?*

There is still little agreement as to exactly why the helicity in the numerical dynamos is negative in the north and positive in the south (outside the tangent cylinder). There is, however, a plausible cartoon. This is based on the observation that a rapidly rotating fluid

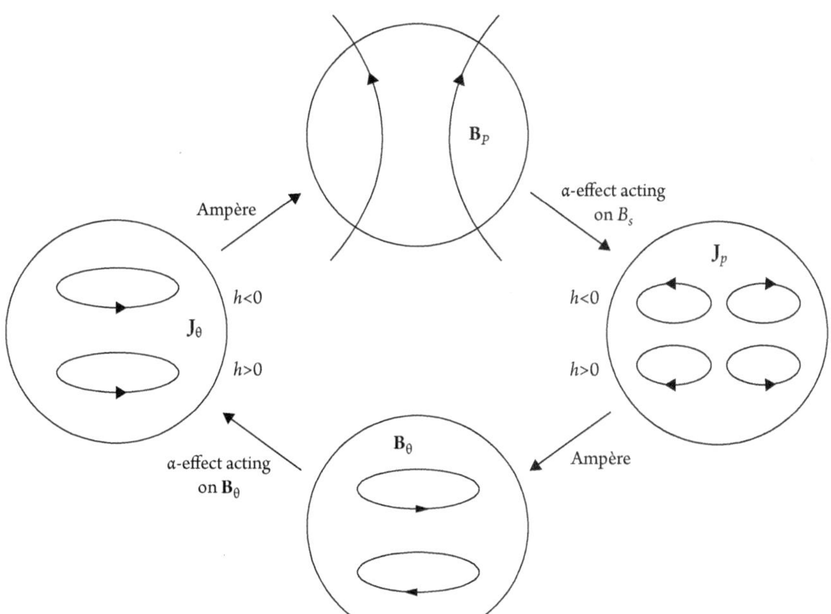


**Figure 21.16** An $\alpha^2$ dynamo with positive (negative) helicity in the south (north).

supports a form of internal wave motion, called *inertial waves*, and these waves disperse helicity in a systematic way. Our discussion follows that of Davidson & Ranjan (2018).

Consider a rapidly rotating fluid in which the Rossby number is small, Ro << 1. Ignoring Lorentz forces, the inviscid equation of motion in the rotating frame is

$$\frac{\partial \mathbf{u}}{\partial t} = 2\mathbf{u} \times \boldsymbol{\Omega} - \nabla \left(\frac{p}{\rho}\right), \tag{21.46}$$

where the irrotational centrifugal force has been absorbed into the pressure term. The corresponding vorticity equation is

$$\frac{\partial \boldsymbol{\omega}}{\partial t} = 2(\boldsymbol{\Omega} \cdot \nabla)\mathbf{u}, \tag{21.47}$$

and the operator $\nabla \times (\partial/\partial t)$ applied to (21.47) yields the wave-like equation

$$\frac{\partial^2}{\partial t^2}(\nabla^2 \mathbf{u}) + 4(\boldsymbol{\Omega} \cdot \nabla)^2 \mathbf{u} = \mathbf{0}. \tag{21.48}$$

This allows for plane waves of the form $\mathbf{u} = \hat{\mathbf{u}} \exp[j(\mathbf{k} \cdot \mathbf{x} - \varpi t)]$, which gives the dispersion relationship

$$\varpi = \pm \frac{2\mathbf{k} \cdot \boldsymbol{\Omega}}{k}, \quad k = |\mathbf{k}|. \tag{21.49}$$

Clearly, inertial waves are restricted to the frequency range $0 \leq \varpi \leq 2\Omega$. Their group velocity, which is the velocity at which wave packets disperse from a localised disturbance, is then

$$\mathbf{c}_g = \pm 2k^{-3}\mathbf{k} \times (\boldsymbol{\Omega} \times \mathbf{k}) = \pm 2k^{-3}[k^2\boldsymbol{\Omega} - (\mathbf{k} \cdot \boldsymbol{\Omega})\mathbf{k}]. \tag{21.50}$$

Note that the wave packets with the lowest frequency have the largest group velocity. Note also that

$$\mathbf{c}_g \cdot \boldsymbol{\Omega} = \pm 2k^{-3}[k^2\Omega^2 - (\mathbf{k} \cdot \boldsymbol{\Omega})^2], \tag{21.51}$$

which tells us that the positive (negative) sign in (21.49) corresponds to wave energy propagating upward (downward).

Crucially, inertial waves are helical. To see why, consider the dispersion relationship (21.49) combined with (21.47). This yields $\hat{\boldsymbol{\omega}} = \mp |\mathbf{k}|\, \hat{\mathbf{u}}$, where $\hat{\boldsymbol{\omega}}$ is the amplitude of the vorticity. We conclude that the vorticity and velocity fields are parallel and in phase. Evidently, inertial waves have the maximum possible helicity, with the positive sign in (21.51) corresponding to negative helicity and vice versa. Thus, when wave packets disperse from a localised disturbance, those wave packets that travel upward carry negative helicity, while wave packets that travel downward carry positive helicity. This phenomenon is illustrated by the simple computation shown in Figure 21.17, where a patch of turbulence in a rapidly rotating fluid emits inertial waves. The left-hand image is the initial condition and the right-hand one corresponds to $\Omega t = 6$. Green indicates positive helicity and red negative helicity,

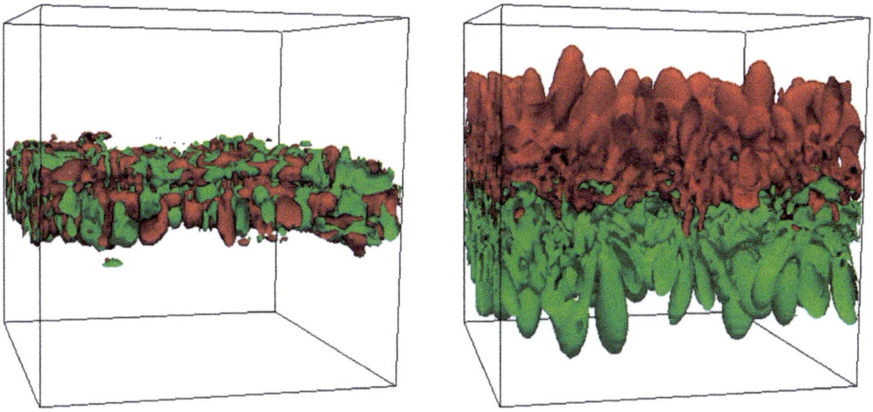

**Figure 21.17** Inertial-wave packets emerge from a patch of turbulence at Ro = 0.1. Red is negative helicity and green is positive helicity. (From P. A. Davidson. *Turbulence in Rotating, Stratified and Electrically Conducting Fluids.* Cambridge University Press, 2013.)

and it is clear that the upward (downward)-propagating wave packets possess negative (positive) helicity.

Next, consider an isolated blob of buoyant material of scale $\ell$ sitting in the outer core. The normalised density perturbation $b = \delta\rho/\rho$ is taken to be small. Adopting local Cartesian coordinates with $z$ pointing to the north and $x$ radially outward, we have

$$\frac{\partial \mathbf{u}}{\partial t} = 2\mathbf{u} \times \mathbf{\Omega} - \nabla\left(\frac{p}{\rho}\right) + b\mathbf{g}, \quad \mathbf{g} = -g\hat{\mathbf{e}}_x. \tag{21.52}$$

Now, $\delta\rho$ is advected on the slow timescale of $\mathbf{u}$, while inertial waves evolve on the fast timescale of $\Omega^{-1}$, so we may take $\delta\rho$ as quasi-steady as far as the inertial waves are concerned. Treating $\delta\rho$ as constant, the operator $\nabla^2(\partial/\partial t)$ applied to (21.52) yields

$$\frac{\partial^2}{\partial t^2}(\nabla^2\mathbf{u}) + (2\mathbf{\Omega}\cdot\nabla)^2\mathbf{u} = (2\mathbf{\Omega}\cdot\nabla)(\mathbf{g}\times\nabla b). \tag{21.53}$$

Evidently our buoyant blob acts as a source of inertial waves. It turns out that these waves propagate at a speed of $\Omega\ell$ in the $\pm\mathbf{\Omega}$ directions, taking the form of columnar wave packets that carrying negative helicity to the north and positive helicity to the south. The dispersion pattern is discussed in Davidson & Ranjan (2018). It consists of a pair of columnar vortices above and below the buoyant blob in the form of a cyclone ($u_\theta > 0$) adjacent to an anti-cyclone ($u_\theta < 0$). This pattern is shown in Figure 21.18 and is reminiscent of the velocity distribution seen in the alternating cyclone–anticyclone pairs observed in the numerical simulations of the geodynamo (Figure 21.18(c)).

Now it so happens that, in the numerical simulations of the geodynamo, the flux of buoyancy from the inner core to the mantle is biased towards the equatorial regions, and so there is a statistical excess of buoyant anomalies in that region. Each of those anomalies will

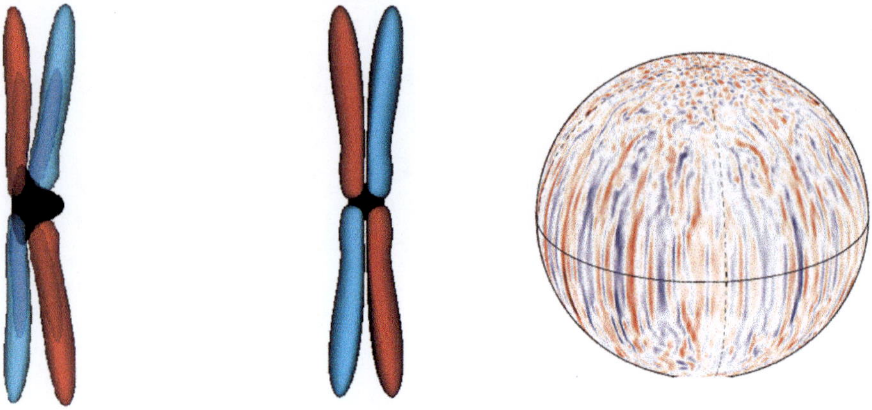


**Figure 21.18** (a) A buoyant blob emits inertial waves at Ro = 0.1. The shading is blue for $u_z < 0$ and red for $u_z > 0$. (b) As for (a), with Ro = 0.01. (c) Radial velocity in the cyclone–anticyclone pairs in a dynamo simulation. (From P. A. Davidson & A. Ranjan. Planetary dynamos driven by helical waves—II. *Geophys. J. Int.* **202**, 1646–1662 (2015).)

disperse inertial wave packets in a manner similar to that shown in Figures 21.17 and 21.18, and the suggestion is that this might explain the observed segregation of helicity above and below the equator. Of course, there are many complications, such as the dissipation of waves through Ohmic losses, the reflection of waves at the mantle and the influence of the magnetic field on the dispersion relationship.

This concludes our all too brief discussion of planetary dynamos. There are still many gaps in our understanding, and so this remains an active area of research. Readers seeking more details might try Moffatt (1978), Jones (2011) or Davidson (2017).

## 21.4 Stellar MHD

### *21.4.1 The Structure of the Sun*

We now provide a brief overview of stellar MHD. The discussion is entirely qualitative. The sun is composed mostly of hydrogen and helium. It has a radius $R_\odot = 6.96 \times 10^5$ km and an equatorial rotation period of ~26 days. With a Rossby number in the outer convective regions of around Ro ~ 0.2-1.0, its interior is not dominated by the Coriolis force in the way that the earth is, although rotation does still play a crucial role.

The interior of the sun is usually divided into three zones, as shown in Figure 21.19. The *core* is the location of the nuclear fusion. It has a radius $\sim 0.25R_\odot$, is of high density and contains around half of the sun's mass. The *radiative zone* that surrounds the core is quiescent, rotates as a rigid body, and extends out to a radius of $4.9 \times 10^5$ km. The outermost

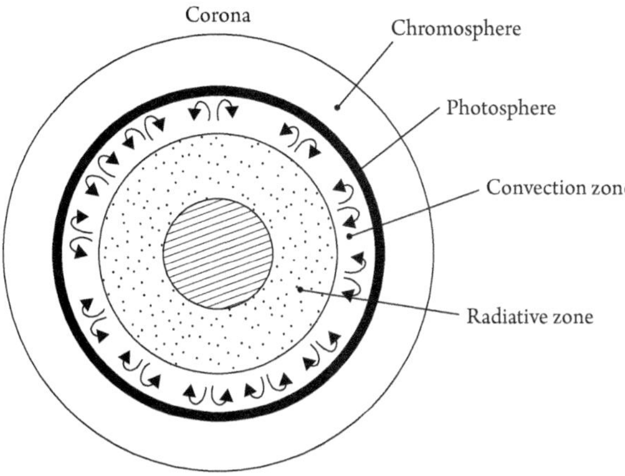


**Figure 21.19** The structure of the sun. The depths of the photosphere and chromosphere are greatly exaggerated for clarity. (Reproduced from P. A. Davidson. *An Introduction to Magnetohydrodynamics*, 2nd ed. Cambridge University Press, 2017.)

region is called the *convection zone* and is dynamically unstable, transporting heat out to the *photosphere* (the solar surface) by turbulent convection. The dynamic nature of this outer layer is evident from the granular appearance of the photosphere, which is reminiscent of Bénard convection. Although the convection zone is $2 \times 10^5$ km deep, it contains only 2% of the sun's mass. Rotation in the convection zone is non-uniform, and so this layer is a prolific generator of an east–west magnetic field through the $\Omega$-effect, a magnetic field that then forms coronal flux loops and sunspots.

Let us consider this differential rotation in a little more detail. The equator rotates faster than the poles at the photosphere, while the radiative zone rotates uniformly at some intermediate rate. Helioseismology shows that the transition from differential rotation in the convection zone to rigid-body rotation in the radiation zone occurs in a thin spherical layer at the base of the convection zone, known as the *tachocline*. This intense shear layer generates an east–west field via the $\Omega$-effect and, based on observations of the magnetic field in sunspots (which has its origins in the tachocline), this east–west field is thought to be of the order of 3000 G. By way of contrast, the dipole field at the poles is a mere 10 G.

By convention, the solar atmosphere is also divided into three regions. The photosphere is a thin layer of relatively dense material with a mean temperature of around 5800 K. Above this lies the chromosphere, which is 2000 km deep and has a temperature that rises steadily with altitude, up to a maximum of 20 000 K. The outermost layer is called the corona, and on entering this region there is a dramatic rise in temperature, from 20 000 K at the top of the chromosphere to $10^6$ K at the base of the corona. The corona has no clear upper boundary, ultimately turning into the solar wind that spirals radially outward from the sun. The unexpectedly high temperature of the corona has been the subject of much speculation and is possibly due to *nano-flares*, which are associated with rapid reconnection of coronal flux loops. These flux loops arch up into the corona from the convection zone and there is a continual jostling of their foot-points in the turbulent photosphere. This causes the flux loops to stretch and intensify in the corona and occasionally the coronal field lines snap back to a state of lower magnetic energy, generating heat and triggering a flare.

### *21.4.2 Sunspots, the 22-Year Solar Cycle and the Solar Dynamo*

Let us now turn to sunspots, the observation of which has offered many clues to events deep within the interior of the sun. The east–west flux tubes generated at the tachocline by the $\Omega$-effect have an internal pressure that is lower than the ambient, the difference being the magnetic pressure $B^2/2\mu$. This makes them mutually buoyant, and every now and then a flux tube becomes unstable and breaks loose from the tachocline. Some of the detached flux loops disintegrate while others remain coherent as they float up through the convection zone and into the corona (Figure 21.20). Sunspots are the foot-points of these flux tubes in the photosphere. They usually occur in pairs and appear relatively dark because the intense magnetic field in a spot suppresses convection, cooling the surface. Since they are associated with east–west fields, sunspot pairs are roughly aligned with a line of latitude. They rotate with the surface of the sun and, of course, the leading and following spots have opposite signs of $\mathbf{B} \cdot d\mathbf{S}$, where $d\mathbf{S}$ is part of the solar surface.

That part of the photosphere which is associated with sunspots varies on a regular 11-year cycle. At sunspot minimum new spots appear first at latitudes of around $\pm 30°$. The number of spots then rises, particularly in the equatorial regions, until finally the sunspots fade away and the cycle restarts. The east–west field linking sunspot pairs is observed to have the same orientation at all points in the northern and southern hemispheres, but this orientation reverses as we move from one hemisphere to the other. This tells us that the east–west field in the tachocline is unidirectional in each hemisphere and changes sign at the equator, which is consistent with an azimuthal field being swept out from a dipole field by differential rotation (Figure 20.13). Interestingly, the magnetic field linking sunspot pairs in any one hemisphere reverses from one 11-year cycle to the next, indicative of a 22-year periodic variation of the subsurface east–west field. This, in turn, is associated with a 22-year oscillation of the sun's

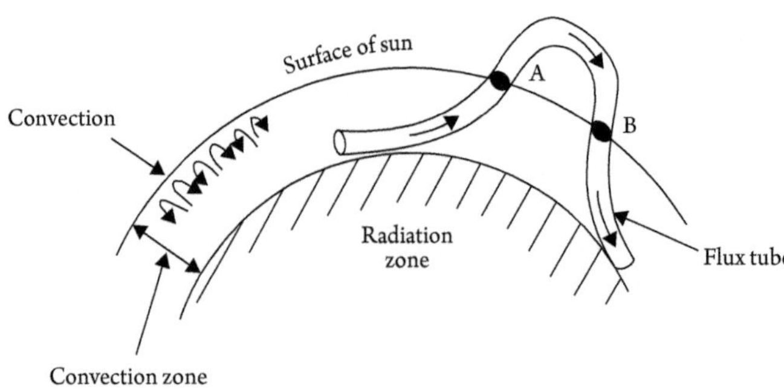


**Figure 21.20** A buoyant flux tube rises through the convection zone and into the corona. Sunspots form at A and B in the photosphere. (Reproduced from P. A. Davidson. *Turbulence in Rotating, Stratified and Electrically Conducting Fluids.* Cambridge University Press, 2013.)

dipole field, which reverses at sunspot maximum. Such oscillations are typical of an $\alpha$-$\Omega$ dynamo (Moffatt, 1978).

The natural decay time for a dipole field in the sun is around $10^{10}$ years, which is roughly the age of the sun itself. So it is not inconceivable that the solar magnetic field is the relic of some galactic field that was trapped in the sun at the time of its formation. However, the periodic 22-year variation of the sun's magnetic field suggests that theories based on a fossil relic are implausible and that the dipole is maintained by dynamo action within the convection zone. Perhaps it is worth noting that the dynamo theories that have been developed for the earth and the sun are very different. For example, in the core of the earth velocities are less than a millimetre per second, and as a result $R_m$ is at most a few hundred, while in the convection zone velocity fluctuations are of the order of a kilometre per second, which yields $R_m \sim 10^8$. Moreover, the prolific east–west field generation in the tachocline means that the solar dynamo is almost certainly of the $\alpha$-$\Omega$ type, whereas numerical simulations of the geodynamo suggest that the earth has an $\alpha^2$ dynamo.

Our current understanding of magnetic activity in the convection zone is still limited. One area of uncertainty is the fate of the buoyant flux tubes that break loose from the tachocline. While stability analysis has revealed the conditions under which such flux tubes can rise up into the convection zone, it is still far from clear what determines the ultimate fate of these flux tubes and the conditions under which they make it all the way to the corona. The nature of the solar $\alpha$-$\Omega$ dynamo is also far from settled. While the $\Omega$-effect in the tachocline is clear cut, the precise manner in which the $\alpha$-effect manifests itself remains a matter of considerable debate.

### *21.4.3 The Solar Atmosphere: Prominences, Flares and Coronal Mass Ejections*

As already noted, the atmosphere of the sun is threaded with vast magnetic flux tubes that arch up from the photosphere into the corona (Figure 21.21). Some of these flux tubes are associated with *prominences*, which are vast tubular arches around $10^5$km in length and perhaps $10^4$ km thick (Figure 21.22). Inside a typical prominence there is a billion tons of cold chromospheric gas, along with a helical magnetic field that spirals along the arch. Prominences are criss-crossed by thinner flux tubes that also arch up from the photosphere. Some of these thin flux tubes lie below a prominence, providing a *magnetic cushion* that counters the weight of the chromospheric gas. Other flux tubes form *magnetic arcades*, which overlie a prominence and push downwards. Typically an X-like saddle point forms below a prominence in those field lines that separate the helical field within the prominence from the underlying cushion and overlying arcade.

Prominences are often stable and survive for many weeks. However, every now and then a rearrangement of the magnetic field surrounding a prominence can trigger an *eruptive prominence* that rises rapidly up through the corona, disappearing into the solar wind within an hour or so. Eruptive prominences are often accompanied by solar flares and the mass that is expelled from the sun in this way is an example of a *coronal mass ejection*. The energy that propels the prominence upward is stored magnetic energy, which is liberated by magnetic reconnection as the local magnetic field lines snap back to a lower-energy state. The largest

**Figure 21.21** Flux tubes in the corona. (Courtesy of the TRACE Project, NASA.)

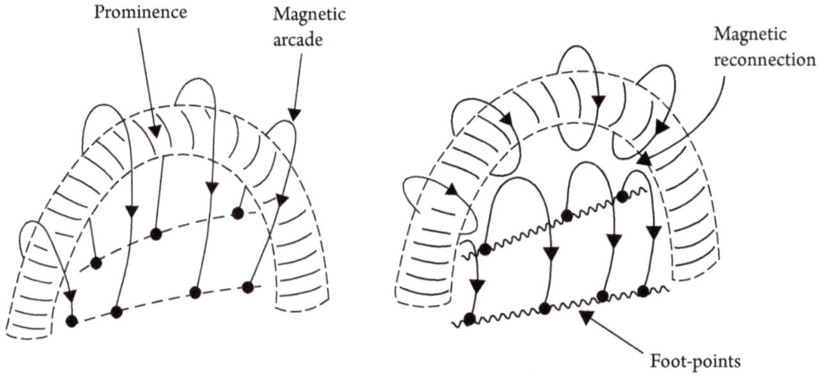


**Figure 21.22** A cartoon of a two-ribbon flare. (Reproduced from P. A. Davidson. *Turbulence in Rotating, Stratified and Electrically Conducting Fluids.* Cambridge University Press, 2013.)

flares may release around $10^{25}$ J and are rare events, while smaller flares are more common, occurring every other day at sunspot maximum.

The largest flares are called *two-ribbon flares* and are thought to occur in the following way. Consider a prominence that is supported by a magnetic cushion and has a magnetic arcade above it. These flux tubes are anchored in the photosphere and so their foot-points are subject to its turbulent motion. Typically this photospheric turbulence causes stretching of the coronal field lines and so their magnetic energy and tension increase with time. If there is a build-up of magnetic pressure in the magnetic cushion below a prominence, then it can start to rise, causing flux tubes of opposite polarity to be pushed together at the X-point below the prominence. Magnetic reconnection can then occur, allowing the arcade field lines to pinch off, as shown in Figure 21.22. When this occurs, the downward force associated with the overlying flux tubes is removed and so the prominence is propelled upward by the pressure in the underlying magnetic cushion. During magnetic reconnection, transient electric fields accelerate charged particles down the arcade field lines to the photosphere, creating two energetic ribbons of H$\alpha$ emission: hence the name 'two-ribbon flare'.

Flares are sometimes accompanied by a coronal mass ejection (CME). The causes of CMEs are still debated, but many are associated with eruptive prominences. Whatever their cause, CMEs are vast events, with some 5–50 billion tons of coronal material expelled into the solar wind. Both flares and CME activity follow the 11-year solar cycle, being most active at sunspot maximum, although the exact relationship between the two phenomena is not well understood. In any event, flares and CMEs are dramatic events, releasing mass and energy that enhances the solar wind. At sunspot maximum the concentration of particles in the solar wind increases, as does the velocity of the wind, and one or two days after a large flare, the earth is buffeted by magnetic storms.

This concludes our brief discussion of magnetic activity in the solar interior and atmosphere. The nature of the solar dynamo and the associated 22-year solar cycle remains poorly understood, as does the relationship among prominences, magnetic reconnection, flares and CMEs. Readers interested in more details of the magnetic interior might wish to consult Hughes, Rosner & Weiss (2007) or Davidson (2013), while Cravens (1997) provides a measured introduction to the solar atmosphere.

### *21.4.4 The Solar Wind*

In 1958 Eugene Parker noted that the solar corona is too hot to stay gravitationally bound to the sun, and so he suggested that the outer regions of the corona must spiral radially outward from the sun in the form of a continuous plasma wind. The occasional ejection of coronal plasma from the sun had long been recognised as a feature of terrestrial magnetic storms, but Parker's suggestion of a *continuous wind* was at first controversial. However, the existence of such a plasma wind was confirmed the following year by the Soviet *Luna* satellites.

The solar wind is extremely tenuous, consisting of a collision-free ionised hydrogen plasma with a number density of a few protons per cm$^3$. It flows out along open magnetic field lines, sweeping past the earth at a speed of $\sim$400 km/s. The sound speed in the wind near the earth is around 70 km/s, and so the wind is supersonic as it spirals through the solar system. As the wind approaches the earth a shock wave forms, the wind slows down and the geomagnetic field is swept back into a region called the magnetosphere (Figure 21.23). The boundary between the wind and earth's field is called the magnetopause.

It is interesting to consider why Parker concluded that the corona cannot maintain a static atmosphere. In order to keep the discussion simple, we shall ignore the coronal magnetic field and take the corona to be spherically symmetric. In a static corona bound to the sun, heat is transferred through the corona by conduction only, and it turns out that the thermal conductivity of such a plasma, $\kappa$, is proportional to the plasma temperature raised to the

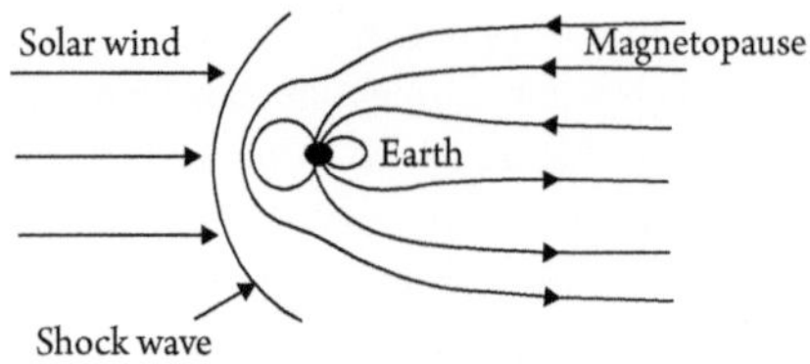


**Figure 21.23** The interaction of the solar wind with the earth's magnetic field. The geomagnetic field is confined to the magnetosphere. (Reproduced from P. A. Davidson. *Turbulence in Rotating, Stratified and Electrically Conducting Fluids.* Cambridge University Press, 2013.)

power of 5/2, say $\kappa = \kappa_0 T^{5/2}$. The net heat flux out through a sphere of radius $r$ within the corona is then independent of radius and equal to

$$\dot{Q} = -4\pi r^2 (\kappa_0 T^{5/2}) \frac{dT}{dr}. \tag{21.54}$$

Let $T_c$ be the temperature at the base of the corona, $r = R_c$, and note that $T << T_c$ for $r >> R_c$. Since $\dot{Q}$ is independent of $r$, (21.54) integrates to give $T/T_c = (R_c/r)^{2/7}$ (see Exercise 21.5). We now combine this temperature profile with the ideal gas law $p = \rho R_g T$ and with the static force balance

$$\frac{dp}{dr} = -\frac{GM_{\odot}\rho}{r^2}, \tag{21.55}$$

where $R_g$ is the gas constant, $G$ the universal gravitational constant and $M_{\odot}$ the solar mass. After a little algebra, this gives the pressure distribution in a static corona as

$$\frac{p}{p_c} = \exp\left\{-\frac{7GM_{\odot}}{5R_g T_c R_c}\left[1 - \left(\frac{R_c}{r}\right)^{5/7}\right]\right\}, \tag{21.56}$$

where $p_c$ is the pressure at the base of the corona. This predicts a pressure in the outer parts of the corona of $p_\infty \approx 10^{-8}\,\mathrm{N/m^2}$, whereas the pressure in the interstellar medium is closer to $10^{-14}\,\mathrm{N/m^2}$. Evidently a static corona is not possible and so Parker concluded that the coronal plasma must spiral radially outward from the sun.

Of course, all of this ignores the coronal magnetic field. In fact, magnetic fields play an important role in guiding the coronal plasma through the corona and into the solar wind. At the base of the corona the plasma velocity is small and **B** large, and so inertia is weak and the plasma just follows the magnetic field lines. At large radius, however, the plasma velocity is large and **B** weak, and so the interplanetary field follows the plasma. Here the magnetic field lines are dragged radially through the solar system by the wind while being deformed into a spiral as their foot-points rotate with the sun (Figure 21.24). The distinctive spiralled structure of the interplanetary field is known as the *Parker spiral,* since Parker showed that the lines of force form an Archimedean spiral.

The spherical surface that more or less divides the region of strong magnetic fields and weak inertia from that of strong inertia and weak fields is sometimes called the *source surface.*

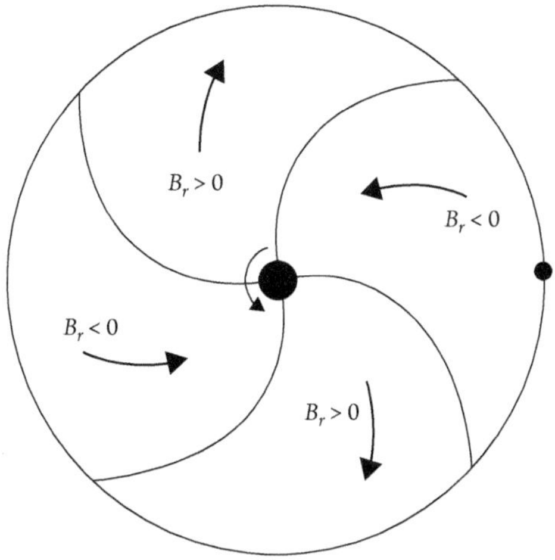


**Figure 21.24** The Parker spiral. (Reproduced from P. A. Davidson. *An Introduction to Magnetohydrodynamics*, 2nd ed. Cambridge University Press, 2017.)

It has a radius that is a few multiples of the solar radius and it is natural to classify the source surface as the point where the solar wind begins. The solar wind has become a very active area of research, in part because measurements made in the wind provide a natural test bed for various theories of MHD turbulence.

## EXERCISES

21.1 Consider the stability of the plasma column shown in Figure 21.2. Allow for a uniform toroidal magnetic field $B_T$ trapped within the plasma, as well as a poloidal field $B_P$ surrounding the column. Let $\Phi = B_T \pi a^2$ be the toroidal flux and $I = 2\pi a B_P/\mu_0$ the total current passing along the column. Show that, if the radius of the column, $a$, locally decreases by $\delta a$, then $B_P$ and $B_T$ rise by$\delta B_P = B_P\, \delta a/a$ and $\delta B_T = 2B_T\, \delta a/a$, respectively. Hence confirm that the inward magnetic pressure increases by $\delta p_m = (B_P^2 - 2B_T^2)\, \delta a/\mu_0 a$, and so the plasma is stable to sausage-mode disturbances provided that $B_T^2 > B_P^2/2$.

21.2 Consider small-amplitude Alfvén waves travelling along a uniform field $\mathbf{B}_0$. Show that the linearised momentum equation written in terms of Maxwell stresses is

$$\frac{\partial \mathbf{u}}{\partial t} = \frac{1}{\rho\mu_0}(\mathbf{B}_0 \cdot \nabla)\mathbf{b} - \nabla\left(\frac{p + p_m}{\rho}\right), \quad \nabla^2(p + p_m) = 0.$$

Explain why the pressure term, including the magnetic pressure, can be dropped from the linearised momentum equation.

21.3 Use the data in Table 21.1 to show that the net azimuthal current flowing in the core of the earth must exceed 2 GA.

21.4 Let us return to Cowling's neutral-point argument of § 21.3.5. Apply Ampère's circuital law to a field line of characteristic radius $\varepsilon$ surrounding the neutral point N to deduce $B\varepsilon \sim \mu_0 J\varepsilon^2$. Now combine this with Ohm's law $\mathbf{J} = \sigma\mathbf{u} \times \mathbf{B}$ to give $u\varepsilon/\lambda \sim 1$ and show that a contradiction arises as $\varepsilon \to 0$.

21.5 Show that (21.54) integrates to give $(T/T_c)^{7/2} = 1 - k(1 - R_c/r)$, where $k$ is positive and proportional to the heat flux. Now show that, if $T << T_c$ for $r >> R_c$, then $k = 1$ and we obtain $T/T_c = (R_c/r)^{2/7}$.

## REFERENCES


Boyd, T. J. M. & Sanderson J. J., 2003, *The Physics of Plasmas*. Cambridge University Press.

Christensen, U. R., 2011, Geodynamo models: tools for understanding properties of Earth's magnetic field. *Phys. Earth Planet. Inter.* **187**, 157–169.

Cowling, T. G., 1934, The magnetic field of sunspots. *Mon. Not. R. Astron. Soc.* **94**, 39–48.

Cravens, T. E., 1997, *Physics of Solar System Plasmas*. Cambridge University Press

Davidson, P. A., 2013, *Turbulence in Rotating, Stratified and Electrically Conducting Fluids*. Cambridge University Press.

Davidson, P. A., 2016, Dynamos driven by helical waves: scaling laws for numerical dynamos and for the planets. *Geophys. J. Int.* **207**, 680–690.

Davidson P. A., 2017, *An Introduction to Magnetohydrodynamics*, 2nd ed. Cambridge University Press.

Davidson, P. A. & Ranjan, A., 2018, Are planetary dynamos driven by helical waves? *J. Plasma Phys.* **84**(3), 735840304.

Hughes, D. W., Rosner, W. & Weiss, N. O. (eds.), 2007, *The Solar Tachocline*. Cambridge University Press.

Jones, C. A., 2011, Planetary magnetic fields and fluid dynamos. *Annu. Rev. Fluid Mech.*, **43**, 583–614.

Moffatt, H. K., 1978, *Magnetic Field Generation in Electrically Conducting Fluids*. Cambridge University Press.

Parker, E. N., 1955, Hydromagnetic dynamo models. *Astrophys. J.* **122**, 293–314.

# 22

# An Introduction to Special Relativity

Nothing puzzles me more than time and space; and yet nothing troubles me less, as I never think about them.

Charles Lamb, 1810

## 22.1 On the Curious Nature of Faraday's Flux Law

We noted at the start of Chapter 10 that Faraday's flux law is a remarkable result when viewed from a strictly classical point of view. Consider Figure 22.1, in which a wire loop moves relative to the poles of a magnet. On the left the magnet is stationary and the loop moves with speed $u$, while on the right we have changed frame of reference so that the loop is stationary and the magnet moves at the same speed but in the opposite direction. In either case a current is induced in the wire and this must be the same current.

Now, when the loop moves, we interpret the induced current to be a consequence of the Lorentz force $\mathbf{f} = q\mathbf{u} \times \mathbf{B}$ acting on the free charges in the wire, $\mathbf{u}$ being the velocity of the free charges. This is expressed in terms of an electromotive force, defined as the force acting on a unit test charge, $\mathbf{E} + \mathbf{u} \times \mathbf{B}$, integrated around the loop:

$$\text{EMF} = \oint \mathbf{u} \times \mathbf{B} \cdot d\mathbf{r} = -uBw, \tag{22.1}$$

where $w$ is the width of the wire. Of course, we may also write (22.1) in terms of Faraday's flux law

$$\text{EMF} = -\frac{d\Phi}{dt}, \tag{22.2}$$

where $\Phi$ is the magnetic flux through the loop. Note that any electric field that may be present in Figure 22.1(a) cannot contribute to the EMF in (22.2), because the magnetic field is steady, and so $\nabla \times \mathbf{E} = 0$ and $\oint \mathbf{E} \cdot d\mathbf{r} = 0$.

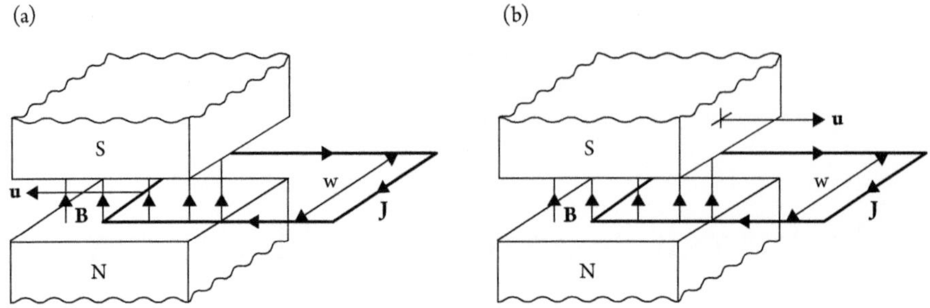


**Figure 22.1** Current induced in a wire loop by the relative motion of a magnet and the loop. (a) The magnet is stationary and the loop moves. (b) The loop is stationary and the magnet moves.

However, when the loop is held stationary and the magnet is pushed from left to right, we have $q\mathbf{u} \times \mathbf{B} = 0$, and so we interpret the induced current to be a consequence of the differential form of *Faraday's law*,

$$\nabla \times \mathbf{E} = -\frac{\partial \mathbf{B}}{\partial t}, \tag{22.3}$$

in which an induced electric field $\mathbf{E}$ drives the current. This too can be written in terms of Faraday's flux rule:

$$\text{EMF} = \oint \mathbf{E} \cdot d\mathbf{r} = \int_S \nabla \times \mathbf{E} \cdot d\mathbf{S} = -\int_S \frac{\partial \mathbf{B}}{\partial t} \cdot d\mathbf{S} = -\frac{d\Phi}{dt}. \tag{22.4}$$

So the same flux rule applies irrespective of whether the current is due to the Lorentz force or Faraday's differential law, yet from a classical perspective these are *distinct physical laws*. For Einstein, this was too much of a coincidence! (See also Exercise 22.5.)

## 22.2 A Re-examination of Space and Time

### 22.2.1 *Space and Time According to Newton and Galileo*

Let us take a step back and reconsider the classical view of space and time, as passed down by Newton and Galileo. As Galileo pointed out, no mechanical experiment can be performed that is capable of determining the absolute velocity of an observer. For example, we might consider, as Galileo did, a laboratory in the hold of a ship that is moving with constant velocity. If there are no windows in the hold, then there is no mechanical experiment that can be done that would determine the speed of the ship, or even indicate that it is in motion. Why is that? Suppose that we have two *inertial* reference frames $S$ and $S'$ that move relative to each other with a *constant* velocity $\mathbf{v}$. We suppose that, in each frame of reference, distance is measured using a Cartesian network of rigid metre sticks that move with that frame, while time in a given frame is measured by clocks that are distributed throughout space and rigidly attached to the metre sticks in that frame. Moreover, all clocks in a given frame are

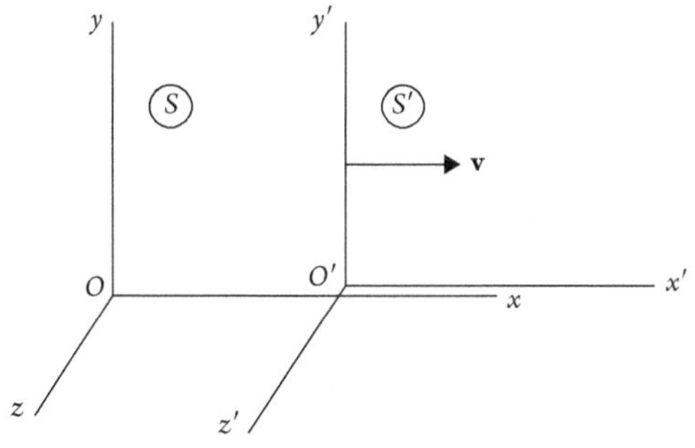


**Figure 22.2** Two inertial reference frames $S$ and $S'$.

synchronised so that, within that frame, all observers can agree what the time it is. Without any loss of generality, we may take $S'$ to be moving in the positive $x$ direction relative to $S$, as shown in Figure 22.2, and we synchronise the clocks in the two frames so that $t = t' = 0$ when the origins of the two coordinate systems $O$ and $O'$ are coincident.

If observers in the two frames measure a given event at locations $(x, y, z)$ and $(x', y', z')$, and at times $t$ and $t'$, then a *Galilean transformation* gives us the apparently obvious result that

$$x' = x - vt,\ y' = y,\ z' = z,\ t' = t. \tag{22.5}$$

Note, in particular, that we need not distinguish between $t$ and $t'$, so when we differentiate $x'$ with respect to either $t$ or $t'$ we get $\mathbf{u}' = \mathbf{u} - \mathbf{v}$, and, on differentiating once again, $\mathbf{a}' = \mathbf{a}$. So, if $m$ is the mass of an object that is subject to a force $\mathbf{F}$, then we have $\mathbf{F} = m\mathbf{a} = m\mathbf{a}'$, and consequently Newton's second law is obeyed in both reference frames. Since the laws of motion are identical in both frames, we cannot determine the absolute speed of either frame of reference by doing a mechanical experiment within that frame. Indeed, as far as mechanics is concerned, it makes no sense to talk of an absolute velocity, only the *relative* velocity between two frames.

Let us now consider wave motion from a classical point of view. Suppose we have a source of acoustic waves in frame $S$ and that the air is *stationary* in that frame of reference. If the source is located near the origin $O$, then wave fronts will travel out along the positive $x$ axis at the *acoustic speed a* in the frame $S$. The same wave fronts measured in frame $S'$, in which the air is moving backwards at velocity $\mathbf{v}$, will travel at $a - v$. Moreover, if the source moves along the $x$ axis in frame $S$, then this makes no difference to the observed speed of the wave fronts in either frame, since acoustic waves always travel at the same speed, $a$, relative to *still* air. This tells us that, when dealing with acoustic waves, we have a preferred inertial frame of reference: that in which the air is stationary.

### 22.2.2 *Maxwell, the Aether and a Crisis in the Newtonian Paradigm*

We now turn from acoustic waves to electromagnetic waves. When Maxwell first pointed out that light was an electromagnetic wave, travelling through free space at speed $c = 1/\sqrt{\mu_0 \varepsilon_0}$, physicists automatically assumed that there must be some form of medium through which these waves travelled. After all, mechanical waves, such as acoustic waves or gravity waves on the surface of a pond, require a medium through which to travel, a medium that supplies the inertia and restoring forces associated with all mechanical waves. So physicists returned to an idea favoured by the early Greeks, that of a *luminiferous aether* which permeates all space. This aether was invoked by nineteenth-century physicists strictly for the purposes of facilitating the propagation of electromagnetic waves. If such an aether exists, it follows from an analogy with mechanical waves that there is a preferred inertial frame of reference—that in which the aether is stationary and wave fronts travel at speed $c$. In all other frames of reference the wave fronts would travel at a different speed.

So, if we buy into the idea of an aether, we should be able to detect the motion of the earth through the aether by recording the speed of light in the earth's frame of reference and then comparing it with $c = 1/\sqrt{\mu_0 \varepsilon_0}$. After all, the earth orbits the sun at around 30 km/s. Although the speed of light is $10^4$ times larger, such an experiment is possible, if difficult. Famously, just such an experiment was undertaken by Michelson and Morley in 1887 using an optical interferometer. This, and all subsequent experiments, produced a null result—the measured speed of light is independent of the velocity of the earth and of the velocity of the source of the light. Various proposals were put forward to explain this, such as the notion that the earth drags the aether along with it, but these various explanations were either inconsistent with the experimental data or else simply implausible. Einstein, already uneasy with the role played by different frames of reference in the classical account of electromagnetic induction, was the first to fully embrace the null results of the Michelson–Morley experiment.

### 22.2.3 *Einstein's Postulates*

In 'On the Electrodynamics of Moving Bodies' Einstein introduces the special theory of relativity with the following statement (Einstein, 1905):

> Maxwell's electrodynamics—as usually understood at the present time—when applied to moving bodies leads to asymmetries which do not appear to be inherent in the phenomena. Take, for example, the reciprocal electrodynamic action of a magnet and a conductor. The observable phenomenon here depends only on the relative motion of the conductor and the magnet, whereas the customary view draws a sharp distinction between the two cases in which either one or the other of these bodies is in motion....
>
> Examples of this sort, together with unsuccessful attempts to discover any motion of the earth relative to the 'light medium', suggest that the phenomena of electrodynamics as well as of mechanics possess no properties corresponding to the idea of absolute rest. They suggest rather that... the same laws of electrodynamics and optics will be valid for all frames of reference for which the equations of mechanics hold good. We shall raise this conjecture (the purport of which will hereafter be called the 'Principle of Relativity') to the status of a postulate, and also introduce another postulate... that light is always propagated in empty space with a definite speed $c$ which is independent of the state of motion of the emitting body. These two postulates suffice for the attainment of a simple and consistent theory of the electrodynamics of moving bodies based on Maxwell's theory for stationary bodies. The introduction of a 'luminiferous aether' will prove superfluous.

We now explore this statement. We might summarise Einstein's two postulates as follows (though note Exercise 22.4):

(1) *The principle of relativity*: The laws of physics are the same in all inertial reference frames.

(2) *The universal speed of light*: The speed of light in vacuum is the same for all inertial observers.

The first of these postulates extends Galileo's observations about relativity in mechanical systems to include the laws of electrodynamics and then raises this conjecture to the status of a general law of physics. The second postulate is more radical. Having let go of the need for an aether, and armed with the results of the Michelson–Morley experiment, it is proposed that the observed speed of light is universal and independent of the speed of an inertial observer. This is in direct contradiction with our intuitive notions about the behaviour of mechanical waves, such as acoustic waves, and rests heavily on the conviction that there is no aether to which light waves are tied.

Einstein's postulates have a number of surprising consequences. The first and most urgent of these is that the Galilean transformation law (22.5) is in general incorrect for velocities comparable with $c$, and needs to be replaced by the so-called *Lorentz transformation*. But what can possibly be wrong with (22.5), which is so intuitively obvious? As we shall see, the key problem is that time is a much more subtle quantity than had hitherto been assumed. For example, two events that appear to be simultaneous in one inertial frame are not, in general, simultaneous when viewed from another frame.

### 22.2.4 *The Lorentz Transformation*

Let us return to the two inertial reference frames introduced in §22.2.1, $S$ and $S'$, which move relative to each other with a constant velocity $\mathbf{v}$ directed along the $x$ axis, as shown in Figure 22.2. Our aim is to use Einstein's postulates, along with symmetry, to derive a relationship between two records of the same event, $(x,y,z,t)$ and $(x',y',z',t')$, that is more general than (22.5) but reverts to the Galilean transformation for $|\mathbf{v}| << c$.

Let us suppose that the two sets of coordinates and times are related by

$$\begin{aligned}
x' &= a_{xx}x + a_{xy}y + a_{xz}z + a_{xt}t,\\
y' &= a_{yx}x + a_{yy}y + a_{yz}z + a_{yt}t,\\
z' &= a_{zx}x + a_{zy}y + a_{zz}z + a_{zt}t,\\
t' &= a_{tx}x + a_{ty}y + a_{tz}z + a_{tt}t,
\end{aligned}$$

where the coefficients $a_{pq}$ are constants. The reason for choosing a linear, rather than a non-linear, relationship is that the latter implies that the *arbitrary* choice of origin plays a role in the transformation of the length of a moving object or the time interval between two events. For example, the expression $x' = a_{xx}x$ would have the length of an object in the

$x$ direction transform as $\Delta x' = a_{xx}(x_2 - x_1) = a_{xx}\,\Delta x$, in which the origin of coordinates plays no role, whereas $x' = a_{xx}x^2$ would have length transform as $\Delta x' = a_{xx}\,\Delta x(x_2 + x_1)$, in which the origin of coordinates *does* play a role. If space is homogeneous, then such a situation is unacceptable, and the choice of a linear transformation is justified.

Next we note that the symmetry of the configuration shown in Figure 22.2 suggests that $y' = y, z' = z$, exactly as for the Galilean transformation (22.5). As noted by several authors, there is a simple thought experiment that confirms this. Suppose we place a sheet of glass with its lower edge on the $x$ axis and lying in the $x$–$y$ plane. In the frame $S$ we paint a horizontal blue line on the glass at a height of $y = 1$ m, as measured in frame $S$. We now supply a paintbrush and a metre stick to someone moving with frame $S'$ and ask them to paint a red line at the height of $y' = 1$ m, as measured using the metre stick. If the rule was that the length of an object (the metre stick) contracts perpendicular to $\mathbf{v}$ when viewed from a moving frame, then the observer in frame $S$ would see the red line below the blue line. However, to an observer in frame $S'$, it is the metre stick in $S$ that is moving, and so they would expect to see the blue line below the red line. Both observers cannot be correct, and so perpendicular lengths cannot contract or expand.

We can make one additional simplification using symmetry. The reflectional symmetry of Figure 22.2 demands that $t'$ cannot change if $y \to -y$ or $z \to -z$, and so we have $a_{ty} = a_{tz} = 0$. Finally, consider the motion of the origin $O'$ along the $x$ axis. Since we have synchronised the clocks in the two frames so that $t = t' = 0$ when the origins of the two coordinate systems are coincident, the positions of $O'$ in the frames $S$ and $S'$ are $x = vt$ and $x' = 0$, respectively. It follows from the $x$ transformation that

$$0 = a_{xx}vt + a_{xy}y + a_{xz}z + a_{xt}t,$$

and hence $a_{xx}v + a_{xt} = 0$ and $a_{xy} = a_{xz} = 0$. So our transformation equations reduce to

$$x' = a_{xx}(x - vt), \quad y' = y, \quad z' = z, \quad t' = a_{tx}x + a_{tt}t. \tag{22.6}$$

Note that (22.6) is perfectly consistent with the Galilean transformation (22.5), provided that we choose $a_{xx} = a_{tt} = 1$ and $a_{tx} = 0$. This is just as well, since we have yet to deploy Einstein's second postulate, and so we are still within the realms of classical mechanics. So let us now invoke the second postulate. Suppose that at $t = t' = 0$, when the origins of the two coordinate systems are coincident, a flashbulb is set off at the common origin. Then observers in either frame of reference see light travelling at the same speed, $c$, and so the resulting spherical wave front satisfies

$$x^2 + y^2 + z^2 = (ct)^2, \tag{22.7}$$

$$x'^2 + y'^2 + z'^2 = (ct')^2. \tag{22.8}$$

Substituting (22.6) into these equations yields the Lorentz transformation equations for the same event recorded in the two different frames of reference:

$$\boxed{\begin{aligned} x' &= \frac{x - vt}{\sqrt{1 - v^2/c^2}}, \quad y' = y, \quad z' = z, \\ t' &= \frac{t - vx/c^2}{\sqrt{1 - v^2/c^2}}. \end{aligned}} \tag{22.9}$$

Evidently, the imposition of the second postulate has introduced a non-trivial coupling between space and time, which can *no longer be regarded as independent entities*. The inverse transformation may be obtained from (22.9) through rearrangement, or by exchanging primed and unprimed variables while replacing $v$ by $-v$. Either way, we find

$$\boxed{\begin{aligned} x &= \frac{x' + vt'}{\sqrt{1 - v^2/c^2}}, \quad y = y', \quad z = z', \\ t &= \frac{t' + vx'/c^2}{\sqrt{1 - v^2/c^2}}, \end{aligned}} \tag{22.10}$$

where the ubiquitous pre-factor

$$\gamma = \frac{1}{\sqrt{1 - v^2/c^2}}$$

is called the *Lorentz factor*. Note that, for $v << c$, these expressions revert to a Galilean transformation. Note also that $v > c$ is forbidden by this transformation, so that no two inertial frames can have a velocity difference greater than $c$.

There are many surprises in store in these transformation equations. Most notably, they require that events viewed from moving frames of reference seem to take longer, which is known as *time dilation*, and that moving objects appear to shorten, which is known as the *Lorentz–FitzGerald contraction*. But perhaps the most profound implication is known as the *relativity of simultaneity*: that is to say, two events that appear to be simultaneous in one inertial frame are not, in general, simultaneous events when viewed from another frame. Let us explore these concepts one at a time.

### 22.2.5 *The Downfall of Simultaneity*

Consider two events that occur simultaneously in frame $S$ at locations $x_1$ and $x_2$. For example, suppose flashbulbs go off at the same instant at $x_1$ and $x_2$. Then, according to (22.9), an observer in frame $S'$ sees the bulbs go off at different times, with a time difference of

$$t'_2 - t'_1 = -\frac{v(x_2 - x_1)/c^2}{\sqrt{1 - v^2/c^2}}. \tag{22.11}$$

So events that occur simultaneously in one inertial frame need not appear to be simultaneous to an observer in another inertial frame. Just as there is no notion of absolute velocity, so

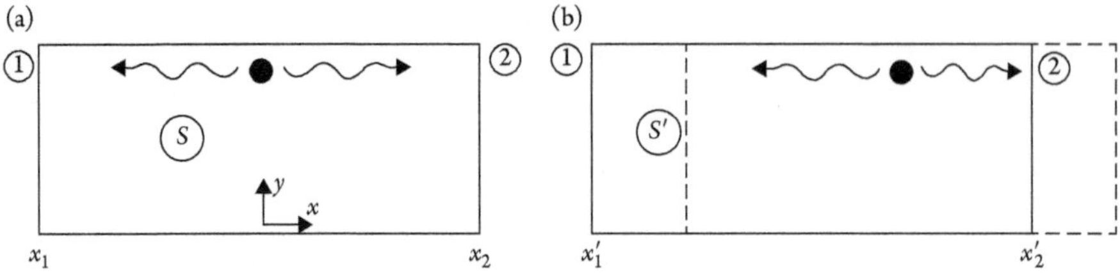


**Figure 22.3** Light travels from a flashbulb to the walls of a warehouse. (a) Events as seen by an observer in the warehouse. (b) Events as seen by an observer driving past.

we have lost the notion of absolute simultaneity. According to (22.11), if the frame $S'$ moves from $x_1$ towards $x_2$ ($\mathbf{v}$ points in the same direction as $\mathbf{x}_2 - \mathbf{x}_1$), then the observer in $S'$ sees the bulb at $x_2$ going off first. Conversely, if the frame $S'$ moves from $x_2$ towards $x_1$, then the observer sees the bulb at $x_1$ going off first.

This is so counterintuitive, it is worth taking a moment see if we can understand how it comes about. Let us consider a different situation. Suppose a flashbulb sits at the centre of a large warehouse (frame $S$) and is set off at $t = 0$. Meanwhile, you are in a car traveling along an adjacent road from left to right at a speed $v$. You are aligned with the centre of the warehouse just as the bulb goes off and observe the light travel from the flash to the walls of the warehouse. Whereas an observer standing under the flashbulb in the warehouse would detect the light from the flash reaching the left and right walls at the same instant, you observe something different. Let us label the light hitting wall 2 on the right of Figure 22.3 as event 2 and light hitting wall 1 on the left as event 1. Then for the observer in the warehouse these two events are simultaneous. For you, however, they are not. The point is that while the light travels from the bulb to the walls, the warehouse has moved in your frame of reference. By the time it strikes wall 2, that wall has moved to the left and so in your frame of reference the light has less distance to travel. Conversely, the light propagating to wall 1 has more distance to travel. Now, the key point is that, according to postulate (2), light waves observed from your frame of reference travel at the same speed $c$ as for the observer in the warehouse. So you observe event 2 before event 1, exactly as predicted by (22.11).

This demise of simultaneity has far-reaching consequences. For example, observers in different reference frames will measure different time intervals between events and so clocks in different inertial frames cannot stay synchronised, as we now discuss.

### 22.2.6 *Time Dilation and the Lorentz-FitzGerald Contraction*

Consider now two events that occur at different times but at the same location in frame $S$, for example the ticking of a metronome. Let the time between ticks be labelled $\Delta t_{\text{rest}}$, because the metronome is at rest in frame $S$. (Frame $S$ is called the *rest frame* of the metronome.) Then, according to (22.9), an observer in frame $S'$ sees the time period between ticks as

$$\Delta t' = \Delta t_{\text{moving}} = \frac{\Delta t_{\text{rest}}}{\sqrt{1 - v^2/c^2}}, \tag{22.12}$$

which we label as $\Delta t_{\text{moving}}$ because frame $S'$ is moving *relative to the metronome*. This time period is evidently longer than that measured in the rest frame, a phenomenon that is called *time dilation*. We conclude that the shortest time period is that measured in the rest frame, while a longer time interval is measured in any other inertial frame.

Some authors paraphrase time dilation by saying that *moving clocks run slow*, a phrase that can be misleading because it begs the question, moving relative to what? In any event, this phrase is intended to convey the following. Suppose our metronome ticks every second in its rest frame $S$, $\Delta t_{\text{rest}} = 1$ s, but our moving observer in frame $S'$ records the time between ticks to be two seconds, $\Delta t_{\text{moving}} = 2$ s. If our moving observer has his own identical metronome ticking every second, he will think that the metronome in frame $S$, which moves relative to him, is ticking more slowly than expected. To the observer in $S'$ it appears that time runs more slowly in frame $S$. Of course, because there is no such thing as absolute motion, the observer in frame $S$ experiences something similar. He sees his metronome in frame $S$ ticking every second, but records the time between ticks for the metronome in frame $S'$ to be two seconds, in accordance with (22.10). So, from his perspective, time is running more slowly in frame $S'$. Both observers are perfectly correct, as seen from their particular vantage points.

We can illustrate time dilation if we return to the flashbulb going off in a warehouse. This time we ask how long it takes for the light to strike the floor below the bulb, as illustrated in Figure 22.4. If the bulb is a height $h$ above the floor, an observer within the warehouse will measure this time as $\Delta t_{\text{rest}} = h/c$. However, you are driving past at a speed $v$, and, by the time the light strikes the floor, the centre of the warehouse has moved to the left by a distance $v\,\Delta t_{\text{moving}}$ in your frame of reference. So, in frame $S'$, the light has to travel the greater distance $\sqrt{h^2 + (v\,\Delta t_{\text{moving}})^2}$ before reaching the floor. We now invoke postulate (2), which demands that light observed in frame $S'$ also travels at speed $c$. We conclude that

$$\Delta t_{\text{moving}} = \frac{\sqrt{h^2 + (v\,\Delta t_{\text{moving}})^2}}{c} = \frac{\sqrt{(c\,\Delta t_{\text{rest}})^2 + (v\,\Delta t_{\text{moving}})^2}}{c}, \tag{22.13}$$

which is indeed longer than $\Delta t_{\text{rest}}$. In fact, on rearrangement, (22.13) yields (22.12), exactly as required by the Lorentz transformation.

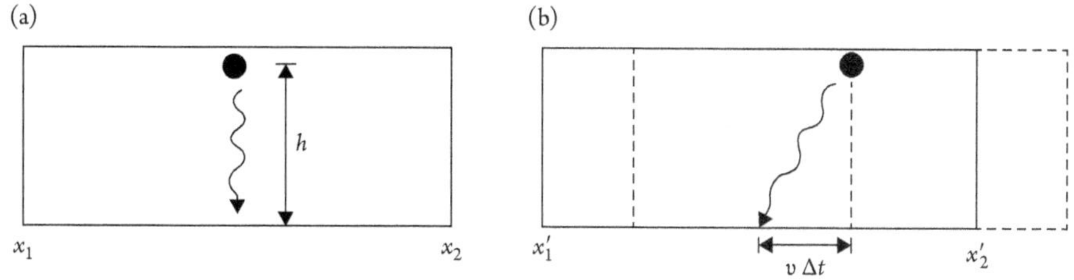


**Figure 22.4** Light travels from a flashbulb to the floor below. (a) Events as seen by an observer in the warehouse. (b) Events as seen by an observer driving past.

If time increments can change from one inertial frame to another, what about length? Consider a rod at rest in frame $S$, aligned with the $x$ axis and running from $x_1$ to $x_2 = x_1 + \ell_{\text{rest}}$. Evidently, $\ell_{\text{rest}}$ is the length of the rod in its rest frame. We now consider the length of the rod as measured by an observer in frame $S'$, being careful to evaluate $x'_1$ and $x'_2$ at the same time, $t'$, in $S'$. Equation (22.10) then tells us that, with $t'_2 = t'_1$,

$$x_2 - x_1 = \ell_{\text{rest}} = \frac{x'_2 - x'_1}{\sqrt{1 - v^2/c^2}} = \frac{\ell_{\text{moving}}}{\sqrt{1 - v^2/c^2}},$$

where $\ell_{\text{moving}}$ is the length of the rod as measured in the frame of reference *moving relative to the rod*. Evidently, the length measured in $S'$ is

$$\ell_{\text{moving}} = \ell_{\text{rest}}\sqrt{1 - v^2/c^2}. \tag{22.14}$$

So we conclude that the length of a rod measured in a frame of reference moving relative to the rod is always shorter than that measured in the rod's rest frame. This is known as the *Lorentz–FitzGerald contraction*, and some authors paraphrase this by saying that *objects moving relative to the observer's frame of reference appear shortened.* Note, however, that a moving object is shortened only along the direction of its motion.

### 22.2.7 The Relativistic Transformation of Velocity

Since both the length and time intervals associated with an event, or sequence of events, vary depending on the frame of reference from which those events are viewed, it is natural to suppose that the observed velocity of an object will vary from one frame to another, and in a way that is more subtle than a simple Galilean transformation. To see that this is indeed the case, let us rewrite (22.9) in terms of increments:

$$dx' = \frac{dx - v\,dt}{\sqrt{1 - v^2/c^2}}, \quad dy' = dy, \quad dz' = dz, \quad dt' = \frac{dt - v\,dx/c^2}{\sqrt{1 - v^2/c^2}}.$$

This tells us how to calculate changes in $d\mathbf{x}'$ and $dt'$ in frame $S'$ that arise from a change in position of an object in frame $S$ of $d\mathbf{x}$ in the time interval $dt$. Dividing through by $dt$ yields

$$\frac{dx'}{dt} = \frac{u_x - v}{\sqrt{1 - v^2/c^2}}, \quad \frac{dy'}{dt} = u_y, \quad \frac{dz'}{dt} = u_z, \quad \frac{dt'}{dt} = \frac{1 - vu_x/c^2}{\sqrt{1 - v^2/c^2}}.$$

We now use the expression for $dt'/dt$ to rewrite $dx'/dt$ as $u'_x = dx'/dt'$, and so on for the other components. We find that an object that has velocity components $u_x, u_y, u_y$ in frame $S$ has components

$$u'_x = \frac{u_x - v}{1 - vu_x/c^2}, \quad u'_y = \frac{u_y\sqrt{1 - v^2/c^2}}{1 - vu_x/c^2}, \quad u'_z = \frac{u_z\sqrt{1 - v^2/c^2}}{1 - vu_x/c^2} \tag{22.15}$$

in frame $S'$. Note that we recover the Galilean transformation of velocity in the limit of $v << c$. Moreover, if $\mathbf{u} = u_x\hat{\mathbf{e}}_x = c\hat{\mathbf{e}}_x$, so that the object being tracked in frame $S$ is a wave front, then (22.15) tells us that $\mathbf{u}' = u'_x\hat{\mathbf{e}}'_x = c\hat{\mathbf{e}}'_x$, as demanded by postulate (2). That is to say, the wave front propagates at the same speed in both frames of reference.

## 22.3 Space–Time, Relativistic Invariants and Four-Vectors

A scalar that takes the same value in all inertial frames is called a *relativistic invariant*. Familiar examples are $c$, by virtue of the second postulate, and electric charge $q$. A third example is, by virtue of (22.7) and (22.8),

$$x^2 + y^2 + z^2 - (ct)^2 = x'^2 + y'^2 + z'^2 - (ct')^2 = 0. \tag{22.16}$$

In fact, (22.16) may be generalised to the invariant,

$$I = (\Delta x)^2 + (\Delta y)^2 + (\Delta z)^2 - (c\,\Delta t)^2 = d^2 - c^2\tau^2, \tag{22.17}$$

which is known as the *invariant space-time interval*. Here $d$ is the distance between two points and $\tau = \Delta t$ the time lapse between two events at these points. (The proof that (22.17) is indeed an invariant rests on the use of (22.10) and is outlined in Exercise 22.1.)

Clearly $I$ is a measure of the separation of two events in space and time. It is instructive to group pairs of events according to their separation as follows:

(i) equal separation in space and time, $I = 0$, which is called *light-like separation*;
(ii) primarily separated in time, $I < 0$, which is called *time-like separation*;
(iii) primarily separated in space, $I > 0$, which is called *space-like separation*.

Two events for which $I = 0$ can be connected by a light signal, the simplest example being a flashbulb going off at the origin and any point on the subsequent wave front that propagates radially outward from the origin. If two events are time-like, then $|d| < c|\tau|$. For such events it is possible to perform a Lorentz transformation to a frame in which the events are coincident in space but separated in time. (Such a frame moves at a speed $v = d/\tau$ relative to the original frame—see Exercise 22.2.) By way of contrast, two events that are space-like can always be transformed to a frame in which they are simultaneous, but their spatial separation remains finite (see e.g. French, 1968). This distinction between time-like and space-like events helps with the notion of *causality* in special relativity, a notion that is somewhat thrown into disarray by the relativity of simultaneity. In particular, two events that are time-like may have a *causal link* between them while two events that are space-like *cannot* have a causal link, because their ordering depends on the inertial frame. (See Exercise 22.3 and also French (1968).)

When working in *space–time*, which is what we are now discussing, it is often helpful to introduce the so-called *four-vector*. The idea is that space and time are so intertwined in special relativity that it is natural to think of an event as being labelled by the four *space–time coordinates*, $(x,y,z,ct)$, rather than by the three spatial coordinates $(x,y,z)$ and a time $t$. This, in turn, leads to the idea of four-vectors, which we can think of as a generalisation of conventional vectors to include the fourth dimension of time.

Now, the prototype three-dimensional vector is the position vector $\mathbf{x} = (x,y,z)$, and we demand that the components of a conventional vector transform under a rotation of the coordinate system in exactly the same way as the components of the position vector. In this way, a vector remains fixed in space as the orientation of the coordinate system changes, so that vectors and the vector laws of nature transcend coordinate systems, being valid for any orientation of a coordinate system. In an analogous way, we demand that a four-vector has the special property that it transforms between inertial reference frames in the exactly same way as $(x,y,z,ct)$. In fact, $(x,y,z,ct)$ *is* the prototype four-vector, and we shall meet several other examples of four-vectors in due course. The idea is that, if we can write the laws of electrodynamics in terms of four-vectors, then it will become more evident that they apply unaltered in any inertial frame of reference.

In order to manipulate four-vectors, we need to devise a kind of four-dimensional vector calculus, and to make that calculus look similar to that in three dimensions, we shall find it convenient to rewrite our *space–time coordinate* in a slightly different way, namely

$$\mathbf{r} = (x,y,z,jct) = (\mathbf{x},jct).$$

(Note that other conventions are in common use that avoid a complex time coordinate. See, for example, Feynman, Leighton & Sands (1964).) Let us see how this works out. In three dimensions the scalar product of two vectors $\mathbf{a}$ and $\mathbf{b}$ is independent of translations and rotations of the coordinate system used to describe them. We say that $\mathbf{a}\cdot\mathbf{b} = a_i b_i$ is an invariant, of which $\mathbf{a}^2 = \mathbf{a}\cdot\mathbf{a}$ is a special case. By analogy, we shall require that the scalar product of two four-vectors, say $\mathbf{a} = (a_x, a_y, a_z, a_t)$ and $\mathbf{b} = (b_x, b_y, b_z, b_t)$, be invariant under rotations of the coordinate system *and* under a change from one inertial frame to another. That is,

$$\mathbf{a}\cdot\mathbf{b} = a_\mu b_\mu = a'_\mu b'_\mu = \mathbf{a}'\cdot\mathbf{b}',$$

of which a special case is $a_\mu a_\mu = a'_\mu a'_\mu$. (Whenever the subscript $\mu$ occurs twice, this implies a summation over *four* indices rather than the usual three.) The prototype four-vector is $\mathbf{r} = (x,y,z,jct)$, and (22.16) confirms that

$$\mathbf{r}^2 = \mathbf{r}\cdot\mathbf{r} = x^2 + y^2 + z^2 - (ct)^2$$

is indeed an invariant. Thus any set of four quantities that transform from one inertial frame to another in the same way as the prototype $\mathbf{r} = (x,y,z,jct)$ is classified as a four-vector, and each four-vector has associated with it a quadratic invariant analogous to $\mathbf{r}^2$. Moreover, using transformation (22.10), it is readily confirmed that four-vectors do indeed satisfy $a_\mu b_\mu = a'_\mu b'_\mu$.

The four-vector analogue of the three-dimensional gradient operator $\nabla$ is now defined as

$$\nabla_\mu = \left(\frac{\partial}{\partial x}, \frac{\partial}{\partial y}, \frac{\partial}{\partial z}, \frac{1}{jc}\frac{\partial}{\partial t}\right) = \left(\nabla, \frac{1}{jc}\frac{\partial}{\partial t}\right). \tag{22.18}$$

This has the property that

$$\nabla_\mu(\mathbf{r}^2) = 2\mathbf{r},$$

just like its counterpart in three dimensions. It follows that $\nabla_\mu(\mathbf{r}^2)$ is a four vector, and indeed it turns out that $\nabla_\mu$ operating on any scalar invariant yields a four-vector. The four-dimensional divergence of $\mathbf{r}$ is $\nabla_\mu \cdot \mathbf{r} = 4$, which is equivalent to the three-dimensional divergence, $\nabla \cdot \mathbf{x} = 3$. It may be shown that the divergence of any four-vector is also a scalar invariant, having the same value in all coordinate systems that differ by a Lorentz transformation. (See, for example, Feynman, Leighton & Sands (1964) for a discussion of the properties of gradients and divergences associated with $\nabla_\mu$.)

Finally, the scalar product of $\nabla_\mu$ with itself yields the wave-like operator

$$\nabla_\mu^2 = \nabla^2 - \frac{1}{c^2}\frac{\partial^2}{\partial t^2}, \tag{22.19}$$

which is called the *d'Alembertian*. We shall return to the d'Alembertian in Chapter 23, where we shall see how to write the equations of electrodynamics in terms of four-vectors and the operator $\nabla_\mu$.

## 22.4 Relativistic Dynamics

### 22.4.1 *Proper Time, Proper Velocity and Relativistic Momentum*

So far we have discussed only kinematics and so we now turn to dynamics. Let us start with some notation. Time measured in the rest frame of some object that undergoes a sequence of events, say a ticking metronome, is called *proper time*, and (22.12) tells us that time intervals measured in the rest frame are shorter than those measured by any other observer. (Griffiths (2013) speculates that the use of 'proper' here is a corruption of the French *propre* for 'own'.) Note that proper time is a relativistic invariant.

For reasons that are not immediately obvious, it is useful to introduce a hybrid quantity called the *proper velocity*. Consider an object that moves with velocity $\mathbf{u} = d\mathbf{x}/dt$ relative to an observer's frame, with $d\mathbf{x}$ and $dt$ both measured in the observer's frame. Now suppose that $t_{\text{prop}}$ is the proper time for the object. We define the proper velocity to be $\mathbf{u}_{\text{prop}} = d\mathbf{x}/dt_{\text{prop}}$ or, using (22.12), $\mathbf{u}_{\text{prop}} = \mathbf{u}/\sqrt{1-u^2/c^2} = \gamma\mathbf{u}$, where $\gamma$ is now based on the velocity of the object. Since $t_{\text{prop}}$ is a relativistic invariant, $\mathbf{u}_{\text{prop}}$ must transform in the same way as $d\mathbf{x}$, which is the Lorentz transformation. It follows that

$$\mathbf{u}_{\text{four}} = \frac{d\mathbf{r}}{dt_{\text{prop}}} = (\mathbf{u}_{\text{prop}}, j\gamma c) = \frac{(\mathbf{u}, jc)}{\sqrt{1-u^2/c^2}} \tag{22.20}$$

is a four-vector, called the *four-velocity*. It has the relativistic invariant

$$\mathbf{u}_{\text{four}}^2 = \gamma^2(\mathbf{u}^2 - c^2) = -c^2.$$

One reason for introducing the proper velocity is that its transformation properties are simpler than those of ordinary velocity, which transforms according to (22.15). A more important reason is that in relativity the momentum of a particle of mass $m$ and velocity $\mathbf{u}$ is defined as

$$\mathbf{p} = m\mathbf{u}_{\text{prop}} = \frac{m\mathbf{u}}{\sqrt{1 - u^2/c^2}}. \tag{22.21}$$

This is called the *relativistic momentum vector* and it turns out that, when summed over many particles, it is conserved for an isolated system in all inertial frames. (We shall not pause to prove this statement, but see, for example, French (1968) for a discussion of relativistic momentum.) Note that the mass $m$ here is a *relativistic invariant,* the same in all inertial frames. Beware, however, that some older texts define $m/\sqrt{1 - u^2/c^2}$ to be a *relativistic mass,* and then call $m$ the *rest mass.* We shall *not* introduce the concept of relativistic mass; for us rest mass *is* simply mass. Note also that it is critical not to confuse the terms *invariant* and *conserved.* Conservation is a dynamical statement about a quantity, whereas invariance is a kinematic property that simply asserts that the quantity in question takes the same value in all inertial frames.

Clearly $\mathbf{p}$ is part of the four-vector

$$\mathbf{p}_{\text{four}} = \frac{(m\mathbf{u}, jmc)}{\sqrt{1 - u^2/c^2}}, \tag{22.22}$$

which is called the *momentum four-vector,* or *energy–momentum four-vector.* Since $\mathbf{p}$ is conserved for an isolated system in *all* inertial frames, and the various components of $\mathbf{p}_{\text{four}}$ transform in accordance with the Lorentz transformation as we move from one inertial frame to another, we conclude that the fourth component, $mc/\sqrt{1 - u^2/c^2}$, must also be conserved. This, in turn, tells us that the energy-like quantity

$$E = \frac{mc^2}{\sqrt{1 - u^2/c^2}} \tag{22.23}$$

is conserved in all inertial frames for an isolated system. (Remember that $c$ is a relativistic invariant by virtue of the second postulate.) We now explore the dynamical significance of the conservation of $E$, starting with Newton's second law.

### 22.4.2 *Newton's Second Law and Relativistic Energy*

Newton's second law, written in terms of force and momentum, remains valid in special relativity provided that we use the relativistic momentum. Thus we have

$$\mathbf{F} = \frac{d\mathbf{p}}{dt} = \frac{d}{dt}\frac{m\mathbf{u}}{\sqrt{1 - u^2/c^2}}, \tag{22.24}$$

for a particle of mass $m$. Note that, in the absence of an external force, the relativistic momentum is conserved, as claimed above. Note also that, as $u$ approaches $c$, the momentum of a particle diverges, $|\mathbf{p}| \to \infty$, which tells us that particles of *finite mass* cannot move at the speed of light. (Of course photons, which have zero mass, *do* travel at the speed of light, so the caveat of a finite mass is not entirely superfluous.)

It is instructive to rewrite (22.23) in a different way. We start by introducing the *relativistic kinetic energy*, defined as

$$T = mc^2 \left( \frac{1}{\sqrt{1 - u^2/c^2}} - 1 \right). \tag{22.25}$$

Note that this reverts to the conventional definition, $T = mu^2/2$, for low speeds. Next we define the *total relativistic energy* to be

$$\boxed{E = \frac{mc^2}{\sqrt{1 - u^2/c^2}},} \tag{22.26}$$

from which we obtain

$$\boxed{\begin{array}{c} E \qquad = mc^2 + \quad T. \\ (\text{total energy} = mc^2 + \text{kinetic energy}) \end{array}} \tag{22.27}$$

In any isolated system the total relativistic energy is conserved, as demanded by (22.23), and so we may think of $mc^2$ as also being a form of energy. It is called the *rest energy* of the particle.

We have finally arrived at the most famous expression in physics: $E_{\text{rest}} = mc^2$. Meanwhile, our energy equation (22.27) reads $E = E_{\text{rest}} + T$. In short, both mass and kinetic energy contribute to the energy of a particle, and while $E$ is conserved, $m$ need not be. Rather, mass can be converted into kinetic energy, or, more strictly, rest energy can be converted into kinetic energy.

In summary, then, *the conserved properties of an isolated system are its relativistic momentum and total relativistic energy.* Let us now return to the momentum four-vector

$$\mathbf{p}_{\text{four}} = \frac{(m\mathbf{u}, jmc)}{\sqrt{1 - u^2/c^2}}.$$

We have already associated the spatial components of this four-vector with the relativistic momentum $\mathbf{p}$ and we now see that the temporal component is just the total relativistic energy (divided by $c$), or, more precisely, $\mathbf{p}_{\text{four}} = (\mathbf{p}, jE/c)$. Thus all four components of $\mathbf{p}_{\text{four}}$ are conserved in all inertial frames. This four-vector has associated with it an invariant (a quantity that is the same in all inertial frames) given by

$$\mathbf{p}_{\text{four}}^2 = \mathbf{p}^2 - E^2/c^2 = \frac{m^2u^2}{1 - u^2/c^2} - \frac{m^2c^2}{1 - u^2/c^2} = -m^2c^2, \tag{22.28}$$

reminding us that mass (or rest mass in the old coinage) is a relativistic invariant. The expression (22.28) is usually rewritten as

$$E^2 = \mathbf{p}^2 c^2 + m^2 c^4. \tag{22.29}$$

* * *

With this briefest of introductions to special relativity, we are now ready to return to electrodynamics and consider electric and magnetic fields from a relativistic point of view.

## EXERCISES

22.1 Use (22.10) to substitute for $\Delta x$ and $\Delta t$ in $I = (\Delta x)^2 + (\Delta y)^2 + (\Delta z)^2 - (c\,\Delta t)^2$ and hence show that $I$ is indeed a relativistic invariant.

22.2 Consider two events $A$ and $B$ with the same $y$ and $z$ coordinates but separated by a distance $d$ in the $x$ direction, $x_A < x_B$, and by a time lapse $\tau$, $t_A < t_B$. The invariant space–time interval is then $I = d^2 - c^2\tau^2$. Consider a *time-like* event in which $d < c\tau$. Confirm that there exists an inertial frame in which the two events occur at the same point and that this frame moves at a speed $v = d/\tau$ relative to the original frame.

22.3 As in Exercise 22.2, consider two events $A$ and $B$ with the same $y$ and $z$ coordinates but separated by a distance $d$ in the $x$ direction and by a time lapse $\tau$. Such events, and the way they transform upon a change of inertial frame, are often represented as points in a *space–time diagram*, as shown in Figure 22.5. Event $A$ sits at the origin and event $B$ is at $x = d$ and $t = \tau$. Confirm that time-like events lie within the *light cone* $x^2 = (ct)^2$, whereas space-like events lie outside the light cone. Sketch the locus of points $x^2 - (ct)^2 = I =$ constant for a space-like event and $(ct)^2 - x^2 = |I| =$ constant for a time-like event. Show that a suitable change of frame of reference can eliminate the spatial separation of two time-like events, or make two space-like events appear simultaneous. Also, confirm that the temporal ordering of space-like events depends on the inertial frame, and so they cannot have a causal link.

22.4 We have stated postulate (2) as 'the speed of light in vacuum is the same for all inertial observers'. In his original 1905 paper Einstein phrased it differently: 'the speed of light in vacuum is independent of the state of motion of the emitting body'. Show that these two statements amount to the same thing.

22.5 In his autobiographical notes Einstein recounts '... a paradox upon which I had already hit at the age of sixteen. If I pursue a beam of light with the velocity $c$, I should observe such a beam as an electromagnetic field constant in time, periodic in space. However, there seems to exist no such thing, neither on the basis of experience, nor according to Maxwell's equations.' Confirm that, as Einstein notes, Maxwell's equations in vacuum do not support static spatial oscillations.

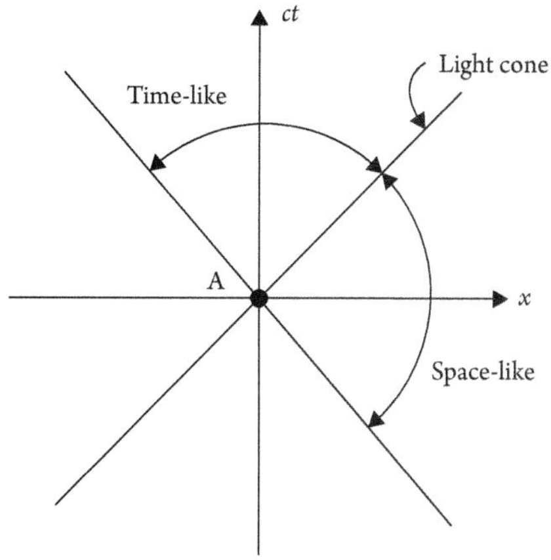


**Figure 22.5** Space–time diagram.

22.6 A constant force **F** is applied to a particle of mass $m$. If the particle starts from rest, show that its velocity at time $t$ is

$$\mathbf{u} = \frac{\mathbf{F}t/m}{\sqrt{1 + (Ft/m)^2/c^2}}.$$

Now find the position of the particle as a function of time.

## REFERENCES

Einstein, A., 1905, On the electrodynamics of moving bodies. In: *The Collected Papers of Albert Einstein*, Volume 2. Princeton University Press, 1990.

Feynman R. P., Leighton R. B. & Sands M., 1964, *The Feynman Lectures on Physics*, Volume II. Addison-Wesley.

French, A. P., 1968, *Special Relativity*. Chapman & Hall.

Griffiths, D. J., 2013, *Introduction to Electrodynamics*, 4th ed. Pearson.

# 23

# Electromagnetism and Special Relativity

> Numerous investigators have endeavoured to solve the question [of the aether], and fresh experiments have been undertaken. What result will they give? I shall take care not to risk a prophecy which might be falsified between the day this book is ready for the press and the day on which it is placed before the public.
>
> *Henri Poincaré* (*La Science et l'Hypothèse*, 1902)

Poincaré wrote *Science and Hypothesis* some three years before Einstein's *Annus Mirabilis*, and some historians of science believe that this book was a source of inspiration for Einstein.

We now consider the implications of relativity for electrodynamics. Our primary aim is to establish how the various fields transform, especially **E** and **B**. Let us start, however, with charge and current density.

## 23.1 The Four-Current Density and Charge Conservation in Relativistic Form

### 23.1.1 *The Transformation of Charge, Volume and Charge Density*

It is an experimental observation that electrical charge is a relativistic invariant, the same in all inertial frames. Charge density, on the other hand, does vary from one frame to another, because volume elements contract. That is, if a charge $q$ occupies a volume element $dV$ in one frame, say $S$, and volume $dV'$ in a second frame $S'$, then the ratio of the charge densities in the two frames is $\rho_e'/\rho_e = dV/dV'$.

To fix our thoughts, suppose that in frame $S$ we have some charge that moves with a constant velocity $\mathbf{u}$. Now consider a second frame $S'$ that moves relative to $S$ with a constant speed $v$ directed along the $x$ axis. We consider first the case where $\mathbf{u} = u_x\hat{\mathbf{e}}_x$, so that the velocity of the charge in frame $S$ is parallel to the relative movement of the two frames (Figure 23.1). The velocity components of the charge measured in frame $S'$ are then, from (22.15),

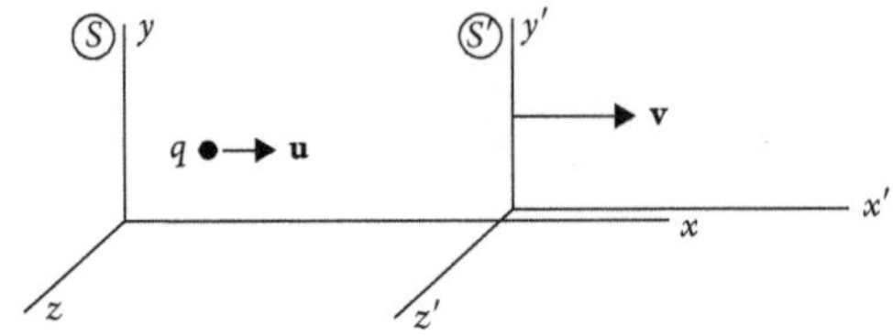


**Figure 23.1** A charge $q$ moves with velocity $\mathbf{u} = u_x\hat{\mathbf{e}}_x$ relative to the inertial reference frame $S$, while frame $S'$ moves relative to $S$ with a constant speed $v$, also along the $x$ axis.

$$u'_x = \frac{u_x - v}{1 - vu_x/c^2}, \quad u'_y = 0, \quad u'_x = 0. \tag{23.1}$$

Now consider a small volume element $dV$ that moves with the charge and is cubic in its rest frame, of side $\ell_0$. Then the length of the volume element is shortened in the $x$ direction when viewed from frames $S$ and $S'$ by factors of $\sqrt{1 - u_x^2/c^2}$ and $\sqrt{1 - u_x'^2/c^2}$, respectively. So the volume elements in frames $S$ and $S'$ are

$$dV = \ell_0^3\sqrt{1 - u_x^2/c^2}$$

and

$$dV' = \ell_0^3\sqrt{1 - u_x'^2/c^2} = \ell_0^3\frac{\sqrt{c^2 - v^2}\sqrt{c^2 - u_x^2}}{c^2 - vu_x},$$

the ratio of which is

$$\frac{dV'}{dV} = \frac{\sqrt{1 - v^2/c^2}}{1 - vu_x/c^2}. \tag{23.2}$$

Although we have assumed here that the charge moves in the $x$ direction, i.e. parallel to the relative motion of the two frames, in fact it is readily confirmed that the same result holds when $\mathbf{u}$ points in any direction relative to $\mathbf{v} = v\hat{\mathbf{e}}_x$. This is discussed in Exercise 23.1. In effect, this is because the Lorentz transformation leaves lengths normal to the relative motion of the two inertial frames unchanged. We conclude that the ratio of the charge density in the two inertial frames is, in general,

$$\frac{\rho'_e}{\rho_e} = \frac{1 - vu_x/c^2}{\sqrt{1 - v^2/c^2}}. \tag{23.3}$$

### 23.1.2 *The Four-Current Density and Charge Conservation Revisited*

Let us now consider the current density associated with this moving charge. In frame $S$ the current density is $\mathbf{J} = \rho_e\mathbf{u}$, while in frame $S'$ we have $\mathbf{J}' = \rho'_e\mathbf{u}'$. The first thing to note is that we may rewrite (23.3) as

$$\rho'_e = \frac{\rho_e - vJ_x/c^2}{\sqrt{1 - v^2/c^2}}. \tag{23.4}$$

Moreover, using (23.3) and (22.15) to substitute for $\rho_e'/\rho_e$ and $u_x'/u_x$ in an expression for $J_x'/J_x$ yields

$$\frac{J_x'}{J_x} = \frac{\rho_e'}{\rho_e}\frac{u_x'}{u_x} = \frac{1 - vu_x/c^2}{\sqrt{1 - v^2/c^2}}\frac{1 - v/u_x}{1 - vu_x/c^2} = \frac{1 - v/u_x}{\sqrt{1 - v^2/c^2}} = \frac{1 - v\rho_e/J_x}{\sqrt{1 - v^2/c^2}},$$

from which

$$J_x' = \frac{J_x - v\rho_e}{\sqrt{1 - v^2/c^2}}. \tag{23.5}$$

Similarly, applying (23.3) and (22.15) to the transverse components of $\mathbf{J}$, we find, for example,

$$\frac{J_y'}{J_y} = \frac{\rho_e'}{\rho_e}\frac{u_y'}{u_y} = \frac{1 - vu_x/c^2}{\sqrt{1 - v^2/c^2}}\frac{\sqrt{1 - v^2/c^2}}{1 - vu_x/c^2} = 1,$$

and so the transverse components transform as $J_y' = J_y, J_z' = J_z$.

In summary, then, the charge and current densities transform as

$$\begin{aligned} J_x' &= \frac{J_x - v\rho_e}{\sqrt{1 - v^2/c^2}}, \quad J_y' = J_y, \quad J_z' = J_z, \\ \rho_e' &= \frac{\rho_e - vJ_x/c^2}{\sqrt{1 - v^2/c^2}}. \end{aligned} \tag{23.6}$$

Comparing this with the Lorentz transformation

$$x' = \frac{x - vt}{\sqrt{1 - v^2/c^2}}, \quad y' = y, \quad z' = z, \quad t' = \frac{t - vx/c^2}{\sqrt{1 - v^2/c^2}},$$

we conclude that

$$\mathbf{J}_{\text{four}} = (J_x, J_y, J_z, jc\rho_e) = (\mathbf{J}, jc\rho_e) \tag{23.7}$$

is a four-vector that transforms just like $\mathbf{r} = (x, y, z, jct)$. This is the *four-current density*.

Let us now reintroduce the four-gradient operator (22.18):

$$\nabla_\mu = \left(\frac{\partial}{\partial x}, \frac{\partial}{\partial y}, \frac{\partial}{\partial z}, \frac{1}{jc}\frac{\partial}{\partial t}\right) = \left(\nabla, \frac{1}{jc}\frac{\partial}{\partial t}\right). \tag{23.8}$$

The four-dimensional divergence of $\mathbf{J}_{\text{four}} = (\mathbf{J}, jc\rho_e)$ is evidently

$$\nabla_\mu \cdot \mathbf{J}_{\text{four}} = \nabla \cdot \mathbf{J} + \frac{\partial \rho_e}{\partial t}, \tag{23.9}$$

and so we conclude that charge conservation can be written in the particularly compact form

$$\nabla_\mu \cdot \mathbf{J}_{\text{four}} = 0. \tag{23.10}$$

Evidently, in space–time, the principle of charge conservation simply comes down to the statement that $\mathbf{J}_{\text{four}}$ has zero divergence. Now, in §22.3 we noted that the four-divergence of a four-vector is a relativistic invariant, and so (23.10) tells us that if charge conservation holds in one frame, then it holds in all inertial frames. Moreover, since $\mathbf{J}_{\text{four}}$ is a four-vector, it follows that

$$\mathbf{J}_{\text{four}}^2 = \mathbf{J}^2 - (c\rho_e)^2 \tag{23.11}$$

is a relativistic invariant, having the same value in all inertial frames. These two results illustrate the utility of four-vector notation.

## 23.2 The Four-Potential

Let us now turn to electrodynamics written in terms of potentials. In Chapter 13 we noted that, provided we adopt the Lorenz gauge

$$\nabla \cdot \mathbf{A} = -\frac{1}{c^2}\frac{\partial \varphi}{\partial t}, \tag{23.12}$$

the scalar and vector potentials are governed by the inhomogeneous wave equations

$$\nabla^2 \varphi - \frac{1}{c^2}\frac{\partial^2 \varphi}{\partial t^2} = -\frac{\rho_e}{\varepsilon_0}, \tag{23.13}$$

$$\nabla^2 \mathbf{A} - \frac{1}{c^2}\frac{\partial^2 \mathbf{A}}{\partial t^2} = -\mu_0 \mathbf{J}. \tag{23.14}$$

Now, the scalar product of $\nabla_\mu$ with itself yields the wave-like operator

$$\nabla_\mu^2 = \nabla^2 - \frac{1}{c^2}\frac{\partial^2}{\partial t^2}, \tag{23.15}$$

which we called the d'Alembertian. Evidently, our inhomogeneous wave equations are more compact when written in terms of $\nabla_\mu^2$:

$$\nabla_\mu^2 \mathbf{A} = -\mu_0 \mathbf{J}, \quad \nabla_\mu^2 (\varphi/c) = -\mu_0 (c\rho_e). \tag{23.16}$$

Moreover, on adding these equations, we find

$$\nabla_\mu^2 (\mathbf{A}, j\varphi/c) = -\mu_0 \mathbf{J}_{\text{four}}, \tag{23.17}$$

and since the right-hand side transforms like a four-vector, so must the left. However, the operator $\nabla_\mu^2$ does not change as we transform from one inertial frame to another, and so we conclude that the term on the left, $(\mathbf{A}, j\varphi/c)$, is also a four-vector. We call this four-vector the *four-potential*, $\mathbf{A}_{\text{four}} = (\mathbf{A}, j\varphi/c)$, and our two wave equations merge into a single equation in space–time,

$$\nabla_\mu^2 \mathbf{A}_{\text{four}} = -\mu_0 \mathbf{J}_{\text{four}}. \tag{23.18}$$

Not only is this compact, but we also know that it holds for any inertial frame, since it is written entirely in four-vector form. This is indicative of the fact that electrodynamics is indeed invariant under a Lorentz transformation, as noticed by Lorentz in the 1890s.

Note that the four-dimensional divergence of $\mathbf{A}_{\text{four}}$ is

$$\nabla_\mu \cdot \mathbf{A}_{\text{four}} = \left(\nabla, \frac{1}{jc}\frac{\partial}{\partial t}\right) \cdot (\mathbf{A}, j\varphi/c) = \nabla \cdot \mathbf{A} + \frac{1}{c^2}\frac{\partial \varphi}{\partial t}, \tag{23.19}$$

and so the Lorenz gauge expressed in four-vector form is simply $\nabla_\mu \cdot \mathbf{A}_{\text{four}} = 0$. In summary, then, we might write the equations of electrodynamics in invariant form as

$$\boxed{\begin{aligned} \nabla_\mu^2 \mathbf{A}_{\text{four}} &= -\mu_0 \mathbf{J}_{\text{four}}, \\ \nabla_\mu \cdot \mathbf{A}_{\text{four}} &= 0,\ \nabla_\mu \cdot \mathbf{J}_{\text{four}} = 0. \end{aligned}} \tag{23.20}$$

These are elegant in the sense that they are compact, but ultimately they contain no more information than Maxwell's equations.

## 23.3 The Four-Potential for a Charge Moving at Constant Velocity

We shall now exploit the fact that $\mathbf{A}_{\text{four}} = (\mathbf{A}, j\varphi/c)$ is a four-vector to deduce the scalar and vector potentials for a moving charge directly from Coulomb's law of electrostatics. The key point is that $\mathbf{A}_{\text{four}}$, being a four-vector, transforms as

$$A'_x = \frac{A_x - u\varphi/c^2}{\sqrt{1 - u^2/c^2}}, \quad A'_y = A_y, \quad A'_z = A_z, \tag{23.21}$$

$$\varphi' = \frac{\varphi - uA_x}{\sqrt{1 - u^2/c^2}}, \tag{23.22}$$

for two inertial frames in relative motion along the $x$ axis (Figure 23.2). We can use this transformation law to get the electromagnetic field of a moving charge, as we now show.

Suppose a charge $q$ moves relative to a frame $S$ with a constant velocity $\mathbf{u} = u\hat{\mathbf{e}}_x$. Now consider a second frame of reference, $S'$, in which the charge $q$ is stationary, and suppose that the charge sits at the origin in this frame. Then Coulomb's law tells us that

$$\varphi' = \frac{q}{4\pi\varepsilon_0\sqrt{x'^2 + y'^2 + z'^2}}, \quad \mathbf{A}' = 0. \tag{23.23}$$

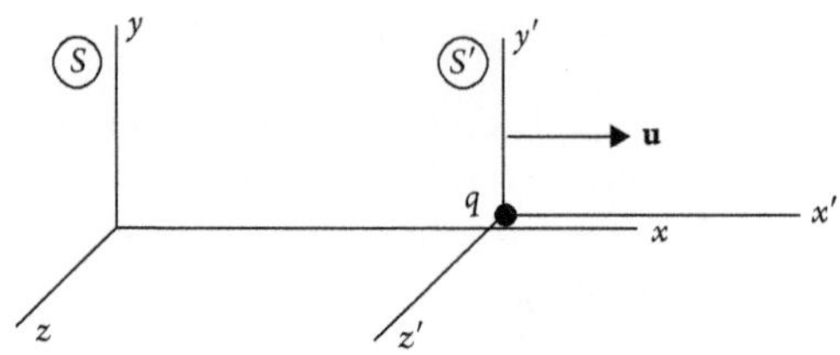


**Figure 23.2** Frame $S'$ moves relative to $S$ with constant speed $u$ along the $x$ axis. A charge $q$ sits at the origin in the reference frame $S'$.

The inverse transformations to (23.21) and (23.22) are, using (22.10),

$$A_x = \frac{A'_x + u\varphi'/c^2}{\sqrt{1-u^2/c^2}}, \quad A_y = Ay', \quad A_z = A'_z,$$

$$\varphi = \frac{\varphi' + uA'_x}{\sqrt{1-u^2/c^2}},$$

and so, viewed from the frame $S$, the potentials are

$$A_x = \frac{u/c^2}{\sqrt{1-u^2/c^2}} \frac{q}{4\pi\varepsilon_0\sqrt{x'^2+y'^2+z'^2}}, \qquad A_y = 0, \quad A_z = 0, \tag{23.24}$$

$$\varphi = \frac{1}{\sqrt{1-u^2/c^2}} \frac{q}{4\pi\varepsilon_0\sqrt{x'^2+y'^2+z'^2}}. \tag{23.25}$$

It is natural to use the Lorentz transformation (22.9) to rewrite these potentials in terms of $(x,y,z)$, rather than $(x',y',z')$,

$$x' = \frac{x-ut}{\sqrt{1-u^2/c^2}}, \quad y = y', \quad z = z',$$

and this gives us

$$\varphi = \frac{q/\sqrt{1-u^2/c^2}}{4\pi\varepsilon_0\sqrt{\left[(x-ut)/\sqrt{1-u^2/c^2}\right]^2 + y^2 + z^2}}, \tag{23.26}$$

$$A_x = u\varphi/c^2, \quad A_y = 0, \quad A_z = 0. \tag{23.27}$$

Notice that (23.26) is identical to the expression we obtained for $\varphi$ using the Liénard–Wiechert potentials in §17.2.4, as it must be. Notice also that (23.26) and (23.27) give the potentials at time $t$ for a charge that passed through the origin at time $t = 0$ and currently

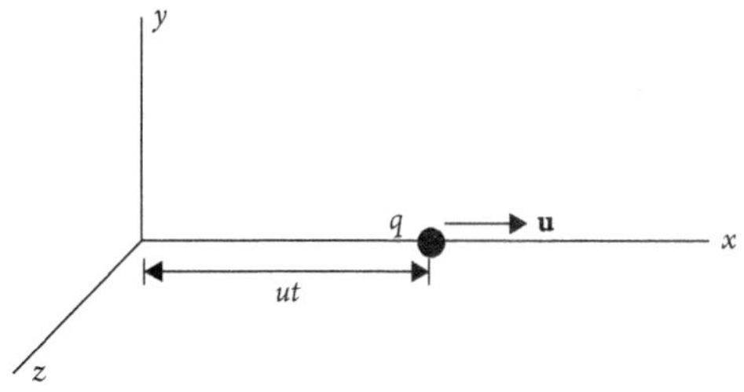


**Figure 23.3** A charge $q$ moves relative to $S$ with constant speed $u$ along the $x$ axis.

sits at the location $\mathbf{x} = \mathbf{u}t$, as shown in Figure 23.3. However, as is clear from the derivation given in §17.2.4, the potentials at location $\mathbf{x}$ at time $t$ were really established by the charge at the earlier retarded time, when the charge was located further up the $x$ axis.

Given the potentials $\mathbf{A}$ and $\varphi$, we can now obtain the $\mathbf{E}$ and $\mathbf{B}$ fields in the usual way from

$$\mathbf{E} = -\nabla\varphi - \frac{\partial \mathbf{A}}{\partial t}, \qquad \mathbf{B} = \nabla \times \mathbf{A}. \tag{23.28}$$

These fields are discussed in §17.2.4. For $\mathbf{E}$ we find

$$\mathbf{E}(\mathbf{x},t) = \frac{q}{4\pi\varepsilon_0} \frac{(1-u^2/c^2)(\mathbf{x}-\mathbf{u}t)}{|\mathbf{x}-\mathbf{u}t|^3\,(1-u^2\sin^2\theta/c^2)^{3/2}}, \tag{23.29}$$

where $\theta$ is the angle between $\mathbf{x} - \mathbf{u}t$ and $\mathbf{u}$, while $\mathbf{B}$ is given by

$$\mathbf{B} = \frac{1}{c^2}\mathbf{u} \times \mathbf{E}. \tag{23.30}$$

These fields are shown in Figure 23.4. The electric field lines are radial, while the magnetic field lines encircle the charge in the azimuthal direction, neither of which is much of a surprise. What is surprising, however, is that the $\mathbf{E}$-lines point radially outward from the *instantaneous* position of the charge, and not the retarded position. This truly remarkable result is discussed in Chapter 17.

## 23.4 How the E and B Fields Transform

It remains to determine how the **E** and **B** fields transform from one inertial frame to another. There are several ways of doing this, but we shall follow Purcell (2011), who gives a characteristically simple derivation.

### *23.4.1 Some Simple Examples*

Suppose we have two large flat parallel plates carrying equal and opposite surface charge densities, $\pm\sigma_e$. The plates are stationary in frame $S$ and lie parallel to the $x$–$z$ plane, as

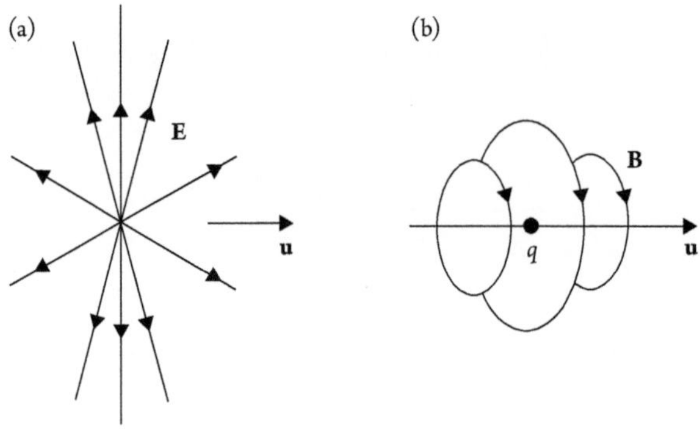


**Figure 23.4** The electromagnetic field of a charge moving with uniform velocity **u**: (a) the electric field; (b) the magnetic field.

shown in Figure 23.5(a). The electric field between the plates points in the $y$ direction and, according to Gauss' law, $\mathbf{E} = (\sigma_e/\varepsilon_0)\hat{\mathbf{e}}_y$. Now consider a second frame $S'$ that moves relative to $S$ with velocity $\mathbf{v} = v\hat{\mathbf{e}}_x$. Since charge is invariant, but lengths parallel to $\mathbf{v}$ contract when viewed from the moving frame, we conclude that the surface charge density viewed from frame $S'$ is $\sigma_e' = \sigma_e/\sqrt{1 - v^2/c^2}$, in accordance with (23.3). Since Gauss' law applied in frame $S'$ demands that $\mathbf{E}' = (\sigma_e'/\varepsilon_0)\hat{\mathbf{e}}_y$, we have

$$\mathbf{E}' = \mathbf{E}/\sqrt{1 - v^2/c^2}. \tag{23.31}$$

Now suppose that the plates are rotated to align with the $y$–$z$ plane, as shown in Figure 23.5(b). This time there is no contraction of the plates when viewed from the frame $S'$ and so $\sigma_e' = \sigma_e$. It follows that $\mathbf{E}' = \mathbf{E} = (\sigma_e/\varepsilon_0)\hat{\mathbf{e}}_x$. So we conclude that, in the absence of a magnetic field, the electric field transforms as

$$\mathbf{E}'_\perp = \mathbf{E}_\perp/\sqrt{1 - v^2/c^2}, \quad \mathbf{E}'_\parallel = \mathbf{E}_\parallel, \tag{23.32}$$

where the subscripts indicate components perpendicular and parallel to the velocity $\mathbf{v} = v\hat{\mathbf{e}}_x$.

How might this change if there is a magnetic field present? Consider again the configuration shown in Figure 23.5(a), only this time we let the plates translate with velocity $\mathbf{u} = u\hat{\mathbf{e}}_x$ as measured in the frame $S$. In this frame there is now a surface current density $\mathbf{K} = \sigma_e\mathbf{u}$, where $\sigma_e$ is the surface charge density as measured in frame $S$. (This charge density is related to the rest charge on the plates by $\sigma_e = \sigma_{\text{rest}}/\sqrt{1 - u^2/c^2}$.) As before, we have $\mathbf{E} = (\sigma_e/\varepsilon_0)\hat{\mathbf{e}}_y$, only now Ampère's law gives us a magnetic field in frame $S$ of $\mathbf{B} = \mu_0 K\hat{\mathbf{e}}_z = \mu_0\sigma_e u\hat{\mathbf{e}}_z$, as shown in Figure 23.6.

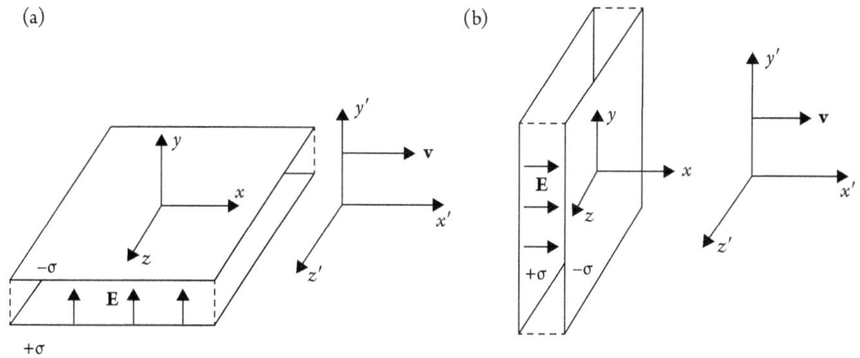


**Figure 23.5** Two large charged plates have an electric field between them. They are viewed from the rest frame $S$ and also from a frame $S'$ that moves with velocity $\mathbf{v} = v\hat{\mathbf{e}}_x$ relative to $S$: (a) the electric field is perpendicular to $\mathbf{v}$; (b) the electric field is parallel to $\mathbf{v}$.

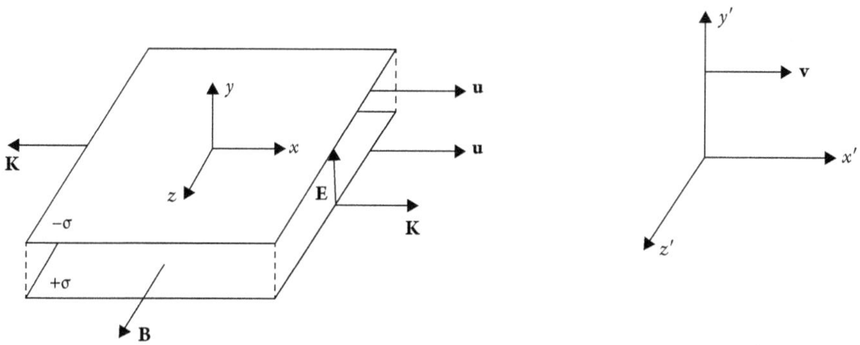


**Figure 23.6** Two charged plates in the $x$–$z$ plane move with velocity $\mathbf{u} = u\hat{\mathbf{e}}_x$ as measured in frame $S$. The moving charge constitutes a surface current $\mathbf{K} = \sigma_e\mathbf{u}$ and so, in addition to the field $\mathbf{E} = (\sigma_e/\varepsilon_0)\hat{\mathbf{e}}_y$, there is a magnetic field $\mathbf{B} = \mu_0 K\hat{\mathbf{e}}_z$.

Once again, we consider how things look from a frame $S'$ that moves relative to $S$ with velocity $\mathbf{v} = v\hat{\mathbf{e}}_x$. The surface charge density $\sigma'_e$ can be found from (23.3) and is evidently

$$\sigma'_e = \frac{\sigma_e - v\sigma_e u/c^2}{\sqrt{1 - v^2/c^2}} = \frac{\sigma_e - vK/c^2}{\sqrt{1 - v^2/c^2}}. \tag{23.33}$$

This then gives the electric field in frame $S'$ as

$$E'_y = \frac{\sigma'_e}{\varepsilon_0} = \frac{\sigma_e/\varepsilon_0 - v\mu_0 K}{\sqrt{1 - v^2/c^2}} = \frac{E_y - vB_z}{\sqrt{1 - v^2/c^2}}. \tag{23.34}$$

Moreover, the surface current density in $S'$ can be found from (23.6),

$$K' = \frac{K - v\sigma_e}{\sqrt{1 - v^2/c^2}} = \frac{1}{\mu_0} \frac{\mu_0 K - vE_y/c^2}{\sqrt{1 - v^2/c^2}},$$

from which we can find the magnetic field in frame $S'$:

$$B_z' = \mu_0 K' = \frac{B_z - vE_y/c^2}{\sqrt{1 - v^2/c^2}}. \tag{23.35}$$

In summary, then, the fields in this geometry transform as

$$E_y' = \frac{E_y - vB_z}{\sqrt{1 - v^2/c^2}}, \quad B_z' = \frac{B_z - vE_y/c^2}{\sqrt{1 - v^2/c^2}}. \tag{23.36}$$

Now consider the situation shown in Figure 23.7, in which the two plates lie parallel to the $x$–$y$ plane and move with the same velocity $\mathbf{u} = u\hat{\mathbf{e}}_x$, as measured in frame $S$. As before, the moving charges give rise to surface currents $\mathbf{K} = \pm\sigma_e\mathbf{u}$ and so there is a magnetic field $\mathbf{B} = -\mu_0 K\hat{\mathbf{e}}_y = -\mu_0\sigma_e u\hat{\mathbf{e}}_y$ in addition to the electric field $\mathbf{E} = (\sigma_e/\varepsilon_0)\hat{\mathbf{e}}_z$. If we now repeat exactly the same steps that led us to (23.36), we find the transformation laws for this rotated geometry are

$$E_z' = \frac{E_z + vB_y}{\sqrt{1 - v^2/c^2}}, \quad B_y' = \frac{B_y + vE_z/c^2}{\sqrt{1 - v^2/c^2}}. \tag{23.37}$$

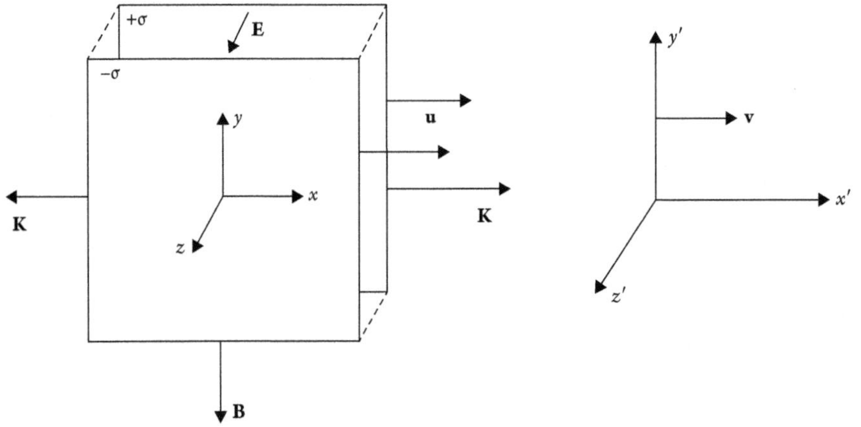


**Figure 23.7** Two charged plates in the $x$–$y$ plane move with velocity $\mathbf{u} = u\hat{\mathbf{e}}_x$ as measured in frame S. The moving charges give rise to a surface current $\mathbf{K} = \sigma_e\mathbf{u}$ and so there is a magnetic field $\mathbf{B} = -\mu_0 K\hat{\mathbf{e}}_y$ in addition to the field $\mathbf{E} = (\sigma_e/\varepsilon_0)\hat{\mathbf{e}}_z$.

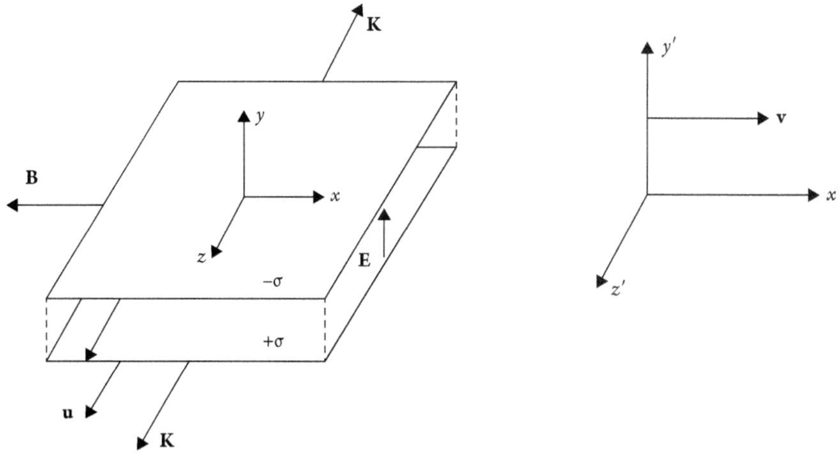


**Figure 23.8** Two charged plates in the $x - z$ plane move with a velocity $\mathbf{u} = u\hat{\mathbf{e}}_z$ as measured in frame $S$. The moving charge constitutes a surface current $\mathbf{K} = \sigma_e \mathbf{u}$ and so, in addition to the field $\mathbf{E} = (\sigma_e/\varepsilon_0)\hat{\mathbf{e}}_y$, there is a magnetic field $\mathbf{B} = -\mu_0 K\hat{\mathbf{e}}_x$.

The final step it to return to the geometry of Figure 23.6, only this time the plates are moved in the $z$ direction with velocity $\mathbf{u} = u\hat{\mathbf{e}}_z$. The surface current density is also in the $z$ direction and consequently the magnetic field in this case is in the minus $x$ direction, as shown in Figure 23.8. So, in addition to the electric field $\mathbf{E} = (\sigma_e/\varepsilon_0)\hat{\mathbf{e}}_y$, there is a magnetic field $\mathbf{B} = -\mu_0 K\hat{\mathbf{e}}_x = -\mu_0\sigma_e u\hat{\mathbf{e}}_x$. However, (23.6) tells us that for this geometry $K' = K$, and so $B'_x = B_x$.

### 23.4.2 *The General Transformation Laws*

Let us now collate all of the transformation rules for $\mathbf{E}$ and $\mathbf{B}$ from these various geometries. They are

$$E'_x = E_x, \quad E'_y = \frac{E_y - vB_z}{\sqrt{1 - v^2/c^2}}, \quad E'_z = \frac{E_z + vB_y}{\sqrt{1 - v^2/c^2}}, \tag{23.38}$$

$$B'_x = B_x, \quad B'_y = \frac{B_y + vE_z/c^2}{\sqrt{1 - v^2/c^2}}, \quad B'_z = \frac{B_z - vE_y/c^2}{\sqrt{1 - v^2/c^2}}, \tag{23.39}$$

where in all cases the frame $S'$ moves relative to $S$ with velocity $\mathbf{v} = v\hat{\mathbf{e}}_x$. Now, these transformation rules were derived for certain specific charge distributions moving in particular ways. However, the way the fields transform from frame to frame cannot depend on the nature of the sources that set up those fields, and so we conclude that these transformation rules are in fact quite general.

The most convenient way to write (23.38) and (23.39) is

$$\boxed{\mathbf{E}'_{\parallel} = \mathbf{E}_{\parallel}, \qquad \mathbf{E}'_{\perp} = \frac{\mathbf{E}_{\perp} + \mathbf{v} \times \mathbf{B}}{\sqrt{1 - v^2/c^2}},} \tag{23.40}$$

$$\boxed{\mathbf{B}'_{\parallel} = \mathbf{B}_{\parallel}, \qquad \mathbf{B}'_{\perp} = \frac{\mathbf{B}_{\perp} - \mathbf{v} \times \mathbf{E}/c^2}{\sqrt{1 - v^2/c^2}},} \tag{23.41}$$

where the subscripts indicate directions parallel and perpendicular to the velocity $\mathbf{v} = v\hat{\mathbf{e}}_x$ with which the frame $S'$ moves relative to $S$. Note that, if $\mathbf{E} = 0$ in frame $S$, then (23.40) and (23.41) reduce to

$$\mathbf{E}' = \frac{\mathbf{v} \times \mathbf{B}}{\sqrt{1 - v^2/c^2}}, \quad \mathbf{B}'_{\perp} = \frac{\mathbf{B}_{\perp}}{\sqrt{1 - v^2/c^2}},$$

from which

$$\mathbf{E}' = \mathbf{v} \times \mathbf{B}'. \tag{23.42}$$

Similarly, if $\mathbf{B} = 0$ in frame $S$, then we have

$$\mathbf{E}'_{\perp} = \frac{\mathbf{E}_{\perp}}{\sqrt{1 - v^2/c^2}}, \quad \mathbf{B}' = -\frac{\mathbf{v} \times \mathbf{E}/c^2}{\sqrt{1 - v^2/c^2}},$$

from which

$$\mathbf{B}' = -\mathbf{v} \times \mathbf{E}'/c^2. \tag{23.43}$$

This last expression lies behind the equation (23.30) for the magnetic field of a moving point charge.

Equations (23.42) and (23.43) demonstrate in a rather stark manner how the **E** and **B** fields are two sides of the same coin, and it was this conviction that led Einstein more or less directly to the special theory of relativity. (See the quotation at the start of Chapter 10.) We have come full circle.

* * *

This completes our all too brief introduction to relativistic electrodynamics. Of course, the subject is vast and we have barely scratched the surface. Readers searching for a more leisurely introduction could do worse than consult Feynman, Leighton & Sands (1964), while Zangwill (2013) offers a more advanced treatment. Perhaps we shall leave the last word to Arnold Sommerfeld, who ends the preface to his classic text on electrodynamics with the following observation:

> The wonderful simplicity and beauty of the Maxwell equations, which is most striking in their relativistic formulation for vacuum, lead to the conviction that these equations, along with the equations of gravitation, are the demonstration of an all-inclusive world geometry.
>
> Arnold Sommerfeld, 1948

One cannot but help think that Faraday, Maxwell and Heaviside would all have approved of this conclusion.

## EXERCISES

23.1 Use the velocity transformation rules (22.15) to show that (23.2) generalises to

$$\frac{dV'}{dV} = \frac{\sqrt{1-\mathbf{u}'^2/c^2}}{\sqrt{1-\mathbf{u}^2/c^2}} = \frac{\sqrt{1-v^2/c^2}}{1-vu_x/c^2},$$

for a charge moving with a constant velocity $\mathbf{u}$ in any direction relative to the $x$ axis.

23.2 Use the transformation laws (23.40) and (23.41) to show that $\mathbf{E}\cdot\mathbf{B}$ is a relativistic invariant. When discussing plane waves in Chapter 15, we found that $\mathbf{E}$ and $\mathbf{B}$ are orthogonal. Show that if this holds in one inertial frame, then it is true in all frames.

23.3 Use the transformation laws (23.40) and (2.41) to show that $\mathbf{E}^2 - c^2\mathbf{B}^2$ is a relativistic invariant. When discussing plane waves in Chapter 15, we found that $|\mathbf{E}| = c|\mathbf{B}|$. Show that if this holds in one inertial frame, then it is true in all frames.

23.4 The earth's magnetic field is predominantly vertical in the polar regions. It is proposed to use this field to measure the speed of an aircraft by measuring the voltage induced in a wire suspended in the aircraft. What orientation should the wire have relative to the aircraft? Estimate the magnitude of the voltage induced in a 1 m wire.

23.5 For $v << c$ the transformation laws (23.40) and (23.41) reduce to

$$\mathbf{E}' = \mathbf{E} + \mathbf{v}\times\mathbf{B}, \quad \mathbf{B}' = \mathbf{B} - \mathbf{v}\times\mathbf{E}/c^2.$$

Discuss under what conditions we may further simplify the second of these to $\mathbf{B}' = \mathbf{B}$.

## REFERENCES

Feynman R. P., Leighton R. B. & Sands M., 1964, *The Feynman Lectures on Physics*, Volume II. Addison-Wesley.

Purcell, E., 2011, *Electricity and Magnetism*, 2nd ed. Cambridge University Press.

Zangwill, A., 2013, *Modern Electrodynamics*. Cambridge University Press.

# Appendix I

# Some Useful Mathematical Relationships

## AI.1 Vector Identities and Theorems

### AI.1.1 Scalar and Vector Triple Products

$$
\begin{aligned}
&(\mathbf{a} \times \mathbf{b}) \cdot \mathbf{c} = (\mathbf{b} \times \mathbf{c}) \cdot \mathbf{a} = (\mathbf{c} \times \mathbf{a}) \cdot \mathbf{b}, \\
&\mathbf{a} \times (\mathbf{b} \times \mathbf{c}) = (\mathbf{a} \cdot \mathbf{c})\mathbf{b} - (\mathbf{a} \cdot \mathbf{b})\mathbf{c}, \\
&\mathbf{a} \times (\mathbf{b} \times \mathbf{c}) + \mathbf{b} \times (\mathbf{c} \times \mathbf{a}) + \mathbf{c} \times (\mathbf{a} \times \mathbf{b}) = 0.
\end{aligned}
$$

### AI.1.2 Grad, Div and Curl in Cartesian Coordinates

$$
\begin{aligned}
&\nabla V = \frac{\partial V}{\partial x}\mathbf{i} + \frac{\partial V}{\partial y}\mathbf{j} + \frac{\partial V}{\partial z}\mathbf{k}, \\
&\nabla \cdot \mathbf{F} = \frac{\partial F_x}{\partial x} + \frac{\partial F_y}{\partial y} + \frac{\partial F_z}{\partial z}, \\
&\nabla \times \mathbf{F} = \left(\frac{\partial F_z}{\partial y} - \frac{\partial F_y}{\partial z}\right)\mathbf{i} + \left(\frac{\partial F_x}{\partial z} - \frac{\partial F_z}{\partial x}\right)\mathbf{j} + \left(\frac{\partial F_y}{\partial x} - \frac{\partial F_x}{\partial y}\right)\mathbf{k}.
\end{aligned}
$$

### AI.1.3 Vector Identities

$$
\begin{aligned}
&\nabla(\varphi\psi) = \varphi\nabla\psi + \psi\nabla\varphi, \\
&\nabla(\mathbf{F} \cdot \mathbf{G}) = (\mathbf{F} \cdot \nabla)\mathbf{G} + (\mathbf{G} \cdot \nabla)\mathbf{F} + \mathbf{F} \times (\nabla \times \mathbf{G}) + \mathbf{G} \times (\nabla \times \mathbf{F}), \\
&\nabla \cdot (\varphi\mathbf{F}) = \varphi\nabla \cdot \mathbf{F} + \mathbf{F} \cdot \nabla\varphi, \\
&\nabla \cdot (\mathbf{F} \times \mathbf{G}) = \mathbf{G} \cdot (\nabla \times \mathbf{F}) - \mathbf{F} \cdot (\nabla \times \mathbf{G}), \\
&\nabla \times (\varphi\mathbf{F}) = \varphi(\nabla \times \mathbf{F}) + (\nabla\varphi) \times \mathbf{F}, \\
&\nabla \times (\mathbf{F} \times \mathbf{G}) = \mathbf{F}(\nabla \cdot \mathbf{G}) - \mathbf{G}(\nabla \cdot \mathbf{F}) + (\mathbf{G} \cdot \nabla)\mathbf{F} - (\mathbf{F} \cdot \nabla)\mathbf{G}, \\
&\nabla \times (\nabla\varphi) = 0, \\
&\nabla \cdot (\nabla \times \mathbf{F}) = 0, \\
&\nabla^2\mathbf{F} = \nabla(\nabla \cdot \mathbf{F}) - \nabla \times (\nabla \times \mathbf{F}).
\end{aligned}
$$

### *AI.1.4 Integral Theorems*

$$\int_V \nabla\cdot\mathbf{F}\,dV = \oint_S \mathbf{F}\cdot d\mathbf{S},$$
$$\int_V \nabla\varphi\,dV = \oint_S \varphi\,d\mathbf{S},$$
$$\int_V \nabla\times\mathbf{F}\,dV = -\oint_S \mathbf{F}\times d\mathbf{S}.$$

$$\int_S \nabla\times\mathbf{F}\cdot d\mathbf{S} = \oint_C \mathbf{F}\cdot d\mathbf{l},$$
$$\int_S \nabla\varphi\times d\mathbf{S} = -\oint_C \varphi\,d\mathbf{l}.$$

### *AI.1.5 A Useful Vector Relationship*

If $\mathbf{\Omega}$ is any constant vector and $\mathbf{J}$ a vector field, then

$$2\mathbf{x}\times(\mathbf{J}\times\mathbf{\Omega}) = (\mathbf{x}\times\mathbf{J})\times\mathbf{\Omega} + \mathbf{J}\cdot\nabla[\mathbf{x}\times(\mathbf{x}\times\mathbf{\Omega})].$$

If $\mathbf{J}$ is solenoidal and has zero normal component at the boundary of the domain $V$, this yields

$$\int_V \mathbf{x}\times(\mathbf{J}\times\mathbf{\Omega})\,dV = \frac{1}{2}\int_V (\mathbf{x}\times\mathbf{J})\,dV\times\mathbf{\Omega}.$$

### *AI.1.6 The Helmholtz Decomposition*

Any vector field $\mathbf{C}$ may be written as the sum of an irrotational and a solenoidal field. The irrotational field may be written in terms of a scalar potential $\varphi$ and the solenoidal field in terms of a vector potential $\mathbf{A}$. The divergence of $\mathbf{A}$ is normally taken to be zero:

$$\mathbf{C} = -\nabla\varphi + \nabla\times\mathbf{A}, \quad \nabla\cdot\mathbf{A} = 0.$$

The two potentials are then solutions of

$$\nabla^2\varphi = -\nabla\cdot\mathbf{C}, \quad \nabla^2\mathbf{A} = -\nabla\times\mathbf{C}.$$

## AI.2 Cylindrical and Spherical Polar Coordinates

In cylindrical polar coordinates $(s,\theta,z)$ the location of a point P is identified by finding the projection of the point on the $x$–$y$ plane and then specifying (i) the vertical distance $z$ from P to the $x$–$y$ plane and (ii) the position of the projection of P in the $x$–$y$ plane using two-dimensional polar coordinates $(s,\theta)$, where $s$ is the distance from the $z$ axis and $\theta$ the azimuthal angle measured from the $x$ axis. This is shown in Figure AI.1.

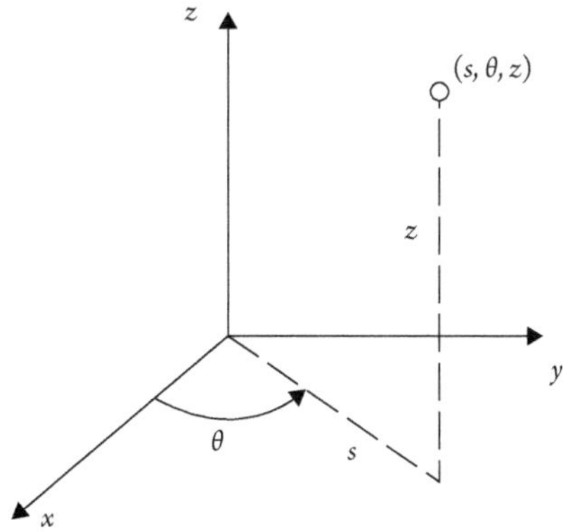


**Figure AI.1** Cylindrical polar coordinates $(s,\theta,z)$.

It is useful to introduce the orthogonal unit vectors $(\hat{\mathbf{e}}_s, \hat{\mathbf{e}}_\theta, \hat{\mathbf{e}}_z)$, where $\hat{\mathbf{e}}_s$ and $\hat{\mathbf{e}}_\theta$ lie in the $x$–$y$ plane and the position vector is $\mathbf{r} = s\hat{\mathbf{e}}_s + z\hat{\mathbf{e}}_z$. A volume element is evidently $dV = ds\,(s\,d\theta)\,dz$, and grad, div, curl and the Laplacian in cylindrical polar coordinates are

$$\nabla V = \frac{\partial V}{\partial s}\hat{\mathbf{e}}_s + \frac{1}{s}\frac{\partial V}{\partial \theta}\hat{\mathbf{e}}_\theta + \frac{\partial V}{\partial z}\hat{\mathbf{e}}_z,$$

$$\nabla \cdot \mathbf{F} = \frac{1}{s}\frac{\partial}{\partial s}(sF_s) + \frac{1}{s}\frac{\partial F_\theta}{\partial \theta} + \frac{\partial F_z}{\partial z},$$

$$\nabla \times \mathbf{F} = \left(\frac{1}{s}\frac{\partial F_z}{\partial \theta} - \frac{\partial F_\theta}{\partial z}\right)\hat{\mathbf{e}}_s + \left(\frac{\partial F_s}{\partial z} - \frac{\partial F_z}{\partial s}\right)\hat{\mathbf{e}}_\theta + \left[\frac{1}{s}\frac{\partial}{\partial s}(sF_\theta) - \frac{1}{s}\frac{\partial F_s}{\partial \theta}\right]\hat{\mathbf{e}}_z,$$

$$\nabla^2 V = \frac{1}{s}\frac{\partial}{\partial s}\left(s\frac{\partial V}{\partial s}\right) + \frac{1}{s^2}\frac{\partial^2 V}{\partial \theta^2} + \frac{\partial^2 V}{\partial z^2},$$

$$\nabla^2 \mathbf{F} = \left(\nabla^2 F_s - \frac{1}{s^2}F_s - \frac{2}{s^2}\frac{\partial F_\theta}{\partial \theta}\right)\hat{\mathbf{e}}_s + \left(\nabla^2 F_\theta - \frac{1}{s^2}F_\theta + \frac{2}{s^2}\frac{\partial F_s}{\partial \theta}\right)\hat{\mathbf{e}}_\theta + (\nabla^2 F_z)\hat{\mathbf{e}}_z.$$

In spherical polar coordinates $(r,\theta,\phi)$, the location of a point P is identified by (i) the distance from the origin to P, which is $r = |\mathbf{r}|$, (ii) the polar angle $\theta$ measured from the positive $z$ axis to the position vector $\mathbf{r}$, as shown in Figure AI.2, and (iii) the azimuthal angle $\varphi$ measured from the $x$ axis to the projection of the point P in the $x$–$y$ plane.

It is useful to introduce the orthogonal unit vectors $(\hat{\mathbf{e}}_r, \hat{\mathbf{e}}_\theta, \hat{\mathbf{e}}_\varphi)$, where the position vector is given by $\mathbf{r} = r\hat{\mathbf{e}}_r$. A volume element is $dV = dr\,(r\,d\theta)(r\sin\theta\,d\phi)$, and grad, div, curl and the Laplacian in spherical polar coordinates are

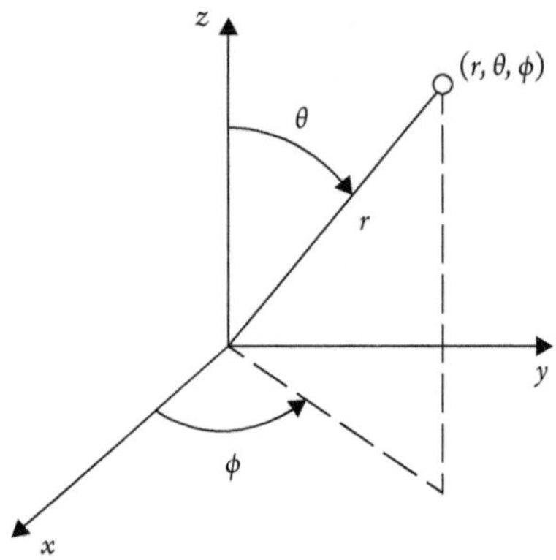


**Figure AI.2** Spherical polar coordinates $(r,\theta,\phi)$.

$$\nabla V = \frac{\partial V}{\partial r}\hat{\mathbf{e}}_r + \frac{1}{r}\frac{\partial V}{\partial \theta}\hat{\mathbf{e}}_\theta + \frac{1}{r\sin\theta}\frac{\partial V}{\partial \phi}\hat{\mathbf{e}}_\phi,$$

$$\nabla\cdot\mathbf{F} = \frac{1}{r^2}\frac{\partial}{\partial r}(r^2F_r) + \frac{1}{r\sin\theta}\frac{\partial(\sin\theta\, F_\theta)}{\partial\theta} + \frac{1}{r\sin\theta}\frac{\partial F_\phi}{\partial\phi},$$

$$\nabla\times\mathbf{F} = \frac{1}{r\sin\theta}\left[\frac{\partial\,(\sin\theta\, F_\phi)}{\partial\theta} - \frac{\partial F_\theta}{\partial\phi}\right]\hat{\mathbf{e}}_r + \frac{1}{r}\left[\frac{1}{\sin\theta}\frac{\partial F_r}{\partial\phi} - \frac{\partial(rF_\phi)}{\partial r}\right]\hat{\mathbf{e}}_\theta$$

$$+\frac{1}{r}\left[\frac{\partial(rF_\theta)}{\partial r} - \frac{\partial F_r}{\partial\theta}\right]\hat{\mathbf{e}}_\phi,$$

$$\nabla^2 V = \frac{1}{r^2}\frac{\partial}{\partial r}\left(r^2\frac{\partial V}{\partial r}\right) + \frac{1}{r^2\sin\theta}\frac{\partial}{\partial\theta}\left(\sin\theta\frac{\partial V}{\partial\theta}\right) + \frac{1}{r^2\sin^2\theta}\frac{\partial^2 V}{\partial\phi^2},$$

$$\nabla^2\mathbf{F} = \left[\nabla^2 F_r - \frac{2F_r}{r^2} - \frac{2}{r^2\sin\theta}\frac{\partial(F_\theta\sin\theta)}{\partial\theta} - \frac{2}{r^2\sin\theta}\frac{\partial F_\phi}{\partial\phi}\right]\hat{\mathbf{e}}_r$$

$$+\left(\nabla^2 F_\theta + \frac{2}{r^2}\frac{\partial F_r}{\partial\theta} - \frac{F_\theta}{r^2\sin^2\theta} - \frac{2\cos\theta}{r^2\sin^2\theta}\frac{\partial F_\phi}{\partial\phi}\right)\hat{\mathbf{e}}_\theta$$

$$+\left(\nabla^2 F_\phi + \frac{2}{r^2\sin\theta}\frac{\partial F_r}{\partial\phi} + \frac{2\cos\theta}{r^2\sin^2\theta}\frac{\partial F_\theta}{\partial\phi} - \frac{F_\phi}{r^2\sin^2\theta}\right)\hat{\mathbf{e}}_\phi.$$

## AI.3 Legendre Functions

When there is azimuthal symmetry in a problem, Laplace's equation written in terms of spherical polar coordinates reduces to

$$\nabla^2 V = \frac{1}{r^2}\frac{\partial}{\partial r}\left(r^2\frac{\partial V}{\partial r}\right) + \frac{1}{r^2\sin\theta}\frac{\partial}{\partial\theta}\left(\sin\theta\frac{\partial V}{\partial\theta}\right) = 0.$$

If we look for solutions of $\nabla^2 V = 0$ in terms of separation of variables, $V(r,\theta) = R(r)P(\theta)$, this yields

$$\frac{1}{R}\frac{d}{dr}\left(r^2\frac{dR}{dr}\right) = \nu(\nu+1), \qquad \frac{1}{P\sin\theta}\frac{d}{d\theta}\left(\sin\theta\frac{dP}{d\theta}\right) = -\nu(\nu+1),$$

for some constant $\nu$. The general solution to the first of these equations is

$$R = Ar^{\nu} + Br^{-(\nu+1)},$$

for two constants $A$ and $B$. Writing $x = \cos\theta$, the second equation above becomes *Legendre's equation,*

$$(1-x^2)\frac{d^2P}{dx^2} - 2x\frac{dP}{dx} + \nu(\nu+1)P = 0,$$

whose bounded solutions in $-1 < x \leq 1$ are known as the *Legendre functions of the first kind of order* $\nu$, denoted by $P_\nu(x)$. The Legendre functions are regular at $x = 1$ and for $|x| < 1$, but singular at $x = -1$ unless $\nu$ is an integer. For problems in which $x = -1$ is excluded from the domain, solutions of the form $V = Ar^{\nu}P_\nu(\cos\theta)$ are physically acceptable, provided $\nu > 0$ to avoid a singularity at the origin. When $x = -1$ is included in the domain, then $\nu$ must be taken to be a positive integer, $\nu = n$, to avoid a singularity at $x = -1$. The functions $P_n(x)$ are known as the *Legendre polynomials* and are

$$\begin{aligned}
P_0(x) &= 1,\\
P_1(x) &= x,\\
P_2(x) &= (3x^2-1)/2,\\
P_3(x) &= (5x^3-3x)/2,\\
P_4(x) &= (35x^4-30x^2+3)/8,\\
&\vdots
\end{aligned}$$

(These polynomials are, by convention, normalised such that $P_n(1) = 1$.) Using superposition, we see that *one* solution to $\nabla^2 V = 0$ in spherical polar coordinates is

$$V(r,\theta) = \sum_{n=0}^{\infty}(A_n r^n + B_n r^{-(n+1)})P_n(\cos\theta).$$

In fact, the orthogonality and completeness of the Legendre polynomials on the interval $-1 \leq x \leq 1$ (which we have not proved) means that this is *the* general solution for problems with azimuthal symmetry.

## AI.4 The Delta Function

The *delta function* in one dimension, usually denoted by $\delta(x)$, is a tall, narrow blip centred on the origin whose integral is equal to unity. Formally, if the blip has a characteristic width $\Delta$ and a maximum height $h$, then we consider the limit as $\Delta \to 0$ and $h \to \infty$ as shown in Figure AI.3, with the limit taken such that the integral of the function remains equal to unity. In this limit the generalised function $\delta(x)$ has the defining characteristics

$$\delta(x) = 0, \quad x \neq 0; \quad \delta(0) = \infty; \quad \int_{-\infty}^{\infty} \delta(x)\, dx = 1.$$

Consider, for example, the Gaussian

$$f(x) = \frac{1}{\Delta\sqrt{\pi}} \exp\left(-\frac{x^2}{\Delta^2}\right),$$

whose integral is unity. We could define $\delta(x)$ to be $f(x)$ in the limit as $\Delta \to 0$, although of course there are many other candidate functions that would do equally well.

The delta function (sometimes called the *impulse function*) is closely related to the *Heaviside function* $H(x)$, which is the step function

$$H(x) = \begin{cases} 0, & x < 0, \\ 1, & x > 0. \end{cases}$$

Evidently $H$ is the integral of the delta function, $H(x) = \int_{-\infty}^{x} \delta(x')\, dx'$, and by implication $\delta(x)$ is the derivative of $H(x)$. However, to give this derivative a clear meaning, it is helpful to start by differentiating a smooth function, such as the integral of the Gaussian defined above, and then consider the limiting process as $\Delta \to 0$. Heaviside, who was quick to see

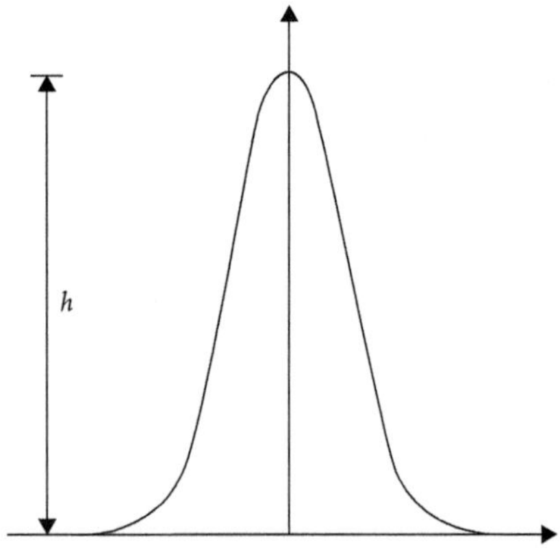


**Figure AI.3** A delta function is obtained as $h \to \infty$ subject to unit area under the curve.

the power of the delta function, was characteristically casual in this matter. As noted by G. Temple (1981),

> ...the outstanding feature of Heaviside's creative imagination was the calm assurance with which he differentiated the step function $H(x)$ to give the impulse function $\delta(x)$.

The *three-dimensional* delta function $\delta(\mathbf{x})$ is a narrow blip in space centred on the origin and whose *volume integral* is unity. It may be thought of as the product of three one-dimensional delta functions, $\delta(\mathbf{x}) = \delta(x)\delta(y)\delta(z)$. If the blip is centred on the point $\mathbf{x}_0$, rather than the origin, then the delta function is written as $\delta(\mathbf{x} - \mathbf{x}_0)$. This has the sifting property,

$$\int g(\mathbf{x})\delta(\mathbf{x} - \mathbf{x}_0)\, d\mathbf{x} = g(\mathbf{x}_0).$$

# Appendix II

## SI Versus Gaussian Units

Privately, I use units that get rid of the $4\pi$'s completely, and then, for publication, liberally season with $4\pi$'s to suit the taste of British Association unit-fed readers.

Heaviside, 1885

In this book we have adopted the *Système International* (SI units). These units are widely used in undergraduate courses in physics and engineering, and so are appropriate for any textbook. However, some physicists still prefer the older system of *Gaussian units*. The advantage of Gaussian units is that the electric and magnetic fields are given the same dimensions, which seems entirely appropriate, since one field can transmute into the other upon a change of frame of reference. So, in the Gaussian system, the two fields **E** and **B** play a more symmetric role in electromagnetic phenomena, and consequently this system is often preferred by those studying, say, electromagnetic radiation. The disadvantages of the Gaussian system are that needless $4\pi$'s creep into Maxwell's equations and the units of charge and current are inconveniently small, being $3 \times 10^9$ smaller than the coulomb or ampere. So Gaussian units are cumbersome for problems in electrical engineering and magnetohydrodynamics. Needless to say, Heaviside had an opinion about the 'eruption' of $4\pi$'s into Maxwell's equations. He did not like it, referring to it as absurd, arbitrary, injurious, irrational, mischievous, obtrusive, perverse, preposterous, ridiculous, stupid, unfortunate, unnecessary and unspeakable! (according to G. A. Campbell, 1933).

There are two differences between Gaussian and SI units. The first is trivial: in Gaussian units mechanical quantities are measured in grams, centimetres and seconds, which does not change the dimensions of any physical quantity, merely their numerical value. The second difference is more profound: in the Gaussian system we assign different values to the constants of proportionality in the laws of mutual attraction or repulsion between two static charges and between two parallel wires carrying current. Let us see how this works. Coulomb's law for two charges $q_1$ and $q_2$, separated by a distance $r$, can be written as

$$f_e = k_e \frac{q_1 q_2}{r^2} = q_1 E_2,$$

**Table AII.1** Maxwell's equations in SI and Gaussian form

| Law | SI | Gaussian |
|---|---|---|
| Gauss | $\nabla \cdot \mathbf{E} = \frac{\rho_e}{\varepsilon_0}$ | $\nabla \cdot \mathbf{E} = 4\pi\rho_e$ |
| Faraday | $\nabla \times \mathbf{E} = -\frac{\partial \mathbf{B}}{\partial t}$ | $\nabla \times \mathbf{E} = -\frac{1}{c}\frac{\partial \mathbf{B}}{\partial t}$ |
| | $\nabla \cdot \mathbf{B} = 0$ | $\nabla \cdot \mathbf{B} = 0$ |
| Ampère–Maxwell | $\nabla \times \mathbf{B} = \mu_0\left(\mathbf{J} + \varepsilon_0\frac{\partial \mathbf{E}}{\partial t}\right)$ | $\nabla \times \mathbf{B} = \frac{4\pi}{c}\mathbf{J} + \frac{1}{c}\frac{\partial \mathbf{E}}{\partial t}$ |

**Table AII.2** Conversion factors for moving from SI to Gaussian units, where $2.998 \times 10^8$ is the speed of light in m/s

| Quantity | Symbol | SI unit | Gaussian unit | Conversion factor |
|---|---|---|---|---|
| Mass | $m$ | kilogram | gram | $10^3$ |
| Length | $l$ | metre | centimetre | $10^2$ |
| Time | $t$ | second | second | 1 |
| Charge | $Q$ | coulomb | esu | $10 \times (2.998 \times 10^8) \approx 3 \times 10^9$ |
| Current | $I$ | ampere | esu/second | $10 \times (2.998 \times 10^8) \approx 3 \times 10^9$ |
| Electric field | $\mathbf{E}$ | newton/coulomb | centimetre·(dyne)$^{1/2}$ | $10^4/(2.998 \times 10^8) \approx 10^{-4}/3$ |
| Magnetic field | $\mathbf{B}$ | tesla | gauss | $10^4$ |

which defines $E_2$, the magnitude of the electric field created by $q_2$. Similarly, the magnitude of the Lorentz force per unit length between two parallel wires separated by a distance $d$ and carrying currents $I_1$ and$I_2$ takes the form

$$f_m = k_m \frac{2I_1 I_2}{d} = \frac{I_1 B_2}{\alpha},$$

which defines $B_2$, the magnetic field caused by current $I_2$. In SI units we take $k_e = 1/4\pi\varepsilon_0$ and $k_m = \mu_0/4\pi$, giving $k_e/k_m = c^2$. Moreover, the units of $\mathbf{B}$ in SI are defined by taking $\alpha = 1$, so that $B_2 = f_m/I_1$.

In the Gaussian system, by contrast, we take $k_e = 1$ and $k_m = 1/c^2$, which leaves $k_e/k_m = c^2$ unchanged, as it must be, since this ratio involves only mechanical units and is independent of the unit of charge. In the Gaussian system Coulomb's law is then

$$f_e = \frac{q_1 q_2}{r^2} = q_1 E_2 \qquad \text{(Gaussian)},$$

which defines a new unit of charge, called the *electrostatic unit* (or esu), with the dimensions of length×(force)$^{1/2}$. Moreover, in the Gaussian system the expression for the force between two current-carrying wires is written using $\alpha = c$, rather than $\alpha = 1$,

$$f_m = \frac{1}{c^2}\frac{2I_1 I_2}{d} = \frac{I_1 B_2}{c} \qquad \text{(Gaussian)},$$

so that the magnetic field $B_2 = cf_m/I_1$ is inflated by a factor of $c$ relative to SI units. With these changes in dimensions, the Lorentz force in SI,

$$\mathbf{f} = q(\mathbf{E} + \mathbf{u}_q \times \mathbf{B}) \qquad \text{(SI)},$$

takes the form

$$\mathbf{f} = q\left(\mathbf{E} + \frac{\mathbf{u}_q}{c} \times \mathbf{B}\right) \qquad \text{(Gaussian)}.$$

Note that the magnetic and electric fields now have the same dimensions, which is the primary merit of Gaussian units. The simplest way to rewrite Maxwell's equations in Gaussian form is to set $\varepsilon_0 \to 1/4\pi$ and $\mu_0 \to 4\pi/c^2$ and then rescale $\mathbf{B}$ by dividing by $c$. The results are shown in Table AII.1. The conversion factors for moving from SI to Gaussian units are given in Table AII.2, where the numerical value in Gaussian units is obtained from the SI value by multiplying by the conversion factor.

# Appendix III

## The Pioneers of Electromagnetism: Brief Historical Sketches

The seventy years from 1820 to 1890 witnessed a fundamental and permanent revolution in our understanding of electromagnetism, the principal players being Ampère, Faraday, Maxwell, Heaviside and Hertz. By way of a footnote, we now offer brief historical sketches of these pioneers. For readers of an historical bent, a few suggestions for further reading are provided at the end of this appendix. Of course, our understanding of electrodynamics did not start on one day and finish on another, so the choice of 1820–1890 is a tad arbitrary and leads to the unfortunate exclusion of Cavendish, Coulomb, Young and Fresnel. I hope readers will forgive these omissions.

### AIII.1 Ampère and the Birth of Electrodynamics

Although many see Ampère's work on electromagnetism as heralding the birth of electrodynamics, and his name will forever be associated with one of Maxwell's equations, Ampère spent no more than six years devoted to this subject. He was a true polymath, dividing up his time between chemistry, where he competed with Avogadro, psychology, astronomy, magnetism, philosophy, logic and biology. But his greatest contribution to science is probably his discovery of the relationship between electric currents and magnetism. Sadly, his scientific success contrasts somewhat with the tragedy that beset his personal life.

André-Marie Ampère was born in 1775 in Lyon during the French Enlightenment, a progressive time during which Voltaire and Rousseau promoted science and logic over the preaching of the Catholic Church. His parents were of comfortable means and encouraged their son to learn as much as he could from their extensive library, which suited the young Ampère very well. Inspired by his father, Ampère read ferociously and started to excel in mathematics. However, the French Revolution soon brought turbulence to France, claiming his father's life by the guillotine in 1793. This marked the start of a particularly dark period for Ampère, as a reign of revolutionary terror swept across the country. Nevertheless, he persisted with his studies in mathematics, first becoming a tutor in mathematics in Lyon and then, in 1802, a professor at the École Centrale in Bourg-en-Bresse. Shortly after, tragedy

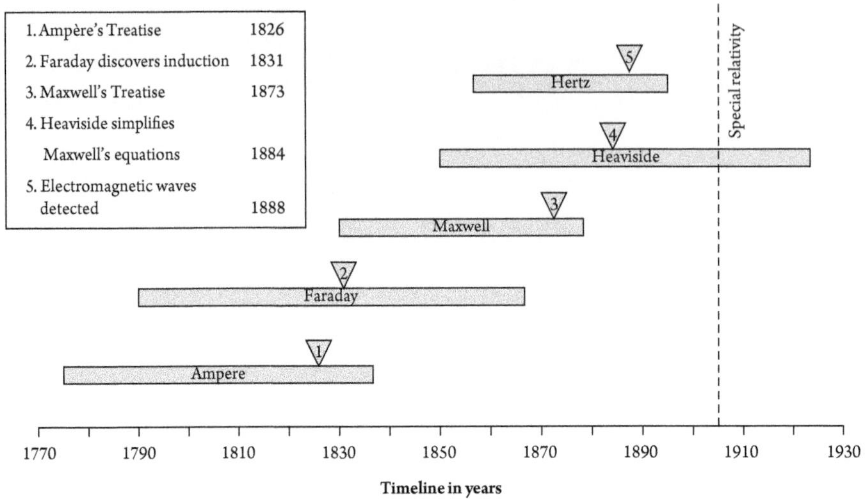


struck for the second time with the premature death of his young wife after less than four years of marriage. A distraught Ampère looked to Paris for a new start in life.

In 1804 Ampère joined the newly created École Polytechnique in Paris as a tutor. Here his reputation quickly spread and in 1809, despite his lack of formal qualifications, he was promoted to a full professorship in mathematics. During this period his interests diversified to encompass chemistry and philosophy. He married once more, but this second marriage rapidly ran into trouble, leaving him with an infant daughter to care for, along with a young son from his first marriage.

Ampère's interest in physics was sparked in 1820 when news reached Paris of the experiments of the Danish physicist Hans Christian Oersted, who had shown that a current flowing in a wire could deflect a compass needle. There followed six highly productive years of research for Ampère, during which he mounted an experimental campaign to explore the relationship between the hitherto distinct disciplines of electricity and magnetism. Ampère discovered that a wire carrying electric current will attract or repel an adjacent wire also carrying a current, the attraction or repulsion depending on whether the currents are parallel or antiparallel. He emphasized that, although this attractive force resembled a magnetic force, no magnets are required. He then went on to explore the relationship between current and magnetism, work that would eventually lead to the law that now bears his name.

At first Ampère's experimental results, and his theoretical interpretation of them, met with considerable scepticism, verging on hostility, with Laplace mocking his abilities as an experimentalist. (See the quotation at the start of Chapter 7.) Perhaps Ampère's fiercest critic was Jean-Baptiste Biot, who, along with his collaborator Felix Savart, was in direct competition with Ampère. However, as Ampère refined his ideas, they gradually won greater acceptance. Ampère's studies culminated in 1826 with the publication of his famous treatise *Memoir on the Mathematical Theory of Electrodynamic Phenomena, Uniquely Deduced from Experience*—the work that coined the term *electrodynamics*. As Maxwell later put it:

> The experimental investigation by which Ampère established the law of the mechanical action between electric currents is one of the most brilliant achievements in science. The whole theory and experiment seems as if it had leaped, full grown and full armed, from the brain of the 'Newton of Electricity'.

From 1827 onward Ampère's interests drifted towards philosophy and biology, and he rarely looked back to the subject that would eventually become his foremost scientific legacy. Nevertheless, his career flourished and honours continued to be bestowed upon him. Tragedy struck for the third and final time when, at the age of 61, he caught a severe case of pneumonia. The years had taken their toll and he died in Marseilles on 10 June 1836. His remains were later moved to Montmartre in Paris.

## AIII.2 Faraday, Prince of Experimentalists

Michael Faraday was born on the outskirts of London on 22 September 1791, the third son of the local blacksmith. His schooling was primitive, consisting of little more than some elementary reading, writing and arithmetic, and at the age of thirteen, the same year Ampère joined the École Polytechnique in Paris, he left school to find employment as an errand boy to a bookbinder, one George Riebau. He later became Riebau's apprentice and took to reading the books he was binding, including the 1797 edition of *Encyclopaedia Britannica*. This revealed an exciting new world to the young Faraday, who seemed particularly interested in the articles on chemistry and the emerging topic of electricity.

The turning point came in 1812 when one of Riebau's customers gave Faraday tickets to see four lectures to be given at the Royal Institution by the eminent chemist Sir Humphry Davy. Faraday was enthralled by the lectures, took copious notes, later added illustrations, and then bound his notes. Keen to pursue a life in science, he sent the bound notes to Davy, who was then the director of the Royal Institution. Davy was flattered and impressed by Faraday's boldness. The following year he offered Faraday a position as a laboratory assistant at the institution, with the grand salary of one pound a week. In 1813 Faraday moved into rooms at the Royal Institution, where he subsequently lived and worked for almost half a century. It would not be long before the laboratory assistant rose to become the most celebrated experimental physicist in the world, outshining his mentor Davy.

In 1813 Davy embarked on an extended scientific tour of Europe, taking Faraday along as his assistant. However, Davy's valet did not wish to go and so the young Faraday was asked to double up as scientific assistant and valet. Victorian England was a class-ridden society and Davy's wife treated Faraday abysmally, forcing him to eat with the servants and to travel outside their coach. This was a humiliation that Faraday found hard to take, although the tour did at least allow him access to some of Europe's most eminent scientists.

Faraday soon developed a passionate interest in chemistry, electricity and magnetism. In chemistry, he discovered benzene, worked on the liquefaction of gases, established the laws of electrolysis, laid the foundations for the chemistry of plasmas, discovered *diamagnetism* and *paramagnetism* (terms he coined), and pioneered the development of special alloy steels, presenting his friends with gifts of non-rusting razors. However, his work on electricity and magnetism was to have a more profound impact, starting with the discovery of electromagnetic rotations in 1821 and culminating in his discovery of electromagnetic induction in 1831. He also made contributions to electrostatics (introducing the Faraday cage), electrical discharges in plasmas, the relationship between light and electromagnetism, and the influence of magnetic fields on spectral lines (later called the Zeeman effect).

However, perhaps the most important contribution of Faraday to electromagnetism was his introduction of the concept of the *field*, particularly his insistence that electric and magnetic fields lay at the heart of all electromagnetic phenomena. Before Faraday, mathematicians were confident that the laws of electrostatics and magnetism could be formulated in terms of action at a distance. Thus, when Ampère discovered that two wires carrying currents attracted or repelled each other, it seemed natural to describe this force in terms of some kind of inverse square law, as in Newton's law of gravitation. But Faraday had a completely different vision, in which the medium between the wires plays a crucial role. In Faraday's mind, a wire that carries a current introduces a field into the space surrounding it, and this field exists whether or not a second wire is present. When the second wire is introduced, it experiences a force by virtue of the field. Moreover, for Faraday, such a field is not just some abstract mathematical intermediary: it is real, possessing energy, for example.

Perhaps this uneasy standoff between Faraday and the mathematicians is best summed up by James Clerk Maxwell in the preface to his *Treatise on Electricity and Magnetism*:

> Before I began the study of electricity I resolved to read no mathematics on the subject till I had first read through Faraday's Experimental Researches in Electricity. I was aware that there was supposed to be a difference between Faraday's way of conceiving phenomena and that of the mathematicians, so that neither he nor they were satisfied with each other's language . . . As I proceeded with the study of Faraday, I perceived that his method of conceiving the phenomena was also a mathematical one, though not exhibited in the conventional form of mathematical symbols . . . For instance, Faraday, in his mind's eye, saw lines of force traversing space where the mathematicians saw centres of force attracting at a distance: Faraday sought the seat of the phenomena in real actions going on in the medium, they were satisfied that they had found it in a power of action at a distance . . . When I translated what I considered to be Faraday's ideas into mathematical form, I found that in general the results of the two methods coincided, so that the same phenomena were accounted for, and the same laws of action deduced by both methods, but that Faraday's methods resembled those in which we begin with the whole and arrive at the parts by analysis, while the ordinary mathematical methods were founded on the principle of beginning with the parts and building up the whole by synthesis. I also found that several of the most fertile methods of research discovered by the mathematicians could be expressed much better in terms of the ideas derived by Faraday than in their original form . . .
>
> If by anything I have written I may assist any student in understanding Faraday's modes of thought and expression, I shall regard it as the accomplishment of one of my principle aims—to communicate to others the same delight which I have found myself in reading Faraday's 'Researches'.
>
> James Clerk Maxwell (1873)

Faraday was a devout Christian, belonging to a sect called the Sandemanians, whose roots lay in the Scottish Presbyterian church. Like the Quakers, the Sandemanians praised simple living and hard work, and opposed the accumulation of wealth. His religion taught Faraday the importance of humility (he would later decline a knighthood) and of public service. Davy had already instituted regular public lectures on science at the Royal Institution, and when Faraday took over as director he championed these public lectures with renewed vigour, becoming famed for his extraordinary eloquence. These lectures did a great deal to popularise science in Victorian England, and indeed the Royal Institution public lectures on science continue to this day, honouring the memory of Faraday.

Faraday's efforts did not go unrewarded, and he was elected to over 60 learned societies over his lifetime, including the Royal Society in 1824, an election that Davy opposed. He was also offered a knighthood, which he turned down on religious grounds. However, unremitting service to the cause of science progressively took its toll on his health. In 1835 increasing lapses of memory and severe headaches caused him to take several months' rest in Switzerland. Work then resumed for a while until, in 1840, he had a complete mental and physical breakdown. He returned to Switzerland with his wife, Sarah, only this time he stayed for the better part of a year. Faraday managed a successful return to science and continued working up until the middle of the 1850s, at which point he realised that his mental faculties were declining. He then slowly withdrew from science, and from the scientific community, turning down invitations to become President of the Royal Society and of the Royal Institution.

Faraday finally stopped giving popular science lectures in 1861. Shortly after, he and his wife moved out of the Royal Institution to a house in Hampton Court that Queen Victoria had made available to them. Michael Faraday died in his seventy-fifth year, on 25 August 1867. At his own request, he was laid to rest in a simple grave in Highgate Cemetery, with a funeral that was devoid of any pomp or ceremony. He had no children.

It is said that Einstein kept a picture of Faraday on his study wall, next to one of Maxwell. Certainly, along with Maxwell, he was one of the most influential nineteenth-century scientists, helping pave the way for modern physics.

## AIII.3 Maxwell: The End of One Scientific Epoch and the Beginning of Another

James Clerk Maxwell was a mathematical physicist who is renowned not only for the equations of electrodynamics that now bear his name, but also for his contribution to the kinetic theory of gases. His work on electrodynamics was strongly influenced by Faraday's views and culminated in Maxwell predicting that light was an electromagnetic wave. As emphasised by Einstein, Maxwell was responsible for establishing once and for all the primacy of Faraday's fields in theoretical physics, which was an essential precursor to the development of modern physics. Maxwell is considered by many scientists to be the third most influential physicist of all time, after Newton and Einstein.

Maxwell was born in Edinburgh on 13 June 1831 to a middle class family of comfortable means. His parents married when they were in their thirties and, while James was still young, settled into a 1500-acre estate in Galloway, southwest Scotland. There they had overseen the building of Glenlair House, which was to become the family home. The scenic countryside of Galloway provided the backdrop to James' happy childhood, where he soon showed himself to be a remarkable child. He was curious about everything that moved, bombarding his parents with questions from the age of three onwards, usually prefixed with the phrase 'What's the go o' that?'.

Like Ampère before him, Maxwell read ferociously as a young child, and by age eight he could recite long passages from Shakespeare, Milton and the Bible. His parents had originally planned for Maxwell to be schooled at home, but after his mother's premature death (Maxwell was only eight at the time) this proved increasingly difficult. Eventually he was sent off to the Edinburgh Academy. Here the ten-year-old Maxwell was at first ridiculed by the other boys on account of his unusual clothes, which had been personally fashioned by his father. After a period of some isolation he eventually started to make friends, which included Peter Guthrie Tait, who would go on to become an influential mathematical physicist in his own right. While still at school, Maxwell had his first paper published. It was on the geometry of curves and was read before the Royal Society of Edinburgh by Professor James Forbes, the fourteen-year-old Maxwell being deemed too young to present it himself.

Maxwell enrolled at Edinburgh University when he was sixteen, where his teachers included Sir William Hamilton (on logic) and James Forbes (on physics). He had two more papers published by the Royal Society of Edinburgh before graduating and moving down to Cambridge in October 1850. Here Maxwell flourished at Trinity College under the tutorship of the formidable William Hopkins, who had tutored both Cayley and Stokes. Hopkins later described Maxwell as 'the most extraordinary man he had met within his whole range of experience; it appeared impossible for him to think incorrectly on physical subjects.' In 1854 Maxwell graduated second top of the Mathematics Tripos, just behind the exceptional Edward Routh, and in October 1855 he was elected a fellow of Trinity College, an unusual honour for one so young.

It would have been easy for Maxwell to be seduced by his comfortable surroundings at Trinity and Cambridge. However, in 1856 his old tutor in Edinburgh, James Forbes, urged him to apply for a Chair in Natural Philosophy (physics) that had just become vacant at Marischal College, Aberdeen. Concerned by his father's failing health, Maxwell applied for, and was appointed to, the chair, although sadly his father died before the appointment was announced. Maxwell was only twenty-five, and immersed himself in his new job, reforming the physics syllabus, lecturing 15 hours a week and introducing laboratory classes for the students. While at Aberdeen he married Katherine Dewar, the daughter of the Principal of Marischal College. They lived in Aberdeen during term time but spent their summers in Glenlair, which Maxwell had inherited after his father's death. In 1857 the topic for the prestigious Adams Prize was announced: it was to be the nature of Saturn's rings. Maxwell devoted himself to the subject for almost two years and was eventually awarded the prize in 1859 for his essay 'On the stability of the motion of Saturn's rings'. Here he showed that the rings must be composed of numerous small particles, rather than being a continuum. The Astronomer Royal, George Airy, commented that the essay was one of the most remarkable applications of mathematics to astronomy he had seen. Maxwell's tenure at Marischal College came to an abrupt end when the college merged with the adjacent King's College to form Aberdeen University, Maxwell's chair being one of the casualties of the merger.

Aberdeen's loss was London's gain when Maxwell moved to a Chair of Natural Philosophy at King's College, London. The five years he spent there, 1860–1865, were to be his most productive, with his being awarded the Royal Society's Rumford Medal in 1860 for his work on colour and then being elected to the Royal Society in 1861. It was during this period that Maxwell developed a keen interest in electromagnetism, possibly as a result of regular meetings with Michael Faraday at the Royal Institution. Despite a forty-year age difference, the two men became friends and developed a strong sense of mutual respect. In a series of papers Maxwell began to assemble a mathematical theory of electromagnetism that would later form the backbone of his *Treatise on Electricity and Magnetism*, and at the heart of this theory lay Faraday's contentious fields. After introducing the displacement current as a correction to Ampère's law, Maxwell went on to show that light was an electromagnetic wave. The pivotal work is probably his 1864 paper 'A dynamical theory of the electromagnetic field', in which Maxwell writes:

> The theory I propose may therefore be called a theory of the Electromagnetic Field because it has to do with the space in the neighbourhood of the electric or magnetic bodies and it may be called a Dynamical Theory because it assumes that in the space there is matter in motion by which the observed electromagnetic phenomena are produced.

Notice that Maxwell starts by placing Faraday's fields before the reader. When Maxwell presented his ideas to the Royal Society in 1864, they were met with bewilderment: no-one knew what to make of this radical departure from orthodoxy. Even the mighty Kelvin was deeply suspicious. Instead of the fanfare that might have been anticipated, Maxwell's efforts were met with a stony silence, a silence that must have caused Maxwell to reflect.

In 1865 Maxwell resigned his chair at King's College and returned with Katherine to their home in Galloway. In part this was because he wanted to look after the family estate of Glenlair, and in part because he wanted to develop his new electromagnetic theory into an extended treatise, a treatise that might present the theory in a more measured and complete form. That treatise appeared in 1873, but the scientific community remained deeply suspicious, clinging to the old notions of action at a distance. It was not until a young Oliver Heaviside picked up a copy of Maxwell's treatise in a public library in Newcastle that the tide began to turn. As we shall see, the tenacious Heaviside championed and simplified Maxwell's theory up to the point where the scientific community could no longer ignore it.

As for Maxwell, he returned to Cambridge in 1871 when he was appointed as the first Cavendish Professor of Physics. Here he set about editing Henry Cavendish's research papers and introducing laboratory classes for students, just as he had done earlier in Aberdeen. He died in Cambridge of abdominal cancer on 5 November 1879. He was forty-eight years of age.

## AIII.4 Heaviside: The Maverick Maxwellian

Given Oliver Heaviside's contributions to mathematical physics and electrical engineering, as well as his colourful if ultimately tragic life, it is a little surprising that he is not more widely known. There are, however, some excellent historical sketches, such as the biography by Nahin, which delves into many aspects of Heaviside's life and work, and Crowe's *History of Vector Analysis*, which focuses on Heaviside's role in the great quaternion versus vector calculus feud. Both books are scholarly and entertaining, and both are well worth a read.

Heaviside was born in 1850 to a lower middle class family in an infamously deprived area of London. Scarlet fever left him partially deaf from the age of eight and, although he studied all of his adult life, he received no formal education after the age of sixteen. His first break came when the highly respectable electrical scientist and entrepreneur Sir Charles Wheatstone decided to marry his cook, who turned out to be Heaviside's aunt. Wheatstone had established the first telegraph system in Britain and, with his help, Heaviside, now aged eighteen, got a well-paid job as a telegraph operator with a Danish company. This was to be his first and last job, but it left him with a passionate interest in the fledgling technology of telegraphy and the deeper mysteries of electricity. The company had just laid its first sub-oceanic cable and, like other telegraph companies of the time, was encountering all manner of technical problems. Electrical engineering barely existed as a science at this stage, yet Heaviside quickly gained a reputation for fixing the problems, and within two years he was promoted to chief operator and transferred from Denmark to Newcastle. He was also educating himself in his free time, and at age 22 he submitted his first scientific article to the renowned *Philosophical Magazine*, a paper that caught the attention of both Kelvin and Maxwell. An epiphany occurred the following year in a public library in Newcastle when Heaviside discovered Maxwell's newly published *Treatise on Electricity and Magnetism* (1873). For the first time he saw the possibility of a scientific explanation for the multitude of mysterious electrical phenomena he had been wrestling with. His destiny was now set. The following year he quit his job, returned to his family home in London and devoted the remainder of his life to the understanding of electrical science.

His first task was to try and understand Maxwell's treatise, but he quickly realised that, despite his self-education, he had not yet the mathematical knowledge required, and so set about teaching himself more mathematics. In the meantime, he quickly established a sound scientific basis for understanding how electrical signals travel along transmission lines, publishing a trilogy of outstanding papers in the years 1876–1879. These were to form the backbone of transmission line theory, even to this day. His new insights led directly to novel methods for minimising the attenuation and distortion of signals. Unfortunately, these were slow to be adopted in the UK because of Heaviside's endless battles with his nemesis William Preece, an ambitious and influential employee of the British Post Office who eventually rose to become its Engineer-in-Chief and who Nahin describes as 'an utter blockhead'. Despite this, Heaviside maintained an interest in telegraphy all of his life and his proposals were at least pursued in the USA. Sadly, he did not patent many of his ideas (although he did patent the coaxial cable) and later, in the 1890s, AT&T and the American physicist and astute self-publicist Michael Pupin took advantage of this to file their own patents. Eventually AT&T bought out Pupin's patent for $445 000 (tax-free), while Heaviside received nothing. This shoddy treatment of Heaviside so outraged the MIT mathematician Norbert Wiener that he sent a detailed film proposal to Orson Welles in 1941, in which the hero was Heaviside whilst AT&T and Pupin were cast as the villains (see Nahin, 2002).

Throughout the 1880s and 1890s Heaviside concentrated on electrodynamics and applied mathematics. His two major contributions to mathematics were vector calculus, which he and Josiah Willard Gibbs developed simultaneously yet independently, and the *operational calculus,* which provides a method of converting linear differential equations into algebraic equations. The operational calculus was, for some time, widely adopted by electrical engineers, but it was criticised by mathematicians for its lack of rigor. It has since been superseded by the related method of Laplace transforms. His work on developing vector calculus was, however, of more lasting value. As with his study of telegraphy, he met considerable opposition from establishment figures, in this case most notably from the Scottish mathematical physicist Peter Guthrie Tait. Tait was determined that physicists adopt the quaternions to manipulate vector fields and, unlike Preece, he was no blockhead and widely respected. The battle with Tait was high-spirited and prolonged, and it took over twenty years for the Gibbs–Heaviside system of vector calculus to be widely accepted. Once established, though, it quickly swept aside quaternions and remains to this day the workhorse of electrodynamics and fluid mechanics. Since we discuss the great quaternion versus vector calculus debate in Chapter 1, we shall not dwell on it here. Rather, we urge readers to consult Crowe (1994), who provides a delightfully humorous account of the Heaviside–Tait feud, in which both protagonists indulged in colourful invective.

Although best known for his work on telegraphy, Heaviside's most profound contribution to science was in electrodynamics. In 1884 he stripped down and greatly simplified Maxwell's analysis, using his newly developed vector calculus to encapsulate the core of Maxwell's new ideas into just four partial differential equations; the four 'Maxwell's equations' we know today. As Sir Edmund Whittaker put it in 1928:

> The great service which Heaviside now rendered to science was... to base the theory on what he called the 'duplex' equations... which modern writers generally call 'Maxwell's equations'—though they are not found in Maxwell's *Treatise,* and the modern writers have in fact copied them from Heaviside.

In the process Heaviside independently discovered the Poynting energy flux, initially publishing in *The Electrician* in June 1884, ahead of Poynting, and then in more detail in January 1885. In 1889 he calculated the electromagnetic force acting on a moving charge, correcting an error in an earlier paper by J. J. Thomson. He also calculated the change in mass of charges moving at relativistic speeds, well before Einstein's special relativity, again building on the work of Thomson, who had already done the calculation for slowly moving charges. In 1902, along with Arthur Kennelly, he predicted the existence of an ionised layer in the upper atmosphere, and in the same year he calculated the electromagnetic fields and radiation associated with accelerated charges moving at relativistic speeds, work he extended in 1904. Until recently, this latter calculation was not well known, and it has been attributed to several other physicists, including Richard Feynman. All-in-all, it is hard to disagree with George FitzGerald's claim, made in 1893, that, more than any other nineteenth-century physicist,

Heaviside was responsible for developing and exploring the consequences of Maxwell's treatise (see the quotation in the preface to the present book).

His theoretical work on electromagnetism led to him being elected a Fellow of the Royal Society in 1891, with Kelvin being one of the signatories on the testimonial. This was an extraordinary honour for the forty-one-year old maverick who had spent most of his life at odds with the scientific establishment. At last the stubbornly independent outsider had the recognition he deserved. In 1922, three years before his death, he was awarded the newly established Faraday Medal by the IEE. This final honour seems particularly appropriate, since both men were of humble origins, received little formal education, were characterised by a striking independence of thought and had a profound influence on electrodynamics. Heaviside disliked honours, but seemed reconciled, perhaps even pleased, to accept this one.

Heaviside was impoverished throughout most of his life, relying on the goodwill of his family and the £40 per year he was paid for his publications in the trade journal *The Electrician*. His pride prevented him from accepting charitable offers from the Royal Society, but as old age crept up on him he was finally persuaded by his friend and fellow Maxwellian, George FitzGerald, to accept a Civil List pension of £120 per year. By this time Heaviside was living alone in a rundown house in Devon, having acquired a reputation as an eccentric recluse, shunning all social contact. He never married and in his later years suffered from poor heath, including bad eyesight. He died, alone, on 3 February 1925.

## AIII.5 Hertz: Maxwell Vindicated

> It was as though scales fell from my eyes when I read Hertz's great paper... Here Maxwell's equations, purified by Heaviside and Hertz, were made the axioms and the starting point of the theory. The totality of electromagnetic phenomena is derived from them systematically by deduction.
>
> A. Sommerfeld, 1948

> Now all has been blended into one theory, the main equations of which can be written on a page of a pocket notebook. That we have got so far is due in the first place to Maxwell, and next to him to Heaviside and Hertz.
>
> H. A. Lorentz, 1922

Heinrich Hertz and Oliver Heaviside, together with George FitzGerald and Oliver Lodge, constituted a small group of nineteenth-century physicists who would eventually become known as the Maxwellians. They were among the first to realise the true significance of Maxwell's ideas and were determined to explore, simplify and popularise this radical new theory. In this regard, it might be said that Hertz accomplished for German physics what Heaviside had done for British science.

Hertz was born in 1857 in Hamburg to a well-respected middle-class family. Most of his schooling was at a strict but progressive private school, after which he moved briefly to the Johanneum Gymnasium in Hamburg to take his final exams. He then began the study of engineering, first at Dresden Polytechnic and then briefly in Munich. However, he soon realised that his heart was really in physics and so he made an abrupt change in direction, eventually ending up in Berlin under the watchful eye of the great Helmholtz. Teacher and student became close friends and there followed a very fruitful collaboration, with Hertz demonstrating his extraordinary skills as an experimental physicist, wining a prestigious gold medal from the University of Berlin. When, in 1879, Helmholtz proposed to the Berlin Academy of Sciences that a prize be offered for the experimental testing of Maxwell's theory, he encouraged Hertz to enter the competition. However, Hertz declined, concerned that the task was too ambitious, and chose instead to complete his doctoral thesis on the topic of electromagnetic induction. Dr Hertz was then appointed as assistant professor in Helmholtz's laboratory, a position he held until 1883.

There followed a two-year period in Kiel, where Hertz specialised in mathematical physics. It was here that, in 1884, he first published his thoughts on simplifying Maxwell's new theory. Sometime later Hertz realised that he and Heaviside had been kindred spirits, embarked on similar journeys. He wrote, with great modesty,

> I have been led to endeavour for some time past to sift Maxwell's formulae and to separate their essential significance from the particular form in which they first happened to appear. The results at which I have arrived are set forth in the present paper. Mr. Oliver Heaviside has been working in the same direction ever since 1885. From Maxwell's equations he removes the same symbols as myself; and the simplest form which these equations thereby attain is essentially the same as that which I arrive. In this respect, then, Mr. Heaviside has the priority.

In 1885, when he was only twenty-eight years old, Hertz secured a full professorship at the University of Karlsruhe. On moving to Karlsruhe, which had excellent laboratories, he returned to his first love of experimental physics. Here, perhaps emboldened by his theoretical studies of electrodynamics at Kiel, he finally took up Helmholtz's suggestion to embark on an experimental examination of Maxwell's equations. These years at Karlsruhe were to prove the most productive for Hertz, culminating in 1888 with his now famous experimental confirmation of the existence of long-wavelength electromagnetic waves in free space. In particular, he showed that they could be reflected and refracted, just like light. Finally Maxwell's prediction of electromagnetic waves, and by implication the field theory of Faraday and Maxwell, was established beyond any doubt. With characteristic modesty, Hertz said at the time: *I do not think that the wireless waves I have discovered will have any practical application*. Of course, in 1901 Marconi was to prove the converse.

At first the impact of the 1888 papers in Germany was somewhat muted, since physicists there were still routed in the action-at-a-distance theories of Weber and Neumann. The British Maxwellians, by contrast, quickly heralded the findings as a great triumph, praising Hertz for his ingenuity and welcoming him into their ranks. Indeed, in some ways, it might be argued that the significance of Hertz's findings found their way back to Germany via Great Britain.

Hertz moved to Bonn in 1889, where he continued to work on Maxwell's equations, while developing an interest in mechanics. However, his health began to take a turn for the worse and this triggered all manner of medical interventions, including surgery, none of which helped. It turns out that he had a malignant bone condition. On 3 December 1893 he sent off his long-awaited manuscript *The Principles of Mechanics* to his publisher, and on 7 December he gave the final lecture of the year to his students. A few weeks later, on 1 January 1894, he died of a blood-vessel inflammation resulting from a failing immune system. He was only 36 years of age.

## REFERENCES AND FURTHER READING


Readers interested in more details of Ampère's tragic yet inspiring life might consult

Hofmann, J. R., 1996, *Andre-Marie Ampère: Enlightenment and Electrodynamics*. Cambridge University Press.

James, I., 2004, *Remarkable Physicists*. Cambridge University Press.

The latter also covers Cavendish, Coulomb, Young, Oersted, Ampère, Ohm, Faraday, Maxwell and Helmholtz, but not Hertz or Heaviside.

A great deal has been written about Faraday. Two excellent biographies are

Thomas, J. M., 1991, *Michael Faraday and the Royal Institution*. Institute of Physics Publishing.

Forbes, N. & Mahon, B., 2014, *Faraday, Maxwell and the Electromagnetic Field*. Prometheus Books.

Forbes & Mahon also cover Maxwell, as does

Harman, P. M., 1998, *The Natural Philosophy of James Clerk Maxwell*. Cambridge University Press.

Two excellent accounts of Heaviside's work are

Nahin P. J., 2002, *Oliver Heaviside: the Life, Work and Times of an Electrical Genius of the Victorian Age*. Johns Hopkins University Press.

Crowe, M. J., 1994, *A History of Vector Analysis*. Dover.

Readers interested in more details of Hertz's tragically short life might consult

Buchwald, J. Z., 1994, *The Creation of Scientific Effects: Heinrich Hertz and Electric Waves*. University of Chicago Press.

Finally, a historical overview of the development of electrodynamics is

Darrigol, O., 2000, *Electrodynamics from Ampère to Einstein*. Oxford University Press.

# INDEX